Nonlinear Solid Mechanics

SOLID MECHANICS AND ITS APPLICATIONS
Volume 160

Aims and Scope of the Series

The fundamental questions arising in mechanics are: *Why?*, *How?*, and *How much?* The aim of this series is to provide lucid accounts written by authoritative researchers giving vision and insight in answering these questions on the subject of mechanics as it relates to solids.

The scope of the series covers the entire spectrum of solid mechanics. Thus it includes the foundation of mechanics; variational formulations; computational mechanics; statics, kinematics and dynamics of rigid and elastic bodies: vibrations of solids and structures; dynamical systems and chaos; the theories of elasticity, plasticity and viscoelasticity; composite materials; rods, beams, shells and membranes; structural control and stability; soils, rocks and geomechanics; fracture; tribology; experimental mechanics; biomechanics and machine design.

The median level of presentation is the first year graduate student. Some texts are monographs defining the current state of the field; others are accessible to final year undergraduates; but essentially the emphasis is on readability and clarity.

For other titles published in this series, go to
www.springer.com/series/6557

Adnan Ibrahimbegovic

Nonlinear Solid Mechanics

Theoretical Formulations and Finite Element Solution Methods

Springer

Adnan Ibrahimbegovic
École Normale Superieure
Lab. Mecanique et Technologie
61 avenue du President
Wilson
94235 Cachan CX
France
ai@lmt.ens-cachan.fr

This is a translation from the French book, *Mécanique non linéaire des solides déformables : formulation théorique et résolution numérique par éléments finis*, published by Hermes Science - Lavoisier Paris in 2006, which was nominated for the Roberval Award for University Textbooks in French.

ISSN 0925-0042
ISBN 978-90-481-8490-3 e-ISBN 978-90-481-2331-5
DOI 10.1007/978-90-481-2331-5
Springer Dordrecht Heidelberg London New York

Printed on acid-free paper

Springer is part of Springer Science+Business Media (www.springer.com)

To my family

Foreword

It is with great pleasure that I accepted invitation of Adnan Ibrahimbegovic to write this preface, for this invitation gave me the privilege to be one of the first to read his book and allowed me to once again emphasize the importance for our discipline of solid mechanics, which is currently under considerable development, to produce the reference books suitable for students and all other researchers and engineers who wish to advance their knowledge on the subject.
The solid mechanics has closely followed the progress in computer science and is currently undergoing a true revolution where the numerical modelling and simulations are playing the central role. In the industrial environment, the 'virtual' (or the computing science) is present everywhere in the design and engineering procedures. I have a habit of saying that the solid mechanics has become the science of modelling and in that respect expanded beyond its traditional frontiers. Several facets of current developments have already been treated in different works published within the series 'Studies in mechanics of materials and structures'; for example, modelling heterogeneous materials (Besson et al.), fracture mechanics (Leblond), computational strategies and namely LATIN method (Ladevèze), instability problems (NQ Son) and verification of finite element method (Ladevèze-Pelle). To these (French) books, one should also add the work of Lemaitre-Chaboche on nonlinear behavior of solid materials and of Batoz on finite element method.
The book of Adnan Ibrahimbegovic also deals with nonlinear solid mechanics, but with the unique approach of the author: each question is examined from all different facets pertaining to either mechanics, mathematics or computations with a special attention to the finite element methods. It is the main strength of this book to provide as complete as possible answer to each question. Such an exhaustive approach is also characteristic of list of different topics studied that count among them some of the main difficulties of modern mechanics: damage theory, localization and failure, discrete models, multi-physics, multi-scale, parallelism etc. Only omission are the issues of verification and validation, which are just mentioned. Quite naturally, this work

is marked by the author's research, demonstrating a thorough understanding of the modern mechanics and a number of essential personal contributions. Collecting in a single book such a broad knowledge and know-how led inevitably to restructuring the presentation of the nonlinear solid mechanics and merging successfully the European and North-American schools, for the greatest benefits of readers. Even if all points of view presented in the book will not necessarily be shared by everybody, together they provide an unusually illuminating and lively image of the subject.

Cachan, August 2006

Pierre LADEVEZE
Professor at Ecole Normale Supérieure de Cachan
EADS Foundation Chair 'Advanced Computational Structural Methods'

Professor Adnan Ibrahimbegovic belongs to a select group of active researchers and developers of the 'finite element method technology' applied to solving the nonlinear problems in mechanics of solids and structures with complex constitutive behavior under static and dynamic loading; the latter includes a wide spectrum that spans from theoretical modelling to validation tests and passing through all numerical implementation aspects. If this kind of developments have been more numerous in eighties and nineties, they have become more rare in 2006 because of the maturity that the discipline has reached as the result of research works of a very active international community of computational mechanics (which is organized within the *International Association of Computational Mechanics*), with this author as its active and established member, as well as the presence of fairly complete and efficient commercial software packages. Nevertheless, there still exist significant needs for improvement, not only in terms of theoretical formulations (variational formulations) and numerical implementation (choice of discrete models and consistent approximations), but also in terms of solution technics of discretized problems, which are nowadays highly nonlinear and non-stationary with different coupling conditions, and without forgetting the programming aspects since the final development stage is inevitably the corresponding software product.

The book 'Nonlinear solid mechanics: Theoretical formulations and finite element solution methods' reflects the rich international teaching experience to master and doctoral students, as well as the joint collaborative research with a number of renown institutions in Europe and North-America (University of California at Berkeley, Swiss Federal Institute of Technology in Lausanne, Compiègne University of Technology in France, Laval University in Quebec, Ecole Normale Supérieure in Cachan, University of Ljubljana, Technical University of Braunschweig). With this detailed and original work, Adnan Ibrahimbegovic offers to master and doctoral students, and also to researchers and to computational software developers, the results and compilation of a number of research works.

This book deals with the analysis of deformable solids accounting for:

– Nonlinear elastic constitutive behavior, plasticity and viscoplasticity with damage, both for small and large deformations

– Contact conditions between deformable solids and the presence of inertia terms.

The theoretical and numerical aspects are pertinent to:

– Hu-Washizu variational principles and associated mixed finite element approximations and incompatible mode elements

– Time-integration schemes for inelastic constitutive models and nonlinear dynamics

– Thermomechanical coupling (also for elastoplasticity) and micro–macro approach

– Geometric and material instabilities.

We can think that the author will eventually publish one more book, dealing with structural mechanics models of rods, plates and shells, given his numerous contributions in that domain.

I wish that the present book meets the success equal to the international reputation of the author.

Compiègne, September 2006

Jean-Louis BATOZ
Professor at Compiègne University of Technology

Preface

The roots of this book go back to my doctoral studies at the University of California at Berkeley, from 1986 to 1989, funded by a Fullbright Grant. The UC Berkeley in general, and Structural Engineering, Mechanics and Materials Division in particular, provided an excellent study and research environment, with the opportunities to exchange the ideas with some extraordinary talented people from all over the world. Subsequently, I had a good fortune to stay on for another couple of years as a post-doc with my Berkeley mentors, Professor Edward L. Wilson and Professor Robert L. Taylor, which allowed me to explore a very wide variety of topics. The same good fortune was my subsequent research appointment at the Swiss Federal Institute of Technology in Lausanne at Structural and Continuum Mechanics Laboratory, directed by Professor François Frey, who granted me complete freedom to carry on with further explorations.

The work on the book started in 1994 with my first Professor appointment in France at the Compiègne University of Technology, continued from 1999 at the Ecole Normale Supérieure of Cachan in France, and kept gradually evolving as the result of a very fruitful interaction with graduate and undergraduate students, my doctoral students and colleagues of both faculties, in Cachan and in Compiègne. The final contents of the book was finally decided while preparing the IPSI course on Computational Solid and Structural Mechanics, which was taught several times in France, and more recently in Germany and in Italy, together with my colleague Robert L. Taylor, Professor at the University of California at Berkeley, to the audience of engineers coming from a number of prominent European companies, university teachers, researchers and graduate students. The first part of my sabbatical leave from ENS-Cachan in 2005 allowed me to finally converge with this long project. The work on French version of the book [115], which was published in 2006 within the collection of graduate textbooks in mechanics edited by Professor Paul Germain and Professor Pierre Ladevèze, was completed during my stay at the Swiss Federal Institute of Technology in Lausanne, as well as the subsequent stay at the University of Ljubljana, Slovenia. The work on the present English version of the book started during the second leg of my sabbatical leave that I spent at the Technical University of Braunschweig, Germany, which was made possible by the financial support of the Alexander von Humboldt Foundation through the Research Award in Technical Mechanics for scientists with internationally recognized qualifications.

Cachan, August 2008

Adnan Ibrahimbegovic
Professor at Ecole Normale Supérieure de Cachan

Acknowledgements

The funding for visiting professors at the Swiss Federal Institute of Technology in Lausanne, NATO fellowship for stay at the University of Ljubljana in Slovenia, and Humboldt Award with TU Braunschweig as my host institution, as well as the kind hospitality of my colleagues, Professor François Frey, Professor Bostjan Brank and Professor Hermann G. Matthies, are hereby gratefully acknowledged.
I would like to thank in particular all those who contributed to improving the book contents, either by generously supplying the results for a number of illustrative numerical simulations, or by participating in what seemed as never ending task of proofreading: some of my former doctoral students, Delphine Brancherie, Jean-Baptiste Colliat, Damijan Markovic, Norberto Dominguez, Guillaume Hervé, Lotfi Chorfi, Said Mamouri, Mazen Almikdad and Fadi Gharzeddine, and my colleagues, Catherine Knopf-Lenoir-Vayssade, Pierre Villon, François Frey, Blaise Rebora, Luc Davenne and Germaine Néfussi. Some of my present doctoral students have also contributed to this English edition, among them: Anna Kucerova, Sergiy Melnyk, Martin Hautefeuille, Amor Boulkertous, Christophe Kassiotis and Pierre Jehel. Thanks are also due to my senior colleagues, Professor Pierre Ladevèze and Professor Jean-Louis Batoz, for kindly accepting to prepare the preface for the French edition of the book, which I took the liberty to translate into English.
Last but not least, I thank my spouse Nita Ibrahimbegovic for her precious support.

Contents

Chapter 1
Introduction

1.1 Motivation and objectives

If motion is a crucial characteristic of life, then mechanics, the science which studies motion, stands at the center of life science studies. Perhaps there is not a single scientific discipline which has affected all aspects of modern society (e.g. transportation, urbanization, environment reconstruction or energy production) nor contributed to technological developments more than mechanics. The industrial applications of mechanics have further increased presently, especially those dealing with nonlinear problems. The latter are problems of great complexity which almost invariably require the use of numerical models as the only ones capable of providing a sufficiently reliable interpretation of the underlying phenomena. The first reason for the current increase in the number of nonlinear problems we deal with, pertains to the high level of competitiveness among industries, resulting with the need to identify any potential weakness of a proposed product not only under standard regime but also under extreme conditions, and to provide the optimal design with greatest economies. One example, rather representative of this kind of nonlinear problems, is the optimization of products of metal forming procedures, such as extrusion (see Figure 1.1). This is made possible by a numerical modelling approach capable of dealing with large deformations, inelastic constitutive behavior and frictional contact phenomena. Other studies of this kind pertain to structural behavior under extreme conditions, either brought by aggressive environment (stormy winds, earthquakes or tsunamis) or man-made incidents (explosions or fires). These studies aim to verify the safety margins with respect to the ultimate limit load design. One such study, currently underway in our research group at the Laboratory of Mechanics and Technology (LMT) in Cachan, deals with the impact of a large commercial aircraft on a massive structure. Here, we have to employ numerical models capable of representing large plastic deformations of the aircraft, the cracking of reinforced concrete structure under impact, the frictional contact of

A. Ibrahimbegovic, *Nonlinear Solid Mechanics: Theoretical Formulations and Finite Element Solution Methods*, Solid Mechanics and its Applications 160,

the plane sliding over the impacted zone, as well as integrating the high-frequency-dominated dynamic responses (Figure 1.2).

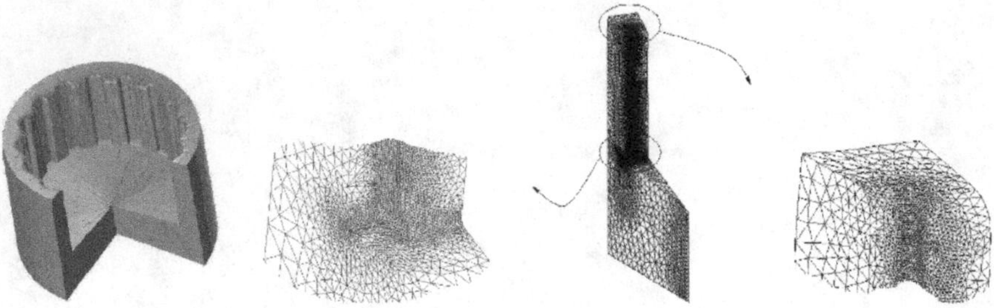

Fig. 1.1 Numerical modelling of metal forming process of extrusion.

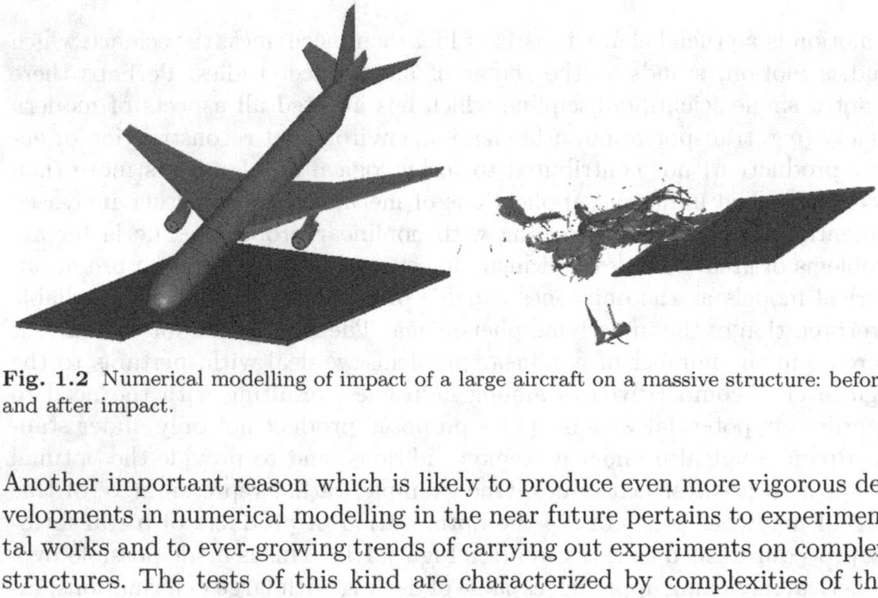

Fig. 1.2 Numerical modelling of impact of a large aircraft on a massive structure: before and after impact.

Another important reason which is likely to produce even more vigorous developments in numerical modelling in the near future pertains to experimental works and to ever-growing trends of carrying out experiments on complex structures. The tests of this kind are characterized by complexities of the structural assemblies chosen for testing, and also by a heterogeneous stress state, which in turn implies heterogenous structural response for inelastic materials. Unlike standard testing procedure on a single structural component, in the tests on a complex structure we rarely have sufficient numbers of measurements. We have to rely instead on numerical models in order to support the measurements, and to make the identification procedure robust. Our research team at LMT-Cachan has been involved in numerical modelling of tests on structures at very different scales. To start, we can deal with scales expressed in microns, or even nanometers for nanoindentation tests (see Figure 1.3), where one seeks to identify the details of the microstructure of a given material. We can also devise numerical models to support identification procedure for the structural tests at mesoscale. The case in point are standard 3-point bending tests that are routinely carried out for quasi-brittle materials, either in statics and in dynamics (see Figure 1.4). The test results interpretation based upon the mesoscale numerical models accounting for distribution of heterogeneities can provide (e.g. see [55]) the natural explanation of the size effect indicating the dominant damage mechanism for a particular specimen size. Finally, we can also provide numerical models improving interpretation of tests on structures at macro scale with (almost) real size. The examples of this kind concern shaking table tests (see Figure 1.5) or pseudo-dynamic tests with a reaction wall, which are carried out in order to explore the dynamic behavior of the complete structural assembly.

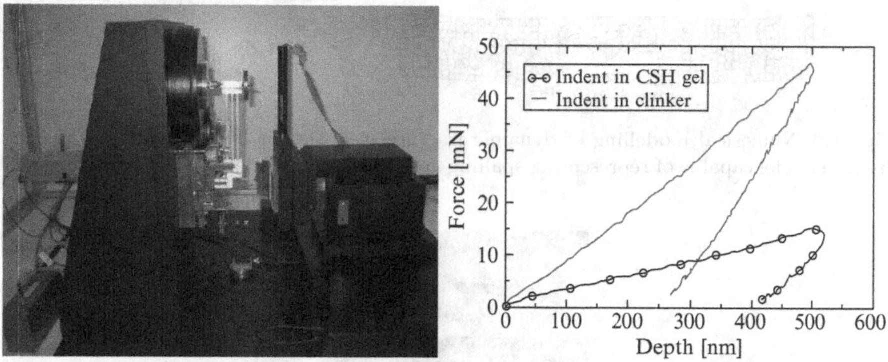

Fig. 1.3 Numerical modelling of nanoindentation tests for cement based composites.

The current tendencies in computational mechanics applications are strongly oriented towards interdisciplinary problems, where a great number of phenomena are led to interact with mechanics. One example of this kind is numerical modelling of the tests for fire-resistance of masonry structures in which our research team at LMT-Cachan was involved (see Figure 1.6). This study required not only a thorough knowledge of nonlinear mechanics but also of the phenomena pertinent to heat transfer and radiation, as well as to thermomechanics coupling. For this kind of problems, more generally referred to as multi-physics problems, we have to enlarge the studies in mechanics to include the coupling with phenomena studied by other scientific disciplines, and thus provide better understanding of the mechanical properties evolution or better quantification of the external loading. The complexity of multi-physics problems is such that the numerical models are often the only ones capable of providing their successful solution.
Therefore, the role of a specialist in mechanics is becoming quite complicated. Mechanics has practically become the science of modelling, or, in other words, it has become both the science and the art of selecting the most appropriate mathematical model for representing mechanics phenomena and their interaction with other phenomena which were traditionally studied in physics,

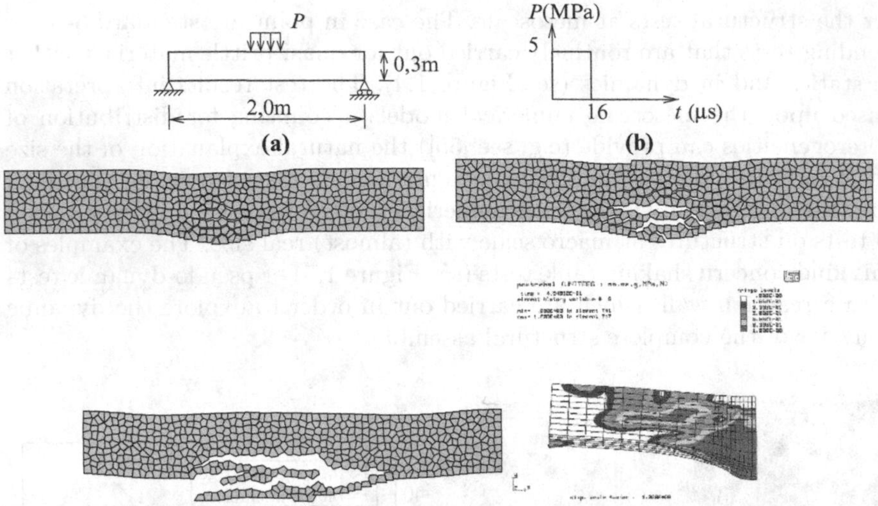

Fig. 1.4 Numerical modelling of dynamic fracture in a 3-point bending test, using the discrete model capable of representing spalling and continuum model with element erosion.

Fig. 1.5 Numerical modelling of shaking table tests for shear-wall structures.

biology, chemistry or nuclear sciences. In that respect, the first and most important task for a specialist in mechanics is that of validation. In validation, one seeks to confirm the right choice of the model with respect to the phenomena which must be represented, or in other words, one seeks "the right equations" to be solved. This can be done by carrying out the necessary experiments to identify the governing mechanisms of the relevant phenomena and their evolution, providing a sound basis for model construction. Most of the problems studied in this book are nonlinear, and extremely difficult to solve other than by numerical modelling and the finite element method. For that reason, the second important role of a specialist in computational mechanics is that of verification, testing the quality of the discrete approximation

and the numerical methods available in a given computer code. Here we seek to estimate the error (e.g. see [159]) of the finite element model with respect to the exact solution of the relevant equations, and furnish an acceptable quality of the numerical results.
It is not possible to equate the role of a specialist in computational mechanics to that of a mathematician, although applied mathematics does provide the essential contribution of rigor needed for using the computer methods. Few complex problems of nonlinear computational mechanics can be solved by a direct application of general solution methods proposed by applied mathematics, since the resulting solution procedure is rarely optimal, and can often fail. A much better result can be obtained by taking into account the mechanics basis of the problem in order to improve the conditioning and computational robustness of the method.

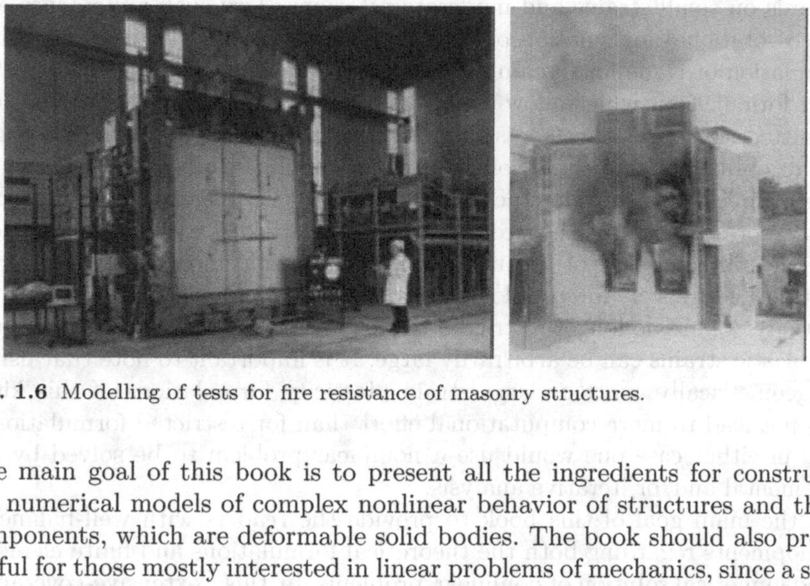

Fig. 1.6 Modelling of tests for fire resistance of masonry structures.

The main goal of this book is to present all the ingredients for constructing numerical models of complex nonlinear behavior of structures and their components, which are deformable solid bodies. The book should also prove useful for those mostly interested in linear problems of mechanics, since a sure way to obtain a sound theoretical formulation of a linear problem is through the consistent linearization of a more general nonlinear problem. The case in point is the linear formulation of shallow shell structures given in [123].
The same point of view that all the problems of mechanics are nonlinear, and that any linear approximation (for example, Hooke's law) has only limited range of applicability, has been adopted by the pioneers of nonlinear mechanics, such as Euler and Bernoulli. They boldly proceeded along such a path and managed to obtain some remarkable solutions of nonlinear problems in mechanics, such as the elastica or the Euler–Bernoulli beam. In fact, it is only the undeniable success of Cauchy's linear theory for continuum mechanics which made us forget that the Euler–Bernoulli theory of beams was originally developed in a geometrically nonlinear or even geometrically exact setting. The main reason for the success of Cauchy's linear theory of

mechanics lay in its ability to furnish a number of results of practical value in a situation where only analytic solutions were available. The revival of interest in nonlinear problems was prompted by the use of computers and has accelerated with ever increasing computational power. Important contributions have been made to bring about the present state of knowledge in nonlinear mechanics. On theoretical side, advances were made by mathematicians, with prominent works of Lyapunov or Poincaré on stability of motion and the work of Truesdell and Noll on the general nonlinear theory of continuum mechanics.[1] Unlike those works, this book is concerned with both qualitative and quantitative methods. We seek to obtain not only the theoretical formulations of nonlinear problems in mechanics, but also their numerical solutions. Our approach to nonlinear problems presented herein in general differs from the usual one. Usually, nonlinear problems are approached by trying to adapt tools developed for linear problems, under restrictive hypothesis on small strains and moderate rotations. Two such approaches are the co-rotational formulation of Argyris for statics, and the floating frame formulation of Kane for dynamics. Instead, we develop herein geometrically exact formulations which allow us to remove all the restrictive hypotheses on permitted size of displacements, strains or rotations. Furthermore, we show how to obtain the numerical solution of such nonlinear problems even when their complexity is considerable. For example, we no longer need an approximate representation of the geometrically exact theory with the restriction that the rotations are of second order, and we can allow for (very) large rotations. Moreover, for problems with material nonlinearities, such as in plasticity, we are no longer restricted to small elastic strains, but both elastic and plastic strains can be arbitrarily large. It is important to note that using the geometrically exact or large strain plasticity formulations of this kind does not lead to more computational effort than for restricted formulations, since in either case one would use a nonlinear problem to be solved by an incremental and/or iterative analysis.

It is the main goal of this book to provide the readers with well-balanced developments regarding both the theoretical formulations and finite element based numerical solution of nonlinear problems. In this "extensive-coverage" this book differs from those of my colleagues from LMT-Cachan (Ladevèze [156] or Lemaître and Chaboche [168]), who consider the theoretical formulation of nonlinear problems in solid mechanics with no discussion of numerical implementation issues, and also those of my former colleagues from UT-Compiègne (Batoz and Dhatt [23] or Dhatt and Touzot [66]), who consider

[1] We note in passing that even though the work of Truesdell and Noll [257] has presented quite a rigorous theoretical formulation of nonlinear continuum mechanics of solids in fully nonlinear setting, none of the pertinent aspects of numerical modelling was considered; The latter was probably the main reason for which most of theoretical developments proposed therein never reached a widespread use in computational mechanics, and, in particular, neither the choice of general curvilinear coordinates nor constitutive models for materials with long memory proved sufficiently convenient basis for numerical solution.

only the finite element method. For the reason of providing a well-balanced presentation of both theoretical formulation and numerical implementation, this book should also be of interest to English speaking mechanics community, despite a number of available books. Its main advantage is in a thorough treatment of nonlinear mechanics problems and a very complete coverage of all the main research topics of current interest in computational solid mechanics.

1.2 Outline of the main topics

Our main challenge is to present all the ingredients of the theoretical and numerical aspects of computational solid mechanics and to review its vast bibliography, within a reasonable number of pages. To achieve this goal, we have selected a typical model problem in nonlinear mechanics as the driving force for the detailed developments. The selected model problem is a boundary value problem, where we consider the response of a solid deformable body occupying a bounded domain under applied loading. In any boundary value problem in nonlinear mechanics, we are only interested in computing three fields: displacements, strains and stresses, which are in agreement with prescribed mechanical properties, applied external loading and the conditions imposed on the boundary. These three fields are defined with three (groups of) equations. The latter consist of: (i) kinematics equations, which describe the dependence of the strain field on the displacement field; (ii) constitutive equations, linking the stress field to the strain field; (iii) equilibrium equations, relating the stress field and the external loading.
Any one of these three equations can become the main source of nonlinearity in a boundary value problem in mechanics, as discussed in Chapters 2, 3 and 4, respectively. More precisely, in Chapter 2 we discuss the first class of boundary value problems dealing with elastostatics, characterized by a unique relationship between the stress and strain fields. We develop in parallel the linear and nonlinear versions of such boundary value problem. This allows us to discuss each phase of the solution procedure in detail. We start with the strong form of a boundary value problem, then proceed to either the weak form or its variational formulation, followed by the computation of the approximate solution by using the finite element method. The finite element discretization procedure delivers the final product in terms of a set of linear algebraic equations, for linear elasticity, or nonlinear algebraic equations, for the nonlinear elasticity case. We will explain how to solve a set of linear algebraic equations by using Gauss elimination, and how to exploit the same method in solving a set of nonlinear algebraic equations by an incremental analysis and Newton's iterations. These model problems of linear and nonlinear elasticity allow us to discuss several important aspects of finite element technology: isoparametric elements, the parent element, numerical integration, the patch test and finite element assembly procedures. We also show

how to improve the performance of isoparametric elements in the presence of a constraint (e.g. incompressibility) on the configuration space, by using either assumed field approximations (e.g. B-bar method for assumed strain or Pian and Sumihara method for assumed stress approximations) or enhanced strain approximation (e.g. the method of incompatible modes). Each method provides significant result improvement with respect to the standard isoparametric interpolations, which is illustrated by means of numerical examples. These methods are currently the most frequently used in dealing with industrial applications.

In Chapter 3, we study the nonlinear boundary value problems where the main source of nonlinearity arises from inelastic constitutive behavior. In this case we need not only the total deformation field in order to obtain the stress values, but also the internal variables. Typical examples of internal variables employed by inelastic constitutive models are: plastic deformations, damage compliance, hardening variable etc. The main role of the internal variable is to describe the result of a given loading program as experienced by a single component (or a particle) of the structure with inelastic constitutive behavior. We note that within a complex structural assembly under heterogeneous stress field, with resulting heterogeneous mechanical properties, the relationship between the given loading program and the evolution of the internal variables in each structural component is hardly ever linear due to the internal stress redistributions among different components. The latter implies that it remains next to impossible to control the evolution of internal variables produced by a particular loading program. For that reason, the interpretation of experimental results and the identification procedure under a heterogeneous stress field are very difficult to devise. Since this class of boundary value problems in very important for a vast majority of industrial applications of nonlinear solid mechanics, the presentation of Chapter 3 is made the most complete. We first present the details of constructing different refinements for two basic inelastic constitutive models: (i) plasticity, which can account for the presence of irreversible (plastic) deformation upon unloading, and (ii) damage, which is capable of representing the modification of the elastic response during the fracture process. We then explain how to couple these two models of inelastic behavior producing a coupled plasticity-damage model, capable of further extending model representation of different inelastic phenomena observed experimentally. A detailed list of topics discussed in Chapter 3 is as follows. We begin with a brief overview of continuum thermomechanics, by introducing different thermodynamic potentials suitable for replacing the strain energy of a hyperelastic material, with either the internal energy, free energy of Helmholtz, enthalpy and Gibbs potential. We show that the choice of a particular potential implies selecting the most suitable couple of independent fields, between strain or stress on one side and between temperature or entropy on the other. The framework of continuum thermomechanics provides a sound definition of the dissipation, produced either by heat conduction or by evolution of internal variables.

The latter is referred to as local or intrinsic dissipation, which appears in a boundary value problem dealing with inelastic constitutive behavior. It is important to note that the last definition also remains valid in purely mechanical framework obtained when the thermal effects ignored. We first study an illustrative boundary value in plasticity, with appropriate internal variables for either 1D or 3D case. We show that the main ingredients of the finite element solution procedure, such as isoparametric elements or numerical integration, constructed previously for elasticity will directly apply to plasticity. In fact, the main advantage of the finite element approach is even more striking in plasticity (or any other inelastic constitutive model), since the values of internal variables are required only at the numerical integration points. We show that the final product of the finite element method for the case of plasticity is a set of differential–algebraic equations, with the algebraic part pertaining to the equilibrium equations and differential equations describing the evolution of internal variables. We show that a very efficient solution procedure can be built for such a system by using the operator split method. The latter separates the global phase of computations for constructing solution of equilibrium equations from a local phase of computation pertaining to evolution of internal variables by the chosen time-integration scheme. The local computational phase is carried out in the inner loop. Having illustrated all the main steps of the solution procedure for the standard plasticity model (for both 1D and 3D case), we show the necessary extensions related to different refinements of plasticity model. Among them we present: plasticity with nonlinear isotropic hardening (e.g. the saturation hardening for limit load computations), plasticity model with kinematic hardening (models capturing the Bauschinger effect in cyclic behavior), plasticity model with strain rate sensitivity (or viscoplasticity), multi-surface plasticity (e.g. models representing different behaviors in compression and in tension), and the plasticity model with nonlinear elastic response (models characterizing behavior of soils or concrete). The improvements obtained by these refinements of plasticity model are illustrated by the numerical examples. We also show in this chapter that the framework of thermodynamics and operator solution procedure proposed for plasticity can be extended to the damage constitutive model, if we accept a somewhat non-standard choice of internal variable in terms of damage compliance tensor. Such a formulation of continuum damage will inherit all the computational advantages granted to plasticity models. For example, we can easily compute the tangent elastodamage modula capable of ensuring the quadratic convergence rate of Newton's iterative method. Moreover, we can also provide different refinements in a similar way as for plasticity, and thus increase the predictive capabilities of the basic damage model. We discuss multi-surface damage model refinement that is illustrated in terms of an anisotropic damage model, which combines the Kachanov-like volumetric damage representing micro-cracks with the surface damage mechanism representing a macro-crack. We draw an important conclusion in the course of this development, that both issues of theoretical formulation and finite element implementation ought to be settled in order to provide the optimal result for a particular inelastic constitutive model. Finally, we also show how to combine the constitutive models of plasticity and of damage into a coupled plasticity-damage constitutive model. The coupled model will inherit all the predictive features of two constituents, without introducing any modification to any of them. The same approach can also be applied to yet more complex constitutive models.

In Chapter 4 we study the boundary value problems where the main source of nonlinearity pertains to the large displacements and large deformations. Here we find the problems from structural mechanics, dealing with models such as beams, plates and shells undergoing very large displacements and rotations, yet only small strains. In this class of problems we also find the models of constitutive behavior of materials built to sustain very large deformations, such as the rubber. In either case, we can not count upon the hypothesis that the displacement gradients remain small, and thus we can no longer leave the developments within the framework of geometrically linear theory and use the infinitesimal strain measure. In fact, in geometrically nonlinear framework there exists theoretically an infinite number of strain measures. Therefore, our first goal is to derive some of those practically used, and to show that all the strain measures can be related to one another by exploiting the polar decomposition of the deformation gradient. Moreover, we show that the geometric nonlinearity also implies that the equilibrium equations are nonlinear. Namely, the Cauchy or true stress is not the best choice for solving a boundary value problem in geometrically nonlinear solid mechanics, for it acts in the (unknown) deformed configuration. In large displacements setting, deformed and (known) initial configurations can be quite different, so that one can no longer simplify the computation of derivatives featuring in the stress equilibrium equations as done for the geometrically linear case. We derive both the strong and weak forms of equilibrium equations in a deformed configuration obtained with large displacements, rotations and strains. By transforming these stress equilibrium equations to the initial configuration, we obtain the two stress tensors of Piola-Kirchhoff, the first as the natural replacement of the Cauchy stress tensor in the strong form of equilibrium equations written in the initial configuration, and the second Piola-Kirchhoff stress as the same kind of replacement but in the weak form. We can identify yet other stress tensors through work-conjugate pairing with different large strain measures, which are very useful for defining a particular constitutive model in large strain regime. Namely, having such a wide selection of work-conjugate stress–strain pairs, represents an important advantage for the development of the constitutive models at large strains. The latter is by far more difficult task than the one carried out at small strains, simply because it is much larger. Namely, we not only have to ensure consistency of the constitutive model with the small strain case in the limit of strains approaching zero, but also to make sure that the constitutive relations obey the frame invariance and remain mechanically sound at very large strains.

The latter can formally be specified in terms of poly-convexity conditions imposed on the strain energy of a hyperelastic constitutive model. We elaborate in particular on quite probably the most efficient computational framework for development of the constitutive models at large strains, defined in terms of the principal directions of the stress and strain tensors. One development of this kind is presented in detail for the case of finite deformation plasticity model, where one assumes the multiplicative decomposition of deformation gradient into elastic and plastic components ($F^e F^p$). Further enhancement of computational efficiency for this kind of plasticity model can be made by using a convenient choice of the logarithmic strain measure. This allows us to separate the geometric and material nonlinearities, and to keep the constitutive model stress computation practically the same as for the small deformation case.

In Chapter 5, we study the boundary value problems where the main source of nonlinearity pertains to constantly changing boundary conditions, yet referred to as contact problems. In any problem of this kind, the motion of the deformable body is no longer unrestrained since it can come in contact with a rigid obstacle (unilateral contact), with another deformable body (bilateral contact) or with two parts of the same body coming into contact (self-contact). We start with the studies of the simplest possible case of unilateral contact in 1D setting, where the contact zone reduces to a single point. Even the simplest contact problem of this kind will allow us, nonetheless, to introduce three main methods for dealing with contact constraint on the configuration space. More precisely, we present the Lagrange multipliers method, which enforces the contact constraint exactly but at the expense of inconvenient computational procedure, the penalty method, which is more convenient computationally but provides only an approximation to contact constraint, and finally the augmented Lagrangian method, which successfully combines the best features of both of the first two. We also show that the computational procedures proposed for 1D case remain practically the same when dealing with 2D and 3D contact problems. However, the latter pose additional difficulty of having to search for contact zone. One such solution procedure for a 2D case is discussed herein in terms of 'slide line' algorithm. At the end of this chapter we also present the mortar element approach which one can use in order to ensure a good accuracy for computed contact pressure and stress.

In conclusion, Chapters 2–5 present all the potential sources of nonlinearity in a boundary value problem in solid mechanics. The tacit hypothesis in any of those cases is that the loading changes slowly, so that the inertial effects need not be taken into account. If that hypothesis is no longer valid, we have to place our studies within the framework of dynamics, and further deal with initial-boundary value problems. This is the main topic of Chapter 6. In an initial-boundary value problem, we still compute the fields of displacement, strain and stress, but any such field has to be defined not only with its spatial but also its temporal variation. Moreover, the equilibrium equations are replaced by equations of motion,[2] where inertial 'forces' are accounted for, and where one also has to define the initial conditions at the start of motion. The solution of this problem can also be obtained by using the finite element method, combined with the time-integration schemes that are directly applied to the standard form of equations of motion in terms of the second-order differential equations. We present both explicit and implicit time-integration schemes, suitable for high and low rate dynamics applications, respectively. We also show how to ensure the robustness of the implicit schemes in a long-term computation, by imposing the conservation of motion properties, such as linear and angular momentum balance or energy conservation. Moreover, we also show how to improve the stress computation accuracy of the implicit schemes, by introducing the numerical dissipation of high frequency modes. We show that all this can be accomplished in a fully nonlinear framework where the high frequency modes keep changing, simply by using the algorithmic constitutive equations. A more detailed illustration is given for finite deformation plasticity and contact. A number of other problems in nonlinear dynamics are addressed in [131].

Having presented the methods for dealing with problems in dynamics, we revisit in Chapter 7 the thermomechanics coupling within the context of dynamics. In other words, we deal in this chapter with initial-boundary value problems in nonlinear thermodynamics and we show how to obtain their solutions by using the finite element method and time-integration schemes. We note in passing that the chosen thermodynamics problem belongs to a wide class of coupled multi-physics problems of practical interest (such as hydromechanical coupling or coupling of mechanics with physical chemistry), which possess the same structure of governing equations and which can be solved with the methods proposed herein. We first choose a direct application of the time-integration schemes presented in previous chapters, in particular the Newmark scheme for mechanics and the mid-point scheme for thermal part, and show that such an approach can hardly be the most efficient due to inherent lack of symmetry of the direct solution of the coupled problem. We therefore turn to the operator split methodology solving the mechanics and thermal part separately, and reestablishing in the process the symmetry of each sub-problem. The way the coupling or the communication between the two sub-problems is done is important. In isothermal case, where one integrates the equations of motion for mechanics component at fixed temperature, the application of implicit, unconditionally stable time-integration schemes in each sub-problem is not sufficient to guarantee the unconditional stability of the operator split solution scheme. The best way to recover the unconditional stability in such a case is by using the adiabatic split, where the mechanics sub-problem is solved at fixed entropy, so that the temperature has to be updated in each sub-problem. An illustrative development for the

[2] In terminology of d'Alembert, the equations of motion are called dynamics equilibrium equations.

case of inelastic behavior with internal variables, which is presented in detail, pertains to the finite deformation thermoplasticity model.
In a boundary value problem in nonlinear solid mechanics, which arises when computing the ultimate limit load of a complex structure, we often have to deal with the instability phenomena. This is the main topic of Chapter 8. The instability phenomena are related to critical equilibrium states, which are characterized by an unproportionately large response to a small perturbation, that leads to inability of reestablishing equilibrium once the perturbation is removed. The linearized form of the equilibrium equations at the critical state shows that the tangent stiffness matrix becomes singular. Within the framework of large displacement gradients we can encounter so-called geometric instabilities, where the stress-dependent, geometric part of the stiffness matrix can introduce matrix-singularity at the critical state. In order to avoid the corresponding risks, it is important to develop the methods capable of detecting the presence of instabilities and the related singularity of the stiffness matrix. We show how to construct one such method for fully nonlinear case with large displacements and large deformations, which is able to detect the critical equilibrium states with the quadratic convergence rate. If the risk of presence of critical equilibrium states is acceptable, we can then use the method of arc-length in order to completely trace the post-critical response. The geometric instability typically occurs for compressed structural components only. The so-called material instability phenomena can also occur for component in tension, when its ultimate stress resistance is defeated. The material instability phenomena are accompanied by softening, where the increase of strain leads to decrease of stress. The corresponding failure mode for such a component is such that the problems of this kind are often referred to as localization problems. The latter implies that the material instability due to softening will change the strain profile and produce the zones with high strain gradients and large inelastic strains. For that reason, the material instability phenomena require not only a sound theoretical formulation of the constitutive models providing the appropriate interpretation of localized inelastic strains, but also an enhanced finite element approximation capable of representing the localized strain profile. We show that the solution procedure for material instability phenomena can be placed within the framework of incompatible mode method, by reinterpreting the enhanced strain field in terms of inelastic deformation. The computational robustness of such a solution procedure is granted by the operator split method.
In Chapter 9 we examine several issues in a currently very active research domain of multiscale modelling of inelastic behavior of materials (e.g. see [118]). The main motivation for multiscale modelling approach is to simplify construction of a physically sound explanation for the mechanisms governing inelastic material behavior. The latter is achieved by accounting for typical heterogeneities or the microstructure of a given material to the finest (affordable) scales, much smaller than a typical scale of the structure. The fine scale is usually referred to as microscale, although the corresponding length can

be of the order of nanometer, micron or centimeter, depending upon not only a particular material but also the chosen goal for our model. The multiscale modelling approach implies that the given material has to be treated as a structure, and that we ought to solve a boundary value problem in order to obtain such an interpretation. Only than can we carry on with the solution of a structural problem at the macroscale, where such material was employed. We first present the multiscale approach to nonlinear material behavior which is equivalent to classical homogenization for linear problems, where only the weak coupling between microscale and macroscale is assumed. This means that two scales are sufficiently separated to allow for the computational procedure to be split in two separate phases. We then present the strong coupling of the scales, where such a split in not allowed, and where one has to adapt the computational procedure to the constant communication between two scales. The key notion of the coupling between two scales is discussed in detail, showing both the displacement-based coupling (which provides the equivalent result to upper or Voigt bound in homogenization theory) and force-based coupling (equivalent to lower or Reuss bound). In order to provide the most general representation of the microstructure, the finite element method is employed at both scales. We show in particular the enhancements of the strain field which allow that the convenient structured mesh can be employed at microscale. We note in closing that the multiscale modelling approach has had the most success with helping develop a more sound identification procedure for experiments carried out on a structural component. Even with ever increasing computational resources, it is very hard to imagine that the multiscale approach, with a detailed, microscale representation of strain and stress field throughout the structural domain, can be employed for a complex structure. One should seek in that case a judicious approach combining the multiscale models for a sub-domain of particular interest and phenomenological models elsewhere. Therefore, the phenomenological models will still play an important role for quite some time to come.

1.3 Further studies recommendations

The book was first and foremost written for students, engineers and expert users of different software products in nonlinear computational solid mechanics, as well as for those in teaching and research seeking to further enhance their understanding of the theoretical formulations which are the basis for development of the solution methods employed in computer codes. What is presented in this book is roughly the current state of the art of the finite element modelling of materials and structures, with the main difficulties (and their solutions) stemming from different industrial applications in mechanical, aerospace or civil engineering and material science. The prerequisites for studying the book are kept to a minimum, and the good part of the presented discussion ought to be accessible to the experts coming from other fields, who are seeking the necessary background for interacting with researchers from computational solid mechanics.

Several courses which I am teaching at École Normale Supérieure in Paris suburb of Cachan, one of the 'Grandes Écoles' of the French system of higher education, pertain to the subjects studied in this book. Among them first of all there are three courses taught to the last year of Master of Engineering students in Graduate Program 'Mechanics and Technology' entitled: 'Numerical models in multi-physics and computational procedures', where we discuss the formulation and the finite element solution for nonlinear mechanics and coupled problems issued from different industrial applications in engineering; 'Integrity and instability of structures under extreme loading conditions', where different methods are presented for dealing with ultimate load computations, geometric and material instabilities, and 'Dynamics of structures and earthquake engineering', where we study different problems in high-rate and low-rate nonlinear dynamics. This book is also used for the course I am teaching to Master students in Applied Mathematics entitled 'Numerical methods in nonlinear solid mechanics', where we cover all different sources of nonlinearities in solid mechanics in 1D setting, and present different ways of improving the conditioning of the system to be solved and the best choice of the solution method, very much inspired by the mechanics formulation of a particular problem.

For further studies of the ideas in computational solid mechanics presented in this book, one can consult a number of different books each specialized in a particular domain. For example, for further studies of the physical aspects, one can start with the books of Bornert et al. [36], François et al. [80], Hill [103], Krajcinovic [153], Lemaitre and Chaboche [168], Lubliner [174], Maugin [185], Prager [224] for nonlinear solid mechanics, accompanied by the books in thermodynamics, such as Atkins [14], Ericksen [74], Kondepudi and Prigogine [152]. The theoretical formulation of nonlinear mechanics can further be elaborated by studying some of the classical works of Germain [85], Green and Zerna [93], Sokolnikoff [243], Truesdell and Noll [257], Truesdell and Toupin [258], as well as the more recent books on the subject, such as Chadwick [46], Duvaut [71], Gurtin [96], Ladevèze [156], Ogden [209], or yet several more specialized works, as those of Argyris et al. [6], Bazant and Cedolin [25], Clough and Penzien [53], Géradin and Rixen [84], Nguyen [202], Rougée [230], as well as a more mathematically oriented works, such as Abraham and Marsden [1], Antman [4], Arnold [12], Choquet-Bruhat and deWitt-Morette [47], Ciarlet [50], Duvaut and Lions [72], Glowinski and Le Tallec [89], Goldstein [90], Lanczos [160], Marsden and Hughes [182]. For more detailed studies of the finite element methods, one can consult the books of Bathe [19], Batoz and Dhatt [23], Belytschko et al. [28], Brezzi and Fortin [41], Ciarlet [49], Crisfield [60], Dhatt and Touzot [66], Hughes [111], Johnson [144], Laursen [164], Owen and Hinton [214], Simo and Hughes [238], Strang and Fix [249], Washizu [260], Wriggers [264], Zienkiewicz and Taylor [271].

The ideas from applied mathematics regarding the different methods presented herein can further be explored in the books of Bathe and Wilson [20], Ciarlet [51], Dahlquist and Bjork [62], Gear [83], Dennis and Schnabel [65], Golub and Van Loan [91], Hairer and Wanner [97], Hirsch and Smale [107], Kelley [148], Luenberger [175], Parlett [219], Strang [248]. The pertinent issues of the functional analysis, which allow a deeper understanding of the approximations developed herein, are addressed in books of Brezis [40], Clarke [52], Ekeland and Temam [73], Hildebrand [102], Lang [161], Oden [205], Oden and Demkowicz [207], Temam [254], Troutman [256].

1.4 Summary of main notations

We first lay down the general rules for distinguishing different notations employed in this book. We use Latin alphabet letters to denote the scalar fields in a boundary value problem in 1D context, which is a frequently used vehicle for introducing the new ideas at the beginning of each chapiter. In 2D and 3D setting, the same fields are denoted by bold face Latin letters, with lower case letters reserved for vectors and upper case for second order tensors. This notation convention is slightly modified for Greek letters, where the long tradition is respected by denoting several second order tensors by lower case Greek letters, such as $\boldsymbol{\sigma}$ and $\boldsymbol{\epsilon}$ for the stress and the strain tensors, respectively. Higher order tensors (fortunately, we do not have to go higher than order three of four) are denoted by calligraphic letters. In order to avoid the risk of possible confusion, occasionally we also state the tensor equation in index notation, which refers explicitly to tensor components. The index notation is also useful when it comes to computations, since the value of any tensor has to be obtained component by component. The computed results are stored in a matrix, exploiting the possibility to reduce the number of stored tensor components for any symmetric tensor and thus increase computational efficiency. We use the sans serif fonts for this matrix notation and try to enforce the convention where the lower case letters are used to denote the vectors (one column matrices) and upper case to denote matrices.

Table of symbols

Symbol	Interpretation in solid mechanics	Equation
A	Cross-section of a bar	(2.148)
$\mathbf{A}^{ep}$	Elastoplastic acoustic tensor	(8.51)
$\mathbf{a}$	Acceleration vector	(6.14)
$\mathbf{B}$	Left Cauchy–Green deformation tensor	(4.30)
B^e	Element strain-displacement matrix	(2.292)
$(\mathbf{b})$ b	Distributed external loading	(2.3)
c	Specific heat capacity coefficient	(3.9)
$\mathcal{C}$	Elasticity tensor (fourth order)	(2.206)

Table of symbols

Symbol	Interpretation in solid mechanics	Equation
$\mathbf{C}$	Right Cauchy–Green deformation tensor	(4.29)
C	Elastic constitutive matrix	(2.296)
$D \equiv D_{loc}$	Local dissipation	(3.75)
D_{cond}	Dissipation by conduction	(3.12)
$\boldsymbol{\mathcal{D}}$	Compliance tensor (fourth order)	(3.315)
$\mathbf{D}$	Rate of deformation tensor	(4.138)
d_a	Total displacement vector (nodal values)	(2.291)
$d_a(t)$	Nodal value of temperature	(3.20)
E	Young's modulus	(2.10)
$\mathbf{E}$ (E_{ij})	Green–Lagrange deformation tensor (components)	(4.31)
$e(\epsilon, s)$	Internal energy density	(7.1)
$\mathbf{e}_i$	Unit base vector	(2.184)
$\mathbf{e}_i^{\varphi}$	Unit base vector in deformed configuration	(4.3)
$\mathbf{F}$	Deformation gradient	(4.16)
$\mathbf{f}$ (f_a^{ext})	Equivalent external force vector (component a)	(2.69)
f^e	Element external force vector	(2.151)
f_a^{int}	Internal force vector (component a)	(2.76)
$\mathsf{f}^{int,e}$	Element internal force vector	(2.160)
G(u;w)	Weak form of equilibrium equations	$(W1)$
G^e	Strain-incompatible modes element matrix	(2.326)
$H^1(\Omega)$	Hilbert functional space	(2.40)
H	Hamiltonian potential in dynamics	(6.21)
h^e	Length of finite element	(2.61)
h	Time step	(3.36)
$\bar{h}$	Imposed flux	$(S3)$
$\boldsymbol{\mathcal{I}}$	Forth order identity tensor	(3.174)
$\mathbf{I}$	Second order identity tensor	(3.174)
J	Determinant of deformation gradient	(4.66)
$j(\xi)$	Jacobian of isoparametric transformation	(2.155)
K (or K^u)	Stiffness matrix	(2.72)
K (or K^θ)	Conductivity matrix	(3.23)
K^e	Element stiffness matrix	(2.147)
K	Isotropic hardening modulus	(3.112)
k	Conductivity coefficient	(3.13)
$L_2(\Omega)$	Space of square-integrable functions	(2.41)
L	Lower triangular matrix ($\mathsf{K} = \mathsf{LU}$)	(2.100)
$\mathbf{L}(\mathbf{x})$	Velocity gradient	(4.136)
L^e	Element connectivity matrix	(2.175)
l	Bar length	(2.1)
M (or M^u)	Mass matrix in dynamics	(6.33)

Table of symbols

M (or M^θ)	Heat capacity matrix	(3.23)
$\mathbf{m}_i$	Principal vector of deformation tensor (deformed config.)	(4.231)
$N_a(x)$	Shape function	(2.61)
$N_a^e(\xi)$	Element shape function	(2.140)
$\mathbf{n}_i$	Principal vector of deformation tensor (initial config.)	(4.231)
$\mathbf{n}$	Exterior normal unit vector	(2.7)
$\mathbf{P}$	First Piola-Kirchhoff stress tensor	(4.47)
p	Pressure	(2.339)
p_c	Contact pressure	(5.7)
q_i	Heat flux	(3.5)
q	Stress-like isotropic hardening variable	(3.118)
$\mathbf{R}$	Rotation tensor	(4.26)
r	Heat source	(3.5)
$\mathbf{S}$	Second Piola-Kirchhoff stress tensor	(4.66)
s	Entropy	(3.5)
T	Kinetic energy	(6.22)
t	Time (or pseudo-time)	(2.113)
$\bar{\mathbf{t}}$	Imposed traction	(2.280)
U	Upper triangular matrix ($\mathsf{K} = \mathsf{LU}$)	(2.100)
$\mathbf{U}(\mathbf{x})$	Right stretch tensor	(4.26)
u	Incremental displacement superposed on deformed config.	(2.124)
$\bar{u}$	Imposed displacement	(2.18)
$\mathbf{V}(\mathbf{x})$	Left stretch tensor	(4.28)
$\mathbf{v}$	Velocity vector	(4.132)
$W(\epsilon)$	Strain energy density	(2.25)
$\mathbf{W}$	Skew-symmetric tensor	(4.154)
w	Virtual displacement	(2.16)
x	Position (vector)	(2.1)
$\mathbf{y}$	Position of rigid obstacle in contact	(5.48)
$\mathbf{Z}$	Back-strain tensor spatial description	(4.325)
z_i	Principal value of back-strain tensor	(4.314)
α	Coefficient of thermal expansion	(7.37)
$\boldsymbol{\alpha}^e$	Element incompatible mode parameters	(2.326)
$\boldsymbol{\beta}$	Stress interpolation parameters	(3.393)
Γ	Domain boundary	(2.2)
Γ_u	Dirichlet boundary (imposed displacement)	(2.6)
Γ_σ	Neumann boundary (imposed stress)	(2.5)
$\boldsymbol{\Gamma}$	Virtual Green–Lagrange deformation tensor	(4.61)
$\boldsymbol{\gamma}$	Virtual infinitesimal deformation tensor	(2.311)

Table of symbols

Symbol	Interpretation in solid mechanics	Equation
δ_{ij}	Kronecker symbol	(2.184)
$\boldsymbol{\epsilon}$	Infinitesimal deformation tensor	(2.239)
ζ	Isotropic hardening variable	(3.77)
η	Viscosity parameter	(3.88)
θ	Temperature	(3.6)
ι	Truncation error for time integration	(3.45)
$\boldsymbol{\kappa}$	Kinematic hardening back-strain tensor	(3.227)
λ_t (λ_i)	(principal) stretch	(4.37)
λ	Lamé parameters	(2.262)
μ	Shear modulus	(2.262)
ν	Poisson's ratio	(2.262)
$\boldsymbol{\nu}$	Normal to plasticity surface	(3.196)
$\Xi(\zeta)$	Isotropic hardening potential	(3.295)
$\boldsymbol{\Xi}$	Back-strain deformation tensor material description	(4.308)
$\boldsymbol{\xi}$ (ξ_i)	Natural coordinates of isoparametric element	(2.289)
Π	Total potential energy	(2.28)
ρ	Mass density	(6.8)
$\boldsymbol{\sigma}$	Cauchy stress tensor (real stress)	(2.247)
σ_y	Limit of elasticity in plasticity	(3.99)
σ_f	Limit of elasticity in damage	(3.294)
$\boldsymbol{\tau}$	Kirchhoff stress tensor	(4.58)
$\boldsymbol{\Upsilon}$	Back-stress tensor in material description	(4.310)
$\boldsymbol{\upsilon}$	Back-stress tensor in spatial description	(3.225)
ϕ	Thermodynamics potential of enthalpy	(3.67)
$\boldsymbol{\varphi}(\mathbf{x})$	Point-wise mapping in large deformation	(4.2)
χ	Gibbs thermodynamics potential	(3.70)
ψ	Helmholtz free energy	(3.62)
Ω	Domain (in a boundary value problem)	(2.1)
$\boldsymbol{\omega}$	Infinitesimal rotation tensor	(2.239)

Chapter 2
Boundary value problem in linear and nonlinear elasticity

Chapter outline: We start with the boundary value problems in elasticity, where the slow rate of external loading application will allow us to neglect the inertia or dynamics effects, and let us deal with the statics problems. Two types of elastic constitutive behavior are considered, linear elastic behavior described by Hooke's law and nonlinear elasticity with a unique (nonlinear) relationship between stress and strain. We study how to develop the theoretical formulation of these problems in terms of their strong and weak forms, and how to construct their numerical solutions by using the methods of Galerkin, Ritz and finite elements. This kind of approach leads us directly to the most costly phase of the solution process: solving a set of linear algebraic equations for linear elasticity or nonlinear algebraic equations for nonlinear elasticity. A detailed analysis of the computational cost is presented for Gauss elimination, perhaps the most frequently used direct method for solving a set of linear algebraic equations. This information on the solution cost remains pertinent to solving a set of nonlinear algebraic equations, because the latter can be reduced to a repetitive task of solving the corresponding set of linear equations by using incremental analysis or Newton's iterations. We start with 1D problems, presenting their variational formulations, and discussing briefly the issues of solution existence and uniqueness. The same 1D framework allows us to explain a number of important ingredients of finite element 'technology': isoparametric elements, numerical integration or the conditions to guarantee the quality of a finite element approximation. In the second part of the chapter we examine the theoretical formulation, with strong and weak forms, and the numerical implementation of 2D and 3D boundary value problems in elasticity. The governing equations are first stated in tensorial notation, which gives them the shortest possible format and which has the advantage of showing the analogy with 1D case. The finite element technology is then reexamined again in 2D and 3D context. It turns out that, unlike the 1D case, isoparametric interpolations are insufficient to achieve an optimal element performance, and that one also needs different enhancements, such as those brought by the method of assumed strains or the method of incompatible modes.

A. Ibrahimbegovic, *Nonlinear Solid Mechanics: Theoretical Formulations and Finite Element Solution Methods*, Solid Mechanics and its Applications 160,

2.1 Boundary value problem in elasticity with small displacement gradients

In this chapter we first present a couple of illustrative boundary value problems in 1D linear and nonlinear elasticity at small displacement gradients, along with their solution by numerical methods, such as the methods of Galerkin and Ritz or yet finite elements. The 1D framework is chosen to let us present these basic ideas clearly, without having to worry about the complexities of the tensor calculus that a more general case would have imposed. Moreover, the 1D framework still allows us to illustrate the finite element solution procedure and its different phases all of which we find again in the 2D and 3D settings.

2.1.1 Domain and boundary conditions

Seeking to develop the theoretical formulation and the finite element solution procedure for a boundary value problem in 1D elasticity, implies finding the variations of displacement, deformation and stress fields of an elastic deformable body in one direction, with respect to the loading which is applied in the same direction. It is assumed that in the plane which is perpendicular to that direction the solution will remain constant. Hence, a convenient choice of the frame allows us to reduce this problem to one-dimensional, or in other words to the model of a bar. For the solution being constant in each section, we can further consider with no loss of generality an elastic bar of a unit cross section ($A = 1$). The domain $\bar{\Omega}$ of the bar can therefore be reduced to an interval of the x axis

$$\bar{\Omega} := \{x \,|\, 0 \leq x \leq l\} \tag{2.1}$$

where l is the length of the bar. In (2.1) above, we use the symbol ':=' to indicate the definition for the quantity on the left hand side of the equation, as explicitly defined with the expression on the right hand side. In particular, the domain $\bar{\Omega}$ is defined by the standard notation to indicate a set, where we place between curvy brackets a typical element of the set, denoted herein as 'x', followed, after '|', by the statement about common properties for all the members of that set. We also use the symbol ':=' to define a function, like a rule for specifying the relationship between two fields.
For the boundary value problems in statics, studied in this chapter, we consider that the loading is applied sufficiently slowly so that the inertia dynamic effects can be neglected. Therefore, the time variation of the loading is not important but only its final value, and its variation in space. In order to specify the spatial variation of loading, we have to distinguish between the interior of the domain denoted as Ω and its boundary Γ, the ends of the

bar. The latter implies that $\Omega \bigcup \Gamma = \bar{\Omega}$ and $\Omega \bigcap \Gamma = \emptyset$, where '$\bigcup$' and '$\bigcap$' indicate the union and the intersection of two sets and $\emptyset$ is the empty set; for the domain defined in (2.1), we can write

$$\Omega := \{x \mid 0 < x < l\}\ ;\quad \Gamma := \{0, l\} \tag{2.2}$$

The first kind of external loading are volume forces, or the force defined in Ω, which reduces to distributed loading along the bar axis in the 1D case

$$b : \Omega \mapsto \mathbb{R} \tag{2.3}$$

where $\mathbb{R}$ is the set of real numbers. In (2.3), for a given function such as b, we state explicitly the domain of definition, indicating where all possible values of its argument are placed (here equal to Ω) and the co-domain where its values are placed (here $\mathbb{R}$). An alternative to denote the same function is by specifying the corresponding evaluation rule which provides the image $b(x) \in \mathbb{R}$ for any value of argument $x \in \Omega$, where '$\in$' denotes that the element is 'in' a given set. For example, we can specify the evaluation rule for linearly distributed loading increasing from zero with slope equal to 10 by

$$b(x) := 10x \tag{2.4}$$

The second type of loading is a surface force or traction, which is defined by a function

$$\bar{t} : \Gamma_\sigma \mapsto \mathbb{R} \tag{2.5}$$

The surface traction is applied on the boundary Γ_σ, and it contributes to force boundary conditions. As indicated in (2.2), in 1D setting the boundary Γ consists of only two points at the ends of the bar domain Ω; for that reason, this class of problems is often referred to as two-point boundary value problems. In order to be able to guarantee the uniqueness of the displacement field in a two-point boundary value problem by eliminating the rigid body translation, we must impose the displacement value at least one of two boundary points; this can be written:

$$\bar{u} : \Gamma_u \mapsto \mathbb{R} \tag{2.6}$$

where $\bar{u}$ is the imposed (known) value of the displacement. For example, if the left end of the bar is fixed in order to eliminate the rigid body translation, we can write for the displacement boundary $\Gamma_u := \{0\}$, with the zero imposed displacement $\bar{u} = 0$. It follows for such an example that the force boundary is at the right end of the bar with $\Gamma_\sigma := \{l\}$, where one should impose the traction force $\bar{t}$. Imposed traction will fix the value of the stress at the free end, since according to the Cauchy principle we can compute the stress vector component at the free end by multiplying the stress by the exterior unit normal (here equal to 1)

$$t^n(l) := \sigma(l) \underbrace{n}_{=1} = \bar{t} \tag{2.7}$$

The boundary conditions of imposed displacement on Γ_u are called the Dirichlet, whereas the conditions of imposed traction on Γ_σ are referred to as the Neumann boundary conditions. In any well-posed boundary value problem, the complete boundary is split between the Dirichlet and von Neumann boundary $\Gamma_u \bigcup \Gamma_\sigma = \Gamma$, such that the two never intersect $\Gamma_u \bigcap \Gamma_\sigma = \emptyset$. In other words, if we want to ensure that a boundary value problem is well posed, we must not impose both displacement and traction at the same boundary. The main goal of the solution procedure for a boundary value problem is to find the displacement field $u(x)$ resulting from applied volume forces $b(x)$, imposed traction $\bar{t}$ and imposed displacement $\bar{u}$. The resulting displacement field can be computed starting either from strong or weak form of the boundary value problem, which are described in the next section. It is important to note that not only the displacement field is of interest, but also the strain $\epsilon(x)$ and stress field $\sigma(x)$. In fact, the latter are often more important than the displacement for verifying the risk of damage or fracture.

2.1.2 *Strong form of boundary value problem in 1D elasticity*

Any boundary value problem in solid mechanics consists of three (groups of) equations governing the kinematics, the equilibrium and the constitutive behavior. One way to define the boundary value problem is through its strong form, where a local, point-wise form ($\forall x \in \bar{\Omega}$) of each of three equations is stated. One such development is presented in this section for the boundary value problem in 1D elasticity at small displacement gradients, for both linear and nonlinear constitutive behavior (see Figure 2.1).

Kinematics: This equation defines the relationship between the displacement field, $u(x)$, and the strain field, $\epsilon(x)$. In geometrically linear theory the axial deformation (or dilatation) is defined by the first derivative of the displacement field

$$\epsilon(x) = \frac{du(x)}{dx} \; ; \; \forall x \in \bar{\Omega} \tag{2.8}$$

In a 1D setting, this choice is simple to justify: The deformation $\epsilon(x)$ can not depend directly on displacement field $u(x)$, or else a rigid body displacement (defined with $u(x) = cst.$ for the 1D case) would affect the strain field. This leads to the deformation is defined through the first derivative of displacement field, which eliminates the influence of any rigid body mode.[1] The hypothesis of small displacement gradients that is used in geometrically linear theory, with $|\frac{du}{dx}| \ll 1$, implies that the deformation is infinitesimal, with $\epsilon \ll 1$. The same hypothesis implies that the deformed length of an infinitesimal segment

[1] We note in passing that we could also add the second derivative of displacement field to strain definition, such as done in the second gradient theory, which can be of interest for dealing with strain localization problems.

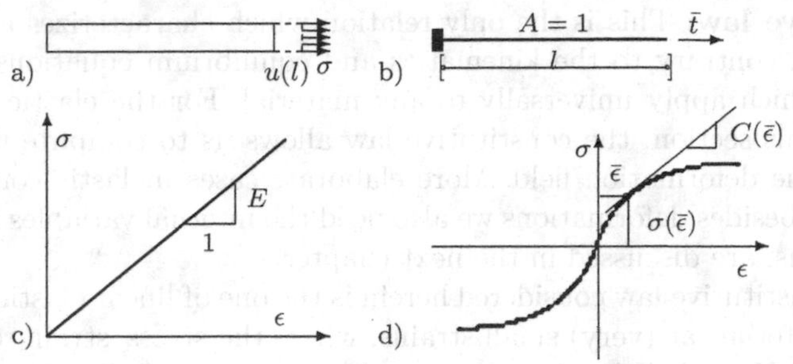

Fig. 2.1 Boundary value problem in 1D elasticity: (a) initial configuration (solid line) and deformed configuration (dashed line) producing free-end displacement $u(l)$ under uniform stress field σ, (b) 1D model, (c) constitutive law of linear elasticity, (d) constitutive law of nonlinear elasticity.

dx from the initial configuration remains approximately the same, since the mapping into deformed configuration is defined according to

$$\left(1+\underbrace{\frac{du}{dx}}_{\ll 1}+\underbrace{\frac{1}{2}\frac{d^2u}{dx^2}dx}_{\approx 0}\right)dx \approx dx$$

Therefore, one can keep the coordinates defined in the initial configuration for any operation where such an infinitesimal segment would enter, such as differentiation or integration in the deformed configuration. Furthermore, in this manner the infinitesimal strain field which is really applied in the deformed configuration can be represented in the initial configuration and so can the stress field.

Equilibrium: This is the first equation which benefits from the hypothesis on small displacement gradient theory, resulting with the corresponding linear form. Namely, by isolating an infinitesimal segment from the deformed configuration, we can identify its position by two end coordinates from the initial configuration, say x and $x+dx$ (where $x, x+dx \in \Omega$). In order to maintain the equilibrium of this isolated segment under volume force $b(x)dx$, we apply the Cauchy stress at each end, with $\sigma(x+dx)$ applied at the right and $-\sigma(x)$ applied at the left end of the segment. Each of these stresses represents the action which (the rest of) the bar exerts on the segment. By further using Taylor's formula to express $\sigma(x+dx)=\sigma(x)+\frac{d\sigma(x)}{dx}dx+O(dx)$, neglecting the higher order terms[2] and dividing through[3] by dx, we can obtain the local form of the force equilibrium equation

$$\frac{d\sigma(x)}{dx}+b(x)=0\ ;\ \forall x\in\bar{\Omega} \tag{2.9}$$

[2] For an infinitesimal segment, one can neglect higher order terms: $dx^2, dx^3 \ldots \ll dx$.

[3] Recall that dx is infinitesimal, but not equal to 0.

Constitutive law: This is the only relation which characterizes a particular material, contrary to the kinematics and equilibrium equations in (2.8) and (2.9) which apply universally to any material. For the elastic bar considered in this section, the constitutive law allows us to compute the stress field from the deformation field. More elaborate cases inelastic constitutive laws, where besides deformations we also need the internal variables for stress computations, are discussed in the next chapter.
The first constitutive law considered herein is the one of linear elasticity, valid for most materials at (very) small strains, where the stress–strain relation is described by Hooke's law

$$\sigma(x) = E(x)\,\epsilon(x)\ ;\ \forall x \in \bar{\Omega} \tag{2.10}$$

where $E(x)$ is Young's modulus. For a bar made of homogeneous elastic material, Young's modulus is constant, $E(x) = E = cst.$
By combining results in (2.8) and (2.10), the stress can be directly expressed as a function of displacement

$$\tilde{\sigma}(u) := E\frac{du}{dx} \tag{2.11}$$

By introducing this relation in (2.9), we can combine the results on kinematics, constitutive law and equilibrium into a single equation

$$\frac{d}{dx}\left(E\frac{du}{dx}\right) + b = 0\ ;\ \forall x \in \bar{\Omega} \implies E\frac{d^2u}{dx^2} + b = 0\ ;\ \text{if } E = cst. \tag{2.12}$$

The last result expressed the equilibrium equation in terms of displacement field. This expression from mathematics point of view is a second order differential equation which can be solved by integrating twice. Moreover, the two conditions needed for eliminating two constants of integration are precisely the boundary conditions chosen in the previous section.

In summary, one can write:

Strong form of boundary value problem in 1D linear elasticity

Given: $b : \Omega \mapsto \mathbb{R}, \bar{u} : \Gamma_u \mapsto \mathbb{R}, \bar{t}^n : \Gamma_\sigma \mapsto \mathbb{R}$
Find: $u : \bar{\Omega} \mapsto \mathbb{R}$, such that:
$\frac{d}{dx}(E\frac{du}{dx}) + b = 0$ in Ω (S1)
$u = \bar{u}$ on Γ_u
$t^n := E\frac{du}{dx}n = \bar{t}$ on Γ_σ

The second constitutive law we will study is that of nonlinear elasticity, where the stress–strain relationship is no longer linear as in (2.10). For nonlinear elasticity, the Cauchy stress is a nonlinear function of infinitesimal strain:

$$\sigma(x) = \hat{\sigma}(x, \epsilon(x)) \tag{2.13}$$

We assume that such a nonlinear relation is single-valued, and sufficiently smooth to define at each strain the corresponding value of tangent modulus

$$C(\epsilon) := \frac{d\hat{\sigma}(\epsilon)}{d\epsilon} \tag{2.14}$$

By combining the results in (2.13) and (2.8), we can thus express the stress as a nonlinear function of the displacement field

$$\tilde{\sigma}(u) := \hat{\sigma}(\frac{du}{dx}) \implies \tilde{\sigma} \equiv \hat{\sigma} \circ \hat{\epsilon}\ ;\ \hat{\epsilon}(u) := \frac{du}{dx} \tag{2.15}$$

where the symbol $\equiv$ indicates the equivalence, and the symbol '$\circ$' indicates the composition of mappings. In order to further clarify the result in (2.15), produced by the composition of mappings (2.8) and (2.13) to express the stress in terms of displacement, we recall that a function is interpreted simply as evaluation rule independent of the choice of argument, so that $\hat{\sigma}(\epsilon)$ and $\hat{\sigma}(\frac{du}{dx})$ are the same functions. Unlike (2.11), the resulting stress–displacement relationship is no longer linear. However, one can still use (2.15) to write the equilibrium equation as a nonlinear function of displacement field:

Strong form of boundary value problem in 1D nonlinear elasticity

Given: $b : \Omega \mapsto \mathbb{R}, \bar{u} : \Gamma_u \mapsto \mathbb{R}, \bar{t}^n : \Gamma_\sigma \mapsto \mathbb{R}$ Find: $u : \bar{\Omega} \mapsto \mathbb{R}$, such that $\frac{d}{dx}\hat{\sigma}(\frac{du}{dx}) + b = 0$ in Ω $u = \bar{u}$ on Γ_u $t^n := \hat{\sigma}(\frac{du}{dx})n = \bar{t}$ on Γ_σ

$(S2)$

We can easily obtain the analytic solution for the strong form of 1D linear elasticity defined in $(S1)$ by integrating twice with respect to 'x' and by using the boundary conditions to eliminate the integration constants. It is more difficult to compute the same integrals for the the nonlinear elasticity case defined in $(S2)$. Moreover, even for a linear case it is much harder to obtain the strong form solution of an elasticity problem in 2D or 3D setting, where a lack of regularity of domain Ω can also impose lack of regularity for the solution of the strong form. For that reason, we turn to the weak form for this kind of problem, which reduces the solution regularity requirements significantly.

2.1.3 *Weak form of boundary value problem in 1D elasticity and the principle of virtual work*

The word 'weak' in this formulation of a boundary value problem pertains to the way in which we now verify the equilibrium equation, no longer at each point $x \in \bar{\Omega}$, but as an integral with a chosen weighting function, $w(x)$. In order to define the weak form, the weighting functions ought to be differentiable and also verify:

$$\mathbb{V}_0 := \{w : \bar{\Omega} \mapsto \mathbb{R} \mid [w]_{\Gamma_u} = 0\} \tag{2.16}$$

The last condition implies that any weighting function has to be zero on the boundary Γ_u, where we impose the displacement value $[u]_{\Gamma_u} = \bar{u}$. This kind of condition is referred to as kinematic admissibility, as we describe in the next section.

In order to establish the relationship between the strong and weak forms of a boundary value problem, we start with the displacement field $u(x)$, satisfying the strong form, which produces the stress field $\sigma(x)$. With such a stress field, the equilibrium equation in (2.9) is satisfied with the sum of all forces equal to zero $\forall x \in \Omega$, and will remain so even if multiplied by a weighting function $w(x)$ and integrated over the domain Ω. In other words, for the solution of the strong form $u(x)$ it also holds that $0 = -\int_\Omega w(\frac{d\sigma}{dx} + b)\, dx$. We can now integrate the first term of this integrand by parts[4] and use the kinematic admissibility requiring that the weighting function remains zero on the Dirichlet boundary, which provides the final result

$$\begin{aligned}
0 &= -\int_\Omega w\left(\frac{d\sigma}{dx} + b\right) dx \\
&= \int_\Omega \left(\frac{dw}{dx}\sigma - wb\right) dx - \int_\Omega \frac{d}{dx}(w\sigma) dx \\
&= \int_\Omega \left(\frac{dw}{dx}\sigma - wb\right) dx - [w\sigma n]_\Gamma \\
&= \int_\Omega \left(\frac{dw}{dx}\sigma - wb\right) dx - [w\bar{t}\,]_{\Gamma_\sigma}
\end{aligned} \tag{2.17}$$

Finally, introducing in the last expression the kinematics and constitutive equations in terms of stress–displacement relationship in (2.11), we can obtain:

Weak form of boundary value problem in 1D linear elasticity

$$\begin{array}{l}
\text{Given: } b : \Omega \mapsto \mathbb{R}, \bar{u} : \Gamma_u \mapsto \mathbb{R}, \bar{t}^n : \Gamma_\sigma \mapsto \mathbb{R} \\
\text{Find: } u \in \mathbb{V}, \text{such that } \forall w \in \mathbb{V}_0 \\
0 = G(u; w) := \int_\Omega \left(\frac{dw}{dx} E \frac{du}{dx} - wb\right) dx - [w\bar{t}\,]_{\Gamma_\sigma}
\end{array} \tag{W1}$$

[4] Integrating by parts allows us to obtain $\int_\Omega w \frac{d\sigma}{dx}\, dx = [w\sigma]_\Gamma - \int_\Omega \frac{dw}{dx} \sigma dx$.

where $\mathbb{V}$ represents a set of test functions, collecting all the sufficiently differentiable candidates for the solution of the weak form

$$\mathbb{V} := \{u : \bar{\Omega} \mapsto \mathbb{R} \mid [u]_{\Gamma_u} = \bar{u}\} \tag{2.18}$$

The comparison of the last result with that in (2.16), shows that the only difference between the functions in $\mathbb{V}$ and in $\mathbb{V}_0$ concerns the boundary conditions on the imposed displacements; each weighting function $w(x)$ takes zero value on Γ_u regardless of imposed displacement value, whereas any test function has to verify the Dirichlet boundary condition with $[u]_{\Gamma_u} = \bar{u}$. The development presented in (2.17) shows that any solution of the strong form will also be the solution of the weak form.
By using very much the same procedure as the one just described, except for employing the constitutive relationship of nonlinear elasticity in (2.15), we can obtain

Weak form of boundary value problem in 1D nonlinear elasticity

Given: $b : \Omega \mapsto \mathbb{R}, \bar{u} : \Gamma_u \mapsto \mathbb{R}, \bar{t}^n : \Gamma_\sigma \mapsto \mathbb{R}$
Find: $u \in \mathbb{V}$, such that $\forall w \in \mathbb{V}_0$
$G(u; w) := \int_\Omega \{\frac{dw}{dx}\hat{\sigma}(\frac{du}{dx}) - wb\}\, dx - [w\bar{t}\,]_{\Gamma_\sigma} = 0$

$(W2)$

Comparison of the strong against weak forms of elasticity, for either $(S1)$ against $(W1)$ for linear or $(S2)$ versus $(W2)$ for nonlinear case, let us conclude that the weak form is less demanding with respect to the solution regularity. The weak form does not require that the stress field be differentiable, nor to satisfy the local equilibrium equations $\forall x \in \bar{\Omega}$. However, if the stress field verifying $(W1)$ and $(W2)$, happens to be differentiable with $\frac{d\sigma}{dx}$; $\forall x \in \Omega$, the integration by parts in (2.17) can be reversed to show that the solution of the weak form will also satisfy the strong form of elasticity:

$$\begin{aligned} 0 &= \int_\Omega \left(\frac{dw}{dx}\sigma - wb\right) dx - [w\bar{t}\,]_{\Gamma_\sigma} \\ &= \int_\Omega \left(-w\frac{d\sigma}{dx} - wb\right) dx + [w\sigma n]_\Gamma - [w\bar{t}\,]_{\Gamma_\sigma} \\ &= \int_\Omega \{-w\left(\frac{d\sigma}{dx} + b\right)\}\, dx + [w(\sigma n - \bar{t}\,)]_{\Gamma_\sigma} \end{aligned} \tag{2.19}$$

The last result is obtained by imposing the kinematic admissibility of the weighting function with $[w]_{\Gamma_\sigma} = 0$. By imposing further that the weighting function outside Dirichlet boundary remains arbitrary, we can conclude from the last expression that

$$\frac{d\sigma}{dx} + b = 0 \; ; \; \forall x \in \Omega$$
$$\sigma n = \bar{t} \; ; \; \forall x \in \Gamma_\sigma$$

The first of these results is precisely the strong form of the equilibrium equation, whereas the second is the boundary condition on von Neumann boundary. We thus conclude that a solution of the weak form, which happens to be sufficiently regular for local stress derivative computation, will also satisfy the strong form. Furthermore, it is not necessary to enforce the imposed traction on the Neumann boundary in the weak form. The latter can be recovered naturally from the weak form, and, for that reason, it is also referred to as the natural boundary condition. On the other hand, the Dirichlet boundary condition on imposed displacement is essential for defining both strong and weak form, and is thus referred to as the essential boundary condition (e.g. Stakgold [246]).

Principle of virtual work: The expressions $(W1)$ and $(W2)$ are known as the statement of the principle of virtual work, and the weighting function $w(x)$ is also called the virtual displacement. The virtual displacement is an imaginary displacement field, which is considered to be sufficiently small so that we can define the virtual deformation to be infinitesimal and equal to $\frac{dw}{dx}$. Moreover, the virtual displacements are assumed to be kinematically admissible, taking the zero value on the Dirichlet boundary and satisfying any other kinematic constraint. This kind of displacement is imagined being imposed on the given deformed configuration and the equilibrium is tested through comparing the work produced by both internal and external forces acting in that configuration. The principle of virtual work simply states that such virtual works of external and internal forces should remain the same. For a bar, the virtual work of internal forces can be computed by accounting for the work of the only non-zero stress component $\sigma(x)$ according to

$$G^{int}(u;w) := \int_\Omega \frac{dw}{dx}\tilde{\sigma}(u)\,dx \tag{2.20}$$

where $\frac{dw}{dx}$ is the infinitesimal virtual strain and $\tilde{\sigma}(u) \equiv \hat{\sigma}(\frac{du}{dx})$ is the stress field computed from the displacement $u(x)$ in the given deformed configuration. The latter is computed from (2.11) for 1D linear elasticity or from (2.15) for nonlinear elasticity case. The virtual work of external forces is computed by accounting for contributions of volume force $b(x)$ and surface traction $\bar{t}$, leading to for the case of a 1D bar

$$G^{ext}(w) := \int_\Omega wb\,dx + [w\bar{t}]_{\Gamma_\sigma} \tag{2.21}$$

The virtual work principle can thus be written:

$$G^{int}(u;w) = G^{ext}(w) \implies G(u;w) := G^{int}(u;w) - G^{ext}(w) = 0 \tag{2.22}$$

which, according to the results in (2.20) and (2.21), is identical to the weak form of the boundary value problem defined in ($W1$) or ($W2$).

2.1.4 Variational formulation of boundary value problem in 1D elasticity and principle of minimum potential energy

For hyperelastic constitutive behavior, the stress can be computed as the derivative of the strain energy density with respect to strain. In such a case, the weak forms in ($W1$) and ($W2$) can also be derived from a variational formulation. The variational formulation provides the necessary condition for the minimum of a functional, which is a function of another function. In solid mechanics, the functional to be minimized is the total potential energy of a deformable solid body $\Pi(u(x))$; the latter is a function of the displacement field $u(x)$, which in turn is a function of position $x \in \Omega$. We show in this section that the displacement field which satisfies the strong form, not only satisfies the weak form of the boundary value problem pertaining to a given deformed configuration, but also minimizes the total potential energy in that configuration.

We next provide the details on how to construct the total potential energy functional for the chosen model problem of 1D elasticity. The total potential energy consists of the strain energy and the energy of external forces. The former can be computed by integrating the strain energy density $W(x, \epsilon(x))$ along the bar, which leads us to

$$\Pi^{int}(u(x)) := \int_{\Omega} W(x, \epsilon(x))\, dx \tag{2.23}$$

For hyperelasticity or Green's elasticity, the strain energy plays the role of the potential for the Cauchy stress computation

$$\sigma(x) = \frac{dW(x, \epsilon)}{d\epsilon} \tag{2.24}$$

The existence of a potential makes hyperelasticity more advantageous for computational purposes than hypoelasticity or Cauchy elasticity, where the stress is a function of strain that not necessarily derives from a potential. The constitutive relationship of linear elasticity, expressed by Hooke's law in (2.10), is one example of a hyperelastic model, where the strain energy density is defined as a quadratic form in terms of strain

$$W(x, \epsilon(x)) = \frac{1}{2}\epsilon(x)\, E(x)\, \epsilon(x) \tag{2.25}$$

The factor $\frac{1}{2}$ in the last expression arises from the gradual increase of deformation from zero to final value under the applied load:

$$W(\epsilon) := \int_0^{\epsilon} \underbrace{\hat{\sigma}(\varepsilon)}_{E\varepsilon} \, d\varepsilon = \frac{1}{2} \underbrace{E\epsilon}_{\sigma} \, \epsilon \tag{2.26}$$

We suppose that the external are applied to the initial configuration with their final values, so that no factor is needed in computing the energy of external forces:

$$\Pi^{ext}(u(x)) := \int_0^u (-\int_{\Omega} b \, dx - [\bar{t}]_{\Gamma_\sigma}) \, du = \int_{\Omega} (-ub) \, dx - [u\bar{t}]_{\Gamma_\sigma} \tag{2.27}$$

We note that the last expression in (2.27) can be considered as the external force potential. The total potential energy is therefore a sum of the strain energy and the potential energy of external forces, which can be written:

$$\Pi(u(x)) := \int_{\Omega} (W(x, \epsilon(x)) - ub) \; dx - [u\bar{t}]_{\Gamma_\sigma} \tag{2.28}$$

We next show that the total potential energy functional reaches the minimum value in a deformed configuration, which implies that the deformed configuration is in equilibrium and furthermore that the corresponding equilibrium state is stable. In order to compute the minimum of the functional of total potential energy in (2.28), we have to explore the ideas from a branch of applied mathematics called calculus of variations (e.g. Hildebrand [102], Troutmann [256]), which has found a number of fruitful applications in mechanics (e.g. Goldstein [90], Lanczos [160]). An easy way to grasp the conditions for minimum of a functional is by comparison with the conditions we have to meet for the minimum of a function. Recall that the minimum of a function $f(x)$ is reached at the point $\bar{x}$, where the first Frechet derivative takes the zero value and the second Frechet derivative remains positive at that point

$$\bar{x} = arg\{min f(x)\} \;\Leftrightarrow\; \begin{cases} \frac{df(\bar{x})}{dx} = 0 \\ \frac{d^2 f(\bar{x})}{dx^2} > 0 \end{cases} \tag{2.29}$$

When spelling the conditions for the minimum of a functional, at the equilibrium 'point' $u(x)$ we can say that its first derivative should take the zero value and its second derivative should remain positive. While this statement is true, there still remains a technical difficulty concerning the definition of the Frechet derivatives, which will not necessarily apply to a functional. However, there is yet another definition of derivative, called the directional or the Gâteaux derivative (e.g. Lang [161] or Troutman [256], p. 44), which can be adapted to handle the problem on hands. In order to point out the difference with respect to the classical Frechet derivative, we write the Gâteaux derivative for a function $f(x)$ in the direction of dx

$$D_{dx}f(\bar{x}) := \lim_{\varepsilon \mapsto 0} \frac{f(\bar{x} + \varepsilon\, dx) - f(\bar{x})}{\varepsilon} \tag{2.30}$$

Under hypothesis of solution regularity, the last expression gives the same result as the Frechet derivative in (2.29). Note however that the Gâteaux derivative does not require any norm on the argument. Therefore, the Gâteaux derivative formalism can easily be applied to a more general task of computing the minimum of a functional. We next carry out this computation for the total potential energy functional of 1D elasticity. Let us assume that $u(x)$ is the solution for the displacement field producing a particular deformed configuration in stable equilibrium state, where the total potential energy $\Pi(u(x))$ reaches the minimum (equivalent to $\bar{x}$ minimizing $f(x)$). By making use of a kinematically admissible perturbation $w(x)$ with respect to this (yet unknown) displacement field, we can represent all other candidates for minimizing the total potential energy. Any admissible candidate, denoted as $u_\varepsilon(x)$, can be parameterized in terms of a small parameter[5] ε:

$$u_\varepsilon(x) = u(x) + \varepsilon\, w(x)\ ;\ u(x), u_\varepsilon(x) \in \mathbb{V}\ ;\ w(x) \in \mathbb{V}_0 \tag{2.31}$$

The last result shows that $u_\varepsilon(x)$ is kinematically admissible if the variation $w(x)$ takes zero value on the Dirichlet boundary Γ_u. Moreover, any kinematically admissible solution candidate ought to be placed in the neighborhood of the true solution, hence the variation $w(x)$ should be small enough to allow the infinitesimal strain computation. We thus conclude that the variation $w(x)$ must obey the same kinematic admissibility conditions as those enforced upon the virtual displacement field. In fact, we will next show that these two fields are the same, since the first variation of the total potential energy functional is identical to the weak form of the equilibrium equations. Namely, the first variation of the total potential energy functional $\Pi(u(x))$ can be computed by the Gâteaux or directional derivative in the direction of the variation $w(x)$; by the analogy with the corresponding expression for Gâteaux derivative in (2.30), we can obtain:

$$D_w \Pi(u) := \lim_{\varepsilon \mapsto 0} \frac{\Pi(u + \varepsilon w) - \Pi(u)}{\varepsilon}\ ;\ \forall w \in \mathbb{V}_0 \tag{2.32}$$

We note in passing that some authors denote the first variation of the functional $\Pi(u(x))$ with the symbol '$\delta\Pi$', also denoting the variation as '$\delta u(x)$'. This kind of notation can be confusing, since the delta symbol has been used to denote two different things, the variation of a functional in '$\delta\Pi$' and the name chosen for the variation field $\delta u(x)$.

For the chosen fields $u(x)$ and $w(x)$, the directional derivative can be interpreted as the Frechet derivative of the function $g(\varepsilon) := \Pi(u + \varepsilon w)$ with respect to the parameter ε. By assuming that such a function is smooth

[5] Parameter ε should not to be confused with the stain field $\epsilon(x)$!

and differentiable in the neighborhood of the true solution, corresponding to $\varepsilon = 0$, we can easily compute its Frechet derivative by making use of the chain rule

$$D_w \Pi(u(x)) := \frac{d}{d\varepsilon}[\Pi(u(x)+\varepsilon w(x))]_{\varepsilon=0} = \int_\Omega \left(\frac{dw}{dx} \underbrace{\frac{\partial W(\cdot)}{\partial \epsilon}}_{\hat{\sigma}(\frac{du}{dx})} - wb \right) dx - [w\bar{t}]_{\Gamma_\sigma} \tag{2.33}$$

The last result confirms that the first Gâteaux derivative, or the first variation, of the total potential energy functional is the same as the weak form of equilibrium equation

$$\frac{d}{d\varepsilon}[\Pi(u(x)+\varepsilon w(x))]_{\varepsilon=0} \equiv G(u; w) \tag{2.34}$$

We can there conclude that the displacement field that satisfies the weak form of equilibrium equation also renders the total potential energy functional stationary, with $\frac{d}{d\varepsilon}[\Pi(u(x)+\varepsilon w(x))]_{\varepsilon=0} = 0$.

Furthermore, the weak form can be integrated by parts for a sufficiently regular solution, to finally show that such a displacement field also satisfies the strong form of equilibrium equation as well as the natural boundary condition

$$0 = \int_\Omega -w\left(\frac{d\sigma}{dx} + b\right) dx + [w(\sigma n - \bar{t}]\Big|_{\Gamma_\sigma} \implies \begin{cases} \frac{d\sigma}{dx} + b = 0 \; ; \; \forall x \in \Omega \\ \sigma n = \bar{t} \; ; \; \forall x \in \Gamma_\sigma \end{cases} \tag{2.35}$$

This result is also known as the fundamental lemma of calculus of variations, and the resulting equations of the strong form are called the Euler–Lagrange equations.

Minimum of the total potential energy: Is an important results that characterize a stable equilibrium state with a positive value of the second variation of the total potential energy, $D_w^2 \Pi(u(x)) > 0$. The latter can by computed as the second Gâteaux derivative, or by repeating twice the directional derivative computation of the total energy functional:

$$\begin{aligned} D_w^2 \Pi(u) &:= D_w[D_w \Pi(u)] = \frac{d}{d\varepsilon}[D_w \Pi(u+\varepsilon w))]|_{\varepsilon=0} \\ &= \int_\Omega \frac{dw}{dx} C \frac{dw}{dx}\, dx \end{aligned} \tag{2.36}$$

In the last expression, $C = \frac{d\hat{\sigma}}{d\epsilon} = \frac{d^2 W(\cdot)}{d\epsilon^2}$ is the elastic tangent modulus, which will be equal to Young's modulus E for linear elastic case. We further assume the stable constitutive behavior, where increasing value of strain is accompanied with a stress increase, which implies that $C > 0$ (note that

$E > 0$ is always satisfied). With this hypothesis in hand, the result in (2.36) will further imply that the second Gâteaux derivative of the total potential energy will remain positive $D_w^2 \Pi(u) > 0$ or zero $D_w^2 \Pi(u) = 0$. However, the latter case is excluded in the presence of Γ_u boundary, which implies placing a support on at least one end of the bar to eliminate the rigid body motion:

$$D_w^2 \Pi(u) = 0 \text{ only if : } \{ \tfrac{dw}{dx} = 0 \Leftrightarrow w = cst. \& [w]_{\Gamma_u} = 0 \implies w = 0 \tag{2.37}$$

We thus conclude that the total potential energy in any adjacent state produced by a kinematically admissible perturbation $w(x)$ will be superior to the total potential energy in the equilibrium state. Namely, by taking into account the results stating that $D_w \Pi(u) = 0$ and $D_w^2 \Pi(u) > 0$, the Taylor series formula can be used to show that

$$\Pi(u + \varepsilon w) - \Pi(u) = \underbrace{G(u; w)}_{D_w \Pi(u)=0} + \underbrace{\int_\Omega \frac{dw}{dx} C \frac{dw}{dx}\, dx}_{D_w^2 \Pi(u) > 0} > 0 \, ; \; \forall w \in \mathbb{V}_0 \tag{2.38}$$

Therefore, any equilibrium state characterized by the minimum of total potential energy is stable. In other words, any kinematically admissible perturbation will increase the total potential energy and ensure that the equilibrium is re-established once the perturbation is removed.
In summary, the variational formulation of a boundary value problem allows us to compute its solution from the conditions of the minimum of the total potential energy. This can be stated in equivalent form to (2.29):

$$u(x) = \arg \left\{ \min_{u_\varepsilon \in \mathbb{V}} \Pi(u_\varepsilon) \right\} \implies \left\{ \begin{array}{l} D_w \Pi(u) = 0 \\ D_w^2 \Pi(u) > 0 \end{array} \right. \tag{2.39}$$

2.2 Finite element solution of boundary value problems in 1D linear and nonlinear elasticity

2.2.1 Qualitative methods of functional analysis for solution existence and uniqueness

The variational formulation of a boundary value problem is not only useful for finding the numerical solution of the problem, but also for studying the qualitative solution properties, such as its existence and uniqueness. In this section, we present a detailed discussion of this kind for linear and nonlinear cases of 1D elasticity.
We start with a boundary value problem in 1D linear elasticity whose variational formulation is defined in (2.28). The quadratic form of the strain energy in (2.25) will require that any admissible solution candidate, or any

test function, $u \in \mathbb{V}$ must have a sufficient regularity as specified by Hilbert functional space (e.g. see Brezis [40] or Oden and Demkowicz [207])

$$H^1(\Omega) := \left\{ u | u \in L_2(\Omega), \frac{du}{dx} \in L_2(\Omega) \right\} \tag{2.40}$$

In the last expression, $L_2(\Omega)$ denotes the space of square-integrable functions

$$L_2(\Omega) := \{u | \int_\Omega u^2 \, dx < \infty\} \tag{2.41}$$

By taking into account the expression for the total potential energy in (2.28), we can conclude that any test function $u(x)$ from Hilbert space in (2.40) will provide a finite value of internal energy. In other words, we can solve a boundary value problems in linear elasticity for the kind of loading and boundary conditions that produce a bounded strain field.[6] The same regularity requirements are imposed upon the variation $w(x)$, along with the kinematic admissibility condition, which can formally be specified as

$$H^1_0(\Omega) := \left\{ w | w \in L_2(\Omega), \frac{dw}{dx} \in L_2(\Omega) | w|_{\Gamma_u} = 0 \right\} \tag{2.42}$$

For a study of solution uniqueness, a couple of interesting properties of the variational formulation will prove useful. First, the variational equation, which is equivalent to the weak form of equilibrium equations, is composed of a linear functional[7] $G^{ext}(w)$ and a bilinear (or linear in each argument) functional $G^{int}(u; w)$,

$$0 = D_w \Pi(u) := G^{int}(u; w) - G^{ext}(w) \tag{W1'}$$

For linear elasticity, with constitutive relation defined in (2.10), the bilinear form is symmetric with respect to exchange of arguments u and w

$$G^{int}(u; w) = G^{int}(w; u) \tag{2.43}$$

Furthermore, the minimum of total potential energy in (2.39) implies that the bilinear functional is positive definite: for any kinematically admissible variation $w(x)$

$$G^{int}(w; w) \geq 0 \;\&\; G^{int}(w; w) = 0 \implies w = 0 \tag{2.44}$$

Therefore, we can use the bilinear form for constructing so called energy norm for measuring the difference between two solutions. With these results

---

[6] Note in particular that such requirement precludes any discontinuity in the displacement field, which would have produced the Dirac delta function in the strain field since the latter is not square-integrable.

[7] If c_1 and c_2 are constants, we can write $G^{ext}(c_1 w_1 + c_2 w_2) = c_1 G^{ext}(w_1) + c_2 G^{ext}(w_2)$.

in hands, we can prove the uniqueness by contradiction: first suppose that there could be two different solutions, $u(x)$ and $u^*(x)$, each satisfying the weak form of the boundary value problem in 1D linear elasticity. We pick a particular choice of variation equal to difference between these two solutions, $w(x) = u^*(x) - u(x)$, which automatically ensures the kinematic admissibility of the variation (since $[u^* - u]_{\Gamma_u} = 0$). By taking into account that both functionals $G^{ext}(w)$ and $G^{int}(u; w)$ are linear in w, this particular choice of w results with

$$\left.\begin{array}{l} G^{int}(u^* - u, u) - G^{ext}(u^* - u) = 0 \\ G^{int}(u - u^*, u^*) - G^{ext}(u - u^*) = 0 \end{array}\right\} \Longrightarrow G^{int}(u^* - u; u^* - u) = 0 \quad (2.45)$$

We can thus conclude that there can not be any difference between two solutions in the energy norm. For linear elasticity, this directly implies the uniqueness of the strain field (as well as the uniqueness of the stress field). The displacement field is made unique only by a judicious choice of the boundary conditions, which precludes any rigid body motion

$$\begin{array}{l} \frac{du}{dx} - \frac{du^*}{dx} = 0 \Longrightarrow \epsilon(x) = \epsilon^*(x)\ ,\ \forall x \in \bar{\Omega} \\ u - u^* = 0 \Longrightarrow u(x) = u^*(x)\ ,\ \forall x \in \bar{\Omega} \end{array} \quad (2.46)$$

The solution existence for linear elasticity can be shown for a sufficiently regular loading, such as, for example, a distributed loading b(x) with finite discontinuities. For this loading case, the solution of the strong form in ($S1$) would require a differentiable function $u(x) \in C^1$. The weak form is less demanding for the same load case, since it also accepts the functions in Hilbert space $u(x) \in H^1(\Omega)$ for the displacement field solution. The latter implies that any continuous function, which is not necessarily continuously differentiable, will be acceptable for representing the displacement field satisfying the weak form. For any such function we can construct the norm in Hilbert space (see Brezis [40], p. 121)

$$\| u \|_{H^1} = [\| u \|_{L_2}^2 + \| \frac{du}{dx} \|_{L_2}^2]^{1/2}\ ;\ \| u \|_{L_2} = \left[\int_\Omega u^2\, dx \right]^{1/2} \quad (2.47)$$

The existence of the solution can be proved for any external load which is sufficiently regular, say $b(x) \in L_2(\Omega)$, by appealing to the Lax-Milgram theorem (see Brezis [40], p. 84). We need to show that the bilinear form $G^{int}(\cdot, \cdot)$ is both continuous and coercive, so that a bounded loading will also produce the bounded displacement field.
The continuity of the bilinear form $G^{int}(\cdot, \cdot)$ implies that there is a constant c_{max} that allows to bound the strain energy from above

$$G^{int}(u; u) \leq c_{max} \| u \|_{H^1}^2 \quad (2.48)$$

By direct comparison of the expressions in (2.20) and in (2.48), we can easily confirm the continuity of strain energy for 1D linear elasticity, with the constant c_{max} equal to maximum value of Young's modulus in a heterogeneous bar

$$c_{max} = E_{max} \equiv \max_{\forall x \in \Omega} E(x) \tag{2.49}$$

The coercivity of the bilinear form $G^{int}(\cdot, \cdot)$ implies that there is a constant c_{min} that bounds the strain energy from below

$$G^{int}(u; u) \geq c_{min} \parallel u \parallel^2_{H^1} \tag{2.50}$$

The comparison between (2.20) and (2.48) shows that the coercivity condition requires to bound the solution by its derivative. This can easily be done for the case of 1D elasticity with homogeneous Dirichlet boundary conditions, $u|_{\Gamma_u} = 0$. In this case, the solution and its derivative are related through $u(x) = \int_0^x \frac{du}{dx}\, dx$. By using furthermore the Cauchy-Schwartz inequality, we can show that

$$u^2(x) = \left[\int_0^x 1 \frac{du}{dx}\, dx \right]^2 \leq \int_0^x dx \int_0^x \left(\frac{du}{dx} \right)^2 dx \leq x \int_0^l \left(\frac{du}{dx} \right)^2 dx \tag{2.51}$$

By integrating the last expression for an elastic bar of length l with fixed ends, we can obtain the Poincaré inequality,[8] which confirms that the solution indeed remains bounded by its first derivative

$$\int_\Omega u^2(x)\, dx \leq \frac{l^2}{2} \int_\Omega \left(\frac{du}{dx} \right)^2 dx \tag{2.52}$$

In the view of the definition of the energy norm in $H_1(\Omega)$, the last result will help us to conclude:

$$\parallel u \parallel^2_{H^1} \leq \left(\frac{l^2}{2} + 1 \right) \int_\Omega \left(\frac{du}{dx} \right)^2 dx \leq \frac{l^2 + 2}{2E_{min}} G^{int}(u, u) \tag{2.53}$$

The latter provides an explicit value of the coercivity constant

$$c_{min} = 2E_{min}/(l^2 + 2) \; ; \; E_{min} = \min_{\forall x \in \Omega} E(x) \tag{2.54}$$

In conclusion, the continuity and coercivity based bounds on the strain energy functional in 1D linear elasticity are sufficient to prove the solution existence for any sufficiently regular loading. In other words, the finite energy which is supplied to a bar by the external loading, results with the finite strain energy stored in the bar, producing bounded stress and strain, as well as bounded displacement field.

---

[8] In 3D case, we replace Poincaré's inequality by Korn's inequality; (e.g. see Duvaut and Lions [72], p. 110).

The analysis of the solution existence and uniqueness can also be carried out for a boundary value problem in 1D nonlinear elasticity. In this case, the Hilbert space ought to be replaced with Sobolev space (e.g. see Oden [205])

$$W^{1,p}(\Omega) := \left\{ u | u \in L^p(\Omega), \frac{du}{dx} \in L^p(\Omega) \right\} \tag{2.55}$$

where

$$L^p(\Omega) := \left\{ u | [\int_\Omega |u|^p \, dx]^{1/p} < \infty \right\} \tag{2.56}$$

The solution existence for 1D nonlinear elasticity can be shown for a sufficiently regular loading and hyperelastic constitutive law. Namely, we can again show the continuity and coercivity of the internal energy functional

$$\int_\Omega \tilde{W}\left(\frac{du}{dx}\right) dx \le c_{max} \parallel u \parallel_{W^{1,p}} \; ; \; \int_\Omega \tilde{W}\left(\frac{du}{dx}\right) dx \ge c_{min} \parallel u \parallel_{W^{1,p}} \tag{2.57}$$

where the constants c_{min} and c_{max} are defined for each particular choice of the strain energy potential $\hat{W}(\cdot)$. We also have to adapt the choice of the norm to this particular solution space $W^{1,p}(\Omega)$ (see Oden [205], p. 35). We note in passing that the norm in the Sobolev space for $p = 2$, which is denoted $\parallel u \parallel_{W^{1,2}}$, is identical to Hilbert space norm $\parallel u \parallel_{H^1}$.

2.2.2 Approximate solution construction by Galerkin, Ritz and finite element methods

It can be quite difficult, even impossible, to obtain the exact analytic solution for a boundary value problem where one is faced with high complexities regarding the loading, the constitutive behavior or the domain. For any such problems, we therefore should rather construct an approximate solution by using the finite element method. We note in passing that solving the boundary value problems in 1D linear elasticity is certainly the simplest task, which will not necessarily require the finite element method. Nonetheless, this provides convenient starting point for illustrating all the steps of the finite element solution procedure, which remain the same for more complex problems.
Historically, the finite element method finds its roots in the approximation methods of Galerkin and Ritz. In either of them, one constructs an approximate solution for displacement field as a linear combination of a finite number of chosen approximation functions, yet called shape functions. In Ritz method (see [229]), we introduce such displacement field representation into the variational formulation in (2.34), whereas in Galerkin method (see [82]) the displacement field representation is introduced into the weak form in

$(W1)$. In hyperelastic case, the weak form and the variational formulation of a boundary value problem are completely equivalent, and the methods of Galerkin and Ritz will lead to the same result. In a more general case than hyperelasticity where there is no potential, one can still define the weak form and use the Galerkin method to find the corresponding approximate solution. Such a solution will always be a linear combination of the chosen shape functions, with the corresponding contributions obtained in the final step from so-called Galerkin equation.
The quality of the approximate solution for the methods of Galerkin and Ritz depends upon the choice of the shape functions. This choice can not be fully arbitrary. Each shape function ought to remain kinematically admissible, which implies the sufficient smoothness as specified by the Hilbert space $H^1(\Omega)$ for linear elasticity (or by the Sobolev space $W^{1,p}(\Omega)$ for nonlinear hyperelastic case), as well as verification of the essential condition on imposed displacement on Dirichlet boundary. It can be quite difficult to find the right candidates for the shape functions in 2D or 3D problems on a complex domain Ω. It is precisely for that reason – providing the advantage of splitting a complex domain into simple sub-domains – that the finite element method has been proposed (see [58] or [259]). More precisely, by using the finite element method one again seeks an approximate solution to a given boundary value problem in the form of a linear combination of shape functions, which is constructed by patching together local approximations defined in each sub-domina. The main point is that the shape functions are now defined independently in each sub-domina, or each finite element Ω^e. The resulting approximations for the real and virtual displacement fields, which we denote as $u^h(x)$ and $w^h(x)$ respectively, must remain kinematically admissible

$$u^h \in \mathbb{V}^h \ ; \ \ w^h \in \mathbb{V}_0^h \tag{2.58}$$

The latter implies that both fields are still sufficiently regular (for example, in linear elasticity $u^h, w^h \in H^1(\Omega)$), and that each verifies a particular form of the displacement boundary condition

$$u^h|_{\Gamma_u} = \bar{u} \ ; \ \ w^h|_{\Gamma_u} = 0 \tag{2.59}$$

Superscript 'h' on both displacement fields is a reminder of discretization. It also indicates the quality of discrete approximation. For example, for 1D domain Ω of elastic bar of length l (see Figure 2.2), represented by 2-node sub-domains or 2-node elements, one can take h to be the length of a typical element among the total number n_{elem} of finite elements in which the bar is sub-divided.

$$\Omega = \bigcup_{e=1}^{n_{elem}} \Omega^e \ ; \ \Omega = [0, l] \ ; \ \Omega^e = [x_a, x_{a+1}] \ ; \ 0 \le x_a < x_{a+1} \le l \tag{2.60}$$

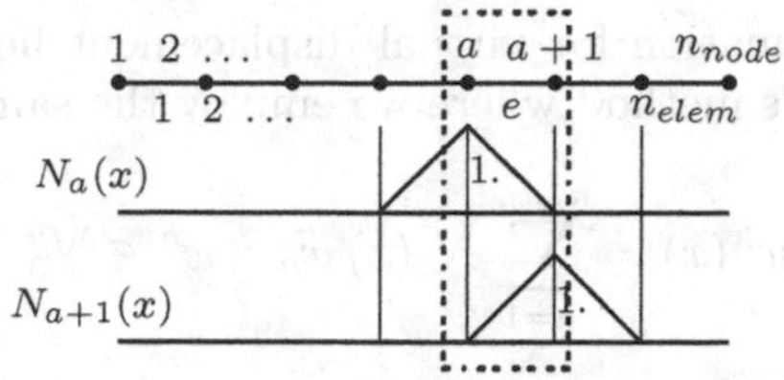

Fig. 2.2 Displacement field discrete approximation constructed by using 2-node finite elements and the shape functions in the form of linear polynomials in each element.

The 1D element boundaries, or the points that separate two neighboring elements, are called the finite element nodes. In finite element terminology, the collection of all the finite elements and nodes is referred to as the finite element mesh. Basically, the finite element mesh is just the selected representation of a given domain Ω in terms of sub-domains Ω^e, $e = 1, 2 \ldots n_{elem}$. The simplest discrete approximation for the displacement field can be constructed by using the shape functions as linear polynomials in the sub-domains surrounding a particular node that remain equal to zero outside that region. For example, the shape function '$N_a(x)$' for one such node 'a' can be written:

$$N_a(x) := \begin{cases} (x_{a+1} - x)/h^e \; ; \; x \in [x_a, x_{a+1}] \\ (x - x_{a-1}/h^{e-1} \; ; \; x \in [x_{a-1}, x_a] \\ 0 \; ; \; otherwise \end{cases} \tag{2.61}$$

The last expression results with the shape function derivatives which are constant in each sub-domain

$$\frac{dN_a(x)}{dx} := \begin{cases} -1/h^e \; ; \; x \in [x_a, x_{a+1}] \\ 1/h^{e-1} \; ; \; x \in [x_{a-1}, x_a] \\ 0 \; ; \; otherwise \end{cases} \tag{2.62}$$

The shape functions of this kind are sufficiently regular with respect to the weak form requirement: $N_a \in H^1(\Omega)$. The discrete approximation for the displacement field can thus be constructed according to

$$u^h(x) = \sum_{a=1}^{n_{node}} N_a(x)\, u_a \tag{2.63}$$

where u_a are the displacement nodal values. In order to ensure the kinematic admissibility of the displacement field approximation in (2.63), the nodal displacements on Dirichlet boundary have to be equal to the imposed displacement

$$\forall a \in \Gamma_u \implies u_a = \bar{u} \tag{2.64}$$

The discrete approximation for virtual displacement field can also be constructed by Galerkin's method where we employ the same shape functions

$$w^h(x) = \sum_{a=1}^{n_{node}} N_a(x)\, w_a \; ; \; w^h \in \mathbb{V}_0^h \tag{2.65}$$

where we choose zero nodal values w_a on Dirichlet boundary

$$\forall a \in \Gamma_u \Longrightarrow w_a = 0 \tag{2.66}$$

The last result enforces the kinematic admissibility of the virtual displacement field in (2.65). In this manner, we reduce the task to computing only the nodal values of the real displacements for all free nodes $a = 1, 2, \ldots, n_{eqs}$ in the chosen finite element mesh. The computation of this kind is carried out with the weak form of the boundary value problem. First, we will consider linear elasticity case. By introducing into the weak form in $(W1)$ the chosen approximations (2.65) and (2.63) for real and virtual displacement fields, along with the corresponding strain approximations,

$$\begin{aligned} \frac{dw^h(x)}{dx} &= \sum_{a=1}^{n_{node}} \frac{dN_a(x)}{dx}\, w_a \\ \frac{du^h(x)}{dx} &= \sum_{a=1}^{n_{node}} \frac{dN_a(x)}{dx}\, u_a \end{aligned} \tag{2.67}$$

we can obtain so-called Galerkin equation:

$$D_{w^h} \Pi(u^h) = 0 \Leftrightarrow \sum_{a=1}^{n_{node}} w_a \left(\sum_{b=1}^{n_{node}} K_{ab} u_b - f_a^{ext} \right) = 0 \tag{G1}$$

In $(G1)$ above, K_{ab} are the components of the stiffness matrix, which can be written:

$$K_{ab} = \int_\Omega \frac{dN_a(x)}{dx}\, E(x)\, \frac{dN_b(x)}{dx}\, dx \tag{2.68}$$

whereas f_a^{ext} are the components of equivalent external nodal load, which is defined as

$$f_a^{ext} = \int_\Omega N_a(x)\, b(x)\, dx + [N_a(x)\bar{t}]_{\Gamma_\sigma} \tag{2.69}$$

The kinematic admissibility will reduce Galerkin equation in $(G1)$ to only n_{eqs} terms, by discarding the contribution of the nodes placed on Dirichlet boundary

$$\forall c \in [n_{eqs} + 1, n_{node}] \Rightarrow w_c|_{\Gamma, u} = 0 \tag{2.70}$$

By considering furthermore that the virtual displacements of free nodes are arbitrary, the Galerkin equation $(G1)$ gives rise to a set of linear algebraic equations with nodal values of real displacement field as unknowns

$$\sum_{b=1}^{n_{eqs}} K_{ab} u_b = f_a^{ext} - \sum_{c=n_{eqs}+1}^{n_{node}} K_{ac} \bar{u}_c \; ; \; \forall a = 1, 2, \ldots, n_{eqs} \tag{2.71}$$

This set of equations can be written in matrix notation

$$\mathsf{K}\,\mathsf{u} = \mathsf{f} \tag{2.72}$$

where K is the stiffness matrix, u is the displacement vector and f is the equivalent nodal load vector

$$\mathsf{K} = \begin{bmatrix} K_{11} & K_{12} & \ldots & K_{1n_{eqs}} \\ K_{21} & K_{22} & \ldots & K_{2n_{eqs}} \\ \vdots & \vdots & & \vdots \\ K_{n_{eqs}1} & K_{n_{eqs}2} & \cdots & K_{n_{eqs}n_{eqs}} \end{bmatrix} ; \; \mathsf{u} = \begin{bmatrix} u_1 \\ u_2 \\ \vdots \\ u_{n_{eqs}} \end{bmatrix} ; \; \mathsf{f} = \begin{bmatrix} f_1 \\ f_2 \\ \vdots \\ f_{n_{eqs}} \end{bmatrix} ;$$

$$f_a = f_a^{ext} - \textstyle\sum_{c=n_{eqs}+1}^{n_{node}} K_{ac} \bar{u}_c \tag{2.73}$$

In conclusion, we have shown that the final product of the proposed solution procedure for a boundary value problem in 1D linear elasticity is a set of linear algebraic equations. It is important to note that all the different phases of the presented solution procedure carry over to a more complex case than 1D linear elasticity, and furthermore the same kind of the final product in terms of a system of linear algebraic equations will be found throughout the book. For that reason, we will next dedicate a complete section to an important task of solving a set of linear algebraic equations, by using the method of Gauss elimination.

With the nodal displacement values in hand, we can easily complete the computations by providing through (2.63) the displacement field discrete approximation throughout the domain ($\forall x \in \Omega$). The strain field approximation can also be obtained by making use of the result for shape function derivatives in (2.62). The latter provides the constant value of strain in each element

$$\epsilon^h|_{\Omega^e} = (u_{e+1} - u_e)/h^e \tag{2.74}$$

The corresponding approximate solution for the stress field in linear elasticity is obtained by multiplying these strains by Young's modulus

$$\sigma^h|_{\Omega^e} = E(u_{e+1} - u_e)/h^e \tag{2.75}$$

In closing this section, we briefly present the final result of applying the proposed finite element solution procedure to a boundary value problem in 1D nonlinear elasticity. By introducing the displacement field and strain field approximations in (2.65), (2.63) and (2.67) into the weak form of the problem defined in ($W2$), we can obtain again the Galerkin equation

$$G(u^h; w^h) = 0 \Leftrightarrow \sum_{a=1}^{n_{node}} w_a \left(f_a^{int} - f_a^{ext} \right) = 0 \tag{G2}$$

where f_a^{ext} and f_a^{int} are components of external and internal force vector, respectively. The former remains the same as already presented in (2.69) for the linear case, whereas the latter can be written:

$$f_a^{int} = \int_\Omega \frac{dN_a(x)}{dx} \hat{\sigma} \left(\underbrace{\sum_{b=1}^{n_{node}} \frac{dN_b(x)}{dx} u_b}_{du^h/dx} \right) dx \tag{2.76}$$

By considering that virtual displacement nodal values w_a on boundary Γ_u ought to be equal to zero, and that the remaining nodal values can be chosen arbitrarily, we can obtain from Galerkin equation ($G2$) a set of nonlinear algebraic equations with nodal displacements as unknowns

$$\hat{f}_a^{int}(u_b) - f_a^{ext} = 0 \; ; \; a, b = 1, 2, \ldots, n_{eqs} \tag{2.77}$$

We can write the same equations in matrix notation

$$\hat{\mathsf{f}}^{int}(\mathsf{u}) = \mathsf{f}^{ext} \tag{2.78}$$

where f^{int} is the internal force vector

$$\hat{\mathsf{f}}^{int}(\mathsf{u}) = \begin{bmatrix} \hat{f}_1^{int}(\mathsf{u}) \\ \vdots \\ \hat{f}_{n_{eqs}}^{int}(\mathsf{u}) \end{bmatrix} \tag{2.79}$$

We can therefore conclude that the final product of the proposed solution procedure for a boundary value problem in 1D nonlinear elasticity is a set of nonlinear algebraic equations. In a subsequent section, we will discuss two ways for computing the solution to this set, with either incremental analysis or Newton's iterative method. We note herein that any such method reduces the main computational task to solving repeatedly a set of linear algebraic equations. Apart this nodal displacements computation, the nonlinear elasticity is not much more demanding than the linear case. Namely, the strain field approximation is again constant in each element, leading to the corresponding element-wise constant stress values

$$\epsilon^h|_{\Omega^e} = (u_{e+1} - u_e)/h^e \;\Rightarrow\; \sigma^h|_{\Omega^e} = \hat{\sigma}(\epsilon^h|_{\Omega^e}) \tag{2.80}$$

2.2.3 Approximation error and convergence of finite element method

2.2.3.1 'Best approximation' property of finite element method

A very important issue which is addressed in this section pertains to the quality of the finite element approximation. The latter can be estimated by the approximation error that is defined as the difference between the finite element solution and exact solutions, '$u^h - u$'. We further show two remarkable features of the finite element solution: i) the approximation error is equal to zero if measured in the norm introduced by the bilinear form $G^{int}(\cdot,\cdot)$, ii) the finite element method gives the best possible approximation in the energy norm with respect to any other approximation method. The last result is referred to as the best approximation property (see Hughes [111], p. 185, Johnson [144], p. 38; Strang and Fix [249]).

It is not difficult to prove the first result. We recall that the exact solution of a boundary value problem $u(x)$ and its finite element approximation $u^h(x)$, both satisfy the weak form of linear elasticity for any kinematically admissible weighting functions $w^h(x) \in \mathbb{V}_0^h$. By linearity of the internal and external energy functionals, we can directly obtain that the difference between these two solutions, or approximation error, will thus satisfy

$$\left.\begin{array}{l} G^{int}(u^h; w^h) = G^{ext}(w^h) \\ G^{int}(u; w^h) = G^{ext}(w^h) \end{array}\right\} \implies G^{int}(u^h - u; w^h) = 0 \tag{2.81}$$

We thus conclude that approximation error $e(x) = u^h(x) - u(x)$ remains orthogonal to each weighting function $w^h \in V_0^h$ when paired together in the strain energy functional, or that there is no error in the energy norm.

The best approximation property can than be proved by contradiction. Let us suppose that there exist another approximate solution, constructed by a different approximation method. Any such solution can be expressed as a kinematically admissible correction of the solution furnished by the finite element method, $u_\varepsilon^h = u^h + \varepsilon w^h$. The energy norm of the approximation error for such a solution can then be expressed

$$\begin{aligned} G^{int}(u_\varepsilon^h - u; u_\varepsilon^h - u) = G^{int}(u^h - u; u^h - u) + 2\varepsilon \underbrace{G^{int}(u^h - u; w^h)}_{=0} \\ + \varepsilon^2 \underbrace{G^{int}(w^h; w^h)}_{>0} > G^{int}(u^h - u; u^h - u) \end{aligned} \tag{2.82}$$

where we have exploited the result in (2.81) and the fact that the strain energy functional in (2.44) is bilinear and positive definite. The last result allows us to conclude that the energy norm of any different solution is superior to the one for the solution produced by the finite element method. In other words, the finite element method provides the smallest error, or the best approximation in energy norm .

2.2.3.2 Lemma of Céa and approximation error

By exploiting the best approximation property of the finite element method, we can further show the convergence of the finite-element-based discrete approximation towards the exact solution. We first recall that the energy norm of the approximation error can be bounded by using the continuity and coercivity of the strain energy functional

$$\begin{aligned} c_{min} \parallel u^h - u \parallel_{H^1}^2 \leq G^{int}(u^h - u; u^h - u) \\ G^{int}(u^h - u; u_\varepsilon^h - u) \leq c_{max} \parallel u^h - u \parallel_{H^1} \parallel u_\varepsilon^h - u \parallel_{H^1} \end{aligned} \tag{2.83}$$

The last result and the best approximation property of the finite element method will thus allow us to draw the following conclusion

$$\parallel u^h - u \parallel_{H^1} \leq \sqrt{\frac{c_{max}}{c_{min}}} \parallel u_\varepsilon^h - u \parallel_{H^1} \tag{2.84}$$

This result, know as the lemma of Céa (see [49]), shows that the approximation error is completely controlled by the chosen approximation space $\mathbb{V}_0^h$, or the quality of the shape functions. For example, the latter that can represent exactly any linear polynomial, will produce the approximation errors in displacement and in strain of the order h^2 and h, respectively

$$\left. \begin{aligned} \parallel u^h - u \parallel_{L_2} \leq c\, h^2 \parallel u \parallel_{H^2} \\ \parallel \frac{du^h}{dx} - \frac{du}{dx} \parallel_{L_2} \leq c\, h \parallel u \parallel_{H^2} \end{aligned} \right. \Longrightarrow \parallel u^h - u \parallel_{H^1} \leq c\, h \;\equiv\; h \parallel u \parallel_{H^2} \tag{2.85}$$

where c is a constant independent of the element size h. In other words, as long as the term $\parallel u \parallel_{H^2}$ proportional to the second derivative remains bounded, the convergence towards the exact solution is enforced by simply decreasing the element size. The same finding is valid for higher order approximations, where we choose the polynomials of order '$k \geq 1$' as the basis of the approximation space $\mathbb{V}_0^h$ to show that

$$\parallel u^h - u \parallel_{H^m} \leq c\, h^r \parallel u \parallel_{H^p} \; ; \text{ with: } r = \min(k + 1 - m, p - m) \tag{2.86}$$

In conclusion, one way to ensure the convergence of the finite element method and approach the exact solution, is by increasing the number of elements in the mesh and thus decreasing each element size. An alternative way with fixed element size, which will also be explored subsequently, is by increasing the order of polynomials in the chosen approximation space by choosing the elements with a larger number of nodes.

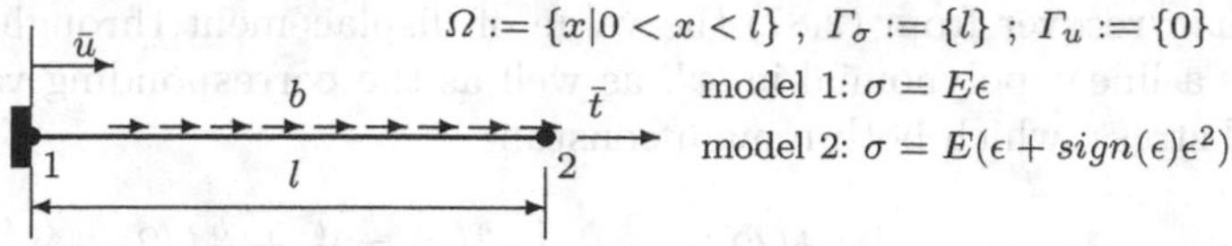

Fig. 2.3 Simple example of a bar model constructed with a single finite element with two nodes.

2.2.3.3 Superconvergence of finite element method for 1D linear elasticity

When solving the boundary value problems in 1D linear elasticity, the finite element method is capable of furnishing even better results than those stated by the bounds. in fact, in this case the finite element method delivers the exact nodal values of displacement. This property is referred to as superconvergence (Hughes [111], p. 188, Zienkiewicz and Taylor [270]).

We further illustrate the superconvergence property along with all the solution phases in the simplest possible context, where the displacement field of a linear elastic bar is represented by a single finite element with two nodes (See Figure 2.3). The bar is of a unit cross-section, with length equal to l and Young's modulus E. We apply a uniformly distributed loading b and a traction force $\bar{t}$ at the right end of the bar, i.e. $\Gamma_\sigma := \{l\}$. The left end of the bar is clamped, $\Gamma_u := \{0\}$, with an imposed displacement value $\bar{u}$.

Kinematically admissible approximations for real and virtual displacement fields are constructed with a single 2-node finite element according to:

$$u^h(0) \equiv u_1 = \bar{u} \Longrightarrow u^h(x) = \bar{u}(1 - \frac{x}{l}) + u_2\frac{x}{l} \; ; \; w^h(x) = w_2\frac{x}{l} \tag{2.87}$$

where u_2 and w_2 are, respectively, the (unknown) real and virtual displacement values at the free node. By introducing these approximations into the weak form in $(W1)$ we obtain

$$\begin{aligned} 0 &= w_2\{\textstyle\int_0^l \frac{d}{dx}(\frac{x}{l})E[\frac{d}{dx}(1-\frac{x}{l})\bar{u} + \frac{d}{dx}(\frac{x}{l})u_2]\,dx - \int_0^l \frac{x}{l}b\,dx - \frac{x}{l}|_x\bar{t}\} \\ &= w_2\{K_{22}u_2 - \tilde{f}_2\} \end{aligned} \tag{2.88}$$

where

$$\begin{aligned} K_{22} &:= \textstyle\int_0^l \frac{d}{dx}(\frac{x}{l})E\frac{d}{dx}(\frac{x}{l})\,dx = \frac{E}{l} \\ \tilde{f}_2 &:= \textstyle\int_0^l \frac{x}{l}b\,dx + \frac{x}{l}|_l\bar{t} - \int_0^l \frac{d}{dx}(\frac{x}{l})E\frac{d}{dx}(1-\frac{x}{l})\,dx\,\bar{u} = \frac{bl}{2} + \bar{t} + \frac{E}{l}\bar{u} \end{aligned} \tag{2.89}$$

For an arbitrary value of virtual displacement w_2, we can obtain from (2.88) a linear algebraic equation with u_2 as the unknown, whose solution can be written:

$$u_2 = \frac{\tilde{f}_2}{K_{22}} = \bar{u} + \frac{\bar{t}l}{E} + \frac{bl^2}{2E} \tag{2.90}$$

We can than recover from (2.87) the value of displacement throughout the domain as a linear polynomial in 'x', as well as the corresponding values of strain and stress, which both remain constant

$$u^h(x) = \bar{u} + \frac{\bar{t} + bl/2}{E} x \longrightarrow \begin{cases} \epsilon^h(x) = \frac{\bar{t}}{E} + \frac{b}{E} l/2 \\ \sigma^h = \bar{t} + bl/2 \end{cases} \tag{2.91}$$

We can also obtain the exact solution of this boundary value problem by integrating with respect to x the strong form in ($S1$), and by using the boundary conditions to eliminate the constant of integration, which results with

$$u(x) = \bar{u} + \frac{tx}{E} + \frac{b(lx - x^2/2)}{E} \tag{2.92}$$

This exact solution for displacement field, which verifies the equilibrium equation locally $\forall x \in \Omega$, is a quadratic polynomial in 'x'. Hence, the latter can not be everywhere the same as the linear polynomial in (2.91) computed as the approximate solution by the finite element method. However, remarkably enough, the finite element approximation of displacement field provides the values that are exact at the nodes

$$\begin{cases} u(0) = \bar{u} \equiv u_1 \\ u(l) = \bar{u} + \bar{t}l/E + bl^2/2E \equiv u_2 \end{cases} \tag{2.93}$$

The finite element nodes are yet referred as the point of superconvergence (see Zienkiewicz and Taylor [270]). This conclusion remains valid for 1D elasticity regardless of the number of elements or nodes we employ in constructing the finite element mesh.

It is interesting to note that the remarkable property of superconvergence for 1D linear elasticity problems is the consequence of the finite element capability to construct the exact solution to the dual problem which can be set to estimate the strain field accuracy. The dual problem of this kind can be formulated as a constrained minimization of the linear displacement functional $Q(u) = \int_0^l \frac{du(x)}{dx}\, dx$, by choosing among the kinematically admissible candidates which satisfy the weak form of equilibrium equations; the corresponding non-constrained minimization problem can then be written:

$$\max_p \min_u L(u,p)\, ;\; L(u,p) = [Q(u) + G^{ext}(p) - G^{int}(u;p)]\, ;\; Q(u) = \int_0^l \frac{du(x)}{dx}\, dx \tag{2.94}$$

The optimality conditions for this problem can be obtained by computing the directional derivatives in the direction of variations $w(x)$ and $q(x)$, respectively

$$
\begin{aligned}
0 = D_w L(u,p) &:= Q(w) - G^{int}(w,p) = \int_0^l \frac{dw(x)}{dx}\, dx - \int_0^l \frac{dw(x)}{dx} E \frac{dp(x)}{dx}\, dx \\
0 = D_q L(u,p) &:= G^{ext}(q) - G^{int}(u;q) = \int_0^l q(x) b(x)\, dx + q(l)\bar{t} \\
&\quad - \int_0^l \frac{dq(x)}{dx} E \frac{du(x)}{dx}\, dx
\end{aligned}
\tag{2.95}
$$

The first of two equations can be written in local form defining the dual function $p(x)$ as the displacement field due to a unit force applied at the free end of the bar

$$
E\frac{dp(x)}{dx} = 1 \Rightarrow p(x) = \frac{1}{E}x \tag{2.96}
$$

Note that such a dual function $p(x)$ can be exactly represented by the chosen 2-node element approximation with linear polynomials, $p(x) \equiv p(x)^h \in \mathbb{V}_0^h$. On the other hand, the chosen finite element model provides only an approximate solution of the displacement field $u^h(x)$ that can not match the exact solution $u(x)$ everywhere. However, when evaluated in terms of the chosen functional $Q(u)$, the difference between exact and approximate solutions, $u(x)$ and $u^h(x)$, disappears due to the best approximation property of the finite element method

$$
\begin{aligned}
Q(u) - Q(u^h) &= G^{int}(u;p) - G^{int}(u^h;p) \\
&= G^{int}(u - u^h;p) = 0
\end{aligned}
\tag{2.97}
$$

For a particular choice of linear form of $Q(\cdot)$, the last result implies that the finite element will provide the exact nodal values of the displacement field

$$
\int_0^l \frac{du(x)}{dx}\, dx = \int_0^l \frac{du^h(x)}{dx}\, dx \Longrightarrow \begin{cases} u(l) = u^h(l) \\ u(0) = u^h(0) \end{cases} \tag{2.98}
$$

In a very similar manner, one can show that the values of the exact solution for strain and stress fields, which are linear polynomials in x, are matched by the finite element approximation at the superconvergence points for strain and stress. The latter, unlike the displacement field superconvergence points, are placed in the center of the element

$$
\left.\begin{aligned}
\epsilon(x) &= \tfrac{\bar{t}}{E} + \tfrac{b}{E}(l - x) \\
\sigma(x) &= \bar{t} + b(l - x)
\end{aligned}\right\} \Longrightarrow \begin{cases} \epsilon(l/2) = \frac{\bar{t}}{E} + \frac{b}{E}l/2 \equiv \epsilon^h(l/2) \\ \sigma(l/2) = \bar{t} + bl/2 \equiv \sigma^h(l/2) \end{cases} \tag{2.99}
$$

In 2D or 3D problems, the points of superconvergence no longer provide the exact solution values. However, the corresponding values are still of higher order of accuracy. In that context, the superconvergence points are yet called Barlow points (see [18]).

2.2.4 Solving a system of linear algebraic equations by Gauss elimination method

The final product of the finite element discretization procedure for a boundary value problem in linear elasticity is a system of linear algebraic equations, with the nodal values of displacements as unknowns. Solving one such system is typically the most costly phase of the total solution procedure, and it is important to find the optimal manner to perform this computation. The same task is of direct interest for practically all the problems treated subsequently, be they nonlinear, non-stationary, coupled multi-physics or multi-scale, since the corresponding solution procedures using incremental analysis, Newton's iterations or a time-integration scheme will finally reduce all these problems to a set of linear algebraic equations.
The solution of a system of linear algebraic equations can be carried out by either iterative or direct methods (see [91]). We discuss the direct methods only, since they are perhaps the only ones to be sufficiently robust in nonlinear problems where the behavior may pass through a number of different regimes, including instabilities and strain localization. Moreover, despite the drawback of the direct methods pertaining to fairly large storage requirements,[9] they offer an important advantage of having the number of operations known in advance.
We present in this section one of the most frequently used direct methods, the method of Gauss elimination. A system of 'n' linear algebraic equations, $\mathsf{K}\mathsf{u} = \mathsf{f}$, is solved by Gauss method by passing through three different solution phases. In the first phase, called triangular decomposition, we replace the stiffness matrix K by a product of two matrices, the first one lower triangular denoted L and the second one upper triangular matrix U. This result can be written explicitly as

$$\mathsf{K} = \mathsf{L}\mathsf{U} \; ; \; \mathsf{L} = \begin{bmatrix} 1 & 0 & \dots & 0 \\ L_{21} & 1 & \dots & 0 \\ \vdots & \vdots & & \vdots \\ L_{n1} & L_{n2} & \cdots & 1 \end{bmatrix} \; ; \; \mathsf{U} = \begin{bmatrix} U_{11} & U_{12} & \dots & U_{1n} \\ 0 & U_{22} & \dots & U_{2n} \\ \vdots & \vdots & & \vdots \\ 0 & 0 & \cdots & U_{nn} \end{bmatrix} \tag{2.100}$$

With this new form of the system matrix, the solution can be obtained easily in two successive phases, which are referred to as forward reduction and back substitution

$$\mathsf{L}\underbrace{\mathsf{U}\mathsf{u}}_{\mathsf{y}} = \mathsf{f} \implies \begin{cases} \mathsf{L}\mathsf{y} = \mathsf{f} \text{ forward reduction phase} \\ \mathsf{U}\mathsf{u} = \mathsf{y} \text{ back substitution phase} \end{cases} \tag{2.101}$$

---

[9] This kind of problem can easily be tackled nowadays by available computational resources, or by using domain decomposition modelling approach as the one presented in the last chapter.

In the forward reduction phase, with the given matrix L and the load vector f, we compute the unknown vector y. By starting this solution procedure from the top, for the given lower triangular form of matrix L we find in the first equation only one unknown component y_1. Having computed this component, we proceed to the second equation where the only unknown will now be y_2, solve for it, and carry on our computations until we reach the last equation which will provide the value of the last unknown component of vector y

$$\mathsf{L}\,\mathsf{y} = \mathsf{f} \;\Leftrightarrow\; \left\{ \begin{aligned} 1\,y_1 = f_1 \;&\Rightarrow\; y_1 = f_1 \\ L_{21}\,y_1 + 1\,y_2 = f_2 \;&\Rightarrow\; y_2 = f_2 - L_{21}y_1 \\ &\ldots\;\ldots \\ L_{n1}\,y_1 + L_{n2}\,y_2 + \ldots + 1\,y_n = f_n \;&\Rightarrow\; y_n = f_n - L_{n1}y_1 \\ &\qquad -L_{n2}y_2 - \ldots - L_{n,n-1}y_{n-1} \end{aligned} \right. \tag{2.102}$$

In the subsequent phase of back substitution, with the given matrix U and computed vector $\mathbf{y}$, we start from the last equation and go backward until we return to the first equation. The special structure of matrix U results with only one unknown displacement component u_n, whose computed value can be used to reduce the number of unknowns to one in the subsequent equation and the same procedure is repeated from one equation to another. As long as the solution is obtained sequentially, we always find a problem with one unknown at the time, which is very easy to solve:

$$\mathsf{U}\,\mathsf{u} = \mathsf{y} \;\Leftrightarrow\; \left\{ \begin{aligned} U_{nn}\,u_n = y_n \;&\Rightarrow\; u_n = y_n/U_{nn} \\ U_{n-1,n-1}\,u_{n-1} + U_{n-1,n}\,u_n = y_{n-1} \;&\Rightarrow\; u_{n-1} = (y_{n-1} \\ &\qquad -U_{n-1,n}u_n)/U_{n-1,n-1} \\ &\ldots\ldots \\ U_{11}\,u_1 + U_{12}\,u_2 + \ldots + U_{1n}\,u_n = y_1 \;&\Rightarrow\; u_1 = (y_1 - U_{12}u_2 \\ &\qquad - \ldots - U_{1n}u_n)/U_{11} \end{aligned} \right. \tag{2.103}$$

We have therefore shown how simple it is to obtain the solution procedure once the stiffness matrix is written in terms of its triangular decomposition in (2.100). In order to provide an efficient manner to perform the triangular decomposition, we exploit two interesting properties of matrix L (see [20]). First we note that matrix L can be written in terms of product of elementary matrices L_i, which will contain a copy of *i-th* column of the matrix L,

$$\begin{aligned} &\mathsf{L} = (\mathsf{I} + \mathsf{L}_1)(\mathsf{I} + \mathsf{L}_2)\ldots(\mathsf{I} + \mathsf{L}_n) \equiv \Pi_{i=1}^{n}(\mathsf{I} + \mathsf{L}_i)\;;\; \mathsf{I} = diag(1,1,\ldots,1)\;; \\ &\mathsf{L}_i = [0, 0, \ldots, \underbrace{\mathsf{l}_i}_{i^{th}\,col.}, \ldots, 0]\;;\; \mathsf{l}_i^T = [\underbrace{0,\ldots,0}_{i\;components}, L_{i+1,i}, \ldots, L_{ni}] \end{aligned} \tag{2.104}$$

Second we can show that the inverse of each elementary matrix L_i can easily be computed, by simply changing the sign of its components

$$(\mathsf{I}+\mathsf{L}_i)^{-1}=\mathsf{I}-\mathsf{L}_i := \begin{bmatrix} 1 & 0 & \cdots & 0 & \cdots & 0 \\ 0 & 1 & \cdots & 0 & \cdots & 0 \\ \vdots & \vdots & & \vdots & & \vdots \\ & & & 1 & & \\ \vdots & \vdots & & -L_{i+1,i} & & \\ \vdots & \vdots & & \vdots & & \vdots \\ 0 & 0 & \cdots & -L_{ni} & \cdots & 1 \end{bmatrix} \tag{2.105}$$

The last two results allow us to write a closed form expression for the inverse of lower triangular matrix L

$$\mathsf{L}^{-1} = (\mathsf{I}-\mathsf{L}_n)\dots(\mathsf{I}-\mathsf{L}_2)(\mathsf{I}-\mathsf{L}_1) \equiv \Pi^{1}_{i=n}(\mathsf{I}-\mathsf{L}_i) \tag{2.106}$$

With this result in hand, we can compute the triangular decomposition of matrix $\mathsf{K}=\mathsf{LU}$ by modifying one row and column at the time. For start, the first row of U and the first column of L are computed, respectively, from the first row and column of matrix K according to

$$[K_{11}, K_{12}, \dots, K_{1n}] = 1\,[U_{11}, U_{12}, \dots, U_{1n}] \Longrightarrow U_{1j} = K_{1j}\,;\; j=1,2,\dots,n$$

$$\begin{bmatrix} K_{11} \\ K_{21} \\ \vdots \\ K_{n1} \end{bmatrix} = U_{11} \begin{bmatrix} 1 \\ L_{21} \\ \vdots \\ L_{n1} \end{bmatrix} \Longrightarrow L_{i1} = K_{i1}/K_{11}\,;\; i=2,3,\dots,n \tag{2.107}$$

Before proceeding to computation of the second row of U and the second column of L, we first obtain the condensed form of matrix K, which is denoted by $\mathsf{K}^{(2)}$, accounting for the modifications of K produced by the first step of triangular decomposition

$$\mathsf{K}^{(2)} = (\mathsf{I}-\mathsf{L}_1)\mathsf{K} = \mathsf{K} - \mathsf{l}_1\mathsf{u}_1^T\,;\; \mathsf{l}_1 = \frac{1}{K_{11}} \begin{bmatrix} 0 \\ K_{21} \\ \vdots \\ K_{n1} \end{bmatrix}\,;\; \mathsf{u}_1 = \begin{bmatrix} K_{11} \\ K_{12} \\ \vdots \\ K_{1n} \end{bmatrix} \tag{2.108}$$

The second row of U and the second column of L can be computed from the corresponding row and column of the condensed matrix $\mathsf{K}^{(2)}$, in precisely the same manner as the first row and column were computed from the original matrix K in (2.107). We can carry on in this manner until the last row and column of K are reached. At each step, a rank-one modification of the previous condensed form of matrix K is used, equivalent to the one in (2.108). This observation allows us to quantify the computational cost of the triangular decomposition in terms of number of operations, which is given in Table 2.1.

In comparison with the computational cost of forward reduction and back substitution (also given in Table 2.1), the cost of triangular decomposition remains largely dominant, proportional to $\frac{2n^3}{3}$ for a matrix K of rank 'n'. If the matrix K is symmetric, $\mathsf{K}^T = \mathsf{K}$, we can cut the number of operations by half, and write the result of triangular decomposition for such a matrix according to:

$$\mathsf{K} = \mathsf{L}\,\mathsf{D}\,\mathsf{L}^T \; ; \; \mathsf{D} = diag(D_1, D_2, \ldots, D_n) \tag{2.109}$$

The computational cost of the triangular decomposition can be reduced even further for a matrix with sparse structure, i.e. the matrix that has a small number of non-zero components which are all grouped around the main diagonal. Namely, we can see from a characteristic step of triangular decomposition in (2.107) that any zero-component in the first row of K will have no influence on condensed matrix $\mathsf{K}^{(2)}$, or any other of subsequent rank-one modifications of the original matrix. Hence, we do not have any reason to store such a component nor use it for further computations. One of the most efficient storage schemes for a spare structure matrix with respect to the Gauss elimination method is so-called 'sky-line' (e.g. see Bathe and Wilson [20], where the columns are stored sequentially in an array, with each column represented in a truncated form starting with the first non-zero component. The number of operation for such a storage scheme, which is proportional to the sum of the squares of number of terms in each column, becomes much smaller than the one for full storage ($\sum_i l_i^2 \ll n^3/3$), especially if the node numbering is optimal.[10] In order to ensure a high computational efficiency, one can process the components of matrix K in the order which accounts for their storage (for example, column-wise) and minimize the data communication overhead. A couple of variants of Gauss elimination method are used for that purpose, such as the methods of Crout or of Cholesky (e.g. see Ciarlet [51], p. 87 or Hughes [111], p. 636).

For a system of linear algebraic equations that can not be solved by a direct method, even with eventual benefits of sparse matrix solvers (see [70], one should turn to iterative methods, such as the methods of Jacobi, of Gauss-Seidle or the conjugate gradient method[11] (e.g. see [20 or 51]). In any iterative method for solving a set of linear algebraic equations, we try to replace the original matrix K by a more convenient form (a diagonal form in Jacobi method, for example) which reduces the storage requirements and renders the solution very easy to find. We thus obtain an iterative solution $\mathsf{u}^{(i)}$, whose accuracy can be checked by the convergence test on a given residual $\mathsf{f} - \mathsf{K}\mathsf{u}^{(i)}$ and eventually be further improved by subsequent iterations.

[10] Normally one should keep the smallest possible difference in node numbers attached to each element; This can be achieved by starting the node numbering from the nodes of periphery, and advancing further into the domain. A number of schemes to accomplish this goal is proposed, most notably the one by Cuthill-McKee [70].

[11] Strictly speaking, the conjugate gradient method is not an iterative method, since it can always converge in 'n' steps; it becomes iterative only if we want to converge faster, with fewer steps.

However, the iterative solvers are in general much less robust than the direct solvers, and a number of modifications are needed to improve their performance (see Ferencz and Hughes [77], Kelley [148]).

Table 2.1 Number of operations for Gauss elimination method used for solving a system of n linear algebraic equations.

Triangular decomposition $\mathsf{K} = \mathsf{L}\,\mathsf{U}$		
Additions	$(n-1)^2 + (n-2)^2 + \ldots + 1^2$	$= \frac{n^3-n}{3}$
Multiplications	$(n-1)^2 + (n-2)^2 + \ldots + 1^2$	$= \frac{n^3-n}{3}$
Divisions	$(n-1) + (n-2) + \ldots + 1$	$= \frac{n(n-1)}{2}$
Forward reduction $\mathsf{L}\mathsf{y} = \mathsf{f}^{ext}$		
Additions	$(n-1) + (n-2) + \ldots + 1$	$= \frac{n(n-1)}{2}$
Multiplications	$(n-1) + (n-2) + \ldots + 1$	$= \frac{n(n-1)}{2}$
Back substitution $\mathsf{U}\,\mathsf{u} = \mathsf{y}$		
Additions	$(n-1) + (n-2) + \ldots + 1$	$= \frac{n(n-1)}{2}$
Multiplications	$(n-1) + (n-2) + \ldots + 1$	$= \frac{n(n-1)}{2}$
Divisions	$1 + 1 + \ldots + 1$	n

2.2.4.1 Condition number of a matrix

The condition number allows us to quantify the sensitivity of the direct or iterative solution procedure for a set of linear algebraic equations, and account for possible solution errors resulting from perturbation on data (for example, from truncation errors) for either the system matrix K or the force vector f. For a given matrix K, the condition number is defined as the ratio of its largest and smallest eigenvalue

$$cond(\mathsf{K}) = \frac{\lambda_{max}}{\lambda_{min}} \; ; \left\{ \begin{array}{c} \mathsf{K}\mathsf{v}_i = \lambda_i \mathsf{v}_i \; ; \; i = 1, 2, \ldots, n \\ \lambda_{max} = \max_{1 \le i \le n} \lambda_i \; ; \; \lambda_{min} = \min_{1 \le i \le n} \lambda_i \end{array} \right. \tag{2.110}$$

The eigenvalues for the stiffness matrix K are always positive (e.g. see [219]), if the matrix is representative of a well-posed boundary value problem or the system where the chosen boundary conditions are sufficient to eliminate the rigid body motion. For such a system, the smallest and the largest eigenvalues correspond to the effective stiffness values for the deformation modes that are the easiest and the most difficult to enforce, respectively. The condition number is therefore always positive. One can show (see Ciarlet [51]) that the condition number of the system matrix K expresses the error bounds for the solution with respect to either loading vector perturbation $\delta\mathsf{f}$ or the perturbation of matrix components $\Delta\mathsf{K}$, which are denoted by $\delta\mathsf{u}$ and $\Delta\mathsf{u}$, respectively. One can write:

$$\frac{\| \delta\mathsf{u} \|}{\| \mathsf{u} \|} \le \underbrace{\| \mathsf{K} \| \| \mathsf{K}^{-1} \|}_{cond(\mathsf{K})} \frac{\| \delta\mathsf{f} \|}{\| \mathsf{f} \|} \tag{2.111}$$

$$\frac{\| \Delta \mathsf{u} \|}{\| \mathsf{u} + \Delta \mathsf{u} \|} \leq \underbrace{\| \mathsf{K} \| \| \mathsf{K}^{-1} \|}_{cond(\mathsf{K})} \frac{\| \Delta \mathsf{K} \|}{\| \mathsf{K} \|} \tag{2.112}$$

The condition number shows that any problem for which $\lambda_{max} \mapsto \infty$ or $\lambda_{min} \mapsto 0$ will be poorly conditioned, with the accuracy of computed result significantly affected by any small error in input data. The systems of equations with $\lambda_{max} \gg \lambda_{min}$ is called "stiff". We note in passing that a number of problems we consider in this book belong to this category. The condition number is equally important for ensuring the robust convergence of iterative methods. Several methods for pre-conditioning (trying to bring the condition number as close as possible to 1) have been proposed for iterative methods (e.g. see [77]).

2.2.5 Solving a system of nonlinear algebraic equations by incremental analysis

By using the finite element method for solving a boundary value problem in 1D nonlinear elasticity, we obtain a system of nonlinear algebraic equations with nodal displacement values as unknowns. We find the same kind of system to solve for practically all other boundary value problems in nonlinear solid mechanics presented in this book, and it is thus very important to explain how to carry on to obtain the solution. In general, it is not possible to obtain a closed form solution to a set of nonlinear algebraic equations, except for the simplest case of a single nonlinear equation. For that reason, we seek an approximate solution to a system of nonlinear algebraic equations by using the incremental analysis of a nonlinear boundary value problem. The principal idea of incremental analysis is very simple: we imagine that the total external load is applied in terms of a number of increments, which are all sufficiently small to permit that the nonlinear problem under consideration be approximated by an equivalent linear problem. The latter will thus replace (in each increment) a set of nonlinear algebraic equations with the linear equations, which can then be solved by Gauss elimination. This kind of approximate solution to a nonlinear problem is constructed through a solution sequence of the equivalent linear problems, and will thus remain linear in each increment.
A simple way to handle the incremental analysis is by introducing so-called pseudo-time parameter[12] denoted as 't', which is used to describe a particular loading program. The set of nonlinear algebraic equations parameterized by pseudo-time can then be written:

$$\hat{\mathsf{f}}^{int}(\hat{\mathsf{d}}(t)) - \hat{\mathsf{f}}^{ext}(t) \; ; \; t \in [0, T] \tag{2.113}$$

[12] The label pseudo-time for parameter 't' implies that it plays the same role as time in describing the loading sequence, except that the problem remains placed within the framework of statics where inertia effects are neglected.

The choice of load increments in a given loading program is then handled through the increments of pseudo-time according to

$$[0, T] = \bigcup_{n=1}^{n_{inc}} [t_n, t_{n+1}] \tag{2.114}$$

where $\mathsf{f}^{ext}(T)$ is the final value of external loading to be applied. In a typical step of the incremental analysis between t_n and t_{n+1}, we can obtain the incremental displacement, denoted by[13] u_{n+1}, as opposed to total displacement, which is denoted by d_{n+1},

$$\mathsf{u}_{n+1} = \mathsf{d}_{n+1} - \mathsf{d}_n \ \ ; \ \mathsf{d}_{n+1} = \mathsf{d}(t_{n+1}) \ ; \ \mathsf{d}_n = \mathsf{d}(t_n) \tag{2.115}$$

The displacement increment is produced by the corresponding load increment, which is denoted by

$$\mathsf{r}_{n+1} = \mathsf{f}_{n+1} - \mathsf{f}_n \ \ ; \ \mathsf{f}_{n+1} = \mathsf{f}(t_{n+1}) \ ; \ \mathsf{f}_n = \mathsf{f}(t_n) \tag{2.116}$$

In the case of proportional loading, which implies that all the external load components are increasing in the same manner, one can write the external load vector as the product of a fixed vector f_0 and a scalar function of pseudo-time $g(t)$. The load increment can then be defined as

$$\mathsf{f}_{n+1} = \mathsf{f}_0 g(t_{n+1}) \ ; \ \mathsf{f}_n = \mathsf{f}_0 g(t_n) \implies \mathsf{r}_{n+1} = \mathsf{f}_0[g(t_{n+1}) - g(t_n)] \tag{2.117}$$

The main goal of the incremental analysis is provide the value of the displacement increment u_{n+1} corresponding to the chosen load increment r_{n+1}. This is accomplished by constructing and solving the equivalent linear problem in each load step. We note that once the incremental displacement is computed, the total displacement can readily be obtained by updating the previous value in time with $\mathsf{d}_{n+1} = \mathsf{d}_n + \mathsf{u}_{n+1}$. For a given nonlinear problem, we can construct the equivalent linear representation by using the consistent linearization procedure (e.g. see Marsden and Hughes [182]). For example, the consistent linearization of a scalar function $g(x)$ can be constructed by keeping in the Taylor series formula only the first two terms; see Figure 2.4 for illustration.

$$Lin[g(\bar{x})] = g(\bar{x}) + \frac{dg(\bar{x})}{dx} u \tag{2.118}$$

where $\bar{x}$ is the value of argument for which the consistent linearization is performed, $\frac{dg(\bar{x})}{dx}$ is the Fréchet derivative of a function to be linearized and u is a small increment.

[13] The chosen incremental displacement notation u remains in agreement with linear analysis explained previously, where one applies only one increment.

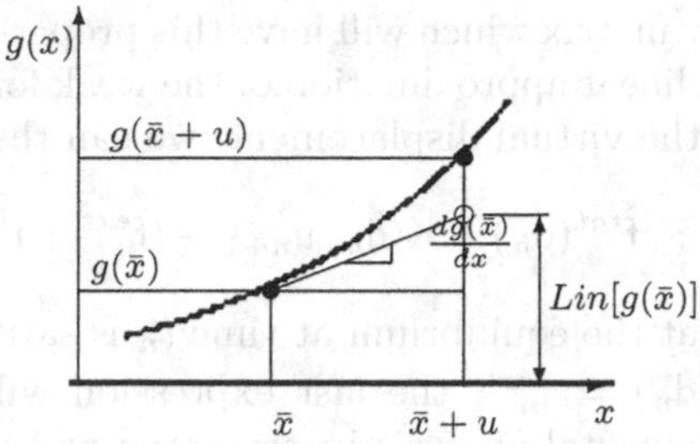

Fig. 2.4 Consistent linear approximation of a scalar function.

The expression for consistent linear approximation of a scalar function $g(x)$ can also be written with the Gâteaux or directional derivative (e.g. see Marsden and Hoffmann [181], p. 349) in the direction of increment u

$$Lin[g(\bar{x})] = g(\bar{x}) + \frac{d}{dt}[g(\bar{x} + t\,u)]\bigg|_{t=0} \tag{2.119}$$

The directional derivative formalism has the advantage for handling a more general case of consistent linearization of a scalar function of vector argument. The latter concerns constructing the consistent linear approximation to the weak form of a boundary value problem at time t_n

$$\begin{aligned} &Lin[G(\mathsf{d}_n;\mathsf{w})] = G(\mathsf{d}_n;\mathsf{w}) + \tfrac{d}{dt}[G(\mathsf{d}_n + t\,\mathsf{u}_{n+1},\mathsf{w})]\bigg|_{t=0} \;; \\ &G(\mathsf{d}_n,\mathsf{w}) := \textstyle\sum_{a=1}^{n_{node}} w_a(\hat{f}_a^{int}(\mathsf{d}_n) - f_{a,n}^{ext}) = 0 \end{aligned} \tag{2.120}$$

For the given value of displacement vector d_n the computation of the Gâteaux derivative in the direction of incremental displacement vector u_{n+1} results with

$$\begin{aligned} \frac{d}{dt}[G(\underbrace{\mathsf{d}_n + t\,\mathsf{u}_{n+1}}_{d_{n,t}},\mathsf{w})]\bigg|_{t=0} &= \sum_{a=1}^{n_{node}} \sum_{b=1}^{n_{node}} w_a \int_\Omega \frac{dN_a(x)}{dx} \underbrace{\frac{\partial\hat{\sigma}(\epsilon^h)}{\partial\epsilon}}_{\hat{C}(\epsilon^h)} \frac{\partial\hat{\epsilon}^h(\mathsf{d}_{n,t})}{\partial d_{b,t}} \frac{\partial d_{b,t}}{dt}\,dx \\ &= \sum_{a=1}^{n_{node}} \sum_{b=1}^{n_{node}} w_a\,\hat{K}_{ab}(\mathsf{d}_n)\,u_{b,n+1} \end{aligned} \tag{2.121}$$

where $\mathsf{K}_n = [\hat{K}_{ab}(\mathsf{d}_n)]$ is the tangent stiffness matrix at pseudo-time t_n. According to Taylor's theorem (see Marsden and Hoffmann [182], p. 341), this tangent stiffness provides the best local approximation of the particular nonlinear problem

$$\lim_{\|\mathsf{u}_{n+1}\|\to 0} \|\,\hat{\mathsf{r}}(\mathsf{d}_{n+1}) - \hat{\mathsf{r}}(\mathsf{d}_n) - \hat{\mathsf{K}}(\mathsf{d}_n)\,\mathsf{u}_{n+1}\,\| \,/\, \|\,\mathsf{u}_{n+1}\,\| \to 0 \tag{2.122}$$

Moreover, K_n is the only matrix which will have this property. With the given result for the consistent linear approximation of the weak form at time t_n and the arbitrary values of the virtual displacement, we can thus obtain that

$$Lin[G(\mathsf{d}_n;\mathsf{w})] = 0 \implies \hat{\mathsf{f}}^{int}(\mathsf{d}_n) + \hat{\mathsf{K}}(\mathsf{d}_n)\,\mathsf{u}_{n+1} - (\mathsf{f}^{ext}_n + \mathsf{r}^{ext}_{n+1}) = 0 \quad (2.123)$$

If we further assume that the equilibrium at time t_n is satisfied (in the weak form sense), with $\hat{\mathsf{f}}^{int}(\mathsf{d}_n) = \mathsf{f}^{ext}_n$, the last expression will imply that the central problem of incremental analysis in each step will reduce to a set of linear algebraic equations with incremental displacements u_{n+1} as unknowns

$$\hat{\mathsf{K}}(\mathsf{d}_n)\,\mathsf{u}_{n+1} = \mathsf{r}_{n+1} \quad (2.124)$$

We can therefore use again Gauss elimination method to solve the linearized problem with tangent stiffness playing the role of the system matrix[14] and obtain the incremental displacements u_{n+1}. The total displacement at subsequent time t_{n+1} can then easily be obtained with displacement update

$$\mathsf{d}_{n+1} = \mathsf{d}_n + \mathsf{u}_{n+1} \quad (2.125)$$

This concludes a typical time step computations, and we can move onto the next step. By means of incremental analysis we replace the true nonlinear problem with a sequence of equivalent linear problems. The computed approximate solution remains linear in each step, and it deviates from the true solution, more and more, with each new step; see Figure 2.5.

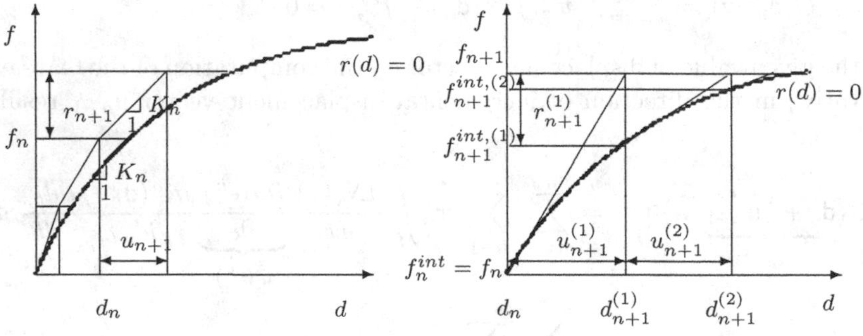

Fig. 2.5 Methods of incremental analysis and iterative corrections to re-establish equilibrium in each time step.

It is important to recognize that we can never have the exact solution to a nonlinear problem with incremental analysis, although the solution accuracy can be improved by taking smaller time steps.

[14] The latter implies that the conditioning of the system can change from step to step of incremental analysis, due to changes in tangent stiffness matrix.

2.2.6 *Solving a system of nonlinear algebraic equations by Newton's iterative method*

Newton's iterative method is used in order to provide the exact solution to a system of nonlinear algebraic equations, or rather to ensure the desired solution accuracy regardless of the pseudo-time increment chosen in incremental analysis. As the main difference from incremental analysis, Newton's solution procedure will not proceed to the next step until the solution is improved and the equilibrium is re-established. More precisely, by starting from the exact solution at time t_n, with $\hat{\mathsf{f}}^{int}(\mathsf{d}_n) = \mathsf{f}_n^{ext}$, we obtain the displacement increment u_{n+1} by using the standard incremental analysis described in the previous section, resulting with: $Lin[\hat{\mathsf{f}}^{int}(\mathsf{d}_n)] = \mathsf{f}_{n+1}^{ext}$. It is clear that the computed solution for displacement vector $\mathsf{d}_{n+1} = \mathsf{d}_n + \mathsf{u}_{n+1}$ can not verify the equilibrium equations at time t_{n+1}, and that the measure of equilibrium violation is given by the residual force, $\hat{\mathsf{r}}(\mathsf{d}_n + \mathsf{u}_{n+1}) = \mathsf{f}_{n+1}^{ext} - \hat{\mathsf{f}}^{int}(\mathsf{d}_n + \mathsf{u}_{n+1})$. One way for improving the incremental analysis solution is to consider this residual as a new increment. We can thus repeat formally the same procedure once again, and thus obtain the first iterative improvement of the solution provided by incremental analysis:

$$\hat{\mathsf{K}}(\mathsf{d}_n + \mathsf{u}_{n+1})\, \mathsf{u}_{n+1}^{(2)} = \mathsf{f}_{n+1} - \hat{\mathsf{f}}^{int}(\mathsf{d}_n + \mathsf{u}_{n+1}) \tag{2.126}$$

The improved value of displacement vector at time t_{n+1} is obtained as the sum of all the computed contributions, resulting with $\mathsf{d}_n + \mathsf{u}_{n+1}^{(1)} + \mathsf{u}_{n+1}^{(2)}$. We can check if the new (improved) value of the displacement vector will reduce the residual force to an acceptable value. If that does not happen, we proceed with additional iterations until the residual becomes sufficiently small. The flowchart of Newton's iterative method combined with incremental analysis is provided in Table 2.2, and the typical results convergence in a single time step is illustrated in Figure 2.5.

2.2.6.1 Quadratic convergence rate of Newton's method

In order to confirm the convergence of Newton's method, we can monitor the reduction of residual force $\| \mathsf{r}_{n+1} \| = \| \mathsf{f}_{n+1}^{ext} - \mathsf{f}_{n+1}^{int} \| \leq tol.$ in Euclidean norm, or any other convenient norm.[15] If the choice of the norm for testing convergence of Newton's method is easy, it is more difficult to choose the convergence tolerance. For example, with the nonlinear response close to the ultimate limit load, where the force–displacement diagram has the tendency to level off, a small residual value can still produce a sizeable iterative correction of incremental displacement. In such a case, it is better to test the

[15] In finite dimensional space all the norms are equivalent (e.g. see [62], and convergence in one norm will imply the convergence in any other norm.

Table 2.2 Incremental/iterative Newton's method for solving a system of n nonlinear algebraic equations.

- For each external load increment $n = 1, 2, \ldots, n_{inc}$

$$\mathsf{d}_{n+1}^{(1)} = \mathsf{d}_n$$

- For each iteration $i = 1, 2, \ldots, n_{iter}$

– Compute iterative correction

$$\hat{\mathsf{K}}(\mathsf{d}_{n+1}^{(i)})\,\mathsf{u}_{n+1}^{(i)} = \mathsf{f}_{n+1}^{ext} - \hat{\mathsf{f}}^{int}(\mathsf{d}_{n+1}^{(i)})$$

– Update displacement vector

$$\mathsf{d}_{n+1}^{(i+1)} = \mathsf{d}_{n+1}^{(i)} + \mathsf{u}_{n+1}^{(i)}$$

– Test convergence

$$\text{IF } \parallel \mathsf{f}_{n+1}^{ext} - \hat{\mathsf{f}}^{int}(\mathsf{d}_{n+1}^{(i+1)}) \parallel \leq tol. \implies \text{next step } n+1 \leftarrow n$$

$$\text{ELSE} \implies \text{next iteration}(i+1) \leftarrow (i)$$

convergence with respect to the computed iterative values of the incremental displacement by using the convergence criterion $\parallel \mathsf{u}_{n+1}^{(i+1)} \parallel \leq tol.$ However, even such strategy is not always safe, especially for computing the nonlinear response of hardening materials, such as rubber, where a small iterative correction of incremental displacement can still produce a large residual force. Therefore, the most general convergence criterion is the one where we test both the residual force and the incremental displacement, by using so-called energy convergence test

$$\mathsf{u}_{n+1}^{(i)} \cdot (\mathsf{f}_{n+1}^{ext} - \hat{\mathsf{f}}^{int}(\mathsf{d}_{n+1}^{(i)})) \leq tol. \tag{2.127}$$

The scalar product between the incremental displacement and the residual force can be computed at a very small supplementary cost with respect to solving a system of linearized equations (the latter is presented in Table 2.1). Namely, once the triangular decomposition of the tangent stiffness is accomplished, we can further obtain

$$\mathsf{u}_{n+1}^{(i)} \cdot (\mathsf{f}_{n+1}^{ext} - \hat{\mathsf{f}}^{int}(\mathsf{d}_{n+1}^{(i)})) = \mathsf{L}^T \mathsf{u}_{n+1}^{(i)} \cdot \mathsf{U}\mathsf{u}_{n+1}^{(i)} \tag{2.128}$$

We show subsequently the quadratic convergence rate of Newton's iterative method (e.g. see [65], p. 89). We can thus remain fairly demanding with respect to the chosen value of tolerance, without an excessive number of iterations and related computational cost. For example, we can require the convergence to $\parallel \mathsf{f}_{n+1}^{int} - \mathsf{f}_{n+1}^{ext} \parallel \leq 10^{-8}$, or $\parallel \mathsf{u}_{n+1}^{(i)} \parallel \leq 10^{-8}$, or yet the equivalent convergence in energy $\mathsf{u}_{n+1}^{(i)} \cdot (\mathsf{f}_{n+1}^{int} - \mathsf{f}_{n+1}^{ext}) \leq 10^{-16}$. By using such a strict convergence criterion, we can eliminate the non-desired round-off error influence on computed solution, which is especially important for inelastic material behavior with path-dependent response where the true path from initial to deformed configuration must be computed. The strict convergence criteria will typically require only a few additional iterations, since the convergence rate of Newton's method is quadratic. We will further illustrate this property for the simplest case where the system reduces to a single nonlinear algebraic equation, with the exact solution denoted $\tilde{u}$,

$$r(u) := f^{ext} - f^{int}(u) \ \& \ r(\tilde{u}) = 0$$

By using Taylor's formula, we can then construct an approximation for the exact solution based upon a given iterative guess $u^{(i)}$

$$0 = r(\tilde{u}) \approx r(u^{(i)}) + (\tilde{u} - u^{(i)}) \frac{\partial r(u^{(i)})}{\partial u} + \frac{1}{2}(\tilde{u} - u^{(i)})^2 \frac{\partial^2 r(\tilde{\xi})}{\partial u^2}$$

where, according to Taylor's theorem (e.g. see Marsden and Hoffmann [181]) we can choose $u^{(i)} \leq \tilde{\xi} \leq \tilde{u}$. By using the result of Newton's scheme, we can simplify the last expression:

$$\begin{aligned} \frac{1}{2}(\tilde{u} - u^{(i)})^2 \left[\frac{\partial^2 r(\tilde{\xi})}{\partial u^2} / \frac{\partial r(u^{(i)})}{\partial u} \right] &= u^{(i)} - [\frac{\partial r(u^{(i)})}{\partial u}]^{-1} r(u^{(i)}) - \tilde{u} \\ &= u^{(i+1)} - \tilde{u} \end{aligned}$$

If the residual is a sufficiently smooth function with bounded second derivative

$$u^{(i+1)} - \tilde{u} \leq c[u^{(i)} - \tilde{u}]^2 \ ; \ c = \sup_{\tilde{\xi}} \left[\frac{\partial^2 r(\tilde{\xi})}{\partial u^2} / \frac{\partial r(u^{(i)})}{\partial u} \right]$$

we can conclude that at each iteration the error is reduced quadratically. In Table 2.3 we give one such example of computations, showing the quadratic convergence rate of Newton's method, where we can see that at each iteration the number of zeros after decimal point is doubled.

2.2.6.2 Line search for increasing robustness of Newton's method

Newton's method allows us to converge towards the exact solution of a boundary value problem in nonlinear solid mechanics for any number of load increments needed to reach the final loading value. In nonlinear elasticity, we could theoretically apply the total load in one increment only, and still obtain the same final solution. Practically, however, we still need the incremental analysis, with the load increments chosen in order to accelerate the convergence of

Table 2.3 Quadratic convergence rate of Newton's iterative method.

itér. (i)	$\| \mathsf{f}_{n+1}^{ext} - \hat{\mathsf{f}}^{int}(\mathsf{u}_{n+1}^{(i)}) \|_2$
1	0.2169324150
2	0.0144234552
3	0.0033014740
4	0.0000214312
5	0.0000000051

Newton's method by reducing the number of initial iterations in each step. The choice of load increments must ensure that we always stay within the radius of convergence of Newton's method. This implies that larger increments can be applied initially when the nonlinear behavior is not very pronounced, whereas smaller increments are needed at a later stage with highly nonlinear response. However, it is in general quite difficult to know what load increments would be too large or too small for ensuring the convergence of Newton's method.

In fact, choosing a large load increment tends to cause convergence difficulties for Newton's method, especially for starting iterates within the step. Hence, the line search procedure will precisely target those initial iterations in order to increase Newton's method robustness. The main idea of line search is quite simple: we assume that the new iterative value of total displacement $\mathsf{d}_{n+1}^{(i+1)} = \mathsf{d}_{n+1}^{(i)} + \mathsf{u}_{n+1}^{(i)}$ is not necessarily optimal, and that there could be other values in the direction specified by iterative displacement increment provided by Newton's method that could give a better result. The problem reduces practically to deciding if the complete increment or only its portion should be used for the iterative displacement update. The displacement update is then written by introducing at each iteration (i) a scaling parameter $s^{(i)}$, which results with

$$\mathsf{d}_{n+1,s}^{(i)} \longleftarrow \mathsf{d}_{n+1}^{(i)} + s^{(i)}\,\mathsf{u}_{n+1}^{(i)} \; ; \quad s^{(i)} \in [0,1] \tag{2.129}$$

The optimal value of parameter $s^{(i)}$ in nonlinear elasticity can be obtained by minimization of the total potential energy

$$p(s^{(i)}) := \Pi(\mathsf{d}_{n+1}^{(i)} + s^{(i)}\,\mathsf{u}_{n+1}^{(i)}) \;\mapsto\; \min \tag{2.130}$$

The optimality condition amounts to the directional derivative computation of the total potential energy at the given value of total displacement $\mathsf{d}_{n+1}^{(i)}$ in the direction of incremental displacement $\mathsf{u}_{n+1}^{(i)}$; this kind of computation renders the corresponding value of parameter $s^{(i)} \in [0,1]$

$$\frac{dp(s^{(i)})}{ds} := \underbrace{\mathsf{u}_{n+1}^{(i)}}_{\frac{\partial \mathsf{d}_{n+1,s}}{\partial s}} \cdot \underbrace{(\mathsf{f}_{n+1}^{ext} - \hat{\mathsf{f}}^{int}(\mathsf{d}_{n+1}^{(i)} + s^{(i)}\,\mathsf{u}_{n+1}^{(i)}))}_{\frac{\partial \Pi}{\partial \mathsf{d}_{n+1}}} = 0 \tag{2.131}$$

The presented optimality condition can also be interpreted as the convergence requirement for enforcing the zero of residual component in the direction of incremental displacement. This kind of interpretation, which remains valid for a more general nonlinear problems where no potential is defined, can formally be written:

$$g(s^{(i)}) := \mathsf{u}_{n+1}^{(i)} \cdot (\mathsf{f}_{n+1}^{ext} - \hat{\mathsf{f}}^{int}(\mathsf{d}_{n+1}^{(i)} + s^{(i)}\, \mathsf{u}_{n+1}^{(i)})) = 0 \tag{2.132}$$

For the known values of $\mathsf{d}_{n+1}^{(i)}$ and $\mathsf{u}_{n+1}^{(i)}$, the optimality condition results with a nonlinear algebraic equation with parameter $s^{(i)}$ as the unknown. By taking into account already computed residual value, it is convenient to use the Regula-Falsi iterative method (e.g. see [62]), which reduces the solution procedure to successive computations of the residual for different values of parameter s

$$\begin{aligned}
&(j) = 1, 2, \ldots \\
&s^{(j+1)} = s^{(j)} \frac{g(0)}{[g(0) - g(s^{(j)})]} \\
&IF\ [g(s^{(j+1)}) = 0]\ (\,or\ \leq 0.5g(0))\ THEN\ s^{(i)} = s^{(j+1)} \\
&ELSEIF\ (j) \longleftarrow (j+1)
\end{aligned} \tag{2.133}$$

We indicate in the last expression that it is not necessary to carry out this computation until convergence, since $g(s^{(i)}) = 0$ will not ensure the global convergence on all residual components. Hence, we can already stop for the iterative value of $s^{(j)}$ reducing the initial residual component by half. As shown by Matthies and Strang [184], one can expect a significant improvement of iterative method robustness from the line search procedure.

2.2.6.3 Quasi-Newton method for reducing computational cost

When solving a system of nonlinear algebraic equations by Newton's method, the most costly phase remains the triangular decomposition of the tangent stiffness matrix at each iteration $\mathsf{K}_{n+1}^{(i)} \longrightarrow \mathsf{LU}$. Therefore, in seeking to reduce the computational cost of Newton's method, the first natural idea is to modify this phase making it less demanding. We are thus led to the quasi-Newton method, where we construct a secant approximation to the tangent stiffness matrix, which has an advantage of being easy to factorize. For a simple problem with only 1 degree of freedom, the secant approximation to the tangent stiffness can be written by using the known displacement and residual values from previous iterations

$$K_{n+1}^{(i)} \approx K_{n+1}^{sec,(i)} := \frac{r_{n+1}^{(i-1)} - r_{n+1}^{(i)}}{d_{n+1}^{(i)} - d_{n+1}^{(i-1)}}\ ;\ r_{n+1}^{(i)} := f_{n+1}^{ext} - \hat{f}^{int}(d_{n+1}^{(i)}) \tag{2.134}$$

This kind of secant approximation can also be presented as the corresponding update of the tangent stiffness from the previous iteration $K_{n+1}^{(i-1)}$, which is carried out by the correction matrix $\Delta K_{n+1}^{(i)}$

$$K_{n+1}^{sec,(i)} = K_{n+1}^{(i-1)} - \underbrace{\frac{r_{n+1}^{(i)}}{d_{n+1}^{(i)} - d_{n+1}^{(i-1)}}}_{\Delta K_{n+1}^{(i)}} \tag{2.135}$$

The same format can also be used for the cases of practical interest for a system of n nonlinear algebraic equations, where the correction matrix will be a rank-one matrix

$$\mathsf{K}_{n+1}^{sec,(i)} = \mathsf{K}_{n+1}^{(i-1)} - \frac{1}{\mathsf{r}_{n+1}^{(i),T}(\mathsf{d}_{n+1}^{(i)} - \mathsf{d}_{n+1}^{(i-1)})} \mathsf{r}_{n+1}^{(i)} \mathsf{r}_{n+1}^{(i),T} \tag{2.136}$$

The main advantage of writing the tangent stiffness in this format, pertains to simplicity of computing the inverse which can be obtained in closed form by using the Sherman-Morisson formula (e.g. see [91])

$$\begin{aligned} \mathsf{K}_{n+1}^{sec,(i),-1} &= \mathsf{K}_{n+1}^{(i-1),-1} + \frac{1}{\beta_{n+1}^{(i-1)}} \left(\mathsf{K}_{n+1}^{(i-1),-1} \mathsf{r}_{n+1}^{(i)}\right) \left(\mathsf{K}_{n+1}^{(i-1),-1} \mathsf{r}_{n+1}^{(i)}\right)^T \; ; \\ \beta_{n+1}^{(i-1)} &= \mathsf{r}_{n+1}^{(i),T}(\mathsf{d}_{n+1}^{(i)} - \mathsf{d}_{n+1}^{(i-1)}) - \mathsf{r}_{n+1}^{(i),T} \mathsf{K}_{n+1}^{(i-1),-1} \mathsf{r}_{n+1}^{(i)} \end{aligned} \tag{2.137}$$

The computational efficiency can thus be significantly increased since the previous iterates on stiffness matrix can be re-utilized to obtain the inverse of $\mathsf{K}_{n+1}^{(i)}$.

One can obtain even better results with respect to computational robustness by using the quasi-Newton method proposed by Broyden-Flechter-Goldfarb-Shanno (BFGS), where the tangent stiffness is obtained as a rank-two update of the previous iterative value (see [42]).

In conclusion, the best starting strategy to be used in solving a system of non-linear algebraic equations is provided by the quasi-Newton method with line search, which provides at small cost the possibility to come within the convergence radius of Newton's method. Once there, no other method can be competitive with Newton's, which is capable of reducing the residual below the tolerance at quadratic convergence rate (as opposed to quasi-Newton method, for which convergence is only quasi-linear, with error reduction power placed between 1 and 2).

2.3 Implementation of finite element method in 1D boundary value problems

In the foregoing discussion, the finite element method was described merely as a convenient manner for constructing a discrete approximation for the solution of a boundary value problem, quite similar to pioneering methods of Galerkin and Ritz. The important difference of the finite element method from its historical predecessors, which is perhaps the main reason for the impressive success that the method has achieved, is in its ability to easily handle rather complex domains. Namely, unlike the methods of Ritz and Galerkin, the finite element method does not need the shape functions that are defined globally in the domain, but only the element-based shape functions with only non-zero values in the element sub-domains connected to a particular node. With the finite element strategy for patching the discrete approximation from element-wise contributions, we gain both in simplicity and in efficiency. Namely, the computation of integrals for stiffness matrix and equivalent nodal load vector can be performed element-wise, and thus limited to only those elements for which a particular shape function takes non-zero values. Moreover, the computations can be standardized for all the elements of the same kind, by referring them to their common parent element, and be carried out easily by means of numerical integration. The element-based approach to integral computations requires a special procedure to account for the contributions of different elements towards the global set of algebraic equations, which is known as the finite element assembly. All those aspects, which constitute the basis of a typical finite element computer program, are presented in detail in this section.

2.3.1 Local or elementary description

In order to construct the discrete approximation for displacement field by the finite element method, we will replace the global description in (2.63) with a local or element-wise description. The latter implies first that we will account only for a portion of any global shape function $N_a(x), a = 1, 2, \ldots, n_{node}$, defined in (2.61), which takes non-zero values in that element. That portion of shape function will be denoted by $N_a^e(\cdot)$. Moreover, in local description we will introduce the change of coordinates, replacing 'x' by so-called natural coordinate 'ξ', which should center the element domain and reduce it to a desired length; see Table 2.4.
The results in Table 2.4 provide the discrete approximations for real and virtual displacement fields in local description, which account for all the nodes of a particular element. This kind of approximation allows in return to rewrite

Table 2.4 Global and local descriptions for 2-node truss-bar element.

#	Ingredient	Global description	Local description
(1)	Domain:	$[x_a, x_{a+1}]$	$[\xi_1, \xi_2\ ]$
(2)	Nodes:	$\{a, a+1\}$	$\{1, 2\}$
(3)	Degrees of freedom:	$\{d_a, d_{a+1}\}$	$\{d_1^e, d_2^e\}$
(4)	Shape functions:	$\{N_a(x), N_{a+1}(x)\}$	$\{N_1^e(\xi), N_2^e(\xi)\}$
(5)	Real displacement field:	$u^h(x) = N_a(x)\, d_a$ $+N_{a+1}(x) d_{a+1}$	$u^h(\xi) = N_1^e(\xi) d_1^e$ $+N_2^e(\xi) d_2^e$
(6)	Virtual displacement field:	$w^h(x) = N_a(x)\, w_a$ $+N_{a+1}(x) w_{a+1}$	$u^h(\xi) = N_1^e(\xi) w_1^e$ $+N_2^e(\xi) w_2^e$

the discrete approximation of the weak form or Galerkin's equations in $(G1)$ and in $(G2)$, by summing over all the elements in the inner loop

$$G(u^h; w^h) := \sum_a \sum_b \sum_{e=1}^{n_{elem}} w_a (K_{ab}^e d_b - f_a^{ext,e}) \qquad (G1e)$$

and

$$G(u^h; w^h) := \sum_a \sum_b \sum_{e=1}^{n_{elem}} w_a (\hat{f}_a^{int,e}(d_b) - f_a^{ext,e}) \qquad (G2e)$$

The latter implies that the integrals in (2.68), (2.69) and (2.76), defining the stiffness matrix, equivalent external load and internal load respectively, are computed by splitting the integration domain Ω into a number of element sub-domains Ω^e to be integrated independently.

In principle, each element sub-domain Ω^e can be different from others, even if they all belong to the same element type (for example, a 2-node truss-bar element). In order to avoid this kind of proliferation of different cases to deal with and standardize the computational task of computing element arrays, we introduce the parent element as the image of the real element corresponding to an interval of natural coordinate, $\xi \in [\xi_1, \xi_2]$. We always choose the same interval for all 1D truss-bar 2-node elements with $\xi_1 = -1$ and $\xi_2 = +1$, which is imposed by requirements of numerical integration procedure used for element integrals computation as elaborated upon subsequently. The change of coordinates between the real and the parent element for a 2-node truss-bar element has to be an affine transformation, which is represented by a linear polynomial

$$x(\xi) = c_1 + c_2 \xi \qquad (2.138)$$

The coefficients c_1 and c_2 can be obtained by equating the values at the ends of interval

$$\left. \begin{array}{r} x(-1) = x_a \\ x(+1) = x_{a+1} \end{array} \right\} \Longrightarrow \left\{ \begin{array}{c} c_1 = \frac{x_a + x_{a+1}}{2} \\ c_2 = \frac{h^e}{2}\ ;\ h^e = x_{a+1} - x_a \end{array} \right. \qquad (2.139)$$

Having defined the coordinate transformation, we can write the local description of the element shape function in (2.61) in terms of natural coordinate ξ according to

$$N_a^e(\xi) := N_a(x(\xi)) = \frac{1}{2}(1 + \xi_a \xi)\,; \quad \xi_a = \begin{cases} -1\,; a = 1 \\ +1\,; a = 2 \end{cases} \tag{2.140}$$

With this result in hand, the affine transformation in (2.138) can also be rewritten:

$$x(\xi)\Big|_{\Omega^e} = \sum_{a=1}^{2} N_a^e(\xi)\, x_a^e \tag{2.141}$$

By choosing so-called isoparametric elements (see Bathe [19], Hughes [111], Zienkiewicz and Taylor [270]), the shape functions for domain representation are also used in constructing the discrete approximations for real and virtual displacement fields

$$u^h(\xi)\Big|_{\Omega^e} = \sum_{a=1}^{2} N_a^e(\xi)\, d_a^e \tag{2.142}$$

$$w^h(\xi)\Big|_{\Omega^e} = \sum_{a=1}^{2} N_a^e(\xi)\, w_a^e \tag{2.143}$$

where d_a^e and w_a^e are nodal values of real and virtual displacement fields, respectively. The discrete approximation of infinitesimal deformation field can then readily be obtained from the displacement approximation in (2.142) resulting with

$$\begin{aligned} \epsilon^h(\xi) := \frac{du^h(\xi)}{dx} &= \sum_{a=1}^{2} \frac{dN_a^e(\xi)}{dx}\, d_a^e \\ &= \frac{dN_a^e}{d\xi} \frac{1}{dx(\xi)/d\xi}\, d_a^e \\ &= \sum_{a=1}^{2} \frac{(-1)^a}{h^e}\, d_a^e \end{aligned} \tag{2.144}$$

where we used the result for the jacobian of affine transformation '$dx/d\xi = h^e/2$', with h^e as the length of real truss-bar element.
The same kind of approximation can also be constructed for virtual deformation field with nodal values of virtual displacements w_a^e replacing those of real displacement d_a^e. We note that both real and virtual deformations remain constant in each element, which also implies (for homogeneous elastic material)

that the stress approximation is constant. By exploiting these results, we can easily obtain element internal force for nonlinear elasticity

$$\mathsf{f}^{int,e} = (f_a^{int,e}(\mathsf{d}^e)) \;\; ; \;\; 1 \leq a \leq 2$$

$$\begin{aligned}
f_a^e(\mathsf{d}^e) &= \int_{\Omega^e} \frac{dN_a(x)}{dx} \hat{\sigma} \left(\sum_{b=1}^{2} \frac{dN_b^e(x)}{dx} d_b^e \right) dx \\
&= \int_{-1}^{1} \frac{dN_a^e(\xi)}{dx} \hat{\sigma} \left(\sum_{b=1}^{2} \frac{dN_b^e(\xi)}{dx} d_b^e \right) j(\xi) d\xi \\
&= \int_{-1}^{1} \frac{(-1)^a}{h^e} \hat{\sigma} \left(\underbrace{\sum_{b=1}^{2} \frac{(-1)^b}{h^e} d_b^e}_{\epsilon^h(\xi)} \right) \frac{h^e}{2} d\xi \\
&= \boxed{(-1)^a \hat{\sigma} \left(\sum_{b=1}^{2} \frac{(-1)^b}{h^e} d_b^e \right)}
\end{aligned} \tag{2.145}$$

The same result can be written in matrix notation, defining the internal force vector for a 2-node truss-bar element

$$\mathsf{f}^{int,e} = \begin{bmatrix} -\sigma \\ \sigma \end{bmatrix} \tag{2.146}$$

In the same manner we can obtain the element tangent stiffness matrix by replacing the previously defined finite element approximations into (2.121)

$$\mathsf{K}^e = [K_{ab}^e] \;\; ; \;\; 1 \leq a, b \leq 2$$

$$\begin{aligned}
K_{ab}^e &= \int_{\Omega^e} \frac{dN_a(x(\xi))}{dx} \hat{C} \left(\sum_{c=1}^{2} \frac{dN_c(x(\xi))}{dx} d_c^e \right) \frac{dN_b(x(\xi))}{dx} dx \\
&= \int_{-1}^{1} \frac{dN_a^e(\xi)}{dx} \hat{C} \left(\underbrace{\sum_{c=1}^{2} \frac{dN_c^e(\xi)}{dx} d_c^e}_{\epsilon^h(\xi)} \right) \frac{dN_b^e(\xi)}{dx} j(\xi) d\xi \\
&= \int_{-1}^{1} \frac{(-1)^a}{h_e} \hat{C} \left(\sum_{c=1}^{2} \frac{(-1)^c}{h^e} d_c^e \right) \frac{(-1)^b}{h^e} \frac{h^e}{2} d\xi \\
&= \boxed{\frac{(-1)^a (-1)^b}{h^e} \hat{C} \left(\sum_{c=1}^{2} \frac{(-1)^c}{h^e} d_c^e \right)}
\end{aligned} \tag{2.147}$$

The element tangent stiffness matrix for a 2-node truss-bar element can also be written in matrix notation according to

$$\mathsf{K}^e = \frac{C}{h^e}\begin{bmatrix} 1 & -1 \\ -1 & 1 \end{bmatrix} \; ; \; (A = 1) \tag{2.148}$$

The element stiffness matrix keeps the same form in the case of linear elasticity with tangent modulus C in the last expression being replaced by Young's modulus E. Finally, when a bar does not have a unit, but an arbitrary cross-section A, the last result provides the valid form of the stiffness matrix if multiplied by A.

2.3.2 Consistence of finite element approximation

The discrete approximation for the displacement field, constructed by the finite element method, should remain consistent and approach the exact solution, when the number of elements increases and the length of each element becomes smaller. The conditions which guarantee the consistence of the finite element approximation can be specified for each finite element or its shape function, according to

$$\begin{aligned} &c1) \;\; N_a^e \in C^1(\Omega^e) \\ &c2) \;\; N_a \in C^o(\Omega) \; ; \; N_a = \bigcup_{e=1}^{n_{elem}} N_a^e \\ &c3) \;\; \sum_{a=1}^{n_{en}} N_a^e = 1 \end{aligned} \tag{2.149}$$

where n_{en} and n_{elem} are, respectively, the number of nodes for each element and the total number of elements chosen in the mesh.

The first condition ensures that the computation of internal force vector or tangent stiffness matrix for each element, defined by integrals in (2.145) and (2.147), can be carried out with no need to further subdivide the element domain of integration Ω^e in order to guarantee the unique expression for integrand. The second condition imposes that the finite-element-based discrete approximation will provide a continuous displacement field approximation from one element to another, allowing only for displacement derivative (or strain) discontinuities. For such a case, the chosen finite element strategy of subdividing the total domain of integration into element sub-domains $\int_\Omega(\cdot) = \sum_e \int_{\Omega^e}$ remains very well suitable.[16] Finally, the third condition in (2.149) is imposed to ensure that the finite element discrete approximation of displacement field is capable of representing the rigid body modes. The

[16] The second condition presents the minimum required regularity, since relaxing this requirement by allowing a discontinuous displacement approximation over finite element boundaries could produce the integrals which are not well defined, such as $\int_\Omega \delta^2 \, dx$, where $\delta(\cdot)$ is the Dirac function, defined according to $\int_\Omega \delta(x - x_d) g(x) \, dx = g(x_d)$.

latter pertains in 1D case to pure translation, where all the nodal values of displacement field would remain constant (say, equal to c_0), resulting with

$$
\begin{aligned}
u^{rb} &= \sum_{a=1}^{2} N_a^e(\xi)\, d_a^e \; ; \; \forall a \in \{1,2\} \Longrightarrow d_a^e = c_0 \\
&= \underbrace{\left(\sum_{a=1}^{2} N_a^e(\xi) \right)}_{=1} c_0
\end{aligned}
\qquad (2.150)
$$

2.3.3 Equivalent nodal external load vector

If trying to standardize the description of the external loading by the finite element approximation, we use again the same element shape functions. Therefore, for a 2-node truss-bar element, we will allow at most a linear variation of distributed external loading, which leads to following equivalent nodal external load vector

$$
\begin{aligned}
\mathsf{f}^e &= (f_a^e) \\
f_a^e &= \int_{\Omega^e} N_a(x) \left(\sum_{b=1}^{2} N_b(x) b_b \right) dx \\
&= \sum_{b=1}^{2} \int_{-1}^{1} N_a^e(\xi) N_b^e(\xi) j(\xi) d\xi \, b_b \\
&= \frac{h^e}{6} \sum_{b=1}^{2} (1 + \delta_{ab}) b_b \; ; \; \delta_{ab} = \begin{cases} 1 \, ; a = b \\ 0 \, ; a \neq b \end{cases}
\end{aligned}
\qquad (2.151)
$$

The same result can also be written in matrix notation as

$$
\mathsf{f}^e = \frac{h^e}{6} \begin{bmatrix} 2b_1 + b_2 \\ b_1 + 2b_2 \end{bmatrix} \qquad (2.152)
$$

By using Taylor's formula for representing the true variation of distributed external load, we can show that the proposed finite element approximation for external load will be of the order of $O((h^e)^2)$, which is comparable with the displacement approximation provided by 2-node finite elements. If this kind of approximation is not acceptable, we ought to employ the finite elements with higher order displacement approximations (quadratic, cubic etc.), which requires introducing truss-bar elements with more than two nodes.

2.3.4 Higher order finite elements

The local description, already presented for 2-node truss-bar element, can easily be extended to higher order isoparametric elements. The domain of the parent element for any such higher order element still remains the same interval in natural coordinate $\xi \in [-1, +1]$, but with $n_{en} > 2$ element nodes located within. The change of coordinates between ξ and x is no longer linear but rather a higher order polynomial which can be written:

$$x(\xi) = \sum_{a=1}^{n_{en}} N_a^e(\xi) x_a^e \; ; \; n_{en} \geq 2 \tag{2.153}$$

where $x_a^e, a = 1, 2, \ldots, n_{en}$ are real element nodal coordinates and $N_a^e(\xi)$ are the corresponding shape functions. The definition of any such shape function $N_a^e(\xi)$ remains the same: it takes a unit value at node a and zero value at all other element nodes (but not equal to zero in-between the nodes). The shape functions of this kind can easily be constructed by using the Lagrange polynomials (e.g. see Hughes [111], p. 126, Zienkiewicz and Taylor [270], p. 119). For a higher order element with n_{en} nodes, we can construct the shape functions as the Lagrange polynomials of order $n_{en} - 1$

$$N_a^e(\xi) := \prod_{b=1, b \neq a}^{n_{en}} \frac{(\xi - \xi_b)}{(\xi_a - \xi_b)} = \frac{(\xi - \xi_1) \ldots (\xi - \xi_{a-1})(\xi - \xi_{a+1}) \ldots (\xi - \xi_{n_{en}})}{(\xi_a - \xi_1) \ldots (\xi_a - \xi_{a-1})(\xi_a - \xi_{a+1}) \ldots (\xi_a - \xi_{n_{en}})} \tag{2.154}$$

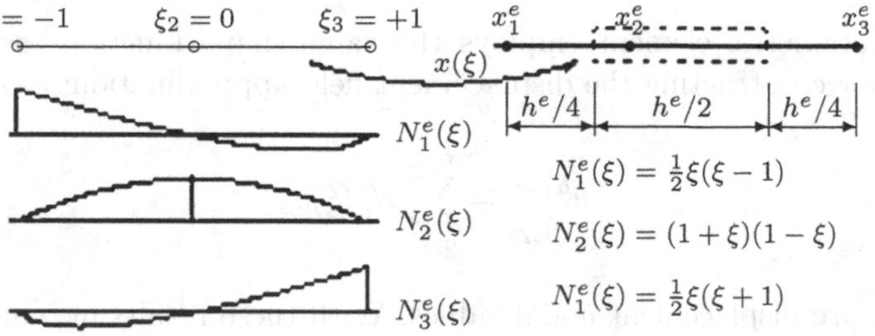

Fig. 2.6 Isoparametric truss-bar element with 3 nodes, its parent element and its shape functions.

For example, as illustrated in Figure 2.6, the shape functions for a 3-node isoparametric element are quadratic polynomials. The jacobian of the coordinate transformation for a higher order element in (2.153) is not a constant, but a polynomial of the order $n_{en} - 2$

$$j(\xi) := \frac{dx(\xi)}{d\xi} = \sum_{a=1}^{n_{en}} \frac{dN_a^e(\xi)}{d\xi} x_a^e \tag{2.155}$$

The same consistency conditions of the finite element discrete approximation as those established in (2.149) will remain valid for higher order elements. These conditions are easy to verify for each particular element of this Lagrangian family. For example, one can readily confirm the capability of a 3-node isoparametric element in Figure 2.6 to represent the rigid body modes by verifying that $\sum_{a=1}^{n_{en}} N_a = 1$. The displacement field continuity between neighboring elements is also apparent, since they share the same nodal value of the displacement. Finally, the condition on strain field continuity within an element, which calls for computation of shape functions derivatives, will require that the coordinate transformation $x(\xi)$ of a higher order isoparametric element remains bijective. By the application of the chain rule, we can readily confirm that such a computational requirement will ask for a positive value the jacobian throughout the element domain $j(\xi) > 0$, $\forall \xi \in [-1, +1]$,

$$\frac{\partial N_a}{\partial x} = \frac{\partial N_a}{\partial \xi} \frac{1}{j(\xi)} \; ; \; N_e^a(x) \in C^1(\Omega^e)$$

For a 3-node isoparametric element, this kind of requirement will limit the acceptable position of the center node within the inner half of the element length; see Figure 2.6 for illustration

$$\begin{aligned} &j(\xi) := h^e/2 + \xi(x_1 - 2x_2 + x_3) > 0; \forall \xi \in [-1, +1] \; ; \; h^e = (x_2^e - x_1^e) \\ &\xi = +1 \implies x_2 < (x_1 + x_3)/2 + h^e/4 \\ &\xi = -1 \implies x_2 > (x_1 + x_3)/2 - h^e/4 \end{aligned} \tag{2.156}$$

The isoparametric element employs the same shape functions as those in (2.153), for constructing the displacement field approximation;

$$u^h \Big|_{\Omega^e} = \sum_{a=1}^{n_{en}} N_a^e(\xi) d_a^e \tag{2.157}$$

where d_a^e are displacement nodal values. With these results in hand, we can easily obtain the discrete approximation for deformation field

$$\epsilon^h(\xi) \Big|_{\Omega^e} = \sum_{a=1}^{n_{en}} B_a^e(\xi) d_a^e \tag{2.158}$$

where $B_a^e(\xi)$ are defined with

$$B_a^e(\xi) := \frac{dN_a(x(\xi))}{dx} = \frac{1}{j(\xi)} \frac{dN_a^e(\xi)}{d\xi} \tag{2.159}$$

The element internal force vector for an isoparametric element with n_{en} nodes can then be written:

$$\mathsf{f}^{int,e} = (f_a^{int,e}) \; ; \; 1 \leq a \leq n_{en}$$
$$f_a^{int,e} = \int_{-1}^{1} B_a^e(\xi)\hat{\sigma}(\epsilon^h(\xi))\, j(\xi) d\xi \tag{2.160}$$

and its tangent stiffness matrix reduces to

$$\mathsf{K}^e = [K_{ab}^e] \; ; \; 1 \leq a, b \leq n_{en}$$
$$K_{ab}^e = \int_{-1}^{1} B_a^e(\xi)\hat{C}(\epsilon^h(\xi))B_b^e(\xi) j(\xi) d\xi \tag{2.161}$$

Finally, by using the higher order isoparametric elements, we can also increase the precision of the element external load vector representation, which can be written:

$$\mathsf{f}^e = (f_a^e) \; ; \; 1 \leq a \leq n_{en}$$
$$f_a^e = \int_{-1}^{1} N_a^e(\xi) \left(\sum_{b=1}^{n_{en}} N_b^e(\xi)\, b_b \right) j(\xi) d\xi \tag{2.162}$$

2.3.5 Role of numerical integration

We could see in the previous section that the higher order elements provide higher order approximations for displacement, strain and stress fields than a simple 2-node truss-bar element. However, the higher order approximations also make the computations of element arrays (element stiffness or internal force vector) quite laborious. For that reason, the computation of the integrals for element arrays components are best carried out by using numerical integration (e.g. see Irons [143]). The main idea of numerical integration is rather simple. First, instead of computing the element integrals exactly, we will approximate the integrand by a polynomial function.[17] This kind of polynomial, denoted as $g(\xi)$, will take the same value as the integrand at the chosen points in the integration domain of the parent element $\xi_l, l = 1, 2.., n_{in}$, referred to as the abscissas of numerical integration. We will then integrate analytically the polynomial integrand $g(\xi)$, and write the final result accordingly

$$\int_{-1}^{1} g(\xi)\, d\xi = \sum_{l=1}^{n_{in}} w_l g(\xi_l) \tag{2.163}$$

where w_l are the weights of numerical integration.

[17] Recall that the approximation of external load components is also written in terms of polynomials.

Among several possibilities to carry out numerical integration (e.g. see Bathe [19], p. 462, Hughes [111], p. 132 or Zienkiewicz and Taylor [270], p. 121), the optimal choice is the Gauss quadrature rule, since it is capable of providing the most accurate result with the smallest number of integration points. More precisely, for Gauss quadrature with n_{in} integration points we consider both the abscissas and weights as free parameters to choose, which allows us to uniquely define an approximating polynomial of order $2n_{in}-1$, which is then integrated exactly. For example, one Gauss quadrature point will integrate exactly any linear polynomial $g(\xi) = c_0 + c_1\xi$

$$\int_{-1}^{1} \underbrace{(c_0 + c_1\xi)}_{g(\xi)} \, d\xi = 2c_0 \equiv 2\,g(0) \tag{2.164}$$

The comparison of last two results in the last expression readily shows that 1 point Gauss quadrature rule should use the abscissa of the matching point in the center of the parent element and the corresponding integration weight equal to 2,

$$n_{in} = 1 \implies \xi_1 = 0 \; ; \; w_1 = 2 \tag{2.165}$$

For higher order approximation polynomials, the abscissas and weights of the numerical integration can also be obtained by comparison with the analytic results obtained for a generic polynomial integrand. For example, with 2 Gauss points $n_{in} = 2$, we can integrate exactly any third order polynomial $g(\xi) = c_0 + c_1\xi + c_2\xi^2 + c_3\xi^3$, where c_0, c_1, c_2, c_3 are constants

$$\int_{-1}^{1} \underbrace{(c_0 + c_1\xi + c_2\xi^2 + c_3\xi^3)}_{g(\xi)} \, d\xi = \sum_{l=1}^{2} w_l g(\xi_l) \tag{2.166}$$

In order to ensure invariance with respect to element node numbering, we impose $-\xi_1 = \xi_2 \equiv \xi$ and $w_1 = w_2 \equiv w$, leading to a reduced form of the last result

$$2c_0 + \frac{2}{3}c_2 = 2w(c_0 + c_2\xi^2) \tag{2.167}$$

The last equality is verified if and only if $w = 1$ and $\xi = \sqrt{3}/3$. Hence, the abscissas and weights of 2 point Gauss quadrature, integrating exactly cubic polynomials, are given as

$$n_{in} = 2 \implies w_1 = w_2 = 1 \; ; \; \xi_2 = -\xi_1 = 1/\sqrt{3} \tag{2.168}$$

The same kind of analysis can be carried out for polynomials of even higher order, with 3 point Gauss quadrature used for integrating the polynomials of order 5, 4 point rule for polynomial of order 7 etc. (see Zienkiewicz and Taylor [270]).

The Gauss quadrature can be used in computation of internal force vector components according to

$$\mathsf{f}^{int,e} = (f_a^{int,e}) \; ; \; 1 \le a \le n_{en}$$
$$f_a^{int,e} = \sum_{l=1}^{n_{in}} B_a(\xi_l)\hat{\sigma}(\epsilon^h(\xi_l))\, j(\xi_l) w_l \tag{2.169}$$

This result illustrates quite well the important role of the numerical integration in reducing the necessary information on stress field variation to Gauss points only. This reduction is especially important for nonlinear constitutive models. Moreover, only the numerical value of any other field at Gauss integration points is finally needed, and the finite element residual computation is thus reduced to algebraic operations that are perfectly suitable for computer implementation. The same cost and data reduction applies to the computation of element tangent stiffness matrix, where we only need the Gauss point values of elastic tangent modulus

$$\mathsf{K}^e = [K^e_{ab}] \; ; \; 1 \le a, b \le n_{en}$$
$$K^e_{ab} = \sum_{l=1}^{n_{in}} B_a(\xi_l)\hat{C}(\epsilon^h(\xi_l))B_b(\xi_l)\, j(\xi_l) w_l \tag{2.170}$$

What is the minimum number of Gauss points to use for computing the components of element residual vector or tangent stiffness matrix, defined by integrals in (2.169) and (2.170), respectively? The answer is easy to provide for 1D problems: the minimum Gauss quadrature rule must be capable of integrating exactly $\int_{-1}^{+1} j(\xi)\, d\xi$. With this choice of the numerical integration, we can ensure the representation of constant deformation and constant stress mode in each element.[18]

A more demanding criterion pertains to a number of Gauss integration points required to ensure the best possible order of convergence for deformation field with respect to the chosen element shape functions. In 1D, this criterion leads to so-called reduced integration rule, with the number of Gauss points being equal to number of element nodes reduced by 1. Hence, the reduced integration implies using 1 point for a 2-node truss-bar element, 2 points for a 3-node element etc. The reduced integration is the best rule to apply in 1D, since those Gauss quadrature points coincide with the points of higher order accuracy for strain and stress fields (see Zienkiewicz and Taylor [270], p. 348). This interesting property was initially pointed out by Barlow [18], and for that reason the reduced integration points are yet referred to as Barlow points.

[18] Verifying that the element can represent a constant stress state is referred to as the 'patch test' (see [19, 111, 270]), which guarantees the finite element method convergence in the limit where all elements become sufficiently small that the stress state in each element approaches a constant.

With 1 Gauss point, reduced integration rule for a 2-node truss-bar element, we can obtain the element internal force vector

$$\begin{aligned} f_a^{int,e} &= 2B_a^e(0)\hat{\sigma}(\epsilon^h(0))\,\frac{h^e}{2} \\ &= (-1)^a\hat{\sigma}(\epsilon^h(0)) \end{aligned} \tag{2.171}$$

as well as its tangent stiffness matrix

$$\begin{aligned} K_{ab}^e &= 2B_a^e(0)\hat{C}(\epsilon^h(0))B_b^e(0)\frac{h^e}{2} \\ &= \frac{(-1)^a(-1)^b}{h^e}\hat{C}(\epsilon^h(0)) \end{aligned} \tag{2.172}$$

These two numerical results of optimal accuracy, turn out to be the same as the analytic results. However, the reduced integration rule is not sufficiently accurate for integrating the external nodal load vector for the same element. Namely, for a linear load variation in a 2-node truss-bar element, 1 point Gauss quadrature rule will lead to a wrong result:

$$\begin{aligned} f_a^{ext,e} &= \sum_{l=1}^{1} 2N_a(0)\left(\sum_{b=1}^{2} N_b(0)\,b_b\right)\frac{h^e}{2} \\ &= \frac{h^e}{4}(b_1+b_2) \end{aligned} \tag{2.173}$$

A simple way to improve this result is by increasing the order of Gauss quadrature rule to 2, which will provide the exact result for external load vector

$$\begin{aligned} f_a^{ext,e} &= [N_a\left(\underbrace{\xi_1}_{-1/\sqrt{3}}\right)\left(\textstyle\sum_{b=1}^{2} N_b(\xi_1)\,b_b\right)\underbrace{w_1}_{1} \\ &\quad + N_a\left(\underbrace{\xi_2}_{1/\sqrt{3}}\right)\left(\textstyle\sum_{b=1}^{2} N_b(\xi_2)\,b_b\right)w_2]\tfrac{h^e}{2} \\ &= \tfrac{h^e}{6}\textstyle\sum_{b=1}^{2}(1+\delta_{ab})b_b \end{aligned} \tag{2.174}$$

This difference between quadrature rule for computing external load vector versus the reduced integration rule used for internal force computation, does not pose any practical problem. On the contrary, we can thus obtain the optimal results for both ingredients of the final set of algebraic equations to be solved. It is important, however, to use the same integration rule for computing the element internal force vector and its tangent stiffness matrix, especially for inelastic constitutive models employing the internal variables, in order to ensure the consistency between the stress and tangent modulus computations.

2.3.6 Finite element assembly procedure

The main advantage of the local (or element) description based upon the parent element pertains to the size reduction of element arrays and the complete standardization of the computational procedure for all finite elements of the same kind. The main advantage becomes the main inconvenience when the results for element arrays should be placed within the structural assembly, since the information of elements positions is not available from local description. The main role of the finite element assembly procedure is to reestablish this connectivity and carry out the summation of local (element) arrays with n_{en} size to provide the set of n_{eqs} (global) structural equations to be solved. For that reason, for each element we define the connectivity matrix L^e, with size $n_{en} \times n_{eqs}$, which provides the relationship between the nodal values of virtual displacement for a particular element w^e and the vector w containing the nodal values of virtual displacement for the chosen finite element mesh.

$$\mathsf{w}^e = \mathsf{L}^e \mathsf{w} \; ; \;\; \mathsf{w}^e = < w_1^e, w_2^e, \ldots, w_{n_{en}}^e >^T \; ; \; \mathsf{w} = < w_1, w_2, \ldots, w_{n_{eqs}} >^T \tag{2.175}$$

The entries in the connectivity matrix depend on the position of a particular element within the finite element mesh. In general, the connectivity matrix L^e is just a Boolean matrix, with each component equal to 0 or 1. The latter is reserved for all the components that correspond to a particular node to which the chosen truss-bar element is attached. In order to illustrate how the entries in the connectivity matrix are chosen, we present in Figure 2.7 a simple example of the finite element mesh for a truss structure composed of either 2-node or 3-node elements. The connectivity matrices for the first mesh with 2-node elements can be written:

$$\mathsf{L}^1 = \begin{bmatrix} 1\,0\,0\,0\,0 \\ 0\,1\,0\,0\,0 \end{bmatrix} ; \mathsf{L}^2 = \begin{bmatrix} 0\,1\,0\,0\,0 \\ 0\,0\,1\,0\,0 \end{bmatrix} ; \mathsf{L}^3 = \begin{bmatrix} 0\,0\,1\,0\,0 \\ 0\,0\,0\,1\,0 \end{bmatrix} ; \mathsf{L}^4 = \begin{bmatrix} 0\,0\,0\,1\,0 \\ 0\,0\,0\,0\,0 \end{bmatrix} \tag{2.176}$$

whereas the connectivity matrices for the mesh with 3-node elements are written:

$$\mathsf{L}^1 = \begin{bmatrix} 1\,0\,0\,0\,0 \\ 0\,1\,0\,0\,0 \\ 0\,0\,1\,0\,0 \end{bmatrix} \; ; \; \mathsf{L}^2 = \begin{bmatrix} 0\,0\,1\,0\,0 \\ 0\,0\,0\,1\,0 \\ 0\,0\,0\,0\,0 \end{bmatrix} \tag{2.177}$$

We note in passing that in each of two cases, the entries of the connectivity matrix for node 5 is equal to 0, since that node is fixed by the support (and the corresponding virtual displacement is equal to 0).

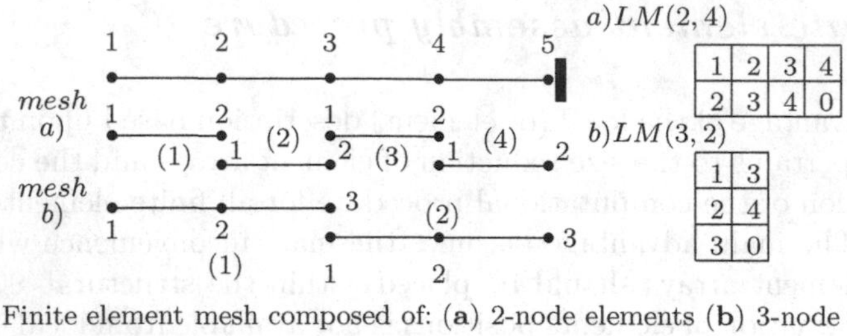

Fig. 2.7 Finite element mesh composed of: (**a**) 2-node elements (**b**) 3-node elements

By exploiting the connectivity matrices connecting the nodal values of virtual displacement of every single element to those of the structure, we can rewrite the discretized weak form (or Galerkin equation) as

$$0 = \sum_{e=1}^{n_{elem}} \mathsf{w}^{e\,T}(\mathsf{f}^{int,e} - \mathsf{f}^{ext,e}) = \sum_{e=1}^{n_{elem}} (\mathsf{L}^e\,\mathsf{w})^T(\mathsf{f}^{int,e} - \mathsf{f}^{ext,e}) \tag{2.178}$$
$$= \mathsf{w}^T \left(\sum_{e=1}^{n_{elem}} \mathsf{L}^{e\,T}\mathsf{f}^{int,e} - \sum_{e=1}^{n_{elem}} \mathsf{L}^{e\,T}\mathsf{f}^{ext,e} \right)$$

The last result implies that the assembly of element contributions to internal and external force vectors for the chosen finite element mesh should be taken into account after multiplication with the transpose of the element connectivity matrix

$$\boxed{\mathsf{f}^{int} = \textstyle\sum_{e=1}^{n_{elem}} \mathsf{L}^{e\,T}\mathsf{f}^{int,e} =: \mathop{\mathbb{A}}_{e=1}^{n_{elem}} \mathsf{f}^{int,e}} \;;$$
$$\boxed{\mathsf{f}^{ext} = \textstyle\sum_{e=1}^{n_{elem}} \mathsf{L}^{e\,T}\mathsf{f}^{ext,e} =: \mathop{\mathbb{A}}_{e=1}^{n_{elem}} \mathsf{f}^{ext,e}} \tag{2.179}$$

The linearized form of this system, or an equivalent linear problem governed by a set of linear algebraic equations, will require the assembly of the tangent stiffness matrix. By taking into account that each element connectivity matrix for connecting the nodal values of real displacement at local and structural level remains the same as the one written for virtual displacement, we obtain

$$\boxed{\mathsf{K} = \textstyle\sum_{e=1}^{n_{elem}} \mathsf{L}^{e\,T}\mathsf{K}^e\mathsf{L}^e =: \mathop{\mathbb{A}}_{e=1}^{n_{elem}} \mathsf{K}^e} \tag{2.180}$$

A more efficient manner to carry out the finite element assembly procedure, first proposed by Wilson [262], pertains to reducing the size of the connectivity matrix from its full format $n_{en} \times n_{eqs}$. For each element $e \in [1, n_{elem}]$, we will write instead a single vector which contains only the node numbers for non-zero entries of the particular element connectivity matrix. As shown in Figure 2.7, all these vectors can be arranged in a single matrix, denoted as LM. The finite element assembly can then be performed according to

$$\begin{aligned} &\forall e \in [1, n_{elem}] \;\&\; \forall a, b \in [1, n_{en}] \Longrightarrow \\ &f^{ext}(LM(a,e)) \longleftarrow f^{ext}(LM(a,e)) + f_a^{ext,e}\;; \\ &f^{int}(LM(a,e)) \longleftarrow f^{int}(LM(a,e)) + f_a^{int,e}\;; \\ &K(LM(a,e), LM(b,e)) \longleftarrow K(LM(a,e), LM(b,e)) + K_{ab}^e \end{aligned} \tag{2.181}$$

It is clear that the efficiency of the finite element assembly is thus improved significantly by operating upon the only contribution terms, non-zero components of each connectivity matrix. The same kind of finite element assembly can easily be extended to 2D or 3D problems, with the only difference regarding more unknowns per node (e.g. Hughes [77], p. 92 or Zienkiewicz and Taylor [270], p. 15 or Bathe [19], p. 28).

2.4 Boundary value problems in 2D and 3D elasticity

The practical engineering problems are mostly represented in terms of 2D or 3D boundary value problems, which require to generalize some of the ingredients of the solution procedure presented for 1D case. For start, the scalar fields of displacement, strain and stress which are sufficient for 1D problems, ought to be replaced by vector and tensor fields, respectively. In the first part of this section we will introduce the notion of a tensor (see [234]) along with the direct or tensorial notation, which allow us to deal efficiently with governing equations of 2D and 3D problems. In order to facilitate comprehension for those unfamiliar with tensor notation, we start with a familiar boundary value problem of linear elasticity, and restate a number of results presented in a more traditional manner by using the index notation in the classical works (e.g. see Duvaut [71], Germain [85], Lubliner [174], Sokolnikoff [243], or Timoshenko [255]).

2.4.1 Tensor, index and matrix notations

Most of physical fields of interest in a 3D boundary value problem in solid mechanics are represented mathematically in terms of vectors and tensors. The latter can be defined in 3D Euclidean space, which is selected herein, in terms of their components with respect to a chosen Cartesian frame. For example, with Cartesian coordinates $\{x, y, z\}$ and unit base vectors $\{\mathbf{i}, \mathbf{j}, \mathbf{k}\}$, we can define the position vector of a particle in 3D Euclidean space by using the classical notation for vectors: $\mathbf{x} = x\mathbf{i} + y\mathbf{j} + z\mathbf{k}$. In order to shorten lengthy expressions, one can use index notation, in which components and base vectors are numbered, such as $\{x, y, z\} \mapsto \{x_1, x_2, x_3\}$ and $\{\mathbf{i}, \mathbf{j}, \mathbf{k}\} \mapsto \{\mathbf{e}_1, \mathbf{e}_2, \mathbf{e}_3\}$, and enforce Einstein summation convention. The latter implies that each index which is repeated two times (so-called dummy index) is varied and summed up in the summation range, with no need to indicate explicitly the summation sign. Therefore, the same position vector can simply be written $\mathbf{x} = x_i\mathbf{e}_i$. The direct or tensor notation is even more compact, where the position vector is simply denoted as $\mathbf{x}$. Hence, with tensor notation we can see the structure of equations, with no need to specify the vector or tensor components, since any such equation should remain valid for an arbitrary choice of reference frame. For that reason, in most of the developments to follow we use the tensor notation, accompanied with index notation only when it is deemed necessary to clarify the details in order to eliminate any confusion. The final form of governing equations that provides the basis for subsequent numerical implementation, is presented in matrix notation with tensor or vector components placed within the corresponding matrix. For example, the position vector can be written in matrix notation as $\mathsf{x} = [x_i] := [x_1, x_2, x_3]^T$.

2.4.1.1 Vectors and tensors

In order to elaborate a bit more on the tensor notation used herein, we first specify the definitions for a number of operations on vectors and tensors in both direct and index notation. We will further consider that any such operation is defined in 3D Euclidean space $\mathbb{E}$ on set of real numbers $\mathbb{R}$.

Scalar product of vectors: to each pair of vectors $\mathbf{x}$ and $\mathbf{y}$ in $\mathbb{E}$ we can attribute their scalar product, $\mathbf{x} \cdot \mathbf{y}$ which is defined in $\mathbb{R}$ by:

$$\boxed{\mathbf{x} \cdot \mathbf{y} = \parallel \mathbf{x} \parallel \parallel \mathbf{y} \parallel \, cos(\widehat{\mathbf{x}, \mathbf{y}})} \tag{2.182}$$

with the following properties ($\forall \mathbf{x}, \mathbf{y}, \mathbf{z} \in \mathbb{E}$, $\forall \alpha, \beta \in \mathbb{R}$)

$$\begin{array}{lll} \mathbf{x} \cdot \mathbf{y} & = \mathbf{y} \cdot \mathbf{x} & \text{commutativity} \\ (\alpha\mathbf{x} + \beta\mathbf{y}) \cdot \mathbf{z} & = \alpha(\mathbf{x} \cdot \mathbf{z}) + \beta(\mathbf{y} \cdot \mathbf{z}) & \text{associativity} \\ \mathbf{x} \cdot \mathbf{x} & = \parallel \mathbf{x} \parallel^2 \, ; \; \mathbf{x} \cdot \mathbf{x} = 0 \implies \mathbf{x} = \mathbf{0} & \text{positive definite} \end{array} \tag{2.183}$$

The last result defines the module or the Euclidian norm of vector $\mathbf{x}$. The geometric interpretation of the scalar product of two vectors, as the product of their moduli and the cosine of the angle between the vectors, confirms that all the possible outcomes of scalar product between two Cartesian base vectors, $\mathbf{e}_i$ and $\mathbf{e}_j$, can be represented by the symbol of Kronecker δ_{ij}

$$\mathbf{e}_i \cdot \mathbf{e}_j = \delta_{ij} := \begin{cases} 1; \text{ si } i = j \\ 0; \text{ si } i \neq j \end{cases} \tag{2.184}$$

By exploiting this result, we can readily obtain an analytic representation of the scalar product of two vectors in terms of their Cartesian components

$$\begin{aligned}\mathbf{x} \cdot \mathbf{y} &= (x_i \mathbf{e}_i) \cdot (y_j \mathbf{e}_j) \\ &= x_i y_j (\mathbf{e}_i \cdot \mathbf{e}_j) \\ &= x_i y_j \delta_{ij} \\ &= x_i y_i \equiv x_1 y_1 + x_2 y_2 + x_3 y_3\end{aligned} \tag{2.185}$$

The Cartesian component of a vector can be obtained as its projection to the corresponding coordinate axis, which can be computed as the scalar product with the corresponding base vector

$$\begin{aligned}\mathbf{e}_i \cdot \mathbf{x} &= \mathbf{e}_i \cdot (x_j \mathbf{e}_j) \\ &= \delta_{ij} x_j \\ &= x_i\end{aligned} \tag{2.186}$$

A typical example where we need vector representation of a particular field is displacement. In 3D case, we have to introduce the displacement vector $\mathbf{u}(\mathbf{x})$ to describe a new placement of the particle whose initial position is defined with vector $\mathbf{x}$. On the other hand, in order to represent deformation and stress fields, we have to employ the second order tensors or simply tensors.[19] A tensor is a linear transformation $\mathbf{A}$ acting on a vector space, which takes one vector $\mathbf{u}$ into another vector $\mathbf{v}$ according to the following rule:

$$\mathbf{v} = \mathbf{A}(\mathbf{u}) \implies \boxed{\mathbf{v} = \mathbf{A}\mathbf{u}} \tag{2.187}$$

As indicated by the last expression, the linearity of operator $\mathbf{A}$ allows us to remove the parentheses. Similarly, the parentheses are no longer needed to indicate the composition of two mappings on a vector $\mathbf{u}$, producing $\mathbf{BAu} := \mathbf{B}(\mathbf{Au}) = \mathbf{Bv}$.

A set of all second order tensors forms a linear space, with the following properties:

$$\begin{array}{lll}(\mathbf{A}+\mathbf{B})\mathbf{u} = \mathbf{Au} + \mathbf{Bu} & \text{addition} \\ (\alpha\mathbf{A})\mathbf{u} \quad = \alpha(\mathbf{Au}) & \text{mult. by scalar} \\ \mathbf{Ou} \quad = \mathbf{0} & \text{zero tensor} \\ \mathbf{Iu} \quad = \mathbf{u} & \text{identity tensor}\end{array} \tag{2.188}$$

By exploiting the rule for computing the Cartesian components of vectors $\mathbf{u}$ and $\mathbf{v}$, as well as the linearity of tensor $\mathbf{A}$, we can restate (2.187) in coordinate representation

$$\begin{aligned}v_i &= \mathbf{e}_i \cdot \mathbf{A}(u_j \mathbf{e}_j) \\ &= \underbrace{\mathbf{e}_i \cdot \mathbf{A}(\mathbf{e}_j)}_{A_{ij} = \mathbf{e}_i \cdot \mathbf{A}\mathbf{e}_j} u_j \\ &= A_{ij} u_j\end{aligned} \tag{2.189}$$

[19] If we refer simply to 'tensors', we will mean the second order tensors; the vectors can thus be considered as the first order tensors.

The last expression provides the definition of the Cartesian components of a tensor $\mathbf{A}$ with respect to base vectors $\mathbf{e}_i$. A very important role in computations is played by the rank-one tensor which is produced by the tensor product of two vectors $\mathbf{u}$ and $\mathbf{v}$,

$$\boxed{(\mathbf{u} \otimes \mathbf{v})\mathbf{w} = (\mathbf{v} \cdot \mathbf{w})\mathbf{u} \; ; \;\; \forall \mathbf{w} \in \mathbb{E}} \tag{2.190}$$

It follows from the last expression that the Cartesian components of the rank-one tensor can be written:

$$\begin{aligned} \mathbf{e}_i \cdot (\mathbf{u} \otimes \mathbf{v})\mathbf{e}_j &= (\mathbf{e}_i \cdot \mathbf{u})(\mathbf{v} \cdot \mathbf{e}_j) \\ &= u_i v_j \end{aligned} \tag{2.191}$$

The tensor products of unit base vectors allow us to write an explicit representation of a tensor in terms of its Cartesian components

$$\boxed{\mathbf{A} = A_{ij}\mathbf{e}_i \otimes \mathbf{e}_j} \; ; \; \mathbf{u} \otimes \mathbf{v} = u_i v_j \mathbf{e}_i \otimes \mathbf{e}_j \; ; \; \mathbf{I} = \delta_{ij}\mathbf{e}_i \otimes \mathbf{e}_j \; ; \tag{2.192}$$

The transpose of a tensor $\mathbf{A}$, denoted as $\mathbf{A}^T$, is a unique tensor satisfying the property: $\mathbf{A}\mathbf{e}_i \cdot \mathbf{e}_j = \mathbf{e}_i \cdot \mathbf{A}^T \mathbf{e}_j$. The Cartesian component representation of $\mathbf{A}^T$ can be written:

$$\mathbf{A}^T = A_{ji}\mathbf{e}_i \otimes \mathbf{e}_j \tag{2.193}$$

Remarks: Role of orthogonal rotation tensor in change of coordinates

1. Active role or orthogonal tensor in change of coordinates: The main advantage of tensor notation is in resulting validity of any tensor equation in different coordinate systems. The choice of a particular coordinate system is made only when it comes to actual computations. The results of any such computation should remain invariant, if the change of coordinates is carried out by an orthogonal rotation tensor. More precisely, the change of coordinates by an orthogonal rotation tensor $\mathbf{R}$ takes a unit base vector $\mathbf{e}_i$ into another unit base vector $\mathbf{e}_i^*$, modifying only the direction and orientation, but not the module of the base vector, which can be written:

$$\mathbf{e}_i^* = \mathbf{R}\mathbf{e}_i \; ; \; \| \mathbf{e}_i^* \| = \| \mathbf{e}_i \| = 1 \implies \begin{cases} \mathbf{R}^T\mathbf{R} = \mathbf{I} \\ \mathbf{R} = R_{ij}\mathbf{e}_i \otimes \mathbf{e}_j \; ; \; R_{ij} = \mathbf{e}_i \cdot \mathbf{e}_j^* \end{cases} \tag{2.194}$$

By exploiting this result on the active role of rotation tensor $\mathbf{R}$ producing new position of unit base vectors, we can compute the vector and tensor components in two bases:

$$\begin{aligned} v_i^* &= \mathbf{e}_i^* \cdot \mathbf{v} & A_{ij}^* &= \mathbf{e}_i^* \cdot \mathbf{A}\mathbf{e}_j^* \\ &= \mathbf{R}\mathbf{e}_i \cdot \mathbf{v} & &= \mathbf{R}\mathbf{e}_i \cdot \mathbf{A}\mathbf{R}\mathbf{e}_j \\ &= \mathbf{e}_i \cdot \mathbf{R}^T\mathbf{v} & &= \mathbf{e}_i \cdot \mathbf{R}^T\mathbf{A}\mathbf{R}\mathbf{e}_j \end{aligned} \tag{2.195}$$

2. Passive role or orthogonal tensor in change of coordinates: the coordinate transformation can be carried out by appealing to the passive role of the

rotation tensor, which transforms directly the Cartesian coordinates of a tensor. The latter is preferred choice and the only one typically presented in classical works on the subject (e.g. Green and Zerna [93], Sokolnikoff [244], Truesdell and Noll [257]). It is easy to see from the last result that such a coordinate transformation is carried out by the transpose of the rotation tensor $\mathbf{R}^T$

$$\begin{aligned} v_i^* &= \mathbf{e}_i \cdot \mathbf{R}^T \mathbf{v} & A_{ij}^* &= \mathbf{e}_i \cdot \mathbf{R}^T \mathbf{A} \mathbf{R} \mathbf{e}_j \\ &= R_{ji} v_j & &= R_{ki} A_{kl} R_{lj} \end{aligned} \tag{2.196}$$

The scalar product of two tensors $\mathbf{A}$ and $\mathbf{B}$ can be defined as the generalization of the notion of scalar product of vectors, according to

$$\mathbf{A} \cdot \mathbf{B} = tr[\mathbf{A}^T \mathbf{B}] := A_{ij} B_{ij} \tag{2.197}$$

where $tr[\cdot]$ is linear transformation defining the trace of a tensor. The trace of a rank-one tensor, obtained as the tensor product of two vectors $(\mathbf{u} \otimes \mathbf{v})$, can be written:

$$tr(\mathbf{u} \otimes \mathbf{v}) = \mathbf{u} \cdot \mathbf{v} \tag{2.198}$$

whereas the trace of a tensor $\mathbf{A}$ is

$$\begin{aligned} \mathbf{I} \cdot \mathbf{A} = tr[\mathbf{A}] &= \delta_{ik} A_{kj} tr[\mathbf{e}_i \otimes \mathbf{e}_j] \\ &= A_{ij} (\mathbf{e}_i \cdot \mathbf{e}_j) \\ &= A_{ij} \delta_{ij} \\ &= A_{ii} \end{aligned} \tag{2.199}$$

This allows us to write the explicit result on scalar product of two tensors $\mathbf{A}$ and $\mathbf{B}$ in the following form

$$\begin{aligned} \mathbf{A} \cdot \mathbf{B} &= A_{ik} B_{ij} tr[\mathbf{e}_k \otimes \mathbf{e}_j] \\ &= A_{ik} B_{ij} \delta_{kj} \\ &= A_{ij} B_{ij} \end{aligned} \tag{2.200}$$

By exploiting these results, we can provide three equivalent manners for writing the scalar product of a tensor $\mathbf{A}$ with a composition of two other tensors $\mathbf{B}$ and $\mathbf{C}$

$$\begin{aligned} \mathbf{A} \cdot (\mathbf{B}\mathbf{C}) &= (\mathbf{B}^T \mathbf{A}) \cdot \mathbf{C} \\ &= (\mathbf{A}\mathbf{C}^T) \cdot \mathbf{B} \end{aligned} \tag{2.201}$$

which is a very important result for the derivation of the weak form of equilibrium equations. The latter also requires the following result stating that the scalar product between a symmetric tensor $\mathbf{S}$ (for which $\mathbf{S}^T = \mathbf{S}$) and a skew-symmetric tensor $\mathbf{W}$ (for which $\mathbf{W}^T = -\mathbf{W}$) is equal to zero

$$\begin{aligned} \mathbf{S} \cdot \mathbf{W} &= \mathbf{S}^T \cdot \mathbf{W}^T \\ &= -\mathbf{S} \cdot \mathbf{W} \end{aligned} \implies \mathbf{S} \cdot \mathbf{W} = 0 \tag{2.202}$$

It then follows that the scalar product between a symmetric tensor $\mathbf{S}$ and an arbitrary tensor $\mathbf{A}$, affects only its symmetric part, denoted as $sym[\mathbf{A}]$; the latter is obtained through the Euclidian decomposition of a tensor into its symmetric and skew-symmetric part

$$\begin{aligned} \mathbf{A}\cdot\mathbf{S} &= sym[\mathbf{A}]\cdot\mathbf{S}\ ;\ sym[\mathbf{A}] = \tfrac{1}{2}(\mathbf{A}+\mathbf{A}^T) \\ \mathbf{A}\cdot\mathbf{W} &= ant\mathbf{A}\cdot\mathbf{W}\ ;\ ant[\mathbf{A}] = \tfrac{1}{2}(\mathbf{A}-\mathbf{A}^T) \end{aligned} \tag{2.203}$$

Higher order tensors: The interpretation of a tensor, or more precisely of the second order tensor, as a linear mapping of a vector into another vector, can be extend to higher order tensors. For example, a third order tensor referred to as the alternator $\boldsymbol{\mathcal{E}}$, produces the skew-symmetric second order tensor $\mathbf{W}$ (for which $\mathbf{W}^T = -\mathbf{W}$) from a vector $\mathbf{w}$

$$\begin{aligned} &\mathbf{W} = \boldsymbol{\mathcal{E}}\mathbf{w}\ ;\ \mathbf{W}^T = -\mathbf{W}\ ; \\ &\boldsymbol{\mathcal{E}} = \mathcal{E}_{ijk}\mathbf{e}_i\otimes\mathbf{e}_j\otimes\mathbf{e}_k\ ;\ \mathcal{E}_{ijk} = \begin{cases} 1\ ;\ ijk = 123, 231, 312 \\ -1\ ;\ ijk = 321, 132, 213 \\ 0\ ;\ \text{otherwise} \end{cases} \end{aligned} \tag{2.204}$$

Vector $\mathbf{w}$, yet referred to as the axial vector of skew-symmetric tensor $\mathbf{W}$, allows to define the vector product. For an arbitrary vector $\mathbf{v}$, we can write

$$\mathbf{w}\times\mathbf{v} := \boldsymbol{\mathcal{E}}\mathbf{w}\mathbf{v} = \mathbf{W}\mathbf{v} \tag{2.205}$$

One example of the fourth order tensor is the elasticity tensor $\boldsymbol{\mathcal{C}}$, producing the second order stress tensor $\mathbf{S}$ from a second order deformation tensor $\mathbf{E}$

$$\boxed{\mathbf{S} = \boldsymbol{\mathcal{C}}\mathbf{E}}\ ;\ \mathbf{S} = S_{ij}\mathbf{e}_i\otimes\mathbf{e}_j\ ;\ \mathbf{E} = E_{kl}\mathbf{e}_k\otimes\mathbf{e}_l\ ;\ \boldsymbol{\mathcal{C}} = \mathcal{C}_{ijkl}(\mathbf{e}_i\otimes\mathbf{e}_j)\otimes(\mathbf{e}_k\otimes\mathbf{e}_l) \tag{2.206}$$

The components of a higher order tensor can also be obtained by the corresponding projections to the base vectors. For example, the components of elasticity tensor can be written:

$$\mathcal{C}_{ijkl} = \mathbf{e}_i\cdot\boldsymbol{\mathcal{C}}(\mathbf{e}_k\otimes\mathbf{e}_l)\mathbf{e}_j \tag{2.207}$$

We could easily carry on to define formally the tensors of order five, six and higher, but (fortunately) none of them is needed for the developments presented herein. We can also extend the definition of a tensor to lower order, by considering vectors to be the tensors of order one and scalars to be the tensors of order zero.

2.4.1.2 Differential calculus with tensors

In this section, we discuss briefly how to generalize some fundamental procedures of differential calculus to vector and tensor fields. Namely, in a 3D boundary value problem in solid mechanics, we have to deal not only with (standard) scalar functions with scalar arguments, but also with vector- or tensor-valued functions with vector arguments. We will denote with $f(\mathbf{x})$, $\mathbf{v}(\mathbf{x})$ and $\mathbf{A}(\mathbf{x})$ scalar, vector and tensor-valued function, respectively, where $\mathbf{x} \in \Omega \subset \mathbb{R}^3$.

For a scalar function of this kind we can define the Gâteaux or directional derivative, by generalizing the notion introduced previously in the finite dimensional context

$$Df(\mathbf{x})[\mathbf{u}] := \frac{d}{d\varepsilon} f(\mathbf{x} + \varepsilon \mathbf{u})\bigg|_{\varepsilon=0} = \frac{\partial f(\mathbf{x})}{\partial x_i} u_i \tag{2.208}$$

where ε is a small parameter for directional derivative computation and $\mathbf{u}$ is a vector fixing the direction in which the derivative is computed. The Gâteaux or directional derivative is also employed for constructing the linear approximation of a function as a part of the consistent linearization procedure often used for solution of nonlinear problems. For example, the linear approximation of a function $f(\mathbf{x})$ can be written by using Taylor's formula for function representation discarding the terms of higher order than chosen increment

$$Lin[f(\mathbf{x})] = f(\mathbf{x}) + Df(\mathbf{x})[\mathbf{u}] \implies |f(\mathbf{x}+\mathbf{u}) - Lin[f(\mathbf{x})]| = O(\| \mathbf{u} \|) \tag{2.209}$$

By appealing to this linearity property, one can often simplify what could be otherwise quite a laborious computation of the direction derivative. For example, the consistent linearization of the determinant of a tensor $\mathbf{A}$ thus reduces to

$$\begin{aligned}
f(\mathbf{A}) &= det\mathbf{A} := \mathcal{E}_{ijk} A_{i2} A_{j3} A_{k1} \implies \\
Df(\mathbf{A})[\mathbf{U}] &= \tfrac{d}{d\varepsilon}[det(\mathbf{A} + \varepsilon \mathbf{U})]\Big|_{\varepsilon=0} \\
&= det(\mathbf{A}) \tfrac{d}{d\varepsilon}[det(\mathbf{I} + \varepsilon\, \mathbf{A}^{-1}\mathbf{U})]\Big|_{\varepsilon=0} \\
&= det(\mathbf{A}) \tfrac{d}{d\varepsilon}[det(\mathbf{I}) + \varepsilon\, tr(\mathbf{A}^{-1}\mathbf{U}) + O(\varepsilon^2)]\Big|_{\varepsilon=0} \\
&= det(\mathbf{A})(\mathbf{A}^{-T} \cdot \mathbf{U})
\end{aligned} \tag{2.210}$$

The procedure for computing the Gâteaux derivative remains the same for a scalar function of tensor argument $f(\mathbf{A})$

$$Df(\mathbf{A})[\mathbf{U}] := \frac{d}{d\varepsilon} f(\mathbf{A} + \varepsilon \mathbf{U})\bigg|_{\varepsilon=0} = \sum_{i,j=1}^{3} \frac{\partial f}{\partial A_{ij}} U_{ij} \tag{2.211}$$

For example, for the trace of a tensor, we can obtain

$$f(\mathbf{A}) = tr[\mathbf{A}] \implies Df(\mathbf{A})[\mathbf{U}] = \frac{d}{d\varepsilon}[\mathbf{I} \cdot (\mathbf{A} + \varepsilon \mathbf{U})]\bigg|_{\varepsilon=0} = \mathbf{I} \cdot \mathbf{U} \equiv tr[\mathbf{U}] \tag{2.212}$$

Finally, for a tensor function of another tensor $\mathbf{A}(\mathbf{B})$ we can compute the directional derivative according to

$$D\mathbf{A}(\mathbf{B})[\mathbf{U}] := \frac{d}{d\varepsilon}[\mathbf{A}(\mathbf{B}+\varepsilon\mathbf{U})]\Big|_{\varepsilon=0} = \frac{\partial A_{ij}}{\partial B_{kl}} U_{kl} \tag{2.213}$$

For example, the fourth order tangent elasticity tensor $\boldsymbol{\mathcal{C}}$ is obtained as the directional derivative of the nonlinear elastic response as

$$\mathbf{S} = \hat{\mathbf{S}}(\mathbf{E}) \implies D\hat{\mathbf{S}}(\mathbf{E})[\mathbf{U}] := \frac{d}{d\varepsilon}[\hat{\mathbf{S}}(\mathbf{E}+\varepsilon\mathbf{U})]\Big|_{\varepsilon=0} = \underbrace{\frac{\partial \hat{S}_{ij}}{\partial E_{kl}}}_{\mathcal{C}_{ijkl}} U_{kl} \tag{2.214}$$

The directional derivative of a product of mappings $\mathbf{f}$ and $\mathbf{g}$, denoted as $\pi(\mathbf{f},\mathbf{g})$, can be computed in the same way as for the classical (or Frechet) derivative computation according to

$$D\pi(\mathbf{f},\mathbf{g})[\mathbf{u}] := \pi(\mathbf{f}, D\mathbf{g}[\mathbf{u}]) + \pi(D\mathbf{f}[\mathbf{u}],\mathbf{g}) \tag{2.215}$$

Several important cases are covered by such a rule, with $\pi(\cdot)$ as bilinear form. For example, the rule applies to a product of a scalar with a vector $\pi(f,\mathbf{v}) := f\mathbf{v}$, or to the scalar product of two vectors $\pi(\mathbf{v},\mathbf{w}) := \mathbf{v}\cdot\mathbf{w}$, or to the tensor product of two vectors $\pi(\mathbf{v},\mathbf{w}) := \mathbf{v}\otimes\mathbf{w}$, or yet to a tensor operating on a vector $\pi(\mathbf{A},\mathbf{v}) := \mathbf{A}\mathbf{v}$.
Another important property of the classical derivative which can be extended to the directional derivative is the chain rule, which applies to dealing with a composition of mappings according to

$$\mathbf{h} = \mathbf{f}\circ\mathbf{g} := \mathbf{f}(\mathbf{g}(\mathbf{x})) \implies D\mathbf{h}(\mathbf{x})[\mathbf{u}] = D\mathbf{f}(\mathbf{g}(\mathbf{x}))\,[D\mathbf{g}(\mathbf{x})[\mathbf{u}]] \tag{2.216}$$

One can also define higher order directional derivatives. For example, the second order directional derivative is the directional derivative of the result obtained by the first directional derivative, which can be written:

$$D^2 f(\mathbf{x})[\mathbf{u},\mathbf{u}] \equiv D(Df(\mathbf{x})[\mathbf{u}])[\mathbf{v}] := \frac{d}{d\varepsilon}\{Df(\mathbf{x}+\varepsilon\mathbf{v})[\mathbf{u}]\} \tag{2.217}$$

This kind of results allow us to write the Taylor series representation of any function with vector argument

$$f(\mathbf{x}+\mathbf{u}) = f(\mathbf{x}) + Df(\mathbf{x})[\mathbf{u}] + D^2 f(\mathbf{x})[\mathbf{u},\mathbf{u}] + O(\parallel \mathbf{u} \parallel^3) \tag{2.218}$$

The directional derivative of a scalar function of vector argument $f(\mathbf{x})$, which can be written in terms of its components:

$$Df(\mathbf{x})[\mathbf{u}] = grad\,[f(\mathbf{x})]\cdot\mathbf{u}\ ;\ grad\, f(\mathbf{x}) = \frac{\partial f(\mathbf{x})}{\partial x_i}\mathbf{e}_i \tag{2.219}$$

can be conveniently rewritten by introducing the gradient operator ∇

$$\nabla = \mathbf{e}_i \frac{\partial}{\partial x_i} \implies grad\, f(\mathbf{x}) = \nabla f(\mathbf{x}) \tag{2.220}$$

By replacing $f(\mathbf{x})$ in the last expression by a component $v_i(\mathbf{x})$ of vector $\mathbf{v}$, we can easily show that the operator ∇ can also be used for writing the gradient of a vector function as the corresponding second order tensor

$$\begin{aligned} grad\,[\mathbf{v}(\mathbf{x})] &= \mathbf{v}(\mathbf{x}) \otimes \nabla \\ &= \tfrac{\partial v_i(\mathbf{x})}{\partial x_j} \mathbf{e}_i \otimes \mathbf{e}_j \end{aligned} \tag{2.221}$$

Similarly, we can define the gradient of a second order tensor $\mathbf{A}$ as the corresponding third order tensor, which can also be written by using the gradient operator

$$grad[\mathbf{A}(\mathbf{x})] = \mathbf{A}(\mathbf{x}) \otimes \nabla := \frac{\partial A_{ij}(\mathbf{x})}{\partial x_k} (\mathbf{e}_i \otimes \mathbf{e}_j) \otimes \mathbf{e}_k \tag{2.222}$$

We could carry on this way for computing the gradients of higher order tensors, but (fortunately) they are not needed in the developments to follow.

2.4.1.3 Integral theorems

The integral theorems concern the relations between the integrals of a vector or tensor field defined in the interior of the chosen domain of definition and its value on the domain boundary. An elementary example of this kind comes from the fundamental integral theorem in 1D case (e.g. see Marsden and Hoffmann [181], p. 209), which allows us to write for any differentiable function $g(x)$ in the interval $[a, b]$ that

$$\int_a^b \frac{dg(x)}{dx} dx = g(b) - g(a) \tag{2.223}$$

By analogy with this result, we can establish in 3D case the Gauss integral theorem (also referred to theorem of Gauss-Green-Ostrogradsky), which states that for a scalar function $f(\mathbf{x})$, defined in a domain Ω with the boundary Γ, we can write

$$\int_\Omega \frac{\partial f(\mathbf{x})}{\partial x_i} d\Omega = \int_\Gamma f(\mathbf{x})\, n_i(\mathbf{x})\, d\Gamma \tag{2.224}$$

The Gauss integral theorem of particular interest for the developments to follow concerns the case when the function $f(\mathbf{x})$ is replaced by a vector field component $v_i(\mathbf{x}) = \mathbf{v}(\mathbf{x}) \cdot \mathbf{e}_i$, which results with

$$\int_\Omega \frac{\partial v_i(\mathbf{x})}{\partial x_i} d\Omega = \int_\Gamma v_i(\mathbf{x}) n_i(\mathbf{x}) d\Gamma \tag{2.225}$$

The last result is yet known as the divergence theorem. By taking into account the following relationship between the divergence and the gradient of a vector field

$$div[\mathbf{v}] := \frac{\partial v_i}{\partial x_i} = tr[grad[\mathbf{v}]] = \mathbf{v} \cdot \boldsymbol{\nabla} \tag{2.226}$$

the divergence theorem can also be written in tensor notation

$$\int_\Omega \underbrace{div[\mathbf{v}(\mathbf{x})]}_{\mathbf{v}\cdot\boldsymbol{\nabla}}\, d\Omega = \int_\Gamma \mathbf{v}(\mathbf{x}) \cdot \mathbf{n}(\mathbf{x})\, d\Gamma \tag{2.227}$$

The linearity of the trace operator $tr[\cdot]$ allows us to restate the last result in an equivalent form

$$\int_\Omega \underbrace{(\mathbf{v}(\mathbf{x}) \otimes \boldsymbol{\nabla})}_{grad\,[\mathbf{v}(\mathbf{x})]}\, d\Omega = \int_\Gamma (\mathbf{v}(\mathbf{x}) \otimes \mathbf{n}(\mathbf{x}))\, d\Gamma \tag{2.228}$$

The divergence theorem also applies to tensor fields $\mathbf{A}(\mathbf{x})$ defining the corresponding vector field $div[\mathbf{A}]$ with

$$\int_\Omega \underbrace{div[\mathbf{A}(\mathbf{x})]}_{\mathbf{A}\boldsymbol{\nabla}}\, d\Omega = \int_\Gamma \mathbf{A}(\mathbf{x})\mathbf{n}(\mathbf{x})\, d\Gamma \;\; ; \;\; \mathbf{A}\boldsymbol{\nabla} := \frac{\partial A_{ij}}{\partial x_j}\mathbf{e}_i \tag{2.229}$$

By analogy with the expression in (2.228), the last result can also be written:

$$\int_\Omega (\mathbf{A}(\mathbf{x}) \otimes \boldsymbol{\nabla})\, d\Omega = \int_\Gamma (\mathbf{A}(\mathbf{x}) \otimes \mathbf{n}(\mathbf{x}))\, d\Gamma \tag{2.230}$$

A summary of different forms of the divergence theorem can be presented in a compact form as

$$\int_\Omega \{grad\,[\mathbf{v}(\mathbf{x})]\,\mathbf{A}^T(\mathbf{x}) + \mathbf{v}(\mathbf{x}) \otimes div[\mathbf{A}(\mathbf{x})]\}\, d\Omega = \int_\Gamma \{\mathbf{v}(\mathbf{x}) \otimes \mathbf{A}(\mathbf{x})\mathbf{n}(\mathbf{x})\}\, d\Gamma \tag{2.231}$$

The last result can easily be verified by exploiting the following identity:

$$\begin{aligned} div\,[(\mathbf{v}(\mathbf{x}) \cdot \mathbf{u})\mathbf{A}^T(\mathbf{x})\mathbf{w}] &= \boldsymbol{\nabla} \cdot \{(\mathbf{v}(\mathbf{x}) \cdot \mathbf{u})\mathbf{A}^T(\mathbf{x})\mathbf{w}\} \\ &= \boldsymbol{\nabla}(\mathbf{v}(\mathbf{x}) \cdot \mathbf{u}) \cdot (\mathbf{A}^T(\mathbf{x})\mathbf{w}) + (\mathbf{v}(\mathbf{x}) \cdot \mathbf{u})(\underbrace{\mathbf{A}(\mathbf{x})\boldsymbol{\nabla}}_{div[\mathbf{A}(\mathbf{x})]} \cdot \mathbf{w}) \\ &= \mathbf{u} \cdot [\underbrace{(\mathbf{v}(\mathbf{x}) \otimes \boldsymbol{\nabla})}_{grad[\mathbf{v}(\mathbf{x})]}\mathbf{A}^T(\mathbf{x}) + (\mathbf{v}(\mathbf{x}) \otimes div[\mathbf{A}(\mathbf{x})])]\mathbf{w} \end{aligned} \tag{2.232}$$

where $\mathbf{u}$ and $\mathbf{v}$ are constant vectors; we can also write

$$(\mathbf{v}(\mathbf{x}) \cdot \mathbf{u})(\mathbf{A}^T(\mathbf{x})\mathbf{w} \cdot \mathbf{n}) = \mathbf{u} \cdot (\mathbf{v}(\mathbf{x}) \otimes \mathbf{A}(\mathbf{x})\mathbf{n}(\mathbf{x}))\mathbf{w} \tag{2.233}$$

By application of the trace operator on both sides of (2.231), we can obtain

$$\int_{\Omega} (grad[\mathbf{v}(\mathbf{x})] \cdot \mathbf{A}(\mathbf{x}) + \mathbf{v}(\mathbf{x}) \cdot div[\mathbf{A}(\mathbf{x})])\, d\Omega = \int_{\Gamma} (\mathbf{v}(\mathbf{x}) \cdot \mathbf{A}(\mathbf{x})\mathbf{n}(\mathbf{x}))\, d\Gamma \tag{2.234}$$

The latter is the key result which is exploited for developing the weak form of a boundary value problem in 2D or 3D case, which is studied next.

2.4.2 Strong form of a boundary value problem in 2D and 3D elasticity

In this section we will generalize the ingredient of the strong form of a boundary value problem, already presented for 1D case, to the case of 3D elasticity. In other words, we develop the equations governing kinematics, equilibrium and constitutive law for 3D elasticity.

2.4.2.1 Linear kinematics and small displacement gradients

We consider a deformable solid body as a collection of particles, where the position of each particle is denoted with $\mathbf{x} \in \Omega$. The first role of the kinematics is to define the deformed or current configuration of the deformable body as the assembly of new positions of all the particles. Each particle in the deformed configuration is identified by its position vector $\mathbf{x}^{\varphi} = \varphi(\mathbf{x})$. For boundary value problems that can be placed in 3D Euclidian space, we can define for each particle $\mathbf{x}$ its displacement vector that takes it into deformed configuration

$$\mathbf{u}(\mathbf{x}) = \mathbf{x}^{\varphi} - \mathbf{x} \tag{2.235}$$

A set of displacement vectors for all the particles represents the displacement field in domain Ω. The second role of the kinematics is to extract deformations or strains from this displacement field. In order to define the strain measure, we ought to know not only the displacement of a given particle but also of the other particles in its local neighborhood. The new position of one such particle, which is initially close to the chosen particle with position vector $\mathbf{x} + d\mathbf{x}$ (where $d\mathbf{x}$ is an infinitesimal vector, with $\| d\mathbf{x} \| \ll 1$), is defined by the displacement vector $\mathbf{u}(\mathbf{x} + d\mathbf{x})$. By assuming that

$$\mathbf{x}^{\varphi} + d\mathbf{x}^{\varphi} = \mathbf{x} + d\mathbf{x} + \mathbf{u}(\mathbf{x} + d\mathbf{x}) \tag{2.236}$$

where $d\mathbf{x}^{\varphi}$ is the placement of the infinitesimal vector $d\mathbf{x}$ in the deformed configuration. By taking into account the definition of the displacement vector in (2.235) and by using the Taylor formula, we can restate the last result:

$$\begin{aligned} d\mathbf{x}^{\varphi} &= d\mathbf{x} + d\mathbf{u}(\mathbf{x} + d\mathbf{x}) - d\mathbf{u}(\mathbf{x}) \\ &\approx [\mathbf{I} + \nabla \mathbf{u}(\mathbf{x})]\, d\mathbf{x} \end{aligned} \tag{2.237}$$

where $\nabla \mathbf{u}(\mathbf{x}) := \boldsymbol{\nabla} \otimes \mathbf{u}$ is the displacement gradient. In geometrically linear theory of solid mechanics studied herein, we assume the hypothesis of small displacement gradients

$$\| \nabla \mathbf{u}(\mathbf{x}) \| \ll 1 \tag{2.238}$$

As indicated in the previous section, we can use the Euclidian decomposition to split the displacement gradient tensor into its symmetric and skew-symmetric part, denoted respectively as $\boldsymbol{\epsilon} = sym[\nabla \mathbf{u}]$ and $\boldsymbol{\omega} = ant[\nabla \mathbf{u}]$,

$$\nabla \mathbf{u} = \boldsymbol{\epsilon} + \boldsymbol{\omega} \; ; \; \boldsymbol{\epsilon} = \frac{1}{2}(\nabla \mathbf{u} + \nabla \mathbf{u}^T) \; ; \; \boldsymbol{\omega} = \frac{1}{2}(\nabla \mathbf{u} - \nabla \mathbf{u}^T) \tag{2.239}$$

We can easily verify that the displacement field, with the symmetric part of the displacement gradient tensor equal to zero, will not modify the norm of any infinitesimal vector in passing from initial to current configuration

$$\begin{aligned} d\mathbf{x}^{\varphi} \cdot d\mathbf{x}^{\varphi} &= [\mathbf{I} + \boldsymbol{\omega}] d\mathbf{x} \cdot [\mathbf{I} + \boldsymbol{\omega}] d\mathbf{x} \\ &= d\mathbf{x} \cdot (\mathbf{I} + \boldsymbol{\omega} + \underbrace{\boldsymbol{\omega}^T}_{-\boldsymbol{\omega}} + \underbrace{\boldsymbol{\omega}^T \boldsymbol{\omega}}_{\approx \mathbf{0}}) d\mathbf{x} \end{aligned} \tag{2.240}$$

In any such case we obtain $\boldsymbol{\epsilon} = \mathbf{0}$, and any infinitesimal vector in the current configuration is produced by a translation (represented with identity tensor $\mathbf{I}$) and/or by an infinitesimal rotation (represented with skew-symmetric tensor $\boldsymbol{\omega}$) of its initial position. For a change in module of infinitesimal vector (or deformation), we must have a non-zero value of $\boldsymbol{\epsilon}$, and we can thus conclude that the symmetric part of the displacement gradient tensor is the appropriate measure of infinitesimal deformation

$$d\mathbf{x}^{\varphi} = \boldsymbol{\epsilon}\, d\mathbf{x} \implies \| d\mathbf{x}^{\varphi} \| \neq \| d\mathbf{x} \| \tag{2.241}$$

By scalar-multiplication of the last expression with base vectors, we can obtain the interpretation for the components of the tensor of infinitesimal deformation: the component ϵ_{ii} or the dilatation, represents the (relative) change of length in direction x_i,

$$\epsilon_{ii} := \mathbf{e}_i \cdot \boldsymbol{\epsilon} \mathbf{e}_j \equiv (dx_i^{\varphi} - dx_i)/dx_i \tag{2.242}$$

the component $\gamma_{ij} = 2\epsilon_{ij} \; ; \; i \neq j$ or the shear deformation, represents the total change of the initially right angle between axes x_i and x_j

$$\gamma_{ij} \equiv 2\epsilon_{ij} := \mathbf{e}_i \cdot \boldsymbol{\epsilon} \mathbf{e}_j \equiv dx_i^{\varphi} \delta_{ij} dx_j^{\varphi} \tag{2.243}$$

The strain tensor can be represented not only by its Cartesian components, but also in terms of its spherical and deviatoric part

$$\epsilon = sph[\epsilon] + dev[\epsilon] \; ; \; sph[\epsilon] = \frac{1}{3} tr[\epsilon] \mathbf{1} \; ; \; dev[\epsilon] = \epsilon - \frac{1}{3} tr[\epsilon] \mathbf{1} \qquad (2.244)$$

Such a decomposition allows us to separate the change of an infinitesimal volume element, described by spherical part of deformation tensor, from the change of shape (without change of volume), which is represented by the deviatoric part. This can be shown by exploiting result in (2.210), which leads to

$$\begin{aligned} \lim_{\|\mathbf{u}\| \mapsto 0} V^{\varphi} &= \lim_{\|\mathbf{u}\| \mapsto 0} \int_{\Omega} det \mathbf{F} \, dV \\ &= \int_{\Omega} tr[\mathbf{I} + sph[\epsilon]] \, dV \\ &= V + \int_{\Omega} tr[\epsilon] \, dV \end{aligned} \qquad (2.245)$$

2.4.2.2 Equilibrium equations in small displacement gradients

The kinematic hypothesis on small displacement gradients leads to yet another important simplification, which concerns the derivation of the equilibrium equations of a deformable body in its deformed configuration. Namely, the linear kinematics allows us to ignore the difference between the coordinates in the initial and deformed configuration in computations of derivatives and integrals featuring in the strong and weak forms of equilibrium equations. In the development of the strong form of equilibrium equations, the linear kinematics hypothesis allows us to parameterize both the volume force $\mathbf{b}(\mathbf{x})$ and the Cauchy (or true) stress $\boldsymbol{\sigma}(\mathbf{x})$ with respect to the coordinates in the initial configuration. The surface force acting on infinitesimal area dA oriented by exterior unit normal $\mathbf{n}$ can also be expressed per unit area in the initial configuration.

The surface force can be related to the traction vector $\mathbf{t}^n(\mathbf{x})$, produced by the Cauchy stress tensor $\boldsymbol{\sigma}$ applied to the unit normal $\mathbf{n}$. This result, first presented by Cauchy to define the second-order stress tensor, is known as the Cauchy principle. The Cauchy principle can be derived by studying equilibrium of all the forces acting on an infinitesimal tetrahedron, yet called Cauchy's tetrahedron, with three facets in the planes of the chosen frame and the fourth facet dA in the plane oriented with unit exterior normal $\mathbf{n} = n_i \mathbf{e}_i$. By taking into account that each of the tetrahedron facets placed in one of the coordinate planes can be computed as the projection of the facet dA to the corresponding plane $dA_i/dA = n_i$, we can write the force equilibrium equation

$$\mathbf{t}^n + \mathbf{t}^{-e_i} n_i + \mathbf{b} \frac{dV}{dA} = \mathbf{0} \qquad (2.246)$$

where dV is the volume of Cauchy's tetrahedron. Furthermore, the principle of action and reaction, that allows to equate the traction forces on both sides of each facets $\mathbf{t}^{-e_i}(\mathbf{x}) = -\mathbf{t}^{e_i}(\mathbf{x})$, and expressions for traction forces in terms of stress tensor components $\sigma_{ij} = \mathbf{t}^{e_i}(\mathbf{x}) \cdot \mathbf{e}_j$, will allow us to obtain the local form of this equilibrium equation for the limit case when the tetrahedron shrinks to a point (with $dV/dA \mapsto 0$),

$$\mathbf{t}^n(\mathbf{x}) = \mathbf{t}^{e_i}(\mathbf{x})n_i \implies t_i^n(\mathbf{x}) = \sigma_{ij}(\mathbf{x})\, n_j(\mathbf{x}) \Leftrightarrow \mathbf{t}^n(\mathbf{x}) = \boldsymbol{\sigma}(\mathbf{x})\,\mathbf{n}(\mathbf{x}) \quad (2.247)$$

This result, known as the Cauchy principle, states that the stress vector $\mathbf{t}^n$ in (2.247), is defined with the stress tensor $\boldsymbol{\sigma}$ and the unit exterior normal vector $\mathbf{n}$.

The Cauchy principle also allows to write the equilibrium equation of the forces acting on an arbitrary sub-domain $\tilde{\Omega} \in \Omega$,

$$\begin{aligned} \mathbf{0} &= \int_{\partial\tilde{\Omega}} \overbrace{\boldsymbol{\sigma}(\mathbf{x})\,\mathbf{n}(\mathbf{x})}^{t^n}\, dA + \int_{\tilde{\Omega}} \mathbf{b}(\mathbf{x})\, dV \\ &= \int_{\tilde{\Omega}} \{div[\boldsymbol{\sigma}(\mathbf{x})] + \mathbf{b}(\mathbf{x})\}\, dV \end{aligned} \quad (2.248)$$

where the divergence theorem (2.229) was used to modify the first term in the last expression. By appealing further to the theorem of localization[20] we can obtain the local form of the force equilibrium equations

$$div[\boldsymbol{\sigma}(\mathbf{x})] + \mathbf{b}(\mathbf{x}) = \mathbf{0} \Leftrightarrow \frac{\partial \sigma_{ij}(\mathbf{x})}{\partial x_j} + b_i(\mathbf{x}) = 0 \ ; \ \forall \mathbf{x} \in \Omega \quad (2.249)$$

These force equilibrium equations should be accompanied by the moment equilibrium equations, which can be written

$$\begin{aligned} \mathbf{0} &= \int_{\Omega} \mathbf{x} \times \mathbf{b}(\mathbf{x})\, dV + \int_{\partial\Omega} \mathbf{x} \times (\boldsymbol{\sigma}(\mathbf{x})\,\mathbf{n}(\mathbf{x}))\, dA \\ &= \int_{\Omega} \mathbf{x} \times \underbrace{[div\boldsymbol{\sigma}(\mathbf{x}) + \mathbf{b}(\mathbf{x})]}_{=\mathbf{0}}\, dV + 2\int_{\Omega} skew\boldsymbol{\sigma}(\mathbf{x})\, dV \end{aligned} \quad (2.250)$$

In deriving the last result we have made use of the following identity

$$\begin{aligned} \mathbf{u} \cdot \int_{\partial\Omega} \mathbf{x} \times \boldsymbol{\sigma}\mathbf{n}\, dA &= \int_{\partial\Omega} (\boldsymbol{\sigma}^T(\mathbf{u} \times \mathbf{x})) \cdot \mathbf{n}\, dA \\ &= \int_{\Omega} div(\boldsymbol{\sigma}^T(\mathbf{u} \times \mathbf{x}))\, dV \\ &= \int_{\Omega} [div\boldsymbol{\sigma} \cdot (\mathbf{u} \times \mathbf{x}) + \boldsymbol{\sigma} \cdot grad(\mathbf{u} \times \mathbf{x})]\, dV \\ &= \mathbf{u} \cdot \int_{\Omega} [\mathbf{x} \times div\boldsymbol{\sigma} + 2\mathbf{s}]\, dV \ ; \ \mathbf{s} \times \mathbf{u} = skew[\boldsymbol{\sigma}]\mathbf{u} \end{aligned} \quad (2.251)$$

---

20 The theorem of localization (see [96], p. 38) states that for a smooth field $\mathbf{v}(\mathbf{x})$ defined in the domain Ω_α, we can obtain the limit case result $\mathbf{v}(\mathbf{x}_0) = \lim_{\alpha \mapsto 0} \int_{\Omega_\alpha} \mathbf{v}(\mathbf{x}_\alpha)\, dV / vol(\Omega_\alpha)$.

which is easy to verify for any constant vector $\mathbf{u}$. The local form of the moment equilibrium equation can thus be written in the form which confirms the symmetry of the stress tensor

$$skew[\boldsymbol{\sigma}] = \mathbf{0} \Longrightarrow \sigma_{ij} = \sigma_{ji} \tag{2.252}$$

Regarding the physical interpretation of the stress tensor components, we can distinguish the axial stress σ_{ii} versus the shear stress components σ_{ij}, with $i \neq j$, which can be computed as

$$\begin{aligned} \mathbf{t}^{e_1} &:= \boldsymbol{\sigma}\mathbf{e}_1 = \sigma_{11}\mathbf{e}_1 + \sigma_{12}\mathbf{e}_2 + \sigma_{13}\mathbf{e}_3 \\ \mathbf{e}_1 \cdot \mathbf{t}^{e_1} &= \mathbf{e}_1 \cdot \boldsymbol{\sigma}\mathbf{e}_1 = \sigma_{11} \\ \boldsymbol{\tau}^{e_1} &:= \mathbf{t}^{e_1} - \sigma_{11}\mathbf{e}_1 = \sigma_{12}\mathbf{e}_2 + \sigma_{13}\mathbf{e}_3 \end{aligned} \tag{2.253}$$

The same kind of stress components, with axial and shear stresses, can be computed for any stress vector $\mathbf{t}(\mathbf{x}, \mathbf{n})$ acting on the plane with any orientation $\mathbf{n}$,

$$\begin{aligned} &\| \mathbf{t} \|^2 = \sigma_{nn}^2 + {\tau^n}^2 \ ; \\ &\sigma_{nn} = \mathbf{n} \cdot (\boldsymbol{\sigma}\mathbf{n}) \ ; \\ &\boldsymbol{\tau}^n = [\mathbf{I} - (\mathbf{n} \otimes \mathbf{n})]\boldsymbol{\sigma}\mathbf{n} = \boldsymbol{\sigma}\mathbf{n} - (\mathbf{n} \cdot \boldsymbol{\sigma}\mathbf{n})\mathbf{n} \end{aligned} \tag{2.254}$$

The axial stress components are work-conjugate to the axial strains, and the shear stresses are conjugate to shear strains. Yet another decomposition, which is useful for identifying work-conjugate pairs, is the one using spherical and deviatoric parts of the stress tensor leading to

$$\begin{aligned} sph[\boldsymbol{\sigma}] &= \tfrac{1}{3} tr[\boldsymbol{\sigma}]\mathbf{1} \Leftrightarrow \tfrac{1}{3}(\sigma_{kk})\delta_{ij}\mathbf{e}_i \otimes \mathbf{e}_j \\ dev[\boldsymbol{\sigma}] &= \boldsymbol{\sigma} - sph[\boldsymbol{\sigma}] \ \Leftrightarrow = [\sigma_{ij} - \tfrac{1}{3}(\sigma_{kk})\delta_{ij}]\mathbf{e}_i \otimes \mathbf{e}_j \end{aligned} \tag{2.255}$$

The spherical part of the stress tensor corresponds to the hydrostatic pressure stress state. It is work-conjugate to the change of volume, which is described by the spherical part of the strain tensor. The deviatoric part of the stress tensor is work-conjugate to the deviatoric part of the strain tensor, which described the isometric change of shape. Therefore, we can write

$$\boldsymbol{\sigma} \cdot \boldsymbol{\epsilon} = \underbrace{sph[\boldsymbol{\sigma}] \cdot sph[\boldsymbol{\epsilon}]}_{p\theta} + \underbrace{dev[\boldsymbol{\sigma}] \cdot dev[\boldsymbol{\epsilon}]}_{\mathbf{s} \cdot \mathbf{e}} \tag{2.256}$$

2.4.2.3 Elastic constitutive law

It is important to note that the equilibrium and kinematics equations, presented in previous sections, always remain the same regardless of the kind of material from which the deformable solid body is composed. The only difference which appears in describing the strong form of a boundary value problem for two solid bodies composed of different materials pertains to the constitutive equations, expressing the governing relation between the stress and strain tensors. In this section, we present only the constitutive equations describing the elastic behavior of material, where the total strain is the only field which must be known to compute the stress field.

In constructing a set of elastic constitutive equations, we can choose between the model of hypoelasticity (or Cauchy's elasticity) and hyperelasticity (or Green's elasticity). In hypoelasticity, the constitutive equations for the stress tensor are defined for each component as an arbitrary function of the total strain tensor components, which can formally be written

$$\sigma_{ij}(\mathbf{x}) = \hat{\sigma}_{ij}(\boldsymbol{\epsilon}(\mathbf{x}), \mathbf{x}) \implies \boldsymbol{\sigma}(\mathbf{x}) = \hat{\boldsymbol{\sigma}}(\boldsymbol{\epsilon}(\mathbf{x}), \mathbf{x}) \tag{2.257}$$

On the other hand, in hyperelasticity we only need to define a potential in terms of strain energy density $W(\boldsymbol{\epsilon}(\mathbf{x}), \mathbf{x})$, which allows to compute the stress tensor components as the partial derivatives with respect to strain components

$$\sigma_{ij} = \frac{\partial W(\cdot)}{\partial \epsilon_{ij}} \Leftrightarrow \boldsymbol{\sigma} = \frac{\partial W(\cdot)}{\partial \boldsymbol{\epsilon}} \tag{2.258}$$

For numerical solution of the boundary value problems in elasticity, the hyperelastic constitutive model is advantageous to the hypoelastic model, because of its ability to provide the guarantees for the solution existence and uniqueness. Yet another advantage of the hyperelasticity concerns a reduced number of parameters to be defined as opposed to hypoelasticity, since the presence of the potential will impose the symmetry of the elasticity tensor. In order to illustrate this property, we consider the case of linear elastic response described by Hooke's law, which can be considered as the hyperelastic constitutive model with the strain energy potential given as a quadratic form in terms of deformation tensor

$$W(\boldsymbol{\epsilon}(\mathbf{x}), \mathbf{x}) = \frac{1}{2}\boldsymbol{\epsilon}(\mathbf{x}) \cdot \boldsymbol{\mathcal{C}}(\mathbf{x})\boldsymbol{\epsilon}(\mathbf{x}) \; ; \; W(\epsilon_{ij}(\mathbf{x}), \mathbf{x}) = \frac{1}{2}\epsilon_{ij}(\mathbf{x})\mathcal{C}_{ijkl}(\mathbf{x})\epsilon_{kl}(\mathbf{x}) \tag{2.259}$$

with $\boldsymbol{\mathcal{C}}(\mathbf{x})$ as the (fourth-order) elasticity tensor. For homogeneous linear elasticity, elasticity tensor will have constant components. The equality of the mixed partial derivatives of the last expression will imply that

$$\frac{\partial^2 W}{\partial \epsilon_{ij} \partial \epsilon_{jk}} = \frac{\partial^2 W}{\partial \epsilon_{jk} \epsilon_{ij}} \Leftrightarrow \mathcal{C}_{ijkl} = \mathcal{C}_{klij} \tag{2.260}$$

which in known as the major symmetry property of the elasticity tensor. We can also count with the minor symmetry of the elasticity tensor implying that

$$\frac{\partial \sigma_{ij}}{\partial \epsilon_{kl}} =: \mathcal{C}_{ijkl} = \mathcal{C}_{jikl} = \mathcal{C}_{ijlk} \; ; \; \mathcal{C}_{ijkl} = (\mathbf{e}_i \otimes \mathbf{e}_j) \cdot \mathcal{C}(\mathbf{e}_k \otimes \mathbf{e}_l) \; ; \; i,j,k,l = 1,2,3 \tag{2.261}$$

which is the consequence of the symmetry of the Cauchy stress tensor and infinitesimal strain tensor. The major and minor symmetries will allow us to reduce the number of independent components of the elasticity tensor from theoretical maximum of $3^4 = 81$ to 21 needed to describe the most general case of anisotropic elasticity. For the simplest case of isotropic linear elastic behavior, the number of independent components of elasticity tensor is reduced to only two. In other words, the elasticity tensor for isotropic linear elastic material can be described by only two parameters. One can thus choose either Lamé's parameters, denoted as $\lambda(\mathbf{x})$ and $\mu(\mathbf{x})$, or make a more usual choice of two parameters like Young's modulus $E(\mathbf{x})$ and Poisson's coefficient $\nu(\mathbf{x})$, which would allow us to write the elasticity tensor components

$$\boxed{\mathcal{C} = \lambda \mathbf{1} \otimes \mathbf{1} + 2\mu \mathbf{I} \;\Leftrightarrow\; \mathcal{C}_{ijkl} = \lambda \delta_{ij}\delta_{kl} + 2\mu \tfrac{1}{2}[\delta_{ik}\delta_{jl} + \delta_{il}\delta_{jk}]} \; ; \quad \lambda = \tfrac{\nu E}{(1+\nu)(1-2\nu)} \; ; \; \mu = \tfrac{E}{2(1+\nu)} \tag{2.262}$$

It is easy to verify that the elasticity tensor in (2.262) above, when multiplied by an infinitesimal rotation (recall that $\omega_{ij} = -\omega_{ji}$) will produce the zero stress. We can thus conclude that the stress in linear elasticity is produced only from the strain tensor

$$\boldsymbol{\sigma}^{rd} := \mathcal{C}\boldsymbol{\omega} = \mathbf{0} = \lambda \underbrace{tr[\boldsymbol{\omega}]}_{=0} \mathbf{1} + 2\mu(\underbrace{\boldsymbol{\omega} + \boldsymbol{\omega}^T}_{=\mathbf{0}}) \;\Longrightarrow\; \boldsymbol{\sigma} = \mathcal{C}\, \nabla \mathbf{u} = \mathcal{C}\, \boldsymbol{\epsilon} \tag{2.263}$$

The stress tensor produced by the strain tensor $\boldsymbol{\epsilon}$ for isotropic linear elastic response can also be written:

$$\boxed{\boldsymbol{\sigma} = \lambda tr(\boldsymbol{\epsilon})\mathbf{1} + 2\mu\boldsymbol{\epsilon} \; ; \; \sigma_{ij} = \lambda \epsilon_{kk}\delta_{ij} + 2\mu\epsilon_{ij}} \tag{2.264}$$

By applying the trace operator on both sides of the last expression, we can obtain yet another set of two parameters defining the isotropic linear elastic response in terms of the bulk modulus K and the shear modulus μ

$$\underbrace{\frac{1}{3}tr[\boldsymbol{\sigma}]}_{p} = \underbrace{\lambda + \frac{2}{3}\mu}_{K} \underbrace{tr[\boldsymbol{\epsilon}]}_{\theta} \; ; \; K = \lambda + \tfrac{2}{3}\mu = \tfrac{E}{3(1-2\nu)}$$
$$dev[\boldsymbol{\sigma}] := \boldsymbol{\sigma} - \tfrac{1}{3}tr[\boldsymbol{\sigma}]\mathbf{1} = 2\mu dev[\boldsymbol{\epsilon}] \tag{2.265}$$

The result in (2.265) above shows that the bulk modulus is parameter which controls the material resistance to the volume change, whereas the shear modulus controls the resistance to the change of shape. The incompressible linear elastic material will have the bulk modulus of an infinite value (or rather an extremely large value compared to shear modulus), which imposes that no change of volume ($tr[\epsilon] = 0$) can take place no matter what is applied loading. Not only incompressible, but also a quasi-incompressible material with $tr[\epsilon] \approx 0$, will have a poor conditioning of the elasticity tensor in (2.262) and thus we should rather use the elastic constitutive law written in the inverse form, where the strain tensor is computed from the stress tensor

$$\boldsymbol{\epsilon} = \frac{1}{E}((1+\nu)\boldsymbol{\sigma} - \nu tr[\boldsymbol{\sigma}]\mathbf{1}) \Leftrightarrow \epsilon_{ij} = \frac{1}{E}((1+\nu)\sigma_{ij} - \nu\sigma_{kk}\delta_{ij}) \quad (2.266)$$

A preferable form of the linear elastic constitutive law for isotropic quasi-incompressible material is the one using the split into spherical and deviatoric parts

$$\begin{aligned} & dev[\boldsymbol{\epsilon}] + \tfrac{1}{3}tr[\boldsymbol{\epsilon}]\mathbf{1} = \tfrac{1}{2\mu}dev[\boldsymbol{\sigma}] + \tfrac{1}{9K}tr[\boldsymbol{\sigma}]\mathbf{1} \\ \Longrightarrow & \boxed{\underbrace{tr[\boldsymbol{\epsilon}]}_{\theta} = \tfrac{1}{K}\underbrace{\frac{1}{3}tr[\boldsymbol{\sigma}]}_{p} \; ; \; dev[\boldsymbol{\epsilon}] = \tfrac{1}{2\mu}dev[\boldsymbol{\sigma}]} \end{aligned} \quad (2.267)$$

For the incompressible case, where $K \mapsto \infty$ and $\nu = 0.5$, only the deviatoric part of the stress tensor is obtained through the constitutive equations in (2.267) above, whereas the spherical part of the strain remains zero. The latter imposes the constraint on the displacement field in the domain, which ought to comply with

$$\begin{aligned} 0 &= \textstyle\int_\Omega \epsilon_{ii}\, dV \\ &= \textstyle\int_\Omega u_{i,i}\, dV \\ &= \textstyle\int_{\partial\Omega} \bar{u}_i n_i\, dA \end{aligned} \quad (2.268)$$

Moreover, the spherical part of the stress tensor or hydrostatic pressure p for an incompressible material is not computed from the constitutive equations, but rather becomes a strain-independent field computed directly from the pressure boundary conditions. The stress tensor can then be written

$$\boldsymbol{\sigma} = \frac{\partial \hat{W}(dev[\boldsymbol{\epsilon}])}{\partial \boldsymbol{\epsilon}} + p\mathbf{1} \quad (2.269)$$

2.4.2.4 2D boundary value problems: plane strain and plane stress

Even though in principle all the boundary value problems can be defined and solved by 3D models, in a number of applications we are interested in reducing the problem to 2D case[21] for either improving the computational efficiency of 3D model or a poor conditioning of the resulting set of equations.

[21] We have already seen that a 1D model can also be useful for providing an interpretation of 3D constitutive behavior.

Two typical 2D models are those of plane strain and plane stress. For either of them, the applied loading and homogeneity of a 3D deformable solid body in one direction (say, direction x_3), will allow to describe the variation of all fields with only two remaining coordinates (x_1 and x_2). We can set equal to zero a number of components of displacement, strain and stress

$$\begin{matrix} \epsilon_{13} = 0 \\ \epsilon_{23} = 0 \end{matrix} \; ; \; \begin{matrix} \sigma_{13} = 0 \\ \sigma_{23} = 0 \end{matrix} \tag{2.270}$$

The 2D plane stain model can be used for a 3D deformable body which is constrained to have zero displacement (and zero strain) in one direction. The zero strain $\epsilon_{33} = 0$, does not imply that the corresponding stress component remains equal to zero. The latter can be computed for the isotropic linear elastic case according to

$$\sigma_{33} = \lambda(\epsilon_{11} + \epsilon_{22}) \tag{2.271}$$

The plane strain case therefore does not change the format of 3D constitutive relations, but simply reduces the number of equations by leaving out the zero strain components. By using Young's modulus E and Poisson's ratio ν, we can write the inverse form of the plane strain constitutive equations

$$\begin{bmatrix} \epsilon_{11} \\ \epsilon_{22} \\ 2\epsilon_{12} \end{bmatrix} = \begin{bmatrix} \frac{1}{E} & -\frac{\nu}{E} & 0 \\ \frac{\nu}{E} & \frac{1}{E} & 0 \\ 0 & 0 & \frac{2(1+\nu)}{E} \end{bmatrix} \begin{bmatrix} \sigma_{11} \\ \sigma_{22} \\ \sigma_{12} \end{bmatrix} \tag{2.272}$$

One example of a 3D deformable body where the 2D plane strain model is appropriate is a very long tunnel with homogeneous mechanical properties and loading conditions along its axis, where the problem can be reduced to a slice of unit thickness under plane strain conditions. The opposite case is the one the plane stress, which concerns a thin, membrane-like 3D deformable body which is loaded only in the membrane plane (say, plane $x_1 - x_2$). By assuming that the stress component in the thickness direction (here, direction of axis x_3) remains equal to zero, we can write $\sigma_{33} = 0$. In the plane stress case, the through-the-thickness deformation ϵ_{33} is not in general equal to zero. The latter can be computed for isotropic linear elastic case with

$$0 = \sigma_{33} := (\lambda + 2\mu)\epsilon_{33} + \lambda(\epsilon_{11} + \epsilon_{22}) \implies \epsilon_{33} = -\frac{\lambda}{\lambda + 2\mu}(\epsilon_{11} + \epsilon_{22}) \tag{2.273}$$

By exploiting the last result, we can obtain a reduced set of the constitutive equations for plane stress case

$$\begin{bmatrix} \sigma_{11} \\ \sigma_{22} \\ \sigma_{12} \end{bmatrix} = \begin{bmatrix} \bar{\lambda} + 2\mu & \bar{\lambda} & 0 \\ \bar{\lambda} & \bar{\lambda} + 2\mu & 0 \\ 0 & 0 & \mu \end{bmatrix} \begin{bmatrix} \epsilon_{11} \\ \epsilon_{22} \\ 2\epsilon_{12} \end{bmatrix} \; ; \; \bar{\lambda} = \frac{2\lambda\mu}{\lambda + 2\mu} \tag{2.274}$$

where λ and μ are Lamé's parameters. An equivalent form of the constitutive equations for the plane stress case can also be written in terms of Young's modulus E and Poisson's ratio ν

$$\begin{bmatrix} \sigma_{11} \\ \sigma_{22} \\ \sigma_{12} \end{bmatrix} = \frac{E}{1-\nu^2} \begin{bmatrix} 1 & \nu & 0 \\ \nu & 1 & 0 \\ 0 & 0 & \frac{1-\nu}{2} \end{bmatrix} \begin{bmatrix} \epsilon_{11} \\ \epsilon_{22} \\ 2\epsilon_{12} \end{bmatrix} \tag{2.275}$$

2.4.3 Weak form of boundary value problem in 2D and 3D elasticity

For constructing numerical solution of a boundary value problem in 3D elasticity, we abandon its strong form in favor of the weak form of the problem, which weakens the interpretation of the equilibrium equations. By introducing the finite element interpolation of the displacement field into the weak form, we reduce the problem to be solved to a system of algebraic equations with nodal values of displacement field as unknowns. Such a displacement-type approximation procedure is fully equivalent to the one explained previously for 1D case, and would thus require only a brief description. However, in a number of applications pertaining to 3D case, the displacement-type finite element approximations can encounter the difficulties in representing the different deformation modes and run into so-called locking phenomena. For that reason, a 3D boundary value problem often requires a new theoretical formulation in terms of so-called mixed weak form, where not only displacement but also the strain and/or the stress field are considered independent, providing more freedom in choosing discrete approximations for dealing with locking phenomena.

Displacement-type weak form

For constructing the displacement-type weak form of a boundary value problem, we choose a virtual displacement field

$$\mathbf{w} = w_i \mathbf{e}_i \ ; \ i = 1, \ldots, n_{sd} \tag{2.276}$$

where n_{sd} is the dimension of space ($n_{sd} = 2$ or 3 for 2D or 3D case, respectively). Each component of the virtual displacement vector w_i should taka a zero value on the Dirichlet boundary Γ_{u_i}, where the corresponding real displacement component is imposed (recall that $\Gamma_i = \Gamma_{u_i} \bigcup \Gamma_{\sigma_i}$, as well as $\Gamma_{u_i} \bigcap \Gamma_{\sigma_i} = 0$, for a well-posed boundary value problem), which can be written:

$$\mathbb{V}_{0i} := \{ w_i : \Omega \mapsto \mathbb{R} \,|\, [w_i]_{\Gamma_{u_i}} = 0 \} \tag{2.277}$$

Similarly, we can define the real displacement vector field

$$\mathbf{u} = u_i \mathbf{e}_i \ ; \ i = 1, \ldots, n_{sd} \tag{2.278}$$

with the components u_i defined with

$$\mathbb{V}_i := \{u_i : \Omega \mapsto \mathbb{R} \mid [u_i]_{\Gamma_{u_i}} = \bar{u}_i\} \tag{2.279}$$

The real and virtual displacement fields verify the weak form of equilibrium equations

$$0 = G(\mathbf{u}; \mathbf{w}) := \int_{\Omega} \nabla^s \mathbf{w} \cdot \hat{\boldsymbol{\sigma}}(\nabla^s \mathbf{u})\, dV - \int_{\Omega} \mathbf{w} \cdot \mathbf{b}\, dV - \int_{\Gamma} \mathbf{w} \cdot \bar{\mathbf{t}}\, dA \tag{2.280}$$

where $\nabla^s \mathbf{w} = sym[\nabla \mathbf{w}]$, with the components $\mathbf{e}_i \cdot \nabla^s \mathbf{w}\, \mathbf{e}_j = \frac{1}{2}(\frac{\partial w_i}{\partial x_j} + \frac{\partial w_j}{\partial x_i})$. By using the integration by parts, it is easy to verify that the solution of the strong form of 3D elasticity is also the solution of the weak form,

$$\begin{aligned} 0 &= -\int_{\Omega} \mathbf{w} \cdot (div[\boldsymbol{\sigma}] + \mathbf{b})\, dV \\ &= \int_{\Omega} \nabla \mathbf{w} \cdot \boldsymbol{\sigma}\, dV - \int_{\Omega} div[\boldsymbol{\sigma} \mathbf{w}]\, dV - \int_{\Omega} \mathbf{w} \cdot \mathbf{b}\, dV \\ &= \int_{\Omega} \nabla^s \mathbf{w} \cdot \boldsymbol{\sigma}\, dV - \int_{\Omega} \mathbf{w} \cdot \mathbf{b}\, dV - \int_{\Gamma} \mathbf{w} \cdot \boldsymbol{\sigma}^T \mathbf{n}\, dA \\ &= \int_{\Omega} \nabla^s \mathbf{w} \cdot \boldsymbol{\sigma}\, dV - \int_{\Omega} \mathbf{w} \cdot \mathbf{b}\, dV - \int_{\Gamma_\sigma} \mathbf{w} \cdot \bar{\mathbf{t}}\, dA \\ &= G(\mathbf{u}; \mathbf{w}) \end{aligned} \tag{2.281}$$

The contrary will be true, with the solution of the weak form verifying the strong form as well, only if such a solution is sufficiently smooth to define the corresponding derivatives. For a hyperelastic material with strain energy density $W(\cdot)$, the weak form in (2.281) is identical to the condition which the displacement field must satisfy to render the minimum of the total potential energy

$$\Pi(\mathbf{u}) := \int_{\Omega} W(\nabla^s \mathbf{u})\, dV - \int_{\Omega} \mathbf{u} \cdot \mathbf{b}\, dV - \int_{\Gamma_\sigma} \mathbf{u} \cdot \bar{\mathbf{t}}\, dA \tag{2.282}$$

The weak form is used as the basis for constructing the finite element approximation. In that sense, we can reduce the number of operations by accounting for the symmetry of stress and strain tensors and replacing the tensor notation by the matrix notation. The weak form in matrix notation can be restated as

$$0 = G(\mathsf{u}^h; \mathsf{w}^h) := \int_{\Omega^h} \hat{\mathsf{e}}(\mathsf{w}^h)^T \tilde{\mathsf{s}}(\mathsf{u})\, dV - \int_{\Omega^h} \mathsf{w}^{hT} \mathsf{b}^h\, dV - \int_{\Gamma_\sigma} \mathsf{w}^{hT} \bar{\mathsf{t}}^h\, dA \tag{2.283}$$

where real and virtual displacement vectors are given as

$$\mathbf{u} \mapsto \mathsf{u}^h = \begin{bmatrix} u_1^h \\ u_2^h \\ u_3^h \end{bmatrix} \;;\; \mathbf{w} \mapsto \mathsf{w}^h = \begin{bmatrix} w_1^h \\ w_2^h \\ w_3^h \end{bmatrix} \tag{2.284}$$

The symmetry of real and virtual strain tensors implies that each has only six independent components, which can be written in terms of a column matrix (or a single index array),

$$\nabla^s \mathbf{w} \mapsto \mathsf{e}(\mathsf{w}^h) = \begin{bmatrix} \frac{\partial w_1^h}{\partial x_1} \\ \frac{\partial w_2^h}{\partial x_2} \\ \frac{\partial w_3^h}{\partial x_3} \\ \left(\frac{\partial w_2}{\partial x_1} + \frac{\partial w_1}{\partial x_2}\right) \\ \left(\frac{\partial w_3}{\partial x_2} + \frac{\partial w_2}{\partial x_3}\right) \\ \left(\frac{\partial w_1}{\partial x_3} + \frac{\partial w_3}{\partial x_1}\right) \end{bmatrix} ; \nabla^s \mathbf{u} \mapsto \mathsf{e}(\mathsf{u}^h) = \begin{bmatrix} \epsilon_{11}^h \\ \epsilon_{22}^h \\ \epsilon_{33}^h \\ 2\epsilon_{12}^h \\ 2\epsilon_{23}^h \\ 2\epsilon_{31}^h \end{bmatrix} = \begin{bmatrix} \frac{\partial u_1^h}{\partial x_1} \\ \frac{\partial u_2^h}{\partial x_2} \\ \frac{\partial u_3^h}{\partial x_3} \\ \left(\frac{\partial u_2}{\partial x_1} + \frac{\partial u_1}{\partial x_2}\right) \\ \left(\frac{\partial u_3}{\partial x_2} + \frac{\partial u_2}{\partial x_3}\right) \\ \left(\frac{\partial u_1}{\partial x_3} + \frac{\partial u_3}{\partial x_1}\right) \end{bmatrix} \tag{2.285}$$

We denoted as $2\epsilon_{ij} = \gamma_{ij}$ so-called engineering shear strains, which regroup the corresponding two strain tensor components which are the same due to symmetry. The symmetry of the stress tensor also implies that only six components are independent; we will write them as a single column matrix

$$\boldsymbol{\epsilon} \cdot \boldsymbol{\sigma} = \epsilon_{ij}\sigma_{ij} \mapsto \mathsf{e}^T\mathsf{s} \Rightarrow \mathsf{s} = \begin{bmatrix} \sigma_{11} \\ \sigma_{22} \\ \sigma_{33} \\ \sigma_{12} \\ \sigma_{23} \\ \sigma_{31} \end{bmatrix} \tag{2.286}$$

The chosen ordering of stress and strain tensors automatically defines the matrix notation for the elasticity tensor, according to

$$\mathcal{C} = \frac{\partial^2 W}{\partial \epsilon^2} \mapsto \mathsf{C} = \begin{bmatrix} \mathcal{C}_{1111} & \mathcal{C}_{1122} & \mathcal{C}_{1133} & \mathcal{C}_{1112} & \mathcal{C}_{1123} & \mathcal{C}_{1131} \\ . & \mathcal{C}_{2222} & \mathcal{C}_{2233} & \mathcal{C}_{2212} & \mathcal{C}_{2223} & \mathcal{C}_{2231} \\ . & . & \mathcal{C}_{3333} & \mathcal{C}_{3312} & \mathcal{C}_{3323} & \mathcal{C}_{3331} \\ . & . & . & \mathcal{C}_{1212} & \mathcal{C}_{1223} & \mathcal{C}_{1231} \\ . & sym. & . & . & \mathcal{C}_{2323} & \mathcal{C}_{2331} \\ . & . & . & . & . & \mathcal{C}_{3131} \end{bmatrix} \tag{2.287}$$

For linear elastic and isotropic case, the elasticity tensor can thus be written:

$$\mathcal{C} \mapsto \mathsf{C} = \lambda 1 1^T + 2\mu \mathsf{I} \; ; \; 1 = \begin{bmatrix} 1 \\ 1 \\ 1 \\ 0 \\ 0 \\ 0 \end{bmatrix} \; ; \; \mathsf{I} = \begin{bmatrix} 1 & 0 & 0 & 0 & 0 & 0 \\ 0 & 1 & 0 & 0 & 0 & 0 \\ 0 & 0 & 1 & 0 & 0 & 0 \\ 0 & 0 & 0 & \frac{1}{2} & 0 & 0 \\ 0 & 0 & 0 & 0 & \frac{1}{2} & 0 \\ 0 & 0 & 0 & 0 & 0 & \frac{1}{2} \end{bmatrix} \tag{2.288}$$

2.5 Detailed aspects of the finite element method

In this section we first present one of the main ingredients in terms of the isoparametric finite elements, which proved very versatile for constructing the finite element approximations in complex 2D and 3D domains. We then show how to enlarge the displacement-base, isoparametric element framework in order to provide a more robust approximation procedure capable of eliminating various locking phenomena, by appealing to the mixed variational formulations. The latter implies that not only displacement field interpolation has to be chosen in each finite element, but also the most appropriate, independent interpolations for stain and stress fields.

2.5.1 Isoparametric finite elements

The isoparametric finite elements are the first choice for constructing the finite element approximations in 2D and 3D domains. Namely, it is possible to standardize the computation even for a very complex domain, by attributing that any real finite element in the constructed mesh derives from a parent element. The latter is defined in the space of natural coordinates (ξ, η, ζ), or simply $\boldsymbol{\xi} = (\xi_1, \xi_2, \xi_3)$, either as a unit square for 2D or a unit cube for 3D case. In other words, in the chosen finite element mesh, the domain of each element is obtained by the corresponding mapping from the parent element domain. In order to provide a clear illustration of this procedure, we consider in Figure 2.8 the 2D case with a 4 node quadrilateral element and its parent element, with the corresponding mapping

$$\left.\begin{array}{l} x^h(\xi,\eta)\Big|_{\Omega^e} = \sum_{a=1}^{4} N_a(\xi,\eta)\, x_a^e \\ y^h(\xi,\eta)\Big|_{\Omega^e} = \sum_{a=1}^{4} N_a(\xi,\eta)\, y_a^e \end{array}\right\} \Longrightarrow \boxed{\mathbf{x}^h(\boldsymbol{\xi})\Big|_{\Omega^e} = \sum_{a=1}^{4} N_a(\boldsymbol{\xi})\, \mathbf{x}^e} \tag{2.289}$$

where $N_a(\xi, \eta)$ are the shape functions. The latter can be written explicitly as

$$N_a(\xi,\eta) = \tfrac{1}{4}(1+\xi_a\xi)(1+\eta_a\eta) \; ; \; \xi_a = \pm 1 \; ; \; \eta_a = \pm 1 \; ; \; a = 1,2,3,4$$
$$\Leftrightarrow \begin{cases} N_1(\xi,\eta) = \tfrac{1}{4}(1+\xi)(1+\eta) \\ N_2(\xi,\eta) = \tfrac{1}{4}(1-\xi)(1+\eta) \\ N_3(\xi,\eta) = \tfrac{1}{4}(1-\xi)(1-\eta) \\ N_4(\xi,\eta) = \tfrac{1}{4}(1+\xi)(1-\eta) \end{cases} \tag{2.290}$$

We can easily verify that the mapping (2.289) for 4-node element Q4 transforms any straight line in the natural coordinate space (for example, take any of the element sides) into the straight line in the physical space.

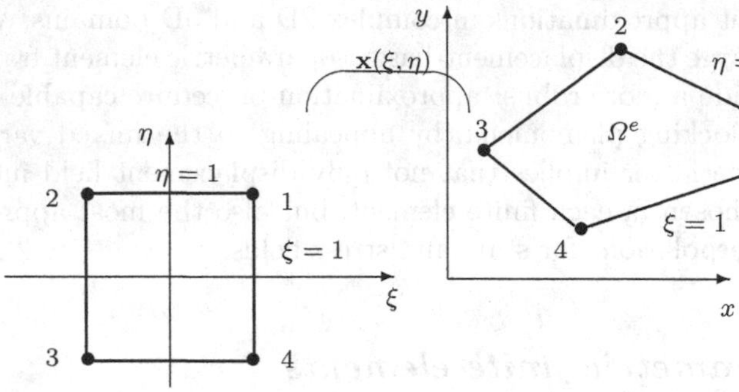

Fig. 2.8 Quadrilateral isoparametric finite element with 4 nodes ($Q4$) and its parent element.

The displacement field of the isoparametric finite element is constructed by employing the same shape functions as those used for the element domain representation

$$u_i^h(\xi,\eta)\Big|_{\Omega^e} = \sum_{a=1}^{4} N_a(\xi,\eta)\, d_a^e \implies \mathbf{u}^h(\xi,\eta)\Big|_{\Omega^e} = \sum_{a=1}^{4} N_a(\xi,\eta)\, \mathbf{d}_a^e \qquad (2.291)$$

where $\mathbf{d}_a^e$ are the nodal displacement values. The corresponding strain field approximation can then easily be obtained as

$$\mathsf{e}^h \mid_{\Omega^e} = \textstyle\sum_{a=1}^{n_{en}} \mathsf{B}_a\, \mathsf{d}_a^e \; ;$$

$$(n_{dm}=2)\ \ \mathsf{B}_a = \begin{bmatrix} \frac{\partial N_a}{\partial x_1} & 0 \\ 0 & \frac{\partial N_a}{\partial x_2} \\ \frac{\partial N_a}{\partial x_2} & \frac{\partial N_a}{\partial x_1} \end{bmatrix} \; ; (n_{dm}=3) \; ; \ \mathsf{B}_a = \begin{bmatrix} \frac{\partial N_a}{\partial x_1} & 0 & 0 \\ 0 & \frac{\partial N_a}{\partial x_2} & 0 \\ 0 & 0 & \frac{\partial N_a}{\partial x_3} \\ \frac{\partial N_a}{\partial x_2} & \frac{\partial N_a}{\partial x_1} & 0 \\ 0 & \frac{\partial N_a}{\partial x_3} & \frac{\partial N_a}{\partial x_2} \\ \frac{\partial N_a}{x_3} & 0 & \frac{\partial N_a}{\partial x_1} \end{bmatrix} \qquad (2.292)$$

By introducing the displacement and strain approximations in (2.291) and (2.292) into the weak form of equilibrium equations, we obtain the Galerkin equation for 3D linear elasticity. Furthermore, by considering that the virtual displacements at free nodes are arbitrary, we arrive at the final product of the finite element discretization procedure in terms of a set of linear algebraic equations, which further defines the displacement vector d to be produced under given external loading f:

$$\mathsf{Kd} = \mathsf{f} \; ; \; \mathsf{K} = \mathop{\mathbb{A}}_{e=1}^{n_{elem}} \mathsf{K}^e \; ; \; \mathsf{f} = \mathop{\mathbb{A}}_{e=1}^{n_{elem}} \mathsf{f}^e \qquad (2.293)$$

The symbol $\mathop{\mathbb{A}}\limits_{e=1}^{n_{elem}}$ in the last equation indicates the finite element assembly procedure of all the element arrays K^e and f^e to be placed within the total stiffness matrix and external load vector, respectively. The finite element assembly for 2D or 3D case is carried out in the same manner as already described for 1D case, except for the need to account for more degrees of freedom at each node. We can thus obtain:

$$\begin{aligned}
&\forall e \in [1, n_{elem}] \;\&\; \forall a, b \in [1, n_{en}] \;\&\; \forall i, j = 1, \ldots, n_{ddl} \Longrightarrow \\
&f^{ext}(LM(p,e)) \longleftarrow \; f^{ext}(LM(p,e)) + \mathbf{e}_i \cdot \mathsf{f}_a^{ext,e} \; ; \; p = (a-1)\, n_{ddl} + i \\
&f^{int}(LM(p,e)) \longleftarrow \; f^{int}(LM(p,e)) + \mathbf{e}_i \cdot \mathsf{f}_a^{int,e} \; ; \; q = (b-1)\, n_{ddl} + j \\
&K(LM(p,e), LM(q,e)) \; \longleftarrow \; K(LM(p,e), LM(q,e)) + \mathbf{e}_i \cdot \mathsf{K}_{ab}^e \mathbf{e}_j
\end{aligned} \tag{2.294}$$

The element load vector is computed by numerical integration. To that end, we will exploit the isoparametric mapping in order to replace the element integration domain by the parent element

$$\begin{aligned}
\mathsf{f}^e = [\mathsf{f}_a^e]\, ; \, \mathsf{f}_a^e &= \textstyle\int_{\Omega^e} N_a(\mathbf{x})\mathbf{b}(\mathbf{x})\, dV \\
&= \textstyle\int_{-1}^{+1} \int_{-1}^{+1} N_a^e(\xi,\eta)\mathbf{b}^e(\xi,\eta)\, j(\xi,\eta)\, d\xi d\eta \\
&= \textstyle\sum_{i=1}^{n_{in}} \sum_{j=1}^{n_{in}} N_a(\xi_i,\eta_j)\, \mathbf{b}^e(\xi_i,\eta_j)\, j(\xi_i,\eta_j) w_i w_j
\end{aligned} \tag{2.295}$$

where $j(\xi,\eta) = \frac{\partial x}{\partial \xi}(\xi,\eta)\frac{\partial y}{\partial \eta}(\xi,\eta) - \frac{\partial x}{\partial \eta}(\xi,\eta)\frac{\partial y}{\partial \xi}(\xi,\eta)$ is the determinant of the Jacobian matrix of the isoparametric coordinate transformation, ξ_i, η_j are abscissas and w_i, w_j and the weights of the numerical integration in ξ and η directions, respectively. The element stiffness matrix is computed in the same manner, leading to

$$\begin{aligned}
\mathsf{K}^e = [\mathsf{K}_{ab}^e]\, ; \, \mathsf{K}_{ab}^e &= \textstyle\int_{\Omega^e} \mathsf{B}_a^T(\mathsf{x})\, \mathsf{C}\mathsf{B}_b(\mathsf{x})\, dV \\
&= \textstyle\int_{-1}^{+1} \int_{-1}^{+1} \mathsf{B}_a^T(\xi,\eta)\mathsf{C}\mathsf{B}_b(\xi,\eta)\, j(\xi,\eta)\, d\xi d\eta \\
&= \textstyle\sum_{i=1}^{n_{in}} \sum_{j=1}^{n_{int}} \mathsf{B}_a^T(\xi_i,\eta_j)\mathsf{C}\mathsf{B}_b(\xi_i,\eta_j)\, j(\xi_i,\eta_j) w_i w_j
\end{aligned} \tag{2.296}$$

The conditions which guarantee that the last computation remain well-defined are known as the consistency conditions of the finite element approximation; the consistency conditions can be written:

Condition 1: $N_a(\xi,\eta) \in C^o(\Omega^h)$ should guarantee the continuity of displacement field between two neighboring elements, with eventual discontinuities limited only to displacement derivatives or deformations. The latter is fully consistent with subdivision of the total domain Ω^h into n_{el} sub-domains Ω^e, for evaluation of integrals (2.283) in the stiffness matrix computation. The isoparametric finite element approximation is precisely designed to verify this condition easily. For example, the displacement field of a 4-node quadrilateral isoparametric element remains a linear polynomial along any of its sides.

Therefore, the displacement compatibility is always ensured between any two neighboring elements, which share these two nodes and the corresponding nodal values of displacement.
Condition 2: $N_a(\xi, \eta) \in C^1(\Omega^e)$ should guarantee that the element arrays computation with integrals in (2.296) would not require any further subdivision of the element domain Ω^e, since the shape function derivatives featuring as matrix B^e components remain continuous functions. We note in passing that the derivatives with respect to x and y for any shape function $N_a(\xi, \eta)$ of isoparametric elements should be computed by using the chain rule and the Jacobian matrix

$$\begin{bmatrix} \frac{\partial}{\partial \xi} \\ \frac{\partial}{\partial \eta} \end{bmatrix} N_a(\xi, \eta) = \begin{bmatrix} \frac{\partial x}{\partial \xi} & \frac{\partial y}{\partial \xi} \\ \frac{\partial x}{\partial \eta} & \frac{\partial y}{\partial \eta} \end{bmatrix} \begin{bmatrix} \frac{\partial}{\partial x} \\ \frac{\partial}{\partial y} \end{bmatrix} N_a(\xi, \eta) \tag{2.297}$$

Each component of the Jacobian matrix is computed at the chosen values of the natural coordinates, typically corresponding to the numerical integration points ξ_i, η_j

$$\begin{aligned} \frac{\partial x(\xi_i,\eta_j)}{\partial \xi} &= \sum_{a=1}^{4} \frac{\partial N_a(\xi_i,\eta_j)}{\partial \xi} x_a^e \\ \frac{\partial y(\xi_i,\eta_j)}{\partial \eta} &= \sum_{a=1}^{4} \frac{\partial N_a(\xi_i,\eta_j)}{\partial \xi} y_a^e \; ; \quad 1 \le i, j \le n_{in} \end{aligned} \tag{2.298}$$

It is now easy to compute the shape function derivative by using the inverse of the Jacobian matrix according to

$$\begin{aligned} \frac{\partial N_a(\xi_i,\eta_j)}{\partial x} &= \left[\frac{\partial N_a(\xi_i,\eta_j)}{\partial \xi} \frac{\partial y(\xi_i,\eta_j)}{\partial \eta} - \frac{\partial N_a(\xi_i,\eta_j)}{\partial \eta} \frac{\partial y(\xi_i,\eta_j)}{\partial \xi} \right] / j(\xi_i, \eta_j) \\ \frac{\partial N_a(\xi_i,\eta_j)}{\partial y} &= \left[-\frac{\partial N_a(\xi_i,\eta_j)}{\partial \xi} \frac{\partial x(\xi_i,\eta_j)}{\partial \eta} + \frac{\partial N_a(\xi_i,\eta_j)}{\partial \eta} \frac{\partial x(\xi_i,\eta_j)}{\partial \xi} \right] / j(\xi_i, \eta_j) \end{aligned} \tag{2.299}$$

In order to obtain the unique result for this inverse, the isoparametric mapping should remain bijective for any value of natural coordinates, placing supplementary requirements on the choice of isoparametric elements. For example, the acceptable distortion for a 4-node quadrilateral element is limited to the forms where not a single interior angle is over 180° (see Figure 2.9 for a counter-example).
Condition 3: $\sum_{a=1}^{n_{en}} N_a(\xi, \eta) = 1$ should guarantee that any isoparametric element is capable of representing a linear displacement field $u(x, y) = c_0 + c_1 x + c_2 y$, with c_0, c_1 and c_2 constants. The latter implies the element capabilities to ensure the exact representation of the rigid body modes, a translation along x, a translation along y, and an infinitesimal rotation around z axis. For node a, with nodal coordinates x_a and y_a, such a rigid body mode should result in

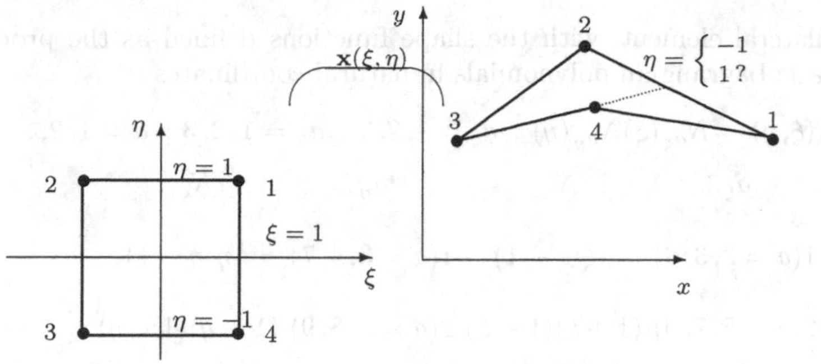

Fig. 2.9 A distorted 4-node isoparametric element, which does not have a bijective mapping with respect to its parent element.

$$u|_{\Omega^e} = \sum_{a=1}^{n_{en}} N_a(\xi,\eta)c_0 + \sum_{a=1}^{n_{en}} N_a(\xi,\eta)c_1 x_a + \sum_{a=1}^{n_{en}} N_a(\xi,\eta)c_2 y_a$$

$$= c_0 \underbrace{\sum_{a=1}^{n_{en}} N_a(\xi,\eta)}_{=1} + c_1 \underbrace{\sum_{a=1}^{n_{en}} N_a(\xi,\eta)x_a}_{=x} + c_2 \underbrace{\sum_{a=1}^{n_{en}} N_a(\xi,\eta)y_a}_{=y} \quad (2.300)$$

which confirms the last convergence condition.

2.5.1.1 Isoparametric finite elements with variable number of nodes

It is very easy to extend the isoparametric interpolations to higher order approximation with the finite elements of Lagrangian family. The shape functions for higher order elements of that family are simply obtained as the products of the chosen order Lagrange polynomials in ξ and in η. For example, quadrilateral 4-node element $Q4$ is the first member of Lagrangian family of finite elements, with the shape functions in (2.290) which can also be defined as the products of linear Lagrange polynomials in natural coordinates

$$N_a(\xi,\eta) = \underbrace{\frac{1}{2}(1+\xi_{a_\xi}\xi)}_{N_{a_\xi}} \underbrace{\frac{1}{2}(1+\eta_{a_\eta}\eta)}_{N_{a_\eta}}; \xi_{a_\xi} = \pm 1; \eta_{a_\eta} = \pm 1; \left\{ \begin{array}{ccccc} a & a_\xi & \xi_{a_\xi} & a_\eta & \eta_{a_\eta} \\ 1 & 2 & +1 & 2 & +1 \\ 2 & 1 & -1 & 2 & +1 \\ 3 & 1 & -1 & 1 & -1 \\ 4 & 2 & +1 & 1 & -1 \end{array} \right. \quad (2.301)$$

Any higher order approximation in the Lagrangian family can be obtained by simply increasing the order of Lagrange polynomials to quadratic, cubic, etc. For example, the first higher order element of Lagrangian family is the 9-node quadrilateral element, with the shape functions defined as the products of quadratic Lagrangian polynomials in natural coordinates

$$N_a(\xi,\eta) = N_{a_\xi}(\xi)N_{a_\eta}(\eta) \; ; \; a_\xi = 1,2,3 \; ; \; a_\eta = 1,2,3 \; ; \; a = 1,2,\ldots,9 \; ;$$

$$\left\{ \begin{array}{llll} a_\xi & N_{a_\xi} & a_\eta & N_{a_\eta} \\ 1(a=2,3,6) & \frac{1}{2}\xi(\xi-1) & 1(a=3,4,7) & \frac{1}{2}\eta(\eta-1) \\ 2(a=5,7,9) & (1+\xi)(1-\xi) & 2(a=6,8,9) & (1+\eta)(1-\eta) \\ 3(a=1,4,8) & \frac{1}{2}\xi(\xi+1) & 3(a=1,2,5) & \frac{1}{2}\eta(\eta+1) \end{array} \right. \quad (2.302)$$

In order to clearly illustrate the difference between first two elements of Lagrangian family, we present in Figure 2.10 the shape function $N_1(\xi,\eta)$ for element $Q4$, and for element $Q9$

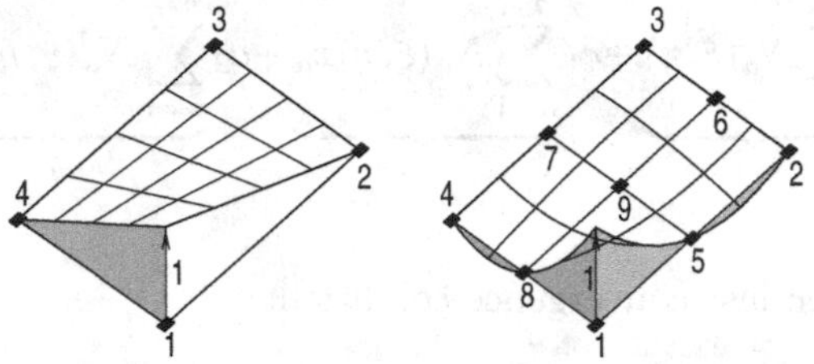

Fig. 2.10 Shape function $N_1(\xi,\eta)$ for isoparametric elements Q4 and Q9.

For a number applications dealing with heterogeneous stress field, where we seek to construct an optimal representation of the displacement or the temperature field, it is advantageous to use the finite element containing the elements $Q9$ in the high gradient sub-domains and the elements $Q4$ where the fields remain nearly homogeneous. In such a case, we should also employ for constructing the mesh the transition elements with variable number of nodes between 4 and 9, which should render two parts of the mesh compatible. We show how to construct the shape function for a 6-node transition element of this kind, which requires the nodes 5 and 8 in order to ensure the compatibility between elements $Q4$ and $Q9$. For example, the final form of the shape function $N_1(\xi,\eta)$ for such element is obtained with the appropriate modification of the original shape function of $Q4$ element, which should ensure the zero value of the modified shape function at all remaining nodes,

$$\begin{aligned} N_1(\xi,\eta) \longleftarrow & \overbrace{\frac{1}{4}(1+\xi)(1+\eta)}^{N_1(\xi,\eta)} \\ & - \underbrace{0.5}_{N_1(0,1)} \underbrace{(1-\xi^2)\frac{1}{2}(1+\eta)}_{N_5(\xi,\eta)} - \underbrace{0.5}_{N_1(1,0)} \underbrace{\frac{1}{2}(1+\xi)(1-\eta^2)}_{N_8(\xi,\eta)} \end{aligned} \quad (2.303)$$

A graphic illustration of this construction leading to the modified shape function is given in Figure 2.11

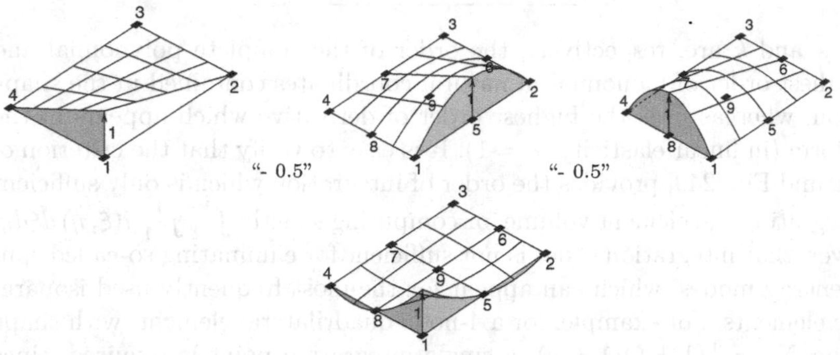

Fig. 2.11 Modified shape function of $Q4$ isoparametric element which is compatible with $Q9$ element.

In the same way we can construct the corresponding modifications for any shape function of the isoparametric element with a variable number of nodes between 5 and 9.

2.5.2 Order of numerical integration

The computation of element arrays is carried out by the numerical integration. For example, a typical component of the stiffness matrix can be computed as

$$\begin{aligned} K^e_{ab} &= \int_{\Omega^e} [\ldots + c_1 \frac{\partial N_a}{\partial x} \frac{\partial N_b}{\partial x} + \ldots] dx dy \\ &= \int_{-1}^{+1} \int_{-1}^{+1} [\ldots + c_1 \frac{\partial N_a(\xi,\eta)}{\partial x} \frac{\partial N_b(\xi,\eta)}{\partial x} + \ldots] j(\xi,\eta)\, d\xi d\eta \\ &= \sum_{i=1}^{n^\xi_{in}} \sum_{j=1}^{n^\eta_{in}} [\ldots + c_1 \frac{\partial N_a(\xi_i,\eta_j)}{\partial x} \frac{\partial N_b(\xi_i,\eta_j)}{\partial x} + \ldots] j(\xi_i,\eta_j)\, w_i w_j \\ &= \sum_{l=1}^{n_{in}} [\ldots + c_1 \frac{\partial N_a(\xi_l,\eta_l)}{\partial x} \frac{\partial N_b(\xi_l,\eta_l)}{\partial x} + \ldots] j(\xi_l,\eta_l)\, w_l \end{aligned} \tag{2.304}$$

where $n_{in} = n^\xi_{in} n^\eta_{in}$ is the total number of integration points and $w_l = w_i w_j$ are the corresponding weights. The number of integration points, or the numerical integration order, has to be selected in agreement with the element polynomial basis. The minimum integration order is obtained according to the criterion proposed by Strang and Fix [249]

$$\boxed{n_{in} \geq \bar{k} + k - 2m} \tag{2.305}$$

where $\bar{k}$ and k are, respectively, the order of the complete polynomial and the highest order of monomial in natural coordinates contained in the shape function, whereas m is the highest order of derivative which appears in the weak form (in linear elasticity $m = 1$). It is easy to verify that the criterion of Strang and Fix [249] provides the order of integration which is only sufficient for integrating the element volume, or computing exactly $\int_{-1}^{1}\int_{-1}^{1} j(\xi,\eta)\, d\xi d\eta$. However, that integration order is not sufficient for eliminating so-called spurious energy modes, which can appear for the most frequently used isoparametric elements. For example, for a 4-node quadrilateral element, with shape function $N_a = \frac{1}{4}(1 \pm \xi)(1 \pm \eta)$, a single integration point is required, since $\bar{k} = 1$, $k = 2$ (the term $\xi\eta$) and $m = 1$ resulting in $n_{in} \geq 1 + 2 - 2 \times 1 = 1$. In other words, according to the Strang and Fix integration rule, the $Q4$ element stiffness matrix is computed by

$$\mathsf{K}^e_{(8\times8)} = \mathsf{B}^T_{(8\times3)}\mathsf{C}_{(3\times3)}\mathsf{B}_{(3\times8)}\, c\ ;\ c = j(0,0)\,4 \tag{2.306}$$

The standard result limiting the rank of a matrix product to the minimum rank of all the matrices in the product confirms that the rank of such a stiffness matrix K^e can not be larger than 3

$$rank(\mathsf{K}^e) = min\{\underbrace{rank(\mathsf{B})}_{=3}, \underbrace{rank(\mathsf{C})}_{=3}\} = 3 \tag{2.307}$$

The matrix with such a rank is not sufficient to guarantee the solution uniqueness even for the simplest elasticity problem with a single $Q4$ element in the mesh, when the minimum number of supports (three supports in 2D case) is used to eliminate the rigid body modes (see Figure 2.12). Namely, a quick check shows that removing three unknowns by support constraints, the stiffness matrix of rank 3 will still leave two remaining displacement vector components (out of five) arbitrary. Moreover, the single point integration rule will provide a completely wrong interpretation to the spurious mode of $Q4$ element shown in Figure 2.12. The spurious mode represents a non-zero deformation everywhere within the element except at the chosen integration point, used for the stiffness matrix computation. Namely, only in the element center the spurious mode results with no change with respect to the element length, neither in x nor in y direction (hence $\epsilon_{11} = \epsilon_{22} = 0$), and the right angle between two axes remains preserved ($\gamma_{12} = 0$), so that the spurious mode is wrongly interpreted as a rigid body mode. This clearly illustrates erroneous interpretation and insufficiency of the Strang and Fix integration rule. A simple remedy for eliminating the spurious mode and computing the element stiffness matrix of sufficient rank is by increasing the order of integration. Hence, the arrays for a 4-node isoparametric element should be computed by $2 \times 2 = 4$ Gauss quadrature points. The same type of analysis shows that the element arrays for a 9-node element should be computed by $3 \times 3 = 9$ numerical points in order to avoid the spurious mode shown in Figure 2.12 (exercise: try to prove this).

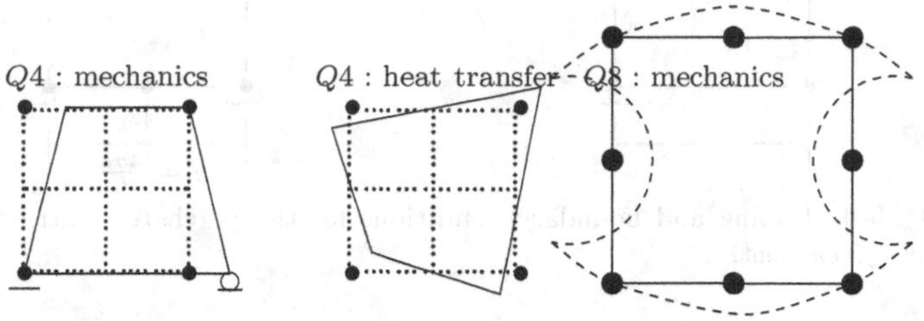

Fig. 2.12 Spurious zero-energy modes for isoparametric elements with four nodes (both mechanics and heat transfer problems) and with eight or nine nodes.

2.5.3 The patch test

With a large variety of isoparametric elements and their spurious modes, we need a convergence test which will confirm definitely that the exact solution can be computed by increasing the number of finite elements in the mesh. One such test is the patch test, initially proposed by Irons [143]. The mechanics basis of the patch test is rather easy to grasp: by refining the mesh in seeking the convergence, each element will become smaller and the stress state in the element will approach the stress state in a point, which remains constant. Hence, in the spirit of the patch test, the convergence is guaranteed if the finite element is capable of exact representation of a constant stress field. Among several possibilities to carry out the patch test (e.g. see [271]), we choose the one which is performed on a single element (see Figure 2.13). We obtain the numerical solution for nodal displacements, with the minimum number of supports imposed to eliminate the rigid body modes, under the equivalent nodal loads computed in agreement with the chosen constant stress field. We can then compare the computed results for displacement, strain and stress against the corresponding analytic solution which can easily be obtained in this case. Although in theory the convergence test of this kind should be concerned with a very small (or vanishing) element, practically the element size will not play any role for a constant stress field solution. We should also choose preferably a more stringent version of the patch test with a distorted element, which does not have a constant Jacobian matrix; see Figure 2.13.

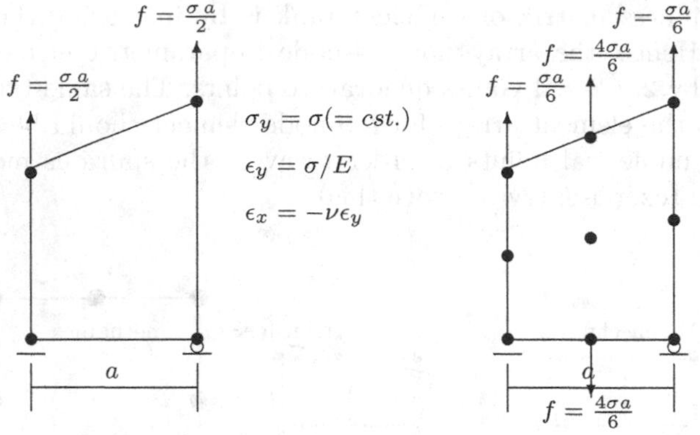

Fig. 2.13 Mesh, loading and boundary conditions for the patch test carried out on a single $Q4$ or $Q9$ element.

The patch test can also be used to check the order of convergence and the optimal choice of the finite element approximation for a particular boundary value problem, or simply the choice of the element type. Choosing the optimal type of finite element is rather a subtle question, which does not only imply the choice between a triangular or a quadrilateral element form, mostly decided by the available tools for mesh preparation, but also the choice of the order of finite element approximation fixed by the number of element nodes. One could think that, each time it is affordable, it should be sufficient to simply increase the order of approximation and the number of element nodes, by say choosing a quadrilateral 9-node element instead of the one with 4 nodes, or continue even further towards a 16-node element in order to obtain an optimal convergence rate. However, such a reasoning is not necessarily true, since an inappropriate use of numerical integration, the element distortion or incompleteness of higher order polynomials in the finite element approximation might significantly impair the performance of isoparametric elements with a high number of nodes. The patch test kind of verification can again be useful in providing the answer to this kind of question. More precisely, one can choose the higher order patch test to verify the capability of a particular element to represent exactly a more complex stress state than constant. For example, a higher order patch test with linear variation of stress field is defined by the pure bending of a beam (see Figure 2.15). A 4-node quadrilateral element can not pass this higher order patch test, contrary to a 9-node quadrilateral element of a rectangular form. In fact, the same test is passed by any 9-node quadrilateral that is produced from its parent element by the 4-node mapping keeping each side straight and placing in its center the mid-side nodes, since this kind of mapping preserves the shape functions as quadratic polynomials not only in ξ and η, but also in x and y. Going further in the same direction, we can find that a 16-node isoparametric element can pass a higher order patch test with a quadratic stress field variation, when it takes a regular (non-distorted) form.

However, these higher order elements (e.g. both 9-node or 16-node) are very sensitive to element distortion, and would loose a great deal of accuracy when distorted. This distortion problem is even more severe in large displacement problems, where even starting with a structured mesh of regular elements is not a sufficient guarantee to avoid the element distortion problems in the deformed configuration, particularly under a heterogeneous stress field. Finally, a higher order finite element approximation would provide an unbalanced approximation with imposing an excessive solution regularity within the element domain, as opposed to the reduced regularity at the element boundaries. Moreover, higher order regularity does not necessarily characterize the exact solution of a boundary value problems dealing with inelastic nonlinear behavior. All these deficiencies of higher order elements are not easy to avoid, and one often prefers to use for nonlinear problems a low order finite element approximation, based on 3-node triangular or 4-node quadrilateral elements, since neither is overly sensitive to element distortion.
The low order finite element approximations do not have this difficulty of providing an excessive solution smoothness, but they might have a new difficulty of not enough smoothness. Fortunately, this is a difficulty we can eliminate. For example, we can improve the behavior of some low order finite elements and construct a complete polynomial in their basis without increasing the number of nodes, by employing the method of incompatible modes or enhanced strains which is described subsequently. Yet another reason for enriching the element basis pertains to so-called locking phenomena, which indicate the element inability to represent certain deformation modes or to accommodate the presence of constraint on displacement field. The locking phenomena appear in a number of practical applications. One example of this kind is a (quasi) incompressible behavior which produces (almost) no change of volume between initial and deformed configurations. The quasi-incompressible behavior is characteristic of rubber, water or metals obeying von Mises plasticity criterion near their ultimate limit state. The locking phenomena can also be dealt with by using the assumed strain method (such as $B-bar$ method), which contrary to the incompatible modes method approximation basis enrichment, is based on judicious reduction of the approximation basis of isoparametric elements which can handle the constraint. The assumed strain method is also described subsequently.

2.5.4 Hu-Washizu (mixed) variational principle and method of incompatible modes

In this section we present the method of incompatible modes or enhanced strains, which allows us to construct an enriched discrete approximation basis for a chosen isoparametric element. The ideas presented herein were first elaborated by the works of Ibrahimbegovic and Wilson [140], Simo and Rifai

[241] and Wilson and Taylor [250]. We first recall the strong form of 3D linear elasticity, where three fields: the displacement $\mathbf{u}$, the deformation $\boldsymbol{\epsilon}$ and the stress $\boldsymbol{\sigma}$, obey three groups of equations

$$
\begin{aligned}
\boldsymbol{\epsilon} = \nabla^s \mathbf{u} &\Leftrightarrow \epsilon_{ij} = \left(\frac{\partial u_i}{\partial x_j} + \frac{\partial u_j}{\partial x_i}\right)/2 \\
\boldsymbol{\sigma} = \boldsymbol{\mathcal{C}}\boldsymbol{\epsilon} &\Leftrightarrow \sigma_{ij} = \mathcal{C}_{ijkl}\epsilon_{kl} \\
div[\boldsymbol{\sigma}] + \mathbf{b} = \mathbf{0} &\Leftrightarrow \frac{\partial \sigma_{ij}}{\partial x_j} + b_i = 0
\end{aligned}
\tag{2.308}
$$

which describe, respectively, the kinematics, the constitutive behavior and the equilibrium at each point $\mathbf{x} \in \Omega$. The displacement-type variational formulation is constructed by replacing the strong form of equilibrium equations with the corresponding weak form, where this equation is no longer enforced point-wise but only in the integral sense. The local enforcement still applies to constitutive and kinematics equations, which allows us to express the stress field directly in terms of displacements thus providing the weak form with only displacement field as unknown. Recall that such a displacement-type variational formulation of linear elasticity requires that the displacement field $\mathbf{u} = u_i \mathbf{e}_i$ belongs to the solution space

$$
\mathbb{V}_i = \left\{ u_i \mid u_i \in H^1(\Omega), u_i = \bar{u}_i(=0)\; on\; \Gamma_{u_i} \right\} \tag{2.309}
$$

The displacement-type variational formulation requires the same regularity for the virtual displacement field $\mathbf{w} = w_i \mathbf{e}_i$, along with the zero value at the Dirichlet boundary

$$
\mathbb{V}_{i,0} = \left\{ w_i \mid w_i \in H^1(\Omega), w_i = 0\; on\; \Gamma_{u_i} \right\} \tag{2.310}
$$

For displacement-type variational formulation, these two spaces also define the real and virtual strain and stress fields through the point-wise use of the kinematics and constitutive equations, respectively. Unlike the choice made for the displacement-type variational formulation, we will now consider that the strain and stress fields are defined independently from the displacement field. Namely, we turn to so-called mixed or Hu-Washizu variational formulation (see [260]), where the weak form is constructed by attributing to all three groups of equations: kinematics, constitutive and equilibrium equations, not a point-wise but only an integral interpretation. Therefore, we ought to choose for mixed variational formulation not only the space of virtual displacements, but also the spaces of virtual stress field $\boldsymbol{\tau} = \tau_{ij} \mathbf{e}_i \otimes \mathbf{e}_j$ and virtual strain field $\boldsymbol{\gamma} = \gamma_{ij} \mathbf{e}_i \otimes \mathbf{e}_j$. For the simplest problem of 3D linear elasticity, we can write:

Mixed or Hu-Washizu variational formulation of 3D linear elasticity
Given: $b_i : \Omega \mapsto \mathbb{R}$, $\bar{t}_i : \Gamma_{\sigma_i} \mapsto \mathbb{R}$, $\bar{u}_i : \Gamma_{u_i} \mapsto \mathbb{R}$
Find: $u_i \in \mathbb{V}_i, \epsilon_{ij} \in \mathbb{E}_{ij}, \sigma_{ij} \in \mathbb{T}_{ij}$
such that $\forall w_i \in \mathbb{V}_{i,0}$, $\forall \gamma_{ij} \in \mathbb{E}_{ij}$, $\forall \tau_{ij} \in \mathbb{T}_{ij}$

$$\begin{aligned}
0 &= G_u(\mathbf{u}, \boldsymbol{\epsilon}, \boldsymbol{\sigma}; \mathbf{w}) := \textstyle\int_\Omega \nabla^s \mathbf{w} \cdot \boldsymbol{\sigma}\, dV - \int_\Omega \mathbf{w} \cdot \mathbf{b}\, dV - \int_{\Gamma_\sigma} \mathbf{w} \cdot \bar{\mathbf{t}}\, dA \\
0 &= G_\sigma(\mathbf{u}, \boldsymbol{\epsilon}, \boldsymbol{\sigma}; \boldsymbol{\tau}) := \textstyle\int_\Omega \boldsymbol{\tau} \cdot (\nabla^s \mathbf{u} - \boldsymbol{\epsilon})\, dV \\
0 &= G_\epsilon(\mathbf{u}, \boldsymbol{\epsilon}, \boldsymbol{\sigma}; \boldsymbol{\gamma}) := \textstyle\int_\Omega \boldsymbol{\gamma} \cdot (\boldsymbol{\mathcal{C}} \boldsymbol{\epsilon} - \boldsymbol{\sigma})\, dV
\end{aligned} \tag{2.311}$$

The corresponding Euler-Lagrange equations of the mixed variational principle in (2.311) above, which can be obtained by the standard procedure of integration by parts (an exercise left to readers), allow to recover the complete set of equations of the strong form in (2.308) providing the sufficient regularity of all the fields. Alternatively, the mixed weak form provides the necessary stationarity condition of the Hu-Washizu functional, which can be written:

$$\Pi(\mathbf{u}, \boldsymbol{\epsilon}, \boldsymbol{\sigma}) = \int_\Omega \frac{1}{2} \boldsymbol{\epsilon} \cdot \boldsymbol{\mathcal{C}} \boldsymbol{\epsilon}\, dV + \int_\Omega \boldsymbol{\sigma} \cdot (\nabla^s \mathbf{u} - \boldsymbol{\epsilon}) dV - \int_\Omega \mathbf{u} \cdot \mathbf{b}\, dV - \int_{\Gamma_\sigma} \mathbf{u} \cdot \bar{\mathbf{t}}\, dA \tag{2.312}$$

where $\mathbf{u}$, $\boldsymbol{\epsilon}$ and $\boldsymbol{\sigma}$ are all independent fields. The regularity of the displacement field in (2.312) above is identical to the one in (2.309) with required kinematic admissibility (for Dirichlet boundary conditions) and continuity of displacement over element boundaries, along with square-integrability of its derivatives $\mathbf{u} \in \mathbb{V}$. The mixed variation formulation is less demanding with respect to the regularity of the stress and strain fields, which need not be continuous but only square-integrable functions

$$\begin{aligned}
\mathbb{E}_{ij} &= \{\epsilon_{ij} \mid \epsilon_{ij} \in L_2(\Omega)\} \\
T_{ij} &= \{\sigma_{ij} \mid \sigma_{ij} \in L_2(\Omega)\}
\end{aligned} \tag{2.313}$$

Moreover, no Dirichlet boundary condition needs to be imposed on stress or strain, hence the same spaces can be used for virtual strain and stress fields, with $\gamma_{ij} \in \mathbb{E}_{ij}$ for strains and $\tau_{ij} \in \mathbb{T}_{ij}$ for stresses. The condition (2.313) on stress and strain fields square-integrability implies that they can be constructed independently from one element to another, allowing for discontinuities between two neighboring elements.

An important goal in choosing the independent strain field in the mixed formulation pertains to enrichment of the corresponding approximation space of a isoparametric element. We can write the enriched strain field representation as the sum of the standard strain field interpolation for the isoparametric element and an enhanced strain, either for real $\tilde{\boldsymbol{\epsilon}}$ or virtual strain $\tilde{\boldsymbol{\gamma}}$

$$\begin{aligned}
\boldsymbol{\epsilon} = \nabla^s \mathbf{u} + \tilde{\boldsymbol{\epsilon}} \quad &\Leftrightarrow \quad \epsilon_{ij} = \tfrac{1}{2}\left(\frac{\partial u_i}{\partial x_j} + \frac{\partial u_j}{\partial x_i}\right) + \tilde{\epsilon}_{ij} \\
\boldsymbol{\gamma} = \nabla^s \mathbf{w} + \tilde{\boldsymbol{\gamma}} \quad &\Leftrightarrow \quad \gamma_{ij} = \tfrac{1}{2}\left(\frac{\partial w_i}{\partial x_j} + \frac{\partial w_j}{\partial x_i}\right) + \tilde{\gamma}_{ij}
\end{aligned} \tag{2.314}$$

By introducing this strain field modification into Hu-Washizu variational formulation, the weak form of the constitutive equations in $(2.311)_3$ can be split into two separate statements, the first related to the displacement variation $\mathbf{w}$ and the second to enhanced strain variation γ

$$\begin{aligned} \int_{\Omega} \nabla^s \mathbf{w} \cdot [-\boldsymbol{\sigma} + \boldsymbol{\mathcal{C}}(\nabla^s \mathbf{u} + \tilde{\epsilon})] \, dV = 0 \\ \int_{\Omega} \tilde{\gamma} \cdot [-\boldsymbol{\sigma} + \boldsymbol{\mathcal{C}}(\nabla^s \mathbf{u} + \tilde{\epsilon})] \, dV = 0 \end{aligned} \tag{2.315}$$

By using the first of these two equations, the weak form of equilibrium equations in $(2.311)_1$ can be recast in the following format:

$$\int_{\Omega} \underbrace{\nabla^s \mathbf{w} \cdot \boldsymbol{\mathcal{C}}(\nabla^s \mathbf{u} + \tilde{\epsilon})}_{\nabla^s \mathbf{w} \cdot \boldsymbol{\sigma}} \, dV - \int_{\Omega} \mathbf{w} \cdot \mathbf{b} \, dV - \int_{\Gamma_\sigma} \mathbf{w} \cdot \bar{\mathbf{t}} \, dA = 0 \tag{2.316}$$

The weak form of kinematics equations in $(2.311)_2$ can also be simplified in the same manner to obtain

$$\int_{\Omega} \boldsymbol{\tau} \cdot \tilde{\epsilon} \, dV = 0 \tag{2.317}$$

It is easy to check that such variational equation derive from the modified Hu-Washizu functional, which can be written:

$$\begin{aligned} \Pi(\mathbf{u}, \tilde{\epsilon}, \boldsymbol{\sigma}) = \int_{\Omega} \left[\tfrac{1}{2} (\nabla^s \mathbf{u} + \tilde{\epsilon}) \cdot \boldsymbol{\mathcal{C}}(\nabla^s \mathbf{u} + \tilde{\epsilon}) - \boldsymbol{\sigma} \cdot \tilde{\epsilon} \right] \, dV \\ - \int_{\Omega} \mathbf{w} \cdot \mathbf{b} \, dV - \int_{\Gamma_\sigma} \mathbf{w} \cdot \bar{\mathbf{t}} \, dA \end{aligned} \tag{2.318}$$

where $\mathbf{u}$, $\tilde{\epsilon}$ and $\boldsymbol{\sigma}$ are independent fields for which we ought to choose the appropriate discrete approximations. The Euler–Lagrange equation corresponding to variation of $\boldsymbol{\sigma}$ leads to the corresponding strong form where the enhanced strain should vanish, $\tilde{\epsilon} = \mathbf{0}$. This result confirms that no improvement is possible beyond the standard definition of strain field as the symmetric part of the displacement gradient with the strong form of the kinematics equation. The weak form, on the contrary, allows us to construct an enhanced deformation field $\tilde{\epsilon}$ capable of providing an enriched deformation field with respect to the standard approximation furnished by an isoparametric finite element.
Any enhanced strain field must satisfy the stress orthogonality conditions in (2.317). We provide subsequently quite a general construction procedure of enhanced strain field, which fulfills this kind of requirement. Without loss of generality the procedure is illustrated for a 2D elasticity case, where only enhancement is constructed for non-zero components. The real and virtual displacement fields in 2D are presented in matrix notation as

$$\mathbf{u} = u_i \mathbf{e}_i \ \mapsto \ \mathsf{u} = [u_i] = \begin{bmatrix} u_1 \\ u_2 \end{bmatrix}, \mathbf{w} = w_i \mathbf{e}_i \ \mapsto \ \mathsf{w} = [w_i] = \begin{bmatrix} w_1 \\ w_2 \end{bmatrix} \tag{2.319}$$

The real and virtual deformation tensor components are also placed each in a column matrix

$$\nabla^s \mathbf{u} \mapsto \hat{\mathsf{e}}(\mathsf{u}) := \begin{bmatrix} \frac{\partial u_1}{\partial x_1} \\ \frac{\partial u_2}{\partial x_2} \\ \frac{\partial u_1}{\partial x_2} + \frac{\partial u_2}{\partial x_1} \end{bmatrix} ; \; \tilde{\epsilon} \mapsto \tilde{\mathsf{e}} = \begin{bmatrix} \tilde{\epsilon}_{11} \\ \tilde{\epsilon}_{22} \\ 2\tilde{\epsilon}_{12} \end{bmatrix}$$

$$\nabla \mathbf{w} \mapsto \hat{\mathsf{e}}(\mathsf{w}) = \begin{bmatrix} \frac{\partial w_1}{\partial x_1} \\ \frac{\partial w_2}{\partial x_2} \\ \frac{\partial w_1}{\partial x_2} + \frac{\partial w_2}{\partial x_1} \end{bmatrix} ; \; \tilde{\gamma} \mapsto \tilde{\mathsf{g}} = \begin{bmatrix} \tilde{\gamma}_{11} \\ \tilde{\gamma}_{22} \\ 2\tilde{\gamma}_{12} \end{bmatrix} \tag{2.320}$$

This enforced the order for storing the stress tensor components and elasticity tensor within the corresponding matrices

$$\boldsymbol{\sigma} \mapsto \mathsf{s} = \begin{bmatrix} \sigma_{11} \\ \sigma_{22} \\ \sigma_{12} \end{bmatrix} ; \; \boldsymbol{\tau} \mapsto \mathsf{t} = \begin{bmatrix} \tau_{11} \\ \tau_{22} \\ \tau_{12} \end{bmatrix} ; \; \boldsymbol{\mathcal{C}} \mapsto \mathsf{C} = \begin{bmatrix} \bar{\lambda} + 2\mu & \bar{\lambda} & 0 \\ \bar{\lambda} & \bar{\lambda} + 2\mu & 0 \\ 0 & 0 & \mu \end{bmatrix} \tag{2.321}$$

where $\bar{\lambda} = \lambda$ for plane strain and $\bar{\lambda} = 2\lambda\mu/(\lambda + 2\mu)$ for plane stress case. Having defined these arrays, we can restate the Hu-Washizu mixed formulation for incompatible mode method in matrix notation according to:

Given: $\mathsf{b}, \bar{\mathsf{t}}, (\bar{\mathsf{u}} = 0)$

Find: $\mathsf{u}^h \in \mathbb{V}^h, \tilde{\mathsf{e}}^h \in \mathbb{E}^h, \mathsf{s}^h \in \mathbb{T}^h$,

such that $\forall \mathsf{w}^h \in \mathbb{V}_0^h, \forall \tilde{\mathsf{g}}^h \in \mathbb{E}^h, \forall \mathsf{t}^h \in \mathbb{T}^h$

$$\begin{aligned} &\int_{\Omega^h} \hat{\mathsf{e}}^T(\mathsf{w}^h) \mathsf{D}(\hat{\mathsf{e}}(\mathsf{u}^h) + \tilde{\mathsf{e}}^h)\, dV - \int_{\Omega^h} \mathsf{w}^{h\,T} \mathsf{b} dV - \int_{\Gamma_\sigma^h} \mathsf{w}^{h\,T} \bar{\mathsf{t}}\, dA = 0 \\ &\int_{\Omega^h} \mathsf{t}^{h\,T} \tilde{\mathsf{e}}^h dV = 0 \\ &\int_{\Omega^h} \mathsf{g}^{h\,T} \left[-\mathsf{s}^h + \mathsf{D}(\hat{\mathsf{e}}(\mathsf{u}^h) + \tilde{\mathsf{e}}^h) \right] dV = 0 \end{aligned} \tag{2.322}$$

where the superscript 'h' denotes the finite-dimensional approximations of the corresponding fields constructed by the finite element method. We consider subsequently an isoparametric finite element with n_{en} as the basis for constructing all the approximations employed by the mixed principle of this kind. First, the standard isoparametric interpolations are used for real and virtual displacement fields

$$\begin{aligned} \mathsf{u}^h \mid_{\Omega^e} &= \textstyle\sum_{a=1}^{n_{en}} N_a^e(\boldsymbol{\xi}) \mathsf{d}_a^e \\ \mathsf{w}^h \mid_{\Omega^e} &= \textstyle\sum_{a=1}^{n_{en}} N_a^e(\boldsymbol{\xi}) \mathsf{w}_a^e \end{aligned} \tag{2.323}$$

where $N_a(\boldsymbol{\xi})$ are shape functions of this element. Discrete approximation of the compatible deformation field is obtained by applying ∇^s operator to this displacement field to obtain

$$\hat{\mathsf{e}}(\mathsf{w}^h)\mid_{\Omega^e} = \sum_{a=1}^{n_{en}} \mathsf{B}_a^e \mathsf{w}_a^e \ ; \ \hat{\mathsf{e}}(\mathsf{u}^h) = \sum_{a=1}^{n_{en}} \mathsf{B}_a^e \mathsf{d}_a^e \ ; \ \mathsf{B}_a = \begin{bmatrix} \frac{\partial N_a}{\partial x_1} & 0 \\ 0 & \frac{\partial N_a}{\partial x_2} \\ \frac{\partial N_a}{\partial x_2} & \frac{\partial N_a}{\partial x_1} \end{bmatrix} \tag{2.324}$$

The enhanced deformation field is constructed by following the proposal of Ibrahimbegovic and Wilson [140], by applying the same ∇^s operator to an incompatible displacement field

$$\tilde{\mathsf{u}}(\boldsymbol{\xi})\mid_{\Omega^e} = \sum_{b=1}^{n_{im}} M_b(\boldsymbol{\xi})\boldsymbol{\alpha}_b \tag{2.325}$$

where $M_b(\xi,\eta)$ are the chosen incompatible model and $\boldsymbol{\alpha}_b$ are the corresponding interpolation parameters defined in each element independently. This produces the following enhanced deformation

$$\tilde{\mathsf{e}}^h\Big|_{\Omega^e} = \sum_{b=1}^{n_{im}} \mathsf{G}_b(\boldsymbol{\xi})\boldsymbol{\alpha}_b^e \ , \ ; \ \mathsf{G}_b = \begin{bmatrix} \frac{\partial M_b}{\partial x_1} & 0 \\ 0 & \frac{\partial M_b}{\partial x_2} \\ \frac{\partial M_b}{\partial x_2} & \frac{\partial M_b}{\partial x_1} \end{bmatrix} \tag{2.326}$$

It is important to note that the reference configuration of the isoparametric element with incompatible modes is still defined only with compatible shape functions, $\mathsf{x}^h \mid_{\Omega^e} = \sum_{a=1}^{n_{en}} N_a(\xi,\eta)\mathsf{x}_a$, which further implies that the derivatives of the incompatible mode shape functions are computed in the same manner as for the chosen isoparametric element whose basis is enriched; therefore, we can write

$$\begin{aligned} \tfrac{\partial M_b}{\partial x}(\xi,\eta) &= [\tfrac{\partial M_b}{\partial \xi}(\xi,\eta)\tfrac{\partial y}{\partial \xi}(\xi,\eta) - \tfrac{\partial M_b}{\partial \eta}(\xi,\eta)\tfrac{\partial y}{\partial \xi}(\xi,\eta)]/j(\xi,\eta) \ ; \\ \tfrac{\partial M_b}{\partial y}(\xi,\eta) &= -\tfrac{\partial M_b}{\partial \xi}(\xi,\eta)\tfrac{\partial x}{\partial \xi}(\xi,\eta) + \tfrac{\partial M_b}{\partial \eta}(\xi,\eta)\tfrac{\partial x}{\partial \xi}(\xi,\eta)]/j(\xi,\eta) \end{aligned} \tag{2.327}$$

where $j(\xi,\eta)$ is the determinant of the Jacobian matrix, which is computed as

$$\begin{aligned} \tfrac{\partial x(\xi,\eta)}{\partial \xi} &= \textstyle\sum_{a=1}^{e} \tfrac{\partial N_a^e(\xi,\eta)}{\xi} x_a^e \ ; \ \tfrac{\partial x(\xi,\eta)}{\partial \eta} = \textstyle\sum_{a=1}^{e} \tfrac{\partial N_a^e(\xi,\eta)}{\eta} x_a^e \ ; \\ \tfrac{\partial y(\xi,\eta)}{\partial \xi} &= \textstyle\sum_{a=1}^{e} \tfrac{\partial N_a^e(\xi,\eta)}{\xi} y_a^e \ ; \ \tfrac{\partial y(\xi,\eta)}{\partial \eta} = \textstyle\sum_{a=1}^{e} \tfrac{\partial N_a^e(\xi,\eta)}{\eta} y_a^e \ ; \\ j(\xi,\eta) &= \tfrac{\partial x(\xi,\eta)}{\partial \xi}\tfrac{\partial y(\xi,\eta)}{\partial \eta} - \tfrac{\partial x(\xi,\eta)}{\partial \eta}\tfrac{\partial y(\xi,\eta)}{\partial \xi} \end{aligned} \tag{2.328}$$

The presence of incompatible modes in defining the discrete approximation of strain field is thus the only difference between an enhanced element and its associated isoparametric element. For that reason, the convergence conditions

for an enhanced element with incompatible modes should first include the corresponding conditions for isoparametric elements, and further require the following:
*Condition 1**: The enhanced deformation should be independent from the deformation field produced from the compatible displacement field of the isoparametric element to which we add the incompatible modes. Considering the proposed manner to produce the enhanced deformation field in (2.326), this condition imposes directly the choice of shape function for compatible displacement field N_a, $a = 1, ..., n_{en}$ versus those for incompatible modes $M_b, b = 1, \ldots, n_{im}$, requiring that they do not belong to the same space

$$N_a \bigcap M_b = \emptyset\,;\, \forall a, b \tag{2.329}$$

Such a requirement can easily be verified if the incompatible modes are chosen as higher order polynomials with respect to the corresponding isoparametric element basis. For example, for a 4-node element with shape functions $N_a(\xi, \eta)$ as linear polynomials in ξ and in η, we choose incompatible modes $M_b(\xi, \eta)$ as quadratic polynomials, whereas for a 9-node element, with shape functions $N_a(\xi, \eta)$ being quadratic, $M_b(\xi, \eta)$ are cubic polynomials.
*Condition 2**: We would like to avoid the presence of stress parameters in the discrete approximation of incompatible mode method. This would allow that the incompatible mode method retains the advantage of stress computation from strain in linear elasticity, which can be generalized with no modification to a more complex constitutive law. For that reason, we have to enforce the element-wise orthogonality between the stress and incompatible mode parameters leading to

$$\begin{aligned} \int_{\Omega^h} \mathsf{t}^{h\,T} \tilde{\mathsf{e}}^h \, dV = 0 \\ \int_{\Omega^h} \tilde{\mathsf{g}}^{h\,T} \mathsf{s}^h dV = 0 \end{aligned} \tag{2.330}$$

*Condition 3**: It is indispensable to make sure that any enhanced strain remains orthogonal to any constant stress within the element and that no work results from coupling between these two fields. This will guarantee convergence of the incompatible mode method in the spirit of the patch test, where the background isoparametric element which serves as the basis for construction of incompatible modes assures the representation of this constant stress field. Hence, the incompatible modes and enhanced strains will contribute only to a coarse mesh representation of a heterogeneous stress field, and will have a negligible contribution to a very fine mesh with the stress field in each element close to a constant. For a constant stress in element denoted as s_c, the orthogonality condition in (2.330) reduces to

$$\left.\begin{aligned} \int_{\Omega^h} \mathsf{t}_c^T \tilde{\mathsf{e}}^h dV = 0 \\ \int_{\Omega^h} \tilde{\mathsf{g}}^{h\,T} \mathsf{s}_c dV = 0 \end{aligned}\right\} \Longrightarrow \int_{\Omega^e} \tilde{\mathsf{G}}_b dV = 0 \tag{2.331}$$

It was shown by Ibrahimbegovic and Wilson [140] that for any incompatible mode strain approximation can be rewritten in the modified form satisfying the patch test condition in (2.331) according to

$$\tilde{\mathsf{e}} = \underbrace{[\mathsf{G} - \frac{1}{\Omega^e}\int_{\Omega^e} \mathsf{G} dV]}_{\tilde{\mathsf{G}}}\boldsymbol{\alpha}^e \; ; \; \tilde{\mathsf{G}}_b = \begin{bmatrix} \frac{\partial \tilde{M}_b}{\partial x} & 0 \\ 0 & \frac{\partial \tilde{M}_b}{\partial y} \\ \frac{\partial \tilde{M}_b}{\partial y} & \frac{\partial \tilde{M}_b}{\partial x} \end{bmatrix} \tag{2.332}$$

The proposed modification of incompatible modes can be carried out independently for each component of matrix G with

$$\begin{aligned} \frac{\partial \tilde{M}_b}{\partial x}(\xi,\eta) &= \frac{\partial M_b}{\partial x}(\xi,\eta) - \tfrac{1}{\Omega^e}\int_{-1}^{1}\int_{-1}^{1}\frac{\partial M_b}{\partial x} j d\xi d\eta \\ \frac{\partial \tilde{M}_b}{\partial y}(\xi,\eta) &= \frac{\partial M_b}{\partial y}(\xi,\eta) - \tfrac{1}{\Omega^e}\int_{-1}^{1}\int_{-1}^{1}\frac{\partial M_b}{\partial y} j d\xi d\eta \\ \Omega^e &= \int_{-1}^{1}\int_{-1}^{1} j \, d\xi d\eta \end{aligned} \tag{2.333}$$

where all the integrals are computed by using the same numerical integration rule as the one used for the background isoparametric element.
With a choice of the incompatible modes verifying the convergence conditions, one can write the matrix form of a single-element-contribution to of the mixed variational principle

$$\begin{aligned} \mathsf{K}^e\mathsf{d}^e + \mathsf{F}^{e\,T}\boldsymbol{\alpha}^e &= \mathsf{f}^e \; ; \; \forall \mathsf{w}^e \\ \mathsf{F}^e\mathsf{d}^e + \mathsf{H}^e\boldsymbol{\alpha}^e &= 0 \; ; \; \forall \boldsymbol{\beta}^e \end{aligned} \tag{2.334}$$

where $\mathsf{K}^e_{ab} = \int_{\Omega^e} \mathsf{B}_a^T \mathsf{D}\mathsf{B}_b dV$, $\mathsf{F}^e_{ab} = \int_{\Omega^e} \tilde{\mathsf{G}}_a^T \mathsf{D}\mathsf{B}_b dV$, $\mathsf{H}^e_{ab} = \int_{\Omega^e} \tilde{\mathsf{G}}_a^T \mathsf{D}\tilde{\mathsf{G}}_b dV$ and $\mathsf{f}^e_b = \int_{\Omega^e} N_a \mathsf{b} dV + \int_{\Gamma^e_\sigma} N_a \bar{\mathsf{t}} dA$.
Before we can proceed towards the finite element assembly accounting for all element contributions, we should first carry out the static condensation (Wilson [263]) in order to eliminate the presence of incompatible mode parameters $\boldsymbol{\alpha}^e$

$$\begin{aligned} \boldsymbol{\alpha}^e &= -\mathsf{H}^{e\,-1}\mathsf{F}^e\mathsf{d}^e \\ \tilde{\mathsf{K}}^e\mathsf{d}^e &= \mathsf{f}^e \; ; \; \tilde{\mathsf{K}}^e = \mathsf{K}^e - \mathsf{F}^{e\,T}\mathsf{H}^{e\,-1}\mathsf{F}^e \end{aligned} \tag{2.335}$$

We thus obtain the reduced form of the stiffness matrix for an element with incompatible modes of the same size as the one for the corresponding isoparametric element. We can then carry on with the standard finite element assembly, leading towards a set of algebraic equations whose solution provides the values of the nodal displacements

$$\mathsf{K}\mathsf{d} = \mathsf{f} \; ; \; \mathsf{K} = \mathop{\mathbb{A}}_{e=1}^{n_{elem}} \tilde{\mathsf{K}}^e \implies \mathsf{d} \tag{2.336}$$

Subsequently, we can select the corresponding element displacements $\mathsf{d}^e = \mathsf{L}^e\mathsf{d}$, which allows to obtain the incompatible mode parameters $\boldsymbol{\alpha}^e$ and carry out the recovery of the total strain field

$$\boldsymbol{\alpha}^e = -\mathsf{H}^{e\,-1}\mathsf{F}^e\mathsf{d}^e \;\Rightarrow\; \mathsf{e}(\boldsymbol{\xi}) = \sum_{a=1}^{n_{en}} \mathsf{B}_a(\boldsymbol{\xi})\mathsf{d}_a^e + \sum_{b=1}^{n_{im}} \tilde{\mathsf{G}}_b(\boldsymbol{\xi})\boldsymbol{\alpha}_b^e \tag{2.337}$$

The element stress field computation can then be performed as dictated by the chosen constitutive equations. The presented procedure applies to any isoparametric element with incompatible modes, regardless of the number of nodes. A couple of elements which were shown to be very robust (see Ibrahimbegovic and Wilson [140]) are a quadrilateral 4-node element with two quadratic incompatible modes $M_1 = 1-\xi^2$ and $M_2 = 1-\eta^2$, and quadrilateral 9-node element with two cubic incompatible modes $M_1 = \xi(1 - \xi^2)$ and $M_2 = \eta(1 - \eta^2)$ (see Figure 2.14). For each of those elements, the main criterion for selecting the incompatible modes pertains to providing the complete polynomial (either quadratic for 4-node or cubic for 9-node element) in the element approximation basis.

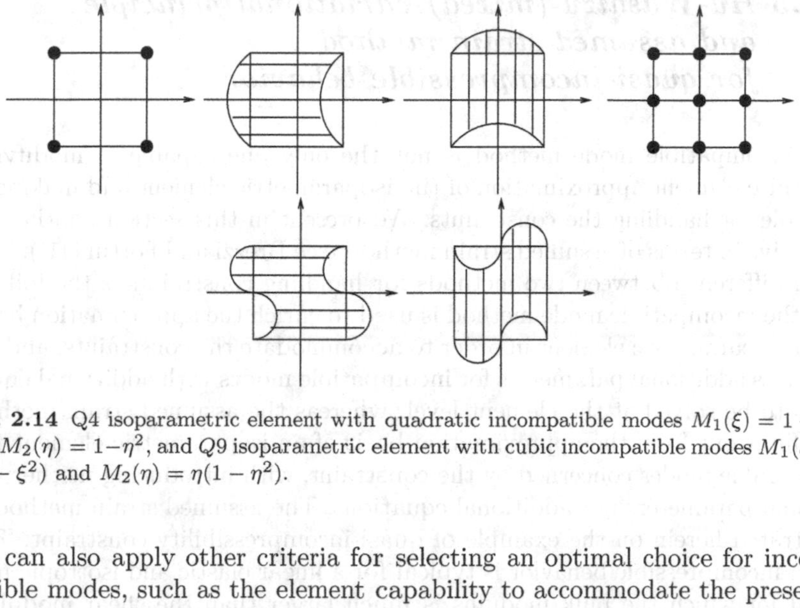

Fig. 2.14 Q4 isoparametric element with quadratic incompatible modes $M_1(\xi) = 1 - \xi^2$ and $M_2(\eta) = 1-\eta^2$, and $Q9$ isoparametric element with cubic incompatible modes $M_1(\xi) = \xi(1 - \xi^2)$ and $M_2(\eta) = \eta(1 - \eta^2)$.

We can also apply other criteria for selecting an optimal choice for incompatible modes, such as the element capability to accommodate the presence of constraint upon the configuration space, such as (quasi-)incompressibility or pure bending constraint. One illustrative example of this kind is given in Figure 2.15, pertaining to the pure bending of a cantilever beam solved with a finite element model containing only four elements. The results for the end displacement are computed for three different types of elements: standard isoparametric $Q4$ element, the same element with two quadratic incompatible modes and mixed element $Q4/P0$ (discussed in detail in the next section), and compared against the exact analytic result for different values of Poisson's ratio. We can see that the standard isoparametric element is lagging well behind the enhanced alternatives both in representing the pure bending mode and for quasi-incompressible case (with $\nu \approx 0.5$).

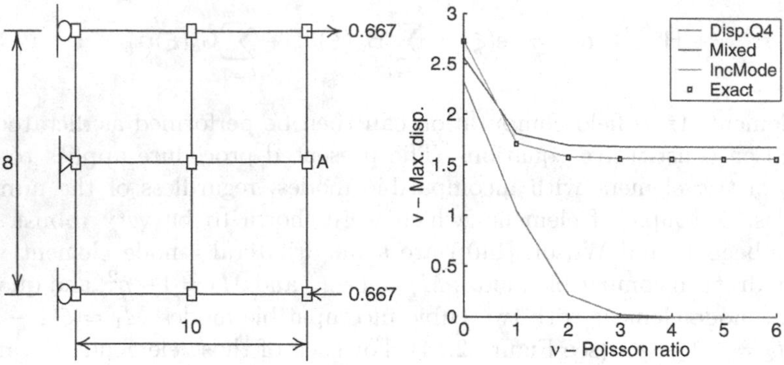

Fig. 2.15 Pure bending of a cantilever: mesh and computed displacement for different finite element approximations.

2.5.5 Hu-Washizu (mixed) variational principle and assumed strain method for quasi-incompressible behavior

The incompatible mode method is not the only one capable of modifying the finite element approximation of the isoparametric element and making it suitable for handling the constraints. We present in this section another alternative in terms of assumed strain method (see Brezzi and Fortin [41]). The main difference between two methods for handling constraints is the following: the incompatible mode method is used to enrich the approximation basis of an isoparametric element in order to accommodate the constraints, and introduces additional parameters for incompatible modes with additional equations to be solved at the element level, whereas the assumed strain method will filter out from the approximation basis of an isoparametric element the undesirable modes concerned by the constraint, with introducing neither additional parameters nor additional equations. The assumed strain method is illustrated herein on the example of quasi-incompressibility constraint. The quasi-incompressible behavior is typical for a linear elastic and isotropic material for which the bulk modulus is much larger than the shear modulus. For the limit case of incompressible behavior $K = \lambda + \frac{2}{3}\mu \mapsto \infty$ $(\lambda \mapsto \infty)$, the constitutive equation concerns only the deviator part of stress and strain with $\boldsymbol{\sigma} - p\mathbf{1} = 2\mu\nabla^s\mathbf{u}$ and the pressure can no longer be obtained from strains but rather from the corresponding boundary conditions. Moreover, the strain and displacement field must respect the incompressibility constraint not allowing any change of volume, which implies that the imposed displacement on Dirichlet boundary must obey the following constraint

$$0 = \int_{\Omega} div[\mathbf{u}]\, dV = \int_{\partial\Omega} \mathbf{u} \cdot \mathbf{n}\, dA = \int_{\Gamma_u} \bar{\mathbf{u}} \cdot \mathbf{n}\, dA \qquad (2.338)$$

The incompressibility constraint can lead to the locking phenomena for a mesh of isoparametric elements, which is especially noticeable for low-order elements. This is illustrated in Figure 2.16 for a mesh of 3-node triangular elements, where nodes near the boundaries can not practically have any admissible displacement.

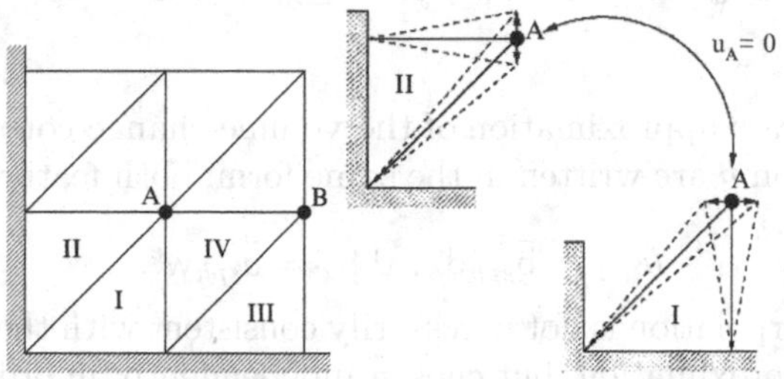

Fig. 2.16 Incompressibility constraint on nodal displacements for a mesh composed of T3-CST elements, which produces the worst-case locking with no admissible displacement values.

A quasi-incompressible behavior with $K \gg \mu$ can also be very difficult, if not impossible, to represent by higher-order isoparametric elements and locking phenomena are still present impairing the rate of convergence. Therefore, in order to avoid such problem, we turn to a formulation where the deviatoric and spherical parts of the strain tensors are separated and can be approximated independently. This can be handled by a mixed Hu-Washizu variational principle introducing an independent variable θ governing the volume change, which is work-conjugate to independent pressure field p. The formulation of this kind, which would remain valid both for quasi-incompressible and the limit case of incompressible behavior, reads

$$\begin{aligned} \Pi(\mathsf{u},\theta,p) := &\int_\Omega \{W(dev[\mathsf{u}]) + U(\theta) + p(tr[\nabla^s \mathsf{u}] - \theta)\}\, dV \\ &- \int_\Omega \mathsf{u}^T \mathsf{b}\, dV - \int_{\Gamma_\sigma} \mathsf{u}^T \bar{\mathsf{t}}\, dA \end{aligned} \tag{2.339}$$

where $W(\cdot)$ and $U(\cdot)$ are two parts of strain energy pertaining to deviatoric and spherical part, respectively. The variational equations obtained from the variational principle in (2.339) above can be written:

$$\begin{aligned} 0 = G_u(\mathsf{u},\theta,p;\mathsf{w}) &:= \int_\Omega \left\{ \mathsf{w}^T \frac{\partial W(\cdot)}{\partial (dev[\nabla^s \mathsf{u}])} + p\, tr[\nabla^s \mathsf{u}] \right\} dV \\ &\quad - \int_\Omega \mathsf{w}^T \mathsf{b}\, dV - \int_{\Gamma_\sigma} \mathsf{w}^T \bar{\mathsf{t}}\, dA \\ 0 = G_\theta(\mathsf{u},\theta,p;\vartheta) &:= \int_\Omega \left(\frac{\partial U(\cdot)}{\partial \theta} - p \right) \vartheta\, dV \\ 0 = G_p(\mathsf{u},\theta,p;q) &:= \int_\Omega (tr[\nabla^s \mathsf{u}] - \theta) q\, dV \end{aligned} \tag{2.340}$$

We choose the finite element approximation for the displacement field by using an isoparametric element with n_{en} nodes, providing the corresponding approximation of the deviatoric and spherical parts of the strain field

$$\begin{aligned} \mathsf{u}\,|_{\Omega^e} &= \textstyle\sum_{a=1}^{n_{en}} N_a \mathsf{d}_a^e = \mathsf{N}\mathsf{d}^e \,;\; dev[\mathsf{u}]\,|_{\Omega^e} = \sum_{a=1}^{n_{en}} \mathsf{B}_{dev}\mathsf{d}^e \,;\; tr[\mathbf{u}]\,|_{\Omega^e} = \mathsf{b}_{sph}\mathsf{d}^e \\ \mathsf{w}\,|_{\Omega^e} &= \textstyle\sum_{a=1}^{n_{en}} N_a \mathsf{w}_a^e = \mathsf{N}\mathsf{w}^e \,;\; dev[\mathsf{w}]\,|_{\Omega^e} = \sum_{a=1}^{n_{en}} \mathsf{B}_{dev}\mathsf{w}^e \,;\; tr[\mathbf{w}]\,|_{\Omega^e} = \mathsf{b}_{sph}\mathsf{w}^e \end{aligned} \tag{2.341}$$

The assumed strain approximation of the volume-change-controlling variable θ and its variation ϑ are written in the same form, both featuring the assume strain matrix $\bar{\mathsf{b}}_{sph}$

$$\theta\,|_{\Omega^e} = \bar{\mathsf{b}}_{sph}\mathsf{d}^e \,;\; \vartheta\,|_{\Omega^e} = \bar{\mathsf{b}}_{sph}\mathsf{w}^e \tag{2.342}$$

This kind of interpolation is not necessarily consistent with the corresponding displacement approximation, but chosen independently in order to avoid the locking. Finally, by considering element-wise independent approximation for pressure p and its variation q, we can rewrite the variational formulation according to

$$\begin{aligned} 0 = G_u(\mathsf{u},\theta,p;\mathsf{w}) &:= \textstyle\int_\Omega \left\{ \frac{\partial W(\cdot)}{\partial(dev[\nabla^s \mathsf{u}])} \mathsf{B}_{dev}\mathsf{w}^e + p\bar{\mathsf{b}}_{sph}\mathsf{w}^e \right\} dV \\ &\quad + \textstyle\int_\Omega p(\mathsf{b}_{sph} - \bar{\mathsf{b}}_{sph})\mathsf{w}^e \, dV - G_{ext} \\ 0 = G_p(\mathsf{u},\theta,p;q) &:= \textstyle\int_{\Omega^e} q(\mathsf{b}_{sph} - \bar{\mathsf{b}}_{sph})\mathsf{d}^e \, dV \;;\; \forall e \in [1, n_{el}] \\ 0 = G_\theta(\mathsf{u},\theta,p;\vartheta) &:= \textstyle\int_{\Omega^e} \left(\frac{\partial U(\cdot)}{\partial \theta} - p \right) \bar{\mathsf{b}}_{sph}\mathsf{w}^e \, dV \;;\; \forall e \in [1, n_{el}] \end{aligned} \tag{2.343}$$

where the following identity $\mathsf{b}_{sph} = \bar{\mathsf{b}}_{sph} + [\mathsf{b}_{sph} - \bar{\mathsf{b}}_{sph}]$ was used. By imposing the element-based orthogonality conditions

$$\begin{aligned} 0 &= \textstyle\int_{\Omega^e} p(\mathsf{b}_{sph} - \bar{\mathsf{b}}_{sph}) \, dV \\ 0 &= \textstyle\int_{\Omega^e} q(\mathsf{b}_{sph} - \bar{\mathsf{b}}_{sph}) \, dV \end{aligned} \tag{2.344}$$

we can practically eliminate the pressure field parameters and reduce the variational formulation in (2.343) to a single equation which can be written:

$$0 = \int_\Omega \left\{ \frac{\partial W(\cdot)}{\partial(dev[\nabla^s \mathsf{u}])} \mathsf{B}_{dev}\mathsf{w}^e + \frac{\partial U(\cdot)}{\partial \theta} \bar{\mathsf{b}}_{sph}\mathsf{w}^e \right\} dV - G_{ext} \tag{2.345}$$

Some of the most popular choices for mixed approximation of quasi-incompressible problems are as follows. We can combine a 4-node quadrilateral isoparametric element displacement interpolation with an element-wise constant pressure p, which produces the assumed strain approximation taking the average value in each element

$$\bar{\mathsf{b}}_{sph} = \frac{1}{\Omega^e} \int_{\Omega^e} \mathsf{b}_{sph} \, dV \tag{2.346}$$

Such a combination produces the element $Q4/P1$ (see Nagtagaal, Park and Rice [197] or Hughes [110]), which remains a very popular choice regardless of checkerboard pressure oscillation it has a tendency of producing.[22] Another element choice providing an improved approximation is a 9-node isoparametric element with bilinear pressure approximation, $Q9/P4$. We should also mention the optimal choice for this kind of approximations, the 9-node element $Q9/P3$, which uses a complete linear polynomial pressure approximation for pressure in physical coordinates x and y defined with three pressure parameters. All these elements are illustrated in Figure 2.17, indicating the displacement nodes with circles and pressure "nodes" with triangles.

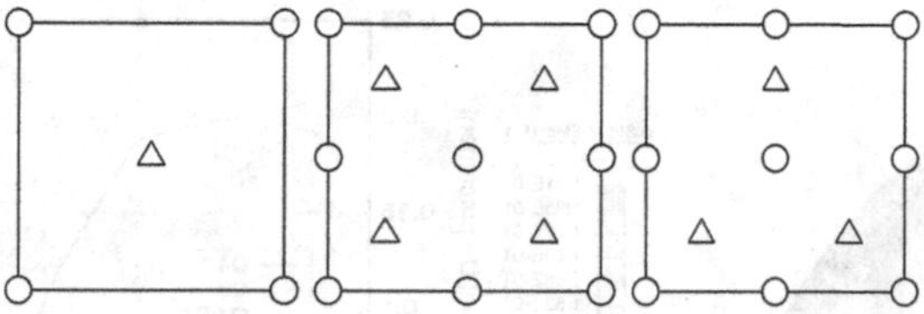

Fig. 2.17 Mixed elements for quasi-incompressible behavior: $Q4/P1$, $Q9/P4$ and $Q9/P3$, where we indicated displacement and pressure nodes.

For any chosen pressure interpolation, we can recast the discrete approximation of this variational principle in $B-bar$ form (see Hughes [110]), replacing the standard strain–displacement matrix by its assumed strain counterpart

$$\bar{\mathsf{b}}_{sph} \mapsto \bar{\mathsf{B}}_{sph} = \frac{1}{3}\begin{bmatrix} \mathsf{b}_{sph} \\ \mathsf{b}_{sph} \\ \mathsf{b}_{sph} \end{bmatrix} \; ; \; \frac{\partial U(\theta)}{\partial \theta} \mapsto \underbrace{\frac{\partial U(\cdot)}{\partial \theta}}_{\bar{p}} 1 = \begin{bmatrix} \bar{p} & \cdot & \cdot \\ \cdot & \bar{p} & \cdot \\ \cdot & \cdot & \bar{p} \end{bmatrix} \tag{2.347}$$

which allows us to write the strain field approximation

$$\nabla^s \mathbf{u} \mapsto \mathsf{e}^h \Big|_{\Omega^e} = (\mathsf{B}^{dev} + \mathsf{B}^{sph})\mathsf{u}^e \equiv \bar{\mathsf{B}}\mathsf{u}^e \tag{2.348}$$

Finally, by taking into account the orthogonality between spherical and deviatoric part, which allows us to write

$$\begin{aligned} &\frac{\partial W(\cdot)}{\partial dev[\nabla^s \mathbf{u}]} \mapsto \mathsf{s}^{dev} = \left(\sigma_{11} - \tfrac{1}{3}tr[\boldsymbol{\sigma}], \sigma_{22} - \tfrac{1}{3}tr[\boldsymbol{\sigma}], \sigma_{33} - \tfrac{1}{3}tr[\boldsymbol{\sigma}], \sigma_{12}, \sigma_{23}, \sigma_{31}\right)^T ; \\ &\frac{\partial U(\cdot)}{\partial \theta} \mapsto \mathsf{p} = (p, p, p, 0, 0, 0)^T ; \\ &\Longrightarrow \mathsf{p}^T \mathsf{B}^{dev} \mathsf{u}^e = 0 \; ; \; \mathsf{s}^{dev\,T} \bar{\mathsf{B}}^{sph} \mathsf{u}^e = 0 \end{aligned} \tag{2.349}$$

[22] The checkerboard pressure oscillation leads to opposite values in neighboring elements (positive versus negative), which can in general be filtered out successfully by taking the average value.

we can rewrite the variational principle in the final form as

$$0 = \mathop{\mathbb{A}}_{e=1}^{n_{elem}} \left\{ \int_{\Omega^e} \underbrace{(\mathsf{s}^{dev} + \mathsf{p})}_{\sigma} \cdot \underbrace{(\mathsf{B}^{dev} + \bar{\mathsf{B}}^{sph})}_{\bar{\mathsf{B}}} \, dV \mathsf{w}^e - G^{ext,e} \right\} \tag{2.350}$$

We briefly present herein an illustrative example of a cylinder under internal pressure, showing to what extent the results obtained by the mixed elements are superior to those obtained by the corresponding isoparametric elements for the case of quasi-incompressible behavior; see Figure 2.18

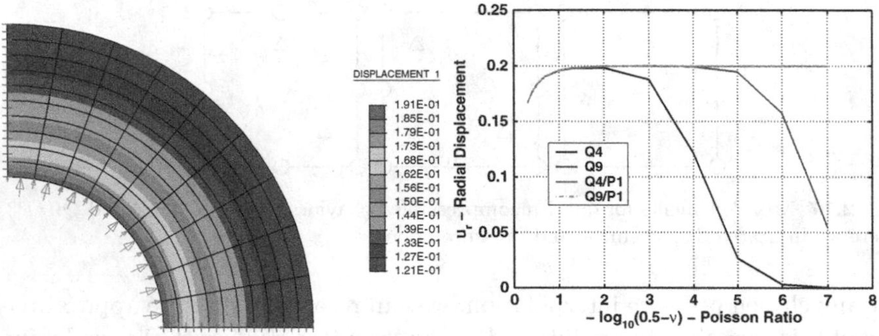

Fig. 2.18 Cylinder under internal pressure: displacement computed by isoparametric and mixed elements.

We note in passing that the elements with incompatible modes are also capable of dealing with quasi-incompressibility problems, and that their results are quite comparable to those obtained by mixed elements specifically designed to handle quasi-incompressibility constraint; see Table 2.5

Table 2.5 Radial displacement of a cylinder under internal pressure computed by mixed elements $Q4/P1$ and $Q9/P3$, and incompatible mode element $Q4 + I2$.

ν	$Q4/P1$	$Q9/P3$	$Q4 + I2$
0.3	1.9053E-01	1.9067E-01	1.9060E-01
0.4	1.9585E-01	1.9600E-01	1.9592E-01
0.45	1.9801E-01	1.9817E-01	1.9808E-01
0.49	1.9950E-01	1.9966E-01	1.9956E-01
0.499	1.9981E-01	1.9997E-01	1.9986E-01
0.4999	1.9984E-01	2.0000E-01	1.9989E-01
0.49999	1.9984E-01	2.0000E-01	1.9990E-01
0.499999	1.9984E-01	2.0000E-01	1.9990E-01
0.4999999	1.9984E-01	2.0000E-01	1.9990E-01

Chapter 3
Inelastic behavior at small strains

In the first part of this chapter, we study an illustrative example for irreversible evolution problem of non-stationary heat transfer under non-homogeneous temperature field. We develop the strong and weak forms of this heat transfer problem, and show how to construct the numerical solution by using the Galerkin and the finite element method. Unlike the systems of algebraic equations as the end results for statics problems studied previously, such a procedure results with a set of ordinary differential equations with respect to time. We show how to compute the numerical solution to this set of equations by using the time-integration schemes. We present the Euler forward and backward schemes, the mid-point scheme and trapezoidal scheme. We recall the essential properties that separate different schemes pertaining to the order of the scheme, as well as its stability and consistence which guarantee the convergence of computations.
We study in the second part of this chapter another important class of problems with irreversible evolutions due to inelastic behavior, where the final value of stress in a particular deformed configuration depends not only on total deformation, but also on the complete deformation trajectory from initial to deformed configuration. We present the fundamental principles of thermodynamics, along with their application to deriving the phenomenological models of inelastic behavior. Among them we present the most frequently used models, such as plasticity, damage as well as the coupled damage–plasticity. The most detailed presentation is given for plasticity model. The increase of model complexities of the presented plasticity models follows the historic path of developments for those models. Namely, we start by presenting the plasticity model governed by the Prandtl-Reuss equations (see Hill [103]). The main ingredients of such model are: an additive decomposition of the total deformation into elastic (reversible) and plastic (irreversible) part, a linear relation (Hooke's law) between the stress and the elastic strain and the normality rule of Levy and St. Venant for plastic flow. We then present different modifications of this classical plasticity model seeking to extend its domain of application, such as nonlinear isotropic hardening for representing

A. Ibrahimbegovic, *Nonlinear Solid Mechanics: Theoretical Formulations and Finite Element Solution Methods*, Solid Mechanics and its Applications 160,

the post-elastic saturation behavior and kinematic hardening of Prager and Ziegler [268] for capturing the Bauschinger effect by shifting the yield stress in load reversal. We also show how to account for rate-strain sensitivity by development of the viscoplasticity models, either the one proposed by Perzyna [222] or the regularized model of Duvaut and Lyons [72]. We also present the nonlinear kinematic hardening model (see Armstrong and Frederick [10] and Chaboche [44]), which is suitable for representing phenomena typical of cyclic loading.

Regarding the numerical implementation of plasticity model, the works of Nguyen and Bui [203] in Europe and of Naghdi and Trapp [195] were among the first to point out that the classical stress-based formulation of plasticity model does not provide the most suitable basis, and that the strain rather than stress should be placed among the state variables. The same discovery on the need to use the strain space formulation of plasticity rather than the one in stress space was also made by experts in numerical methods leading eventually to the return mapping algorithm (e.g., see Krieg and Krieg [154], Wilkins [261]). For the simplest model of perfect plasticity the return mapping algorithm is reduced to the catching-up algorithm devised by convex analysis (e.g., see Nguyen [201] or Moreau [191]). We will show in this chapter how to adapt the return mapping algorithm to more complex models, such as a multi-surface plasticity criterion or the nonlinear elastic response. We will also show how to obtain the consistent elastoplastic tangent modulus in order to ensure the quadratic convergence rate of the return mapping combined with Newton's iterative scheme.

Other essential features of the numerical implementation of classical plasticity models concern the necessary modification which ought to be brought about to isoparametric interpolations to eliminate the locking phenomena (Nagtegaal, Parks and Rice [197]). The latter appears as the consequence of a quasi-incompressibility limitation on strain field in the case of deviatoric plasticity or viscoplasticity models near the ultimate limit state where the plastic deformation component is much larger than its elastic counterpart. Some of the methods proposed in the previous chapter can be used for dealing with this kind of locking phenomena, but not necessarily with the optimal result or with the most general applicability. For example, for the successful application of the *B-bar* method (e.g., see Hughes [110]) to plasticity one is limited to the models where the volumetric and deviatoric parts of deformation tensors can easily be separated. Similarly, the reduced integration method (see Zienkiewicz et al. [272]) can produce the spurious energy modes that can be very damaging for the quality of computed results. We show in this chapter how to apply the incompatible modes method to plasticity, and provide some illustrative results improvement one can expect with such a method.

We also present the damage model, which fits the same format and allows the same kind of refinement as the plasticity model. We finish the chapter with an interesting development showing how to combine the plasticity and damage models into a coupled model inheriting the features of both constituents, without the need to modify the computational procedure for either of them.

3.1 Boundary value problem in thermomechanics

3.1.1 Rigid conductor and heat equation

On the classical example of heat equation, we will explain the first principle of thermodynamics dealing with energy conservation. Without loss of generality, we start with a simple case of 1D rigid conductor (at zero deformation $\epsilon = 0$), with the domain occupying the interval $x \in [0, l]$.
For an isolated conductor that has no exchange with its environment, the internal energy $E(t)$ will remain constant in time. The same applies to any arbitrary segment of the conductor, placed between x_1 and x_2 with $x_1, x_2 \in [0, l]$,

$$E = cst. \implies \frac{dE}{dt} = 0 \tag{3.1}$$

The same result is true for the case of stationary heat transfer, where the heat source from the distributed heat supply $r(x)$ is equal to the outgoing flux q. For a segment placed between x_1 and x_2, the stationary heat transfer implies that

$$\int_{x_1}^{x_2} r\, dx = [q]_{x_1}^{x_2} \Leftrightarrow 0 = Q := \int_{x_1}^{x_2} r\, dx - [q]_{x_1}^{x_2} \implies E = cst. \tag{3.2}$$

The first principle of thermodynamics for a rigid conductor implies that the change of internal energy of a rigid conductor (or its segment between x_1 and x_2) in a non-stationary heat transfer, is proportional to $Q \neq 0$

$$\boxed{\tfrac{d}{dt} E = Q} \tag{3.3}$$

The same result can be restated in local form. Namely, the internal energy is proportional to the total mass of the segment, and it can be written by using the internal energy density e

$$E = \int_{x_1}^{x_2} e(s)\, dx \tag{3.4}$$

The internal energy density allows us to rewrite the first principle of thermodynamics in (3.3) in local form

$$\boxed{\frac{\partial}{\partial t} e(s) = r - \frac{\partial q}{\partial x}} \tag{3.5}$$

We have indicated in (3.4) that the internal energy density is a function of entropy, s. The latter is a state variable parameterizing different states during a non stationary heat transfer. The definition of entropy, as the measure for disorder in statistical mechanics (see Atkins [14] or Kondepudi and Prigogine [152]), is not easy to grasp intuitively. Thus, we often switch to using an alternative state variable in terms of absolute temperature. The temperature is the dual variable to entropy, defined with the internal energy as the potential

$$\boxed{\theta = \frac{\partial e(s)}{\partial s} > 0} \tag{3.6}$$

The last result is also referred to as the zero principle of thermodynamics. We note that the absolute temperature (or simply the temperature) is always considered positive. By exploiting the last two results, we can further obtain the reduced form of the first principle of thermodynamics, which can be written:

$$\boxed{\theta \, \frac{\partial s}{\partial t} = r - \frac{\partial q}{\partial x}} \tag{3.7}$$

The reduced form of the first principle provides a thermodynamics-based definition of entropy at the fixed temperature, $\theta = cst.$, since we can integrate (3.7) to obtain the (reversible) change of entropy between the initial and final states,

$$s_{fin.} = s_{init.} + \int_{init.}^{final} [(r - \frac{\partial q}{\partial x})/\theta] \, dt \tag{3.8}$$

The last result also confirms that the temperature can replace the entropy as the state variable. For a rigid conductor, we can write the internal energy density as a linear function of absolute temperature

$$e = c\,\theta \tag{3.9}$$

where c is the heat capacity coefficient. This definition is consistent with a more general case, where the heat capacity coefficient is defined (see Atkins [14], p. 56) as the partial derivative of the internal energy with respect to temperature

$$c(\theta) := \frac{\partial e}{\partial \theta} \tag{3.10}$$

The absolute temperature also allows to identify the only possibility for spontaneous heat transfer, from a hot sub-domain (with a higher temperature) to cold sub-domains (with lower temperature). This the main result of the second principle of thermodynamics when applied to a rigid conductor under heterogeneous temperature field. More precisely, the second principle of thermodynamics postulates that the heat transfer along a segment placed between x_1 and x_2 results in an increase of entropy that is superior (or equal, for the reversible case) to the corresponding heat supply divided by temperature

$$\boxed{\frac{\partial}{\partial t}\int_{x_1}^{x_2} s\,dx \geq \int_{x_1}^{x_2} \frac{r}{\theta}\,dx - [\tfrac{q}{\theta}]_{x_1}^{x_2}} \tag{3.11}$$

We can also obtain the local form of this result by going to the limit case with $x_2 \mapsto x_1 = x$ and by exploiting the reduced form of the first principle in (3.7); we thus obtain the definition of dissipation produced by heat conduction, D_{cond}

$$\begin{aligned} 0 \leq D_{cond} &:= \overbrace{\theta\,\frac{\partial s}{\partial t} - (r - \frac{\partial q}{\partial x})}^{=0} - \tfrac{1}{\theta} q \tfrac{\partial \theta}{\partial x} \\ &= -\tfrac{1}{\theta} q \tfrac{\partial \theta}{\partial x} \end{aligned} \tag{3.12}$$

The second principle applied to a rigid conductor with non-zero temperature gradient states that the dissipation by conduction remains positive. This confirms the validity of the Fourier law for heat conduction, postulating that the heat flux $q(\theta)$ is proportional to the temperature gradient $\frac{\partial \theta}{\partial x}$, where the coefficient of proportionality is a negative value of coefficient of conductivity k,

$$q = -k\frac{\partial \theta}{\partial x}\ ;\ k > 0 \Leftrightarrow D_{cond} \geq 0 \tag{3.13}$$

The negative sign in (3.13) (with a positive value of k) is in agreement with the experimental observation that the heat flows spontaneously in the opposite direction to the temperature gradient, from hot to cold.
The dissipation by conduction remains equal to zero only for a homogeneous temperature field and only in this case we encounter a reversible processus. We note that a reversible processus in thermodynamics (e.g. see Lubliner [174] or Ericksen [74]), will allow us replace the pseudo-time 't' by '$-t$' without any change of the governing equation or the physical nature of the problem. By replacing the results (3.9) and (3.13) into (3.5), we can rewrite the first principle of thermodynamics for non-stationary case in an equivalent format

$$\boxed{c\tfrac{\partial \theta}{\partial t} = r + \tfrac{\partial}{\partial x}(k\tfrac{\partial \theta}{\partial x})} \tag{3.14}$$

The last result is the classical form of the heat equation. This result shows the irreversible nature of the non-stationary heat flow with a non-homogeneous temperature field, since replacing t by $-t$ will no longer result with the same equation as (3.14).
From mathematics point of view, the non-stationary heat transfer is described by a partial differential equation featuring the temperature field $\theta(x,t)$ and its partial derivatives with respect to space and time. For that reason, in order to obtain the unique solution to this equation, we need to impose not

only the boundary conditions, specifying the value of temperature or its space derivative on the domain boundary, with $\theta|_{\Gamma_\theta} = \bar{\theta}$ or $[qn]|_{\Gamma_q} = \bar{h}$, but also the initial conditions, specifying the temperature field at the time where the heat transfer starts, $\theta(x,0) = \theta_0(x)$. We can thus write:

Strong form of non-stationary heat transfer

$$\boxed{\begin{array}{ll} \text{Given: } r(x,t), \bar{\theta}(t), \bar{h}(t), \theta_0(x) & \\ \text{Find: } \quad \text{temprature } \theta(x,t) \text{ such that} & \\ \qquad c\frac{\partial \theta}{\partial t} + \frac{\partial q}{\partial x} = r \;\; \text{in } \Omega \times]0,T[\text{ (heat eq.)} & \\ \qquad \left.\begin{array}{l} \theta = \bar{\theta} \;\; \text{on } \Gamma_\theta \times]0,T[\\ -q\,n = \bar{h} \text{ on } \Gamma_q \times]0,T[\end{array}\right\} \text{(boundary conds.)} & \\ \qquad \theta(x,0) = \theta_0(x) \;\; x \in \Omega \text{ (initial conds.)} & \\ \qquad \text{where : } \;\; q = -k\frac{\partial \theta}{\partial x} & \end{array}} \qquad (S3)$$

The stationary heat transfer problem can easily be obtained from $(S3)$ by removing the temperature field time-dependence with $\frac{\partial \theta}{\partial t} = 0$. This stationary case is formally equivalent to 1D elasticity problem with the temperature $\theta(x, \bar{t})$ replacing displacement $u(x)$, the heat source $r(x, \bar{t})$ replacing the distributed loading $b(x)$ and finally the heat conduction coefficient k used instead of Young's modulus E. The equivalence of the stationary heat transfer in 1D setting with linear elasticity problem studied in detail in the previous chapter, will allow us to adopt the same kind of finite element approximation for the temperature field. Therefore, the finite element procedure for 1D stationary heat transfer governed by Fourier's law will lead to a set of linear algebraic equations with K as the conductivity matrix, f as the heat source term and d as the unknown nodal values of temperature

$$\mathsf{Kd} = \mathsf{f} \; ; \; \mathsf{K} = \mathop{\mathbb{A}}_{e=1}^{n_{elem}} \mathsf{K}^e \; ; \; \mathsf{f} = \mathop{\mathbb{A}}_{e=1}^{n_{elem}} \mathsf{f}^e$$

The conductivity matrix is obtained by finite element assembly accounting for different element contributions. For a 2-node element the conductivity matrix is the same as the elasticity matrix for a bar, but with conductivity coefficient k replacing Young's modulus E resulting with

$$\boxed{\begin{aligned} K^e_{ab} &= \textstyle\int_{\Omega^e} B_a(x)\, k\, B_b(x)\, A dx \\ &= \textstyle\int_{-1}^{1} B^e_a(\xi)\, kA\, B^e_b(\xi)\, j(\xi) d\xi \\ &= \textstyle\sum_{l=1}^{n_{in}} kA\, B^e_a(\xi_l) B^e_b(\xi_l)\, j(\xi_l) w_l \end{aligned}} \implies \mathsf{K}^e = \frac{kA}{l^e} \begin{bmatrix} 1 & -1 \\ -1 & 1 \end{bmatrix} \qquad (3.15)$$

The nodal values of temperature are obtained as the solution of this set of linear algebraic equations by using the Gauss elimination method in exactly the same manner as described for 1D elasticity. In other words, the stationary heat transfer in 1D setting does not require any new method.

We therefore turn to the finite element solution of a non-stationary case of heat transfer, which allows us to illustrate the key ideas of the solution procedure based upon time integration schemes. We note in passing that such a solution procedure will be of direct interest for dealing with evolution problems of inelastic material behavior described with internal variables.
For the starting point in constructing the finite element solution of a non-stationary heat transfer problem we select the weak form of energy balance equation, rather than its strong form. The main advantage of the weak form concerns a reduced solution regularity, where for a fixed value of time $\bar{t}$ the admissible test functions should belong to

$$\begin{aligned} \mathbb{V}_t = \{\theta(x,t) \,|\, \forall \bar{t} \in [0,T] \Rightarrow \theta(x,\bar{t}) \in H^1(\Omega) \,| \\ \forall \bar{x} \in \Gamma_\theta \Rightarrow \theta(\bar{x},t) = \bar{\theta}(t)\} \end{aligned} \tag{3.16}$$

The same regularity is required from the weighting functions[1] $w(x)$, but they should also take the zero value at the Dirichlet boundary

$$\mathbb{V}_0 = \{w(x) \,|\, w(x) \in H^1(\Omega) \,|\, \forall \bar{x} \in \Gamma_\theta \Rightarrow w(\bar{x}) = 0\} \tag{3.17}$$

Having defined the trial and test solution spaces, we can write:

Weak form of non-stationary heat transfer

Given:	$r(x,t), \bar{h}(t), (\bar{\theta}(t) = 0), \theta_0(x)$	
Find:	$\theta(x,t) \in \mathbb{V}_t$ such that $\forall w \in \mathbb{V}_0$	$(W3)$
$0 = G(\theta; w) :=$	$\int_\Omega w c \frac{\partial \theta}{\partial t}\, dx + \int_\Omega \frac{dw}{dx} k \frac{\partial \theta}{\partial x}\, dx - \int_\Omega w r\, dx - [w\bar{h}]_{\Gamma_q}$	

It is easy to show that the solution of the strong form will always be the solution of the weak form as well. We simply multiply the heat equation by weighting function $w \in \mathbb{V}_0$, integrate over the domain Ω and employ the integration by parts to obtain the final result that is identical to the weak form of the problem

$$\begin{aligned} 0 &= \int_\Omega w[c\frac{\partial \theta}{\partial t} - \frac{\partial}{\partial x}(k\frac{\partial \theta}{\partial x}) - r]\, dx \\ &= \int_\Omega wc\frac{\partial \theta}{\partial t} + \frac{dw}{dx}k\frac{\partial \theta}{\partial x} - wr\, dx - [w\bar{h}]_{\Gamma_q} \end{aligned} \quad \Leftrightarrow \; G(\theta; w) = 0 \tag{3.18}$$

We can not take the inverse path, going from the weak to the strong form of the problem, unless the solution regularity is sufficient to provide the interpretation of the second derivative which appears in $(S3)$.
The finite element discrete approximation is based upon the weak form. First, we choose the finite element interpolation for the weighting functions

[1] Even for a non-stationary heat transfer problem, the weighting functions $w(x)$ remain independent of time, since they are chosen at fixed time value of $\bar{t}$.

$$w^h(x) = \sum_{a=1}^{n_{node}} N_a(x) w_a \tag{3.19}$$

where w_a are the corresponding nodal values and $N_a(x)$ are the shape functions for a particular choice of element. All the elements presented in 1D linear elasticity are suitable for constructing this kind of approximation. The method of separation of variables, where we assume that $\theta^h(x,t) = f_1(x)f_2(t)$, is used to construct the discrete approximation of the temperature field for non-stationary heat transfer problem. This method allows to describe the temperature space variation by using again the same shape functions $N_a(x)$ (as typical of the Galerkin method), while the time variation pertains to nodal values of temperature

$$\theta^h(x,t) = \sum_{a=1}^{n_{node}} N_a(x) d_a(t) \tag{3.20}$$

Such a discrete approximation allows us to obtain easily the partial derivative with respect to time and space, as well as the initial temperature field

$$\begin{aligned} &\tfrac{\partial \theta^h(x,t)}{\partial t} = \textstyle\sum_{a=1}^{n_{node}} N_a(x) \dot{d}_a(t) \ ; \\ &\tfrac{\partial \theta^h(x,t)}{\partial x} = \textstyle\sum_{a=1}^{n_{node}} B_a(x) d_a(t) \ ; \ B_a = \tfrac{dN_a}{dx} \\ &\theta_0^h(x) = \textstyle\sum_{a=1}^{n_{node}} N_a(x) d_{a,0} \end{aligned} \tag{3.21}$$

By exploiting the approximations in (3.19), (3.20) and (3.21), the weak form $(W3)$ for non-stationary heat transfer can be recast in terms of the Galerkin equation

$$\begin{aligned} &G(\theta^h; w^h) = 0 \Leftrightarrow \textstyle\sum_{a=1}^{n} w_a r_a = 0 \ ; \\ &r_a = \textstyle\sum_{b=1}^{n_{node}} [\overbrace{\int_\Omega N_a(x)\, c\, N_b(x)\, dx}^{M_{ab}}]\, \dot{d}_b(t) \\ &\quad + \textstyle\sum_{b=1}^{n_{node}} \underbrace{[\int_\Omega B_a(x)\, k\, B_b(x)\, dx]}_{K_{ab}}\, d_b(t) - \underbrace{\int_\Omega N_a(x) r\, dx - [N_a(x)\bar{h}]_{\Gamma_q}}_{f_a^{ext}} \end{aligned} \tag{3.22}$$

The terms concerning the nodes on the Dirichlet boundary, where the weighting functions take zero value, will drop from the sum in (3.22); by assuming that the Galerkin equation must be verified for an arbitrary choice of nodal values of weighting functions of temperature for all active node with unknown values of real temperature ($w_a \neq 0,\ a = 1, 2, ..., n_{eqs} < n_{node}$), the Galerkin equation will give rise to a first-order system of ordinary differential equations in time

$$\boxed{\begin{aligned} &r_a = 0,\ a = 1, 2, ..., n_{eqs} < n_{node} \implies \\ &\mathsf{M}\,\dot{\mathsf{d}}(t) + \mathsf{K}\,\mathsf{d}(t) = \mathsf{f}^{ext}(t) \ ; \ t \in]0, T[\ ; \ \mathsf{d}(0) = \mathsf{d}_0 \end{aligned}} \tag{3.23}$$

In this equation, $\mathsf{d} = (d_a^e)$ is the vector collecting all the unknown nodal values of temperature, $\dot{\mathsf{d}}$ is its time derivative, K is the conductivity matrix and M the heat capacity matrix. The matrix M can be computed by the standard finite element procedure, making use of the isoparametric interpolations, numerical integration and the finite element assembly procedure

$$\boxed{\begin{aligned} \mathsf{M} &= \mathbb{A}_{e=1}^{n_{elem}} \mathsf{M}^e; \qquad \mathsf{M}^e = [M_{ab}^e] \\ M_{ab}^e &= \int_{\Omega^e} N_a(x)\, c\, N_b(x)\, dx \\ &= \int_{-1}^{1} N_a(\xi)\, c\, N_b(\xi)\, j(\xi) d\xi \\ &= \textstyle\sum_{l=1}^{n_{in}} c\, N_a(\xi_l) N_b(\xi_l)\, j(\xi_l) w_l \end{aligned}} \tag{3.24}$$

For a 2-node bar element, the element heat capacity matrix can be written:

$$\mathsf{M}^e = \frac{cl^e}{6}\begin{bmatrix} 2 & 1 \\ 1 & 2 \end{bmatrix} \tag{3.25}$$

The consistent form of the heat capacity matrix can be replaced by a diagonal form, which is obtained by summing up all the components in a particular raw and placing the result on the main diagonal

$$\mathsf{M}^e = \frac{cl^e}{6}\begin{bmatrix} (2+1) & 0 \\ 0 & (1+2) \end{bmatrix} \mapsto \mathsf{M}^e = \frac{cl^e}{2}\begin{bmatrix} 1 & 0 \\ 0 & 1 \end{bmatrix} \tag{3.26}$$

The advantage of using the diagonal form of the heat capacity matrix for reducing the solution cost of explicit time-integration schemes is discussed in detail subsequently.

Remark :

1. Non-stationary 3D heat transfer: the heat equation governing non-stationary heat transfer for 3D case can also be obtained by the same procedure as the one just described for 1D case. Even in 3D case, the temperature is still a scalar field, but the temperature flux is a vector field. Hence, by replacing $\theta(x,t) \mapsto \theta(\mathbf{x},t)$, and $q(x,t) \mapsto \mathbf{q}(\mathbf{x},t)$, where $\mathbf{x}$ is the position vector, we can obtain the strong form for 3D case:

$$\begin{aligned} c\tfrac{\partial\theta}{\partial t} + div[\underbrace{-k\nabla\theta}_{\mathbf{q}}] &= r \\ \theta|_{\Gamma_\theta} = \bar{\theta}\ ;\ -\mathbf{q}\cdot\mathbf{n}|_{\Gamma_q} &= \bar{h} \end{aligned} \tag{3.27}$$

The latter allows to obtain the weak form of non-stationary heat transfer problem

$$0 = G(\theta; w) := \int_\Omega \{wc\frac{\partial\theta}{\partial t} + \nabla w \cdot k\nabla\theta - wr\}\, dV - \int_{\Gamma_q} w\bar{h}\, dA \tag{3.28}$$

The finite element method can transform this weak form into a system of ordinary differential equations in time for computing the evolutions of nodal values of temperature. This system takes the same form as the system in (3.23) obtained for 1D case, except that the conductivity and heat capacity matrices are now defined as

$$\begin{aligned} \mathsf{M}^e &= [M^e_{ab}] \; ; \; M^e_{ab} = \int_{\Omega^e} N^e_a(\mathsf{x})\, c\, N^e_b(\mathsf{x})\, dV \; ; \\ \mathsf{K}^e &= [K^e_{ab}] \; ; \; K^e_{ab} = \int_{\Omega^e} \mathsf{B}^{e,T}_a(\mathsf{x})\, k\, \mathsf{B}^e_b(\mathsf{x})\, dV \; ; \\ \mathsf{B}^{e,T}_a &= [\tfrac{\partial N^e_a}{\partial x_1}; \tfrac{\partial N^e_a}{\partial x_2}; \tfrac{\partial N_a}{\partial x_3}] \end{aligned} \tag{3.29}$$

3.1.2 *Numerical solution by time-integration scheme for heat transfer problem*

In this section we elaborate upon integrating the system of ordinary differential equations in (3.23), by using the time integration schemes. Both implicit and explicit schemes are proposed. The time integration schemes of this kind are directly applicable to solving the evolution equations for internal variables and mastering a more complex problem of inelastic behavior of materials, which is studied subsequently in this chapter.

We note in passing that one can also obtain the analytic solution for the system of linear differential equation in (3.23), by using the modal superposition method (e.g. see [53, 84 or 111]). Namely, we first have to solve the associated eigenvalue problem, defined by the same system with a zero force vector $\mathsf{f}^{ext} = 0$. The solution of the eigenvalue problem can be written:

$$\begin{aligned} &\mathsf{M}\dot{\mathsf{d}} + \mathsf{K}\mathsf{d} = 0 \; ; \; \mathsf{d}(t) = \textstyle\sum_{i=1}^{n_{eqs}} \mathsf{p}_i exp(-\lambda_i t) \; ; \; \lambda_i > 0 \\ &\Longrightarrow \boxed{\mathsf{K}\mathsf{p}_i = \lambda_i \mathsf{M}\mathsf{p}_i} \; ; \; i = 1, 2, \ldots, n_{eqs} \end{aligned} \tag{3.30}$$

According to the standard result of linear algebra (see Parlett [219]), all the eigenvalues λ_i are positive and real provided that the matrices K and M are positive definite. Moreover, we can count with the orthogonality of eigenvectors p_i with respect to the heat capacity matrix M, which also implies their orthogonality with respect to the conductivity matrix K

$$\mathsf{p}_i^T \mathsf{M} \mathsf{p}_j = \delta_{ij} \; ; \; \mathsf{p}_i^T \mathsf{K} \mathsf{p}_j = \delta_{ij}\lambda_i \; ; \; \delta_{ij} = \begin{cases} 1 \; ; \; i = j \\ 0 \; ; \; i \neq j \end{cases} \tag{3.31}$$

By introducing the modal coordinates $y_i(t)$, and writing the solution as $\mathsf{d}(t) = \sum_{i=1}^{n_{eqs}} \mathsf{p}_i y_i(t)$, which further implies $\dot{\mathsf{d}}(t) = \sum_{i=1}^{n_{eqs}} \mathsf{p}_i \dot{y}_i(t)$, we can transform the system of n-equations in (3.23) into a set of n uncoupled equations, yet called modal equations. Namely, by using the eigenvector orthogonality, we can write

$$\mathsf{p}_i^T [\mathsf{M}\dot{\mathsf{d}} + \mathsf{K}\mathsf{d} = \mathsf{f}^{ext}] \; \Longrightarrow \; \boxed{\dot{y}_i + \lambda_i y_i = l_i(t) \; ; \; l_i(t) = \mathsf{p}_i^T \mathsf{f}^{ext}(t)} \tag{3.32}$$

For each modal equation, we can obtain the analytic solution depending upon a particular form of excitation $l_i(t)$, simply by integrating

$$y_i(t) = y_i(0) + \int_0^t [l_i(s) - \lambda_i y_i(s)]\, ds \tag{3.33}$$

The corresponding solution of the system of coupled equations can then be obtained by modal superposition

$$\mathsf{d}(t) = \sum_{i=1}^{n_{eqs}} \mathsf{p}_i \left\{ y_i(0) + \int_0^t [l_i(s) - \lambda_i y_i(s)]\, ds \right\} \tag{3.34}$$

Unlike analytic solution specifying the value of nodal temperature for any particular time, the numerical schemes provide only the values at the chosen times within the interval of interest, $\forall t \in [0, T]$,

$$0 \leq t_1 \leq t_2 \ldots \leq t_n \leq t_{n+1} \leq \ \ldots \leq T \tag{3.35}$$

The choice of time instants is dictated by time steps. The discrete values of nodal temperatures at the chosen times are computed by a time integration scheme. In order to simplify the architecture of the finite element codes, we use in general a single step schemes,[2] where the computations are carried out one step at the time. Hence, for a typical time step, $h = t_{n+1} - t_n$, we are facing:

Central problem of time integration scheme: computation over a typical step

$$\boxed{\begin{array}{ll} \text{Given: } \mathsf{d}_n = \mathsf{d}(t_n),\ \mathsf{v}_n = \dot{\mathsf{d}}(t_n),\ h = t_{n+1} - t_n & \\ \text{Find: } \mathsf{d}_{n+1} = \mathsf{d}(t_{n+1}),\ \mathsf{v}_{n+1} = \dot{\mathsf{d}}(t_{n+1}) & \\ \text{such that:} \qquad \mathsf{M}\, \mathsf{v}_{n+1} + \mathsf{K}\, \mathsf{d}_{n+1} = \mathsf{f}_{n+1} & \end{array}} \tag{3.36}$$

We will start the study for the simplest case of a single modal equation, which can be written:

$$\dot{d}(t) + \lambda d(t) = l(t)\ ;\ \lambda := K/M > 0 \tag{3.37}$$

The exact analytic solution of this equation can easily be obtained by integrating,

$$d(t_{n+1}) = d(t_n) + \int_{t_n}^{t_{n+1}} [-\lambda d(t) + l(t)]\, dt \tag{3.38}$$

However, given a very large diversity of dynamic excitations, it is practically impossible to store all such solutions in a finite element code. Therefore, we

---

[2] An alternative choice of multi-step schemes, where the results of several time steps are needed for advancing computations, is quite popular in mathematics community, especially for dealing with stiff equations, see [83, 97]. We show subsequently that the system of stiff equations, characterized by a large difference between maximal and minimal eigenvalues, can also be dealt with by using the appropriate modification of single-step schemes.

seek to standardize the computation of this kind for different load cases by using the time integration schemes, even though they will supply only the approximate solution.[3] The key idea in computing the approximate solution is to replace the time variation of dynamic excitation, or integrand in (3.38), by a polynomial expression, and then integrate exactly. This is illustrated in Figure 3.1, where we show the approximations with the constant value, either at time t_n, the time t_{n+1} or yet the time $t_{n+1/2}$, as well as the linear approximation of the integrand. We thus obtain, respectively, the forward and backward Euler schemes, the mid-point rule and the trapezoidal integration scheme.

The first of these scheme, the forward Euler scheme, is based upon the integrand approximation by a constant value at time t_n, which allows us to obtain

$$d_{n+1} = d_n + h\,\overbrace{(-\lambda d_n + l_n)}^{v_n} \tag{3.39}$$

The forward Euler scheme is explicit, since only the known values at time t_n appear on the right hand side. All the other schemes presented in Figure 3.1 are implicit, and require a more elaborate computation since the unknown values appear on both sides of equation. For example, the backward Euler scheme, based upon a constant approximation of the integrand at the end of the step, results with

$$d_{n+1} = d_n + h\,\overbrace{(-\lambda d_{n+1} + l_{n+1})}^{v_{n+1}} \implies d_{n+1} = (d_n + h l_{n+1})/(1 + h\lambda) \tag{3.40}$$

The last two schemes, mid-point scheme and trapezoidal rule, result with an improved approximation for the integrand in (3.38) with respect to Euler schemes. For linear problems, these schemes provide the same final result, which can be written:

$$\begin{aligned} d_{n+1} &= d_n + h\,\overbrace{[\frac{1}{2}(-\lambda d_n - \lambda d_{n+1}) + \frac{1}{2}(l_n + l_{n+1})]}^{v_{n+1/2}=(1/2)(v_n+v_{n+1})} \\ \implies d_{n+1} &= [d_n(1 - \lambda h/2) + h(l_n + l_{n+1})/2]/(1 + h\lambda/2) \end{aligned} \tag{3.41}$$

It is interesting to note that the computational procedure for these four schemes can be presented in a unified manner, by introducing a free parameter $\alpha \in [0, 1]$ that allows to write:

Unified implementation of four schemes – α method

[3] Needless to say, the quality of this approximate solution can always be improved by selecting smaller time steps or higher order schemes.

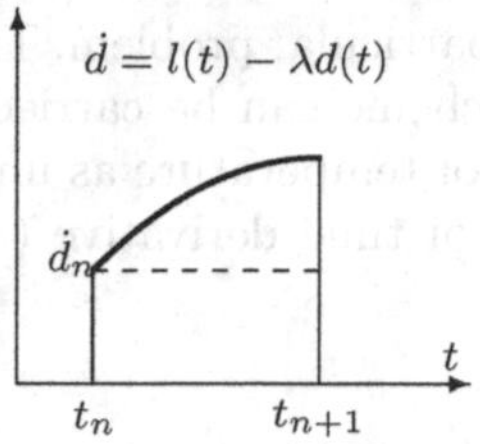

1- forward Euler (explicit) : $d_{n+1} = d_n + h \underbrace{(-\lambda d_n + l_n)}_{v_n}$

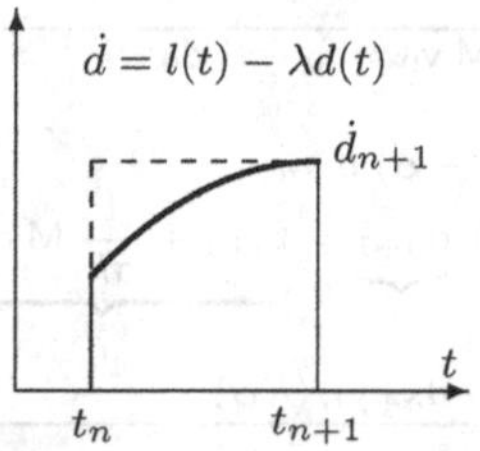

2- backward Euler (implicit) : $d_{n+1} = d_n + h \underbrace{(-\lambda d_{n+1} + l_{n+1})}_{v_{n+1}}$

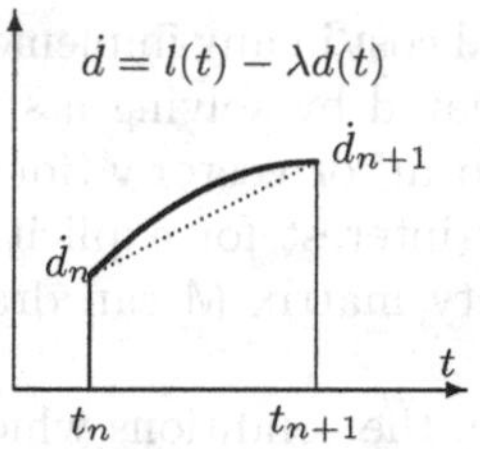

3- trapezoidal rule (implicit) $d_{n+1} = d_n + h \underbrace{[\frac{1}{2}(-\lambda d_n - \lambda d_{n+1}) + \frac{1}{2}(l_n + l_{n+1})]}_{\frac{v_{n+1}+v_n}{2}}$

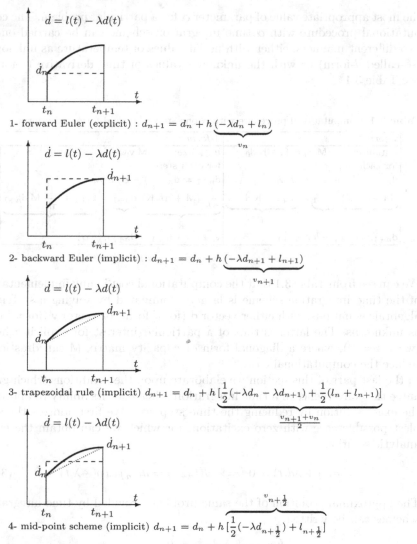

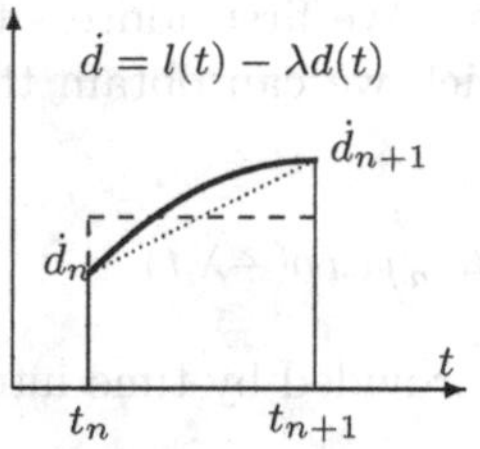

4- mid-point scheme (implicit) $d_{n+1} = d_n + h \overbrace{[\frac{1}{2}(-\lambda d_{n+\frac{1}{2}}) + l_{n+\frac{1}{2}}]}^{v_{n+\frac{1}{2}}}$

Fig. 3.1 Integrand approximations for time-integration schemes for: forward Euler, backward Euler, mid-point scheme and trapezoidal rule: – – exact variation, - - - approximation.

$\mathsf{M}\,\mathsf{v}_{n+1} + \mathsf{K}\,\mathsf{d}_{n+1} = \mathsf{f}_{n+1}$ $\mathsf{d}_{n+1} = \mathsf{d}_n + h\,\mathsf{v}_{n+\alpha}$ $\mathsf{v}_{n+\alpha} = (1-\alpha)\,\mathsf{v}_n + \alpha\,\mathsf{v}_{n+1}$
$\alpha = 0$ forward Euler $\alpha = 1$ backward Euler $\alpha = \frac{1}{2}$ trapezoidal rule (or mid-point scheme)

We can thus ensure a significant versatility of the computer code by implementing all these schemes, and leaving the final choice to the user to define the most appropriate value of parameter α for a particular problem. The computational procedure with α time integration scheme can be carried out in two different manners, either with nodal values of temperature as unknowns (so-called d-form) or with the unknown values of time derivative (v-form); See Table 3.1

Table 3.1 Computational procedure with α one-step scheme: d-form and v-form.

v-*form*	d-*form*
initialize: $\mathsf{M}\,\mathsf{v}_0 = \mathsf{f}_0 - \mathsf{K}\,\mathsf{d}_0$	initialize: $\mathsf{M}\,\mathsf{v}_0 = \mathsf{f}_0 - \mathsf{K}\,\mathsf{d}_0$
for each step:	for each step:
$\tilde{\mathsf{d}}_{n+1} = \mathsf{d}_n + (1-\alpha)\,h\,\mathsf{v}_n$	$\tilde{\mathsf{d}}_{n+1} = \mathsf{d}_n + (1-\alpha)\,h\,\mathsf{v}_n$
$\underbrace{(\mathsf{M} + \alpha h\,\mathsf{K})}_{\tilde{\mathsf{K}}}\ \underbrace{\mathsf{v}_{n+1}}_{\tilde{\mathsf{d}}} = \underbrace{\mathsf{f}_{n+1} - \mathsf{K}\,\tilde{\mathsf{d}}_{n+1}}_{\tilde{\mathsf{f}}}$	$\underbrace{\frac{1}{\alpha h}\,(\mathsf{M} + \alpha h\,\mathsf{K})}_{\tilde{\mathsf{K}}}\ \underbrace{\mathsf{d}_{n+1}}_{\tilde{\mathsf{d}}} = \underbrace{\mathsf{f}_{n+1} + \frac{1}{\alpha h}\,\mathsf{M}\,\tilde{\mathsf{d}}_{n+1}}_{\tilde{\mathsf{f}}}$
$\mathsf{d}_{n+1} = \tilde{\mathsf{d}}_{n+1} + h\,\alpha\,\mathsf{v}_{n+1}$	$\mathsf{v}_{n+1} = (\mathsf{d}_{n+1} - \tilde{\mathsf{d}}_{n+1})/(h\,\alpha)$

We can see from Table 3.1 that the computational cost in any implementation of the time integration scheme is largely dominated by solving a system of algebraic equations, with either vector d (for d-form) or vector v (for v-form) as unknowns. The latter is thus of a particular interest for explicit scheme (with $\alpha = 0$), where a diagonal form of capacity matrix M can drastically reduce the computational cost.
In the last part of this section we elaborate upon the conditions which guarantee the convergence of a time integration scheme, or its ability to approach the exact solution by reducing the time step size. We first address the simplest possible case with zero excitation, for which we can obtain the exact analytic solution

$$\dot{d}(t) + \lambda d(t) = 0 \implies d(t_{n+1}) = d(t_n)\,exp(-\lambda_i t) \tag{3.42}$$

The approximate solution of the same problem provided by time-integration schemes can be written:

$$\begin{cases} v_{n+1} + \lambda\, d_{n+1} = 0 \\ d_{n+1} = d_n + h\, v_{n+\alpha} \\ v_{n+\alpha} = (1-\alpha) v_n + \alpha\, v_{n+1} \end{cases} \tag{3.43}$$

For d-form implementation, this further results with

$$\begin{aligned} (1 + \alpha\, h\, \lambda)\, d_{n+1} &= \tilde{d}_{n+1} \\ &= d_n + h\,(1-\alpha)\, v_n \quad \Big|\ v_n + \lambda\, d_n = 0 \mapsto v_n = -\lambda\, d_n \\ &= [1 - (1-\alpha)\, h\, \lambda]\, d_n \end{aligned}$$

$$\implies \boxed{d_{n+1} = A\, d_n \quad ; \quad A = \frac{1-(1-\alpha)\,h\,\lambda}{1+\alpha\, h\,\lambda}} \tag{3.44}$$

where A is so-called amplification factor. In Figure 3.2 we can see that none of the time integration schemes can give the exact amplification factor.[4] However, a higher quality of approximation of approximation factor does not imply that the scheme is more precise. The precision of a time integration scheme is governed by its truncation error, $\iota(t_n)$, which can be defined by replacing the exact solution into the recurrence equation in (3.44). For example, we can define the truncation error for a more general case with non-zero excitation according to

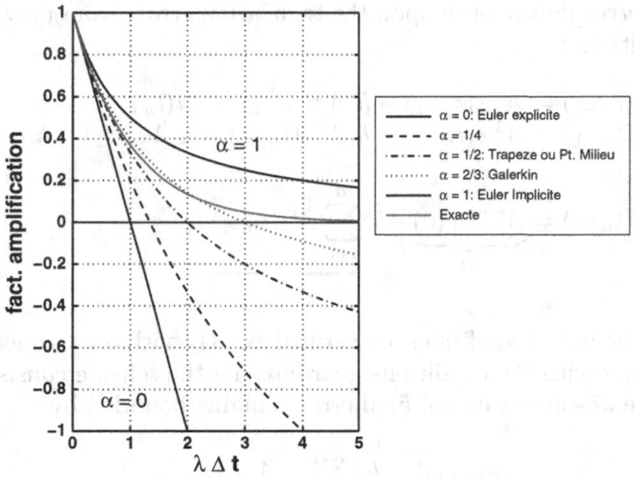

Fig. 3.2 Amplification factor of time integration scheme depending on α.

$$\boxed{d(t_{n+1}) - A\, d(t_n) - l_n = h\iota(t_n)} \tag{3.45}$$

The convergence of a time integration scheme can be proved with the Lax equivalence theorem, enforcing the conditions on the amplification factor and truncation error. According to Lax equivalence theorem, the error produced by integration scheme $e(t_n) = d_n - d(t_n) \mapsto 0$, with $h \mapsto 0$, if the scheme is:

1. Consistent in the sense that the truncation error remains bounded by $|\iota(t_n)| \leq c\, h^k, \forall t \in [0, T]$, where c is a constant independent of time and k is the order of the scheme
2. Stable with absolute value of amplification factor bounded by 1, $|A| \leq 1$

The proof of Lax equivalence theorem is done by induction. In order to describe the propagation of truncation error over a single time step, from $e(t_n) = d_n - d(t_n)$ to $e(t_{n+1}) = d_{n+1} - d(t_{n+1})$, we subtract the result in (3.45) from the recurrence equation defined for a scheme

[4] In fact the best approximation for the amplification factor is obtained by the time integration scheme which can be constructed as the Galerkin linear approximation of the weak form of non-stationary heat transfer equation, resulting with $\alpha = 2/3$.

$$\begin{array}{l} d_{n+1} \quad - A\, d_n \quad - l_n \quad = 0 \\ d(t_{n+1}) - A\, d(t_n) - l_n = h\, \iota(t_n) \\ \hline e(t_{n+1}) - A\, e(t_n) \qquad = -\, h\, \iota(t_n) \end{array}$$

By induction, we can also write the corresponding result for the previous steps; for example, for step $n-1$ we obtain

$$e(t_n) = A\, e(t_{n-1}) - h\, \iota(t_{n-1})$$

By combining the results over successive time steps, we can obtain the final truncation error dependence upon the truncation errors from previous steps, which is written:

$$\begin{array}{l} e(t_{n+1}) = A^2\, e(t_{n-1}) - h\, A\, \iota(t_{n-1}) - h\, \iota(t_n) \\ e(t_{n+1}) = A^3\, e(t_{n-2}) - h\, A^2\, \iota(t_{n-2}) - h\, A\, \iota_{n-1}) - h\, \iota(t_n) \\ \quad \ldots\ldots \\ \Rightarrow \\ e(t_{n+1}) = \underbrace{A^{n+1}\, e(0)}_{=0} - h \underbrace{\sum_{i=0}^{n} A^i \iota(t_{n-i})} \end{array}$$

Given that the initial conditions are verified exactly with zero truncation error $e(0) = 0$ along with the conditions guaranteeing the scheme consistency and stability, the absolute value of final error remains bounded by

$$\begin{aligned} |e(t_{n+1})| &= h\, |\textstyle\sum_{i=0}^{n} A^i\, \iota(t_{n-i})| \\ &\leq h\, \textstyle\sum_{i=0}^{n} \underbrace{|A^i|}_{\leq 1}\, |\iota(t_{n-i})| \\ &\leq h\, \textstyle\sum_{i=0}^{n} \underbrace{|\iota(t_{n-i})|}_{\leq max|\iota(t)|} \\ &\leq \underbrace{(n+1)h}_{t_{n+1}}\, \underbrace{max|\iota(t)|}_{\leq c\, h^k} \\ &\leq t_{n+1}\, c\, h^k \end{aligned}$$

It follows that the convergence is ensured for any scheme of order $k \geq 1$ with

$$\boxed{h \mapsto 0 \Longrightarrow e(t_{n+1}) = O(h^k) \mapsto 0}$$

We can see that the rate of convergence or the order of the scheme 'k' depends directly on truncation error. Therefore, it is important to note that the Taylor formula development allows us to express

$$d(t_{n+1}) = d(t_{n+\alpha}) + (1-\alpha)\, h\dot{d}(t_{n+\alpha}) + \frac{[(1-\alpha)\, h]^2}{2}\, \ddot{d}(t_{n+\alpha}) + O(h^2)$$
$$d(t_n) = d(t_{n+\alpha}) - \alpha\, h\dot{d}(t_{n+\alpha}) + \frac{[(-\alpha)\, h]^2}{2}\, \ddot{d}(t_{n+\alpha}) + O(h^2)$$

which allows us to write the truncation error for all members of α-scheme presented in Figure 3.1 as a function of α parameter

$$\boxed{\iota = (1-2\,\alpha)O(h) + O(h^2)}$$

where $O(h)$ and $O(h^2)$ denote the terms going to zero more quickly than h and h^2, respectively. We can see that two Euler schemes, Euler forward and Euler backward, are both first order, whereas the mid-point scheme and trapezoidal rule are second order.

While two Euler schemes possess the same truncation error, and thus the same accuracy, they are quite different regarding the stability condition in (3.44). The latter requiring $|A| \leq 1$ leads to two conditions to be examined. The first one with $A := (1-(1-\alpha)\, h\, \lambda)/(1+\alpha\, \lambda\, h) < 1$ is always verified $\forall \alpha \in [0,1]$ and $\lambda h > 0$. The verification of the second condition,

$$-1 < \frac{1-(1-\alpha)\, h\, \lambda}{1+\alpha\, \lambda\, h} \Rightarrow \boxed{h < \frac{2}{\lambda\,(1-2\alpha)}}$$

is also always guaranteed for the implicit, backward Euler scheme with $\alpha = 1$ as well as for the mid-point scheme or trapezoidal rule with $\alpha = 1/2$, for any size of the time step. However, the last condition applied to the explicit, forward Euler scheme with $\alpha = 0$ will result with the restriction on the time step needed to preserve the stability, which requires that $h_{max} \leq 2/\lambda$.

Such a stability criterion for explicit Euler scheme can also be derived for a system of equations, resulting with the same restriction applied to all the eigenvalues, $h_i \leq 2/\lambda_i$, $i = 1, 2, \ldots, n_{eqs}$. Clearly, we ought to keep the most stringent of all these requirements $h_{min} \leq 2/\lambda_{max}$, which in turns increases the computational cost for explicit scheme.[5] The most demanding in that respect of time step reduction is the case of a system of stiff equations with $\lambda_{max} \gg \lambda_{min}$. We thus conclude that preserving the stability of the explicit schemes might require much smaller time steps than the implicit schemes. Hence, the choice of the type of scheme, explicit or implicit, in trying to achieve the most efficient computation is far from trivial. The implicit scheme can allow larger time steps from the explicit one, but the computation in each

[5] Application of the time integration scheme to a system of differential equation does not require modal superposition, and we can deal directly with the set of original differential equations. However, for the explicit schemes we would still need an estimate of the maximum eigenvalue, which can be obtained from the corresponding eigenvalue problem set with the heat capacity and conductivity matrix of a single element; this result follows from Cauchy interlace theorem (e.g. see [62]).

step is much more demanding from the corresponding solution obtained by explicit scheme dealing with diagonal form of capacity matrix.
This basic consideration of time integration schemes applied to non-stationary heat transfer problem is further expanded in dealing with a more complex problem of inelastic constitutive behavior of materials with internal variables. As shown subsequently, choosing the most appropriate scheme for integration of evolution of internal variables, as well as elaborating the details of numerical implementation, results with a number of further important differences between the explicit and implicit schemes.

3.1.3 Thermomechanical coupling in elasticity

In the hierarchy of models capable of placing the mechanics and heat transfer problem within the same framework and accounting for the thermomechanical coupling, the simplest one is the model of thermoelasticity. Such a model does not require any supplementary hypothesis with respect to two thermodynamics principles, and it is practically the only one which was not subjected to criticism and disagreements (e.g. see Ericksen [74] or Kondepudi and Prigogine [152], Lubliner [174], Truesdell and Noll [257] or Ziegler [269]) regarding the validity of thermodynamics principles.
For clarity of presentation, we start with a simple 1D problem of thermomechanical coupling, defined in the domain $x \in [0, l]$. We study herein the non-stationary heat transfer problem in an elastic deformable body that also acts as a conductor. The hypothesis of small displacement gradients allows us to retain the same kinematics and equilibrium equations as those already obtained for linear elasticity, since they will remain valid for any constitutive behavior

$$\frac{\partial \sigma}{\partial x} + b = 0 \; ; \; \epsilon = \frac{\partial u}{\partial x} \tag{3.46}$$

The constitutive model of thermoelasticity must be able to account for both mechanics and thermal effects. The mechanics part of the response for thermoelasticity model (with temperature dependent material properties) should provide the stress that remains completely reversible upon unloading. In order to master the thermal response of thermoelasticity model, we ought to replace the hyperelasticity model with a more general potential which fits within the framework of the first principle of thermodynamics. The latter postulates that the rate of increase of internal energy is proportional to the supply of heat Q and mechanical power P. The internal energy in thermoelasticity is a function of both entropy and deformation. This kind of balance can be written for any segment of the bar $x \in [x_1, x_2]$ according to

$$\begin{aligned} &\tfrac{\partial E}{\partial t} = P + Q \; ; \; E = \textstyle\int_{x_1}^{x_2} e(\epsilon, s)\, dx \; ; \\ &P = \textstyle\int_{x_1}^{x_2} b \tfrac{\partial u}{\partial t}\, dx + \left[\sigma \tfrac{\partial u}{\partial t}\right]_{x_1}^{x_2} \; ; \; Q = \textstyle\int_{x_1}^{x_2} r\, dx - [q]_{x_1}^{x_2} \end{aligned} \tag{3.47}$$

By using equilibrium and kinematic equations in (3.46), the last expression gives rise to the local form of the first principle on energy conservation, which can be written:

$$\begin{aligned}\frac{\partial}{\partial t}e(\epsilon,s) &= b\frac{\partial u}{\partial t}+\frac{\partial}{\partial x}\left(\sigma\frac{\partial u}{\partial t}\right)+r-\frac{\partial q}{\partial x}\\ &= \frac{\partial u}{\partial t}\underbrace{\left(\frac{\partial\sigma}{\partial x}+b\right)}_{=0}+\sigma\frac{\partial\epsilon}{\partial t}+r-\frac{\partial q}{\partial x}\\ &= \sigma\frac{\partial\epsilon}{\partial t}+r-\frac{\partial q}{\partial x}\end{aligned} \tag{3.48}$$

For the special case with constant (or zero) value of strain in time (with $\frac{\partial\epsilon}{\partial t}=0$), the contribution of stress power term disappears and the last result is simplified to the local form of the first principle of thermodynamics for a rigid conductor in (3.5); the latter is written by using so-called zero principle of thermodynamics defining the absolute temperature θ

$$\frac{\partial e}{\partial t}|_\epsilon = r-\frac{\partial q}{\partial x}\ \&\ \theta=\left.\frac{\partial e(\epsilon,s)}{\partial s}\right|_\epsilon \implies \theta\dot{s}=r-\frac{\partial q}{\partial x} \tag{3.49}$$

One special case of thermoelasticity is isentropic case, where the entropy remains constant in time ($\frac{\partial s}{\partial t}=0$). In agreement with result in (3.49), the isentropic case for thermoelasticity is identical to the adiabatic case with the zero heat source (with $r-\frac{\partial q}{\partial x}=0$). In this special case, the first principle in (3.48) confirms that the internal energy will play the role of the potential furnishing the stress value as the corresponding derivative with respect to strain, and the thermoelastic case is reduced to hyperelasticity

$$\frac{\partial e(\epsilon,s)}{\partial t}|_s=\sigma\frac{\partial\epsilon}{\partial t}\implies\sigma=\left.\frac{\partial e(\epsilon,s)}{\partial\epsilon}\right|_s\ \Leftrightarrow\ e|_s\equiv W\,;\,\sigma=\frac{dW}{d\epsilon} \tag{3.50}$$

The duality between the strain ϵ and the stress σ remains valid in thermoelasticity, even for the most general case with a heat source and stress power both driving the internal energy change; moreover, we recover the duality between the entropy s and the temperature θ along with a reduced form of the first principle

$$\begin{aligned}&\frac{\partial e}{\partial\epsilon}\frac{\partial\epsilon}{\partial t}+\frac{\partial e}{\partial s}\frac{\partial s}{\partial t}=\sigma\frac{\partial\epsilon}{\partial t}+r-\frac{\partial q}{\partial x}\,;\\ &\implies \boxed{\sigma=\frac{\partial e}{\partial\epsilon}}\ ;\ \boxed{\theta=\frac{\partial e}{\partial s}}\ ;\ \boxed{\theta\dot{s}=r-\frac{\partial q}{\partial x}}\end{aligned} \tag{3.51}$$

The reduced form of the first principle of energy conservation can be used to compute the entropy in a deformed configuration. Namely, at a fixed temperature, the reversible change of entropy in thermoelasticity is given as

$$\begin{aligned} \frac{\partial s}{\partial t}_{reversible} &= r/\theta - \frac{\partial q}{\partial x}/\theta \\ &= \underbrace{\frac{\partial e}{\partial t}}_{c(\theta)\frac{\partial \theta}{\partial t}}/\theta - \sigma\frac{\partial \epsilon}{\partial t}/\theta \end{aligned} \tag{3.52}$$

where the result in (3.51) was used. We can see that both the temperature and the strain time variations will contribute to the rate of change of entropy.

3.1.3.1 Remarks on entropy for perfect gas

The name 'entropy' was first used by Clausius, with the same connotation as label 'caloric' previously used by Carnot. The definitions of entropy in a vast majority of works on thermodynamics often lack clarity and the firm theoretical basis, comparable to the one provided by Cauchy for stress tensor. The only exception to this state of affairs is the unambiguous interpretation of the entropy for perfect gas (e.g. Atkins [14]), which has inspired the result in (3.52). The model of perfect gas, occupying a volume V and subjected to a pressure p is an assembly of particles or molecules, which is constantly in (random) motion whose velocity increases as a function of temperature θ. The particles do not interact, except of eventual trajectory modifications due to collision. The total number of particles does not change, hence the corresponding constitutive relations are proportional to this number, denoted as n. Moreover, by taking into account that the product between the pressure and the volume must remain fixed at constant temperature, we can write explicitly (see Atkins [14]):

$$pV = nR\theta \;;\; cst. = R := kN_A \tag{3.53}$$

where $R = 8.31451\ \mathrm{JK^{-1}mol^{-1}}$ is the constant of perfect gas, which can be obtained in accordance with this molecular model as the product between the Boltzman constant $k = 1.38 \times 10^{-23}\ \mathrm{JK^{-1}}$ (the measure of quantity of energy carried by a particular motion mode) and the Avogadro constant $N_A = 6.025 \times 10^{23}\ \mathrm{mol^{-1}}$ (the number of particles in a mole, the base unit of amount of pure substance in chemistry).

For the application of these results to thermoelasticity, it is important to realize that the expression in (3.53) can also be presented in the form of constitutive law between the pressure p and the volume change $1/V$,

$$p := f(\theta, V, n) = \underbrace{nR\theta}_{K_V(\theta)}\frac{1}{V} \tag{3.54}$$

where $K_V(\theta)$ is the bulk modulus, directly proportional to temperature. With this result in hand we can derive the first principle of thermodynamics for perfect gas:

$$
\begin{aligned}
dQ_V &= dE_V - \overbrace{dP_V}^{-pdV} \\
&= dE_V + \underbrace{p}_{K_V(\theta)/V} dV \\
&= dE_V + nR\theta \tfrac{dV}{V}
\end{aligned}
\tag{3.55}
$$

By making use of the definition of specific heat capacity $C_V(\theta)$ in (3.10), we can write the first term of the expression (3.55) according to

$$
dE_V = \underbrace{\frac{\partial E_V}{\partial \theta}}_{C_V(\theta)} d\theta \tag{3.56}
$$

By exploiting last two results, we can write the reversible change of entropy (at fixed temperature) for perfect gas in the form which is fully equivalent to the corresponding result in (3.52) for thermoelasticity:

$$
\begin{aligned}
dS_{reversible} &:= \tfrac{dQ_V}{\theta} = dE_V/\theta + pdV/\theta \\
&= C_V(\theta)\tfrac{d\theta}{\theta} + nR\tfrac{dV}{V}
\end{aligned}
\tag{3.57}
$$

The last equation pertaining to infinitesimal increase can be integrated to obtain the final value of entropy for perfect gas

$$
\begin{aligned}
S &= S_0 + \int_{t_0}^{t} (C_V(\theta)\tfrac{d\theta}{\theta} + nR\tfrac{dV}{V})\, dt \\
&= S_0 + C_V \ln \tfrac{\theta}{\theta_0} + nR \ln \tfrac{V}{V_0}
\end{aligned}
\tag{3.58}
$$

where $S_0 = S(t_0)$, $V_0 = V(t_0)$ and $\theta_0 = \theta(t_0)$ are, respectively, initial values of entropy, volume and temperature.

3.1.3.2 Remarks on second principle for thermoelasticity

A non-homogeneous temperature field and heat transfer along the bar will restrict that all the heat be transformed into entropy increase. In other words, unlike reversible processes in thermoelasticity at fixed value of temperature, the heat transfer in a non-homogeneous temperature field will be accompanied by dissipation, typical of irreversible processes. Due to such dissipation, the rate of increase of entropy ought to be superior to the source of heat. This is precisely the statement of the second principle of thermodynamics applied to thermoelasticity

$$
\frac{\partial}{\partial t} \int_{x_1}^{x_2} s\, dx > \int_{x_1}^{x_2} (r/\theta)\, dx - [q/\theta]_{x_1}^{x_2} \tag{3.59}
$$

By exploiting the reduced form of the first principle, we will simplify the last result to obtain the corresponding expression for dissipation by conduction, which must remain positive for non-homogeneous temperature field

$$0 < D_{cond} = \overbrace{\theta \frac{\partial s}{\partial t} - r + \frac{\partial q}{\partial x}}^{=0} - \tfrac{1}{\theta} q \tfrac{\partial \theta}{\partial x} \qquad (3.60)$$
$$= -\tfrac{1}{\theta} q \tfrac{\partial \theta}{\partial x}$$

The second principle of thermodynamics should remain in agreement with the experimentally observed result that the spontaneous heat flow is always directed from hot to cold. This can be satisfied with the Fourier law of heat conduction, postulating that the heat flux is proportional to the negative value of gradient of temperature, with the heat conduction $k > 0$ as coefficient of proportionality

$$q = -k \frac{\partial \theta}{\partial x} \; ; \; k > 0 \qquad (3.61)$$

3.1.4 Thermodynamics potentials in elasticity

The potential of internal energy, with the entropy and the deformation as the state variable, is not the only possibility to pursue the development of a theoretical formulation in thermodynamics. We can also use the dual variables, the temperature or the stress, to replace, respectively, the entropy or the strain as the state variables. This also implies an alternative choice of the potential, among those which are presented in this section.

3.1.4.1 Free energy of Helmholtz

Considering the complexity of eventual experiments for measuring the entropy as specified in (3.52), versus the simplicity of measuring the temperature, we often give preference to the temperature as the state variable. To that end, we can use the Legendre transform to exchange the roles between the entropy and the temperature, and thus introduce the free energy of Helmholtz $\psi(\epsilon, \theta)$ according to

$$\psi(\epsilon, \theta) = e(\epsilon, s) - s\theta \qquad (3.62)$$

We can easily show that the free energy of Helmholtz is the thermodynamic potential that allows to define constitutive equations for entropy and stress in terms of temperature and deformation as state variables. More precisely, the computation of partial derivatives of potentials e and ψ in (3.62), allows us to write

$$\frac{\partial}{\partial\theta}[\psi(\cdot) = e(\cdot) - s\theta] \Rightarrow s = -\frac{\partial\psi(\epsilon,\theta)}{\partial\theta}\bigg|_{\epsilon}$$
$$\frac{\partial}{\partial\epsilon}[\psi(\cdot) = e(\cdot) - s\theta] \Rightarrow \underbrace{\frac{\partial\psi(\epsilon,\theta)}{\partial\epsilon}\bigg|_{\theta}}_{isothermal} = \underbrace{\frac{\partial e(\epsilon,s)}{\partial\epsilon}\bigg|_{s}}_{isentrope} =: \sigma \tag{3.63}$$

The last results shows that the stress computation in thermoelasticity for either isothermal state (with $\theta = cst.$) or isentropic state (with $s = cst.$) can be carried out in the same manner as in hyperelasticity. Moreover, for an isentropic state (with $\frac{\partial\theta}{\partial t} = 0$) the free energy can be interpreted as the part of internal energy which is available for producing the mechanical power at fixed temperature:

$$d\psi = de - d(\theta s) = \underbrace{de - \overbrace{\theta ds}^{dQ}}_{dP} - s\overbrace{d\theta}^{=0} \tag{3.64}$$

The choice of free energy potential does not affect the result in (3.51) on the reduced form of the first principle. Namely, by replacing (3.62) into (3.48), we can obtain

$$\begin{aligned} r - \frac{\partial q}{\partial x} &= \frac{\partial e(\epsilon,s)}{\partial t} - \sigma\frac{\partial\epsilon}{\partial t} \\ &= \frac{\partial\psi(\epsilon,\theta)}{\partial t} + \frac{\partial\theta}{\partial t}s + \theta\frac{\partial s}{\partial t} - \sigma\frac{\partial\epsilon}{\partial t} \\ &= \underbrace{\left(\frac{\partial\psi(\epsilon,\theta)}{\partial\epsilon} - \sigma\right)}_{=0}\frac{\partial\epsilon}{\partial t} + \underbrace{\left(\frac{\partial\psi(\epsilon,\theta)}{\partial\theta} + s\right)}_{=0}\frac{\partial\theta}{\partial t} + \theta\frac{\partial s}{\partial t} \end{aligned} \tag{3.65}$$

By taking into account that evolution of each state variable remains independent of others, the last result further allows to conclude

$$\sigma = \frac{\partial\psi(\epsilon,\theta)}{\partial\epsilon}\ ;\ s = -\frac{\partial\psi(\epsilon,\theta)}{\partial\theta}\ ;\ \theta\dot{s} = r - \frac{\partial q}{\partial x} \tag{3.66}$$

3.1.4.2 Enthalpy

The Legendre transform can also be used to exchange the roles between the deformation and the stress, to allow the stress as a state variable

$$\phi(\sigma,s) = e(\epsilon,s) - \sigma\epsilon \tag{3.67}$$

The new potential of enthalpy will allow us to define the constitutive equations for deformation and temperature in terms of stress and entropy

$$
\begin{aligned}
&\frac{\partial}{\partial \sigma}[\phi(\cdot) = e(\cdot) - \sigma\epsilon] \Rightarrow \qquad \epsilon = -\left.\frac{\partial\phi(\sigma,s)}{\partial\sigma}\right|_s \\
&\frac{\partial}{\partial s}[\phi(\sigma,\theta) = e(\epsilon,s) - \sigma\epsilon] \Rightarrow \left.\frac{\partial\phi(\sigma,s)}{\partial s}\right|_\sigma = \left.\frac{\partial e(\epsilon,s)}{\partial s}\right|_\epsilon =: \theta
\end{aligned}
\tag{3.68}
$$

The main advantage of using the stress as one of the state variables is in accommodating a more general form of the constitutive equation with constrains, which can be of interest for practical applications. For example, the inverse of standard constitutive law for elasticity in (3.68) for computing strain from stress remains valid for incompressible cane, contrary to the form in (3.63) which is no longer applicable, since the strain components in the incompressible limit $tr[\epsilon] = 0$ are not independent.
By introducing the result in (3.68) into the first principle of thermodynamics, we can show that the enthalpy corresponds to the part of free energy which is available to produce heating at fixed value of strain. Namely, with $\sigma = cst.$ and $\dot{\sigma} = 0$, we can write:

$$
d\phi = \underbrace{\overbrace{de}^{dQ+dP} - \overbrace{\sigma d\epsilon}^{dP}}_{dQ} - \epsilon \overbrace{d\sigma}^{=0}
\tag{3.69}
$$

3.1.4.3 Free enthalpy of Gibbs

We conclude the list of thermodynamics potentials with the free enthalpy of Gibbs, where the Legendre transform allows us to exchange the places between the entropy and the temperature to obtain the corresponding modification of enthalpy

$$
\chi(\sigma,\theta) = \phi(\sigma,s) - s\theta
\tag{3.70}
$$

The Gibbs potential provides the constitutive equations for deformation and entropy in terms of stress and temperature

$$
\begin{aligned}
&\frac{\partial}{\partial \theta}[\chi(\cdot) = \phi(\cdot) - s\theta] \Rightarrow \ s = -\left.\frac{\partial\chi(\sigma,\theta)}{\partial\theta}\right|_\sigma \\
&\frac{\partial}{\partial \sigma}[\chi(\cdot) = \phi(\cdot) - s\theta] \Rightarrow \left.\frac{\partial\chi(\sigma,\theta)}{\partial\sigma}\right|_\theta = \left.\frac{\partial\phi(\sigma,s)}{\partial\sigma}\right|_s =: -\epsilon
\end{aligned}
\tag{3.71}
$$

3.1.5 Thermodynamics of inelastic behavior: constitutive models with internal variables

3.1.5.1 Second principle of thermodynamics and local dissipation

The role of the second principle of thermodynamics is particularly important for inelastic behavior, in directing the thermomechanical coupling by

limiting a spontaneous exchange between the mechanical and thermal energies. In that sense, we can no longer count with the conservation principle and energy balance without accounting for energy dissipation characteristic of inelastic behavior. This can be illustrated by a simple experiment, where a solid block is posed freely on the inclined plane and stabilized by frictional resistance. If we supply heat to the system, by heating the block, it will not start to slide and the thermal energy will not be transformed into mechanical energy. On the other hand, if we apply mechanical power (of external forces) starting the frictional sliding of block, the frictional resistance will stop the sliding, and, more importantly, produce the heating of the surface of block in contact with the plane. We thus conclude that the mechanical energy can be spontaneously transformed into thermal energy in an irreversible processus of frictional sliding, with the corresponding heat loss. This example illustrates the dissipation and the heat loss of the system (here the solid block) in contact with its environment.
More generally, the dissipation can also be produced inside a system (or a solid deformable body with inelastic behavior) submitted to a heterogeneous stress field. The dissipation is accompanied by a heat loss in the part of the system where the irreversible processus takes place. The loss of heat in a irreversible processus implies that the corresponding rate of increase of entropy ought to be superior to the heat supply. This is the statement of the second law of thermodynamics, also expressed in the form of the Clausius-Duhem inequality:

$$\frac{\partial}{\partial t}\int_{x_1}^{x_2} s\,dx > \int_{x_1}^{x_2}\frac{r}{\theta}\,dx - [\frac{q}{\theta}]_{x_1}^{x_2} \tag{3.72}$$

In the limit case with $x_2 \mapsto x_1 \equiv x$, we can obtain the local form of the second principle and define the dissipation $D > 0$, containing the local part D_{loc} and the dissipation by conduction D_{cond}

$$\begin{aligned} 0 < D &:= \theta\tfrac{\partial s}{\partial t} - (r - \tfrac{\partial q}{\partial x}) - \tfrac{1}{\theta}q\tfrac{\partial \theta}{\partial x} \\ &= \underbrace{\theta\frac{\partial s}{\partial t} + \sigma\frac{\partial \epsilon}{\partial t} - \frac{\partial}{\partial t}e}_{D_{loc}}\underbrace{-\frac{1}{\theta}q\frac{\partial \theta}{\partial x}}_{D_{cond}} \end{aligned} \tag{3.73}$$

We note that the dissipation produced by conduction D_{cond} in (3.73) is the same as for the rigid conductor case;

$$\begin{aligned} 0 \le D_{cond} &:= -\tfrac{1}{\theta}\overbrace{q}^{-k\frac{\partial \theta}{\partial x}}\tfrac{\partial \theta}{\partial x} \\ &= \tfrac{k}{\theta}(\tfrac{\partial \theta}{\partial x})^2 \end{aligned} \tag{3.74}$$

The last expression confirms that, as long as the Fourier law is chosen to describe the heat conduction (with $q = -k\frac{\partial \theta}{\partial x}$, where $k > 0$), the dissipation by conduction always remains positive. For that reason, we will only focus upon the local dissipation (Truesdell and Toupin [258]), which characterizes the inelastic behavior

$$0 \leq D_{loc} := \theta \frac{\partial s}{\partial t} + \sigma \frac{\partial \epsilon}{\partial t} - \frac{\partial e}{\partial t} \tag{3.75}$$

By taking into account that $\theta = \frac{\partial e}{\partial s}$ and $\sigma = \frac{\partial e}{\partial \epsilon}$, the last result clearly indicates that the internal energy ought to depend upon more state variables than only the deformation and the entropy. Namely, in order to describe an inelastic behavior and the corresponding irreversible processus, we have to include among the state variables so-called internal variables, capable of representing different phenomena that characterize a particular inelastic behavior. Several examples of this kind are the response sensitivity to deformation rate, the response dependency on deformation trajectory or yet the change of material internal structure during deformation.

The choice of internal variables can be made according to various criteria. For example, the internal variables can represent the physical phenomena observed at a fine scale, which are at the origin of the inelastic behavior. Several examples of this kind are the internal variables characterizing the microstructure internal evolution, such as dislocation pile-up of plastic crystal for metallic materials (e.g. see Lubliner [174]) as well as the micro-cracks (e.g. see Krajcinovic [153] or Lemaitre and Chaboche [168]) for cement-based materials, and yet the internal variables characterizing the coupling between mechanics and physical chemistry (e.g. see Atkins [14]).

More frequently, the internal variables are chosen for the phenomenological model defined at the structural level without taking into account the phenomena at the origin of inelastic behavior (which is impossible at that level), but only the corresponding consequences observed experimentally. Several examples of this kind are the hardening variable specifying the increase of yield stress due to accumulated plastic deformation, damage variable providing a homogenized description of the cracks and the corresponding modification of Young's modulus etc.

Internal variables are yet referred to as the hidden variables, the connotation which might imply difficulties in measuring their values. However, this is not necessarily true, since a large number of phenomenological internal variables (e.g. hardening variables, plastic deformation or damage variable) can be measured, even though this requires a somewhat elaborate loading program including loading and unloading cycles. In fact, the main difficulty with internal variables is not related to measuring them, but rather to controlling their evolution during an experiment.

In the rest of this chapter we are interested only in phenomenological internal variables (the other kind are dealt with in the last chapter) and presenting the thermodynamics foundation of the constitutive models with phenomenological internal variables. In a simple 1D case, any internal variable is represented as a scalar field, ζ_i. In more general 2D and 3D cases, the internal variables can be represented as either scalars, vectors or tensors, without modification of the thermodynamics framework we developed for 1D case. In either case, the internal energy for a constitutive model of inelastic behavior with internal variables can be written as a function of the deformation, entropy and the internal variables

$$e(\epsilon, s, \zeta_i) \tag{3.76}$$

The key assumption which allows us to exploit the thermodynamics framework developed for elasticity with internal energy as the potential, and apply to present case of inelastic constitutive model with internal variables, concerns the accompanied elastic state (see Germain, Nguyen and Suquet [86], Kestin and Rice [149]) with frozen values of internal variables. With such an internal energy potential governing the irreversible processus of inelastic behavior, we can also define the thermodynamic flux q_i, conjugate to internal variable ζ_i

$$\frac{\partial e}{\partial t} = \underbrace{\frac{\partial e}{\partial \epsilon}}_{\sigma} \frac{\partial \epsilon}{\partial t} + \underbrace{\frac{\partial e}{\partial s}}_{\theta} \frac{\partial s}{\partial t} + \sum_i \underbrace{\frac{\partial e}{\partial \zeta_i}}_{q_i} \frac{\partial \zeta_i}{\partial t} \tag{3.77}$$

The hypothesis of the accompanied elastic state and the last result are valid only if the characteristic evolution time for the internal variables is much shorter than the evolution time scale of other state variables: strain and entropy. In this case, the result in (3.77) will further allow to obtain the reduced form of local dissipation in (3.75)

$$\begin{aligned} 0 < D_{loc} &:= \theta \tfrac{\partial s}{\partial t} + \sigma \tfrac{\partial \epsilon}{\partial t} - \tfrac{\partial e}{\partial t} \\ &= \textstyle\sum_i \underbrace{(-\frac{\partial e}{\partial \zeta_i})}_{q_i} \tfrac{\partial \zeta_i}{\partial t} \end{aligned} \tag{3.78}$$

With this result in hand we can also write the local form of the first principle of thermodynamics, which is valid both for the elastic case (with $D_{loc} = 0$) and inelastic case (with $D_{loc} > 0$)

$$\boxed{\theta \tfrac{\partial s}{\partial t} = D_{loc} + r - \tfrac{\partial q}{\partial x}} \quad ; \; D_{loc} := \sum_i q_i \dot{\zeta}_i \geq 0 \tag{3.79}$$

Other thermodynamic potentials besides the internal energy, which were presented in the previous section for thermoelasticity, can also be adapted for inelastic constitutive models with internal variables. For example, we can also introduce the free energy potential by means of Legendre transform which will replace the entropy by the temperature as the state variable

$$\psi(\epsilon, \theta, \zeta_i) = e(\epsilon, s, \zeta_i) - \theta s \tag{3.80}$$

With such a choice of the free-energy potential we will modify neither the local dissipation nor the local form of the first principle of thermodynamics in (3.79)

$$\begin{aligned} 0 < D_{loc} &:= \theta \tfrac{\partial s}{\partial t} - [r - \tfrac{\partial q}{\partial x}] \\ &= \sigma \tfrac{\partial \epsilon}{\partial t} - \tfrac{\partial}{\partial t}\psi(\epsilon,\theta,\zeta_i) - \tfrac{\partial \theta}{\partial t} s \\ &= \underbrace{(-\frac{\partial \psi}{\partial \epsilon} + \sigma)}_{=0} \tfrac{\partial \epsilon}{\partial t} - \underbrace{(\frac{\partial \psi}{\partial \theta} + s)}_{=0} \tfrac{\partial \theta}{\partial t} - \textstyle\sum_i \underbrace{\frac{\partial \psi}{\partial \zeta_i}}_{-q_i} \tfrac{\partial \zeta_i}{\partial t} \\ \Longrightarrow \quad & D_{loc} := \textstyle\sum_i q_i \tfrac{\partial \zeta_i}{\partial t} \geq 0 \; ; \; \sigma = \tfrac{\partial \psi(\epsilon,\theta,\zeta_i)}{\partial \epsilon} \; ; \; s = -\tfrac{\partial \psi(\epsilon,\theta,\zeta_i)}{\partial \theta} \end{aligned} \tag{3.81}$$

We can also choose the Gibbs thermodynamic potential, with stress σ and temperature θ as the state variables accompanied by the internal variables ζ_i. The Gibbs potential will not change the definition of local dissipation, since the Legendre transform $\chi(\sigma,\theta,\zeta_i) = -\psi(\epsilon,\theta,\zeta_i) + \epsilon\,\sigma$ and result (3.81) allow us to write

$$\begin{aligned} 0 \quad & \leq D_{loc} := (\tfrac{\partial \chi}{\partial \sigma} - \epsilon)\tfrac{\partial \sigma}{\partial t} + (\tfrac{\partial \chi}{\partial \theta} - s)\tfrac{\partial \theta}{\partial t} - \textstyle\sum_i \tfrac{\partial \chi}{\partial \zeta_i}\tfrac{\partial \zeta_i}{\partial t} \\ & \Longrightarrow \epsilon = \tfrac{\partial \chi}{\partial \sigma} \; ; \; s = \tfrac{\partial \chi}{\partial \theta} \; ; \; q_i = -\tfrac{\partial \chi}{\partial \zeta_i} \; ; \; D_{loc} := \textstyle\sum_i q_i \tfrac{\partial \zeta_i}{\partial t} \geq 0 \end{aligned} \tag{3.82}$$

In order to complete the description of the inelastic constitutive model, we also need to supply the evolution equation for each internal variable. The latter can be chosen as the first order differential equation in time, very much alike the evolution equation for non-stationary heat transfer, in order to reflect the irreversible nature of the evolution of internal variable. For example, with the state variables of the Gibbs potential, the evolution equation can be written:

$$\boxed{\tfrac{\partial \zeta_i}{\partial t} = f_i(\sigma,\theta,\zeta_i)} \tag{3.83}$$

The last result allows us to explicitly define the local dissipation with respect to the corresponding evolution of stress, temperature and internal variables

$$D := \sum_i q_i \frac{\partial \zeta_i}{\partial t} = \sum_i q_i f_i(\sigma,\theta,\zeta_i) \geq 0 \; ; \; q_i = -\frac{\partial \chi}{\partial \zeta_i} \tag{3.84}$$

Several examples of internal variables describing the inelastic phenomena observed experimentally are presented in the next section.

3.1.6 *Internal variables in viscoelasticity*

The first example of the evolution equation for internal variable is presented herein for the linear viscoelasticity model. The latter is the simplest

model capable of accounting for the rate-dependency of the response. Similar viscoelasticity models are used for representing the phenomena of creep and relaxation, resulting in time evolution of deformation.
The rheological model of linear viscoelasticity (see Figure 3.3) is constructed as the assembly in series between the elastic spring and an assembly in parallel between another elastic spring and a viscous damper. In general, the elasticity moduli of springs can be temperature dependent, $E_0(\theta)$ and $E_1(\theta)$, and the same temperature dependence is assumed for viscosity coefficient $\eta(\theta)$. From the viscoelasticity model of this kind, we can recover the model of Kelvin (with $E_0 = 0$) or the model of Maxwell (with $E_1 = 0$).

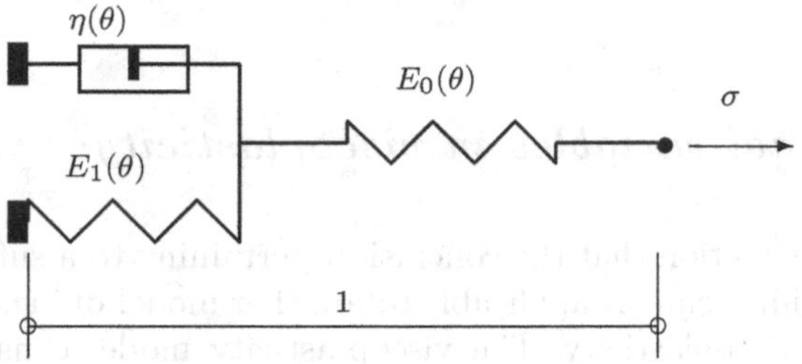

Fig. 3.3 Rheological model of 1D linear viscoelasticity.

By taking into account that the elastic spring on one side and the assembly of spring and viscous damper on another (sharing the same value of stress σ), each contribute to the total deformation, we can postulate the additive decomposition of the deformation. The deformation of the first spring, which is the first to appear in loading and to disappear in unloading, is referred to as the elastic deformation $\epsilon^e = \sigma/E_0$. The deformation of the assembly spring-damper is called viscous ϵ^v. If the latter is chosen as the internal variable, we can write:

$$\boxed{\zeta = \epsilon^v} \quad \Longrightarrow \epsilon = \underbrace{\sigma/E_0(\theta)}_{\epsilon^e} + \underbrace{\zeta}_{\epsilon^v} \tag{3.85}$$

By taking into account that the stress σ in the first spring will be shared between the second spring and the viscous damper, which both have the same viscous deformation, we can obtain

$$\sigma = E_1(\theta)\zeta + \eta(\theta)\dot{\zeta} \implies \boxed{\dot{\zeta} = -\frac{E_1(\theta)}{\eta(\theta)}\zeta + \frac{1}{\eta(\theta)}\sigma =: f(\sigma, \theta, \zeta)} \tag{3.86}$$

The last result represents the evolution equation of the internal variable for linear viscoelasticity model, which complies with the standard format proposed in (3.83). It is possible (see Strang [248], p. 472) to obtain the analytic solution[6] of such an equation:

$$\zeta(t) = \zeta(0)exp(-t/\tau) + \int_0^t \frac{1}{\eta(\theta)} exp[-(t-s)/\tau]\sigma(s)\, ds \; ; \; \tau = \frac{\eta}{E_1} \quad (3.87)$$

The parameter $\tau := E_1(\theta)/\eta(\theta)$ is called the characteristic time of standard viscoelasticity model, capturing the rate of evolution of internal variables with respect to a variation of the stress or the temperature imposed by a particular loading program. We can see from (3.87) that the hypothesis of accompanied elastic state, postulating that the internal variables remain frozen, is justified for the linear viscoelasticity model only if the characteristic time τ remains (much) smaller than the evolution time of stress and temperature.

3.1.7 Internal variables in viscoplasticity

We show in this section that the conclusion pertaining to a sufficiently short characteristic time remains applicable to another model of inelastic behavior referred to as viscoplasticity. The viscoplasticity model considers the viscoplastic deformation component as the internal variable. No modification of internal variable will occur in the elastic domain where the absolute value of stress remains inferior to yield stress. Outside of elastic domain, one has to provide the evolution equation of viscoplastic deformation. The rheological model of 1D viscoplasticity presented in Figure 3.4 consists of an elastic spring mounted in series with the frictional slider with resistance to sliding $\sigma_y(\theta)$ and the viscous damper. For the stress value larger than yield stress, the total deformation can be decomposed into elastic part ϵ^e and viscoplastic part $\epsilon^{vp} = \zeta$. The latter is the chosen internal variable, which corresponds to the damper deformation. The evolution equation of the internal variable can be written:

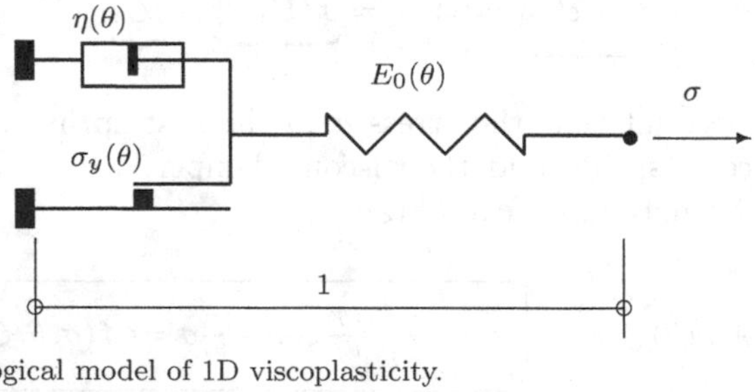

Fig. 3.4 Rheological model of 1D viscoplasticity.

[6] For a first order differential equation $\dot{d}(t) + d(t)/\tau = g(t)$, with $d(0) = d_0$, we can find the exact solution by pre-multiplying by $exp(t/\tau)$ and by integrating directly to obtain $d(t) = d_0 exp(-t/\tau) + \int_0^t exp[-(t-s)/\tau]g(s)\, ds$.

$$\zeta = \begin{cases} 0 & ; |\sigma| \leq \sigma_y \\ \frac{1}{\eta(\theta)}[\sigma - \sigma_y sign(\sigma)] & ; |\sigma| > \sigma_y \end{cases} \tag{3.88}$$

The characteristic evolution time for internal variable is proportional to the extra stress, $[\sigma - \sigma_y sign(\sigma)]$, and inversely proportional to the viscosity parameter $\eta(\theta)$. The characteristic time of evolution for other state variables is directly dependent on the given loading program. This can be illustrated by the relaxation test, where we impose a given value of deformation, which is larger than the elasticity limit $\epsilon_0 > \frac{\sigma_y}{E}$. This deformation is subsequently kept constant in time, leading to the corresponding stress evolution

$$\left.\begin{array}{r} \dot{\sigma} = E(\dot{\epsilon} - \dot{\epsilon}^{vp}) \\ \dot{\epsilon}^{vp} = \frac{1}{\tau}E^{-1}[\sigma - \sigma_y] \end{array}\right\} \Leftrightarrow \left\{\begin{array}{c} \dot{\sigma} + \frac{1}{\tau}\sigma = E\dot{\epsilon} + \frac{1}{\tau}\sigma_y \\ \epsilon = \epsilon(0) > \sigma_y/E \end{array}\right. \tag{3.89}$$
$$\Longrightarrow \sigma(t) = [E\epsilon_0 - \sigma_y]e^{-t/\tau} + \sigma_y \; ; \tau = \eta/E \, ; t \in [0, +\infty)$$

We can note that the final value of stress, produced in relaxation test for the viscoplasticity model, tends towards the limit value of the yield stress σ_y for classical plasticity. We can also note that the relaxation time $\tau = \eta/E$ dictates the rate with which this limit value is attained; see Figure 3.5.

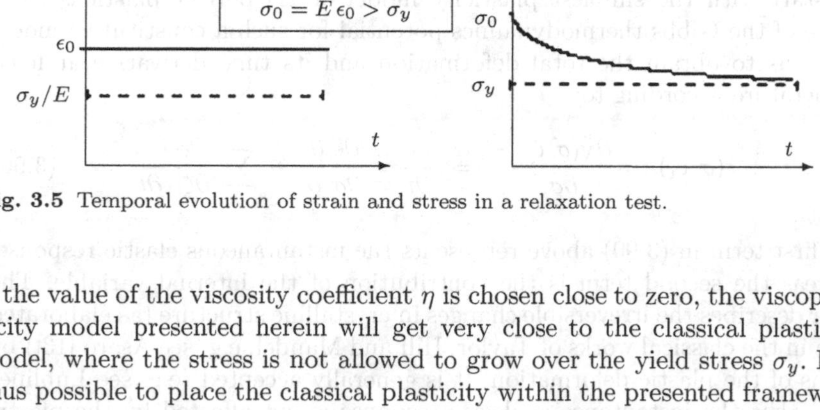

Fig. 3.5 Temporal evolution of strain and stress in a relaxation test.

If the value of the viscosity coefficient η is chosen close to zero, the viscoplasticity model presented herein will get very close to the classical plasticity model, where the stress is not allowed to grow over the yield stress σ_y. It is thus possible to place the classical plasticity within the presented framework of viscoplasticity models, with the zero value of relaxation time.[7] We can also exploit the inverse point of view, where the viscoplasticity model can be considered as a regularized form of the classical plasticity model, which introduces a continuity in the time evolution of the plastic deformation.
In the next section we present a 1D model of perfect plasticity and the plasticity model with linear hardening. In order to simplify the presentation, in

[7] Note that the latter also ensures the validity of the standard thermodynamics framework with accompanying elastic state in dealing with irreversible processes.

the rest of this chapter we will ignore the influence of temperature. The role of temperature will further be explored in the later chapter on thermodynamical coupling.

3.2 1D models of perfect plasticity and plasticity with hardening

The plasticity model is perhaps the most well understood basic example of a constitutive model with internal variables which is already capable of describing several observed phenomena of inelastic behavior. Namely, the plasticity model discussed in this section can account for the presence of irreversible plastic deformation upon complete unloading, as well as provide the representation of change in resistance outside of elastic domain and evolution of plasticity threshold as a function of accumulated plastic deformation. These ideas are first illustrated in 1D framework, and then generalized to 3D case.

3.2.1 1D perfect plasticity

We start with the simplest plasticity model of 1D perfect plasticity. The choice of the Gibbs thermodynamics potential for such a constitutive model allows us to obtain the total deformation and its time derivative at fixed temperature according to:

$$\epsilon(\sigma, \zeta_i) := \frac{\partial \chi(\sigma, \zeta_i)}{\partial \sigma} \implies \frac{\partial \epsilon}{\partial t} = \frac{\partial \epsilon}{\partial \sigma}\frac{\partial \sigma}{\partial t} + \sum_i \frac{\partial \epsilon}{\partial \zeta_i}\frac{\partial \zeta_i}{\partial t} \tag{3.90}$$

The first term in (3.90) above represents the instantaneous elastic response, whereas the second term is the contribution of the internal variable. The latter describes the irreversible changes in crystalline structure (as elaborated upon in the classical works of Taylor, Hill and Mandel, e.g. see Asaro [13]) by means of the plastic deformation. It is generally accepted (e.g. see Lubliner [172]) that the instantaneous elastic response is not affected by the plastic deformation. This observation can formally be written:

$$\frac{\partial}{\partial \zeta_i}\frac{\partial \epsilon}{\partial \sigma} = 0 \tag{3.91}$$

The results in (3.90) and (3.91) can be used to confirm the validity of the additive decomposition of the total deformation into elastic part ϵ^e and plastic part ϵ^p. We can also accept that the former depends on stress but not on internal variables, whereas the latter is only a function of the internal variable.

$$\boxed{\epsilon(\sigma, \zeta_i) = \epsilon^e(\sigma) + \epsilon^p(\zeta_i)} \tag{3.92}$$

In fact, for 1D model of perfect plasticity, we use only one internal variable, which is equal to the plastic deformation

$$\epsilon^p = \zeta_1 \tag{3.93}$$

The equivalence between the perfect plasticity and dry friction in 1D framework allows us to consider the plastic deformation as the frictional sliding. We can thus construct an illustrative rheological model of 1D perfect plasticity, which consists of an elastic spring and a frictional device mounted in series; see Figure 3.6. The equivalence with plasticity imposes that the spring elastic constant is equal to Young's modulus E and that the sliding resistance is equal[8] to elasticity limit σ_y.

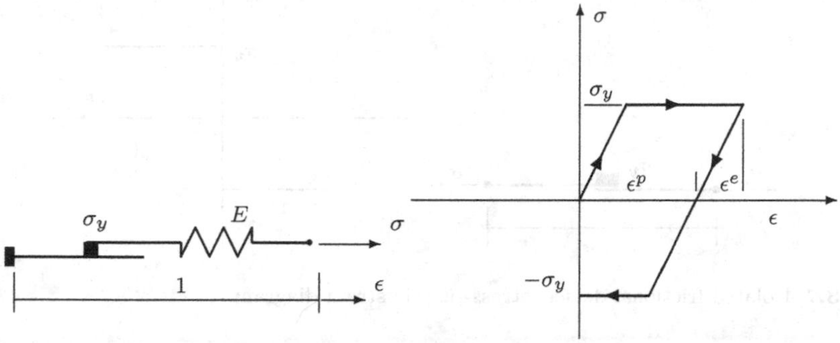

Fig. 3.6 Rheological model of 1D perfect plasticity as simple spring-frictional device and its stress–strain diagram considering both loading and unloading.

The key observation that allows to easily provide the stress-strain diagram presented in Figure 3.6, pertains to computing the stress value in each component of the chosen rheological model directly from equilibrium equation. For a traction value applied at free end, the stress remains constant (and equal to applied traction) in each component, the spring and the frictional device. The elastic spring is always activated first providing the same elastic response, both in loading and unloading. Moreover, the stress value is always defined by the deformation of the elastic spring $\epsilon^e \leq \sigma_y/E$, independent on the amount of frictional sliding (or plastic deformation). In agreement with these observations, we can provide the free energy potential for 1D perfect plasticity

$$\psi(\epsilon, \underbrace{\epsilon^p}_{\zeta_1}) = \frac{1}{2}(\epsilon - \epsilon^p)E(\epsilon - \epsilon^p) \tag{3.94}$$

[8] Note that the frictional model in 1D framework can not have the sliding resistance dependent on contact pressure.

The stress is obtained as the derivative of such a potential with respect to infinitesimal strain

$$\sigma := \frac{\partial \psi}{\partial \epsilon}|_{\epsilon^p} = E\,\underbrace{(\epsilon - \epsilon^p)}_{\epsilon^e} \tag{3.95}$$

We can also interpret the last result as the definition of the elastic strain in (3.92) in terms of stress

$$\epsilon^e = \frac{1}{E}\sigma \tag{3.96}$$

The stress value in perfect plasticity can not be larger than the elasticity limit σ_y, neither in loading nor in unloading. Moreover, the stress value has to stay fixed and equal to σ_y during active plastic slip. The plastic slip can also be represented by the corresponding rheological model of isolated frictional device (see Figure 3.7), leading to a further simplification of stress-plastic strain diagram.

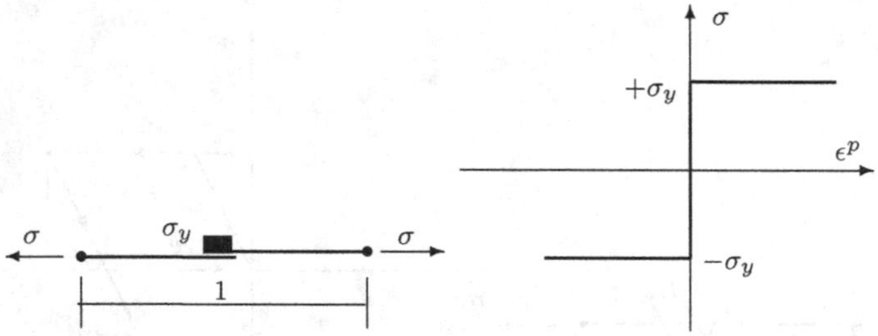

Fig. 3.7 Isolated frictional device: stress-plastic strain diagram.

The plastic slip can be described by means of the plastic multiplier γ, which is considered to take only non-negative values requiring that $\gamma > 0$ and $\frac{\partial \gamma}{\partial t} > 0$; we can thus write

$$\begin{aligned} \frac{\partial \epsilon^p}{\partial t} &= \frac{\partial \gamma}{\partial t} \;;\; \text{if } \sigma = \sigma_y > 0 \\ \frac{\partial \epsilon^p}{\partial t} &= -\frac{\partial \gamma}{\partial t} \;;\; \text{if } \sigma = -\sigma_y < 0 \end{aligned} \tag{3.97}$$

The pseudo-time derivative is applied to both side of the last equation, to both ϵ^p and γ, to indicate that the plastic deformation is rate independent. The rate of plastic deformation takes the same sign (positive or negative) as stress, which allows us to write the general form of the evolution equation applicable to both tension and compression

$$\boxed{\frac{\partial \epsilon^p}{\partial t} = \frac{\partial \gamma}{\partial t}\, sign(\sigma)} \;;\; \text{if } |\sigma| = \sigma_y \tag{3.98}$$

The last result is typically recast in terms of plasticity criterion which pertains to zero value of the yield function in the presence of plastic flow

$$\phi(\sigma) := |\sigma| - \sigma_y = 0 \implies \frac{\partial \gamma}{\partial t} > 0 \implies \frac{\partial \epsilon^p}{\partial t} = \frac{\partial \gamma}{\partial t}\, sign(\sigma) \tag{3.99}$$

By taking into account that the derivative of such a yield function is equal to $\frac{\partial|\sigma|}{\partial\sigma} = sign(\sigma)$, the evolution equation for the plastic deformation can also be written by using the yield surface as the corresponding potential

$$\frac{\partial\epsilon^p}{\partial t} = \frac{\partial\gamma}{\partial t}\frac{\partial\phi}{\partial\sigma}\;;\;\frac{\partial\gamma}{\partial t} > 0 \tag{3.100}$$

The plasticity model for which the yield function also plays the role of the plastic deformation potential is called the associative plasticity (e.g. see Hill [103]) or yet the standard material (e.g. see Halphen and Nguyen [98]).
If we want to keep the same form of flow rule for the elastic case where no change of plastic deformation occurs, we must admit that the plastic multiplier can also take the zero value. The latter is reserved for the elastic domain, which is characterized by a negative value of the yield function

$$\phi(\sigma) < 0 \implies \frac{\partial\gamma}{\partial t} = 0 \implies \frac{\partial\epsilon^p}{\partial t} = 0 \tag{3.101}$$

The elastic and plastic case are combined in a single constraint $\phi(\sigma) \leq 0$, which defines the plastic admissibility of stress field. The latter implies that the stress value for perfect plasticity can never have larger absolute value than the chosen elastic limit σ_y.
In summary of these consideration, we can provide the standard form of evolution equation for the plastic deformation, which is valid for both elastic and plastic case, along with the loading/unloading conditions

$$\boxed{\dot\epsilon^p = \dot\gamma\frac{\partial\phi}{\partial\sigma} \equiv \dot\gamma\, sign(\sigma)\;;\;\dot\gamma \geq 0}$$
$$\implies \left.\begin{array}{l}\phi(\sigma) < 0 \implies \dot\gamma = 0\\ \phi(\sigma) = 0 \implies \dot\gamma > 0\end{array}\right\} \Leftrightarrow \boxed{\dot\gamma\phi(\sigma) = 0} \tag{3.102}$$

The same result can also be obtained with a more abstract approach that fits within the thermodynamics framework and does not require any rheological scheme for illustrations. To that end, the free energy for perfect plasticity in (3.94) will allow us identify the negative value of stress as the flux thermodynamically conjugate to plastic deformation

$$\frac{\partial\psi}{\partial\epsilon^p} = -E(\epsilon - \epsilon^p) =: -\sigma \tag{3.103}$$

The last result, the second principle of thermodynamics and the stress constitutive equation will allow to define a compact form of the plastic dissipation for perfect plasticity model:

$$\begin{aligned} 0 \leq D^p &:= \sigma\tfrac{\partial\epsilon}{\partial t} - \tfrac{\partial}{\partial t}\psi(\epsilon,\epsilon^p)\\ &= \underbrace{(\sigma - \frac{\partial\psi}{\partial\epsilon})}_{=0}\tfrac{\partial\epsilon}{\partial t} - \underbrace{\frac{\partial\psi}{\partial\epsilon^p}}_{-\sigma}\tfrac{\partial\epsilon^p}{\partial t}\\ &= \sigma\tfrac{\partial\epsilon^p}{\partial t} \end{aligned} \tag{3.104}$$

We note in passing that the plastic dissipation for perfect plasticity takes the same form as local dissipation in (3.81), with the plastic deformation as internal variable and the negative value of stress as the thermodynamically conjugate flux.
Principle of maximum plastic dissipation: Allows us to obtain the evolution equation for the plastic deformation along with the loading/unloading conditions and. This principle postulates (e.g. see Hill [103] or Lubliner [173]) that among all plastically admissible values of stress, for which $\phi(\sigma) \leq 0$, it is the true stress that delivers the maximum of plastic dissipation. The principle of maximum of plastic dissipation can formally be presented as a constrained minimization problem:

$$\boxed{-D^p(\sigma) = \min_{\phi(\sigma^*)\leq 0} -D^p(\sigma^*)} \tag{3.105}$$

By using the Lagrange multiplier method, we can further transform this constrained minimization problem into the corresponding unconstrained version which can accept any stress field. this leads to min–max problem for the Lagrangian potential L^p, which is written:

$$\boxed{L^p(\sigma, \tfrac{\partial \gamma}{\partial t}) = \max_{\frac{\partial \gamma^*}{\partial t} \geq 0} \min_{\sigma^*} L^p(\sigma^*, \tfrac{\partial \gamma^*}{\partial t})} \quad ; \\ L^p(\sigma, \tfrac{\partial \gamma}{\partial t}) = -D^p(\sigma) + \tfrac{\partial \gamma}{\partial t}\phi(\sigma) \tag{3.106}$$

where the role of Lagrange multiplier is played by the plastic multiplier $\frac{\partial \gamma}{\partial t} \geq 0$. The Kuhn-Tucker optimality conditions (e.g. see Luenberger [175] or Strang [248]) corresponding to the Lagrangian in (3.106) above allow us to obtain the familiar evolution equation for plastic deformation accompanied by the loading/unloading conditions

$$\boxed{\left.\frac{\partial L^p}{\partial \sigma^*}\right|_{\sigma} = 0 \implies \frac{\partial \epsilon^p}{\partial t} = \frac{\partial \gamma}{\partial t}\frac{\partial \phi}{\partial \sigma}} \\ \boxed{\tfrac{\partial \gamma}{\partial t} \geq 0 \; ; \; \phi(\sigma) \leq 0 \; ; \; \tfrac{\partial \gamma}{\partial t}\phi(\sigma) = 0} \tag{3.107}$$

The principle of maximum plastic dissipation remains applicable for 2D and 3D cases, where it is often not possible to provide simple rheological models. This principle always provides the associated evolution equations typical of associative plasticity model (e.g. Hill [103], Lubliner [174]) or standard material (e.g. Nguyen and Halphen [98]). For all the models of this kind the yield function also plays the role of the potential governing plastic flow.
Consistency condition for the perfect plasticity model enforces that the absolute value of stress can never be larger than the elasticity limit σ_y. The

consistency condition also places restriction on possible evolution of the plastic flow. We first note that the plastic admissibility of a stress state satisfying the yield criterion at a given value of pseudo-time $\bar{t}$, will impose that $\phi(\sigma(x,\bar{t})) = 0$. The consistency condition follows from trying to impose the plastic admissibility of stress at subsequent time $\bar{t} + dt$, where dt is an infinitesimal time increment. Namely, we have to ensure that $\phi(\sigma(x,\bar{t}+dt)) \leq 0$. The latter can be computed by exploiting the Taylor series representation

$$\phi(\sigma(x,\bar{t} + dt)) = \underbrace{\phi(\sigma(x,\bar{t}))}_{=0} + \frac{\partial \phi(\sigma(x,\bar{t}))}{\partial t} dt + O(dt) = 0 \tag{3.108}$$

We can conclude from the last expression that the plastic admissibility of stress at subsequent time is decided by the value of the time derivative of yield function at present time

$$\left. \begin{array}{l} \frac{\partial \phi(\sigma(x,\bar{t}))}{\partial t} = 0 \Rightarrow \phi(\sigma(x,\bar{t} + dt)) = 0 \Rightarrow \frac{\partial \gamma(x,\bar{t})}{\partial t} > 0 \\ \frac{\partial \phi(\sigma(x,\bar{t}))}{\partial t} < 0 \Rightarrow \phi(\sigma(x,\bar{t} + dt)) < 0 \Rightarrow \frac{\partial \gamma(x,\bar{t})}{\partial t} = 0 \end{array} \right\} \tag{3.109}$$

The last result, written in compact form: $\frac{\partial \gamma}{\partial t}\frac{\partial \phi}{\partial t} = 0$, is also known as the plastic consistency condition on the stress field. It allows to distinguish between elastic unloading ($\frac{\partial \phi}{\partial t} < 0$ and $\frac{\partial \gamma}{\partial t} = 0$) and plastic loading ($\frac{\partial \phi}{\partial t} = 0$ and $\frac{\partial \gamma}{\partial t} > 0$). The latter is finally the only case where the plastic multiplier takes a non-zero value, which can be computed according to:

$$\begin{array}{l} 0 = \frac{\partial \phi(\sigma(x,t))}{\partial t} \\ \quad = \frac{\partial \phi}{\partial \sigma}\frac{\partial \sigma}{\partial t} \\ \quad = \frac{\partial \phi}{\partial \sigma} E(\frac{\partial \epsilon}{\partial t} - \frac{\partial \epsilon^p}{\partial t}) \\ \quad = \frac{\partial \phi}{\partial \sigma} E \frac{\partial \epsilon}{\partial t} - \frac{\partial \gamma}{\partial t}\frac{\partial \phi}{\partial \sigma} E \frac{\partial \phi}{\partial \sigma} \end{array} \Rightarrow \boxed{\phi = 0 \;\&\; \frac{\partial \phi}{\partial t} = 0 \Rightarrow \frac{\partial \gamma}{\partial t} = \frac{\partial \phi}{\partial \sigma}\frac{\partial \epsilon}{\partial t}} \tag{3.110}$$

By exploiting the last result we can also obtain the relation between the stress rate and deformation rate for the model of perfect plasticity

$$\frac{\partial \sigma}{\partial t} = E(\frac{\partial \epsilon}{\partial t} - \frac{\partial \gamma}{\partial t}\frac{\partial \phi}{\partial \sigma}) = 0 \Longrightarrow \boxed{\frac{\partial \sigma}{\partial t} = \left\{ \begin{array}{l} E\frac{\partial \epsilon}{\partial t} \; ; \; \frac{\partial \gamma}{\partial t} = 0 \\ 0 \; ; \; \frac{\partial \gamma}{\partial t} > 0 \end{array} \right.} \tag{3.111}$$

The last result explicitly shows than once the stress state leaves the elastic domain, the value of stress can no longer increase and it remains constant (equal to σ_y).

3.2.2 1D plasticity with isotropic hardening

The plasticity model with isotropic hardening is a simple modification of the perfect plasticity model, which allows to take into account the increase in plasticity threshold beyond limit of elasticity σ_y as a function of accumulated plastic deformation. The isotropic hardening phenomena are described by an additional internal variable with respect to plastic strain, which is referred to as hardening variable ζ. The plasticity criterion that accounts for increase of plasticity threshold can be written:

$$0 \geq \hat{\phi}(\sigma, \zeta) := |\sigma| - (\sigma_y + K\,\zeta) \tag{3.112}$$

where $K = cst.$ is the hardening modulus. It is important to note that the presence of hardening does not affect the additive decomposition of total strain into elastic and plastic component, nor does it change the elastic response. Therefore, the thermodynamic potential for plasticity with isotropic hardening can be constructed as a simple modification of the perfect plasticity model. For linear isotropic hardening model, we add to this potential a quadratic form in hardening variable

$$\psi(\epsilon, \epsilon^p, \zeta) = \underbrace{\frac{1}{2}(\epsilon - \epsilon^p)\,E\,(\epsilon - \epsilon^p)}_{\psi^e} + \underbrace{\frac{1}{2}\zeta\,K\,\zeta}_{\psi^p} \tag{3.113}$$

For any stress state in the elastic domain, the values of internal variables will not change, and plastic dissipation remains equal to zero. For the plasticity model with isotropic hardening this leads to:

$$\begin{aligned} 0 = D^p_{loc} &:= \sigma \tfrac{\partial \epsilon}{\partial t} - \tfrac{\partial}{\partial t}\psi(\epsilon, \epsilon^p, \zeta) \\ &= \underbrace{(\sigma - \frac{\partial \psi}{\partial \epsilon})}_{=0} \tfrac{\partial \epsilon}{\partial t} - \tfrac{\partial \psi}{\partial \epsilon^p} \underbrace{\frac{\partial \epsilon^p}{\partial t}}_{=0} - \tfrac{\partial \psi}{\partial \zeta} \underbrace{\frac{\partial \zeta}{\partial t}}_{=0} \end{aligned} \tag{3.114}$$

which confirms that the stress is computed in the same manner as for perfect plasticity,

$$\sigma := \frac{\partial \psi}{\partial \epsilon} = E\,(\epsilon - \epsilon^p) \tag{3.115}$$

By assuming the same constitutive relation remains valid for plastically admissible stress on the yield surface, we can obtain the corresponding (positive) value of the plastic dissipation from the second principle of thermodynamics according to:

$$0 < D^p := \sigma \frac{\partial \epsilon^p}{\partial t} + q \frac{\partial \zeta}{\partial t} \tag{3.116}$$

This compact form of plastic dissipation follows from observation that σ and q are the corresponding dual variables, or thermodynamic fluxes conjugate to internal variables ϵ^p and ζ, respectively

$$\frac{\partial \psi}{\partial \epsilon^p} = -E\,(\epsilon - \epsilon^p) =: -\sigma \;;\; \frac{\partial \psi}{\partial \zeta} = K\,\zeta = -q \tag{3.117}$$

For the developments to follow, it is interesting to recast the yield function in terms of dual variables

$$0 \geq \phi(\sigma, q) := |\sigma| - (\sigma_y - \underbrace{q}_{-K\zeta}) \tag{3.118}$$

Such a form of the plasticity criterion does not change the evolution equation of the plastic deformation with respect to perfect plasticity. However, we now also obtain an additional evolution equation for the hardening variable. The evolution equations of this kind are obtained from the principle of maximum plastic dissipation. We will look among all the plastically admissible couples of dual variables (for which $\phi(\sigma^*, q^*) = 0$), for the true solution (σ, q) that will maximize plastic dissipation. This task can be formulated as a constrained minimization problem

$$D^p(\sigma, q) = \max_{\phi(\sigma^*, q^*) \leq 0} [D^p(\sigma^*, q^*)] \tag{3.119}$$

The Lagrange multiplier method can be used to handle constraint and define the corresponding unconstrained minimization problem

$$\boxed{L^p(\sigma, q, \tfrac{\partial \gamma}{\partial t}) = \max_{\frac{\partial \gamma^*}{\partial t} \geq 0} \min_{\forall \sigma^*, q^*} L^p(\sigma^*, q^*, \tfrac{\partial \gamma^*}{\partial t})} \; ; \qquad L^p(\sigma, q, \tfrac{\partial \gamma}{\partial t}) = -D^p(\sigma, q) + \tfrac{\partial \gamma}{\partial t}\, \phi(\sigma, q) \tag{3.120}$$

The Kuhn-Tucker optimality condition for this problem will provide the evolution equations of plastic strain and hardening variable, along with the loading/unloading conditions

$$\boxed{\tfrac{\partial L^p}{\partial \sigma^*}|_\sigma = 0 \Longrightarrow \tfrac{\partial \epsilon^p}{\partial t} = \tfrac{\partial \gamma}{\partial t}\tfrac{\partial \phi}{\partial \sigma} \equiv \tfrac{\partial \gamma}{\partial t}\, sign(\sigma)}$$
$$\boxed{\tfrac{\partial L^p}{\partial q^*}|_q = 0 \Longrightarrow \tfrac{\partial \zeta}{\partial t} = \tfrac{\partial \gamma}{\partial t}\tfrac{\partial \phi}{\partial q} \equiv \tfrac{\partial \gamma}{\partial t}} \tag{3.121}$$
$$\boxed{\tfrac{\partial \gamma}{\partial t} \geq 0 \; ; \; \phi \leq 0 \; ; \; \tfrac{\partial \gamma}{\partial t}\, \phi = 0}$$

We can note that the plasticity model with isotropic hardening provides the same evolution equation for plastic deformation and the same stress constitutive relation as the model of perfect plasticity. Nevertheless, the presence of hardening changes the yield function, introduces the modification of the elastic domain and produces in general different plastic deformation from perfect plasticity for the same loading program.
The hardening variable introduced herein is a phenomenological variable, which seeks to ensure a good agreement with experimental results (e.g. those

obtained from a simple tension test) for evolution of plasticity threshold, without trying to represent any sound explanation of hardening mechanisms. In fact, the comparison of evolution equations in (3.121) readily reveals that the chosen phenomenological hardening variable ζ can be interpreted as the absolute value of accumulated plastic deformation

$$\frac{\partial \zeta}{\partial t} := |\frac{\partial \epsilon^p}{\partial t}| = \frac{\partial \gamma}{\partial t} \underbrace{|sign(\sigma)|}_{=1} \tag{3.122}$$

However, this choice of hardening variable is not unique. We can also take the hardening variable that is proportional to the plastic work (e.g. see Lubliner [174]), which can sometimes be easier to identify. We record here the corresponding form of the evolution equation for the case where the hardening variable is considered to be the plastic work

$$\frac{\partial \zeta}{\partial t} := \sigma \frac{\partial \epsilon^p}{\partial t} = \frac{\partial \gamma}{\partial t} \underbrace{\sigma \; sign(\sigma)}_{|\sigma|} \tag{3.123}$$

We note that such a choice of hardening variable does not fit within the framework of associative plasticity, and thus we will not further explore this possibility.

The appropriate value of the plastic multiplier is again obtained from the consistency condition, which will impose the plastic admissibility of stress. For plastic loading case, this leads to:

$$\begin{aligned} 0 = \tfrac{\partial \phi}{\partial t} &:= \frac{\partial \phi}{\partial \sigma}\frac{\partial \sigma}{\partial t} + \frac{\partial \phi}{\partial q}\frac{\partial q}{\partial t} \\ &= \tfrac{\partial \phi}{\partial \sigma} E \tfrac{\partial \epsilon}{\partial t} - \tfrac{\partial \gamma}{\partial t}(\tfrac{\partial \phi}{\partial \sigma} E \tfrac{\partial \phi}{\partial \sigma} + \tfrac{\partial \phi}{\partial q} K \tfrac{\partial \phi}{\partial q}) \\ &\Longrightarrow \tfrac{\partial \gamma}{\partial t} = \tfrac{\partial \phi}{\partial \sigma} E \tfrac{\partial \epsilon}{\partial t} / (E + K) \end{aligned} \tag{3.124}$$

The last result allows us to express the stress-rate equation for plasticity with linear isotropic hardening according to:

$$\frac{\partial \sigma}{\partial t} = \begin{cases} E \, \frac{\partial \epsilon}{\partial t} \; ; \; \frac{\partial \gamma}{\partial t} = 0 \\ \frac{EK}{E+K} \frac{\partial \epsilon}{\partial t} \; ; \; \frac{\partial \gamma}{\partial t} > 0 \end{cases} \tag{3.125}$$

The response of the hardening plasticity model computed in a closed cycle of loading/unloading in compression and in tension is presented in Figure 3.8. This result illustrates that the instantaneous response in the plastic loading phase will change with respect to the perfect plasticity to allow further increase of stress beyond elasticity limit σ_y. In fact, by exploiting again the equivalence between 1D plasticity and dry friction (at constant normal pressure) we can illustrate the plastic loading response of the hardening plasticity by using the rheological model shown in Figure 3.8. The model is constructed

by connecting in series the elastic spring (with elasticity coefficient E) and the frictional slider (with sliding resistance σ_y) which activates another spring (with coefficient equal to hardening modulus K); see Figure 3.8.
In summary, the governing equations of the plasticity model with linear isotropic hardening can be written as follows:

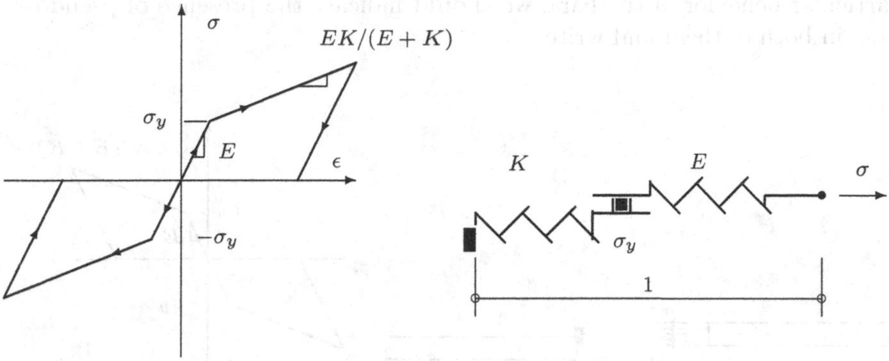

Fig. 3.8 Rheological model of hardening plasticity for plastic loading and stress–strain diagram.

$$\begin{array}{ll} \text{constitutive relation:} & \sigma = E(\epsilon - \epsilon^p) \ ; \ q = -K\zeta \\ \text{yield function:} & \phi(\sigma, q) = |\sigma| - (\sigma_y - q) \\ \text{loading-unloading:} & \frac{\partial \gamma}{\partial t} \geq 0; \ \phi(\sigma) \leq 0 \, ; \ \frac{\partial \gamma}{\partial t} \, \phi(\sigma) = 0 \\ \text{consistency condition:} & \frac{\partial \gamma}{\partial t} \frac{\partial \phi}{\partial t} = 0 \\ \text{plastic flow:} & \frac{\partial \epsilon^p}{\partial t} = \frac{\partial \gamma}{\partial t} \frac{\partial \phi}{\partial \sigma} \equiv \frac{\partial \gamma}{\partial t} \, sign(\sigma) \\ \text{hardening evolution:} & \frac{\partial \zeta}{\partial t} = \frac{\partial \gamma}{\partial t} \frac{\partial \phi}{\partial q} \equiv \frac{\partial \gamma}{\partial t} \end{array} \tag{3.126}$$

3.2.3 Boundary value problem for 1D plasticity

In this section, we will consider the methods of Galerkin and Ritz and the finite element method for solving a boundary value problem for the plasticity model with linear isotropic hardening. We take again the simplest 1D boundary value problem for a bar (see Figure 3.9), submitted to a distributed loading along its axis, imposed displacement on one end and imposed traction on another. The constitutive behavior of the bar is governed by the 1D plasticity model with linear isotropic hardening.
The principal difference with respect to the case of elastic bar studied in the previous chapter is the presence of the plastic deformation field $\epsilon^p(x,t)$, defined at each point x, whose evolution has to be obtained with respect

to the chosen loading program described by pseudo-time parameter t. This dependence of plastic deformation on pseudo-time will propagate to all other variables due to field coupling through governing equations; hence, the stress $\sigma(x,t)$, total strain $\epsilon(x,t)$ and displacement field $u(x,t)$ also become functions of pseudo-time. Therefore, although the kinematics and equilibrium equations remain the same as those in elasticity (since they are independent of a particular behavior of the bar), we should indicate the presence of pseudo--time in both of them and write:

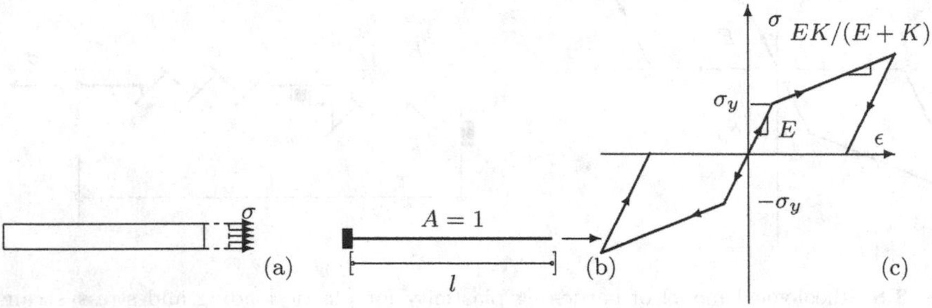

Fig. 3.9 Bar with constitutive model of 1D plasticity with linear isotropic hardening: (a) initial and deformed configurations, (b) 1D model, (c) stress–strain diagram.

$$\epsilon(x,t) = \frac{\partial u(x,t)}{\partial x} \; ; \; \frac{\partial \sigma}{\partial x}(x,t) + b(x,t) = 0 \tag{3.127}$$

The constitutive equations for the bar are those defined for the plasticity model with linear isotropic hardening:

$$\begin{aligned} &\sigma(x,t) = E(\epsilon(x,t) - \epsilon^p(x,t)) \; ; \; q(x,t) = -K\zeta(x,t) \\ &\phi(\sigma,q) := |\sigma| - (\sigma_y - q) \leq 0 \; ; \; \tfrac{\partial \gamma}{\partial t} \geq 0 \; ; \; \tfrac{\partial \gamma}{\partial t}\phi = 0 \\ &\tfrac{\partial \epsilon^p(x,t)}{\partial t} = \tfrac{\partial \gamma(x,t)}{\partial t} sign(\sigma(x,t)) \; ; \; \tfrac{\partial \zeta(x,t)}{\partial t} = \tfrac{\partial \gamma(x,t)}{\partial t} \end{aligned} \tag{3.128}$$

The strong form of the boundary value problem in 1D plasticity consists of the last two groups of equations, accompanied by the boundary conditions pictured in Figure 3.9

$$u(0,t) = \bar{u}(t)(=0) \; ; \; \sigma(l,t) = \bar{t}(t) \tag{3.129}$$

The dependence with respect to pseudo-time will also require to specify the initial conditions on the displacement field and the internal variables. For simplicity, we choose the starting point of a virgin material, where all fields initially take the zero value

$$u(x,0) = 0 \; ; \; \epsilon^p(x,0) = 0 \; ; \; \zeta(x,0) = 0 \tag{3.130}$$

Following in the steps of the finite element procedure applied to elasticity, we will abandon this strong form of 1D plasticity in favor of its weak form as the starting point for constructing the discrete approximations. In this manner we can reduce regularity requirement on the admissible displacement field and be able to find the potential solution candidates more easily. For the weighting functions (or virtual displacement field) we can choose in hardening plasticity the functions of bounded deformation, which belong to the space $S_{BD}(\Omega)$. In 1D, the space of functions of bounded deformation $S_{BD}(\Omega)$ is equivalent to the space of functions with bounded variation $S_{BV}(\Omega)$ (see Temam [254]); the latter can be written:

$$\mathbb{V}_0 := \{w | w : \Omega \mapsto \mathbb{R}, w \in S_{BV}(\Omega), [w(x)]_{\Gamma_u} = 0\} \tag{3.131}$$

The solution candidates are chosen to have the same regularity, as defined by bounded-variation space $S_{BV}(\Omega)$, but contrary to weighting functions (which have zero value at the boundary Γ_u), the test functions must satisfy the imposed displacement boundary condition

$$\mathbb{V} := \{u | u : \Omega \times [0, T] \mapsto \mathbb{R}, u \in S_{BV}(\Omega), [u(x,t)]_{\Gamma_u} = \bar{u}(t)\} \tag{3.132}$$

The weak form can be obtained by integrating the weighted form of equilibrium equation in (3.127) over the domain Ω, while keeping the local form for all other governing equations. By using the standard procedure of integration by parts, we then obtain the following result:

Weak form of 1D plasticity with linear isotropic hardening

$$\begin{array}{l}
\text{Given: } b(x,t),\, \bar{t}(t),\, \bar{u}(t) = 0 \\
\text{Find: } u(x,t) \in \mathbb{V},\, \forall w(x) \in \mathbb{V}_0 \\
\hline
0 = G(u, \epsilon^p, \zeta; w) := \int_\Omega \frac{dw}{dx} \hat{\sigma}(x, t, \epsilon, \epsilon^p, \zeta)\, dx \\
\quad - \int_\Omega w(x) b(x,t)\, dx - [w(x)\bar{t}(t)]_{\Gamma_\sigma} \\
\sigma = E\, (\frac{\partial u}{\partial x} - \epsilon^p)\, ;\; q = -K\, \zeta \\
\phi(\sigma, q) := |\sigma| - (\sigma_y - q) \\
\frac{\partial \gamma}{\partial t} \geq 0\, ;\; \phi \leq 0\, ;\; \frac{\partial \gamma}{\partial t}\, \phi = 0 \\
\frac{\partial \epsilon^p}{\partial t} = \frac{\partial \gamma}{\partial t}\, sign(\sigma)\, ;\; \frac{\partial \zeta}{\partial t} = \frac{\partial \gamma}{\partial t}
\end{array} \tag{W3}$$

3.2.3.1 Galerkin method in plasticity

Galerkin method allows to construct the approximate solution to a boundary value problem in plasticity, which is defined in the weak form ($W3$). At each instant of pseudo-time, such an approximate solution belongs to a chosen finite-dimensional approximation space $u^h(x,t) \in \mathbb{V}^h$. The approximation basis of such space can be constructed by Galerkin method, which employs the shape functions $N_a(x), a = 1, 2, \ldots, n$, defined in the domain Ω. By using the method of separation of variables, we can then express the approximate solution at any time t according to:

$$u^h(x,t) = \sum_{a=1}^{n} N_a(x)\, d_a(t) \in \mathbb{V}^h \subset \mathbb{V} \tag{3.133}$$

where time dependence in $d_a(t)$ is chosen to describe a particular evolution with respect to pseudo-time.

In order to illustrate this kind of approximation, we imagine that the domain Ω of the bar is split by n nodes into $n_{elem} = n-1$ sub-domains (or elements). The simplest choice for the shape functions providing the approximation basis are linear polynomials that take non-zero value only in the elements (or sub-domains) adjacent to a given node

$$N_a(x) := \begin{cases} (x_{a+1} - x)/h^e \; ; \; x \in [x_a, x_{a+1}] \; ; \; h^e = x_{a+1} - x_a \\ (x - x_{a-1})/h^{e-1} \; ; \; x \in [x_{a-1}, x_a] \; ; \; h^{e-1} = x_a - x_{a-1} \\ 0 \; ; \; otherwise \end{cases} \tag{3.134}$$

With this kind of shape function $N_a(x)$, which takes a unit value at the given node a and zero value at all other nodes,

$$N_a(x_b) = \delta_{ab} \; ; \quad \delta_{ab} = \begin{cases} 1 \; ; a = b \\ 0 \; ; a \neq b \end{cases} \tag{3.135}$$

we can identify $d_a(t)$ as the time-evolution of the nodal displacement at node a. For this kind of construction, the quality of approximate solution $u^h(x,t)$ depends directly upon the size of the chosen element, with $h = \min_e h^e$. It is clear that increasing the number of elements and reducing their size will lead to a solution of better quality.

The Galerkin method implies that the same approximation basis is employed to construct the virtual displacement field with

$$w^h(x) = \sum_{a=1}^{n} N_a(x)\, w_a \in \mathbb{V}_0^h \subset \mathbb{V}_0 \tag{3.136}$$

It is important to note that, contrary to real displacement field, the nodal values of virtual displacement field w_a are not a function of pseudo-time, but constants. This is in agreement with the chosen interpolations and the interpretation of virtual displacement field as an imaginary displacement superposed on a particular deformed configuration, with a fixed value of pseudo-time.

According to regularity requirements in (3.132), each shape function should have a bounded derivative. It is easy to see that the shape functions in (3.134) obey this condition since their derivatives are constant in each element. The corresponding approximation of the strain field can then be obtained as:

$$\epsilon^h(x,t) = \sum_{a=1}^{n} \frac{dN_a(x)}{dx} d_a(t) \; ; \qquad (3.137)$$
$$\frac{dN_a(x)}{dx} := \begin{cases} -1/h^e \; ; \; x \in [x_a, x_{a+1}] \; ; \; h^e = x_{a+1} - x_a \\ 1/h^{e-1} \; ; \; x \in [x_{a-1}, x_a] ; \; h^{e-1} = x_a - x_{a-1} \\ 0 \; ; \; otherwise \end{cases}$$

In order to ensure the validity of the additive decomposition of the total deformation field into elastic and plastic component within the discrete approximation framework, the same element-wise constant approximation should be chosen for all the variables which govern plastic flow: plastic deformation, hardening variable and plastic multiplier. We can thus write:

$$\begin{aligned} \epsilon^{ph}(x,t) &= \textstyle\sum_{e=1}^{n_{elem}} H^e(x)\epsilon^p_e(t) \; ; \\ \zeta^h(x,t) &= \textstyle\sum_{e=1}^{n_{elem}} H^e(x)\zeta_e(t) \; ; \\ \tfrac{\partial \gamma}{\partial t}(x,t) &= \textstyle\sum_{e=1}^{n_{elem}} H^e(x)\dot{\gamma}_e(t) \; ; \end{aligned} \qquad (3.138)$$

where $H^e(x)$ is defined with

$$H^e(x) := \begin{cases} 1 \, ; \, x_a \leq x \leq x_{a+1} \\ 0 \, ; \, \text{otherwise} \end{cases} \qquad (3.139)$$

By introducing the chosen approximation into the weak form in $(W3)$, we can obtain the Galerkin equation for 1D plasticity:

$$0 = G(u^h, \epsilon^{ph}, \zeta^h; w_a) := \textstyle\sum_{a=1}^{n} w_a \{ \underbrace{\int_0^l \frac{dN_a(x)}{dx} \hat{\sigma}^h(x,t,\epsilon^h,\epsilon^{ph},\zeta^h)\, dx}_{f_a^{int}(t)}$$
$$- \underbrace{\int_0^l N_a(x)\, b(x,t)\, dx - N_a(l)\bar{t}(t) \}}_{f_a^{ext}(t)}$$
$$[\sigma^h = E(\tfrac{du^h}{dx} - \epsilon^{ph}) \; ; \; q^h = -K\zeta^h]_e \; ; \; \forall e \in [1, n_{elem}]$$
$$\phi(\sigma^h, q^h) := |\sigma^h| - (\sigma_y - q^h) \; ; \; [\phi \leq 0; \dot{\gamma} \geq 0; \dot{\gamma}\phi = 0]_e$$
$$\textstyle\sum_{e=1}^{n_{elem}} H^e(x)[\dot{\epsilon}^p - \dot{\gamma} sign(\sigma)]_e = 0 \; ; \; \sum_{e=1}^{n_{elem}} H^e(x)[\dot{\zeta} - \dot{\gamma}]_e = 0 \qquad (G3)$$

We note that the second group of Galerkin equations for plasticity in $(G3)$ is supplied only to define the pseudo-time evolution of internal variables in each sub-domain with respect to a given loading program. By requiring that the Galerkin equations in $(G3)$ be satisfied for any virtual displacement, we obtain a set of algebraic-differential equations, with the algebraic equations expressing equilibrium and the differential equations governing the pseudo-time evolution of internal variables

$$\boxed{\begin{array}{l} f_a^{int}(\sigma^h(t)) - f_a^{ext}(t) = 0\ ;\ \ \forall a \in [1,n] \\ [\sigma(t) = E(\sum_{a=1}^{n} \frac{dN_a}{dx} d_a(t) - \epsilon^p(t))\ ;\ q(t) = -K\zeta(t)]_e \\ [\phi(\sigma,q) := |\sigma| - (\sigma_y - q)\ ;\ \ \dot{\gamma} \geq 0\ ;\ \phi \leq 0\ ;\ \dot{\gamma}\,\phi = 0]_e \\ [\dot{\epsilon}^p = \dot{\gamma} sign(\sigma)]_e\ ;\ [\dot{\zeta} = \dot{\gamma}]_e\ ;\ \forall e \in [1, n_{elem}] \end{array}} \quad (M3)$$

The solution to the last group of equations corresponding to internal variables evolution in each element, will require the use of a time-integration scheme. Therefore, for chosen value of pseudo-time t_n, the unknowns to be computed will be all nodal values of displacement, as well as the internal variables in each element.

3.2.3.2 Finite element method in plasticity

The finite element method provides the most convenient framework for constructing the discrete approximation for a boundary value problem in plasticity, which allows to reduce the cost for the internal variable computation and standardize the computational procedure in each element. Namely, in the finite element method we will consider any sub-domain Ω^e as an isoparametric element that derives from its parent element. The discrete approximations for element domain is constructed by using the shape functions $N_a^e(\xi)\ ;\ a \in [1, n_{en}]$ depending upon the natural coordinate ξ, and the same basis is applied to construction of real and virtual displacement fields

$$\begin{aligned} x(\xi)\Big|_{\Omega^e} &= \textstyle\sum_{a=1}^{n_{en}} N_a^e(\xi)\, x_a^e\ ; \\ u^h(\xi,t)\Big|_{\Omega^e} &= \textstyle\sum_{a=1}^{n_{en}} N_a^e(\xi)\, d_a^e(t)\ ; \\ w^h(\xi)\Big|_{\Omega^e} &= \textstyle\sum_{a=1}^{n_{en}} N_a^e(\xi)\, w_a^e \end{aligned} \quad (3.140)$$

The shape function $N_a^e(\xi)$ for any isoparametric element with n_{en} nodes is chosen as the Lagrange polynomial of order $n_{en} - 1$, constructed in the parent element domain with $\xi \in [-1, 1]$

$$\begin{aligned} N_a^e(\xi) &:= \textstyle\prod_{b=1,b\neq a}^{n_{en}} \frac{(\xi-\xi_b)}{(\xi_a-\xi_b)} \\ &= \frac{(\xi-\xi_1)\ldots(\xi-\xi_{a-1})(\xi-\xi_{a+1})\ldots(\xi-\xi_{n_{en}})}{(\xi_a-\xi_1)\ldots(\xi_a-\xi_{a-1})(\xi_a-\xi_{a+1})\ldots(\xi_a-\xi_{n_{en}})} \end{aligned} \quad (3.141)$$

By exploiting the chain rule, we can easily obtain the discrete approximation for the total deformation field in each element in terms of natural coordinate:

$$\boxed{\begin{array}{l} \epsilon^h(\xi,t)\Big|_{\Omega^e} := \frac{\partial u^h(\xi,t)}{\partial x} = \sum_{a=1}^{n_{en}} B_a^e(\xi) d_a^e(t) \; ; \\ B_a^e(\xi) = \frac{dN_a^e}{d\xi}\frac{1}{j(\xi)} \; ; \; j(\xi) = \sum_{a=1}^{n_{en}} \frac{dN_a^e(\xi)}{d\xi} x_a^e \end{array}} \tag{3.142}$$

Any element contribution toward the internal force vector can then be computed by using the numerical integration, such as the Gauss quadrature rule for example. By choosing the quadrature rule with n_{in} points, with abscissas ξ_l and integration weights w_l, we can write an explicit form of the element internal force vector

$$\begin{aligned} \mathrm{f}^{int,e} &= [f_a^{int,e}] \; ; \; 1 \leq a \leq n_{en} \\ f_a^{int,e} &= \int_{\Omega^e} B_a^e(x) \hat{\sigma}(\epsilon^h(x,t), \epsilon^{ph}(x,t), \zeta^h(x,t))\, dx \\ &= \int_{-1}^{1} B_a^e(\xi) \hat{\sigma}(\epsilon^h(\xi,t), \epsilon^{ph}(\xi,t), \zeta^h(\xi,t))\, j(\xi)\, d\xi \\ &= \sum_{l=1}^{n_{in}} B_a^e(\xi_l) \hat{\sigma}(\epsilon^h(\xi_l,t), \epsilon^{ph}(\xi_l,t), \zeta^h(\xi_l,t))\, j(\xi_l) w_l \end{aligned} \tag{3.143}$$

The last result illustrates yet another important advantage of the finite element method over other approximation methods (such as Galerkin or Ritz method) in simplifying the implementation of plasticity model. Namely, we are able not only to standardize the computational procedure for element arrays with the choice of isoparametric elements, but also to reduce the computations of internal variables to Gauss quadrature points only. Moreover, with the only values of internal variables needed at ξ_l; $l \in [1, n_{in}]$, we also reduce the computational cost. For example, in the case of the simplest finite element interpolation with 2-node isoparametric elements, we can use a single Gauss point integration, with abscissa $\xi_1 = 0$ and integration weight $w_1 = 2$, and thus recover the same result for internal force vector in (3.143) as obtained previously by the exact integration.

$$\begin{aligned} f_a^{int,e} &= \underbrace{\frac{(-1)^a}{l^e}}_{B^e(0)} \hat{\sigma}(\underbrace{\epsilon(0,t)}_{\sum_{b=1}^{2} \frac{(-1)^b}{l^e} d_b^e(t)}, \epsilon_e^p(\underbrace{0}_{\xi_1},t), \zeta_e(\underbrace{0}_{\xi_1},t)) \underbrace{\frac{l^e}{2}}_{j(0)} \underbrace{2}_{w_1} \\ &= (-1)^a \, \sigma^h(0,t) \; ; \; a \in [1,2] \end{aligned} \tag{3.144}$$

The same advantage of the finite element approach using the numerical integration also applies to computation of the stiffness matrix for an isoparametric element

$$\begin{aligned} \mathsf{K}^e &= [K_{ab}^e] \; ; \; 1 \leq a,b \leq n_{en} \; ; \; e \in [1, n_{elem}] \\ K_{ab}^e &:= \frac{\partial f_a^{int,e}}{\partial d_a^e} \\ &= \int_{\Omega^e} B_a^e(x) \hat{C}^{ep}(\epsilon^h((x,t), \epsilon^{ph}(x,t), \zeta^h(x,t)) B_b^e(x)\, dx \\ &= \int_{-1}^{1} B_a^e(\xi) \hat{C}^{ep}(\epsilon^h(\xi,t), \epsilon^{ph}(\xi,t), \xi^h(\xi,t)) B_b^e(\xi)\, j(\xi)\, d\xi \\ &= \sum_{l=1}^{n_{in}} B_a^e(\xi_l) \hat{C}^{ep}(\epsilon^h(\xi_l,t), \epsilon^{ph}(\xi_l,t), \zeta^h(\xi_l,t)) B_b^e(\xi_l)\, j(\xi_l) w_l \end{aligned} \tag{3.145}$$

where the elastoplastic tangent modulus C^{ep} is obtained only at Gauss quadrature points.

3.2.3.3 Time-integration in plasticity

The finite element method provides an efficient tool for discretization in space or semi-discretization, reducing the boundary value problem in plasticity to a set of differential–algebraic equations. The differential equations are evolution equations of internal variables defined locally at each Gauss quadrature point, whereas the algebraic equations are those defined globally to express the static equilibrium condition at all nodes of the chosen mesh. For a mesh constructed with 2-node isoparametric elements, such a problem can be written:

$$\begin{aligned}
&\mathop{\mathbb{A}}_{e=1}^{n_{elem}} \left(\mathsf{f}^{int,e} - \mathsf{f}^{ext,e}\right) = 0\,;\\
&\mathsf{f}^{int,e} = \left[f_a^{int,e}\right]\,;\; f_a^{int,e} = (-1)^a \hat{\sigma}(\mathsf{d}^e(t), \epsilon_e^p(t), \zeta_e(t))\;;\; a \in [1,2]\\
&[\sigma(t) = E(\textstyle\sum_{a=1}^{2} \frac{(-1)^a}{l^e} d_a(t) - \epsilon^p(t))\;;\; q(t) = -K\zeta(t)]_e\\
&\left[\hat{\phi}(t) = |\sigma(t)| - (\sigma_y - q(t)) \leq 0\right]_e\,;\\
&\left[\dot{\epsilon}^p = \dot{\gamma}\frac{\partial \phi}{\partial \sigma}\right]_e\,;\; \left[\dot{\zeta} = \dot{\gamma}\right]_e\;;\; \forall e \in [1, n_{elem}]\\
&\dot{\gamma}_e(t) = \begin{cases} 0\,;\, \phi < 0\,,\, \phi = 0\,;\, \dot{\phi} < 0 \\ \frac{E sign(\sigma)}{(E+K)} \sum_{a=1}^{2} \frac{(-1)^a}{l^e} \dot{d}_a^e(t)\,;\; \hat{\phi}(t) = 0\,;\, \dot{\hat{\phi}}(t) = 0 \end{cases}
\end{aligned} \tag{3.146}$$

where the local element evolution equations concern the values at each element center. It is not easy to construct the exact solution of the evolution problem in (3.146). The main difficulty concerns the presence of constraint of plastic admissibility of stress with $\phi(\sigma, q) \leq 0$, resulting with a non-smooth evolution of plastic multiplier $\dot{\gamma}(t) \leq 0$ and plastic deformation. For that reason, we seek a numerical solution to this evolution problem by using a time-integration scheme, providing the solution values only for the chosen instants of the pseudo-time

$$[0, T] = \bigcup_{n=1}^{n_{pt}} [t_n, t_{n+1}] \tag{3.147}$$

In order to ensure the simplicity of the computer code architecture, we are interested only in time-integration schemes that construct the approximate solution one step at the time. This allows us to reduce the state variable computation to:

Central problem of computational plasticity in incremental analysis over a typical step $h = t_{n+1} - t_n$

$$\begin{array}{|l|}\hline \text{Given: } [\epsilon(t_n) = \epsilon_n, \epsilon^p(t_n) = \epsilon^p_n, \zeta(t_n) = \zeta_n]_e \;;\; \forall e \in [1, n_{el}] \\ \text{Find: } [\epsilon_{n+1}, \epsilon^p_{n+1}, \zeta_{n+1}] \\ \text{as well as: } [\sigma_{n+1} = E(\epsilon_{n+1} - \epsilon^p_{n+1}) \;;\; q_{n+1} = -K\zeta_{n+1}]_e \\ \text{such that: } [\phi(\sigma_{n+1}, q_{n+1}) \leq 0 \,;\, \dot{\gamma}_{n+1} \geq 0 \,;\, \dot{\gamma}_{n+1}\phi(\sigma_{n+1}, q_{n+1}) = 0]_e \\ \hline \end{array} \tag{3.148}$$

We will discuss the use of two Euler schemes, forward and backward, for plastic flow computation and the solution of central problem.

Forward Euler explicit scheme is the simplest possibility to provide the numerical solution to central problem at time t_{n+1}. We first recall the result that the forward Euler scheme will produce for a typical first-order differential equation

$$\dot{y} = f(y,t); y_n = y(t_n) \implies y_{n+1} = y_n + \int_{t_n}^{t_{n+1}} \underbrace{f(y_n, t_n)}_{f_n} \, dt \equiv y_n + h\, f_n$$

By applying the same scheme to constrained evolution problem for internal variables, we can obtain

$$\begin{aligned} &\dot{\epsilon}^p = \dot{\gamma}\, sign(\sigma) \Rightarrow \epsilon^p_{n+1} = \epsilon^p_n + \gamma_n\, sign(\sigma_n) \\ &\dot{\zeta} = \dot{\gamma} \qquad\quad\;\; \Rightarrow \zeta_{n+1} = \zeta_n + \gamma_n \\ &\qquad\quad \gamma_n := h\,\dot{\gamma}_n = \begin{cases} 0 \;;\; \phi_n < 0 \\ h\frac{E sign(\sigma_n)\,\dot{\epsilon}_n}{E+K} \;;\; \phi_n = 0 \,;\, \dot{\phi}_n = 0 \end{cases} \end{aligned} \tag{3.149}$$

The forward Euler scheme can also be applied to find the solution of equilibrium equation, when the latter is written in rate form. It is easy to see that one obtains in this manner the results which are equivalent to those furnished by standard incremental analysis

$$\begin{aligned} &\dot{\mathsf{d}} = \mathbb{A}_{e=1}^{n_{elem}} [\mathsf{K}^e]^{-1} \mathbb{A}_{e=1}^{n_{elem}} \left(\dot{\mathsf{f}}^e\right) \Rightarrow \underbrace{\mathsf{d}_{n+1} - \mathsf{d}_n}_{\mathsf{u}_{n+1}} = \mathsf{K}_n^{-1} \underbrace{h\dot{\mathsf{f}}_n}_{r_n} \\ &\mathsf{d}_{n+1} = \mathsf{d}_n + \mathsf{u}_{n+1} \end{aligned} \tag{3.150}$$

In the last expression, u_{n+1} is the incremental displacement vector and K_n is the tangent matrix computed at time t_n by the finite element assembly of element contributions. For a mesh of 2-node elements the latter can be written explicitly as

$$\begin{aligned} &\mathsf{K}_n = \mathbb{A}_{e=1}^{n_{elem}} \mathsf{K}^e_n \;;\; \mathsf{K}^e_n = [K^e_{ab,n}] \;; \\ &K^e_{ab,n} = \tfrac{(-1)^a(-1)^b}{l^e} \hat{C}^{ep}_n(\epsilon_n(0), \epsilon^p_n(0), \zeta_n(0)) \;; \\ &C^{ep}_n = \begin{cases} E \;;\; \dot{\gamma}_n = 0 \\ \frac{EK}{E+K} \;;\; \dot{\gamma}_n > 0 \end{cases} \end{aligned} \tag{3.151}$$

The last expression clearly shows that the elastoplastic tangent modulus for the explicit scheme depends upon the value of the plastic multiplier at time t_n. The latter can always be chosen in agreement with the given value of the yield function in (3.149), resulting with $\phi_n < 0 \Rightarrow \gamma_n = 0$ or $\phi_n = 0 \Rightarrow \gamma_n > 0$. The advantage of computational simplicity of explicit scheme can easily be forgotten with the lack of reliability of computed results. Namely, the critical time step size for the Euler explicit scheme can change rapidly with every new tangent stiffness matrix modification (especially for elastic unloading). Moreover, the time step size should also enforce the plastic admissibility of computed stress. In other words, there is no guarantee that independent computations in (3.150) and (3.149), with the former providing displacements (and thus total strains) and the latter providing the values of internal variables, will jointly produce the stress that results with a negative or zero value of the yield function at time t_{n+1}

$$0 \overset{?}{\geq} \phi_{n+1} := |E[\textstyle\sum_{a=1}^{2} \mathsf{B}_a^e(\mathsf{d}_n + \mathsf{u}_{n+1}) - (\epsilon_n + \gamma_n sign(\sigma_n))]| \\ -[\sigma_y + K(\zeta_n + \gamma_n)]) \tag{3.152}$$

One way to remove this deficiency of the explicit scheme is by controlling the time step size to maintain the plastic admissibility of computed stress at time t_{n+1} by making sure that $\phi_{n+1} \leq 0$. The latter imposes an additional restriction upon the critical time step which might be more stringent than the one imposed by the stability of computations for equilibrium equations by the explicit scheme (Recall that the forward Euler scheme will impose $h \leq 2/\lambda_{n,max}$, where $\lambda_{n,max}$ is the maximum eigenvalue of the tangent stiffness matrix K_n). The computational cost can thus be further increased because of the plasticity model time step requirement on admissible stress, which can be imposed by any such local computation. One strategy for controlling the computational cost is by sub-cycling (e.g. see Owen and Hinton [214]), where the time step for the forward Euler scheme is reduced only when integrating local evolution equations, with no reduction of time step used for integrating the equilibrium equations by forward Euler scheme. We can thus improve the efficiency of the explicit scheme for plasticity, but not its reliability in general. More precisely, for a complex loading program, the estimate of the critical time step[9] might not be able to follow frequent loading–unloading regime changes. Therefore, the explicit scheme computation can produce what is sometimes referred to as arrested instability, where the stability of the scheme is lost during several steps, leading to an artificial response amplification, and then restored once the element enters into the plastic regime. In this manner

[9] The critical time step of the explicit scheme can be obtained from the eigenvalues of the element tangent matrix $\lambda_{max}(\mathsf{K}^e)$, which represent according to the Cauchy interlace theorem (e.g. see [20 or 62]) the upper bound for the eigenvalues of the structural tangent matrix $\lambda_{max}(\mathsf{K}) \leq \lambda_{max}(\mathsf{K}^e)$; All the elements ought to be examined at each step, checking if any tangent matrix has changed from one step to another with element entering plastic loading or elastic unloading phase.

we obtain the least desirable result, which is not "sufficiently" wrong for the analyst to notice. The only possibility to restor the reliability of the computations is to turn to using the implicit schemes.

Backward Euler implicit scheme can ensure that the equilibrium equations are satisfied (within a chosen tolerance) and that the computed values of internal variables produce the plastically admissible stress field at each instant of the incremental sequence. In each time step, the computation is reduced to:

Central problem in computational plasticity: incremental-iterative analysis over a typical time step $h = t_{n+1} - t_n$

Given: $[\epsilon_n, \epsilon_n^p, \zeta_n]_{e,l}$; $\forall l \in [1, n_{in}]$, $\forall e \in [1, n_{elem}]$
Find: $[\epsilon_{n+1}, \epsilon_{n+1}^p, \zeta_{n+1}]_{e,l}$
such that: $[\sigma_{n+1} = E(\epsilon_{n+1} - \epsilon_{n+1}^p)]_{e,l}$; $[q_{n+1} = -K\zeta_{n+1}]_{e,l}$
plastically admissible:
$[\dot{\gamma}_{n+1} \geq 0\,;\ \phi(\sigma_{n+1}, q_{n+1}) \leq 0\,;\ \dot{\gamma}_{n+1}\phi_{n+1} = 0]_{e,l}$
and equilibrium satisfied: $\mathop{\mathbb{A}}_{e=1}^{n_{elem}} \left(\mathrm{f}_{n+1}^{int,e} - \mathrm{f}_{n+1}^{ext,e}\right) = 0$; $f_{n+1,a}^{int,e} = (-1)^a \sigma_{n+1}$

(3.153)

The equations to be solved by an implicit scheme in central problem of computational plasticity do not pertain to the same domains: on one side, the equilibrium equations concern the global domain which includes all the elements and nodes, whereas the evolution equations are local since they are limited to a single element and a particular Gauss quadrature point of that element. In seeking the solution efficiency, we will not solve all these equations at once but rather apply the operator split methodology to separate the computation of internal variables from the solution of equilibrium equations. The operator split solution procedure performs two phases of computations sequentially, first global then local phase, with all the computed results from one phase used immediately afterwards. For more detailed explanation, we start with:

I Local phase of operator split computing internal variables: With the best iterative value of the total deformation $\epsilon_{n+1}^{(i)}$, we can start the local phase of the operator split solution procedure. The latter should provide the values of the internal variables at the end of the time step guaranteeing the plastic admissibility of the stress field. This computation is carried out by integrating the evolution equations by implicit backward Euler scheme. We first recall the result provided by the backward Euler scheme for a standard first-order differential equation:

$$\dot{y} = f(y,t)\,;\ y_n = y(t_n) \Longrightarrow y_{n+1} = y_n + \int_{t_n}^{t_{n+1}} \underbrace{f(y_{n+1}, t_{n+1})}_{f_{n+1}}\, dt \equiv y_n + h\, f_{n+1}$$

By using the same scheme to integrate the evolution equations, we can obtain:

$$\begin{aligned}
&\dot{\epsilon}^p = \dot{\gamma}\, sign(\sigma) \Rightarrow \epsilon^p_{n+1} = \epsilon^p_n + \gamma_{n+1}\, sign(\sigma_{n+1})\ ; \\
&\dot{\zeta} = \dot{\gamma} \Rightarrow \zeta_{n+1} = \zeta_n + \gamma_{n+1}\ ; \\
&\sigma_{n+1} = E(\epsilon^{(i)}_{n+1} - \epsilon^p_{n+1})\ ; \\
&\gamma_{n+1} := h\,\dot{\gamma}_{n+1} = \begin{cases} 0\ ;\ \phi(\sigma_{n+1}, q_{n+1}) < 0 \\ > 0\ ;\ \phi(\sigma_{n+1}, q_{n+1}) = 0 \end{cases}
\end{aligned} \tag{3.154}$$

The implicit backward Euler scheme no longer provides the computational simplicity of its explicit counterpart, since the unknown values appear on both side of the equations. The best computational efficiency of the Euler implicit scheme is ensured by a special treatment for the plastic multiplier. Namely, in accordance to (3.154) the latter can either be positive (in the case of plastic step) or zero (in the case of an elastic step), and there are no more than these two cases to check. We start with a simpler case, with so-called elastic trial step, where the plastic multiplier remains equal to zero $\gamma^{trial}_{n+1} = 0$. The internal variables remain frozen during such an elastic trial step, and keep their value from the previous step

$$\gamma^{trial}_{n+1} = 0 \Longrightarrow \begin{cases} \epsilon^{p,trial}_{n+1} = \epsilon^p_n \\ \zeta^{trial}_{n+1} = \zeta_n \end{cases} \tag{3.155}$$

The stress value for the elastic trial step can then easily be computed according to:

$$\sigma^{trial}_{n+1} = E(\epsilon^{(i)}_{n+1} - \epsilon^p_n)\ ;\ q^{trial}_{n+1} = -K\zeta_n \equiv q_n \tag{3.156}$$

which further leads to the trial value of the yield function

$$\phi^{trial}_{n+1} \equiv \phi(\sigma^{trial}_{n+1}, q^{trial}_{n+1}) := |\sigma^{trial}_{n+1}| - (\sigma_y - q_n) \overset{?}{\leq} 0 \tag{3.157}$$

If such a trial value of the yield function is indeed negative or zero, $\phi^{trial}_{n+1} \leq 0$, the elastic trial step is accepted for final and the internal variable values at time t_{n+1} remain the same as those already computed at time t_n. In Figure 3.10, we have illustrated two different possibilities where such an elastic trial step is accepted.
In the opposite, if the trial value of yield function is positive $\phi^{trial}_{n+1} > 0$, we conclude that the current step is not elastic and that the final value of the internal variables at time t_{n+1} are produced by the plastic flow. We thus first compute the positive value of the plastic multiplier $\gamma_{n+1} > 0$, followed by the corresponding evolution of the internal variables that provides the plastically admissible stress at time t_{n+1}, with $\phi(\sigma_{n+1}) = 0$. For 1D plasticity model with isotropic hardening, this task can be presented as a set of three algebraic equations with three unknowns, $\epsilon^p_{n+1}, \zeta_{n+1}, \gamma_{n+1}$

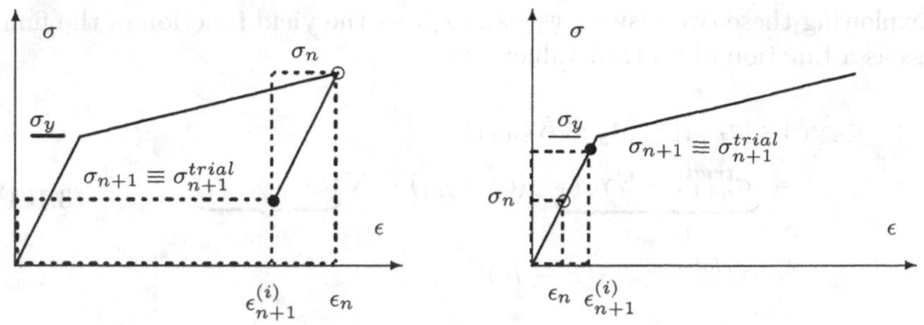

Fig. 3.10 Two possibilities when an elastic trial step of the operator split solution procedure in plasticity is accepted as the final step at t_{n+1}, with no change of internal variables from time t_n.

$$
\begin{aligned}
0 &= -\epsilon^p_{n+1} + \underbrace{\epsilon^p_n}_{\epsilon^{p,trial}_{n+1}} + \gamma_{n+1} sign(\sigma_{n+1}) \\
0 &= -\zeta_{n+1} + \underbrace{\zeta_n}_{\zeta^{trial}_{n+1}} + \gamma_{n+1} \\
0 &= \phi_{n+1} := |\sigma_{n+1}| - (\sigma_y - q_{n+1}) \; ; \\
\sigma_{n+1} &= E(\epsilon_{n+1} - \epsilon^p_{n+1}) \; ; \; q_{n+1} = -K\zeta_{n+1}
\end{aligned}
\tag{3.158}
$$

The system of this kind for local computation in the plastic step is always well posed for hardening plasticity. We seek herein the most efficient solution method, which is capable of exploiting the results already obtained for the elastic trial state. To that end, we first rewrite the plastically admissible value of stress as a correction of the trial stress:

$$
\begin{aligned}
\sigma_{n+1} &= E(\epsilon^{(i)}_{n+1} - \epsilon^p_{n+1}) \\
&= \underbrace{E(\epsilon^{(i)}_{n+1} - \epsilon^p_n)}_{\sigma^{trial}_{n+1}} - E \underbrace{(\epsilon^p_{n+1} - \epsilon^p_n)}_{\gamma_{n+1}\, sign(\sigma_{n+1})} \\
&= \sigma^{trial}_{n+1} - E\gamma_{n+1} sign(\sigma_{n+1})
\end{aligned}
\tag{3.159}
$$

By taking into account that $\gamma_{n+1} > 0$ and $E > 0$, we can conclude that the plastically admissible and trial stresses ought to have the same sign and that their absolute values are related according to:

$$
\begin{aligned}
&[|\sigma_{n+1}| + E\gamma_{n+1}] sign(\sigma_{n+1}) = |\sigma^{trial}_{n+1}| sign(\sigma^{trial}_{n+1}) \\
\Longrightarrow \quad & sign(\sigma_{n+1}) = sign(\sigma^{trial}_{n+1}) \;\Leftrightarrow\; \frac{\partial \phi}{\partial \sigma^{trial}_{n+1}} = \frac{\partial \phi}{\partial \sigma_{n+1}} \\
& |\sigma_{n+1}| + E\gamma_{n+1} = |\sigma^{trial}_{n+1}|
\end{aligned}
\tag{3.160}
$$

By exploiting these two results, we can express the yield function of the final stress as a function of its trial value:

$$\begin{aligned}\phi_{n+1} &:= |\sigma_{n+1}| - (\sigma_y + K\zeta_{n+1}) \\ &= \underbrace{|\sigma^{trial}_{n+1}| - E\gamma_{n+1}}_{|\sigma_{n+1}|} - (\sigma_y - q_n) - K\underbrace{(\zeta_{n+1} - \zeta_n)}_{\gamma_{n+1}} \\ &= \phi^{trial}_{n+1} - \gamma_{n+1}(E+K)\end{aligned} \tag{3.161}$$

which allows us to obtain the positive value of the plastic multiplier

$$\bar{\gamma}_{n+1} = \frac{\phi^{trial}_{n+1}}{E+K} \ (>0) \tag{3.162}$$

With the plastic multiplier in hand, we can perform the corresponding update of the internal variables in (3.158)

$$\epsilon^p_{n+1} = \epsilon^p_n + \bar{\gamma}_{n+1} sign(\sigma^{trial}_{n+1}) \ ; \ \zeta_{n+1} = \zeta_n + \bar{\gamma}_{n+1} \tag{3.163}$$

The result in (3.162) can also be interpreted as the stress radial projection onto admissible domain (see Figure 3.11), in which the total deformation $\epsilon^{(i)}_{n+1}$ is kept fixed, leading to:

$$\sigma_{n+1} = (1 - \frac{E\bar{\gamma}_{n+1}}{|\sigma^{trial}_{n+1}|})\sigma^{trial}_{n+1} \tag{3.164}$$

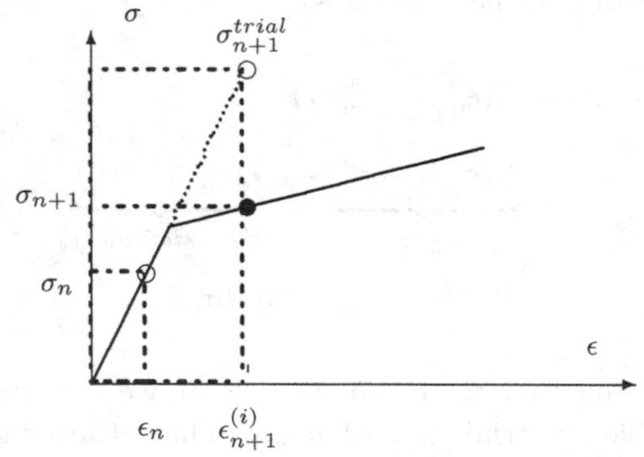

Fig. 3.11 Computation of plastically admissible stress as the radial projection (which passes through the center of elastic domain) of the trial elastic stress keeping the total strain fixed.

An alternative interpretation for plastically admissible stress computation from elastic trial stress is proposed by Moreau [191] and Nguyen [201], which is known as catching-up algorithm. In such an approach we will minimize the distance measured in complementary energy (or free enthalpy) norm between the elastic trail stress and plastically admissible final stress. The latter is thus obtained as the result of this minimization, and the corresponding projection of elastic trial stress to the yield surface. For an associative plasticity model with isotropic hardening, or yet called generalized standard material (Nguyen [201]), we can thus obtain

$$\begin{aligned} \chi(\sigma_{n+1}, q_{n+1}) &= \min_{\phi(\sigma^*, q^*) \leq 0} \{\chi(\sigma^*, q^*)\} \\ \chi(\sigma^*, q^*) &= \tfrac{1}{2}(\sigma_{n+1}^{trial} - \sigma^*) E^{-1} (\sigma_{n+1}^{trial} - \sigma^*) \\ &\quad + \tfrac{1}{2}(q_{n+1}^{trial} - q^*) K^{-1} (q_{n+1}^{trial} - q^*) \; ; \\ q_{n+1}^{trial} &= -K\zeta_n \; ; \; \sigma_{n+1}^{trial} = E(\epsilon_{n+1}^{(i)} - \epsilon_n^p) \end{aligned} \tag{3.165}$$

We note that the requirement of plastic admissibility of stress appears as a constraint equation in the last expression. Such a constraint can be eliminated by introducing the Lagrange multiplier method featuring the corresponding Lagrangian

$$\begin{aligned} L(\sigma_{n+1}, q_{n+1}, \gamma_{n+1}) &= \max_{\gamma^* \geq 0} \min_{\forall \sigma^*, q^*} L(\sigma^*, q^*, \gamma^*) \; ; \\ L(\sigma, q, \gamma) &:= \chi(\sigma, q) + \gamma \phi(\sigma, q) \end{aligned} \tag{3.166}$$

It is easy to check that the Kuhn-Tucker optimality conditions of such a Lagrangian are identical to the radial projection equations given in (3.164)

$$\begin{aligned} 0 &= \tfrac{\partial}{\partial \sigma} L(\sigma_{n+1}, q_{n+1}, \gamma_{n+1}) \implies \sigma_{n+1} = \sigma_{n+1}^{trial} - E\gamma_{n+1} sign(\sigma_{n+1}) \\ 0 &= \tfrac{\partial}{\partial q} L(\sigma_{n+1}, q_{n+1}, \gamma_{n+1}) \implies q_{n+1} = q_n + K\gamma_{n+1} \\ &\gamma_{n+1} \geq 0 \; ; \; \phi(\sigma_{n+1}, q_{n+1}) \leq 0 \; ; \; \gamma_{n+1} \phi(\sigma_{n+1}, q_{n+1}) = 0 \end{aligned} \tag{3.167}$$

II Global phase of operator split computation of equilibrium equations, is carried out, once the local phase of the operator split solution procedure has provided the final values of internal variables, further denoted with $\bar{\epsilon}_{n+1}^p$ and $\bar{\zeta}_{n+1}$, the plastically admissible stress field with $\phi(\sigma_{n+1}, q_{n+1}) \leq 0$ and $\bar{\gamma}_{n+1} \geq 0$, as well as the elastoplastic tangent modulus. The main goal of the global solution phase is to verify the equilibrium equation and to perform, if needed, an iterative correction of the total deformation. The latter can formally be presented as follows:

Given: ϵ_{n+1}^p, ζ_{n+1} (as well as $\sigma_{n+1} = E(\epsilon_{n+1}^{(i)} - \epsilon_{n+1}^p)$, $q_{n+1} = -K\bar{\zeta}_{n+1}$, $\gamma_{n+1}\phi_{n+1} = 0$)

Find: $\epsilon_{n+1}^{(i)} (= \sum_e \mathsf{B}^e \mathsf{d}_{n+1}^{e,(i)})$

which satisfies the equilibrium equations:

$$
\begin{aligned}
0 &\overset{?}{=} \mathsf{r}(\mathsf{d}_{n+1}^{(i)}, \epsilon_{n+1}^{p}, \zeta_{n+1}) := \mathop{\mathbb{A}}_{e=1}^{n_{elem}} \left(\mathsf{f}^{int,e}(\mathsf{d}_{n+1}^{e,(i)}, \epsilon_{n+1}^{p}, \zeta_{n+1}) - \mathsf{f}^{ext,e} \right) \\
\mathsf{f}^{int,e} &= [f_a^{int,e}] \; ; \\
f_a^{int,e} &:= \int_0^{l^e} B_a^e(x) \sigma_{n+1}(\epsilon_{n+1}^{(i)}(x), \bar{\epsilon}_{n+1}^{p}(x), \bar{\zeta}_{n+1}(x))\, dx \; ; \; B_a^e = \frac{dN_a^e}{dx} \\
&= \int_{-1}^{1} B_a^e(\xi) \hat{\sigma}_{n+1}(\epsilon_{n+1}^{(i)}(\xi), \bar{\epsilon}_{n+1}^{p}(\xi), \bar{\zeta}_{n+1}(\xi))\, j(\xi)\, d\xi \\
&= \sum_{l=1}^{n_{in}} B_a^e(\xi_l) \hat{\sigma}_{n+1}(\underbrace{\epsilon_{n+1}^{(i)}(\xi_l)}_{\sum_{b=1}^{n_{el}} B_b^e(\xi_l) d_{b,n+1}^{e,(i)}}, \bar{\epsilon}_{n+1}^{p}(\xi_l), \bar{\zeta}_{n+1}(\xi_l))\, j(\xi_l)\, w_l
\end{aligned}
$$

If the equilibrium equations are not satisfied within the chosen tolerance, we have to perform a new iterative sweep and compute an improved iterative value of the total deformation $\epsilon_{n+1}^{(i+1)}$. When Newton's iterative procedure is used for that computation, we will carry out the consistent linearization of the equilibrium equations, obtain the solution for the corresponding iterative correction of incremental displacement $\mathsf{u}_{n+1}^{(i)}$ and provide the updated value of total deformation

$$
\begin{aligned}
0 = Lin[\mathsf{r}(\mathsf{d}_{n+1}^{(i)}, \cdot)] &:= \mathsf{r}(\mathsf{d}_{n+1}^{(i)}, \cdot) + \mathop{\mathbb{A}}_{e=1}^{n_{elem}} [\mathsf{K}_{n+1}^{e}]\, \mathsf{u}_{n+1}^{(i)} \\
\mathsf{d}_{n+1}^{(i+1)} = \mathsf{d}_{n+1}^{(i)} + \mathsf{u}_{n+1}^{(i)} &\implies \left. \epsilon_{n+1}^{(i+1)} \right|_{\Omega^e} = \mathsf{B}^e \mathsf{d}_{n+1}^{(i+1)}
\end{aligned}
$$

The consistent tangent stiffness matrix for each element K_{n+1}^{e} can easily be obtained by using the corresponding result from the local phase for the elasto-plastic tangent modulus C_{n+1}^{ep}; the corresponding result of this computation can be written:

$$
\begin{aligned}
\mathsf{K}_{n+1}^{e} &= [K_{ab,n+1}^{e}] \; ; \\
K_{ab,n+1}^{e} &= \int_0^{l^e} B_a^e(x) \hat{C}_{n+1}^{ep}(\epsilon_{n+1}^{(i)}(x), \bar{\epsilon}_{n+1}^{p}(x), \bar{\zeta}_{n+1}(x)) B_b^e(x)\, dx \\
&= \int_{-1}^{1} B_a^e(\xi) \hat{C}_{n+1}^{ep}(\epsilon_{n+1}^{(i)}(\xi), \bar{\epsilon}_{n+1}^{p}(\xi), \bar{\zeta}_{n+1}(\xi)) B_b^e(\xi)\, j(\xi)\, d\xi \\
&= \sum_{l=1}^{n_{in}} B_a^e(\xi_l) \hat{C}_{n+1}^{ep}(\epsilon_{n+1}^{(i)}(\xi_l), \bar{\epsilon}_{n+1}^{p}(\xi_l), \bar{\zeta}_{n+1}(\xi_l)) B_b^e(\xi_l)\, j(\xi_l)\, w_l
\end{aligned}
$$

The explicit form of the elastoplastic tangent modulus is obtained in accordance with the results from the local phase. Namely, for an elastic step the tangent modulus is the same as Young's modulus, since:

$$
\sigma_{n+1} = \sigma_{n+1}^{trial} = E(\epsilon_{n+1} - \epsilon_n^p) \implies C_{n+1}^{ep} := \frac{\partial \sigma_{n+1}^{trial}}{\partial \epsilon_{n+1}} = E \tag{3.168}
$$

However, for a plastic step, the tangent modulus can be computed from (3.164) leading to:

$$\begin{aligned} C^{ep}_{n+1} &= \frac{\partial \sigma^{trial}_{n+1}}{\partial \epsilon_{n+1}} - \frac{\partial \gamma_{n+1}}{\partial \epsilon_{n+1}} E sign(\sigma^{trial}_{n+1}) - \gamma_{n+1} E \frac{\partial sign(\sigma^{trial}_{n+1})}{\partial \epsilon_{n+1}} \\ &= E - \frac{\partial |\sigma^{trial}_{n+1}|}{\partial \epsilon_{n+1}} \frac{E sign(\sigma_{n+1})}{E+K} \\ &= E - \frac{E^2 (sign(\sigma^{test}_{n+1}))^2}{E+K} \\ &= \frac{EK}{E+K} \end{aligned} \tag{3.169}$$

The comparison of the last result with the one in (3.125) shows that the tangent modulus for 1D plasticity model with hardening remains the same both for continuum and discrete problem. This is no longer true for 3D case, which is studied next.

3.3 3D plasticity

3.3.1 Standard format of 3D plasticity model: Prandtl-Reuss equations

In this section we develop the standard format of 3D plasticity model described by Prandtl-Reuss equations. We choose for illustration the yield function proposed by von Mises, which describes well the plasticity phenomena in metals. The choice of this kind represents the first practically applicable model of 3D plasticity which was developed by a number of classical works on the subject (e.g. see Hill [103] or Prager [224]). A typical development of plasticity model in these classical works elaborates upon numerous details important only for a particular 3D plasticity criterion, which hides the fact that the structure of governing equations remains the same as in 1D case. Unlike the classical approach, we follow the same development procedure as the one presented for 1D plasticity model, by using the extensions appropriate for 3D case and replacing the scalar fields by vectors or tensors. The material model of this kind is referred to as standard. The standard model of 3D plasticity is characterized by three main ingredients: (i) an additive decomposition of the total deformation into elastic and plastic part, (ii) a quadratic form of the free energy in terms of strains and internal variables, (iii) a quadratic form of yield function in terms of stress and dual variables. The additive decomposition of the total deformation implies that the elastic response is not affected by plastic flow. The quadratic form of free energy in terms of strains implies that the elastic response is linear. The latter allows to ensure the efficiency of the plastic flow computation, which is reduced to

a single equation for computing the plastic multiplier. The quadratic form of the yield function is sufficiently general to accommodate a number of well-known criteria applicable to a wide range of materials, such as von Mises for metals, Drucker-Parger for soils, or anisotropic criterion of Hill [103] capable of accounting for residual stress used in sheet metal forming.
For a boundary value problem in 3D plasticity, the displacement is represented as a vector field, which is a function of both space position '$\mathbf{x}$' and pseudo-time 't'

$$\mathbf{u}(\mathbf{x}, t) = u_i(\mathbf{x}, t)\mathbf{e}_i \; ; \; \mathbf{x} = x_i \mathbf{e}_i \tag{3.170}$$

We consider the case of small displacement gradient, which allows us to use the symmetric part of displacement gradient tensor as the corresponding infinitesimal deformation measure, $\boldsymbol{\epsilon} = \epsilon_{ij}\mathbf{e}_i \otimes \mathbf{e}_j$

$$\begin{aligned} \nabla \mathbf{u} = \tfrac{\partial u_i}{\partial x_j}\mathbf{e}_i \otimes \mathbf{e}_j \; ; \; \parallel \tfrac{\partial u_i}{\partial x_j} \parallel \ll 1 \\ \Longrightarrow \boldsymbol{\epsilon} = \nabla^s \mathbf{u} \Leftrightarrow \epsilon_{ij} = \tfrac{1}{2}\left(\tfrac{\partial u_i}{\partial x_j} + \tfrac{\partial u_j}{\partial x_i}\right) \end{aligned} \tag{3.171}$$

The hypothesis of small displacement gradient also allows to express the equilibrium equations in terms of Cauchy (or true) stress, $\boldsymbol{\sigma} = \sigma_{ij}\mathbf{e}_i \otimes \mathbf{e}_j$, directly in the initial configuration

$$div[\boldsymbol{\sigma}] + \mathbf{b} = \mathbf{0} \Leftrightarrow \frac{\partial \sigma_{ij}}{\partial x_j} + b_i = 0 \tag{3.172}$$

The kinematics and equilibrium equations defined by (3.171) and (3.172), respectively, do not depend on constitutive behavior and remain the same as in the elastic case. We recall that for 3D linear elasticity the constitutive model is governed by Hooke's law which only requires to specify the elasticity tensor $\boldsymbol{\mathcal{C}}$

$$\boldsymbol{\sigma} = \boldsymbol{\mathcal{C}}\boldsymbol{\epsilon} \; ; \; \sigma_{ij} = \mathcal{C}_{ijkl}\epsilon_{kl} \tag{3.173}$$

We can also recall that the elasticity tensor of an isotropic material can be constructed with two parameters only; by choosing Lame's parameters, λ and μ, we can write such an elasticity tensor as:

$$\boldsymbol{\mathcal{C}} = \lambda \mathbf{I} \otimes \mathbf{I} + 2\mu \boldsymbol{\mathcal{I}} \Leftrightarrow \mathcal{C}_{ijkl} = \lambda \delta_{ij}\delta_{kl} + \mu(\delta_{ik}\delta_{jl} + \delta_{il}\delta_{jk}) \tag{3.174}$$

In the presence of plastic deformation, the constitutive relation in (3.173) is no longer featuring the total but the elastic deformation. Namely, by assuming (see Lublimer [172]) the independence of the elastic response on plastic flow, the total deformation can be split additively into elastic $\boldsymbol{\epsilon}^e$ and plastic part $\boldsymbol{\epsilon}^p$,

$$\boxed{\boldsymbol{\epsilon} = \boldsymbol{\epsilon}^e + \boldsymbol{\epsilon}^p} \tag{3.175}$$

By further assuming that the elastic response remains linear, reducing to Hooke's law in (3.173) in the absence of plastic deformation, we can construct the free energy potential as a quadratic form in terms of deformations

$$\boxed{\psi(\boldsymbol{\epsilon},\boldsymbol{\epsilon}^p,\boldsymbol{\xi}) := \tfrac{1}{2}(\boldsymbol{\epsilon}-\boldsymbol{\epsilon}^p)\cdot\boldsymbol{\mathcal{C}}(\boldsymbol{\epsilon}-\boldsymbol{\epsilon}^p)+\tfrac{1}{2}\zeta K\zeta} \tag{3.176}$$

In order to complete the development of this standard material model of plasticity, we choose the yield function as a quadratic form in stress.

$$\boxed{\phi(\boldsymbol{\sigma},q) :=\parallel \boldsymbol{\sigma}\cdot\boldsymbol{\mathcal{P}}\boldsymbol{\sigma}\parallel -(\sigma_y-q)\le 0\ ;\ \dot{\gamma}\ge 0\ ;\ \dot{\gamma}\phi(\boldsymbol{\sigma},q)=0} \tag{3.177}$$

where $\boldsymbol{\mathcal{P}}$ is a fourth-order tensor with constant components. The last expression indicates that the plastically admissible stresses are those resulting with a negative or zero value of yield function.

We can show that these three main ingredients of the standard form of 3D plasticity model are sufficient to completely define the stress tensor computation as well as the internal variables evolution corresponding to the associative plasticity model or generalized standard material model. To that end, with the model ingredients in (3.175), (3.176) and (3.177), we appeal to the second principle of thermodynamics and the principle of maximum plastic dissipation. The final results for 3D case are therefore fully equivalent to those already presented for 1D case, apart the appropriate representation of displacement, strain and stress fields by vectors and tensors, respectively. Namely, by ignoring the thermal effects, we can write the second principle of thermodynamics for 3D plasticity case according to:

$$\begin{aligned} 0\le D &:= \boldsymbol{\sigma}\cdot\dot{\boldsymbol{\epsilon}}-\tfrac{\partial}{\partial t}\psi \\ &= \boldsymbol{\sigma}\cdot\dot{\boldsymbol{\epsilon}}-\tfrac{\partial\psi}{\partial\boldsymbol{\epsilon}}\cdot\dot{\boldsymbol{\epsilon}}-\tfrac{\partial\psi}{\partial\boldsymbol{\epsilon}^p}\cdot\dot{\boldsymbol{\epsilon}}^p-\tfrac{\partial\psi}{\partial\zeta}\cdot\dot{\zeta} \\ &= (\boldsymbol{\sigma}-\tfrac{\partial\psi}{\partial\boldsymbol{\epsilon}})\cdot\dot{\boldsymbol{\epsilon}}-\tfrac{\partial\psi}{\partial\boldsymbol{\epsilon}^p}\cdot\dot{\boldsymbol{\epsilon}}^p-\tfrac{\partial\psi}{\partial\zeta}\cdot\dot{\zeta} \end{aligned} \tag{3.178}$$

In the elastic case, with no change of internal variables, $\dot{\boldsymbol{\epsilon}}^p=\mathbf{0}$ and $\dot{\zeta}=0$, the last result confirms that the plastic dissipation remains equal to zero, and that the stress can be computed as:

$$D=0\,;\ \dot{\boldsymbol{\epsilon}}^p=\mathbf{0}\,;\ \dot{\zeta}=0\implies \boldsymbol{\sigma}:=\frac{\partial\psi}{\partial\boldsymbol{\epsilon}}=\boldsymbol{\mathcal{C}}(\boldsymbol{\epsilon}-\boldsymbol{\epsilon}^p) \tag{3.179}$$

By assuming that such a stress computation remains valid in the plastic case, and by introducing the thermodynamic fluxes conjugate to internal variables,

$$\boldsymbol{\sigma}:=-\frac{\partial\psi}{\partial\boldsymbol{\epsilon}^p}\ ;\ q:=-\frac{\partial\psi}{\partial\zeta} \tag{3.180}$$

we can obtain the final expression for the plastic dissipation according to:

$$\begin{aligned} 0<D^p &:= \overbrace{(\boldsymbol{\sigma}-\frac{\partial\psi}{\partial\boldsymbol{\epsilon}})}^{=0}\cdot\dot{\boldsymbol{\epsilon}}\overbrace{-\frac{\partial\psi}{\partial\boldsymbol{\epsilon}^p}}^{\sigma}\cdot\dot{\boldsymbol{\epsilon}}^p\overbrace{-\frac{\partial\psi}{\partial\zeta}}^{q}\cdot\dot{\zeta} \\ &= \boldsymbol{\sigma}\cdot\dot{\boldsymbol{\epsilon}}^p+q\cdot\dot{\zeta} \end{aligned} \tag{3.181}$$

We assume that the plastic case corresponds to the stress state giving the zero value of the yield function. The evolution of internal variables for plastic case can then be obtained from the principle of maximum plastic dissipation. We postulate that, among all plastically admissible stress states (for which $\phi(\boldsymbol{\sigma}, q) \leq 0$), we choose the one which renders the maximum of plastic dissipation:

$$\boxed{D^p(\boldsymbol{\sigma}, q) = \max_{\phi(\boldsymbol{\sigma}^*, q^*) \leq 0} [D^p(\boldsymbol{\sigma}^*, q^*)]} \tag{3.182}$$

By the Lagrange multiplier method, this problem of computing the maximum under the corresponding constraint can be transformed into an unconstrained minimization problem:

$$\begin{aligned} &\boxed{L^p(\boldsymbol{\sigma}, q, \dot{\gamma}) = \max_{\dot{\gamma}^*>0} \min_{\forall \boldsymbol{\sigma}^*, q^*} L^p(\boldsymbol{\sigma}^*, q^* \dot{\gamma}^*)} \; ; \\ &L^p(\boldsymbol{\sigma}, q, \dot{\gamma}) := -D^p(\boldsymbol{\sigma}, q) + \dot{\gamma}\phi(\boldsymbol{\sigma}, q) \; ; \; \dot{\gamma} \geq 0 \end{aligned} \tag{3.183}$$

The Kuhn-Tucker optimality conditions of the Lagrangian in (3.183) will provide the corresponding evolution equations for internal variables:

$$\begin{aligned} 0 &= \frac{\partial L^p}{\partial \boldsymbol{\sigma}} = -\dot{\boldsymbol{\epsilon}}^p + \dot{\gamma}\frac{\partial \phi}{\partial \boldsymbol{\sigma}} \\ 0 &= \frac{\partial L^p}{\partial q} = -\dot{\zeta} + \dot{\gamma}\frac{\partial \phi}{\partial q} \end{aligned} \tag{3.184}$$

along with the loading/unloading conditions:

$$\dot{\gamma} \geq 0 \; ; \; \phi(\boldsymbol{\sigma}, q) \leq 0 \; ; \; \dot{\gamma}\phi = 0 \tag{3.185}$$

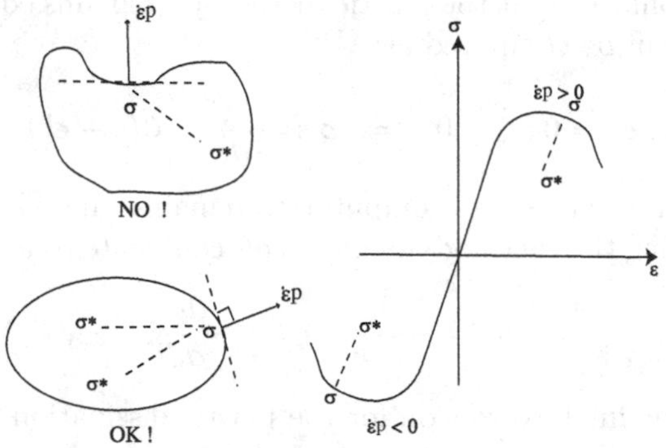

Fig. 3.12 Model of associative plasticity or standard material, which obeys the principle of maximum plastic dissipation, will impose: (i) convexity of yield surface, (ii) normality of plastic flow.

The principle of maximum plastic dissipation is applicable to the model of associative plasticity or yet called standard material model. There are two consequences of this principle that characterize any associative plasticity model: (i) the convexity of the yield surface in stress space, (ii) the normality of the plastic flow with respect to the yield surface. An illustration of these two important properties is given in Figure 3.12.
The plastic multiplier value can be computed from the consistency condition, requiring that starting from a stress state on the yield surface we recover subsequently a plastically admissible state, by either plastic loading or elastic unloading. more precisely, by using the Taylor formula to express the yield function value at the subsequent time '$t + dt$', we can write:

$$\phi(\boldsymbol{\sigma}(t), q(t)) \equiv \bar{\phi}(t) = 0 \; ; \; 0 \overset{?}{\geq} \phi(t+dt) \approx \underbrace{\bar{\phi}(t)}_{=0} + \dot{\bar{\phi}}(t)\, dt \tag{3.186}$$

By taking into account that the plastic admissibility imposes that the yield function should not be positive, we can obtain a detailed form of the loading/-unloading conditions:

$$\begin{cases} \phi > 0 & \text{plastically inadmissible at time } t \\ \phi = 0 \, ; \, \dot{\phi} > 0 & \text{plastically inadmissible at time } t+dt \\ \hline \phi < 0 & \dot{\gamma} = 0 \text{ elastic loading / unloading} \\ \phi = 0 \, ; \, \dot{\phi} < 0 \; \dot{\gamma} = 0 & \text{elastic unloading} \\ \phi = 0 \, ; \, \dot{\phi} = 0 \; \dot{\gamma} = 0 & \text{neutral loading} \\ \hline \phi = 0 \, ; \, \dot{\phi} = 0 \; \dot{\gamma} > 0 & \text{plastic loading} \end{cases} \tag{3.187}$$

We can see from the last expression that the plastic loading is the only case which results with a positive value of the plastic multiplier, which can be computed from:

$$\begin{aligned} 0 = \dot{\phi} &= \frac{\partial \phi}{\partial \boldsymbol{\sigma}} \cdot \dot{\boldsymbol{\sigma}} + \frac{\partial \phi}{\partial q} \cdot \dot{q} \\ &= \frac{\partial \phi}{\partial \boldsymbol{\sigma}} \cdot \boldsymbol{C} \dot{\boldsymbol{\epsilon}} - \dot{\gamma} \left(\frac{\partial \phi}{\partial \boldsymbol{\sigma}} \cdot \boldsymbol{C} \frac{\partial \phi}{\partial \boldsymbol{\sigma}} + \frac{\partial \phi}{\partial q} \cdot K \frac{\partial \phi}{\partial q} \right) \\ \Longrightarrow \quad \dot{\gamma} &= \frac{\partial \phi}{\partial \boldsymbol{\sigma}} \cdot \boldsymbol{C} \dot{\boldsymbol{\epsilon}} / \left(\frac{\partial \phi}{\partial \boldsymbol{\sigma}} \cdot \boldsymbol{C} \frac{\partial \phi}{\partial \boldsymbol{\sigma}} + \frac{\partial \phi}{\partial q} \cdot K \frac{\partial \phi}{\partial q} \right) \end{aligned} \tag{3.188}$$

Having established this result, we can finally express the constitutive relation in the form connecting the stress rate with the strain rate according to:

$$\dot{\boldsymbol{\sigma}} = \begin{cases} \boldsymbol{C} \dot{\boldsymbol{\epsilon}} \; ; \; \dot{\gamma} = 0 \\ [\boldsymbol{C} - (\boldsymbol{C} \frac{\partial \phi}{\partial \boldsymbol{\sigma}}) \otimes (\boldsymbol{C} \frac{\partial \phi}{\partial \boldsymbol{\sigma}}) / (\frac{\partial \phi}{\partial \boldsymbol{\sigma}} \cdot \boldsymbol{C} \frac{\partial \phi}{\partial \boldsymbol{\sigma}} + \frac{\partial \phi}{\partial q} \cdot K \frac{\partial \phi}{\partial q})] \dot{\boldsymbol{\epsilon}} \; ; \; \dot{\gamma} > 0 \end{cases} \tag{3.189}$$

We can note that the governing equations of 3D plasticity model presented in (3.175) to (3.189), as well as the corresponding steps in the model development, are fully equivalent to those previously presented for 1D case, apart the appropriate use of the tensors to replace the stress and strain scaler fields from 1D case. The von Mises plasticity model which is presented in the next section, is only one of the possible models which fit within the presented framework.

3.3.2 J2 plasticity model with von Mises plasticity criterion

The hypothesis of Tresca, postulating that the principal shear stress which reaches a given elasticity limit will activate the plastic flow, turned out to be applicable to a large number of metals. The plasticity criterion of von Mises is based on essentially the same hypothesis[10] that the plastic flow does not pertain to change of volume but only the change of form for any infinitesimal volume element. For that reason, only the deviatoric part of the stress tensor, work-conjugate to the change of shape, is used to construct the yield function for von Mises plasticity criterion

$$\begin{aligned} &\phi(\boldsymbol{\sigma}, q) :=\parallel dev[\boldsymbol{\sigma}] \parallel - \sqrt{\tfrac{2}{3}}(\sigma_y - q) \leq 0 \\ &\parallel dev[\boldsymbol{\sigma}] \parallel := \sqrt{dev[\boldsymbol{\sigma}] \cdot dev[\boldsymbol{\sigma}]} \ ; \ dev[\boldsymbol{\sigma}] = \boldsymbol{\sigma} - \tfrac{1}{3}(tr[\boldsymbol{\sigma}])\mathbf{I} \end{aligned} \tag{3.190}$$

where σ_y is the elasticity limit and q is the stress-like internal variable of isotropic hardening which handles the evolution of plasticity threshold during the plastic flow. The von Mises plasticity criterion can also be written in coordinate-independent format in terms of the second invariant J_2 of the deviatoric part of stress tensor, and it is thus often referred to as $J2$-plasticity,

$$\begin{aligned} &\phi(\boldsymbol{\sigma}, q) := \sqrt{J_2} - \sqrt{\tfrac{2}{3}}(\sigma_y - q) \leq 0 \\ &(dev[\boldsymbol{\sigma}])^3 + J_2\,(dev[\boldsymbol{\sigma}]) + J_3\mathbf{I} = \mathbf{0} \ ; \\ &J_2 =\parallel dev[\boldsymbol{\sigma}] \parallel \ ; \ J_3 = det(dev[\boldsymbol{\sigma}]) \end{aligned} \tag{3.191}$$

The factor $\sqrt{\frac{2}{3}}$ in front of the limit of elasticity is needed if the latter is identified from a simple tension test. Namely, if we define the yielding for such a case when the only non-zero stress component reaches the value σ_y, we will find it necessary to scale the von Mises stress by $\sqrt{\frac{2}{3}}$,

$$\begin{aligned} &\boldsymbol{\sigma} = \begin{bmatrix} \sigma_y & 0 & 0 \\ 0 & 0 & 0 \\ 0 & 0 & 0 \end{bmatrix} ; \ dev[\boldsymbol{\sigma}] = \begin{bmatrix} \frac{2\sigma_y}{3} & 0 & 0 \\ 0 & -\frac{\sigma_y}{3} & 0 \\ 0 & 0 & -\frac{\sigma_y}{3} \end{bmatrix} \\ &\Longrightarrow \parallel dev[\boldsymbol{\sigma}] \parallel = \sqrt{(\tfrac{2}{3}\sigma_y)^2 + 2(-\tfrac{1}{3}\sigma_y)^2} = \sqrt{\tfrac{2}{3}}\sigma_y \end{aligned} \tag{3.192}$$

[10] In fact, it is easy to verify that for the stress state of pure shear the Tresca and von Mises plasticity criteria will predict the same plastic flow.

It is important to note that the scaling factor can change when using other kind of tests for identifying the limit of elasticity; for example, if the yielding is produced in a pure shear state (which is carried out by a torsional test on a thin-walled cylinder), the scaling factor turns out to be equal to $\sqrt{2}$

$$\boldsymbol{\sigma} = \begin{bmatrix} 0 & \sigma_y & 0 \\ \sigma_y & 0 & 0 \\ 0 & 0 & 0 \end{bmatrix} = dev[\boldsymbol{\sigma}] \Longrightarrow \| \, dev[\boldsymbol{\sigma}] \, \| = \sqrt{2}\sigma_y \tag{3.193}$$

Therefore, one has to specify not only the value of limit of elasticity, but also the kind of test with which that value was obtained.
The factor $\sqrt{\frac{2}{3}}$ also appears in the von Mises yield function (3.190) placed in front of the stress-like hardening variable. In this manner, the principle of maximum plastic dissipation will produce the following evolution equation for the strain-like hardening variable:

$$\dot{\zeta} := \dot{\gamma}\frac{\partial \phi}{\partial q} = \dot{\gamma}\sqrt{\frac{2}{3}} \tag{3.194}$$

The last result is in agreement with the interpretation of the strain-like hardening variable as the equivalent plastic deformation, which can be defined from the plastic deformation tensor as

$$\dot{\zeta}(t) := \sqrt{\frac{2}{3}} \, \| \, \dot{\boldsymbol{\epsilon}}^p(t) \, \| \tag{3.195}$$

We should also take into account the incompressibility of the plastic deformation, which is imposed upon associative plasticity model by the von Mises plasticity criterion

$$\dot{\boldsymbol{\epsilon}}^p = \dot{\gamma} dev[\boldsymbol{\sigma}] / \, \| \, dev[\boldsymbol{\sigma}] \, \| = \dot{\gamma}\boldsymbol{\nu} \Longrightarrow \| \, \dot{\boldsymbol{\epsilon}}^p \, \| = \dot{\gamma} \underbrace{\| \, \boldsymbol{\nu} \, \|}_{=1} \tag{3.196}$$

With these results in hand, we can easily show that the factor $\sqrt{\frac{2}{3}}$ is needed to provide the appropriate interpretation for the plastic deformation in a simple tension test, where the corresponding plastic strain tensor components can be written:

$$\dot{\boldsymbol{\epsilon}}^p = \begin{bmatrix} \dot{\epsilon}^p & 0 & 0 \\ 0 & -\frac{1}{2}\dot{\epsilon}^p & 0 \\ 0 & 0 & -\frac{1}{2}\dot{\epsilon}^p \end{bmatrix} \Longrightarrow \sqrt{\frac{2}{3}} \, \| \, \dot{\boldsymbol{\epsilon}}^p \, \| = \dot{\epsilon}^p \tag{3.197}$$

The incompressibility constraint on plastic deformation (with $tr[\boldsymbol{\epsilon}^p] = 0$) implies that the components of this tensor are not independent, which represents yet another reason for giving preference to stress tensor components in defining the von Mises plasticity criterion.
The plastic multiplier for von Mises criterion can be obtained by enforcing the plastic consistency condition. We can simplify the corresponding result in

(3.188), which is obtained for the standard material format of 3D plasticity model, to obtain:

$$\dot{\gamma} = \frac{1}{1 + K/3\mu}\boldsymbol{\nu} \cdot \dot{\boldsymbol{\epsilon}} \tag{3.198}$$

In Figure 3.13, we give an illustration of loading/unloading criteria for the von Mises plasticity criterion (with no hardening). The latter can be represented as a circle in the deviatoric plane, the plane which is perpendicular to the hydrostatic axis defined with $\sigma_{kk} = cst.$. Within this frame, in accordance with the result in (3.198), the plastic loading case imposes the radial direction of the plastic deformation rate,

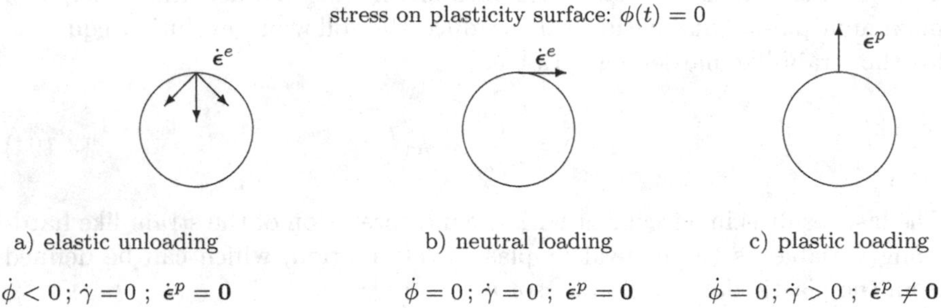

Fig. 3.13 von Mises yield surface (a circle in the deviatoric plane) and three different possibilities to define the deformation rate, for a plastically admissible stress on the yield surface.

Having established the result in (3.198), we can write the stress rate constitutive equations for von Mises plasticity model, which clearly shows that only the deviatoric part of the stress tensor is affected by plastic flow:

$$\dot{\boldsymbol{\sigma}} = \begin{cases} \boldsymbol{\mathcal{C}}\dot{\boldsymbol{\epsilon}}\,;\; \dot{\gamma} = 0 \\ [\boldsymbol{\mathcal{C}} - \frac{2\mu}{1+K/3\mu}\boldsymbol{\nu} \otimes \boldsymbol{\nu}]\dot{\boldsymbol{\epsilon}}\,;\; \dot{\gamma} > 0 \end{cases} \tag{3.199}$$

3.3.3 Implicit backward Euler scheme and operator split for von Mises plasticity

In this section, we will outline the main steps in applying the implicit backward Euler scheme to von Mises plasticity model. While the procedure remains the same as the one described in 1D case, the end result concerning the elastoplastic tangent modulus is not. Namely, contrary to 1D case, the consistent form of the elastoplastic tangent modulus for the discrete problem is no longer the same as the one for continuum problem in (3.199).

The implicit Euler scheme reduces the internal variable computation in each time step to:

Central problem of implicit Euler scheme applied to J_2 plasticity - 3D case

Given: $\boldsymbol{\epsilon}_n = \boldsymbol{\epsilon}(t_n)$, $\boldsymbol{\epsilon}^p_n = \boldsymbol{\epsilon}^p(t_n)$, $\zeta_n = \zeta(t_n)$, $h = t_{n+1} - t_n$

Find: $\boldsymbol{\epsilon}_{n+1}$, $\boldsymbol{\epsilon}^p_{n+1}$, ζ_{n+1}

such that the plastic admissibility and the weak form of equilibrium equations are verified:

$\gamma_{n+1} \geq 0\,;\ \phi_{n+1} \leq 0\,;\ \gamma_{n+1}\phi_{n+1} = 0$

$\mathbb{A}_{e=1}^{n_{el}} [\mathsf{f}^{e,int}_{n+1} - \mathsf{f}^{ext}_{n+1}] = 0$

The operator split allows us to carry out the solution procedure in an efficient manner, by separating the local phase of computation, which imposes the plastic admissibility of stress, from the global phase of computation, which will impose equilibrium. The stress tensor changes in each phase of computations, either through modification of internal variables in the local phase or through modification of the total deformation in the global phase.

We start the computation with the local phase, in which we compute the internal variables producing the plastically admissible stress in accordance with the given (best) iterative value of the total deformation $\boldsymbol{\epsilon}^{(i)}_{n+1}$. This can be written as follows:

I Local phase of plastic flow computation

Given: $\boldsymbol{\epsilon}^p_n$, ζ_n and $\boldsymbol{\epsilon}^{(i)}_{n+1}$

Find: $\boldsymbol{\epsilon}^p_{n+1}$, ζ_{n+1}

such that: $\gamma_{n+1} \geq 0\,;\ \phi_{n+1} \leq 0\,;\ \gamma_{n+1}\phi_{n+1} = 0$

The first attempt to find the solution leads to an elastic trial step, which is obtained by assuming that the plastic multiplier takes zero value and that the internal variables remain fixed in this step:

$$\phi^{trial}_{n+1} < 0\,;\ \gamma^{trial}_{n+1} = 0 \Longrightarrow \begin{array}{l} \boldsymbol{\epsilon}^{p,trial}_{n+1} = \boldsymbol{\epsilon}^p_n \\ \zeta^{trial}_{n+1} = \zeta_n \end{array} \tag{3.200}$$

We can thus readily compute the trial values of stress

$$\boldsymbol{\sigma}^{trial}_{n+1} = \boldsymbol{\mathcal{C}}(\boldsymbol{\epsilon}_{n+1} - \boldsymbol{\epsilon}^p_n)\ ;\ \bar{q}^{trial}_{n+1} = \bar{q}_n := -K\zeta_n \tag{3.201}$$

If the trial value of yield function is indeed not positive, the trial state is accepted for final, and the internal variables will not change their values with respect to the previous step.

$$\begin{array}{l} 0 \geq \phi^{trial}_{n+1} := \parallel dev[\boldsymbol{\sigma}^{trial}_{n+1}] \parallel - \sqrt{\tfrac{2}{3}}(\sigma_y - q^{trial}_{n+1}) \\ \boldsymbol{\epsilon}^p_{n+1} = \boldsymbol{\epsilon}^p_n\ ;\ \zeta_{n+1} = \zeta_n \end{array} \tag{3.202}$$

In the opposite case producing a positive trial value of the yield function, $\phi^{trial}_{n+1} > 0$, we know that the step is plastic in fact. We thus have to find the true (positive) value of plastic multiplier $\gamma_{n+1} > 0$, which will re-establish the plastic admissibility of stress with $\phi_{n+1} = 0$. This value of plastic multiplier will also provide the corresponding new values of plastic deformation and hardening variable:

$$\begin{aligned} \boldsymbol{\epsilon}^p_{n+1} &= \boldsymbol{\epsilon}^p_n + \gamma_{n+1}\boldsymbol{\nu}_{n+1} \; ; \; \boldsymbol{\nu}_{n+1} = dev[\boldsymbol{\sigma}_{n+1}]/ \parallel dev[\boldsymbol{\sigma}_{n+1}] \parallel \\ \zeta_{n+1} &= \zeta_n + \gamma_{n+1}\sqrt{\tfrac{2}{3}} \end{aligned} \tag{3.203}$$

In order to ensure efficiency in the plastic step computations, we will try to exploit the result obtained in elastic trial step. For example, the final value of the yield function can be obtained as a modification of the trial value

$$\begin{aligned} \boldsymbol{\sigma}_{n+1} &= \mathcal{C}(\boldsymbol{\epsilon}_{n+1} - \boldsymbol{\epsilon}^p_{n+1}) \\ &= \mathcal{C}(\boldsymbol{\epsilon}_{n+1} - \boldsymbol{\epsilon}^p_n) - \gamma_{n+1}\mathcal{C}\boldsymbol{\nu}_{n+1} \\ &= \boldsymbol{\sigma}^{trial}_{n+1} - \gamma_{n+1}\mathcal{C}\boldsymbol{\nu}_{n+1} \end{aligned} \tag{3.204}$$

The deviatoric part of the final stress can also be written as the appropriate modification of the trial stress, which allows us to obtain a couple of important auxiliary results:

$$\underbrace{dev[\boldsymbol{\sigma}_{n+1}]}_{\boldsymbol{\nu}_{n+1}\|dev[\boldsymbol{\sigma}_{n+1}]\|} \quad = \quad \underbrace{dev[\boldsymbol{\sigma}^{trial}_{n+1}]}_{\boldsymbol{\nu}^{trial}_{n+1}\|dev[\boldsymbol{\sigma}^{trial}_{n+1}]\|} \quad -\gamma_{n+1}2\mu\boldsymbol{\nu}_{n+1} \Longrightarrow$$

$$\begin{aligned} &\boxed{\boldsymbol{\nu}_{n+1} = \boldsymbol{\nu}^{trial}_{n+1}} \\ &\boxed{\parallel dev[\boldsymbol{\sigma}_{n+1}] \parallel = \parallel dev[\boldsymbol{\sigma}^{trial}_{n+1}] \parallel - \gamma_{n+1}2\mu} \end{aligned} \tag{3.205}$$

The first auxiliary result confirms that we keep the same normal to the yield surface in the trial and in the final state. The plastic flow can thus be obtained as the radial projection (in the deviatoric plane) from the trial state onto the yield surface. For that reason, this algorithm is also referred to as the radial return mapping. The second auxiliary result provides the relation between trial and final value of von Mises stress, which allows to express the yield function in terms of its trial value:

$$\begin{aligned} 0 &= \phi_{n+1} \\ &:= \parallel dev[\boldsymbol{\sigma}_{n+1}] \parallel - \sqrt{\tfrac{2}{3}}(\sigma_y + K\zeta_{n+1}) \\ &= \parallel dev[\boldsymbol{\sigma}^{trial}_{n+1}] \parallel - 2\mu\gamma_{n+1} - \sqrt{\tfrac{2}{3}}[\sigma_y + K(\zeta_n + \gamma_{n+1}\sqrt{\tfrac{2}{3}})] \\ &= \underbrace{\parallel dev[\boldsymbol{\sigma}^{trial}_{n+1}] \parallel - \sqrt{\frac{2}{3}}(\sigma_y + K\zeta_n)}_{\phi^{trial}_{n+1}} - \gamma_{n+1}[2\mu + 2K/3] \end{aligned} \tag{3.206}$$

We can then easily obtain the true value of the plastic multiplier:

$$\boxed{\gamma_{n+1} = \frac{\phi^{trial}_{n+1}}{2\mu+2K/3}} \qquad (3.207)$$

The corresponding graphic illustration of the return mapping algorithm computation is given in Figure 3.14.

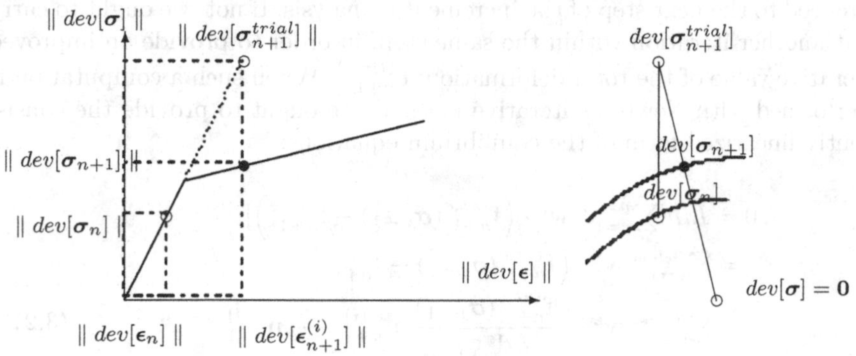

Fig. 3.14 Computation of final stress from elastic trial state by radial return mapping algorithm, where total deformation is kept fixed.

II Global phase of solving equilibrium equations

Having computed the internal variables for each integration point which ensures the plastic admissibility of stress for given iterative value of total deformation $\epsilon^{(i)}_{n+1}$, we can now turn to the equilibrium equations. The problem to be solved can be defined as follows:

Given: $\bar{\epsilon}^p_{n+1}, \bar{\zeta}_{n+1}$,

as well as the stress $\sigma_{n+1} = \mathcal{C}(\epsilon^{(i)}_{n+1} - \bar{\epsilon}^p_{n+1})$; $q_{n+1} = -K\bar{\zeta}_{n+1}$

which are plastically admissible with $\phi(\sigma_{n+1}, q_{n+1}) \leq 0$; $\gamma_{n+1} \geq 0$; $\gamma_{n+1}\phi_{n+1} = 0$

Test: equilibrium equations at time t_{n+1}

$$\mathbf{r}(\epsilon^{(i)}_{n+1}, \bar{\epsilon}^p_{n+1}, \bar{\zeta}_{n+1}) := \mathop{\mathbb{A}}_{e=1}^{n_{el}} (\mathbf{f}^{int,e}(\sigma_{n+1}) - \mathbf{f}^{ext,e}) \stackrel{?}{=} \mathbf{0} \Longrightarrow \qquad (3.208)$$

$$\begin{array}{l} \text{IF } \| \mathbf{r}^{(i)}_{n+1} \| \leq tol.\text{: next step } n \leftarrow n+1 \\ \text{ELSE : next iteration } (i) \leftarrow (i+1) \end{array}$$

We recall that the internal force vector for each element is computed by numerical integration, by using the appropriate rule for each element type.

Such a computation requires the values of internal variables only at numerical integration point, with the abscissae $\boldsymbol{\xi}_l$,

$$\begin{aligned}\mathbf{f}^{int,e}(\boldsymbol{\sigma}_{n+1}) &:= \int_{\Omega^e} \mathbf{B}^{e\,T} \hat{\boldsymbol{\sigma}}_{n+1}(\boldsymbol{\epsilon}_{n+1}^{(i)}, \boldsymbol{\epsilon}_{n+1}^{p}, \zeta_{n+1})\, dV \\ &= \sum_{l=1}^{n_{in}} \mathbf{B}^{e\,T}(\boldsymbol{\xi}_l) \hat{\boldsymbol{\sigma}}(\underbrace{\boldsymbol{\epsilon}_{n+1}^{(i)}}_{\mathbf{B}^e \mathbf{d}_{n+1}^{e\,(i)}}, \boldsymbol{\epsilon}_{n+1}^{p}, \zeta_{n+1})|_{\boldsymbol{\xi}_l}\, j(\boldsymbol{\xi}_l)\, w_l\end{aligned} \tag{3.209}$$

If the equilibrium equations are satisfied within specified tolerance, we can proceed to the next step of the incremental analysis. If not, we ought to carry out another iteration within the same step, in order to provide an improved iterative value of the total deformation, $\boldsymbol{\epsilon}_{n+1}^{(i)+1)}$. When such a computation is performed with Newton's iterative method, we ought to provide the consistently linearized form of the equilibrium equations:

$$\begin{aligned}0 &= Lin[\sum_{e=1}^{n_{elem}} \mathbf{w}^e \cdot \left(\mathbf{f}_{n+1}^{int,e}(\boldsymbol{\sigma}_{n+1}) - \mathbf{f}_{n+1}^{ext,e}\right)] \\ &:= \sum_{e=1}^{n_{elem}} \mathbf{w}^e \cdot \left(\mathbf{f}_{n+1}^{int,e}(\boldsymbol{\sigma}_{n+1}) - \mathbf{f}_{n+1}^{ext,e}\right) \\ &\quad + \sum_{e=1}^{n_{elem}} \mathbf{w}^e \cdot \underbrace{\frac{\partial \mathbf{f}_{n+1}^{int,e}(\boldsymbol{\sigma}_{n+1})}{\partial \mathbf{d}_{n+1}^e}}_{\mathbf{K}_{n+1}^{e,(i)}} \mathbf{u}_{n+1}^{e,(i)} \;\mapsto\; \mathbf{u}_{n+1}^{e,(i)} \\ \mathbf{d}_{n+1}^{e,(i+1)} &= \mathbf{d}_{n+1}^{e,(i)} + \mathbf{u}_{n+1}^{e,(i)} \implies \boldsymbol{\epsilon}_{n+1}^{(i+1)} = \mathbf{B}^e\, \mathbf{d}_{n+1}^{e,(i+1)}\end{aligned} \tag{3.210}$$

where $\mathbf{K}_{n+1}^{e,(i)}$ is the element tangent stiffness matrix. The latter is computed for a particular finite element by using the same rule of numerical integration with abscise $\boldsymbol{\xi}_l$ as the one used for the element internal force computation,

$$\begin{aligned}\mathbf{K}_{n+1}^{e\,(i)} &= \int_{\Omega^e} \mathbf{B}^{e\,T} \mathcal{C}_{n+1}^{ep,(i)} \mathbf{B}^e\, dV \;;\; \mathcal{C}_{n+1}^{ep,(i)} := \frac{\partial \boldsymbol{\sigma}_{n+1}}{\partial \boldsymbol{\epsilon}_{n+1}^{(i)}} \\ &= \sum_{l=1}^{n_{in}} \mathbf{B}^{e\,T}(\boldsymbol{\xi}_l) \mathcal{C}_{n+1}^{ep,(i)}\Big|_{\boldsymbol{\xi}_l} \mathbf{B}^e(\boldsymbol{\xi}_l)\, j(\boldsymbol{\xi}_l) w_l\end{aligned} \tag{3.211}$$

The tangent elastoplastic modulus $\mathcal{C}_{n+1}^{ep,(i)}$ in the last expression ought to be computed in the consistent manner with respect to the stress computation in the local phase. In particular, for an elastic step the consistent elastoplastic tangent modulus is the same as the elasticity tensor, since the elastic trial step provides the final value of stress

$$\boldsymbol{\sigma}_{n+1} = \boldsymbol{\sigma}_{n+1}^{trial} := \mathcal{C}(\boldsymbol{\epsilon}_{n+1}^{(i)} - \boldsymbol{\epsilon}_n^p) \implies \mathcal{C}_{n+1}^{ep,(i)} := \frac{\partial \boldsymbol{\sigma}_{n+1}}{\partial \boldsymbol{\epsilon}_{n+1}^{(i)}} = \mathcal{C} \tag{3.212}$$

For a plastic step, the consistent elastoplastic tangent modulus is computed as the corresponding modification of the elasticity tensor. Namely, by taking into account the final value of stress tensor for plastic step defined in (3.204) and (3.205), we can define:

$$
\begin{aligned}
\mathcal{C}^{ep,(i)}_{n+1} &:= \frac{\partial \boldsymbol{\sigma}_{n+1}}{\partial \boldsymbol{\epsilon}^{(i)}_{n+1}} \\
&= \frac{\partial \boldsymbol{\sigma}^{trial}_{n+1}}{\partial \boldsymbol{\epsilon}_{n+1}} - \boldsymbol{\nu}^{trial}_{n+1} \otimes 2\mu \frac{\partial \bar{\gamma}_{n+1}}{\partial \boldsymbol{\epsilon}_{n+1}} - 2\mu \bar{\gamma}_{n+1} \frac{\partial \boldsymbol{\nu}^{trial}_{n=1}}{\partial \boldsymbol{\epsilon}_{n+1}} \\
&= \underbrace{\mathcal{C} - \frac{2\mu}{1+K/3\mu} \boldsymbol{\nu}^{trial}_{n+1} \otimes \boldsymbol{\nu}^{trial}_{n+1}}_{\mathcal{C}^{ep}} - 2\mu \gamma_{n+1} \frac{\partial \boldsymbol{\nu}^{trial}_{n+1}}{\partial \boldsymbol{\epsilon}^{(i)}_{n+1}}
\end{aligned}
\tag{3.213}
$$

The last result indicates that the elastoplastic tangent modulus for the discrete problem, which is consistent with respect to the implicit Euler scheme computation of stress, is not the same as the one in (3.199) defined for the continuum problem. The difference between the two tangent moduli, which can be written explicitly as

$$
-2\mu \bar{\gamma}_{n+1} \frac{\partial \boldsymbol{\nu}^{trial}_{n+1}}{\partial \boldsymbol{\epsilon}_{n+1}} = -\frac{(2\mu)^2 \bar{\gamma}_{n+1}}{\| \boldsymbol{\eta}^{trial}_{n+1} \|} [\mathcal{I} - \boldsymbol{\nu}_{n+1} \otimes \boldsymbol{\nu}_{n+1} - \frac{1}{3} \mathbf{I} \otimes \mathbf{I}] \tag{3.214}
$$

will disappear only for vanishing time step $h \mapsto 0 \Rightarrow \gamma_{n+1} \mapsto 0$. However, for a large time it is very important to employ the consistent (rather than continuum) elastoplastic tangent modulus in order to ensure the quadratic convergence rate of Newton's method.

In conclusion, the linearized form of the discrete problem in $J2$ plasticity, which should replace the one in (3.199), can be written:

$$
d\boldsymbol{\sigma}_{n+1} = \begin{cases} \mathcal{C} d\boldsymbol{\epsilon}_{n+1} \ ; \ \bar{\gamma}_{n+1} = 0 \\ [\mathcal{C} - \frac{2\mu}{1+K/3\mu} \boldsymbol{\nu}^{trial}_{n+1} \otimes \boldsymbol{\nu}^{trial}_{n+1} - \frac{(2\mu)^2 \bar{\gamma}_{n+1}}{\| \boldsymbol{\eta}^{trial}_{n+1} \|} [\mathcal{I} \\ -\boldsymbol{\nu}_{n+1} \otimes \boldsymbol{\nu}_{n+1} - \frac{1}{3} \mathbf{I} \otimes \mathbf{I}]] d\boldsymbol{\epsilon}_{n+1} \ ; \ \bar{\gamma}_{n+1} > 0 \end{cases} \tag{3.215}
$$

3.3.4 Finite element numerical implementation in 3D plasticity

Most of the main steps of finite element numerical implementation already described for 3D elasticity will directly carry over to 3D plasticity. First, we can reduce the number of independent components and computational cost by exploiting the symmetry of both total and plastic strain tensors and of stress tensor, and store their components into one-column matrices:

$$\nabla^s \mathbf{w} \mapsto \mathsf{e}(\mathsf{w}^h) = \begin{bmatrix} \frac{\partial w_1^h}{\partial x_1} \\ \frac{\partial w_2^h}{\partial x_2} \\ \frac{\partial w_3^h}{\partial x_3} \\ (\frac{\partial w_2}{\partial x_1} + \frac{\partial w_1}{\partial x_2}) \\ (\frac{\partial w_3}{\partial x_2} + \frac{\partial w_2}{\partial x_3}) \\ (\frac{\partial w_1}{\partial x_3} + \frac{\partial w_3}{\partial x_1}) \end{bmatrix} ; \; \mathsf{s} = \begin{bmatrix} \sigma_{11} \\ \sigma_{22} \\ \sigma_{33} \\ \sigma_{12} \\ \sigma_{23} \\ \sigma_{31} \end{bmatrix} ;$$

$$\nabla^s \mathbf{u} \mapsto \mathsf{e}(\mathsf{u}^h) = \begin{bmatrix} \frac{\partial u_1^h}{\partial x_1} \\ \frac{\partial u_2^h}{\partial x_2} \\ \frac{\partial u_3^h}{\partial x_3} \\ (\frac{\partial u_2}{\partial x_1} + \frac{\partial u_1}{\partial x_2}) \\ (\frac{\partial u_3}{\partial x_2} + \frac{\partial u_2}{\partial x_3}) \\ (\frac{\partial u_1}{\partial x_3} + \frac{\partial u_3}{\partial x_1}) \end{bmatrix} ; \; \boldsymbol{\epsilon}^p \mapsto \mathsf{e}^p = \begin{bmatrix} \epsilon_{11}^p \\ \epsilon_{22}^p \\ \epsilon_{33}^p \\ 2\epsilon_{12}^p \\ 2\epsilon_{23}^p \\ 2\epsilon_{31}^p \end{bmatrix} \tag{3.216}$$

The chosen numbering for the stress and strain tensor components imposes the corresponding matrix representation of the elastoplastic modulus which can be written:

$$\mathcal{C}^{ep} \mapsto \mathsf{C}^{ep} = \begin{bmatrix} \mathcal{C}_{1111}^{ep} & \mathcal{C}_{1122}^{ep} & \mathcal{C}_{1133}^{ep} & \mathcal{C}_{1112}^{ep} & \mathcal{C}_{1123}^{ep} & \mathcal{C}_{1131}^{ep} \\ \cdot & \mathcal{C}_{2222}^{ep} & \mathcal{C}_{2233}^{ep} & \mathcal{C}_{2212}^{ep} & \mathcal{C}_{2223}^{ep} & \mathcal{C}_{2231}^{ep} \\ \cdot & \cdot & \mathcal{C}_{3333}^{ep} & \mathcal{C}_{3312}^{ep} & \mathcal{C}_{3323}^{ep} & \mathcal{C}_{3331}^{ep} \\ \cdot & \cdot & \cdot & \mathcal{C}_{1212}^{ep} & \mathcal{C}_{1223}^{ep} & \mathcal{C}_{1231}^{ep} \\ \cdot & sym. & \cdot & \cdot & \mathcal{C}_{2323}^{ep} & \mathcal{C}_{2331}^{ep} \\ \cdot & \cdot & \cdot & \cdot & \cdot & \mathcal{C}_{3131}^{ep} \end{bmatrix} \tag{3.217}$$

The element internal force vector and its tangent stiffness matrix can also be cast in matrix notation, and computed by exploiting the Gauss quadrature to obtain:

$$\begin{aligned} \mathbf{f}^{int,e} &\mapsto \mathsf{f}^{int,e} = \textstyle\sum_{l=1}^{n_{in}} [\mathsf{B}^{e,T} \mathsf{s} j]_{\boldsymbol{\xi}_l} \, w_l \; ; \\ \mathbf{K}^e &\mapsto \mathsf{K}^e = \textstyle\sum_{l=1}^{n_{in}} [\mathsf{B}^{e,T} \mathsf{C}^{ep} \mathsf{B}^e j]_{\boldsymbol{\xi}_l} \, w_l \end{aligned} \tag{3.218}$$

where $\boldsymbol{\xi}_l$ and w_l are the abscise and weights of the chosen numerical integration rule. The strain–displacement matrix B^e in the last expression allows us to compute the strain tensor components from the given nodal values of displacement

$$\mathsf{e}(\mathsf{u}^h) = \mathsf{B}^e \, \mathsf{d}^e \tag{3.219}$$

For an isoparametric finite element, we can write an explicit form of matrix B^e for either 3D or 2D case

$$\mathbf{B}^e \mapsto \mathsf{B}^e = [\mathsf{B}_1, \mathsf{B}_2, \ldots, \mathsf{B}_{n_{en}}] \; ;$$

$$\mathsf{B}_a = \begin{bmatrix} \frac{\partial N_a}{\partial x_1} & 0 & 0 \\ 0 & \frac{\partial N_a}{\partial x_2} & 0 \\ 0 & 0 & \frac{\partial N_a}{\partial x_3} \\ \frac{\partial N_a}{\partial x_2} & \frac{\partial N_a}{\partial x_1} & 0 \\ 0 & \frac{\partial N_a}{\partial x_3} & \frac{\partial N_a}{\partial x_2} \\ \frac{\partial N_a}{\partial x_3} & 0 & \frac{\partial N_a}{\partial x_1} \end{bmatrix} ; \; \mathsf{B}_a = \begin{bmatrix} \frac{\partial N_a}{\partial x_1} & 0 \\ 0 & \frac{\partial N_a}{\partial x_2} \\ \frac{\partial N_a}{\partial x_2} & \frac{\partial N_a}{\partial x_1} \end{bmatrix} \tag{3.220}$$

We indicate in Figure 3.15 all the values of state variables which ought to be computed in a 4-node quadrilateral element $Q4$, along with the position within the element domain where these computations are performed. We note in passing that the plane strain case requires that the plastic deformation component ϵ_{33}^p be computed and stored, since the plane strain constraint applies to total deformation only ($\epsilon_{33} = 0$) without affecting the value of the corresponding elastic strain component, which is computed from the stress component $\sigma_{33} \neq 0$. These remarks apply not only to isoparametric strain field approximation, but also to any other kind of approximations, such as those provided by $B - bar$ method or the method of incompatible modes.

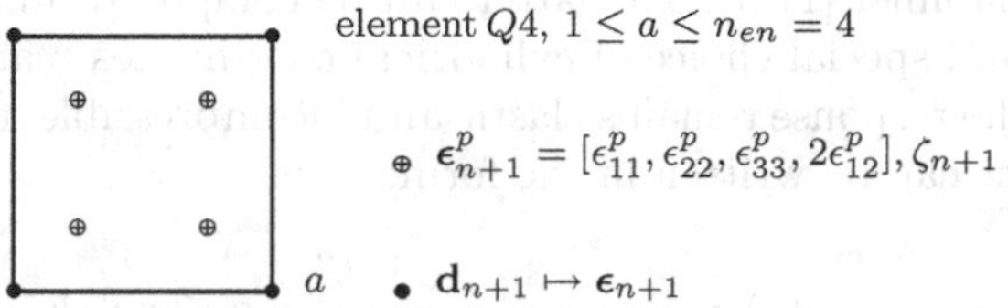

Fig. 3.15 State and internal variables computed within element domain for a 4-node quadrilateral element $Q4$, with displacement computed at nodes and internal variables at Gauss points.

3.3.4.1 Numerical examples of locking problem in $J2$ plasticity

The von Mises or $J2$ plasticity criterion imposes the incompressibility of plastic deformation. In ultimate limit load computation, where the plastic deformation is a dominant deformation component (much larger than elastic deformation), the $J2$ model will produce a quasi-incompressible deformation field (with $tr[\epsilon] \approx 0$). The latter can not be correctly represented by the standard isoparametric approximations, which leads to so-called locking phenomena. In order to accommodate quasi-incompressibility constraint within the discrete approximation framework, the isoparametric elements ought to be enhanced by using the methods presented in the previous chapter, either $B - bar$ or incompatible mode method. The method of incompatible modes, for example, can relieve the quasi-incompressibility locking phenomena with a single incompatible mode $M(\xi, \eta) = (\xi^2 + \eta^2)$ added to a 4-node isoparametric element (e.g. see Gharzeddine and Ibrahimbegovic [124]). The enhancement of discrete approximation provided by either of two methods pertains only to the total deformation field, with no change to internal variable, stress and elastoplastic tangent modulus computations as previously described for 3D plasticity and isoparametric element interpolations.

In this section we present the results of two numerical simulations for J2 plasticity, obtained by either isoparametric or incompatible mode approximations. The main objective of these examples is to illustrate the quasi-incompressibility locking phenomena which plague the isoparametric elements and the capability of incompatible mode method to eliminate them. In order to quantify the kind of result improvement that the method of incompatible modes can provide, we have selected the examples where the analytic solution can also be obtained.

3.3.4.2 Cylinder under internal pressure

The elastoplastic cylinder submitted to internal pressure is one of the rare examples in 3D plasticity where the analytic solution is available (e.g. see Hill [103], Lubliner [174]). The latter can be computed due to problem axial symmetry and special choice of cylindrical coordinates r and θ.

As long as the response remains elastic and incompressible, the cylinder radial displacement can be written in the form:

$$u(r) = \frac{3p}{2E[(b/a)^2 - 1]} \frac{b^2}{r} \; ; \; a \leq r \leq b$$

where a and b are, respectively, the internal and external cylinder radius, E is Young's modulus and p is internal pressure. The numerical value chosen for computations are $E = 1{,}000$ and $p = 10$. We also choose to approximate the cylinder incompressible behavior by a quasi-incompressible behavior, by setting the value of Poisson's coefficient to $\nu = 0.4999$.

Due to symmetry, the plane strain model considered in computation is constructed for only a quarter of cylinder, as shown in Figure 3.16.

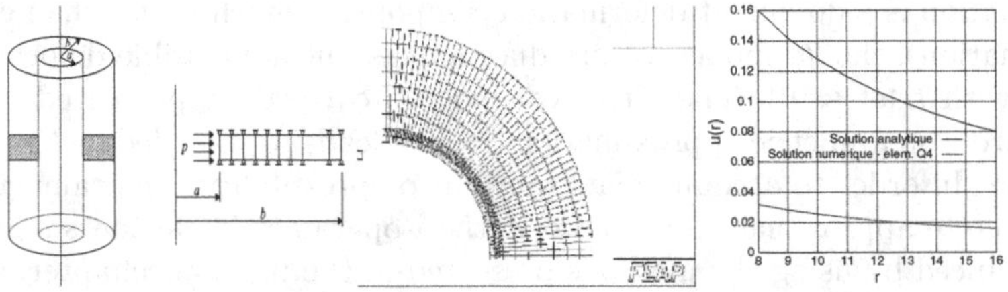

Fig. 3.16 Cylinder under internal pressure: problem definition, FE mesh for plane strain model and radial displacement for elastic case.

The comparison of the analytic result for radial displacement with the corresponding numerical result computed by the isoparametric element shown in Figure 3.16 indicates that the isoparametric elements suffer from severe locking problem already in the elastic case. This is a sufficient reason to turn to other interpolations capable of enhancing the isoparametric element performance, especially since this locking problem becomes even more severe for the case of von Mises plasticity. We thus employ the elements with incompatible modes, which have demonstrated a very good performance for the elastic case (see the previous chapter). We would thus like to show that the incompatible mode elements can provide the same enhanced performance in computing the elastoplastic response.

In the second phase of computation we consider the elastoplastic response, which appear when the stress reaches the limit of elasticity $\sigma_y = 20$. An increase in pressure starts spreading the plastic zone in the internal part of the cylinder, which propagates until the radius $r = c \leq b$. The remaining part of the cylinder $c < r < b$ remains elastic. By using the plasticity criterion of Tresca, we can obtain the relation between the value of applied pressure p and the position of frontier c between plastic and elastic part:

$$p = k(1 - \frac{c^2}{b^2} + ln\frac{c^2}{a^2})$$

where k is the elasticity limit for plasticity criterion of Tresca. For the same case, we can also obtain (see Hill [103], Lubliner [174]) the analytic solution for the stress field components, both in the elastic zone ($c < r < b$)

$$\sigma_r = -k[\frac{c^2}{r^2} - \frac{c^2}{b^2}] \; ; \; \sigma_\theta = k[\frac{c^2}{r^2} + \frac{c^2}{b^2}] \; ; \; \sigma_z = k\frac{c^2}{b^2}$$

as well as in the plastic zone ($a < r < c$)

$$\sigma_r = -k[1 - \frac{c^2}{b^2} + ln\frac{c^2}{r^2}] \; ; \; \sigma_\theta = k[1 + \frac{c^2}{b^2} - ln\frac{c^2}{r^2}] \; ; \; \sigma_z = k[\frac{c^2}{b^2} - ln\frac{c^2}{r^2}]$$

For incompressible elastic response (with $\nu = 0.5$), the von Mises plasticity criterion coincides with the one proposed by Tresca coincide, if the chosen value of limit of elasticity is equal to $k = \sigma_y/\sqrt{3}$. Therefore, with the corresponding value of yield stress we can exploit the analytic solution obtained with Tresca criterion, and evaluate the quality of the results computed for von Mises plasticity criterion. All the numerical results for stress components and radial displacement presented in Figure 3.17 are obtained by using the incompatible mode approximations; these results are in excellent agreement with the analytic solution.

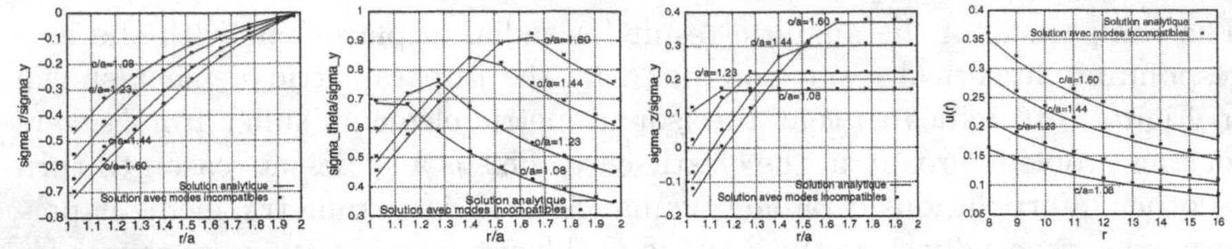

Fig. 3.17 Cylinder under internal pressure: : distribution of radial, hoop and axial stress, as well as radial displacement for plastic case.

3.3.4.3 Double edge notched specimen

The example dealing with the limit load computation of a double edge notched specimen, proposed by Nagtegaal, Parks and Rice [197], was the first to draw attention to quasi-incompressibility locking phenomena exhibited by the isoparametric elements in $J2$ plasticity. The locking phenomena appear for plasticity criterion of von Mises imposing the incompressibility of plastic deformation, in the case where the plastic deformation is much larger than the elastic deformation.

The double edge notched specimen geometry is taken as: length equal to $l =$ 16, width equal to $w = 10$, and the ligament width $b = 2$; see Figure 3.18. The constitutive model considers elastic-perfectly-plastic behavior governed by von Mises plasticity criterion, with yield stress $\sigma_y = 0.243$, Young's modulus $E = 70$ and Poisson's coefficient $\nu = 0.3$. The analytic solution for the limit load in this case is given as $\sigma_{lim} = 2.97\sigma_y$.

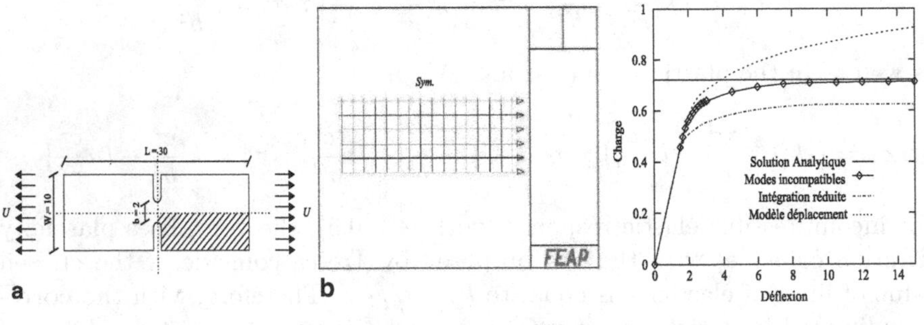

Fig. 3.18 Double edge notched specimen: problem definition, FE mesh for a quarter of specimen, and normalized force–displacement diagram $(E\delta/\sigma_y w)$.

By exploiting the symmetry of the problem, we can construct the finite element model for only a quarter of the specimen, by using the FE mesh with 5×15 elements (see Figure 3.18). The computations are performed under displacement control, in order to eliminate the convergence difficulties of the force-driven incremental analysis near the limit load. We can see in Figure 3.18 that the isoparametric element based computation can not capture the limit load, and provides instead overly stiff response which keeps increasing. On the other hand, the elements with incompatible modes can deliver the exact result. We have also shown in Figure 3.18 that the isoparametric elements employing the reduced integration (with one Gauss quadrature point for isoparametric element $Q4$), can also eliminate the locking problem, but can not provide the same quality of computed results as the elements with incompatible modes (ever in the case where the zero-energy mode is suppressed by the supports).

3.4 Refined models of 3D plasticity

The standard plasticity model with linear isotropic hardening and a particular choice of von Mises plasticity criterion have allowed us to illustrate all the main steps of the operator split methodology for both explicit and implicit integration schemes. In this section, we discuss the eventual modification of the operator split solution procedure for more refined models of 3D plasticity. The refined plasticity models are developed for representation of the inelastic behavior mechanism for a large class of materials, not only metals, but also alloys, concrete, soils or composites. In fact, we can quickly realize that the number of potential applications of refined plasticity models is practically unlimited, which makes it very difficult to elaborate the details for all of them. Instead of providing such a catalog-like presentation of existing plasticity criteria, we rather choose to present how to introduce any desired modification for enhancing predictive capabilities of the standard plasticity model. The list of potential modifications is made as comprehensive as possible. More precisely, we show how to add: nonlinear isotropic hardening, kinematic hardening, dependence of response on rate of deformation with viscoplasticity model, multi-surface plasticity criterion and nonlinear elastic response. In this manner we hope to assist the readers with the development of a new plasticity model suitable for their particular needs.

3.4.1 Nonlinear isotropic hardening

It is often observed in simple tension tests that the hardening behavior of the specimen is nonlinear. One example of this kind, which is capable of describing the stress increase until the limit value denoted as σ_∞, is referred to as the saturation hardening

$$q = \hat{q}(\zeta) := -(\sigma_\infty - \sigma_y)[1 - exp(-\beta\zeta)] \tag{3.221}$$

In (3.221) above, β is the hardening parameter that governs the rate with which the saturation is achieved. The hardening modulus of nonlinear isotropic hardening model is no longer constant, but a function of the accumulated plastic deformation

$$\hat{K}(\zeta) := -\frac{d\hat{q}(\zeta)}{d\zeta} \tag{3.222}$$

In this case, the operator split solution procedure described for the standard plasticity model is just slightly modified. Namely, it is no longer possible to obtain the direct solution for the plastic multiplier, such as the one given in (3.207), but rather use a local iterative procedure. The latter can be reduced to a single nonlinear equation with plastic multiplier as unknown:

$$\begin{aligned} 0 &= \hat{\phi}_{n+1}(\gamma_{n+1}) \\ &:= \| \, dev[\boldsymbol{\sigma}_{n+1}] \, \| - \sqrt{\tfrac{2}{3}}(\sigma_y - q_{n+1}) \\ &= \| \, dev[\boldsymbol{\sigma}^{trial}_{n+1}] \, \| - \sqrt{\tfrac{2}{3}}(\sigma_y - q_n) - 2\mu\gamma_{n+1} + \sqrt{\tfrac{2}{3}}(q_{n+1} - q_n) \\ &= \phi^{trial}_{n+1} - 2\mu\gamma_{n+1} - \sqrt{\tfrac{2}{3}}(q_n - \hat{q}(\zeta_n + \sqrt{\tfrac{2}{3}}\gamma_{n+1})) \end{aligned} \tag{3.223}$$

It is easy to see that the solution of the last equation provides the correct value of plastic multiplier that ensures the plastic admissibility of stress with the zero value of the yield function. This iterative solution for plastic multiplier can be carried out by Newton's method, starting with the first iterative guess which is provided by the elastic trial state with $\gamma^{trial}_{n+1} = 0$ and $\phi^{trial}_{n+1} > 0$

$$\begin{aligned} &\gamma^{(1)}_{n+1} = 0 \; ; \; \bar{\phi}^{(1)}_{n+1} = \phi^{trial}_{n+1} \\ &(j) = 1, 2, \ldots \\ &\quad 0 = \hat{\phi}^{(j)}_{n+1} + \frac{d\hat{\phi}_{n+1}(\gamma^{(j)}_{n+1})}{d\gamma_{n+1}} \Delta\gamma^{(j)}_{n+1} \; ; \; \frac{d\hat{\phi}(\cdot)}{d\gamma_{n+1}} = 2\mu + \hat{K}(\zeta^{(j)}_{n+1}) \\ &\quad \gamma^{(j+1)}_{n+1} = \gamma^{(j)}_{n+1} + \Delta\gamma^{(j)}_{n+1} \\ &\quad \zeta^{(j+1)}_{n+1} = \zeta_n + \sqrt{\tfrac{2}{3}}\gamma^{(j+1)}_{n+1} \\ &IF \; \phi(\gamma^{(j+1)}_{n+1}) \neq 0 \; (j+1) \longleftarrow (j) \\ &ELSE \; \bar{\zeta} = \zeta^{(j+1)}_{n+1} \end{aligned} \tag{3.224}$$

It is important to provide the robustness of this local iterative procedure and make sure it will always converge. To that end, we note that as long as the nonlinear isotropic hardening is described by a convex function, such as the one typical of saturation hardening in (3.221), the convergence of the iterative procedure starting from $\gamma^{test}_{n+1} = 0$ is guaranteed. In general, we need several Newton's iterations to converge to the plastically admissible value of stress. An illustrative representation of this kind of computation, showing several intermediate iterative values of stress is given in Figure 3.19. It is important to note that during the local iterations at each quadrature points the value of the total deformation is kept fixed.

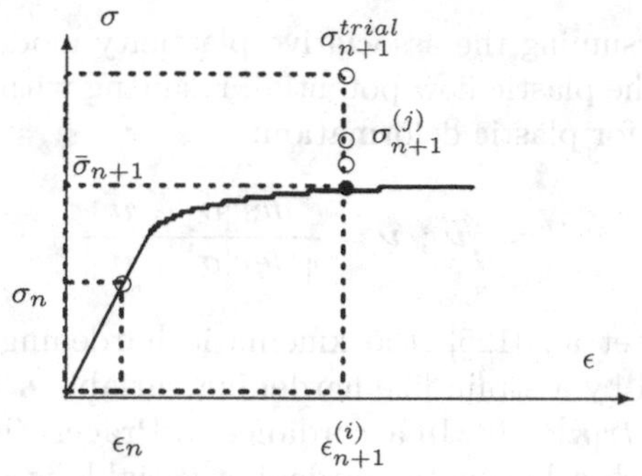

Fig. 3.19 Computing plastically admissible stress for 1D plasticity model with nonlinear isotropic hardening, where we indicate intermediate iterative values of stress as well as the final, plastically admissible value of stress.

The nonlinear isotropic hardening model also requires a small modification of the consistent elastoplastic tangent modulus written in (3.215). Namely, the value of the hardening modulus is no longer constant, and it has to be replaced with the last value used in local iteration $\hat{K}(\bar{\zeta}_{n+1})$, as defined in (3.222).

3.4.2 Kinematic hardening

Another modification of von Mises plasticity model, which will make it applicable to cyclic loading, pertains to kinematic hardening. Such a hardening model ought to be introduced for representing the Bauschinger effect, observed in experiments with cyclic loading. Namely, in a typical cycle with the load reversal (from compression to tension or vice versa), the plasticity threshold value is typically reduced from the previous value. In order to explain the Bauschinger effect, we can assume that the evolution of plasticity threshold in the plastic loading phase will also displace the center of the elastic domain with respect to previous plastic state. For example, the hardening model proposed by Prager-Ziegler [268] represents the kinematic hardening phenomena for von Mises yield function, where the size of the elastic domain remains fixed, as a circle with fixed radius in the deviatoric plane, but with the position of its center changing. We use a deviatoric stress tensor, referred to as the back-stress, to trace the evolution of the elastic domain center during the plastic flow. By denoting the back-stress as deviatoric tensor $\boldsymbol{v}$ (with $tr[\boldsymbol{v}] = 0$), we can recast the von Mises plasticity criterion for kinematic hardening case:

$$0 \geq \phi(\boldsymbol{\sigma}, \boldsymbol{v}) := \parallel dev[\boldsymbol{\sigma}] + \boldsymbol{v} \parallel - \sqrt{\frac{2}{3}}\sigma_y \tag{3.225}$$

Furthermore, by assuming the associative plasticity model, we can take the same function for the plastic flow potential, resulting with the corresponding evolution equation for plastic deformation:

$$\dot{\boldsymbol{\epsilon}}^p = \dot{\gamma}\boldsymbol{\nu} \; ; \; \boldsymbol{\nu} = \frac{dev[\boldsymbol{\sigma}] + \boldsymbol{v}}{\| \, dev[\boldsymbol{\sigma}] + \boldsymbol{v} \, \|} \tag{3.226}$$

In Ibrahimbegovic et al. [125], the kinematic hardening model of Prager-Ziegler is described by a strain-like hardening variable $\boldsymbol{\kappa}$ in terms of a deviatoric tensor (with $tr[\boldsymbol{\kappa}] = 0$). In accordance to Prager-Ziegler hypothesis, it is assumed that such a kinematic hardening variable is proportional to the plastic deformation tensor, which can be written:

$$\dot{\boldsymbol{\kappa}} = \dot{\gamma}\boldsymbol{\nu} =: \dot{\gamma}\frac{\partial \phi}{\partial \boldsymbol{v}} \tag{3.227}$$

The last result shows that the back-stress is thermodynamically conjugate to the chosen internal variable of kinematic hardening. The flow potential characterizing the evolution of this hardening variable is the same as the yield function in (3.225). Therefore, the proposed kinematic hardening modification of the standard plasticity model will remain placed within the framework of the associative plasticity or generalized standard material (e.g. see Nguyen and Halphen [98]). By denoting the kinematic hardening modulus with H, we can write the free energy potential as a quadratic form:

$$\psi(\boldsymbol{\epsilon}, \boldsymbol{\epsilon}^p, \boldsymbol{\kappa}) = \frac{1}{2}(\boldsymbol{\epsilon} - \boldsymbol{\epsilon}^p) \cdot \boldsymbol{\mathcal{C}}(\boldsymbol{\epsilon} - \boldsymbol{\epsilon}^p) + \frac{H}{3}\boldsymbol{\kappa} \cdot \boldsymbol{\kappa} \tag{3.228}$$

The latter allows us to obtain the linear form of the constitutive relation between the kinematic hardening variable and back-stress

$$\boldsymbol{v} := -\frac{\partial \psi(\cdot)}{\partial \boldsymbol{\kappa}} = -\frac{2H}{3}\boldsymbol{\kappa} \tag{3.229}$$

The operator split procedure that is already presented for isotropic hardening case remains practically the same for the present case of kinematic hardening, apart the tensorial nature of the internal variable. In the plastic step, we can carry out the direct computation of plastic multiplier:

$$\gamma_{n+1} = \frac{\phi_{n+1}^{trial}}{2\mu + 2H/3} \tag{3.230}$$

The graphic illustration of this computation resulting with the plastically admissible stress is presented in Figure 3.20

In the global phase of the operator split method for computing a new iterative guess for displacement vector, we will need the consistent elastoplastic tangent modulus. For kinematic hardening model of von Mises plasticity, the elastoplastic modulus can be written:

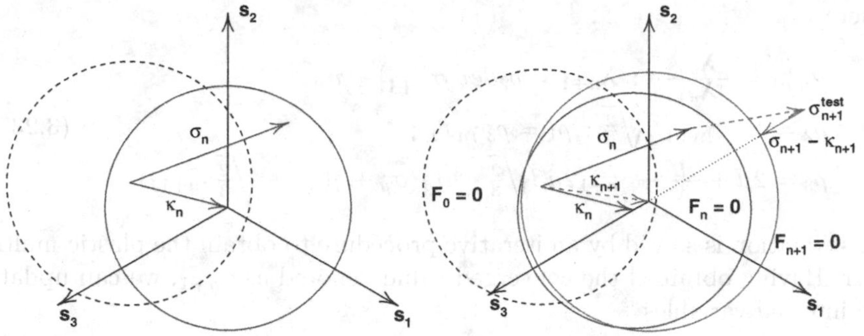

Fig. 3.20 Plastically admissible stress computation for kinematic hardening model presented in deviatoric plane.

$$d\boldsymbol{\sigma}_{n+1} = \begin{cases} \mathcal{C} d\boldsymbol{\epsilon}_{n+1} \; ; \; \bar{\gamma}_{n+1} = 0 \\ [\mathcal{C} - \frac{2\mu}{1+H/3\mu} \boldsymbol{\nu}_{n+1}^{trial} \otimes \boldsymbol{\nu}_{n+1}^{trial} - \frac{(2\mu)^2 \bar{\gamma}_{n+1}}{\| dev[\boldsymbol{\sigma}_{n+1}^{trial}] + \boldsymbol{\upsilon}^{trial} \|} [\mathcal{I} \\ -\boldsymbol{\nu}_{n+1} \otimes \boldsymbol{\nu}_{n+1} - \frac{1}{3} \mathbf{I} \otimes \mathbf{I}]] d\boldsymbol{\epsilon}_{n+1} \; ; \; \bar{\gamma}_{n+1} > 0 \end{cases} \quad (3.231)$$

There exist other models of kinematic hardening, much more elaborate than the simple linear model proposed by Prager-Ziegler. One such model is a non-linear kinematic hardening model (e.g. see Armstrong and Frederick [10]) or Chaboche [44], [45]), which is particularly useful for representing the ratchetting effect which occur in cyclic loading between non-symmetric plasticity thresholds in compression and in tension. The nonlinear kinematic hardening model no longer belongs to the class of associative plasticity models, since the evolution equation for back-stress assumes that the rate of back-stress is proportional non only to the plastic deformation but also to the current value of the back-stress

$$\dot{\boldsymbol{\upsilon}} = -\frac{2H}{3} \dot{\boldsymbol{\epsilon}}^p + \beta \dot{\zeta} \boldsymbol{\upsilon} \; ; \; tr[\boldsymbol{\upsilon}] = 0 \quad (3.232)$$

where β ia a chosen material parameter. For a non-associative plasticity model of this kind, the plastic flow potential is not the same as the yield function in (3.225). Despite this increase in complexity, it is still possible (see Chorfi and Ibrahimbegovic [48]) to ensure the efficiency of the operator split methodology for the hardening model of this kind. Namely, by using implicit Euler scheme, we can obtain:

$$\boldsymbol{\upsilon}_{n+1} = \boldsymbol{\upsilon}_n + \gamma_{n+1} \left(-\frac{2H}{3} \boldsymbol{\nu}_{n+1} + \beta \sqrt{\frac{2}{3}} \boldsymbol{\upsilon}_{n+1} \right) \quad (3.233)$$

The local phase of the plastic flow computation can again be reduced to a single nonlinear equation representing the yield function, with the plastic multiplier as unknown:

$$0 = \phi_{n+1} = \| \boldsymbol{\lambda}_{n+1} \| - \rho_1 - \sqrt{\frac{2}{3}} (\sigma_y + \hat{q}(\zeta_{n+1})) \quad (3.234)$$

where

$$\boldsymbol{\nu}_{n+1} = \frac{\boldsymbol{\lambda}_{n+1}}{\|\boldsymbol{\lambda}_{n+1}\|} \; ; \; \boldsymbol{\lambda}_{n+1} = \rho_2 dev[\boldsymbol{\sigma}^{tr}_{n+1}] + \boldsymbol{\upsilon}_n$$
$$\rho_2 = 1 - \gamma_{n+1}\bar{\beta}\sqrt{\tfrac{2}{3}} \; ; \; \rho_1 = \rho_3\gamma_{n+1} \; ; \tag{3.235}$$
$$\rho_3 = 2\mu + \tfrac{2H}{3} - \gamma_{n+1}2\mu\bar{\beta}\sqrt{\tfrac{2}{3}} - \bar{\beta}\tfrac{2}{3}(\sigma_y + \hat{q}(\zeta_n + \sqrt{\tfrac{2}{3}}\gamma_{n+1}))$$

This equation is solved by an iterative procedure to obtain the plastic multiplier. Having obtained the converged value denoted as $\bar{\gamma}_{n+1}$, we can update the internal variables

$$\boldsymbol{\upsilon}_{n+1} = \frac{1}{1 - \sqrt{\frac{2}{3}}\beta\bar{\gamma}_{n+1}}[\boldsymbol{\upsilon}_n - \bar{\gamma}_{n+1}\frac{2H}{3}\boldsymbol{\nu}_{n+1}] \tag{3.236}$$

We also have to compute the consistent tangent elastoplastic modulus, needed for the global phase of operator split procedure. It is important to note that the nonlinear kinematic hardening is a non-associative plasticity model, which produces a non-symmetric form of the elastoplastic tangent modulus:

$$\begin{aligned}\mathbf{C}^{ep\,(i)}_{n+1} &:= \left.\frac{\partial\boldsymbol{\sigma}^{(i)}_{n+1}}{\partial\boldsymbol{\epsilon}_{n+1}}\right|_{\boldsymbol{\epsilon}^{(i)}_{n+1},\bar{\boldsymbol{\epsilon}}^{p}_{n+1},\bar{\zeta}_{n+1},\bar{\boldsymbol{\kappa}}_{n+1}} \\ &= \mathbf{C} - \omega_1(\boldsymbol{\mathcal{I}} - \tfrac{1}{3}\mathbf{I}\otimes\mathbf{I}) - \omega_2(\boldsymbol{\nu}_{n+1}\otimes\boldsymbol{\nu}_{n+1}) \\ &\quad -\omega_3(dev[\boldsymbol{\sigma}^{tr}_{n+1}]\otimes\boldsymbol{\nu}_{n+1})\end{aligned} \tag{3.237}$$

with:

$$\omega_1 = \bar{\gamma}_{n+1}\frac{(2\mu)^2\rho_2}{\|\boldsymbol{\lambda}_{n+1}\|} \; ; \; \omega_3 = \bar{\gamma}_{n+1}\frac{(2\mu)^2\rho_2}{\rho_4\|\boldsymbol{\lambda}_{n+1}\|}\bar{\beta}\sqrt{\tfrac{2}{3}}$$
$$\omega_2 = \frac{\rho_2(2\mu)^2}{\rho_4\|\boldsymbol{\lambda}_{n+1}\|}\left[\|\boldsymbol{\lambda}_{n+1}\| - \bar{\gamma}_{n+1}(\rho_4 + \bar{\beta}\sqrt{\tfrac{2}{3}}tr(dev[\boldsymbol{\sigma}^{tr}_{n+1}]\,\boldsymbol{\nu}_{n+1}))\right] \tag{3.238}$$

The values of $\boldsymbol{\lambda}_{n+1}$ and ρ_2 are defined in $(3.235)_1$, whereas ρ_4 is defined by the following relation:

$$\rho_4 = \rho_3 + \gamma_{n+1}\bar{\beta}\sqrt{\frac{2}{3}}(2\mu + \frac{2}{3}\bar{k}'_{n+1}) + \frac{2}{3}\bar{k}'_{n+1} - \bar{\beta}\sqrt{\frac{2}{3}}tr(dev[\boldsymbol{\sigma}^{tr}_{n+1}]\,\boldsymbol{\nu}_{n+1}) \tag{3.239}$$

where ρ_3 is given in (3.235). For nonlinear kinematic hardening model of Armstrong-Frederick, it is important to take the non-symmetric form of the consistent elastoplastic tangent modulus in order to ensure the quadratic convergence of Newton's method used for global solution phase. An illustrative example which confirms this statement is given next.

3.4.2.1 Thick cylinder under cyclic loading

We consider in this example a thick cylinder with internal radius of 30 mm and external radius of 60 mm. The cylinder is built of elastoplastic material with Young's modulus $E = 210{,}000[\mathrm{N/mm}^2]$, Poisson's ratio $\nu = 0,3$ and limit of elasticity $\sigma_y = 300\,[\mathrm{N/mm}^2]$. The nonlinear kinematic hardening model is represented with Armstrong-Frederick hardening model, with hardening parameters $H = 30{,}000\,[\mathrm{N/mm}^2]$ and $\beta = 60$.
The cylinder is submitted to a cyclic loading, which is driven by imposed displacement $|d| = 0{,}437$ mm (see Figure 3.21). We construct 2D model under plane strain conditions and exploit the symmetry to limit the study to a quarter of cylinder. The FE mesh is constructed by using 36 finite elements $Q4$ with incompatible modes; see Figure 3.21.
The analysis is carried out under displacement control at the inner radius of cylinder. The time step is kept constant and equal to $h = 0.05$ s. The typical convergence rate for Newton's method is represented in Table 3.2.

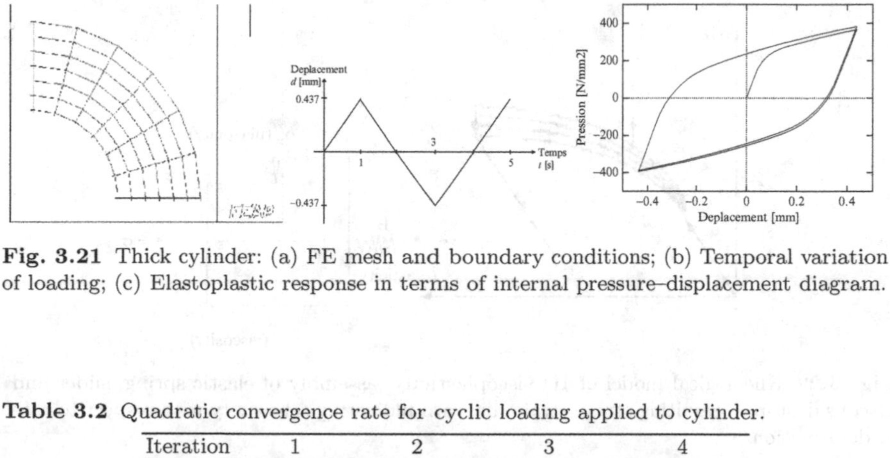

Fig. 3.21 Thick cylinder: (a) FE mesh and boundary conditions; (b) Temporal variation of loading; (c) Elastoplastic response in terms of internal pressure–displacement diagram.

Table 3.2 Quadratic convergence rate for cyclic loading applied to cylinder.

Iteration	1	2	3	4
Residual	$2{,}049 \times 10^4$	$1{,}596 \times 10^1$	$1{,}205 \times 10^{-3}$	$1{,}614 \times 10^{-8}$

3.4.3 Plasticity model dependent on rate of deformation or viscoplasticity

It has been observed experimentally that the elastoplastic behavior can be sensitive to the rate of deformation, in the sense that a rapidly applied loading producing higher rate of deformation can also lead to an increase of resistance in plastic phase. This kind of behavior can be represented with a plasticity model dependent on rate of deformation, or viscoplasticity (e.g.

see Duvaut and Lyons [72] or Perzyna [222], which can be constructed as the appropriate modification of standard plasticity model. As the main difference from plasticity, the viscoplasticity model will remove the constraint on plastic admissibility of stress. We will accept not only the stress state for which $\phi(\boldsymbol{\sigma}) \leq 0$, but also all other values which can result with a so-called excessive stress producing a positive value of yield function $\phi(\boldsymbol{\sigma}) > 0$. In agreement with standard plasticity model, we postulate that no modification of irreversible deformation will occur within the elastic domain where $\phi(\boldsymbol{\sigma}) < 0$, and that only the excessive stress will produce the change of viscoplastic deformation. Furthermore, excessive stress and rate of viscoplastic deformation are related through the viscosity coefficient η, a material parameter to be chosen. This will result with a rate dependent viscoplastic deformation. A rheological model of 1D viscoplasticity is presented in Figure 3.22, as an assembly in series of an elastic sprig (with coefficient E) and a parallel assembly of a slider (with resistance σ_y) and viscous dash-pot (with viscosity coefficient η).

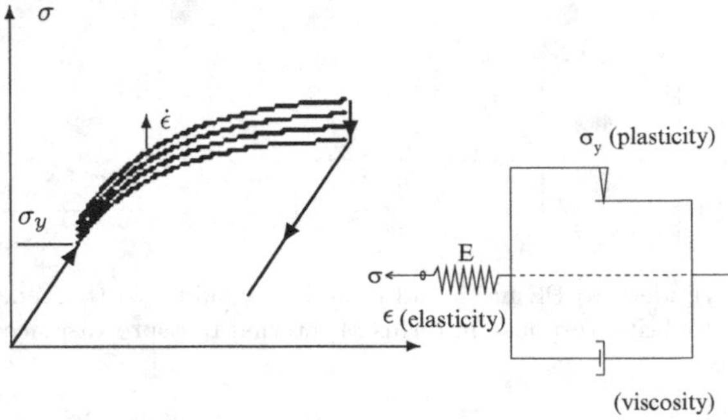

Fig. 3.22 Rheological model of 1D viscoplasticity: assembly of elastic spring, slider and viscous dash-pot; resulting stress–strain diagram indicating sensitivity with respect to rate of deformation.

The deformation of elastic spring, which will always appear first either in loading or unloading, is called elastic deformation, and the remaining part of total deformation is referred to as the viscoplastic deformation. We can thus keep the first hypothesis of standard plasticity model on the additive decomposition of the total deformation into elastic and viscoplastic part

$$\epsilon = \epsilon^e + \epsilon^{vp} \tag{3.240}$$

Moreover, it is again the elastic deformation that will decide the value of stress. We can thus keep the same internal energy potential as to the one defined for plasticity, but with plastic deformation replaced by viscoplastic deformation

$$\psi(\boldsymbol{\epsilon}, \boldsymbol{\epsilon}^{vp}) = \frac{1}{2}(\boldsymbol{\epsilon} - \boldsymbol{\epsilon}^{vp}) \cdot \boldsymbol{\mathcal{C}}(\boldsymbol{\epsilon} - \boldsymbol{\epsilon}^{vp}) \tag{3.241}$$

The role of the yield function $\phi(\boldsymbol{\sigma})$ in viscoplasticity in terms of indicating the elastic domain with its negative values remains the same as in plasticity

$$\phi(\boldsymbol{\sigma}) :=\| dev[\boldsymbol{\sigma}] \| - \sqrt{\frac{2}{3}}\sigma_y \; ; \; \begin{matrix} \phi(\boldsymbol{\sigma}) < 0 \Rightarrow \dot{\boldsymbol{\epsilon}}^{vp} = \mathbf{0} \\ \phi(\boldsymbol{\sigma}) \geq 0 \Rightarrow \dot{\boldsymbol{\epsilon}}^{vp} \neq \mathbf{0} \end{matrix} \tag{3.242}$$

In fact, the only modification with respect to the standard plasticity model reduces to removing the plastic admissibility constraint on stress and accepting as admissible any positive value of the yield function. In order to illustrate this point clearly, we refer to 1D rheological model in Figure 3.22; we can observe that the excessive stress can only be taken by viscous dash-pot, which allows us to write the corresponding evolution equations for viscoplastic deformation in 1D case:

$$\sigma_{ex} := (|\sigma| - \sigma_y) sign(\sigma) = \eta \dot{\epsilon}^{vp} \;\; \Rightarrow \;\; \dot{\epsilon}^{vp} = \frac{1}{\eta}(|\sigma| - \sigma_y) sign(\sigma) \tag{3.243}$$

An equivalent result can be obtained in 3D case, by taking into account the associative nature of the viscoplasticity model and considering the yield function to be the same as the potential governing the viscoplastic flow:

$$\dot{\boldsymbol{\epsilon}}^{vp} = \frac{< \phi(\boldsymbol{\sigma}) >}{\eta} \frac{\partial \phi(\boldsymbol{\sigma})}{\partial \boldsymbol{\sigma}} \; ; \; \phi(\boldsymbol{\sigma}) :=\| dev[\boldsymbol{\sigma}] \| - \sqrt{\frac{2}{3}}\sigma_y \tag{3.244}$$

where $< \cdot >$ is the Macauley parenthesis, which is defined as:

$$< x >:= (x + |x|)/2 = \begin{cases} x \; ; \; x > 0 \\ 0 \; ; \; x < 0 \end{cases} \tag{3.245}$$

It is interesting to note that such an evolution equation for viscoplastic deformation can formally be written as the corresponding equation for plastic deformation, by using a modified definition of the plastic multiplier

$$\dot{\boldsymbol{\epsilon}}^{vp} = \dot{\gamma} \frac{\partial \phi(\boldsymbol{\sigma})}{\partial \boldsymbol{\sigma}} \; ; \; \dot{\gamma} = \frac{< \phi(\boldsymbol{\sigma}) >}{\eta} \tag{3.246}$$

Hence, both plasticity and viscoplasticity model can be placed within the same framework for computing the evolution of inelastic deformation from the gradient of the yield surface potential scaled by the plastic multiplier. However, the crucial difference between these two models concerns the manner in which this plastic multiplier is computed: either from current value of stress in viscoplasticity or from the constraint condition of plastic consistency for rate-independent plasticity.

Exploiting the equivalence with 1D rheological model is not the only way to obtain the evolution equations for 3D viscoplasticity model. We can obtain the same result form a regularized form of the maximum dissipation principle, which applies to standard 3D plasticity model. To that end, we will consider the 3D viscoplasticity model, characterized by the free energy in (3.241) and the elastic domain definition in (3.242). The viscoplastic dissipation, which pertains to any stress state giving a positive value of the yield function $\phi(\boldsymbol{\sigma}) > 0$, can be defined from the second principle of thermodynamics:

$$0 \leq D^{vp} := (\boldsymbol{\sigma} - \frac{\partial \psi(\cdot)}{\partial \boldsymbol{\epsilon}^e}) \cdot \dot{\boldsymbol{\epsilon}}^e + \boldsymbol{\sigma} \cdot \dot{\boldsymbol{\epsilon}}^{vp} \implies \begin{cases} \boldsymbol{\sigma} = \frac{\partial \phi(\cdot)}{\partial \boldsymbol{\epsilon}} \\ D^{vp} := \boldsymbol{\sigma} \cdot \dot{\boldsymbol{\epsilon}}^{vp} \end{cases} \tag{3.247}$$

We will then look for the stress that will maximize the viscoplastic dissipation. In principle, all the stress values are admissible, but those outside the elastic domain are penalized by an additional term $P(\cdot)$ directly proportional to the penalty factor $1/\eta$ (with η as the viscosity coefficient)

$$\begin{array}{l} \min_{\phi(\boldsymbol{\sigma}^*)=0}\{-D^p(\boldsymbol{\sigma}^*)\} \Leftrightarrow \min_{\forall \boldsymbol{\sigma}^*}[-D^{vp}_\eta(\boldsymbol{\sigma}^*)] \; ; \\ D^{vp}_\eta[\boldsymbol{\sigma}] = \boldsymbol{\sigma} \cdot \dot{\boldsymbol{\epsilon}}^{vp} + \frac{1}{\eta} P(\phi(\boldsymbol{\sigma})) \end{array} \tag{3.248}$$

We replace in this manner the constrained minimization problem, which pertains to the principle of maximum dissipation for standard plasticity, by a penalty version of the same problem, where the plastic admissibility constraint is relaxed and handled by the penalty method. We make the simplest choice of the penalty term in terms of a quadratic functional (see Figure 3.23) placing higher penalty on the stress states which are further outside the elastic domain.

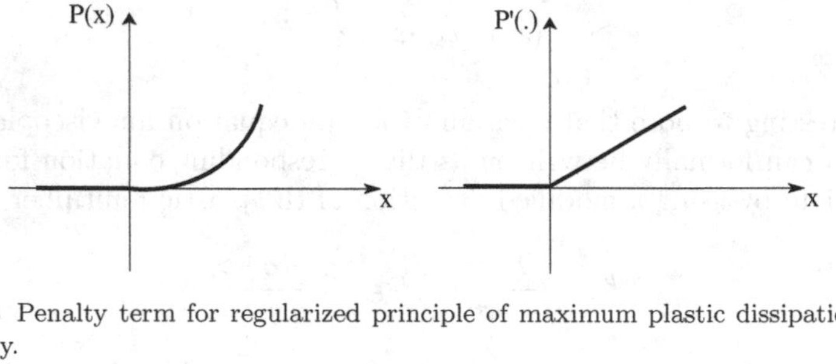

Fig. 3.23 Penalty term for regularized principle of maximum plastic dissipation in viscoplasticity.

$$P(x) = \begin{cases} \frac{1}{2}x^2 \; ; \; x \geq 0 \\ 0 \; ; \; x < 0 \end{cases} \implies \frac{d}{dx} P'(x) = < x > = \begin{cases} x \; ; \; x \geq 0 \\ 0 \; ; \; x < 0 \end{cases} \tag{3.249}$$

The Kuhn-Tucker optimality conditions for the minimization problem defined in (3.248) will lead to the same evolution equation as the one presented in (3.244),

$$\mathbf{0} = \frac{\partial D_\eta^{vp}}{\partial \boldsymbol{\sigma}} := -\dot{\boldsymbol{\epsilon}}^{vp} + \frac{<\phi>}{\eta}\frac{\partial \phi}{\partial \boldsymbol{\sigma}} \tag{3.250}$$

The operator split solution procedure for viscoplasticity remains practically the same as the one already presented for standard plasticity model. The only difference concerns the local phase of computation for the plastic multiplier, where the corresponding value for viscoplasticity is computed directly from the trial value of stress

$$\gamma_{n+1} := \frac{h}{\eta} \underbrace{<\phi_{n+1}>}_{<\phi_{n+1}^{trial}>-2\mu\gamma_{n+1}} \implies \gamma_{n+1} = \frac{<\phi_{n+1}^{trial}>}{2\mu + \eta/h} \tag{3.251}$$

With the computed value of plastic multiplier, we can proceed to computing the final value of stress from its elastic trial value, by using the update which is formally the same as in plasticity:

$$\boldsymbol{\sigma}_{n+1} = \boldsymbol{\sigma}_{n+1}^{trial} - 2\mu\gamma_{n+1}\boldsymbol{\nu}_{n+1}^{trial} \tag{3.252}$$

This computational approach, which fully exploits the formal equivalence with the plasticity model computation, is not the most suitable choice for viscoplasticity with a very small value of viscosity parameter. In the case $\eta \mapsto 0$, the finite rate of viscoplastic deformation will impose that $\phi(\boldsymbol{\sigma}) \mapsto 0$, and the proposed viscoplasticity model will recover the same kind of response as the standard plasticity model. For handling this limit case, an alternative approach proposed by Duvaut and Lions [72] is more suitable, where the viscoplasticity model computation are considered as the appropriate modification of the result delivered by standard plasticity due to stress relaxation. We can thus recast the evolution equation for viscoplastic deformation in terms of stress

$$\left.\begin{array}{c} \dot{\boldsymbol{\sigma}} = \boldsymbol{\mathcal{C}}(\dot{\boldsymbol{\epsilon}} - \dot{\boldsymbol{\epsilon}}^{vp}) \\ \dot{\boldsymbol{\epsilon}}^{vp} = \frac{1}{\tau}\boldsymbol{\mathcal{C}}^{-1}[\boldsymbol{\sigma} - \boldsymbol{\sigma}_\infty] \end{array}\right\} \Leftrightarrow \dot{\boldsymbol{\sigma}} + \frac{1}{\tau}\boldsymbol{\sigma} = \boldsymbol{\mathcal{C}}\dot{\boldsymbol{\epsilon}} + \frac{1}{\tau}\boldsymbol{\sigma}_\infty \tag{3.253}$$

where $\tau = \eta/2\mu$ is the relaxation period and $\boldsymbol{\sigma}_\infty = \sigma_y \mathbf{n}$ is the stress relaxation limit which would have been produced by the standard plasticity model for the same deformation. The stress computation can thus be carried out by integrating the last equation by the implicit Euler scheme:

$$\begin{array}{c} \boldsymbol{\sigma}_{n+1} = \boldsymbol{\sigma}_n + \boldsymbol{\mathcal{C}}\Delta\boldsymbol{\epsilon}_{n+1} - \frac{h}{\tau}\boldsymbol{\sigma}_{n+1} + \frac{h}{\tau}\boldsymbol{\sigma}_{\infty_{n+1}} \\ \implies \boldsymbol{\sigma}_{n+1} = \frac{1}{1+\frac{h}{\tau}}\boldsymbol{\sigma}_{n+1}^{trial} + \frac{\frac{h}{\tau}}{1+\frac{h}{\tau}}\boldsymbol{\sigma}_{\infty_{n+1}} \end{array} \tag{3.254}$$

The final value of stress in viscoplasticity can be expressed as a linear combination of the trial elastic value and the corresponding stress which would have been furnished by standard plasticity for the same deformation, with weighting factors computed as a function of the time step and the relaxation period. The consistent viscoplastic tangent modulus can be computed as the corresponding linear combination, which can be written:

$$\begin{aligned}\mathcal{C}^{evp}_{n+1} := \frac{\partial \boldsymbol{\sigma}_{n+1}}{\partial \boldsymbol{\epsilon}_{n+1}} &= \frac{\partial \boldsymbol{\sigma}^{trial}_{n+1}}{\boldsymbol{\epsilon}_{n+1}} \frac{1}{1+\frac{h}{\tau}} + \frac{\frac{h}{\tau}}{1+\frac{h}{\tau}} \frac{\partial \boldsymbol{\sigma}_{n+1}}{\partial \boldsymbol{\epsilon}_{n+1}} \\ &= \frac{1}{1+h/\tau} \mathcal{C} + \frac{h/\tau}{1+h/\tau} \mathcal{C}^{ep}_{n+1}\end{aligned} \tag{3.255}$$

where $\mathcal{C}^{ep}$ is the elastoplastic tangent modulus for standard plasticity model. The last result clearly shows that the viscoplasticity model will provide a stiffer response from the standard plasticity model, since at the same level of loading the former produces a smaller value of inelastic deformation. This is illustrated in a numerical example presented in Figure 3.24, where a cyclic loading is applied to a plate with a hole. The plate is of square shape, with each side equal to $a = 100$, and the hole radius is equal to $r = 10$. The maximum value of applied loading equal to 465 is reached in 1 s, and the same rate is kept in unloading.

3.4.4 Multi-surface plasticity criterion

The plasticity criterion expressed by a yield function $\phi(\boldsymbol{\sigma}) = 0$ can be interpreted as a smooth surface in the stress space. With respect to such an interpretation, it is possible to imagine yet another modification of great practical interest – the plasticity criterion represented by a non-smooth surface (e.g. a yield surface with corners). The latter is formally equivalent to representing a given plasticity criterion in terms of a number of intersecting (smooth) yield surfaces $\phi_i(\boldsymbol{\sigma}) = 0$; $i = 1, 2, \ldots, m$. Whatever interpretation one chooses, we have to deal with the main novelty of an exterior normal to yield surface and the corresponding plastic flow that are no longer unique at each and every point on the yield surface.

The first use of multisurface plasticity criterion is to eliminate the eventual drawbacks of an existing smooth surface model. One example of this kind is the modified version of the Drucker-Prager plasticity criterion, the smooth surface generalization of Mohr-Coulomb criterion applicable to soil materials where cohesion and friction are main inelastic behavior mechanisms; the usual modification consists of adding a supplementary surface or so-called 'cap', in order to avoid an excessive values of stress under hydrostatic pressure loading (e.g. see DiMaggio and Sander [67] or Hoffstetter, Simo and Taylor [108]). We note in passing that it might be possible to bring such a modified model within the framework of standard models, by using a modified form of the cap which shares the same exterior normal with the Drucker-Prager surface at the intersection points (see Dolarevic and Ibrahimbegovic [68]).

Perhaps the most important reason to resort to a multisurface plasticity criterion is in its capability for representing several deformation modes and attribute to each one the most appropriate yield surface governing the

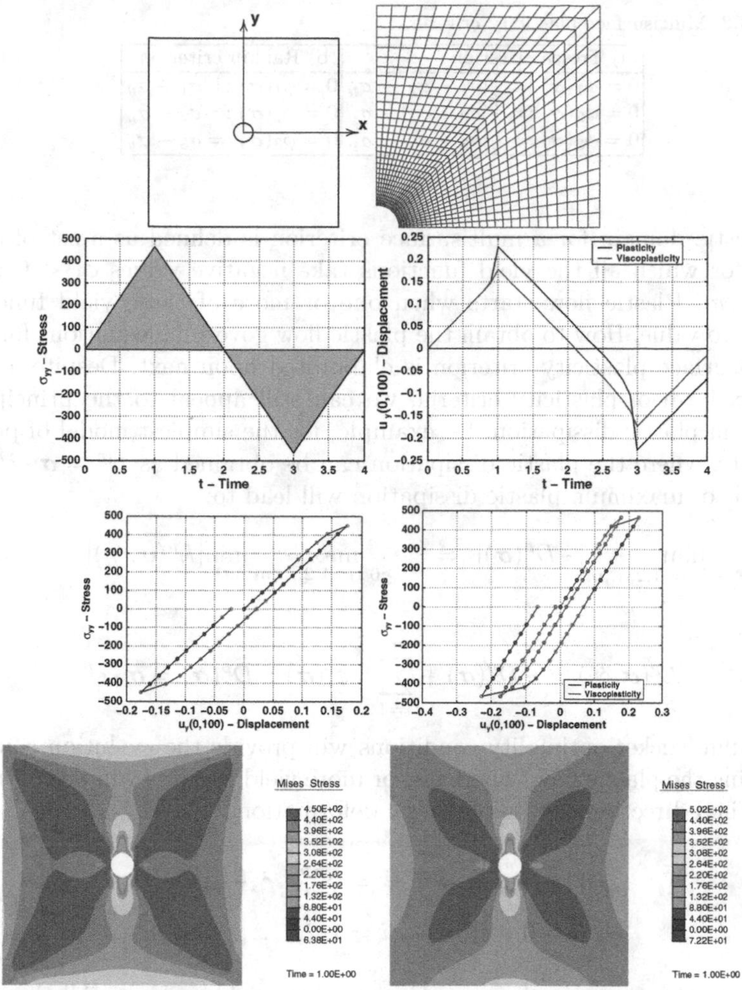

(a) plasticity (b) viscoplasticity

Fig. 3.24 Plate with a hole: response comparison computed by standard plasticity and viscoplasticity model.

inelastic behavior and plastic flow. We can then leave it to such a multisurface criterion to account for eventual interaction of different inelastic mechanisms, which can occur in a given loading program. One example of multisurface plasticity criterion is the one proposed by Tresca, as a historic precursor to von Mises criterion used for metals (e.g. see Hill [103]), where one postulates that the plastic flow starts when any principal value of shear stress reaches the limit of elasticity. Another example of this kind is the multisurface plasticity criterion of Rankin, where the yield functions will place a chosen elasticity limit on each principal value of stress, be it in traction or in compression; we refer to Table 3.3 for exact form of these two multisurface criteria.

Table 3.3 Multisurface plasticity criteria.

(a) Tresca criterion	(b) Rankin criterion
$0 = \phi_1(\boldsymbol{\sigma}) := \frac{1}{2}\lvert\sigma_1 - \sigma_2\rvert - \sigma_y$	$0 = \phi_1(\boldsymbol{\sigma}) := \sigma_1 - \sigma_y$
$0 = \phi_2(\boldsymbol{\sigma}) := \frac{1}{2}\lvert\sigma_2 - \sigma_3\rvert - \sigma_y$	$0 = \phi_2(\boldsymbol{\sigma}) := \sigma_2 - \sigma_y$
$0 = \phi_3(\boldsymbol{\sigma}) := \frac{1}{2}\lvert\sigma_3 - \sigma_1\rvert - \sigma_y$	$0 = \phi_3(\boldsymbol{\sigma}) := \sigma_3 - \sigma_y$

The elastic domain for a multisurface criterion is defined as a set of stress values for which all the yield functions take negative values $\phi_i < 0$; $i = 1, 2, \ldots, m$. Plastic flow starts when one or more of these yield functions reach zero value. How to obtain the plastic flow governing equations for such a multisurface plasticity criterion is elaborated upon next. Despite a more complex form of plasticity criteria, we can still appeal to the principle of maximum plastic dissipation; for example, for the simplest model of perfect plasticity, where the plastic dissipation can be obtained as $D^p = \boldsymbol{\sigma} \cdot \dot{\boldsymbol{\epsilon}}^p$, the principle of maximum plastic dissipation will lead to:

$$\min_{\phi_i(\boldsymbol{\sigma}) \leq 0\,;\, i=1,2,\ldots,m} [-D^p(\boldsymbol{\sigma})] \Leftrightarrow \max_{\dot{\gamma}_i \geq 0\,;\, i=1,2,\ldots,m} \min_{\forall \boldsymbol{\sigma}} [L^p(\boldsymbol{\sigma}, \dot{\gamma}_i)] \; ; \qquad (3.256)$$

where

$$L^p(\boldsymbol{\sigma}, \dot{\gamma}_i) = -D^p(\boldsymbol{\sigma}) + \sum_{i=1}^{m} \dot{\gamma}\, \phi_i(\boldsymbol{\sigma}) \; ; \; D^p(\boldsymbol{\sigma}) = \boldsymbol{\sigma} \cdot \dot{\boldsymbol{\epsilon}}^p$$

The Kuhn-Tucker optimality conditions will provide the evolution equation governing the plastic flow, where one or more yield surfaces will be contributing to final direction of rate of plastic deformation

$$\begin{aligned} 0 = \frac{\partial L^p(\cdot)}{\partial \boldsymbol{\sigma}} \implies \dot{\boldsymbol{\epsilon}}^p = \textstyle\sum_{i=1}^{p} \dot{\gamma}_i \frac{\partial \phi_i(\cdot)}{\partial \boldsymbol{\sigma}} \; ; \\ \dot{\gamma}_i > 0 \; ; \; \phi_i = 0 \; ; \; i = 1, 2, \ldots, p \leq m \end{aligned} \qquad (3.257)$$

where p is the corresponding number of active yield surfaces. All the active surfaces will take zero value for the final, plastically admissible stress state. Recall again that none of the yield surfaces can take a positive value, so that all the remaining, or "inactive" surfaces should remain with a negative value:

$$\dot{\gamma}_i = 0 \; ; \; \phi_i < 0 \; ; \; i = p + 1, \ldots, m \qquad (3.258)$$

The evolution equation in (3.257) is often attributed to Koiter [150]. It states that the direction of the plastic flow at each intersection point between different yield surfaces is obtained as a linear combination of exterior normals to each of the intersecting surfaces, each one scaled accordingly by its own Lagrange multiplier. The corresponding value of each Lagrange multiplier is obtained from the consistency condition, which requires that any stress state on yield surface satisfying (3.257) at time 't' be followed at subsequent time by another plastically admissible stress state. The consistency condition for the case of plastic step will then lead to a system of equations with plastic multipliers $\dot{\gamma}_j$ as unknowns:

$$\begin{aligned} 0 = \dot{\phi}_i &:= \textstyle\sum_{i=1}^{p} \frac{\partial \phi_i}{\partial \boldsymbol{\sigma}} \cdot \dot{\boldsymbol{\sigma}} \\ &= \textstyle\sum_{i=1}^{p} [\frac{\partial \phi_i}{\partial \boldsymbol{\sigma}} \cdot \boldsymbol{\mathcal{C}} \dot{\boldsymbol{\epsilon}} - \sum_{j=1}^{p} \dot{\gamma}_j \frac{\partial \phi_i}{\partial \boldsymbol{\sigma}} \cdot \boldsymbol{\mathcal{C}} \frac{\partial \phi_j}{\partial \boldsymbol{\sigma}}] \\ &\Longrightarrow \textstyle\sum_{j=1}^{p} \underbrace{\frac{\partial \phi_i}{\partial \boldsymbol{\sigma}} \cdot \boldsymbol{\mathcal{C}} \frac{\partial \phi_j}{\partial \boldsymbol{\sigma}}}_{G_{ij}} \dot{\gamma}_j = - \frac{\partial \phi_i}{\partial \boldsymbol{\sigma}} \cdot \boldsymbol{\mathcal{C}} \dot{\boldsymbol{\epsilon}} \; ; \; \forall i \in [1, p] \end{aligned} \tag{3.259}$$

The system of equations in (3.259) will have a unique solution if the matrix of the system is invertible; the latter is the case only if all the active surfaces are linearly independent. The solution can be written:

$$\dot{\gamma}_j = -G_{ij}^{-1} (\frac{\partial \phi_i}{\partial \boldsymbol{\sigma}} \cdot \boldsymbol{\mathcal{C}} \dot{\boldsymbol{\epsilon}}) \tag{3.260}$$

where G_{ij}^{-1} is the inverse of the matrix G_{ij}. By exploiting the last result, we can obtain the stress rate constitutive equations, which can be written:

$$\dot{\boldsymbol{\sigma}} = \boldsymbol{\mathcal{C}}^{ep} \dot{\boldsymbol{\epsilon}} \; ; \; \boldsymbol{\mathcal{C}}^{ep} := \begin{cases} \boldsymbol{\mathcal{C}} \; ; \; \forall \dot{\gamma}_i = 0 \, ; \, i = 1, 2, \ldots, m \\ \boldsymbol{\mathcal{C}} - \sum_i \sum_j G_{ij}^{-1} \boldsymbol{\mathcal{C}} \frac{\partial \phi_i}{\partial \boldsymbol{\sigma}} \otimes \boldsymbol{\mathcal{C}} \frac{\partial \phi_j}{\partial \boldsymbol{\sigma}} \; ; \; \forall \dot{\gamma}_i, \dot{\gamma}_j > 0 \end{cases} \tag{3.261}$$

The time integration of the last equation is carried out by using the implicit backward Euler scheme. The central problem in multisurface plasticity for advancing the computation of the state variables over a typical time step, is again simplified by the operator split solution procedure to global and local phase. In multisurface plasticity the latter implies solving for unknown values of plastic multipliers for all active yield surfaces. In terms of interpretation given by catching up algorithm, this problem is reduced to minimizing the complementary energy functional specifying the projection of elastic trial stress on the plastically admissible domain:

$$\chi(\boldsymbol{\sigma}) = \min_{\phi_i(\sigma^*) \leq 0 \, i=1,2,\ldots,p} \chi(\boldsymbol{\sigma}^*) \; ; \; \chi(\boldsymbol{\sigma}^*) := \frac{1}{2} (\boldsymbol{\sigma}^{trial} - \boldsymbol{\sigma}^*) \cdot \boldsymbol{\mathcal{C}}^{-1} (\boldsymbol{\sigma}^{trial} - \boldsymbol{\sigma}^*) \tag{3.262}$$

This constrained minimization problem can further be transformed into its unconstrained counterpart, by appealing to the method of Lagrange multipliers

$$\begin{aligned} &L(\boldsymbol{\sigma}, \gamma_i) = \max_{\gamma_i^* \geq 0} \min_{\forall \sigma^*} L(\boldsymbol{\sigma}^*, \gamma_i) \; ; \\ &L(\boldsymbol{\sigma}, \gamma_i^*) = \chi(\boldsymbol{\sigma}) + \textstyle\sum_{i=1}^{p} \gamma_i \phi_i(\boldsymbol{\sigma}) \; ; \; i = 1, 2, \ldots, p \leq m \end{aligned} \tag{3.263}$$

where p is the number of active surfaces. The optimality conditions for the minimization problem in (3.263) will result with a system of equations with stress components and plastic multipliers as unknowns:

$$\begin{aligned} \mathbf{0} &= \frac{\partial L}{\partial \boldsymbol{\sigma}} := \frac{\partial \chi}{\partial \boldsymbol{\sigma}} + \sum_i \gamma_i \frac{\partial \phi_i}{\partial \boldsymbol{\sigma}} \\ 0 &= \frac{\partial L}{\partial \gamma_i} := \phi_i \; ; \; i = 1, 2, \ldots, p \leq m \end{aligned} \tag{3.264}$$

The solution of system in (3.264) above can be obtained by Newton's iterative method, where at each iteration we solve the corresponding set of linearized equations:

$$\begin{bmatrix} \left(\frac{\partial^2 \chi}{\partial \boldsymbol{\sigma}^2} + \sum_{i=1}^{p} \gamma_i \frac{\partial^2 \phi_i}{\partial \boldsymbol{\sigma}^2}\right) & \frac{\partial \phi_1}{\partial \boldsymbol{\sigma}} & .. & \frac{\partial \phi_m}{\partial \boldsymbol{\sigma}} \\ \frac{\partial \phi_1}{\partial \boldsymbol{\sigma}} & 0 & .. & 0 \\ .. & .. & .. & .. \\ \frac{\partial \phi_p}{\partial \boldsymbol{\sigma}} & 0 & .. & 0 \end{bmatrix} \begin{pmatrix} \Delta \boldsymbol{\sigma}^{(k)} \\ \Delta \gamma_1^{(k)} \\ .. \\ \Delta \gamma_p^{(k)} \end{pmatrix} = - \begin{pmatrix} \frac{\partial L}{\partial \boldsymbol{\sigma}} \\ \phi_1 \\ .. \\ \phi_m \end{pmatrix} \tag{3.265}$$

and carry on with the corresponding update of the stress and plastic multipliers:

$$\Longrightarrow \begin{aligned} \boldsymbol{\sigma}_{n+1}^{(k+1)} &= \boldsymbol{\sigma}_{n+1}^{(k)} + \Delta \boldsymbol{\sigma}^{(k)} \\ \gamma_1^{(k+1)} &= \gamma_1^{(k)} + \Delta \gamma_1^{(k)} \\ &.. \\ \gamma_p^{(k+1)} &= \gamma_p^{(k)} + \Delta \gamma_p^{(k)} \end{aligned} \tag{3.266}$$

After the convergence of this computation, we also have to obtain the consistent elastoplastic tangent modulus. For the present case of multisurface plasticity, the latter can be written:

$$d\boldsymbol{\sigma}_{n+1} = \boldsymbol{\mathcal{C}}^{ep} d\boldsymbol{\epsilon}_{n+1} \; ; \; \boldsymbol{\mathcal{C}}^{ep} = \boldsymbol{\Xi} - \sum_{i,j} G_{ij}^{-1} \boldsymbol{\Xi} \frac{\partial \phi_i}{\partial \boldsymbol{\sigma}} \otimes \boldsymbol{\Xi} \frac{\partial \phi_j}{\partial \boldsymbol{\sigma}} \tag{3.267}$$

were $\boldsymbol{\Xi} = [\frac{\partial^2 \chi}{\partial \boldsymbol{\sigma}^2} + \sum_i \dot{\gamma}_i \frac{\partial^2 \phi}{\partial \boldsymbol{\sigma}^2}]^{-1}$. It is only the final converged solution to (3.264) that will confirm if a particular surface indeed remains active (with a positive value of plastic multiplier) or not. If we obtain a negative value of plastic multiplier, it is the sure sign that the particular surface should not have been considered as active. The choice of iterative strategy and the number of active surfaces are very important for robustness of the computations in local phase. In general, the number of active surfaces should not be changed during iterative procedure, since such a strategy can prevent convergence.

3.4.4.1 Saint-Venant multisurface plasticity criterion in deformation space

The only manner to guarantee the robustness of computations in local phase of the operator split procedure for a multisurface plasticity criterion is brought by a detailed analysis of the proposed criterion that allows to decide in advance which yield surfaces will finally remain active. One such development is presented in this section for a 2D form of Saint-Venant plasticity criterion. This kind of multisurface criterion allows us to define the plasticity threshold for each principal value of elastic deformation, which leads to plasticity criterion in strain space defined by two surfaces:

$$\begin{aligned} 0 &\geq \phi_1^\epsilon(\boldsymbol{\epsilon}^e) := \epsilon_1^e - \epsilon_y \\ 0 &\geq \phi_2^\epsilon(\boldsymbol{\epsilon}^e) := \epsilon_2^e - \epsilon_y \end{aligned} \qquad (3.268)$$

If we suppose that the elastic response is isotropic, the principal directions of the elastic strain tensor and the stress tensor will coincide,

$$\boldsymbol{\epsilon}^e = \sum_{i=1}^{2} \epsilon_i^e \mathbf{n}_i \otimes \mathbf{n}_i \ ; \ \boldsymbol{\sigma} = \sum_{i=1}^{2} \sigma_i \mathbf{n}_i \otimes \mathbf{n}_i \qquad (3.269)$$

By assuming moreover the linear elastic response, we can express the constitutive response in terms of principal values of stress and strain, by using the Lame parameters, $\bar{\lambda}$ and μ, for example

$$\begin{bmatrix} \epsilon_1^e \\ \epsilon_2^e \end{bmatrix} = \frac{1}{4\mu(\bar{\lambda}+\mu)} \begin{bmatrix} \bar{\lambda}+2\mu & -\bar{\lambda} \\ -\bar{\lambda} & \bar{\lambda}+2\mu \end{bmatrix} \begin{bmatrix} \sigma_1 \\ \sigma_2 \end{bmatrix} \qquad (3.270)$$

where

$$\bar{\lambda} = \begin{cases} \lambda \ ; \ \text{plane strain} \\ 2\lambda\mu/(\lambda+2\mu) \ ; \ \text{plane stress} \end{cases}$$

With these results in hand, we can express the Saint-Venant multisurface plasticity criterion directly in stress space, in terms of the principal values of stress tensor:

$$\begin{aligned} 0 &\geq \phi_1(\boldsymbol{\sigma}) := \tfrac{\bar{\lambda}+2\mu}{2\mu}\sigma_1 - \tfrac{\bar{\lambda}}{2\mu}\sigma_2 - \sigma_y \\ 0 &\geq \phi_2(\boldsymbol{\sigma}) := -\tfrac{\bar{\lambda}}{2\mu}\sigma_1 + \tfrac{\bar{\lambda}+2\mu}{2\mu}\sigma_2 - \sigma_y \end{aligned} \qquad (3.271)$$

where we have chosen the reference value of the elasticity limit obtained from a bi-axial tension test with:

$$\sigma_1 = \sigma_2 = \sigma \implies 0 \geq \phi_1(\sigma) \equiv \phi_2(\sigma) := \sigma - \sigma_y \qquad (3.272)$$

For the multisurface Saint-Venant plasticity criterion, we are able to construct a clear graphic interpretation for the consistency condition in (3.259) , which allows us to decide upfront if only one or both yield surfaces will remain active. To that end, we first note that in the case where both surfaces are active, the plastic flow equation provides corresponding components of plastic deformation rate $\dot{\boldsymbol{\epsilon}}^p$ in the reference frame with base vectors $\frac{\partial \phi}{\partial \boldsymbol{\sigma}}$,

$$\dot{\boldsymbol{\epsilon}}^p = \sum_{i=1}^{2} \dot{\gamma}_i \mathbf{g}_i \ ; \ \begin{cases} \mathbf{g}_1 := \frac{\partial \phi_1}{\partial \boldsymbol{\sigma}} = \frac{\bar{\lambda}+2\mu}{2\mu}\mathbf{n}_1 \otimes \mathbf{n}_1 - \frac{\bar{\lambda}}{2\mu}\mathbf{n}_2 \otimes \mathbf{n}_2 \\ \mathbf{g}_2 := \frac{\partial \phi_2}{\partial \boldsymbol{\sigma}} = -\frac{\bar{\lambda}}{2\mu}\mathbf{n}_1 \otimes \mathbf{n}_1 + \frac{\bar{\lambda}+2\mu}{2\mu}\mathbf{n}_2 \otimes \mathbf{n}_2 \end{cases} \qquad (3.273)$$

The last result allows us to provide the appropriate interpretation of the consistency condition in (3.259), as the equality of the strain rate tensor components,[11] between the total deformation $\dot{\boldsymbol{\epsilon}}$ and the plastic deformation $\dot{\boldsymbol{\epsilon}}^p$, which are expressed in the reference frame with base vectors $\boldsymbol{\mathcal{C}}\frac{\partial \phi}{\partial \boldsymbol{\sigma}}$; this leads to:

$$\boldsymbol{\mathcal{C}}\frac{\partial \phi_i}{\partial \boldsymbol{\sigma}} \cdot \underbrace{\sum_{j=1}^{2} \dot{\gamma}_j \frac{\partial \phi_j}{\partial \boldsymbol{\sigma}}}_{\dot{\boldsymbol{\epsilon}}^p} = \boldsymbol{\mathcal{C}}\frac{\partial \phi_i}{\partial \boldsymbol{\sigma}} \cdot \dot{\boldsymbol{\epsilon}} \; ; \; i \in [1,2] \; ; \; \begin{cases} \boldsymbol{\mathcal{C}}\frac{\partial \phi_1}{\partial \boldsymbol{\sigma}} := 2(\bar{\lambda}+\mu)\mathbf{n}_1 \otimes \mathbf{n}_1 \\ \boldsymbol{\mathcal{C}}\frac{\partial \phi_2}{\partial \boldsymbol{\sigma}} = 2(\bar{\lambda}+\mu)\mathbf{n}_2 \otimes \mathbf{n}_2 \end{cases} \tag{3.274}$$

The plastic consistency condition for the Saint-Venant criterion in the case when two surfaces are active will impose that the elastic deformation components should not change in the direction $\mathbf{n}_i \otimes \mathbf{n}_i$. This conclusion is in complete agreement with the chosen Saint-Venant yield surfaces in strain space defined in (3.268); we refer to Figure 3.25 for the illustration of the elastic domain M^- and two reference frames used for computing the components of stress or different strain tensors. In the same figure, we denoted by M^+ the domain where both components of $\dot{\boldsymbol{\epsilon}}^p$ are positive imposing that both yield surfaces are declared active initially by an elastic trial step, as opposed to the domain Γ^+ where two surfaces are revealed as active finally at the end of the local phase of computation. We note that any elastic trial stress state which is placed in the region denoted as (A), which represents the difference between Γ^+ and M^+, would send a false signal of only one surface being active.

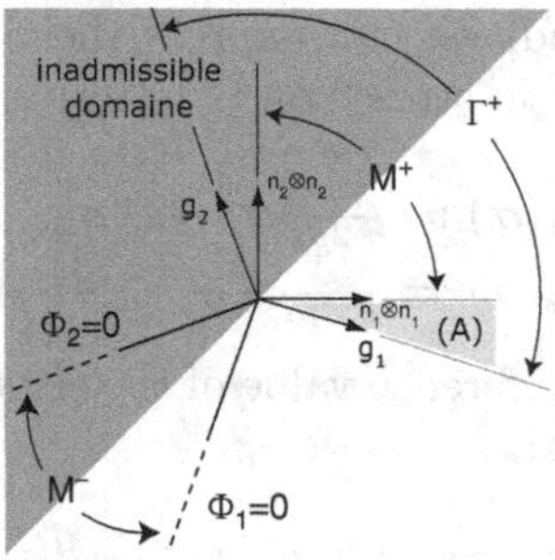

Fig. 3.25 Multisurface Saint-Venant plasticity criterion in principal directions of elastic strain and stress, with the admissible domain ($\epsilon_1^e \geq \epsilon_2^e$).

We can employ the Saint-Venant plasticity criterion to represent the different behavior in compression and in tension, which is characteristic of brittle materials, such as concrete or rocks. Namely, according to this plasticity criterion

[11] The consistency condition for the stress state on the yield surface for perfect plasticity requires that the complete total strain increment be transformed into the plastic strain increment.

the limit of elasticity for a simple tension test will not be the same as the elasticity limit in compression, with the ratio between the two limits equal to $1 + 2\mu/\bar{\lambda}$,

$$\begin{aligned} \sigma_1 = \sigma > 0 \; ; \; \sigma_2 = 0 &\Longrightarrow \phi_1(\sigma_1) := \tfrac{\bar{\lambda}+2\mu}{2\mu}\sigma_1 - \sigma_y \\ \sigma_1 = 0 \; ; \; \sigma_2 = -\sigma < 0 &\Longrightarrow \phi_1(\sigma_2) := -\tfrac{\bar{\lambda}}{2\mu}\sigma_2 - \sigma_y \end{aligned} \tag{3.275}$$

The multisurface Saint-Venant plasticity criterion can be adapted (see Colliat, Ibrahimbegovic and Davenne [56]) for representing the inelastic micro-cracking mechanisms in compression and in traction (see Figure 3.26), by using the appropriate value of fracture energy for each of these two modes, G_c or G_t. The fracture energies G_c and G_t are defined as the amount of energy which must be input in order to drive the stress to zero in compression and in tension, respectively; by interpolating between these two values, we can obtain a modified form of Saint-Venant plasticity criterion where the plasticity threshold evolution is governed by a softening internal variable:

$$\hat{q}(\zeta) := \sigma_y[1 - exp(-\frac{\sigma_y}{G_f}\zeta] \; ; \; G_f = \frac{G_c + G_t}{2} - \frac{G_c - G_t}{2} tanh(tr[\epsilon^e]) \tag{3.276}$$

We note that the mechanism of inelastic deformation with micro-cracking will be the same in each case (see Figure 3.26), but that the corresponding amount of energy can be quite different because of a typically large difference in number of cracks.

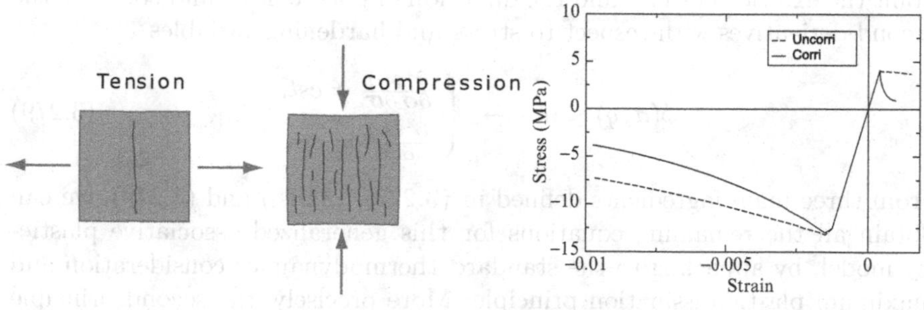

Fig. 3.26 Mechanisms of micro-cracking in compression and in tension, represented by multisurface Saint-Venant plasticity criterion, and resulting stress–strain diagram with G_f corrected (solid line) and uncorrected (dashed-line).

3.4.5 Plasticity model with nonlinear elastic response

The last modification of the standard plasticity model that we describe in this section pertains to its ability to account for a nonlinear elastic response.

This feature can be very useful for modelling the elastoplastic behavior of materials like soils of concrete, as well as a number of other materials in the large strain regime. Even if the elastic response is now assumed to be nonlinear, it is still independent on plastic flow, which again leads to the additive decomposition of the total deformation into elastic and plastic part:

$$\boldsymbol{\epsilon} = \boldsymbol{\epsilon}^e + \boldsymbol{\epsilon}^p \tag{3.277}$$

Moreover, the strain energy can again be expressed as a function of elastic deformation only, but contrary to all other refined plasticity models we no longer assume this strain energy to be a quadratic form in elastic strain. In order to ensure a robust performance of Newton's iterative method in the global phase of the operator split solution procedure, we will also postulate that the second derivative of the strain energy can be computed for any value of elastic strain to define the tangent elasticity tensor. The development of the present plasticity model remains within the framework of generalized standard materials and we can also admit a nonlinear isotropic hardening model with the corresponding hardening modulus which is not constant:

$$\psi(\boldsymbol{\epsilon}^e, \zeta) \longrightarrow \begin{cases} \frac{\partial^2 \psi(\cdot)}{\partial \boldsymbol{\epsilon}^e \partial \boldsymbol{\epsilon}^e} \neq cst. \\ \frac{\partial^2 \psi(\cdot)}{\partial \zeta^2} \neq cst. \end{cases} \tag{3.278}$$

The final model ingredient we ought to choose is the yield function; for generality, the latter can no longer be reduced to a quadratic form in stress like for all other plasticity models studied in the foregoing. However, we suppose the yield surface to be sufficiently smooth in order to be able to define at each point the exterior normal (and the direction of plastic flow) and compute the second derivatives with respect to stress and hardening variables

$$\phi(\boldsymbol{\sigma}, q) \leq 0 \longrightarrow \begin{cases} \frac{\partial^2 \phi(\cdot)}{\partial \boldsymbol{\sigma} \partial \boldsymbol{\sigma}} \neq cst. \\ \frac{\partial^2 \phi(\cdot)}{\partial q^2} \neq cst. \end{cases} \tag{3.279}$$

From three main ingredients defined in (3.277), (3.278) and (3.279), we can obtain all the remaining equations for this generalized associative plasticity model, by appealing to the standard thermodynamics consideration and maximum plastic dissipation principle. More precisely, the second principle of thermodynamics allows to obtain not only the explicit form of constitutive equations but also the reduced form of the plastic dissipation, which can be written:

$$\begin{aligned} 0 \leq D &:= \boldsymbol{\sigma} \cdot \dot{\boldsymbol{\epsilon}} - \dot{\psi}(\boldsymbol{\epsilon}^e) \\ &= (\boldsymbol{\sigma} - \tfrac{\partial \phi(\cdot)}{\partial \boldsymbol{\epsilon}^e}) \cdot \dot{\boldsymbol{\epsilon}}^e + \boldsymbol{\sigma} \cdot \dot{\boldsymbol{\epsilon}}^p - \tfrac{\partial \phi(\cdot)}{\partial \zeta} \dot{\zeta} \\ &\Longrightarrow \begin{cases} \boldsymbol{\sigma} = \frac{\partial \psi(\cdot)}{\partial \boldsymbol{\epsilon}^e} \\ q = -\frac{\partial \psi(\cdot)}{\partial \zeta} \\ 0 \leq D^p := \boldsymbol{\sigma} \cdot \dot{\boldsymbol{\epsilon}}^p + q\dot{\zeta} \end{cases} \end{aligned} \tag{3.280}$$

With these results in hand, we can further obtain the evolution equations of the internal variables and the corresponding loading/unloading conditions simply by appealing to the principle of maximum plastic dissipation

$$\max_{\dot\gamma \geq 0} \min_{\forall \boldsymbol{\sigma}} [-D^p(\boldsymbol{\sigma}) + \dot\gamma \phi(\boldsymbol{\sigma})] \implies \begin{array}{c} \dot{\boldsymbol{\epsilon}}^p = \dot\gamma \frac{\partial \phi}{\partial \boldsymbol{\sigma}} \\ \dot\zeta = \dot\gamma \frac{\partial \phi}{\partial q} \\ \dot\gamma \geq 0\,;\ \phi(\boldsymbol{\sigma}) \leq 0\,;\ \dot\gamma \phi(\boldsymbol{\sigma}) = 0 \end{array} \tag{3.281}$$

The plastic consistency condition, imposing that the stress evolution from the yield surface remains in agreement with the yield surface subsequent evolution, will allow to compute the positive value of plastic multiplier $\dot\gamma > 0$:

$$\begin{aligned} 0 = \dot\phi(\cdot) &:= \frac{\partial \phi}{\partial \boldsymbol{\sigma}} \cdot \dot{\boldsymbol{\sigma}} + \frac{\partial \phi}{\partial q} \dot q \\ &= \frac{\partial \phi}{\partial \boldsymbol{\sigma}} \cdot \frac{\partial^2 \psi}{\partial \boldsymbol{\epsilon}^e \partial \boldsymbol{\epsilon}^e} \dot{\boldsymbol{\epsilon}}^e - \frac{\partial \phi}{\partial q} \frac{\partial^2 \psi}{\partial \zeta} \dot\zeta \\ &= \frac{\partial \phi}{\partial \boldsymbol{\sigma}} \cdot \frac{\partial^2 \psi}{\partial \boldsymbol{\epsilon}^e \partial \boldsymbol{\epsilon}^e} \dot{\boldsymbol{\epsilon}} - \dot\gamma \left(\frac{\partial \phi}{\partial \boldsymbol{\sigma}} \cdot \frac{\partial^2 \psi}{\partial \boldsymbol{\epsilon}^e \partial \boldsymbol{\epsilon}^e} \frac{\partial \phi}{\partial \boldsymbol{\sigma}} + \frac{\partial \phi}{\partial q} \frac{\partial^2 \psi}{\partial \zeta} \frac{\partial \phi}{\partial q} \right) \\ \implies \dot\gamma &= \frac{\left(\frac{\partial \phi}{\partial \boldsymbol{\sigma}} \cdot \frac{\partial^2 \psi}{\partial \boldsymbol{\epsilon}^e \partial \boldsymbol{\epsilon}^e} \dot{\boldsymbol{\epsilon}} \right)}{\left(\frac{\partial \phi}{\partial \boldsymbol{\sigma}} \cdot \frac{\partial^2 \psi}{\partial \boldsymbol{\epsilon}^e \partial \boldsymbol{\epsilon}^e} \frac{\partial \phi}{\partial \boldsymbol{\sigma}} + \frac{\partial \phi}{\partial q} \frac{\partial^2 \psi}{\partial \zeta} \frac{\partial \phi}{\partial q} \right)} \end{aligned} \tag{3.282}$$

With this result in hand, we can also write stress rate constitutive equation, which will be nonlinear both in elastic ($\dot\gamma = 0$) and in plastic phase ($\dot\gamma > 0$):

$$\dot{\boldsymbol{\sigma}} = \begin{cases} \frac{\partial^2 \psi}{\partial \boldsymbol{\epsilon}^e \partial \boldsymbol{\epsilon}^e} \dot{\boldsymbol{\epsilon}}\ ;\ \dot\gamma = 0 \\ \left[\frac{\partial^2 \psi}{\partial \boldsymbol{\epsilon}^e \partial \boldsymbol{\epsilon}^e} - \frac{1}{\left(\frac{\partial \phi}{\partial \boldsymbol{\sigma}} \cdot \frac{\partial^2 \psi}{\partial \boldsymbol{\epsilon}^e \partial \boldsymbol{\epsilon}^e} \frac{\partial \phi}{\partial \boldsymbol{\sigma}} + \frac{\partial \phi}{\partial q} \frac{\partial^2 \psi}{\partial \zeta} \frac{\partial \phi}{\partial q} \right)} \left(\frac{\partial^2 \psi}{\partial \boldsymbol{\epsilon}^e \partial \boldsymbol{\epsilon}^e} \frac{\partial \phi}{\partial \boldsymbol{\sigma}} \right) \right. \\ \left. \otimes \left(\frac{\partial^2 \psi}{\partial \boldsymbol{\epsilon}^e \partial \boldsymbol{\epsilon}^e} \dot{\boldsymbol{\epsilon}} \frac{\partial \phi}{\partial \boldsymbol{\sigma}} \right) \right] \dot{\boldsymbol{\epsilon}}\ ;\ \dot\gamma > 0 \end{cases} \tag{3.283}$$

Time integration of the evolution equation for internal variables can carried out by using the implicit backward Euler scheme, leading to:
Central problem in plasticity with nonlinear elastic response
Given: $\boldsymbol{\epsilon}_n$, $\boldsymbol{\epsilon}^e_n$, ζ_n, $h = t_{n+1} - t_n$
Find: $\boldsymbol{\epsilon}_{n+1}$, $\boldsymbol{\epsilon}^e_{n+1}$, ζ_{n+1}
such that: $\dot\gamma_{n+1} \geq 0\ ;\ \phi_{n+1} \leq 0\ ;\ \dot\gamma_{n+1}\phi_{n+1} = 0$
$G(\cdot|_{n+1}; \mathbf{w}) = 0$
The main difference from the standard plasticity model (with the von Mises plasticity criterion) is the choice of elastic rather than plastic deformation as one of the state variables. As shown in Ibrahimbegovic, Gharzeddine and Chorfi [125], the choice of elastic rather than plastic deformation as a state variable is more suitable for the operator split solution procedure, which is described next.
The local phase of the operator split solution procedure concerns the computation of internal variables and plastically admissible stresses in accordance

with the best iterative value of the total deformation, $\boldsymbol{\epsilon}_{n+1}^{(i)} = \boldsymbol{\epsilon}_n + \Delta\boldsymbol{\epsilon}_{n+1}^{(i)}$. The computations are started by assuming that the step remains elastic, with $\dot{\gamma}^{trial} = 0$ and no change of internal variables, which leads to:

$$\left.\begin{array}{l}\boldsymbol{\epsilon}_{n+1}^{e,trial} = \boldsymbol{\epsilon}_n^e + \Delta\boldsymbol{\epsilon}_{n+1}^{(i)} \\ \zeta_{n+1}^{trial} = \zeta_n\end{array}\right\} \Longrightarrow \left\{\begin{array}{l}\boldsymbol{\sigma}_{n+1}^{trial} = \dfrac{\partial\psi(\boldsymbol{\epsilon}_{n+1}^{trial},\cdot)}{\partial\boldsymbol{\epsilon}^e} \\ q_{n+1}^{trial} = -\dfrac{\partial\psi(\cdot,\zeta_{n+1}^{trial})}{\partial\zeta}\end{array}\right. \tag{3.284}$$

If the trial value of the yield function is positive, $\phi(\boldsymbol{\sigma}_{n+1}^{trial}) > 0$, we ought to correct this trial state. We thus have to compute the positive value of the plastic multiplier, $\gamma_{n+1} > 0$, which will establish the plastic admissibility of the stress and ensure that $\phi_{n+1} = 0$; the latter can formally be accomplished as a part of the solution to a set of nonlinear algebraic equations:

$$\begin{bmatrix} \boldsymbol{\epsilon}_{n+1}^e - \boldsymbol{\epsilon}_n^e - \Delta\boldsymbol{\epsilon}_{n+1}^{(i)} + \gamma_{n+1}\dfrac{\partial\phi(\cdot)}{\partial\boldsymbol{\sigma}}_{n+1} \\ \phi(\boldsymbol{\sigma}_{n+1}, q_{n+1}) \\ -\zeta_{n+1} + \zeta_n + \gamma_{n+1}\dfrac{\partial\phi(\cdot)}{\partial q_{n+1}} \end{bmatrix} = \mathbf{0} \tag{3.285}$$

Unlike the standard plasticity model, the plasticity model with nonlinear elastic response does not allow that the solution of the system in (3.285) above be reduced to a single scalar equation ($\hat{\phi}(\gamma_{n+1}) = 0$), followed by a simple update of the internal variables; the iterative sweeps ought to be performed on all equations in (3.285) simultaneously, and Newton's method can be employed providing the sufficient smoothness that allows computation of the second derivatives of functions $\phi(\cdot)$ in (3.278) and $\psi(\cdot)$ in (3.279),

$$\begin{bmatrix} [\frac{\partial^2\psi}{\partial\boldsymbol{\epsilon}^e\partial\boldsymbol{\epsilon}^e}]^{-1} + \gamma_{n+1}\frac{\partial^2\phi}{\partial\boldsymbol{\sigma}\partial\boldsymbol{\sigma}} & \frac{\partial\phi}{\partial\boldsymbol{\sigma}} & \mathbf{0} \\ \frac{\partial\phi}{\partial\boldsymbol{\sigma}} & 0 & \frac{\partial\phi}{\partial q} \\ 0 & \frac{\partial\phi}{\partial q} & [\frac{\partial^2\psi}{\partial\zeta^2}]^{-1} + \gamma_{n+1}\frac{\partial^2\phi}{\partial q^2} \end{bmatrix}_{n+1}^{(k)} \begin{bmatrix} \Delta\boldsymbol{\sigma}_{n+1}^{(k)} \\ \Delta\gamma_{n+1}^{(k)} \\ \Delta q_{n+1}^{(k)} \end{bmatrix}$$
$$= \begin{bmatrix} -\boldsymbol{\epsilon}_{n+1}^e + \boldsymbol{\epsilon}_n^e + \Delta\boldsymbol{\epsilon}_{n+1}^{(i)} + \gamma_{n+1}\frac{\partial\phi}{\partial\boldsymbol{\sigma}}_{n+1} \\ \phi_{n+1} \\ -\zeta_{n+1} + \zeta_n + \gamma_{n+1}\frac{\partial\phi}{\partial q_{n+1}} \end{bmatrix} \tag{3.286}$$

The subsequent iterative values can then be obtained with simple updates:

$$\begin{aligned} \boldsymbol{\sigma}_{n+1}^{(k+1)} &= \boldsymbol{\sigma}_{n+1}^{(k)} + \Delta\boldsymbol{\sigma}_{n+1}^{(k)} \\ \gamma_{n+1}^{(k+1)} &= \gamma_{n+1}^{(k)} + \Delta\gamma_{n+1}^{(k)} \\ q_{n+1}^{(k+1)} &= q_{n+1}^{(k)} + \Delta q_{n+1}^{(k)} \end{aligned} \tag{3.287}$$

Having obtained the convergence of this iterative procedure, we can advance to the global phase of the operator split, which allows us to obtain an improved iterative value of total deformation, $\epsilon_{n+1}^{(i+1)} = \epsilon_n + \Delta\epsilon_{n+1}^{(i+1)}$. This is done by solving the linearized system of equilibrium equations, and updating the total deformation field. With the value of residual already available from stress computation in local phase, the other key result needed pertains to the consistent tangent elastoplastic modulus, which is obtained as the derivative of stress with respect to the total strain. The tangent modulus value depends on the manner in which the stress is computed in the local phase, and it is by far the easiest to complete those computations in the local phase, even though the results might finally not be needed in the global phase if the equilibrium equations happens to be satisfied and no further update of total deformation field is needed. Therefore, at the convergence of the local computation phase, we write a new linearized form of the system (3.285), which is computed with the converged value of internal variables and a new increment of total deformation:

$$\begin{bmatrix} [\frac{\partial^2\psi}{\partial\epsilon^e_{n+1}\partial\epsilon^e_{n+1}}]^{-1} + \gamma_{n+1}\frac{\partial^2\phi}{\partial\boldsymbol{\sigma}_{n+1}\partial\boldsymbol{\sigma}_{n+1}} & \frac{\partial\phi}{\partial\boldsymbol{\sigma}_{n+1}} & \mathbf{0} \\ \frac{\partial\phi}{\partial\boldsymbol{\sigma}_{n+1}} & 0 & \frac{\partial\phi}{\partial q_{n+1}} \\ 0 & \frac{\partial\phi}{\partial q_{n+1}} & [\frac{\partial^2\psi}{\partial\zeta^2_{n+1}}]^{-1} + \gamma_{n+1}\frac{\partial^2\phi}{\partial q^2_{n+1}} \end{bmatrix} \begin{bmatrix} \Delta\boldsymbol{\sigma}_{n+1}^{(k)} \\ \gamma_{n+1} \\ q_{n+1} - q_n \end{bmatrix} = \begin{bmatrix} -\Delta\epsilon_{n+1}^{(i)} \\ 0 \\ 0 \end{bmatrix} \tag{3.288}$$

The static condensation (see Wilson [263]) then allows to reduce the size of the last system and reduce it to a single equation expressing the relation between the strain and stress increment:

$$\begin{aligned} &[\Xi_{n+1}^{-1} + \frac{1}{\hat{K}(\zeta_{n+1})}\frac{\partial\phi}{\partial\boldsymbol{\sigma}_{n+1}} \otimes \frac{\partial\phi}{\partial\boldsymbol{\sigma}_{n+1}}]\,\Delta\boldsymbol{\sigma}_{n+1} = \Delta\epsilon_{n+1}^{(i+1)} \\ &\Xi_{n+1}^{-1} = [\frac{\partial^2\psi}{\partial\epsilon^e_{n+1}\partial\epsilon^e_{n+1}}]^{-1} + \gamma_{n+1}\frac{\partial^2\phi}{\partial\boldsymbol{\sigma}_{n+1}\partial\boldsymbol{\sigma}_{n+1}}\ ; \\ &\hat{K}(\zeta_{n+1}) = \frac{\partial\phi}{\partial q}[(\frac{\partial^2\psi}{\partial\zeta^2_{n+1}})^{-1} + \gamma_{n+1}\frac{\partial^2\phi}{\partial q^2_{n+1}}]\frac{\partial\phi}{\partial q_{n+1}} \end{aligned} \tag{3.289}$$

We obtain in this manner the consistent tangent elastoplastic compliance tensor. The Sherman-Morison formula can then be used to compute the closed-form result for the inverse of this compliance tensor, which exploits its special form as a rank-one update, leading to the tangent elastoplastic modulus:

$$\boldsymbol{C}^{ep}_{n+1} = \Xi_{n+1} - \frac{1}{\frac{\partial\phi}{\partial\boldsymbol{\sigma}_{n+1}}\cdot\Xi\frac{\partial\phi}{\partial\boldsymbol{\sigma}_{n+1}} + \hat{K}(\zeta_{n+1})}\Xi_{n+1}\frac{\partial\phi}{\partial\boldsymbol{\sigma}_{n+1}} \otimes \Xi_{n+1}\frac{\partial\phi}{\partial\boldsymbol{\sigma}_{n+1}} \tag{3.290}$$

The tangent elstoplastic modulus of this kind should be able to provide the quadratic convergence rate of the global set of equilibrium equations.

3.5 Damage models

The basic hypothesis of plasticity models that the elastic response remains the same in loading and unloading is no longer valid if the elastic response is affected by inelastic deformation. This kind of phenomena are indeed observed in loading/uloading cycles of brittle materials, such as ceramics, glass or concrete, where the inelastic behavior pertains to cracks which will also modify the elastic response in the unloading phase with respect to the elastic response of virgin material in loading.
It is the damage model (e.g. see François, Pineau and Zaoui [80], Krajcinovic [153], Lematre and Chaboche [168] and Maugin [185] for classical works, or [3, 31] for more recent works) which is capable to reproduce this kind of inelastic behavior.

3.5.1 1D damage model

The first damage model was introduced by Kachanov in 1958, as a simple idea for representing the cracking phenomena by a continuum mechanics model. The proposed model has exploited a single internal variable of damage, with the zero value representing the virgin material with no cracks which can further increase to 1 for completely damaged material. For a one-dimensional stress state (see Figure 3.27), one can provide the internal damage variable interpretation as the ratio between the damaged cross-section area $\bar{A}$ with respect to the nominal cross-section of the virgin specimen A. Namely, one can define the effective stress $\bar{\sigma}$, as the true stress acting in undamaged material with:

$$\left.\begin{array}{l} A - \bar{A} = (1-d)A \ ; \ 0 \leq d \leq 1 \\ f := \sigma A = \bar{\sigma}(A - \bar{A}) \end{array}\right\} \Longrightarrow \boxed{\bar{\sigma} = \sigma/(1-d)} \qquad (3.291)$$

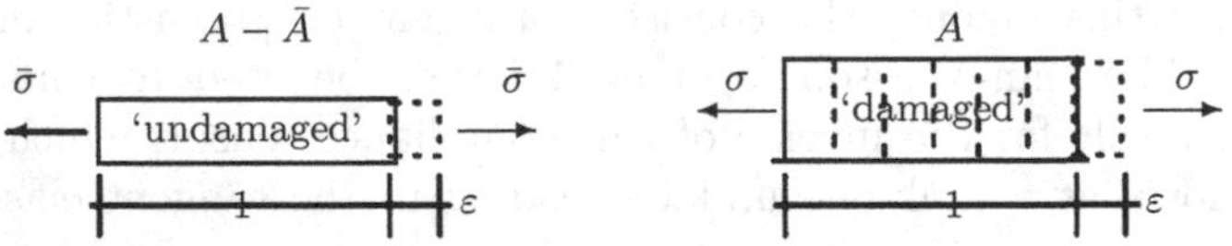

Fig. 3.27 1D damage model – effective stress interpretation.

By the hypothesis of equivalent deformation, the strain response of damaged material is considered the same to the response of undamaged material, when the nominal stress σ is replaced by the effective stress $\bar{\sigma}$; this can be written:

$$\epsilon := \frac{\bar{\sigma}}{E} = \frac{\sigma}{(1-d)E} \Longrightarrow \sigma = (1-d)E\epsilon \tag{3.292}$$

Such a formulation of the damage model is developed in detail by Lematre and Chaboche [168]. In this section, we will carry out the developments of an alternative approach for constructing a damage model ingredients, which is more suitable for generalizing the damage model to an anisotropic case where all the components of the elasticity tensor are not necessarily modified in the same manner by damage. The model is based on the idea proposed by Ortiz [211] where for the internal damage variable we chose the compliance D of a damaged state.[12] The constitutive relation for a damaged state can then be written in the inverse form of (3.292), which can be written:

$$\epsilon = D\sigma \ ; \ D \in [\frac{1}{E}, \infty) \tag{3.293}$$

With such a choice of internal variable, we can place the damage model development in the same framework as already developed for plasticity, with the principal model ingredients which concern the internal energy and damage criterion. For clarity, we start with a simple 1D framework, which allows us to illustrate all the pertinent phases in model formulation and numerical solution of the corresponding boundary value problems. We will show later that the same solution phases apply to 3D framework.
We choose the damage criterion to prescribe the admissible values of stress in the sense of damage with:

$$0 \geq \phi(\sigma, q) = |\sigma| - (\sigma_f - q) \tag{3.294}$$

where σ_f is the corresponding elasticity limit that indicates the first cracking and q is a stress-like hardening variable which handles the damage threshold evolution. Similar to plasticity model, the stress producing a positive value of damage function in (3.294) is considered inadmissible. Any negative value of the damage function $\phi(\boldsymbol{\sigma}, q) < 0$ corresponds to the elastic domain, where no change of the internal damage variable is possible. The damage is produced (resulting with the evolution of internal damage variables) for zero value of damage function $\phi(\boldsymbol{\sigma}, q) = 0$.
The internal energy of such a damage model can formally be written:

$$\psi(\epsilon, D, \zeta) = \frac{1}{2}\epsilon D^{-1}\epsilon + \Xi(\zeta) \ ; \ \Xi(\zeta) = \frac{1}{2}\zeta K \zeta \tag{3.295}$$

where ζ is the internal variable which controls hardening. The three main ingredients given in (3.293), (3.294) and (3.295), respectively, are sufficient to obtain all other equations governing the chosen damage model by using the procedure fully equivalent to the one already presented for plasticity model.

---

[12] For a virgin material, the compliance D is equal to the inverse of elasticity modulus E.

We only need the second principles of thermodynamics and the principle of maximum dissipation.
In order to define the damage dissipation, we use the Legendre transform to exchange the roles between the stress and deformation and introduce the complementary energy potential:

$$\begin{aligned}\chi(\sigma, D, \zeta) &= \sigma\epsilon - \psi(\epsilon, D, \zeta) \\ &= \tfrac{1}{2}\sigma D\sigma - \Xi(\zeta)\end{aligned} \tag{3.296}$$

With this result in hand, we can exploit the second principle of thermodynamics to obtain the explicit form of the damage model dissipation:

$$\begin{aligned}0 \leq D^d &:= \sigma\dot{\epsilon} - \dot{\psi} \\ &= \sigma\dot{\epsilon} - \sigma\dot{\epsilon} - \dot{\sigma}\epsilon + \dot{\chi}(\sigma, D, \zeta) \\ &= \dot{\sigma}(-\epsilon + D\sigma) + \tfrac{1}{2}\sigma\dot{D}\sigma - \tfrac{d\Xi}{d\zeta}\dot{\zeta}\end{aligned} \tag{3.297}$$

For an elastic process, the existing values of internal variables will remain fixed, $\dot{D} = 0$ and $\dot{\zeta} = 0$, and the damage dissipation will be equal to zero, $D^d = 0$. We can thus obtain the appropriate form of constitutive equations for damage model:

$$\epsilon = D\sigma \implies \sigma = D^{-1}\epsilon \; ; \; q = -\frac{d\Xi}{d\zeta} \tag{3.298}$$

By assuming that the same constitutive equations remain valid for an inelastic process, we can use the result of second principle in (3.297) to define the damage dissipation:

$$0 < D^d = \frac{1}{2}\sigma\dot{D}\sigma + q\dot{\zeta} \tag{3.299}$$

We will complete this development constructing the evolution equations of associative damage model, placed in the framework of generalized standard materials, by assuming the maximum of damage dissipation. In other words, among all admissible values of stress in the sense of the chosen damage criterion, we choose the one which will maximize the damage dissipation. This can formally be cast as the following minimization problem:

$$\begin{aligned}&\min_{\phi(\cdot)\leq 0}[-D^d(\sigma, q^d)] \;\Leftrightarrow\; \max_{\dot{\gamma}>0}\min_{\forall(\sigma,q)}[L^d(\sigma, D, \zeta)] \; ; \\ &L^d(\sigma, D, \zeta, \dot{\gamma}) = -D^d(\sigma, D, \zeta) + \dot{\gamma}\phi(\sigma, q)\end{aligned} \tag{3.300}$$

The Kuhn-Tucker optimality conditions for the chosen damage Lagrangian can then provide the evolution equations for the internal variables:

$$\begin{aligned}0 = \tfrac{\partial L}{\partial \sigma} &:= -D\sigma + \dot{\gamma}\tfrac{\partial \phi}{\partial \sigma} \implies \dot{D} = \dot{\gamma}\tfrac{1}{\sigma}\tfrac{\partial \phi}{\partial \sigma} \\ 0 = \tfrac{\partial L}{\partial q} &:= -\dot{\zeta} + \dot{\gamma}\tfrac{\partial \phi}{\partial q} \implies \dot{\zeta} = \dot{\gamma}\tfrac{\partial \phi}{\partial q}\end{aligned} \tag{3.301}$$

The optimality conditions will also include the corresponding loading/-unloading criteria:

$$\dot{\gamma} \leq 0 \ ; \ \phi \leq 0 \ ; \ \dot{\gamma}\phi = 0 \tag{3.302}$$

Finally, by enforcing the stress admissibility at the subsequent time to damage activation, we recover the consistency condition, which is very much equivalent to the one in plasticity. The consistency condition and the following auxiliary result:

$$1 = DD^{-1} \implies \dot{D}^{-1} = -D^{-1}\dot{D}D^{-1} \tag{3.303}$$

will allow us to compute the final value of damage multiplier, $\dot{\gamma} > 0$

$$\begin{aligned} 0 = \dot{\phi} &:= \tfrac{\partial \phi}{\partial \sigma}\dot{\sigma} + \tfrac{\partial \phi}{\partial q}\dot{q} \\ &= \tfrac{\partial \phi}{\partial \sigma}(D^{-1}\dot{\epsilon} + \dot{D}^{-1}\epsilon) - \tfrac{\partial \phi}{\partial q}K\dot{\zeta} \quad \implies \dot{\gamma} = \frac{sign(\sigma)D^{-1}\dot{\epsilon}}{D^{-1}+K} \\ &= \tfrac{\partial \phi}{\partial \sigma}D^{-1}\dot{\epsilon} - \dot{\gamma}[\tfrac{\partial \phi}{\partial \sigma}D^{-1}\tfrac{\partial \phi}{\partial \sigma} + K] \end{aligned} \tag{3.304}$$

With this result in hand, we can easily obtain stress rate constitutive equations for the standard damage model:

$$\dot{\sigma} = \begin{cases} D^{-1}\dot{\epsilon} \ ; \ \dot{\gamma} = 0 \\ \frac{D^{-1}K}{D^{-1}+K}\dot{\epsilon} \ ; \ \dot{\gamma} > 0 \end{cases} \tag{3.305}$$

The graphic illustration of the result in (3.305), which is presented in Figure 3.28, shows clearly two main differences of this kind of damage model from standard plasticity models; the first one concerns the elastic response of the damage model that no longer remains the same in loading and unloading, and the second concerns the absence of any irreversible deformation upon complete unloading. We also note that using a linear hardening law for the damage model results with a nonlinear relationship between the stress and strain. In fact, it is precisely the difference between the nonlinear response in loading and linear response in unloading that is at the origin of dissipation for the damage model.

The computation of internal variables evolution for 1D damage model can be carried out in fully equivalent manner to the one already explained in detail for 1D plasticity. We thus only present here the outline of the main results. The incremental analysis and the implicit backward Euler scheme can reduce the computation over a typical time step to:

Central problem of incremental/iterative analysis for 1D damage model

Given: $\epsilon_n = \epsilon(t_n)$, $D_n = D(t_n)$, $\zeta_n = \zeta(t_n)$, and $h = t_{n+1} - t_n$

Find: ϵ_{n+1}, D_{n+1}, ζ_{n+1}

such that: $\sigma_{n+1} = D_{n+1}^{-1}\epsilon_{n+1}$, $q_{n+1} = -\frac{d\Xi(\zeta_{n+1})}{d\zeta_{n+1}}$

$$\mathop{\mathbb{A}}_{e=1}^{n_{el}} [\mathsf{f}_{n+1}^{e,int} - \mathsf{f}_{n+1}^{e,ext}] = 0 \ ; \ f_{a,n+1}^{e,int} = \int_{l^e} B_a^e \sigma_{n+1}\, dx$$

$$\dot{\gamma}_{n+1} \geq 0 \ ; \ \phi(\sigma_{n+1}, q_{n+1}) \leq 0 \ ; \ \dot{\gamma}_{n+1}\phi(\sigma_{n+1}, q_{n+1}) = 0$$

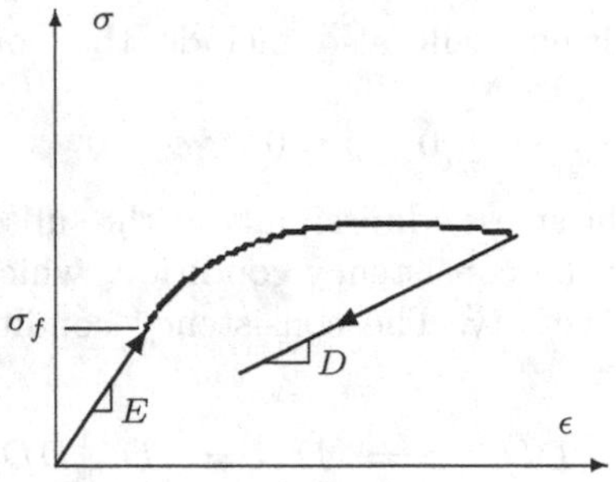

Fig. 3.28 Stress–strain diagram for 1D damage model in loading–unloading cycle.

We use the operator split solution procedure for this central problem in order to separate the global computation of equilibrium equation from local computation providing admissible values of stress in the sense of the chosen damage criterion. The local phase of operator split computation is carried out at each Gauss quadrature point, for the given iterative guess of the total deformation $\epsilon_{n+1}^{(i)}$. We start by assuming the elastic step, which allows us to compute the following trial values of stress:

$$\gamma^{trial} = 0 \implies \begin{array}{l} D_{n+1}^{trial} = D_n \\ \zeta_{n+1} = \zeta_n \end{array} \implies \begin{array}{l} \sigma_{n+1}^{trial} = D_n^{-1} \epsilon_{n+1}^{(i)} \\ q_{n+1}^{d,trial} = -K\zeta_n \end{array} \tag{3.306}$$

We can thus readily obtain the corresponding trial value of the damage function:

$$\implies \phi_{n+1}^{trial} = |\sigma_{n+1}^{trial}| - (\sigma_f - q_{n+1}^{trial}) \tag{3.307}$$

If this value is negative or zero, $\phi_{n+1}^{trial} \leq 0$, the trial is considered as successful and the present step is indeed elastic (with either loading or unloading), with no need to further modify the given values of computed stress or internal variables.
In the opposite case, for a positive value of the damage function $\phi_{n+1}^{d,trial} > 0$, the elastic trial step is not admissible, and we have to compute the true (positive) value of damage multiplier $\dot{\gamma}_{n+1} > 0$ and reestablish the admissibility of stress with $\phi_{n+1} = 0$. This will also produce new values of internal variables computed from evolution equations in (3.301) by the implicit Euler scheme:

$$\begin{array}{ll} D_{n+1} &= D_n + \gamma_{n+1} \frac{1}{|\sigma_{n+1}|} \\ \zeta_{n+1} &= \zeta_n + \gamma_{n+1} \end{array} \tag{3.308}$$

The final, admissible value of stress can also be written as the corresponding modification of its trial value:

$$\begin{aligned} \epsilon_{n+1}^{(i)} &:= D_{n+1}\sigma_{n+1} = [D_n + \gamma_{n+1}\tfrac{1}{|\sigma_{n+1}|}]\sigma_{n+1} \\ \Longrightarrow \underbrace{D_n^{-1}\epsilon_{n+1}^{(i)}}_{\sigma_{n+1}^{trial}} &=: [1 + \gamma_{n+1}\tfrac{1}{D_n|\sigma_{n+1}|}]\sigma_{n+1} \end{aligned} \tag{3.309}$$

The last result allows us to draw the following conclusions: the sign of trial and final stress remains the same (more generally, the direction of damage flow remains the same) and the absolute values of trial and final stress can be related through the damage multiplier according to:

$$\begin{aligned} |\sigma_{n+1}^{trial}| &= |\sigma_{n+1}| + \tfrac{\gamma_{n+1}}{D_n} \\ sign(\sigma_{n+1}) &= sign(\sigma_{n+1}^{trial}) \end{aligned} \tag{3.310}$$

Having established these two results, we can express the damage function for final stress as the corresponding modification of its trial value, which provides the closed-form result for the damage multiplier:

$$\begin{aligned} 0 = \phi_{n+1} &:= |\sigma_{n+1}| - (\sigma_f - q_{n+1}) \\ &= \underbrace{|\sigma_{n+1}^{trial}| - (\sigma_f - q_n)}_{\phi_{n+1}^{trial}} - \gamma_{n+1}(1/D_n + K) \\ \Longrightarrow \gamma_{n+1} &= \tfrac{\phi_{n+1}^{trial}}{1/D_n + K} \end{aligned} \tag{3.311}$$

Having computed the internal variables in the local phase of the operator split solution procedure, we proceed to the global phase to check the equilibrium. If the equilibrium equations are not verified, we will carry out a new iterative sweep providing an iterative improvement of the incremental displacement $\mathsf{u}_{n+1}^{(i)}$, along with an improved value of total strain $\epsilon_{n+1}^{(i+1)}$. If Newton's iterative method is employed for global iterations, we ought to obtain the consistently linearized form of the equilibrium equations:

$$\begin{aligned} &\mathop{\mathbb{A}}_{e=1}^{n_{elem}} \mathsf{K}_{n+1}^{e,(i)} \mathsf{u}_{n+1}^{(i)} = \mathsf{f}_{n+1}^{ext} - \mathop{\mathbb{A}}_{e=1}^{n_{elem}} \mathsf{f}_{n+1}^{e,(i)} \; ; \; K_{ab,n+1}^{e,(i)} = \int_{l^e} B_a^e C_{n+1}^{ed,(i)} B_b \, dx \\ &\mathsf{d}_{n+1}^{(i+1)} = \mathsf{d}_{n+1}^{(i)} + \mathsf{u}_{n+1}^{(i)} \end{aligned} \tag{3.312}$$

The main new ingredient which is needed for computing the linearized form of equilibrium equations is the tangent elastodamage modulus. The latter is obtained as the derivative of stress with respect to total deformation, which should be obtained in the local phase of computations after computing the admissible value of stress:

$$C^{ed}_{n+1} := \frac{\partial \sigma_{n+1}}{\partial \epsilon_{n+1}} = \begin{cases} 1/D_n \, ; \, \gamma_{n+1} = 0 \\ \frac{K/D_n}{1/D_n+K} \, ; \, \gamma_{n+1} > 0 \end{cases} \tag{3.313}$$

The new iterative value of incremental displacement allows us to update the displacement field $\mathsf{d}^{(i+1)}_{n+1} = \mathsf{d}^{(i)}_{n+1} + \mathsf{u}^{(i)}_{n+1}$, and then compute the new iterative value of the total deformation:

$$\epsilon^{(i+1)}_{n+1}|_{l^e} = \sum_{a=1}^{2} B^e_a \, d^{e,(i+1)}_{a,n+1} \tag{3.314}$$

We can then return to the local phase and start the new iterative cycle of the operator split solution scheme.

3.5.2 3D damage model

In the construction of a 3D damage model, we will find the same ingredient and we will have to go through the same phases of development as those just described for 1D standard damage model. However, the 3D case also requires a careful treatment of a number of modifications of the standard damage model that are not possible in 1D case, such as the multisurface damage model.

The internal damage variable for 3D damage model is chosen as the fourth order compliance tensor, $\boldsymbol{\mathcal{D}}$. For a virgin material, $\boldsymbol{\mathcal{D}}$ is equal to the inverse of elasticity tensor $\boldsymbol{\mathcal{C}}$, $\boldsymbol{\mathcal{D}} = \boldsymbol{\mathcal{C}}^{-1}$. The compliance tensor components are modified by the damage process, in order to represent the cracking produced by different damage mechanisms. By assuming that all the cracks remain inactive in unloading, we can write:

$$\nabla^s \mathbf{u} =: \boldsymbol{\epsilon} = \boldsymbol{\mathcal{D}} \boldsymbol{\sigma} \tag{3.315}$$

which allows to confirm that the damage model does not produce any residual deformation upon unloading. The same result also allows to construct the complementary energy potential as a quadratic form in terms of stress:

$$\chi(\boldsymbol{\sigma}, \boldsymbol{\mathcal{D}}, \zeta) := \frac{1}{2} \boldsymbol{\sigma} \cdot \boldsymbol{\mathcal{D}} \boldsymbol{\sigma} - \Xi(\zeta) \tag{3.316}$$

where $\Xi(\cdot)$ is the corresponding hardening potential. The damage criterion for standard model is written in terns of stress components:

$$0 \geq \phi(\boldsymbol{\sigma}, q) := \| \boldsymbol{\sigma} \|_D - (\sigma_f - q) \; ; \; \| \boldsymbol{\sigma} \|_D = [\boldsymbol{\sigma} \cdot \boldsymbol{\mathcal{P}} \boldsymbol{\sigma}]^{1/2} \tag{3.317}$$

where $\boldsymbol{\mathcal{P}}$ is a fourth-order tensor with constant components, $\|\cdot\|_D$ is the weighted norm for stress tensor, σ_f is the limit of elasticity indicating the first cracking and q is the internal variable which controls the evolution of the damage threshold.
Any negative value of the damage function $\phi(\cdot) < 0$ corresponds to the elastic domain where internal variables remain frozen, whereas a zero value of the damage function will indicate the presence of damage and evolution of compliance tensor. This produces the damage dissipation, which ought to be in agreement with the second thermodynamics principle according to:

$$\begin{aligned} 0 \leq D^d &:= \boldsymbol{\sigma} \cdot \dot{\boldsymbol{\epsilon}} + \tfrac{\partial}{\partial t}[\chi(\boldsymbol{\sigma}, \boldsymbol{\mathcal{D}}, q) - \boldsymbol{\sigma} \cdot \boldsymbol{\epsilon}] \\ &= \dot{\boldsymbol{\sigma}}(-\boldsymbol{\epsilon} + \boldsymbol{\mathcal{D}}\boldsymbol{\sigma}) + \tfrac{1}{2}\boldsymbol{\sigma} \cdot \dot{\boldsymbol{\mathcal{D}}}\boldsymbol{\sigma} + q\dot{\zeta} \ ; \ q = -\tfrac{d\Xi}{d\zeta} \end{aligned} \tag{3.318}$$

In an elastic process, with no change of internal variables and zero damage dissipation, the last result allows us to confirm the constitutive relationship between stress and strain tensors in (3.315)

$$\left.\begin{array}{r} \dot{\boldsymbol{\mathcal{D}}} = \mathbf{0} \\ \dot{\zeta} = 0 \\ D^d = 0 \end{array}\right\} \implies \boldsymbol{\epsilon} = \boldsymbol{\mathcal{D}}\boldsymbol{\sigma} \tag{3.319}$$

By assuming that the same relation remains valid in an inelastic process, we can obtain from (3.318) a reduced form of the damage dissipation:

$$0 < D^d := \frac{1}{2}\boldsymbol{\sigma} \cdot \dot{\boldsymbol{\mathcal{D}}}\boldsymbol{\sigma} + q\dot{\zeta} \tag{3.320}$$

The principle of maximum of damage dissipation can then be exploited to select among all admissible candidates in the sense of proposed damage criterion, the stress which will maximize the damage dissipation in (3.320) above. The principle of maximum damage dissipation can be cast as a constrained minimization problem and, by means of the Lagrange multiplier method, further recast as the unconstrained minimization problem:

$$\begin{aligned} &\min_{\phi(\sigma,q)\leq 0}[-D^d] \implies \max_{\dot{\gamma}\geq 0} \min_{\forall(\sigma,q)} L^d(\boldsymbol{\sigma}, q, \dot{\gamma}) \ ; \\ &L^d(\boldsymbol{\sigma}, q, \dot{\gamma}) := -D^d(\boldsymbol{\sigma}, q) + \dot{\gamma}\phi(\boldsymbol{\sigma}, q) \end{aligned} \tag{3.321}$$

By assuming that the weighted norm for stress tensor $\| \boldsymbol{\sigma} \|_P$ in (3.317) is a homogeneous function of degree one, we can further write:

$$\frac{\partial}{\partial \boldsymbol{\sigma}} \| \boldsymbol{\sigma} \|_D \cdot \boldsymbol{\sigma} = \| \boldsymbol{\sigma} \|_D \tag{3.322}$$

By exploiting the last result, it is possible to provide the explicit form of the evolution equations from the corresponding Kuhn-Tucker optimality conditions for minimization problem in (3.321),

$$
\begin{aligned}
0 = \tfrac{\partial L^d}{\partial \boldsymbol{\sigma}} \cdot d\boldsymbol{\sigma} &:= -\boldsymbol{\sigma} \cdot \dot{\boldsymbol{\mathcal{D}}} d\boldsymbol{\sigma} + \dot{\gamma}(\tfrac{\partial \phi}{\partial \boldsymbol{\sigma}} \cdot d\boldsymbol{\sigma})(\boldsymbol{\sigma} \cdot \tfrac{\partial \phi}{\partial \boldsymbol{\sigma}}) \tfrac{1}{\|\boldsymbol{\sigma}\|_P} \\
&\implies \dot{\boldsymbol{\mathcal{D}}} = \dot{\gamma} \tfrac{1}{\|\boldsymbol{\sigma}\|_P} \tfrac{\partial \phi}{\partial \boldsymbol{\sigma}} \otimes \tfrac{\partial \phi}{\partial \boldsymbol{\sigma}} \\
0 = \tfrac{\partial L^d}{\partial q} &:= -\dot{\zeta} + \dot{\gamma} \tfrac{\partial \phi}{\partial q} \implies \dot{\zeta} = \dot{\gamma} \tfrac{\partial \phi}{\partial q}
\end{aligned}
\tag{3.323}
$$

These equations are accompanied by loading/unloading conditions, which are also obtained as a part of the Kuhn-Tucker optimality conditions:

$$
\dot{\gamma} \geq 0 \; ; \; \phi \leq 0 \; ; \; \dot{\gamma}\phi = 0 \tag{3.324}
$$

From the evolution equation for compliance tensor, we can also obtain the corresponding evolution equation of its inverse, by appealing to the following identity:

$$
\boldsymbol{\mathcal{I}} = \boldsymbol{\mathcal{D}} \boldsymbol{\mathcal{D}}^{-1} \implies \dot{\boldsymbol{\mathcal{D}}}^{-1} = -\boldsymbol{\mathcal{D}}^{-1} \dot{\boldsymbol{\mathcal{D}}} \boldsymbol{\mathcal{D}}^{-1} \tag{3.325}
$$

By exploiting the last result, and making use of consistency condition to eliminate all the stress state resulting with a positive value of the damage function $\phi(\cdot) > 0$, we can obtain the value of the Lagrange multiplier according to:

$$
\begin{aligned}
0 = \dot{\phi} &:= \tfrac{\partial \phi}{\partial \boldsymbol{\sigma}} \cdot \dot{\boldsymbol{\sigma}} + \tfrac{\partial \phi}{\partial q} \dot{q} \\
&= \tfrac{\partial \phi}{\partial \boldsymbol{\sigma}} \cdot \boldsymbol{\mathcal{D}}^{-1} \dot{\boldsymbol{\epsilon}} - \tfrac{\partial \phi}{\partial \boldsymbol{\sigma}} \cdot \boldsymbol{\mathcal{D}}^{-1} \dot{\boldsymbol{\mathcal{D}}} \boldsymbol{\mathcal{D}}^{-1} - \tfrac{d^2 \Xi}{d\zeta^2} \dot{\zeta} \\
&= \tfrac{\partial \phi}{\partial \boldsymbol{\sigma}} \cdot \boldsymbol{\mathcal{D}}^{-1} \dot{\boldsymbol{\epsilon}} - \dot{\gamma}(\tfrac{d^2 \Xi}{d\zeta^2} + \tfrac{\partial \phi}{\partial \boldsymbol{\sigma}} \cdot \boldsymbol{\mathcal{D}}^{-1} (\tfrac{\partial \phi}{\partial \boldsymbol{\sigma}} \otimes \tfrac{\partial \phi}{\partial \boldsymbol{\sigma}}) \boldsymbol{\sigma} \tfrac{1}{\|\boldsymbol{\sigma}\|_D} \\
&\implies \dot{\gamma} = \frac{\frac{\partial \phi}{\partial \boldsymbol{\sigma}} \cdot \boldsymbol{\mathcal{D}}^{-1} \dot{\boldsymbol{\epsilon}}}{\frac{\partial \phi}{\partial \boldsymbol{\sigma}} \cdot \boldsymbol{\mathcal{D}}^{-1} \frac{\partial \phi}{\partial \boldsymbol{\sigma}} + \frac{d^2 \Xi}{d\zeta^2}}
\end{aligned}
\tag{3.326}
$$

With this result in hand, we can also write the constitutive equation for the stress rate, and define the tangent elastodamage modulus for the continuum problem:

$$
\dot{\boldsymbol{\sigma}} = \boldsymbol{\mathcal{C}}^{ed} \dot{\boldsymbol{\epsilon}} \; ; \; \boldsymbol{\mathcal{C}}^{ed} = \begin{cases} \boldsymbol{\mathcal{D}}^{-1} \; ; \; \dot{\gamma} = 0 \\ \boldsymbol{\mathcal{D}}^{-1} - \frac{1}{(\frac{\partial \phi}{\partial \boldsymbol{\sigma}} \cdot \boldsymbol{\mathcal{D}}^{-1} \frac{\partial \phi}{\partial \boldsymbol{\sigma}} + \frac{d^2 \Xi}{d\zeta^2})} \boldsymbol{\mathcal{D}}^{-1} \frac{\partial \phi}{\partial \boldsymbol{\sigma}} \otimes \boldsymbol{\mathcal{D}}^{-1} \frac{\partial \phi}{\partial \boldsymbol{\sigma}} \end{cases} \tag{3.327}
$$

3.5.2.1 Rate-sensitive damage model

It is also possible to choose an alternative form of the damage model, which is capable of taking into account rate-sensitivity of the response. To that end, it is sufficient to use a penalized form of the maximum dissipation principle, and replace the constrained minimization problem in (3.321) by:

$$
\min_{\forall(\boldsymbol{\sigma}, q)} [-D^d(\boldsymbol{\sigma}, q) + P(\phi(\boldsymbol{\sigma}, q))] \tag{3.328}
$$

With a quadratic form of the penalty term defined as:

$$P(\phi(\boldsymbol{\sigma}, q)) = \begin{cases} \frac{1}{2\eta}\phi^2 \; ; \; \phi > 0 \\ 0 \; ; \; \phi \leq 0 \end{cases} \tag{3.329}$$

we can further obtain the evolution equation of the internal variables for rate-sensitive damage model:

$$\begin{aligned} \dot{\mathcal{D}} &= \frac{<\phi>}{\eta} \frac{1}{\|\boldsymbol{\sigma}\|_P} \frac{\partial \phi}{\partial \boldsymbol{\sigma}} \otimes \frac{\partial \phi}{\partial \boldsymbol{\sigma}} \\ \dot{\zeta} &= \frac{<\phi>}{\eta} \end{aligned} \tag{3.330}$$

In the last expression, symbol $< \cdot >$ denotes again the Macauley parenthesis, which will filter out only the positive value of its argument. By introducing the damage multiplier $\dot{\gamma} := \frac{<\phi>}{\eta}$, the evolution equation for rate-sensitive damage model in (3.330) can be recast in an equivalent format to the corresponding one in (3.323) for the 3D standard damage model. This further allows to provide a unified implementation of the 3D standard damage model and its rate-sensitivity counterpart.

3.5.3 Refinements of 3D damage model

As presented in the previous section, the development of the 3D standard damage model is carried out in the same manner as presented previously for standard plasticity. The equivalent development to plasticity will also apply to any further refinement of the standard damage model. Namely, besides the rate-sensitivity of the damage model that we already discussed in the previous section, we can introduce other refinements of the standard damage model to increase its predictive capabilities. In the list of all possible refinements (which are in one-to-one correspondence to those introduced for plasticity), we will only elaborate upon a couple of them: a simple isotropic Kachanov-Lemaitre type of damage, and a novel anisotropic damage that can illustrate the damage model capabilities in representing different phases of localized failure.

3.5.4 Isotropic damage model of Kachanov

The standard 3D damage model, presented in the previous section, allows us to recover the isotropic damage model of Kachanov, which employs only one scalar damage variable. The latter is obtained by using the isotropic damage criterion, where the cracking will affect all the components of the compliance tensor in exactly the same manner. The isotropic damage criterion of Kachanov can be written:

$$0 \geq \phi(\boldsymbol{\sigma}, q) := \underbrace{(\boldsymbol{\sigma} \cdot \boldsymbol{\mathcal{D}}^e \boldsymbol{\sigma})^{1/2}}_{\|\boldsymbol{\sigma}\|_{D^e}} - \frac{1}{\sqrt{E}}(\sigma_f - q) \tag{3.331}$$

where $\boldsymbol{\mathcal{D}}^e$ is the elastic compliance tensor, equal to inverse of the elasticity tensor, $\boldsymbol{\mathcal{D}}^e = \boldsymbol{\mathcal{C}}^{-1}$. The isotropic damage criterion allows us to write the evolution equation in a simplified form:

$$\left.\begin{aligned} 0 &= -\boldsymbol{\sigma} \cdot \dot{\boldsymbol{\mathcal{D}}}\, d\boldsymbol{\sigma} + \dot{\gamma} \frac{\partial \phi}{\partial \boldsymbol{\sigma}} \cdot d\boldsymbol{\sigma} \\ &= -\boldsymbol{\sigma} \cdot \dot{\boldsymbol{\mathcal{D}}}\, d\boldsymbol{\sigma} + \dot{\gamma} \frac{\boldsymbol{\mathcal{D}}^e \boldsymbol{\sigma}}{\|\boldsymbol{\sigma}\|_{D^e}}\, d\boldsymbol{\sigma} \end{aligned}\right\} \Longrightarrow \dot{\boldsymbol{\mathcal{D}}} = \frac{\dot{\gamma}}{\| \boldsymbol{\sigma} \|_{D^e}} \boldsymbol{\mathcal{D}}^e \tag{3.332}$$

We can introduce a new definition of the damage multiplier:

$$\dot{\bar{\mu}} = \frac{\dot{\gamma}}{\| \boldsymbol{\sigma} \|_{D^e}} \tag{3.333}$$

leading to a very simple evolution equation for damage internal variables. For such a simple evolution equation we can compute the analytic result for the corresponding evolution of the compliance tensor, starting from the virgin state:

$$\dot{\boldsymbol{\mathcal{D}}} = \dot{\bar{\mu}} \boldsymbol{\mathcal{D}}^e \Longrightarrow \boldsymbol{\mathcal{D}}(t) = [1 + \bar{\mu}(t)] \boldsymbol{\mathcal{D}}^e \; ; \; \mu(0) = 0 \tag{3.334}$$

The same result can be recast in the usual format used for isotropic damage model of Kachanov (e.g. see Lemaitre and Chaboche [168]), by introducing a new definition of the damage variable $d \in [0, 1]$:

$$d = \frac{\bar{\mu}}{1 + \bar{\mu}} \; ; \; d \in [0, 1] \Longrightarrow \boldsymbol{\mathcal{D}}^{-1} = (1 - d)\boldsymbol{\mathcal{C}} \Leftrightarrow \boldsymbol{\sigma} = (1 - d)\boldsymbol{\mathcal{C}} \boldsymbol{\epsilon} \tag{3.335}$$

3.5.4.1 Anisotropic damage model with multisurface criterion

Most of the 3D damage criteria of interest for practical applications are of multi-surface type. They can be written:

$$\phi_k(\boldsymbol{\sigma}, q) \leq 0 \; ; \; k = 1, 2, \ldots, m \tag{3.336}$$

where each damage surface $\phi_k(\cdot)$ is chosen to describe a particular damage mechanism. For example, the need to use this type of multisurface damage model is imposed by geomaterials, where one ought to account for a significant difference between the behavior in tension and in compression, or yet for the crack opening versus the frictional sliding along the cracks mouth. The latter is represented by a 2D anisotropic damage model, which makes use of the crack opening displacement $\bar{\bar{u}}_n$ and the sliding displacement over the crack mouth $\bar{\bar{u}}_m$. The two-surface damage model of this kind can account for mode

I cracking in traction and mode II in shear. The corresponding driving forces are the normal and tangential traction components on the crack surface Γ_c. We can thus write:

$$
\begin{aligned}
0 \geq \phi_1(\boldsymbol{\sigma}, q) &:= \mathbf{n} \cdot \underbrace{\boldsymbol{\sigma}\mathbf{n}}_{\mathbf{t}_{\Gamma_c}} - (\sigma_f - q) \\
0 \geq \phi_2(\boldsymbol{\sigma}, q) &:= \mathbf{m} \cdot \underbrace{\boldsymbol{\sigma}\mathbf{n}}_{\mathbf{t}_{\Gamma_c}} - (\sigma_s - \tfrac{\sigma_s}{\sigma_f} q)
\end{aligned}
\tag{3.337}
$$

where $\mathbf{n}$ is external unit normal vector on the crack, $\mathbf{m}$ is the unit tangent vector, σ_f and σ_s are the values of elasticity limit in mode I and mode II, respectively, whereas q is the hardening variable which controls the evolution of damage threshold. For representing the final fracture phase, one actually assumes the softening behavior with exponential law:

$$
q = \sigma_f [1 - exp(-\frac{\beta}{\sigma_f} \zeta)] \tag{3.338}
$$

where the value of parameter β is chosen in accordance with the fracture energy that is established experimentally.
The original feature of this damage model is the coupling between two modes, which allows to account not only for sliding resistance reduction due to separation but also for the reduction in cohesive force due to tangential sliding. In order to make the mode coupling fully consistent, the choice of elasticity limits in mode I and II is made so that the full stress reduction is achieved simultaneously.
The evolution equations for internal damage variables can be written:

$$
\begin{aligned}
\dot{\bar{\bar{\boldsymbol{\mathcal{D}}}}} &= \dot{\gamma}_1 \tfrac{1}{\mathbf{n} \cdot \mathbf{t}_{\Gamma_c}} \mathbf{n} \otimes \mathbf{n} + \dot{\gamma}_2 \tfrac{1}{\mathbf{m} \cdot \mathbf{t}_{\Gamma_c}} \mathbf{m} \otimes \mathbf{m} \\
\zeta &= \dot{\gamma}_1 + \dot{\gamma}_2 \tfrac{\sigma_s}{\sigma_f}
\end{aligned}
\tag{3.339}
$$

where $\dot{\gamma}_1$ and $\dot{\gamma}_2$ are the Lagrange multiplier corresponding to surface ϕ_1 and ϕ_2, respectively. The loading/unloading conditions impose that:

$$
\dot{\gamma}_i \geq 0 \; ; \; \phi_i \leq 0 \; ; \; \dot{\gamma}_i \phi_i = 0 \; ; \; i = 1, 2 \tag{3.340}
$$

The consistency conditions which ensure the stress admissibility in the course of evolution with respect to either surface will allow the computation of the corresponding values of damage multipliers. The main difficulty in that sense is to decide the number of active surfaces, between one only or both; for the later case, we can obtain the damage multipliers according to:

$$
\begin{aligned}
0 &= \dot{\phi}_i \\
&:= \tfrac{\partial \phi_i}{\partial \mathbf{t}_{\Gamma_c}} \cdot \bar{\bar{\boldsymbol{\mathcal{D}}}}^{-1} \dot{\bar{\bar{\mathbf{u}}}} - \textstyle\sum_{j=1}^{2} \dot{\gamma}_j \underbrace{[\frac{\partial \phi_i}{\partial \mathbf{t}_{\Gamma_c}} \cdot \bar{\bar{\boldsymbol{\mathcal{D}}}}^{-1} \frac{\partial \phi_j}{\partial \mathbf{t}_{\Gamma_c}} + \frac{\partial \phi_i}{\partial q} \cdot \hat{K}(\zeta) \frac{\partial \phi_j}{\partial q}]}_{G_{ij}} \\
&\Longrightarrow \; \dot{\gamma}_i = \textstyle\sum_{j=1}^{2} G_{ij}^{-1} \bar{\bar{\boldsymbol{\mathcal{D}}}}^{-1} \dot{\bar{\bar{\mathbf{u}}}} \; ; \; i = 1, 2 \; ; \; \hat{K}(\zeta) := -\tfrac{\partial q}{\partial \zeta}
\end{aligned}
\tag{3.341}
$$

With these results in hand, we can easily obtain the stress rate constitutive equations for the damage model of this kind, which can be written:

$$\dot{\mathbf{t}}_{\Gamma_c} = \tilde{\boldsymbol{\mathcal{D}}}^t \dot{\bar{\bar{\mathbf{u}}}} \; ; \tag{3.342}$$

$$\tilde{\boldsymbol{\mathcal{D}}}^t := \begin{cases} \bar{\bar{\boldsymbol{\mathcal{D}}}}^{-1} \; ; \dot{\gamma}_1 = 0 \; ; \dot{\gamma}_2 = 0 \\ [\bar{\bar{\boldsymbol{\mathcal{D}}}}^{-1} - \frac{1}{\mathbf{n} \cdot \bar{\bar{\boldsymbol{\mathcal{D}}}}^{-1} \mathbf{n} + \hat{K}(\zeta)} (\bar{\bar{\boldsymbol{\mathcal{D}}}}^{-1} \mathbf{n}) \otimes (\bar{\bar{\boldsymbol{\mathcal{D}}}}^{-1} \mathbf{n})] \; ; \dot{\gamma}_1 > 0 \; ; \dot{\gamma}_2 = 0 \\ [\bar{\bar{\boldsymbol{\mathcal{D}}}}^{-1} - \frac{1}{\mathbf{m} \cdot \bar{\bar{\boldsymbol{\mathcal{D}}}}^{-1} \mathbf{m} + (\frac{\sigma_s}{\sigma_f})^2 \hat{K}(\zeta)} (\bar{\bar{\boldsymbol{\mathcal{D}}}}^{-1} \mathbf{m}) \otimes (\bar{\bar{\boldsymbol{\mathcal{D}}}}^{-1} \mathbf{m})] \; ; \dot{\gamma}_1 = 0 \; ; \dot{\gamma}_2 > 0 \\ [\bar{\bar{\boldsymbol{\mathcal{D}}}}^{-1} - \sum_{i,j=1}^{2} G_{ij}^{-1} (\bar{\bar{\boldsymbol{\mathcal{D}}}}^{-1} \frac{\partial \phi_i}{\mathbf{t}_{\Gamma_c}}) \otimes (\bar{\bar{\boldsymbol{\mathcal{D}}}}^{-1} \frac{\partial \phi_j}{\partial \mathbf{t}_{\Gamma_c}})] \; ; \dot{\gamma}_1 > 0 \; ; \dot{\gamma}_2 > 0 \end{cases}$$

The solution to the central problem for this damage model is computed by the operator split procedure. The global phase of such a computation will again provide the best iterative value of the total deformation field, along with the corresponding iterative value of the crack opening and sliding, denoted as $\bar{\bar{\mathbf{u}}}_{n+1}^{(i)}$. With these iterative valued available, we can then obtain the corresponding evolution of the internal variables in the local computation phase by using the implicit backward Euler scheme. The local computation is started by assuming the elastic trial step, which furnishes the corresponding value of stress according to:

$$\begin{aligned} & \gamma_{i,n+1}^{trial} = 0 \; ; \; i = 1,2 \implies \bar{\bar{\boldsymbol{\mathcal{D}}}}_{n+1}^{trial} = \bar{\bar{\boldsymbol{\mathcal{D}}}}_n \; ; \; \zeta_{n+1}^{trial} = \zeta_n \\ & \implies \begin{array}{l} \mathbf{t}_{\Gamma_c,n+1}^{trial} = \bar{\bar{\boldsymbol{\mathcal{D}}}}_{n+1}^{test,-1} \bar{\bar{\mathbf{u}}}_{n+1}^{(i)} \\ q_{n+1}^{trial} = q_n \end{array} \end{aligned} \tag{3.343}$$

The trial step solution can be accepted as final if both damage functions take non-positive values

$$\begin{aligned} 0 \geq \phi_1^{trial} &:= \mathbf{n} \cdot \mathbf{t}_{\Gamma_c,n+1}^{trial} - (\sigma_f - q_{n+1}^{trial}) \; ; \\ 0 \geq \phi_2^{trial} &:= \mathbf{m} \cdot \mathbf{t}_{\Gamma_c,n+1} - (\sigma_s - \frac{\sigma_s}{\sigma_f} q_{n+1}^{trial}) \end{aligned} \tag{3.344}$$

On the other hand, if any of the damage functions is positive, we will have to compute the true (positive) value of the corresponding damage multiplier $\gamma_{i,n+1} \geq 0$, which will reestablish admissibility of the final value of stress in the sense of the given damage criterion, $\phi_{i,n+1} \leq 0$, $i = 1, 2$. The final values of the internal variables will also be computed in this case according to:

$$\begin{aligned} & \bar{\bar{\boldsymbol{\mathcal{D}}}}_{n+1} = \bar{\bar{\boldsymbol{\mathcal{D}}}}_n + \gamma_{1,n+1} \frac{1}{\mathbf{n} \cdot \mathbf{t}_{\Gamma_c,n+1}} \mathbf{n} \otimes \mathbf{n} + \gamma_{2,n+1} \frac{1}{|\mathbf{m} \cdot \mathbf{t}_{\Gamma_c,n+1}|} \mathbf{m} \otimes \mathbf{m} \\ & \zeta_{n+1} = \zeta_n + \gamma_{1,n+1} + \gamma_{2,n+1} \frac{\sigma_s}{\sigma_f} \end{aligned} \tag{3.345}$$

By taking into account the last result and two possibilities to express the total deformation in terms of stress, we can obtain the relationship between the trial and the final value of driving traction acting on the crack:

$$\bar{\bar{\mathbf{u}}}_{n+1}^{(i)} = \begin{cases} \bar{\bar{\mathcal{D}}}_n \mathbf{t}_{\Gamma_c,n+1}^{trial} \\ \bar{\bar{\mathcal{D}}}_{n+1} \mathbf{t}_{\Gamma_c,n+1} \end{cases}$$

$$\Longrightarrow \mathbf{t}_{\Gamma_c,n+1} = \mathbf{t}_{\Gamma_c,n+1}^{trial} - \bar{\bar{\mathcal{D}}}_n^{-1} (\gamma_{1,n+1}\mathbf{n} + \gamma_{2,n+1} sign(\mathbf{m} \cdot \mathbf{t}_{\Gamma_c,n+1})\mathbf{m}) \tag{3.346}$$

With these results in hand, we can express the final value of the damage function in terms of its trail value:

$$\begin{aligned} \phi_1(\mathbf{t}_{\Gamma_c,n+1}, q_{n+1}) &= \phi_{1,n+1}^{trial} - \gamma_{1,n+1}(\mathbf{n} \cdot \mathcal{D}_n \mathbf{n}) - (q_{n+1} - q_n) \\ \phi_2(\mathbf{t}_{\Gamma_c,n+1}, q_{n+1}) &= \phi_{2,n+1}^{trial} - \gamma_{2,n+1}(\mathbf{m} \cdot \mathcal{D}_n \mathbf{m}) \\ & sign(\mathbf{t}_{\Gamma_c,n+1}^{trial} \cdot \mathbf{m}) - \tfrac{\sigma_s}{\sigma_f}(q_{n+1} - q_n) \end{aligned} \tag{3.347}$$

For a multisurface damage model of this kind, we ought to consider in total four different possibilities concerning loading/unloading conditions: i) $\gamma_{1,n+1} < 0$ and $\gamma_{2,n+1} < 0$ for an elastic step, ii) $\gamma_{1,n+1} > 0$ and $\gamma_{2,n+1} < 0$ for damage step in mode I, iii) $\gamma_{1,n+1} < 0$ and $\gamma_{2,n+1} > 0$ for damage step in mode II, and iv) $\gamma_{1,n+1} > 0$ and $\gamma_{2,n+1} > 0$ for damage step where both mode I and mode II are active simultaneously. The final computation in the local phase at the converged value of internal variables is carried out to obtain the consistent elastodamage tangent modulus; the latter is computed in agreement with the nature of the step (elastic versus damage) and the number the active surfaces:

$$\begin{aligned} \mathcal{C}^{ed} &:= \tfrac{\partial \mathbf{t}_{\Gamma_c,n+1}}{\partial \bar{\bar{\mathbf{u}}}_{n+1}} \\ &= \begin{cases} \bar{\bar{\mathcal{D}}}_n^{-1} - \sum_{i,j} [\mathcal{G}_{ij,n+1}]^{-1} (\bar{\bar{\mathcal{D}}}_n^{-1} \frac{\partial \phi_i}{\partial \mathbf{t}_{\Gamma_c,n+1}}) \otimes (\bar{\bar{\mathcal{D}}}_n^{-1} \frac{\partial \phi_i}{\partial \mathbf{t}_{\Gamma_c,n+1}}) \\ \bar{\bar{\mathcal{D}}}_n^{-1} - \frac{1}{\frac{\partial \phi_i}{\partial \mathbf{t}_{\Gamma_c,n+1}} \cdot \bar{\bar{\mathcal{D}}}_n^{-1} \frac{\partial \phi_i}{\partial \mathbf{t}_{\Gamma_c,n+1}}} (\bar{\bar{\mathcal{D}}}_n^{-1} \frac{\partial \phi_i}{\partial \mathbf{t}_{\Gamma_c,n+1}}) \otimes (\bar{\bar{\mathcal{D}}}_n^{-1} \frac{\partial \phi_i}{\partial \mathbf{t}_{\Gamma_c,n+1}}) \end{cases} \end{aligned} \tag{3.348}$$

3.5.5 Numerical examples: damage model combining isotropic and multisurface criteria

In this section we present the numerical results computed by using the anisotropic damage model combining an isotropic damage criterion for the continuum part and a multisurface damage criterion for the discrete part. Between two parts, the former represents the fracture process zone and the latter represents the macro-crack. This kind of failure pattern is typical of massive structures, where the fracture process zone (which provides a substantial contribution to the total dissipation) is a precursor to a macro-crack activation. With large number of micro-cracks, the fracture process zone can be represented by a continuum damage model with isotropic criterion that implies a large spectrum of micro-crack orientations. The macro-crack is represented by an embedded discontinuity of fixed orientation and strain profile including the crack opening.

In the finite element framework, this model is implemented in a 3-node triangular element with constant deformation field (the CST element), with its kinematic field enhanced by en embedded displacement discontinuity. The latter provides the possibility to represent the crack opening displacement. In order to illustrate the ability of such a finite element approximation to provide a mesh-insensitive result for computing the solution to a failure problem, we have chosen a simple tension test with two different finite element meshes, a coarse and a fine one. In each mesh, a single element is slightly weakened in order to better control the macro-crack creation; see Figure 3.29.

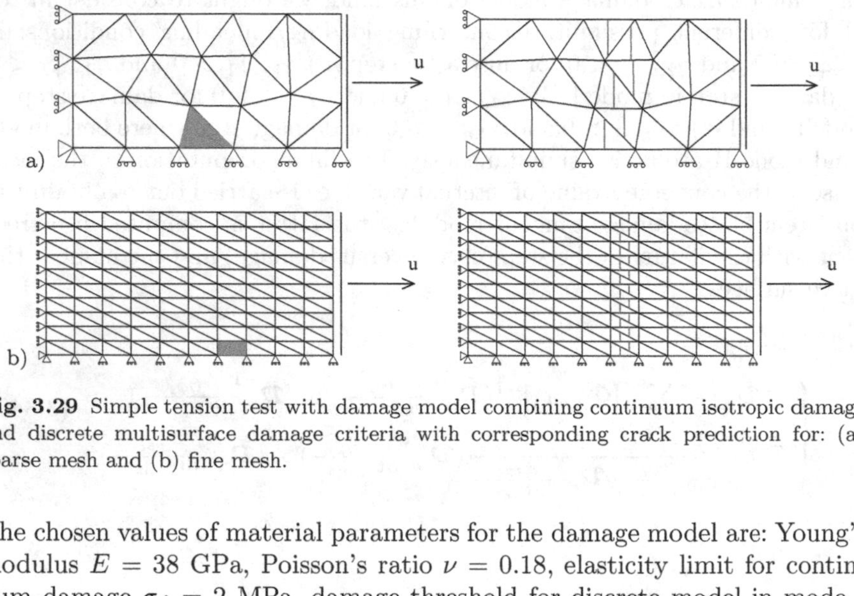

Fig. 3.29 Simple tension test with damage model combining continuum isotropic damage and discrete multisurface damage criteria with corresponding crack prediction for: (a) coarse mesh and (b) fine mesh.

The chosen values of material parameters for the damage model are: Young's modulus $E = 38$ GPa, Poisson's ratio $\nu = 0.18$, elasticity limit for continuum damage $\sigma_f = 2$ MPa, damage threshold for discrete model in mode I $\bar{\sigma}_f = 2.55$ MPa, damage threshold for discrete model in mode II $\sigma_s = 0.3\bar{\sigma}_f$, softening parameter $\beta = 2.55$ MPa/mm. We consider that the macro-crack is produced in the center of the element, in the direction which is perpendicular to the principal stress at the time when it reaches the chosen damage threshold value. The computed macro-cracks are indicated in Figure 3.29 for both coarse and fine mesh. These two meshes predict very different local response features with different crack patterns. The coarse mesh is non-structured and, contrary to fine mesh, it can not provide the macro-crack representation as a straight line as obtained with the fine, structured mesh model. Despite this difference, the global response computed for both meshes in terms of the force–displacement diagram remains practically the same for either of two meshes; see Figure 3.30.

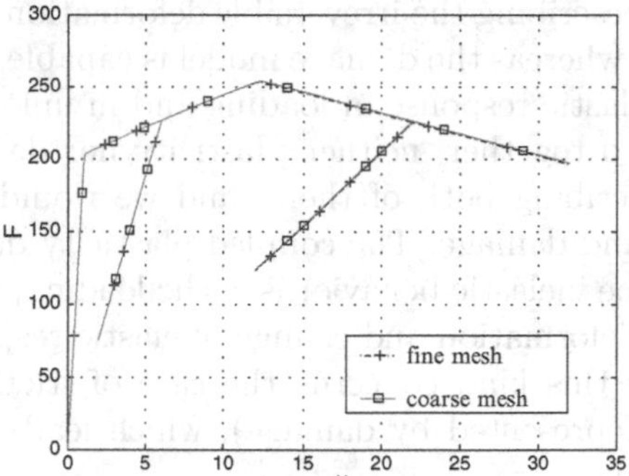

Fig. 3.30 Force–displacement diagram in simple tension test with damage model combining continuum isotropic damage and discrete multisurface damage criterion computed for coarse and fine mesh.

For further illustration of the predictive capabilities of the presented damage model combining continuum isotropic damage and discrete multisurface damage criterion, we have carried out the same traction test under a more complex loading program, which passes through both loading and unloading phases (see Figure 3.31).
The corresponding results are as follows. As long as the only active component is the continuum damage, the loading/unloading cycle is typical of models of this kind. Once we pass the peak resistance and discrete component governing crack opening is activated, the loading/unloading cycle will no longer produce a closed loop with the same deformation value. However, the two model components do remain coupled, which can be seen in restarting the loading phase at the same value of stress which was reached before the unloading started.

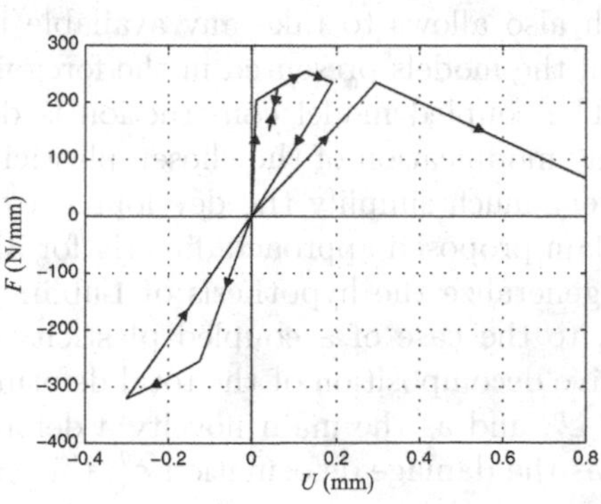

Fig. 3.31 Force–displacement diagram of damage model combining continuum isotropic and discrete multisurface damage criterion obtained during loading/unloading cycle.

3.6 Coupled plasticity-damage model

The constitutive models of plasticity and damage, studied previously, are characterized by distinct mechanisms of inelastic behavior: the plasticity model is capable of describing the irreversible deformation that remains upon complete unloading, whereas the damage model is capable of representing the difference between elastic response in loading and in unloading. If these two mechanisms appeared together, neither plasticity nor damage model would be sufficient for describing both of them, and we would need the coupled model of plasticity and damage. The coupled plasticity-damage model is often needed to describe inelastic behavior is cyclic loading, where we find both irreversible, plastic deformation and change of elastic response in unloading. Another example of this kind concerns the case of frictional sliding along the cracks mouth (represented by damage), which leads to the irreversible (plastic) deformations.

3.6.1 Theoretical formulation of 3D coupled model

In this section we develop a 3D coupled model of inelastic behavior that is capable of accounting for all inelastic mechanisms typical of both plasticity and damage. The vast majority of coupled plasticity-damage models (see Benallal, Billardon and Doghi [32] or Ju [146], Lematre [167], Simo and Ju [239]) employ the plasticity criteria based on the effective stress produced by damage model. The latter implies that the order of computation between two model components is always fixed, with the damage which supplies the stress to plasticity model component.
We advocate a different approach (see [135]) with respect to these previous works, which allows to decouple the computations we carry out on plasticity component with respect to those which are carried out on damage component. We thus obtain the procedure for local computation of internal variables where none of the coupled model components will take a privileged place. This approach also allows to take any available model plasticity or damage (including all the models presented in the foregoing) as the coupled model component. The coupled model construction is done without introducing practically any modification of the chosen plasticity or damage component, which can very much simplify the development of complex coupled models. We will explain proposed approach directly for 3D case.
First of all we will generalize the hypothesis of Lubliner [172], introduced for plasticity model, to the case of a coupled plasticity-damage model, by considering an additive decomposition of the total deformation ϵ into elastic part ϵ^e, plastic part ϵ^p, and as the main novelty a deformation component which is referred to as the damage deformation ϵ^d; this can be written:

$$\epsilon = \epsilon^e + \epsilon^p + \epsilon^d \tag{3.349}$$

The additive decomposition is also assumed for the internal energy potential for the coupled model, where the contribution of each components is accounted for:

$$\psi(\boldsymbol{\epsilon}, \boldsymbol{\epsilon}^d, \boldsymbol{\mathcal{D}}, \zeta^d, \boldsymbol{\epsilon}^p, \zeta^p) = \psi^e(\boldsymbol{\epsilon}^e) + \Xi^p(\zeta^p) + \psi^d(\boldsymbol{\epsilon}^d, D) + \Xi^d(\zeta^d) \qquad (3.350)$$

In the last expression, ζ^d and ζ^p are, respectively, the internal hardening variables for plasticity and damage, whereas $\Xi^p(\zeta^p)$ and $\Xi^d(\zeta^d)$ are the corresponding potentials. In order to illustrate these ideas more clearly, we can select a quadratic form for the internal energy of plastic component and a quadratic form for the complementary energy of the damage component, which allows us to define the internal energy through the Legendre transform:

$$\begin{aligned} &\psi^e(\boldsymbol{\epsilon}^e) = \tfrac{1}{2}\boldsymbol{\epsilon}^e \cdot \boldsymbol{\mathcal{C}}^e \boldsymbol{\epsilon}^e \;;\; \chi^d(\boldsymbol{\sigma}, \boldsymbol{\mathcal{D}}) := \tfrac{1}{2}\boldsymbol{\sigma} \cdot \boldsymbol{\mathcal{D}}\boldsymbol{\sigma} \;; \\ &\psi^d(\boldsymbol{\epsilon}^d, \boldsymbol{\mathcal{D}}) = \boldsymbol{\sigma} \cdot \boldsymbol{\epsilon}^d - \chi^d(\boldsymbol{\sigma}, \boldsymbol{\mathcal{D}}) \end{aligned} \qquad (3.351)$$

In (3.351) above, $\boldsymbol{\mathcal{C}}^e$ is the elasticity tensor and $\boldsymbol{\mathcal{D}}$ is the damage compliance tensor.

We finally postulate that each model component will have its own criterion, which defines the admissible values of stress both for plasticity and damage component:

$$\begin{aligned} 0 \geq \phi^p(\boldsymbol{\sigma}, q^p) &:= \| \boldsymbol{\sigma} \|_P - (\sigma_y - q^p) \\ 0 \geq \phi^d(\boldsymbol{\sigma}, q^d) &:= \| \boldsymbol{\sigma} \|_D - (\sigma_f - q^d) \end{aligned} \qquad (3.352)$$

In (3.352) above, σ_y is the yield stress, σ_f is the fracture stress, q^p and q^d are hardening variables controlling the threshold evolution for plasticity and for damage, respectively. It is important to note that the chosen criteria in (3.352) both pertain to the same stress, which allows to enforce the coupling of two model components.

These main ingredients specified in (3.349)–(3.352) allow to complete the model development by using nothing more than the second principle of thermodynamics and the principles of maximum dissipation for both plasticity and damage. The second principle of thermodynamics applied to the present coupled model of plasticity and damage will lead to:

$$\begin{aligned} 0 \leq D &= \boldsymbol{\sigma} \cdot \dot{\boldsymbol{\epsilon}} - \dot{\psi} \\ &= (\boldsymbol{\sigma} - \tfrac{\partial \psi^e}{\partial \boldsymbol{\epsilon}^e}) \cdot \dot{\boldsymbol{\epsilon}}^e + \overbrace{\boldsymbol{\sigma} \cdot \dot{\boldsymbol{\epsilon}}^p - \frac{\partial \Xi^p}{\partial \zeta^p}\dot{\zeta}^p}^{D^p} \\ &+ \dot{\boldsymbol{\sigma}} \cdot (\boldsymbol{\mathcal{D}}\boldsymbol{\sigma} - \boldsymbol{\epsilon}^d) + \underbrace{\frac{1}{2}\boldsymbol{\sigma} \cdot \dot{\boldsymbol{\mathcal{D}}}\boldsymbol{\sigma} - \frac{\partial \Xi^d}{\partial \zeta^d}\dot{\zeta}^d}_{D^d}, \end{aligned} \qquad (3.353)$$

where D^p and D^d are, respectively, the plastic and the damage dissipation. In an elastic process, the total dissipation remains equal to zero and all the internal variables remain fixed, with $\dot{\boldsymbol{\epsilon}}^p = \mathbf{0}$, $\dot{\zeta}^p = 0$, $\dot{\boldsymbol{\mathcal{D}}} = \mathbf{0}$ and $\dot{\zeta}^d = 0$. This allows to obtain the constitutive relation for stress:

$$\boldsymbol{\sigma} := \frac{\partial \Psi^e}{\partial \boldsymbol{\epsilon}^e} = \boldsymbol{\mathcal{C}}^e \boldsymbol{\epsilon}^e \tag{3.354}$$

as well as the corresponding definition of the damage deformation component:

$$\boldsymbol{\epsilon}^d = \boldsymbol{\mathcal{D}} \boldsymbol{\sigma} \tag{3.355}$$

By assuming that last two results also apply to an inelastic process, the total dissipation can be written in the following form:

$$0 < D = \underbrace{\boldsymbol{\sigma} \cdot \dot{\boldsymbol{\epsilon}}^p + q^p \dot{\zeta}^p}_{D^p} + \underbrace{\frac{1}{2} \boldsymbol{\sigma} \cdot \dot{\boldsymbol{\mathcal{D}}} \boldsymbol{\sigma} + q^d \dot{\zeta}^d}_{D^d}, \tag{3.356}$$

In the last expression, q^p and q^d are the stress-like variables which govern evolution of the plasticity and damage thresholds that can be obtained from given hardening potentials:

$$q^p = -\frac{\partial \Xi^p}{\partial \zeta^p} \; ; \; q^d = -\frac{\partial \Xi^d}{\partial \zeta^d}. \tag{3.357}$$

In order to obtain the corresponding evolution equations for internal variables, we can exploit the principle of maximum dissipation, applied both to plasticity and damage component of the coupled model. For plasticity, we will thus obtain the following result:

$$\begin{aligned} &\min_{\boldsymbol{\sigma}, q^p : \phi^p(\boldsymbol{\sigma}, q^p) = 0} [-\mathcal{D}^p(\boldsymbol{\sigma}, q^p)] \Leftrightarrow \max_{\dot{\gamma}^p > 0} \min_{\forall \boldsymbol{\sigma}, q^p} [L^p(\boldsymbol{\sigma}, q^p, \dot{\gamma}^p)] \\ &L^p = -\mathcal{D}^p(\boldsymbol{\sigma}, q^p) + \dot{\gamma}^p \phi^p(\boldsymbol{\sigma}, q^p) \implies \end{aligned} \tag{3.358}$$

$$\begin{aligned} 0 &= \frac{\partial L^p(\boldsymbol{\sigma}, q^p, \dot{\gamma}^p)}{\partial \sigma} = -\dot{\boldsymbol{\epsilon}}^p + \dot{\gamma}^p \frac{\partial \phi^p}{\partial \boldsymbol{\sigma}} \implies \dot{\boldsymbol{\epsilon}}^p = \dot{\gamma}^p \frac{\partial \phi^p}{\partial \boldsymbol{\sigma}} \\ 0 &= \frac{\partial L^p(\boldsymbol{\sigma}, q^p, \dot{\gamma}^p)}{\partial q^p} = -\dot{\zeta}^p + \dot{\gamma}^p \frac{\partial \phi^p}{\partial q^p} \implies \dot{\zeta}^p = \dot{\gamma}^p \frac{\partial \phi^p}{\partial q^p}, \end{aligned} \tag{3.359}$$

accompanied by the loading/unloading conditions for plastic component:

$$\dot{\gamma}^p \geq 0 \; ; \; \phi^p \leq 0 \; ; \; \dot{\gamma}^p \phi^p = 0 \tag{3.360}$$

In the same manner, we can obtain the evolution equations for the internal variables of the damage component:

$$\begin{aligned} &min_{\sigma, q^d : \phi^d(\sigma, q^d) = 0} [-\mathcal{D}^d(\sigma, q^d)] \Leftrightarrow max_{\dot{\gamma}^d > 0} min_{\forall \sigma, q^d} [L^d(\sigma, q^d), \dot{\gamma}^d] \\ &L^d = -\mathcal{D}^d(\sigma, q^d) + \dot{\gamma}^d \phi^d(\sigma, q^d) \qquad \implies \end{aligned} \tag{3.361}$$

$$0 = \frac{\partial L^d(\boldsymbol{\sigma}, q^d, \dot{\gamma}^d)}{\partial \boldsymbol{\sigma}} = -\dot{\boldsymbol{\mathcal{D}}}\boldsymbol{\sigma} + \dot{\gamma}^d \frac{1}{\| \boldsymbol{\sigma} \|_D} \frac{\partial \phi^d}{\partial \boldsymbol{\sigma}} (\frac{\partial \phi^d}{\partial \boldsymbol{\sigma}} \cdot \boldsymbol{\sigma}) \qquad (3.362)$$

$$\Longrightarrow \dot{\boldsymbol{\mathcal{D}}} = \dot{\gamma}^p \frac{1}{\| \boldsymbol{\sigma} \|_D} \frac{\partial \phi^d}{\partial \boldsymbol{\sigma}} \otimes \frac{\partial \phi^d}{\partial \boldsymbol{\sigma}}$$

$$0 = \frac{\partial L^d(\sigma, q^d, \dot{\gamma}^d)}{\partial q^d} = -\dot{\zeta}^d + \dot{\gamma}^d \frac{\partial \phi^d}{\partial q^d} \Longrightarrow \dot{\zeta}^d = \dot{\gamma}^d \frac{\partial \phi^d}{\partial q^d} \qquad (3.363)$$

along with the corresponding loading/unloading conditions:

$$\dot{\gamma}^d \geq 0 \; ; \; \phi^d \leq 0 \; ; \; \dot{\gamma}^d \phi^d = 0, \qquad (3.364)$$

Finally, we can impose the consistency condition on the stress state requiring that its evolution remains in agreement with the given yield criterion, which allows us to compute the plastic multiplier $\dot{\gamma}^p$. The same kind of consistency condition is imposed on the stress state with respect to the chosen damage criterion, which allows us to obtain the damage multiplier $\dot{\gamma}^d$. More precisely, in a plastic process where the plasticity threshold is reached, we will require that the computed stress remains equal to the plastic threshold evolution for the subsequent moment in time. This reduces to requirement that the time derivative of the yield function remains equal to zero:

$$\begin{aligned} 0 = \dot{\phi}^p &:= \tfrac{\partial \phi^p}{\partial \boldsymbol{\sigma}} \dot{\boldsymbol{\sigma}} + \tfrac{\partial \phi^p}{\partial q^p} \dot{q}^p \\ &= \frac{\partial \phi^p}{\partial \boldsymbol{\sigma}} \cdot \boldsymbol{\mathcal{C}}^e (\dot{\epsilon} - \dot{\epsilon}^d - \dot{\gamma}^p \boldsymbol{\mathcal{C}}^e \frac{\partial \phi^p}{\partial \boldsymbol{\sigma}} - \frac{\partial \phi^p}{\partial q^p} \dot{\gamma}^p \frac{\partial^2 \Xi^p}{\partial \zeta^{p2}} \frac{\partial \phi^p}{\partial q^p}) \\ \Longrightarrow \; \dot{\gamma}^p &= \frac{\frac{\partial \phi^p}{\partial \boldsymbol{\sigma}} \cdot \boldsymbol{\mathcal{C}}^e (\dot{\epsilon} - \dot{\epsilon}^d)}{\frac{\partial \phi^p}{\partial \boldsymbol{\sigma}} \cdot \boldsymbol{\mathcal{C}}^e \frac{\partial \phi^p}{\partial \boldsymbol{\sigma}} + \frac{\partial \phi^p}{\partial q^p} \frac{\partial^2 \Xi^p}{\partial \zeta^{p2}} \frac{\partial \phi^p}{\partial q^p}} \end{aligned} \qquad (3.365)$$

With this result in hand, we can further obtain the stress rate constitutive equation for the plastic component of the coupled model, which can be written:

$$\dot{\boldsymbol{\sigma}} = \boldsymbol{\mathcal{C}}^{ep} (\dot{\epsilon} - \dot{\epsilon}^d) \; ; \; \boldsymbol{\mathcal{C}}^{ep} = \begin{cases} \boldsymbol{\mathcal{C}}^e \; ; \; \dot{\gamma}^p = 0 \\ [\boldsymbol{\mathcal{C}}^e - \frac{\boldsymbol{\mathcal{C}}^e \frac{\partial \phi^p}{\partial \boldsymbol{\sigma}} \otimes \boldsymbol{\mathcal{C}}^e \frac{\partial \phi^p}{\partial \boldsymbol{\sigma}}}{\frac{\partial \phi^p}{\partial \boldsymbol{\sigma}} \cdot \boldsymbol{\mathcal{C}}^e \frac{\partial \phi^p}{\partial \boldsymbol{\sigma}} + \frac{\partial \phi^p}{\partial q^p} \frac{\partial^2 \Xi^p}{\partial \zeta^{p2}} \frac{\partial \phi^p}{\partial q^p}}] \; ; \; \dot{\gamma}^p > 0 \end{cases} \qquad (3.366)$$

The damage multiplier can be obtained in a similar manner. Namely, we will appeal to the damage consistency condition and require that the time derivative of the damage function remains equal to zero, which ensures that the stress remains in agreement with the damage threshold evolution in the subsequent time:

$$\begin{aligned} 0 = \dot{\phi}^d &:= \tfrac{\partial \phi^d}{\partial \boldsymbol{\sigma}} \cdot \dot{\boldsymbol{\sigma}} + \tfrac{\partial \phi^d}{\partial q^d} \dot{q}^d \\ &= \frac{\partial \phi^d}{\partial \boldsymbol{\sigma}} \cdot \boldsymbol{\mathcal{D}}^{-1} \dot{\epsilon}^d - \dot{\gamma}^d \boldsymbol{\mathcal{D}}^{-1} \frac{\partial \phi^d}{\partial \boldsymbol{\sigma}} - \dot{\gamma}^d \frac{\partial \phi^d}{\partial q^d} \frac{\partial^2 \Xi^d}{\partial \zeta^{d2}} \frac{\partial \phi^d}{\partial q^d} \\ \Longrightarrow \; \dot{\gamma}^d &= \frac{\frac{\partial \phi^d}{\partial \boldsymbol{\sigma}} \cdot \boldsymbol{\mathcal{D}}^{-1} \dot{\epsilon}^d}{\frac{\partial \phi^d}{\partial \boldsymbol{\sigma}} \cdot \boldsymbol{\mathcal{D}}^{-1} \frac{\partial \phi^d}{\partial \boldsymbol{\sigma}} + \frac{\partial \phi^d}{\partial q^d} \frac{\partial^2 \Xi^d}{\partial \zeta^{d2}} \frac{\partial \phi^d}{\partial q^d}} \end{aligned} \qquad (3.367)$$

The instantaneous response of the damage component of the coupled model can then be written according to:

$$\dot{\boldsymbol{\sigma}} = \mathcal{C}^{ed}\,\boldsymbol{\epsilon}^{d}\ ;\ \mathcal{C}^{ed} = \begin{cases} \mathcal{C}^{-1}\ ;\ \dot{\gamma}^{d} = 0 \\ [\mathcal{D}^{-1} - \dfrac{\mathcal{D}^{-1}\frac{\partial \phi^{d}}{\partial \boldsymbol{\sigma}} \otimes \mathcal{D}^{-1}\frac{\partial \phi^{d}}{\partial \boldsymbol{\sigma}}}{\frac{\partial \phi^{d}}{\partial \sigma} \cdot \mathcal{D}^{-1}\frac{\partial \phi^{d}}{\partial \boldsymbol{\sigma}} + \frac{\partial \phi^{d}}{\partial q^{d}}\frac{\partial^{2} \Xi^{d}}{\partial \zeta^{d2}}\frac{\partial \phi^{d}}{\partial q^{d}}}]\ ;\ \dot{\gamma}^{d} > 0 \end{cases} \tag{3.368}$$

These evolution equations will remain the same in the case where both components of coupled model, plasticity and damage, are active at the same time. However, in such a case, we need to make sure that the final stress values produced by two components, with either (3.366) or (3.368) lead to the same result. This condition allows us further to obtain the consistent elastoplastic-damage tangent modulus, which can be written:

$$\begin{aligned} \mathcal{C}^{ep}(\dot{\boldsymbol{\epsilon}} - \dot{\boldsymbol{\epsilon}}^{d}) = \mathcal{C}^{ed}\dot{\boldsymbol{\epsilon}}^{d} \Longrightarrow \dot{\boldsymbol{\epsilon}}^{d} = [\mathcal{C}^{ep} + \mathcal{C}^{ed}]^{-1}\mathcal{C}^{ep}\dot{\boldsymbol{\epsilon}} \\ \dot{\boldsymbol{\sigma}} = \mathcal{C}^{epd}\dot{\boldsymbol{\epsilon}}\ ;\ \mathcal{C}^{epd} = \mathcal{C}^{ed}[\mathcal{C}^{ep} + \mathcal{C}^{ed}]^{-1}\mathcal{C}^{ep} \end{aligned} \tag{3.369}$$

The graphic illustration in Figure 3.32 constructed for 1D case allows to obtain a very clear representation of different phases of the computed response for such a coupled model.

We should also note that any refinement trying to extend their predictive capabilities of the plasticity or damage component presented previously, can easily be incorporated within the present coupled model of plasticity and damage; one can add sensitivity to rate of deformation with either viscoplasticity or viscodamage, difference between behavior in compression and in tension with multisurface criteria for plasticity or damage, nonlinear hardening and so on, with only change introduced at the level of a single component (plasticity or damage) and no change to the presented framework.

3.6.2 Time integration of stress for coupled plasticity-damage model

In this section we present the numerical implementation details for the coupled plasticity-damage model. The most important goal is to achieve the kind of implementation which allows that computations on each coupled model component, plasticity or damage, remain independent from one another.

The proposed computational procedure exploits the operator split methodology to separate the global from local computations. Namely, we first use the implicit Euler time integration scheme to carry out the local phase of computations leading to:

Central problem for coupled plasticity-damage model

Given: $\boldsymbol{\epsilon}_n$, $\boldsymbol{\epsilon}^{p}_{n}$, $\boldsymbol{\epsilon}^{d}_{n}$, $\zeta^{p}_{n} = \zeta^{p}_{n}$, $\mathcal{D}_n$, ζ^{d}_{n}, $h \equiv \Delta t = t_{n+1} - t_n$

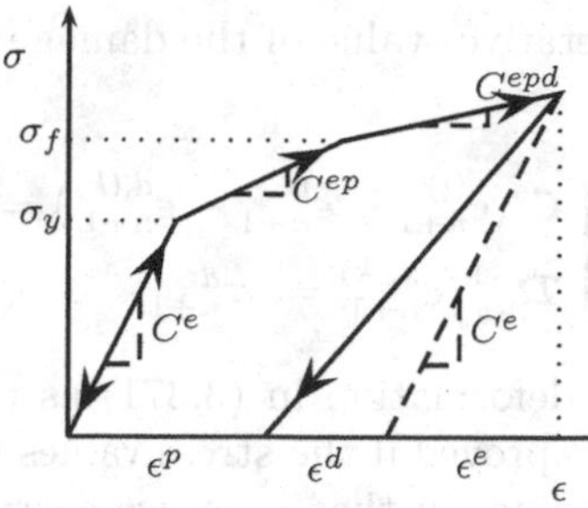

Fig. 3.32 Stress–strain diagram for coupled plasticity-damage model in loading and unloading.

Find: $\boldsymbol{\epsilon}_{n+1}$, $\boldsymbol{\epsilon}^p_{n+1}$, $\boldsymbol{\epsilon}^d_{n+1}$, ζ^p_{n+1}, $\boldsymbol{\mathcal{D}}_{n+1}$, ζ^d_{n+1}
such that:

$$
\begin{aligned}
&A_{e=1}^{n_{el}}[\mathbf{f}^{e,int}_{n+1} - \mathbf{f}^{e,ext}_{n+1}] = \mathbf{0}\ ;\\
&\mathbf{f}^{e,int}_{n+1} = \int_{\Omega^e} \mathbf{B}^{e,T}\boldsymbol{\sigma}(\boldsymbol{\epsilon}_{n+1}, \boldsymbol{\epsilon}^p_{n+1}, \zeta^p_{n+1}, \boldsymbol{\mathcal{D}}_{n+1}, \zeta^d_{n+1}, \boldsymbol{\epsilon}^d_{n+1})\, dV\ ;\\
&\mathbf{f}^{e,ext}_{n+1} = \int_{\Omega^e} \mathbf{N}^{e,T}\mathbf{b}\, dV + \int_{\Gamma^e_\sigma} \mathbf{N}^{e,T}\bar{\mathbf{t}}\, dA\ ;\\
&\dot{\gamma}^p_{n+1} \geq 0,\ \phi^p_{n+1}(\boldsymbol{\sigma}_{n+1}, q^p_{n+1}) \leq 0,\ \dot{\gamma}^p_{n+1}\phi^p_{n+1} = 0\ ;\\
&\dot{\gamma}^d_{n+1} \geq 0,\ \phi^d_{n+1}(\boldsymbol{\sigma}_{n+1}, q^d_{n+1}) \leq 0,\ \dot{\gamma}^d_{n+1}\phi^d_{n+1} = 0.
\end{aligned}
\tag{3.370}
$$

The operator split procedure will first deal with the local phase of computation. For the given best iterative value of total deformation, we will seek the values of internal variables that provide the admissible stress field with respect to the chosen plasticity and damage criteria. This can formally be presented as:

Phase I: Computation of internal variables for each Gauss quadrature point
In the case of traditional damage-plasticity coupled model (e.g. see [32, 146, 167 or 239]), this local phase of the operator split solution procedure for computing internal variables can become significantly more complex than the corresponding computation for either plasticity or damage model component. This complexity is very much eliminated with the present model, since it can directly exploit the standard stress computations for both plasticity and the damage components, with only one additional iterative loop enforcing the stress equality supplied by two model components. More precisely, with the best iterative value of the total deformation, $\boldsymbol{\epsilon}^{(i)}_{n+1}$, supplied with the solution to (linearized) equilibrium equations, we will start the local phase by computing separately the internal variables enforcing stress admissibility with respect to the given plasticity and damage criteria. Initially, we will let each components carry out the standard procedure to supply its own stress value. The results of these computations are denoted as $\boldsymbol{\sigma}^p_{n+1}(\boldsymbol{\epsilon}^{d,(k)}_{n+1})$ for plasticity and $\boldsymbol{\sigma}^d_{n+1}(\boldsymbol{\epsilon}^{d,(k)}_{n+1})$ for damage. Both stress values are considered

as functions of chosen (iterative) value of the damage deformation $\epsilon_{n+1}^{d,(k)}$ that can be explicitly defined:

$$\sigma_{n+1} = \begin{cases} \mathcal{C}^e(\epsilon_{n+1}^{(i)} - \epsilon_{n+1}^p - \epsilon_{n+1}^{d,(k)}) =: \sigma_{n+1}^p \\ \mathcal{D}_{n+1}^{-1}\epsilon_{n+1}^{d,(k)} =: \sigma_{n+1}^d \end{cases} \tag{3.371}$$

We consider the damage deformation in (3.371) as the best iterative value, which should be further improved if the stress values produced by two model components are not the same. In that case, we carry out the next iterative sweep producing a new iterative value of damage deformation $\epsilon_{n+1}^{d,(k+1)}$. At each iteration $(k) = 1, 2, \ldots$, the complete iterative procedure for coupled plasticity-damage model can be described as follows:

Ip. Stress and internal variables computations for plasticity

Given: $\epsilon_{n+1}^{(i)}$, $\epsilon_{n+1}^{d,(k)}$, $\epsilon_n^p = \epsilon_n^p(\cdot, t_n)$, $\zeta_n^p = \zeta_n^p(\cdot, t_n)$

Find: ϵ_{n+1}^p, ζ_{n+1}^p and σ_{n+1}^p, q_{n+1}^p

such that:

$$\dot\gamma_{n+1}^p \geq 0,\ \phi_{n+1}^p(\sigma_{n+1}^p, q_{n+1}^p) \leq 0,\ \dot\gamma_{n+1}^p \phi_{n+1}^p = 0$$

The computation of this kind is the same as the one already described for the constitutive model of plasticity, which starts by the elastic trial state computation assuming that the internal variable values remain frozen, which results with:

$$\begin{aligned} \sigma_{n+1}^{p,trial} &= \mathcal{C}^e(\epsilon_{n+1}^{(i)} - \epsilon_n^p - \epsilon_{n+1}^{d(k)})\ ;\ q_{n+1}^{p,trial} = \hat{q}(\zeta_n^p) \\ \Longrightarrow \phi_{n+1}^{p,trial} &= \| \sigma_{n+1}^{p,trial} \|_P - (\sigma_y - q_{n+1}^{p,trial}) \end{aligned} \tag{3.372}$$

We can stop plasticity component stress computation in the case the trial value of yield function is indeed non-positive with $\phi_{n+1}^{p,trial} \leq 0$. Otherwise, if $\phi_{n+1}^{trial} > 0$, we carry on in order to obtain the plastically admissible stress values:

$$\begin{aligned} \sigma_{n+1}^p &= \mathcal{C}^e(\epsilon_{n+1}^{(i)} - \epsilon_{n+1}^p - \epsilon_{n+1}^{d(k)}) \\ &= \sigma_{n+1}^{p,trial} - \mathcal{C}^e \gamma_{n+1}^p \frac{\partial \phi_{n+1}^p}{\partial \sigma_{n+1}} \end{aligned}$$

$$q_{n+1}^p = \hat{q}^p(\zeta_n^p + \gamma_{n+1}^p \frac{\partial \phi^p}{q_{n+1}^p}) \tag{3.373}$$

where $\gamma_{n+1}^p > 0$ and $\phi_{n+1}^p = 0$.

Id. Stress and internal variables computations for damage

Given: $\epsilon_{n+1}^{(i)}$, $\epsilon_{n+1}^{d,(k)}$, $\mathcal{D}_n$, ζ_n^d

Find: $\mathcal{D}_{n+1}$, ζ_{n+1}^d

such that:

$$\dot\gamma_{n+1}^d \geq 0,\ \phi_{n+1}^d(\sigma_{n+1}, q_{n+1}^d) \leq 0,\ \dot\gamma_{n+1}^d \phi_{n+1}^d = 0$$

The solution to this problem is computed in the same manner as already explained for the damage model. We start by computing the elastic trail step by assuming that the damage compliance remains fixed:

$$\boldsymbol{\sigma}_{n+1}^{d,trial} = \boldsymbol{\mathcal{D}}_n^{-1} \boldsymbol{\epsilon}_{n+1}^{d,(k)} \; ; \; q_{n+1}^{d,trial} = \hat{q}^d(\zeta_n^d) \Longrightarrow \phi_{n+1}^{d,trial} = \| \boldsymbol{\sigma} \|_D - (\sigma_f - q^d) \tag{3.374}$$

The elastic trial stress value for damage is accepted as final if $\phi_{n+1}^{d,trial} \leq 0$, or else, if $\phi_{n+1}^{d,trial} > 0$, we have to correct the computed stress in order to make it admissible with respect to the chosen damage criterion:

$$\begin{aligned}\boldsymbol{\sigma}_{n+1}^{d} &= \boldsymbol{\mathcal{D}}_{n+1}^{-1} \boldsymbol{\epsilon}_{n+1}^{d(k)} \\ &= \boldsymbol{\sigma}_{n+1}^{d,trial} - D_n^{-1} \gamma_{n+1}^d \frac{\partial \phi_{n+1}^d}{\partial \boldsymbol{\sigma}_{n+1}^d}\end{aligned} \tag{3.375}$$

$$q_{n+1}^d = q^d(\zeta_n^d + \gamma_{n+1}^d \frac{\partial \phi_{n+1}^d}{\partial q_{n+1}^d}) \tag{3.376}$$

which results with $\gamma_{n+1}^d > 0$ and $\phi_{n+1}^d = 0$.
It is important to note that the stress computation for two coupled model components can be carried out in parallel, using the same iterative value of the damage deformation $\boldsymbol{\epsilon}_{n+1}^{d,(k)}$. The convergence of the local phase ought to be verified with respect to the right choice of the damage deformation enforcing the same value of stress in each component:

$$0 \stackrel{?}{=} \| \mathbf{r}(\boldsymbol{\epsilon}_{n+1}^{d,(k)}) \| := \| \boldsymbol{\sigma}_{n+1}^{p}(\boldsymbol{\epsilon}_{n+1}^{d,(k)}) - \boldsymbol{\sigma}_{n+1}^{d}(\boldsymbol{\epsilon}_{n+1}^{d,(k)}) \| \tag{3.377}$$

If necessary, we can continue to the next iteration, and provide a modified value of the damage deformation with:

$$\Delta\boldsymbol{\epsilon}_{n+1}^{d,(k)} = [\boldsymbol{\mathcal{C}}_{n+1}^{ep} + \boldsymbol{\mathcal{C}}_{n+1}^{ed}]^{-1} \mathbf{r}(\boldsymbol{\epsilon}_{n+1}^{d,(k)}) \Longrightarrow \boldsymbol{\epsilon}_{n+1}^{d,(k+1)} = \boldsymbol{\epsilon}_{n+1}^{d,(k)} + \Delta\boldsymbol{\epsilon}_{n+1}^{d,(k)} \tag{3.378}$$

The local phase of operator split procedure is completed by computation of the consistent tangent elastoplastic-damage modulus. Namely, upon convergence of the internal variable computation along with the convergence towards the best damage deformation value, we can provide the linearized form of stress supplied by each component of the coupled plasticity-damage model. The latter includes the consistent tangent for both plasticity and damage components, which will jointly provide the consistent tangent for the coupled model that can be written:

$$\begin{aligned} d\boldsymbol{\sigma}_{n+1} &:= \begin{cases} \boldsymbol{\mathcal{C}}_{n+1}^{ep}(d\boldsymbol{\epsilon}_{n+1} - d\boldsymbol{\epsilon}_{n+1}^d) \\ \boldsymbol{\mathcal{C}}_{n+1}^{ed} d\boldsymbol{\epsilon}_{n+1}^d \end{cases} \Longrightarrow \\ d\boldsymbol{\sigma}_{n+1} &= \boldsymbol{\mathcal{C}}_{n+1}^{epd} d\boldsymbol{\epsilon}_{n+1} \; ; \; \boldsymbol{\mathcal{C}}_{n+1}^{epd} = \boldsymbol{\mathcal{C}}_{n+1}^{ed}(\boldsymbol{\mathcal{C}}_{n+1}^{ep} + \boldsymbol{\mathcal{C}}_{n+1}^{ed})^{-1} \boldsymbol{\mathcal{C}}_{n+1}^{ep} \end{aligned} \tag{3.379}$$

For clarity, we present in Table 3.4 the flowchart of the complete computational procedure for this coupled model over a typical time step.

3.6.2.1 Numerical examples: plate with a hole

In this section we present a couple of numerical examples that can illustrate the capabilities of the coupled plasticity-damage model to represent the inelastic behavior mechanisms, which can not be captured by a single component, neither plasticity nor damage.

3.6.2.2 Porous metallic materials

The first example deals with porous metals, whose inelastic behavior is traditionally represented by the Gurson plasticity criterion [95]. The Gurson model is based on the hypothesis that the inelastic behavior of metallic matrix is governed by the von Mises yield criterion, and that the porosity will contribute an additional term proportional to the trace of stress tensor; the most general form of the Gurson criterion can be written:

$$\phi^G(\boldsymbol{\sigma}) := c_1 \parallel dev[\boldsymbol{\sigma}] \parallel + c_2 tr[\boldsymbol{\sigma}] - c_3 \sigma_y$$

where c_1, c_2 and c_3 are chosen constants (see Needleman and Tveegard [199]). With the Gurson model as motivation, we can construct (e.g. see [135]) an equivalent form of coupled plasticity-damage constitutive model. The plasticity component of such a model corresponds to the metallic matrix, whose inelastic behavior is governed by von Mises yield criterion:

$$\phi^p(\boldsymbol{\sigma}, q^p) = \sqrt{dev[\boldsymbol{\sigma}] \cdot dev[\boldsymbol{\sigma}]} - (\sigma_y^p - q^p) \; ; \; \; dev[\boldsymbol{\sigma}] \equiv \boldsymbol{\sigma} - \frac{1}{3} tr[\boldsymbol{\sigma}] \quad (3.380)$$

The damage component of the coupled plasticity-damage represents pore--space filled with a damage-like material with a very low damage threshold, whose behavior is governed by a spherical part of stress tensor:

$$\phi^d(\boldsymbol{\sigma}, q^d) = < tr[\boldsymbol{\sigma}] > - (\boldsymbol{\sigma}_f^d - q^d), \quad (3.381)$$

The Gurson criterion can be reproduced from this kind of coupled plasticity-damage model by a judicious choice of the coupled model parameters. However, the coupled model can provide more advantageous features to the Gurson plasticity model, such as the difference in volume change response in loading and in unloading, as well as the difference of the behavior in traction and in compression. The latter is introduced by using the Macauley bracket $< tr[\boldsymbol{\sigma}] >$:

$$< x > = \begin{cases} x \; ; x \geq 0 \\ 0 \; ; x < 0 \end{cases} . \quad (3.382)$$

Table 3.4 Flowchart of operator split computational procedure for 3D coupled plasticity-damage model.

Given: $\boldsymbol{\epsilon}_n^{(i)}, \boldsymbol{\epsilon}_n^p, \mathbf{D}_n, \zeta_n^p, \zeta_n^d$

Find: $\boldsymbol{\sigma}_{n+1}^{(i)}, \boldsymbol{\epsilon}_{n+1}^{p(i)}, \mathbf{D}_{n+1}^{(i)}, \zeta_{n+1}^{p(i)}$ and $\zeta_{n+1}^{d(i)}$

elastic trial step

$$\boldsymbol{\sigma}_{n+1}^{trial} = \mathbf{C}^e(\boldsymbol{\epsilon}_{n+1}^{(i)} - \boldsymbol{\epsilon}_n^p - \boldsymbol{\epsilon}_{n+1}^{d(k=0)}) = (\mathbf{C}^{e\ -1} + D_n)^{-1}(\boldsymbol{\epsilon}_{n+1}^{(i)} - \boldsymbol{\epsilon}_n^p)$$

$$\boldsymbol{\epsilon}_{n+1}^{d(k=0)} = \mathbf{D}_n \boldsymbol{\sigma}_{n+1}^{trial}$$

$\boldsymbol{\epsilon}_{n+1}^{d(k)}$	$\boldsymbol{\epsilon}_{n+1}^{d(k)}$
$\Downarrow$	$\Downarrow$
plasticity	damage
$\phi_{n+1}^{p,trial} = \phi^p(\boldsymbol{\sigma}_{n+1}^{trial}, q_n^p)$	$\phi_{n+1}^{d,trial} = \phi^d(\boldsymbol{\sigma}_{n+1}^{d}, q_n^d)$
IF $\phi_{n+1}^{p,trial} \leq 0$:	IF $\phi_{n+1}^{d,trial} \leq 0$:
$\boldsymbol{\sigma}_{n+1}^p = \boldsymbol{\sigma}_{n+1}^{trial}$	$\boldsymbol{\sigma}_{n+1}^d = \boldsymbol{\sigma}_{n+1}^{trial}$
$\boldsymbol{\epsilon}_{n+1}^p = \boldsymbol{\epsilon}_n^p$	$\boldsymbol{\mathcal{D}}_{n+1} = \boldsymbol{\mathcal{D}}_n$
$\zeta_{n+1}^p = \zeta_n^p$	$\zeta_{n+1}^d = \zeta_n^d$
$q_{n+1}^p = q_n^p$	$q_{n+1}^d = q_n^d$
ELSEIF $\phi_{n+1}^{p,trial} > 0$:	ELSEIF $\phi_{n+1}^{d,trial} > 0$:
solve	solve
$\phi_{n+1}^p(\boldsymbol{\sigma}_{n+1}^p(\gamma_{n+1}^p), q_{n+1}^p(\gamma_{n+1}^p)) = 0$	$\phi_{n+1}^d(\boldsymbol{\sigma}_{n+1}^d(\gamma_{n+1}^d), q_{n+1}^d(\gamma_{n+1}^d)) = 0$
with	with
$\boldsymbol{\sigma}_{n+1}^p = \boldsymbol{\sigma}_{n+1}^{trial} - \gamma_{n+1}^p \boldsymbol{\mathcal{C}}^e \frac{\partial \phi_{n+1}^p}{\partial \boldsymbol{\sigma}}$	$\boldsymbol{\sigma}_{n+1}^d = \boldsymbol{\sigma}_{n+1}^{trial} - \gamma_{n+1}^d \boldsymbol{\mathcal{D}}_n^{-1} \frac{\partial \phi_{n+1}^d}{\partial \boldsymbol{\sigma}}$
$\zeta_{n+1}^p = \zeta_n^p + \gamma_{n+1}^p$	$\zeta_{n+1}^d = \zeta_n^d + \gamma_{n+1}^d$
$q_{n+1}^p = \hat{q}^p(\zeta_{n+1}^p)$	$q_{n+1}^d = \hat{q}^d(\zeta_{n+1}^d)$
$\boldsymbol{\epsilon}_{n+1}^p = \boldsymbol{\epsilon}_n^p + \gamma_{n+1}^p \frac{\partial \phi_{n+1}^p}{\partial \boldsymbol{\sigma}}$	$\boldsymbol{\mathcal{D}}_{n+1} = \boldsymbol{\mathcal{D}}_n + \frac{\gamma_{n+1}^d}{\phi_{n+1}^d + \sigma_f - q_{n+1}^d} \frac{\partial \phi_{n+1}^d}{\partial \boldsymbol{\sigma}} \otimes \frac{\partial \phi_{n+1}^d}{\partial \boldsymbol{\sigma}}$
$\Downarrow$	$\Downarrow$
$\boldsymbol{\sigma}_{\mathbf{n+1}}^{\mathbf{p}}$	$\boldsymbol{\sigma}_{\mathbf{n+1}}^{\mathbf{d}}$

IF $\boldsymbol{\sigma}_{n+1}^p \neq \boldsymbol{\sigma}_{n+1}^d$:

correct $\boldsymbol{\epsilon}_{n+1}^{d(k)}$

$$\Delta\boldsymbol{\epsilon}_{n+1}^{d(k)} = (\boldsymbol{\mathcal{C}}^{ep} + \boldsymbol{\mathcal{C}}^{ed})^{-1}(\boldsymbol{\sigma}_{n+1}^p - \boldsymbol{\sigma}_{n+1}^d)$$

$$\boldsymbol{\epsilon}_{\mathbf{n+1}}^{\mathbf{d(k+1)}} = \boldsymbol{\epsilon}_{\mathbf{n+1}}^{\mathbf{d(k)}} + \boldsymbol{\Delta\epsilon}_{\mathbf{n+1}}^{\mathbf{d(k)}}$$

Other possible modification can be introduced in order to increase the predictive capabilities of the proposed coupled model, such as two different hardening mechanisms; for example, by choosing saturation hardening for each component, we can write:

$$q^p(\zeta^p) = (\sigma_y^p - \sigma_\infty)(1 - e^{-\beta^p \zeta^p})$$
$$q^d(\zeta^d) = (\sigma_f^d - \sigma_\infty)(1 - e^{-\beta^d \zeta^d}), \tag{3.383}$$

where σ_∞ is the limit stress value for both plasticity and damage component, whereas β^p and β^d are the parameters governing the rate of saturation for plasticity and damage component, respectively.
For the example of a hollow plate composed of porous metallic material, we chose the following parameter values: Young's modulus $E = 240$ GPa, shear modulus $\mu = 92$ GPa, ultimate stress $\sigma_\infty = 210$ MPa, yield stress for plasticity $\sigma_y = 170$ MPa, damage stress $\sigma_f = 170$ MPa, saturation parameter for plasticity $\beta^p = 50$, and saturation parameter for damage $\beta^d = 50$.
The model symmetry allows us to choose only a quarter of the plate; see Figure 3.33. The incremental-iterative analysis is performed under displacement control, imposed at the free-end of the plate. The local iterations are also carried out in order to ensure that the stress remains the same in both components. The typical rate of convergence and computed iterative values of this computation are presented in Table 3.5.

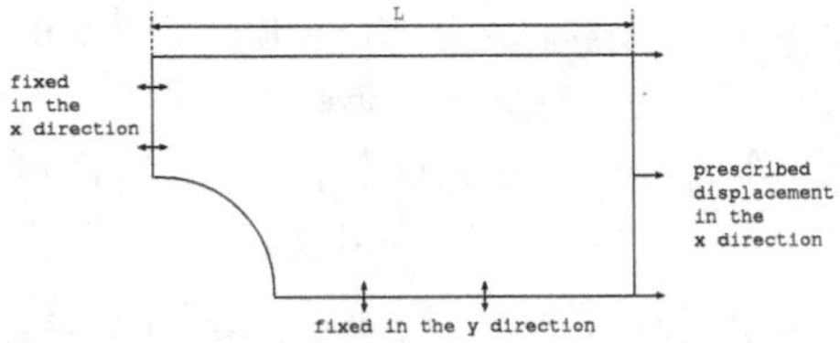

Fig. 3.33 Model geometry and boundary conditions.

Table 3.5 Iterative values for stress computations in plasticity and damage components of coupled plasticity-damage model.

iter.	$\|\boldsymbol{\sigma}^p - \boldsymbol{\sigma}^d\|^2$	ϵ_{11}^d	ϵ_{22}^d	ϵ_{12}^d
1	2.400521 10^{17}	0.01681690	0.006922074	−0.002569930
2	7.003845 10^{12}	0.01614148	0.007330000	−0.002210263
3	3.758989 10^{9}	0.01615158	0.007325925	−0.002293375
4	1.504535 10^{2}	0.01615126	0.007326051	−0.002292093
5	9.815405 10^{-5}	0.01615126	0.007326049	−0.002292127

In Figure 3.34 we present contours of the hardening variable for the von Mises plasticity model and the equivalent values for the coupled plasticity--damage model. We can observe for former the creation of the shear band failure mechanism typical of von Mises plasticity, which is completely absent from the coupled model.

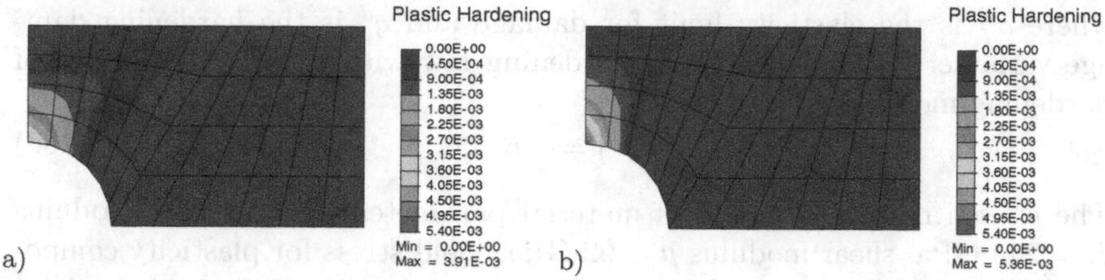

Fig. 3.34 Porous metallic material in traction: contours of plastic hardening variable for (a) plasticity model (b) coupled plasticity-damage model.

3.6.2.3 Concrete in compaction

We take the same specimen from the previous example, but assume now that it is made of concrete and submitted to compressive loading. The coupled plasticity-damage model components are selected in order to provide the representation of concrete in compaction. With no intention of constructing the most elaborate model of this kind, we choose for the plastic component the Drucker-Prager plasticity modle (see Figure 3.35).

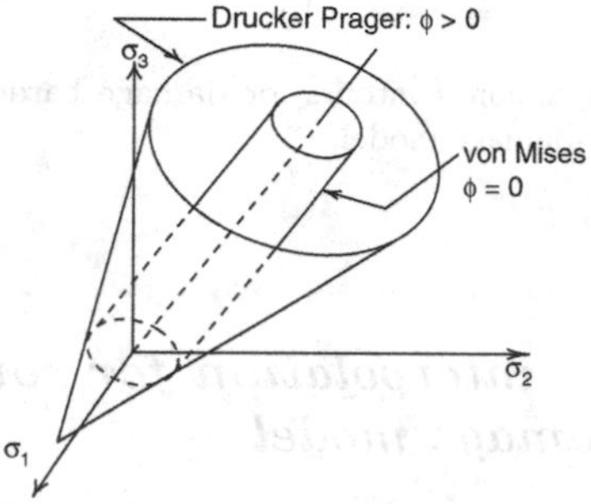

Fig. 3.35 Comparison between von Mises and Drucker-Prager plasticity criteria.

$$\phi^p(\boldsymbol{\sigma}) = \sqrt{dev(\boldsymbol{\sigma}) : dev(\boldsymbol{\sigma})} - tan(\alpha)\,\frac{1}{3}tr(\boldsymbol{\sigma}) - \sqrt{2/3}\,\sigma_f^p \tag{3.384}$$

where $tan(\alpha)$ is the material parameter which can characterize the internal friction. The damage component of the coupled model is supposed to represent the corresponding hardening phenomena due to concrete compaction. For that reason, we choose the damage criterion in terms of the spherical part of stress tensor:

$$\phi^d(\boldsymbol{\sigma}, q^d) =< -tr[\boldsymbol{\sigma}] > -(\sigma_f^d - q^d), \tag{3.385}$$

where σ_f^d is the elasticity limit for damage and q^d is the hardening damage variable. We choose a linear hardening law with a fairly large value of hardening modulus:

$$q^d(\zeta^d) = -K^d \zeta^d \tag{3.386}$$

The chosen numerical values of material parameters are: Young's modulus $E = 240$ GPa, shear modulus $\mu = 92$ GPa, yield stress for plasticity component $\sigma_y = 170$ MPa, internal friction parameter $tan\alpha = 0.6$, limit of elasticity for damage $\sigma_f = 210$ MPa, and damage hardening modulus $K^d = 200$. In order to illustrate the performance of the coupled plasticity-damage model, we compare the contours of damage hardening variable computed with the proposed coupled model along with the damage component only; see Figure 3.36.

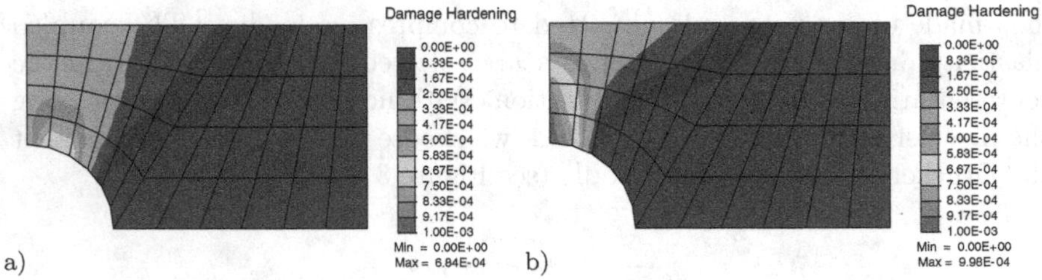

Fig. 3.36 Concrete in compaction: contours for damage hardening variable (a) damage model (b) coupled plasticity-damage model.

3.6.3 Direct stress interpolation for coupled plasticity-damage model

The coupled plasticity-damage model, which is described in the previous section, requires an iterative approach to local computation phase in order to obtain the final value of the damage deformation that will impose the equality of stress in each model component. Such an iterative procedure can be completely eliminated if we abandon the standard displacement-type interpolation in favor of a direct interpolation of stress tensor components.[13] The latter thus requires the presence of stress in the list of state variable, which further leads to a mixed variational principle referred to as Hellinger-Reissner, where independent interpolations can be chosen for displacement and stress fields (e.g. see [128]). At the level of constitutive model, the main difference from the previous approach concerns the use of the complementary energy instead of the internal energy. We can thus write:

[13] In terms of accuracy, the stress interpolation is not necessarily more accurate than the incompatible mode method, and two methods can be considered equivalent (see [35]).

$$\begin{aligned} \chi(\boldsymbol{\sigma}, \boldsymbol{\mathcal{D}}) &= \chi^e(\boldsymbol{\sigma}) + \chi^d(\boldsymbol{\sigma}, \boldsymbol{\mathcal{D}}) \\ \chi^e(\boldsymbol{\sigma}) &= \boldsymbol{\sigma}\boldsymbol{\epsilon}^e - \psi^e(\boldsymbol{\epsilon}^e) \; ; \; \chi^d(\boldsymbol{\sigma}, \boldsymbol{\mathcal{D}}) = \tfrac{1}{2}\boldsymbol{\sigma}\boldsymbol{\mathcal{D}}\boldsymbol{\sigma} \end{aligned} \tag{3.387}$$

All the other equations which define behavior of the coupled model remain the same, including the yield and damage criteria as well as the corresponding evolution equations.

The direct interpolation of stress components will impose that the time integration of stress be modified with respect to the procedure presented in the previous section. We will start again by computing the internal variables at time t_{n+1} for each component of the coupled model by using the implicit Euler scheme:

$$\begin{aligned} \boldsymbol{\epsilon}^p_{n+1} &= \boldsymbol{\epsilon}^p_n + \gamma^p_{n+1} \frac{\partial \phi^p_{n+1}}{\partial \boldsymbol{\sigma}_{n+1}}, \\ \zeta^p_{n+1} &= \zeta^p_n + \gamma^p_{n+1} \frac{\partial \phi^p_{n+1}}{\partial q^p_{n+1}}, \end{aligned} \tag{3.388}$$

$$\begin{aligned} \boldsymbol{\mathcal{D}}_{n+1}\sigma_{n+1} &= \boldsymbol{\mathcal{D}}_n \boldsymbol{\sigma}_{n+1} + \gamma^d_{n+1} \frac{\partial \phi^d_{n+1}}{\partial \boldsymbol{\sigma}_{n+1}}, \\ \zeta^d_{n+1} &= \zeta^d_n + \gamma^d_{n+1}, \end{aligned} \tag{3.389}$$

The computed internal variables should be admissible with respect to the chosen plasticity and damage criteria:

$$\begin{aligned} \gamma^p_{n+1}\phi^p(\boldsymbol{\sigma}_{n+1}, q^p_{n+1}) = 0; \; \gamma^p_{n+1} \geq 0; \; \phi^p_{n+1} \leq 0 \\ \gamma^d_{n+1}\phi^d(\boldsymbol{\sigma}_{n+1}, q^d_{n+1}) = 0; \; \gamma^d_{n+1} \geq 0; \; \phi^d_{n+1} \leq 0. \end{aligned} \tag{3.390}$$

Having completed the local computation phase, we will hold the internal variable values fixed and carry on with the global phase. The latter can be interpreted as enforcing the stationary condition for the Hellinger-Reissner functional, which employs independent displacement and stress fields:

$$\begin{aligned} &\Pi_{comp}(\mathbf{u}_{n+1}, \boldsymbol{\sigma}_{n+1}, \cdot) = \textstyle\int_\Omega \big(-\chi^e(\boldsymbol{\sigma}_{n+1}) - \chi^d(\boldsymbol{\sigma}_{n+1}, \boldsymbol{\mathcal{D}}_{n+1}) \\ &+ \boldsymbol{\sigma}_{n+1} \cdot (\nabla^s \mathbf{u}_{n+1} - \boldsymbol{\epsilon}^p_{n+1})\big) \, dV - \textstyle\int_{\Gamma_\sigma} \mathbf{u}_{n+1} \cdot \bar{\mathbf{t}}_{n+1} \, dA. \end{aligned} \tag{3.391}$$

The corresponding optimality condition for such a functional, leads to the following variational equations:

$$\begin{aligned} 0 &= G_u(\mathbf{u}_{n+1}, \boldsymbol{\sigma}_{n+1}, \cdot; \mathbf{w}) \\ &:= \textstyle\int_\Omega \nabla^s \mathbf{w} \cdot \boldsymbol{\sigma}_{n+1} \, dV - \int_{\Gamma_\sigma} \mathbf{w} \cdot \bar{\mathbf{t}}_{n+1} \, dA \\ 0 &= G_\sigma(\mathbf{u}_{n+1}, \boldsymbol{\sigma}_{n+1}, \cdot; \boldsymbol{\tau}) \\ &:= \textstyle\int_\Omega \boldsymbol{\tau} \cdot \left[\nabla^s \mathbf{u}_{n+1} - \boldsymbol{\epsilon}^p_{n+1} - \boldsymbol{\mathcal{D}}_{n+1}\boldsymbol{\sigma}_{n+1} - \boldsymbol{\mathcal{C}}^{-1}\boldsymbol{\sigma}_{n+1}\right] \, dV \end{aligned} \tag{3.392}$$

We note that the last equation is the weak form of the additive decomposition of the total deformation field into plastic deformation $\boldsymbol{\epsilon}^p_{n+1}$, elastic and damage deformation. The last two are expressed in terms of stress, by using $\boldsymbol{\epsilon}^e_{n+1} := \boldsymbol{\mathcal{C}}^{-1}\boldsymbol{\sigma}_{n+1}$ and $\boldsymbol{\epsilon}^d_{n+1} := \boldsymbol{\mathcal{D}}_{n+1}\boldsymbol{\sigma}_{n+1}$, respectively. Hence, contrary to the coupled model implementation presented in the previous section, the additive decomposition of the total deformation field is not enforced point-wise (at each Gauss quadrature point), but only at the element level. Namely, the stress field is defined independently in each element by using the Pian and Sumihara [223] stress interpolation in terms of five stress interpolation parameters

$$\boldsymbol{\sigma} = \mathbf{S}^e\boldsymbol{\beta} \;\Leftrightarrow\; \begin{cases} \sigma_{11} = \beta_1 + \beta_2\,\xi_2 \\ \sigma_{22} = \beta_3 + \beta_4\,\xi_1 \\ \sigma_{12} = \beta_5 \end{cases} \tag{3.393}$$

The chosen displacement approximation corresponds to the standard isoparametric $Q4$ element

$$u_{n+1}|_{\Omega^e} = \sum_{a=1}^{4} \mathbf{N}^e_a(\xi_1,\xi_2)\mathbf{d}^e_a \;;\; N^e_a(\xi_1,\xi_2) = \frac{1}{4}(1+\xi_{1,a}\xi_1)(1+\xi_{2,a}\xi_2) \tag{3.394}$$

This stress approximation is further transformed from natural to physical coordinates, by using the Jacobian matrix of $Q4$ element; the latter is computed only at the element center (see Zienkiewicz and Taylor [271]), in order to ensure the satisfaction of the patch test.
With the chosen stress interpolation of Pian and Sumihara [223], the last variational equation can be enforced in each element independently, resulting with:

$$\begin{aligned} \mathbf{0} = \mathbf{h}^e \;;\; &\forall e \in [1, n_{el}] \\ := \mathbf{e}_{n+1} - \mathbf{H}_n\boldsymbol{\beta}_{n+1} &- \hat{\mathbf{e}}^p(\boldsymbol{\beta}_{n+1},\zeta^p_{n+1},\gamma^p_{n+1}) - \hat{\mathbf{e}}^d(\boldsymbol{\beta}_{n+1},\zeta^d_{n+1},\gamma^d_{n+1}) \end{aligned} \tag{3.395}$$

In the last expression, we defined:

$$\begin{aligned} \mathbf{e}_{n+1} &= \int_\Omega \mathbf{S}^{e,T}(\mathbf{B}^e\mathbf{u}^e_{n+1} - \boldsymbol{\epsilon}^p_n)\,dV \\ \mathbf{H}_n &= \int_\Omega \mathbf{S}^{e,T}(\boldsymbol{\mathcal{D}}_n + \boldsymbol{\mathcal{C}}^{-1})\mathbf{S}^e\,dV \\ \hat{\mathbf{e}}^p(\boldsymbol{\beta}_{n+1},\zeta^p_{n+1},\gamma^p_{n+1}) &= \int_\Omega \mathbf{S}^{e,T}\gamma^p_{n+1}\frac{\partial\phi^p_{n+1}}{\partial\boldsymbol{\sigma}_{n+1}}\,dV \\ \hat{\mathbf{e}}^d(\boldsymbol{\beta}_{n+1},\zeta^d_{n+1},\gamma^d_{n+1}) &= \int_\Omega \mathbf{S}^{e,T}\gamma^d_{n+1}\frac{\partial\phi^d_{n+1}}{\partial\boldsymbol{\sigma}_{n+1}}\,dV, \end{aligned} \tag{3.396}$$

The discrete approximation of the global equilibrium equations can thus be rewritten as follows:

$$\mathbf{0} = \mathbf{r} := \mathop{\mathbb{A}}_{e=1}^{n_{elem}} (\mathbf{f}^{int,e} - \mathbf{f}^{ext,e}) \; ; \qquad \mathbf{f}^{int,e} = \int_{\Omega^e} \mathbf{B}^{e,T} \mathbf{S}^e \boldsymbol{\beta}_{n+1} \, dV \; ; \; \mathbf{f}^{ext,e} = \int_{\Gamma_\sigma^e} \mathbf{N}^{e,T} \bar{\mathbf{t}} \, dA \tag{3.397}$$

With the chosen Pian-Sumihara interpolations, we have to define three levels of computations: (i) point-wise computation at each Gauss quadrature point for internal variables, (ii) element-wise computation of stress parameters, and (iii) global-level computation of the set of equilibrium equations providing nodal values of displacements. We can still apply the operator split solution procedure to this problem, by solving sequentially the problems i, ii and iii. The flowchart of this solution procedure is given in Tables 3.6, 3.7 and 3.8. Having obtained the converged values of internal variables with iterative procedure presented in Table 3.6, we write again the same linearized form for the case when a new displacement increment is introduced:

$$\begin{bmatrix} -\mathbf{H}_n & \frac{\partial \hat{\mathbf{e}}^p}{\partial \gamma^p} & 0 & \frac{\partial \mathbf{e}^d}{\partial \gamma^d} & 0 \\ \frac{\partial \phi^p}{\partial \boldsymbol{\sigma}} \mathbf{S} & 0 & \frac{\partial \phi^p}{\partial q^p} & 0 & 0 \\ 0 & \frac{\partial \phi^p}{\partial q^p} & (\frac{\partial q^p}{\partial \zeta^p})^{-1} & 0 & 0 \\ \frac{\partial \phi^d}{\partial \boldsymbol{\sigma}} \mathbf{S} & 0 & 0 & 0 & \frac{\partial \phi^d}{\partial q^d} \\ 0 & 0 & 0 & \frac{\partial \phi^d}{\partial q^d} & (\frac{\partial q^d}{\partial \zeta^d})^{-1} \end{bmatrix} \begin{bmatrix} \Delta \boldsymbol{\beta}_{n+1}^{(i)} \\ \gamma_{n+1}^p \\ q_{n+1}^p - q_n^p \\ \gamma_{n+1}^d \\ q_{n+1}^d - q_n^d \end{bmatrix} = \begin{bmatrix} \Delta \mathbf{e}_{n+1}^{(i)} \\ 0 \\ 0 \\ 0 \\ 0 \end{bmatrix} \tag{3.400}$$

Table 3.6 Coupled plasticity-damage model: computation of internal variables.

i) local computation (at Gauss quadrature point) for plasticity and damage internal variables

Given: $\boldsymbol{\beta}_{n+1}^{(i)}$

Find: $\boldsymbol{\epsilon}_{n+1}^p$, ζ_{n+1}^p, $\boldsymbol{\mathcal{D}}_{n+1}$, ζ_{n+1}^d, such that γ_{n+1}^p and γ_{n+1}^d

step 1: compute elastic trial step $\phi^p(\boldsymbol{\beta}_{n+1}^{(i)}, q^p(\zeta_n^p))$; $\phi^d(\boldsymbol{\beta}_{n+1}^{(i)}, q^d(\zeta_n^d)) \overset{?}{\leq} 0$

IF $\phi^{p,d} \leq 0$ then $\gamma_{n+1}^{p,d} = 0$ exit

ELSE *step* 2: plastic correction, to solve

$$\begin{aligned} \boldsymbol{\epsilon}_{n+1}^p &= \boldsymbol{\epsilon}_n^p + \gamma_{n+1}^p \frac{\partial \phi^p}{\partial \boldsymbol{\sigma}_{n+1}} \\ \zeta_{n+1}^p &= \zeta_n^p + \gamma_{n+1}^p \frac{\partial \phi_{n+1}^p}{\partial q_{n+1}^p} \\ 0 &= \phi^p(\boldsymbol{\beta}_{n+1}^{(i)}, q^p(\zeta_{n+1}^p)) \end{aligned} \tag{3.398}$$

$$\begin{aligned} \boldsymbol{\mathcal{D}}_{n+1} \mathbf{S} \boldsymbol{\beta}_{n+1}^{(i)} &= \boldsymbol{\mathcal{D}}_n \mathbf{S} \boldsymbol{\beta}_{n+1}^{(i)} + \gamma_{n+1}^d \frac{\partial \phi^d}{\partial \boldsymbol{\sigma}_{n+1}} \\ \zeta_{n+1}^d &= \zeta_n^d + \gamma_{n+1}^d \frac{\partial \phi_{n+1}^d}{\partial q_{n+1}^d} \\ 0 &= \phi^d(\boldsymbol{\beta}_{n+1}^{(i)}, q^d(\zeta_{n+1}^d)) \end{aligned} \tag{3.399}$$

The static condensation is then used to reduce the size of this system, by keeping only the equations which are needed to carry out the stress computation; see Table 3.7

Table 3.7 Coupled plasticity-damage model: stress computations.

ii) Element level computation

Given: $\mathbf{u}_{n+1}^{(i)}, \gamma_{n+1}^{p}, \boldsymbol{\epsilon}_{n+1}^{p}, \zeta_{n+1}^{p}, \gamma_{n+1}^{d}, \boldsymbol{\mathcal{D}}_{n+1}, \zeta_{n+1}^{d}$

Find: $\boldsymbol{\beta}_{n+1}^{(i+1)}$

step 1: check convergence - IF $\mathbf{h}_{n+1}^{(i)} \leq tol$ sortir

ELSE $\mathbf{h}_{n+1}^{(i)} > tol$ continue with iterations

step 2: compute

$$\begin{aligned} \boldsymbol{\beta}_{n+1}^{(i+1)} &= \boldsymbol{\beta}_{n+1}^{(i)} + \Delta\boldsymbol{\beta}_{n+1}^{(i)} \\ \Delta\boldsymbol{\beta}_{n+1}^{(i)} &= -\left(\frac{\partial \mathbf{h}_{n+1}^{(i)}}{\partial \boldsymbol{\beta}_{n+1}^{(i)}}\right)^{-1} \Delta\mathbf{e}_{n+1}^{(i)} \end{aligned} \tag{3.401}$$

where

$$\frac{\partial \mathbf{h}_{n+1}^{(i)}}{\partial \boldsymbol{\beta}_{n+1}^{(i)}} = -\mathbf{H}_n - \mathbf{P}_{n+1}^{p} - \mathbf{P}_{n+1}^{d}, \tag{3.402}$$

and

$$\begin{aligned} \mathbf{P}_{n+1}^{p} = \int_{\Omega} \mathbf{S}^{e,T} \Bigg[& \gamma_{n+1}^{p} \frac{\partial^2 \phi_{n+1}^{p}}{\partial \boldsymbol{\sigma}_{n+1}^2} - \\ & \frac{\partial \phi_{n+1}^{p}}{\partial \boldsymbol{\sigma}_{n+1}} \left(\frac{\partial \phi_{n+1}^{p}}{\partial q_{n+1}^{p}} \frac{d q_{n+1}^{p}}{d \zeta_{n+1}^{p}} \frac{\partial \phi_{n+1}^{p}}{\partial q_{n+1}^{p}} \right)^{-1} \frac{\partial \phi_{n+1}^{p}}{\partial \boldsymbol{\sigma}_{n+1}} \Bigg] \mathbf{S}^{e} \, dV. \end{aligned} \tag{3.403}$$

$$\begin{aligned} \mathbf{P}_{n+1}^{d} = \int_{\Omega} \mathbf{S}^{e,T} \Bigg[& \gamma_{n+1}^{d} \frac{\partial^2 \phi_{n+1}^{d}}{\partial \boldsymbol{\sigma}_{n+1}^2} - \\ & \frac{\partial \phi_{n+1}^{d}}{\partial \boldsymbol{\sigma}_{n+1}} \left(\frac{\partial \phi_{n+1}^{d}}{\partial q_{n+1}^{d}} \frac{d q_{n+1}^{d}}{d \zeta_{n+1}^{d}} \frac{\partial \phi_{n+1}^{d}}{\partial q_{n+1}^{d}} \right)^{-1} \frac{\partial \phi_{n+1}^{d}}{\partial \boldsymbol{\sigma}_{n+1}} \Bigg] \mathbf{S}^{e} \, dV. \end{aligned} \tag{3.404}$$

Once we have obtained the converged value of stress, we can carry on with the iterative procedure for global equilibrium equations, in order to provide (if needed) an improved iterative guess for total strain field.

Table 3.8 Coupled plasticity-damage model: displacement computations.

iii) Global solution phase for equilibrium equations

Given: $\mathbf{u}^{(i)}_{n+1}$, $\boldsymbol{\beta}_{n+1}$, γ^p_{n+1}, $\boldsymbol{\epsilon}^p_{n+1}$, ζ^p_{n+1}, γ^d_{n+1}, $\boldsymbol{\mathcal{D}}_{n+1}$, ζ^d_{n+1}

step 1 : check convergence - IF $\| \mathbf{r}^{(i)}_{n+1} \| \leq tol$ exit - start new step

ELSE $\| \mathbf{r}^{(i)}_{n+1} \| > tol$ restart new iteration in the same step

step 2

$$\mathbf{u}^{(i+1)}_{n+1} = \mathbf{u}^{(i)}_{n+1} + \Delta\mathbf{u}^{(i)}_{n+1}$$

$$\mathop{\mathbb{A}}_{e=1}^{n_{elem}} \left(\mathbf{K}^{e,(i)}_{n+1} \right) \Delta\mathbf{u}^{(i)}_{n+1} = -\mathbf{r}^{(i)}_{n+1}, \tag{3.405}$$

(3.406)

where

$$\mathbf{K}^{e,(i)}_{n+1} = \int_\Omega \mathbf{B}^{e,T}\mathbf{S}^e \, dV [\mathbf{H}_n + \mathbf{P}^p_{n+1} + \mathbf{P}^d_{n+1}]^{-1} \int_\Omega \mathbf{S}^{e,T}\mathbf{B}^e \, dV \tag{3.407}$$

Chapter 4
Large displacements and deformations

In this chapter we will revisit some of the mechanics models for elastic and inelastic behavior studied previously, when placed within a geometrically nonlinear framework. In other words, we will study the motion of a deformable solid body submitted to large displacements, large rotations and large deformations, which will no longer allow to ignore the difference between the initial configuration at the beginning of a given loading program and the deformed configuration at the end.

The main difficulty in constructing the solutions for this class of problems concerns making the right choice with respect to a large diversity of suitable formulations for large displacement/large strain regime, such as Eulerian, Lagrangian, updated Lagrangian, arbitrary Lagrangian–Eulerian and many others. Choosing a particular formulation implies the corresponding choice of the reference frame, as well as the properly invariant measure of large strain. One way of dealing with a large number of possible formulations is by placing the theoretical development of a large deformation theory within the framework of differential manifolds (e.g. see Abraham, et al. [1] or Choquet-Bruhat and deWitt-Moretti [47]). In such a case, the appropriate transformation of the metric tensor[1] would allow us to easily switch from one to another formulation and the corresponding measure of large strain.[2] Unfortunately, the conceptual clarity of the theoretical formulation of finite deformation elasticity set on a differential manifold (e.g. see Marsden and Hughes [182] or Rougée [230]) does not imply any simplicity of implementation. Namely, we obtain elaborate component form of governing equations in curvilinear coordinates (e.g. see Green and Zerna [93] or Truesdell and Noll [257]), accompanied by a lack of robustness of the numerical models, which

[1] The metric tensor specifies how to measure the distance in the chosen curvilinear frame.

[2] As indicated in Marsden and Hughes [182], any objective strain rate in large deformation regime can be obtained as the Lie derivative of the metric tensor by using the appropriate choice of the operator to carry out the corresponding convective transfer back to the initial configuration (pull-back), followed by the subsequent transfer of time-derivative computation to the deformed configuration (push-forward).

A. Ibrahimbegovic, *Nonlinear Solid Mechanics: Theoretical Formulations and Finite Element Solution Methods*, Solid Mechanics and its Applications 160,

remain plagued by the locking phenomena due to inability of the polynomial basis to properly represent the rigid body modes (e.g. see Fonder and Clough [78]).
For that reason, we stay away in this chapter from the general framework of manifolds, and develop all large strain theoretical formulations in the three-dimensional Euclidean space. The notation we use is similar to the one proposed by Ciarlet [50], with the superscript 'φ' attached to all the fields which are parameterized in the deformed configuration. However, contrary to the choice in [50] (and number of others[3]), we make every effort to distinguish between unknown fields of vectors and tensors and their coordinate representation in terms of matrices, as well as to ensure the clarity as to what configuration the given field belongs to and with respect to which reference frame it is parameterized. The choice of the Euclidean space is sufficient to tackle the vast majority of problems in large deformations related to practical applications. In fact, even some typical exceptions, such as shells (e.g. see Naghdi [193]) that normally require a manifold framework, can significantly be simplified (e.g., Ibrahimbegovic [114] or Ibrahimbegovic and Gruttmann [128]) by making use of the finite element technology and placing the local Cartesian frames at numerical integration points.
The main goal of this chapter is to show how to solve a boundary value problem concerning the deformable solid body in large motion, or in other words how to obtain the corresponding displacement, deformation and stress fields for the case when the displacements are no longer infinitesimal. Even in large displacement regime, the boundary value problem stull contains three groups of equations, governing the kinematics, the equilibrium and the constitutive behavior, accompanied by the boundary conditions that should ensure that the problem solution remains unique.
The main difference from small deformation theory, where we invariably use the infinitesimal deformation, concerns a large (theoretically infinite) number of different possibilities to define the strain measure in large deformation case. Some of the most frequently used strain measures are presented in this chapter. For each of these strain measures, the corresponding kinematics equation, expressing the relation between the displacement and strain, is nonlinear. The latter implies that the equilibrium equations are also nonlinear.[4]
We also present in this chapter how to construct the constitutive laws in large deformations, both for elastic and inelastic case. Constructing a constitutive model for large deformations is by far more demanding task from the similar one for small deformation case, since we have to consider a number of different strain regimes (for both small and large strains). Moreover, we have to ensure the constitutive response invariance in the large strain regime under superposed large rotations and the corresponding change of frame in both

---

[3] The same choice is made by [46], [50], [96], or [209] among others.

[4] One exception to this rule is the Euler buckling as a typical example of linear instability problems, where the kinematics equation is linear, but the equilibrium is established in the deformed configuration by using a nonlinear equation.

initial and deformed configurations. We show that the constitutive models constructed in the space of principal axes of strain tensor can easily fulfill the objectivity requirements, both for elastic and inelastic case. Two important examples of this kind of constitutive models are studied in detail in this chapter: finite deformation elasticity and finite deformation plasticity. The pertinent details of their numerical implementation within the finite element framework are also presented. We evaluate both spatial and material descriptions of proposed constitutive models, which employ the deformed and initial configurations, respectively.

4.1 Kinematics of large displacements

The main role of the kinematics is to study the motion of a solid deformable body, which means to identify the successive configurations of the body during a given loading program parameterized by the pseudo-time t.

4.1.1 Motion in large displacements

Within the framework of 3D Euclidean space, $\mathbb{R}^3$, the configuration of a solid body is considered as an assembly of interacting particles Ω (see Figure 4.1) with each particle identified by its position vector:

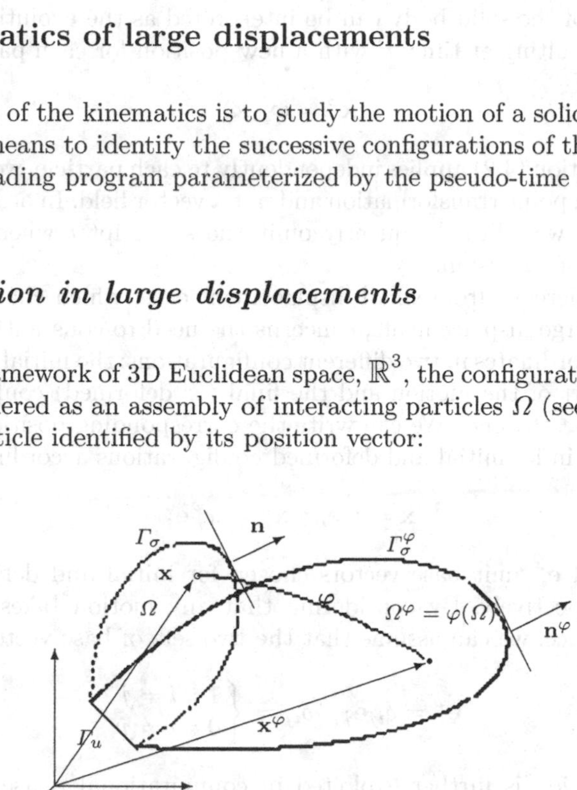

Fig. 4.1 Initial and deformed configurations of solid deformable body in Euclidean space.

$$\mathbf{x} = x_i \mathbf{e}_i \mapsto \mathrm{x} = \begin{bmatrix} x_1 \\ x_2 \\ x_3 \end{bmatrix} \tag{4.1}$$

The set of all particles which constitute the solid body will occupy the domain $\Omega \in \mathbb{R}^3$. The domain Ω is considered bounded with the frontier or boundary $\Gamma = \partial\Omega$, which is sufficiently smooth (at least piece-wise) to define the unit vector of exterior normal $\mathbf{n}$. The complete boundary is divided in two parts, one denoted Γ_u where we prescribe the displacements,[5] and the other denoted Γ_σ where we impose the stress component values. For a well-posed boundary value problem, it is not possible to impose simultaneously both the displacement and stress; hence it must hold that $\Gamma_u \bigcap \Gamma_\sigma = \emptyset$ and $\Gamma_u \bigcup \Gamma_\sigma = \Gamma$. Moreover, in order to be able to compute the unique solution to a boundary value problem, we require that $\Gamma_u \neq \emptyset$ be chosen such that the rigid body motion is eliminated.

The motion of the solid body can be interpreted as the evolution of its configuration, resulting at time 't' with a new position for each particle:

$$\mathbf{x}^\varphi = \varphi_t(\mathbf{x}) \tag{4.2}$$

Since the relation (4.2) applies independently to each particle, we will consider that $\varphi_t(\cdot)$ is a point-transformation and not a vector field. In order to simplify the notation, we will subsequently omit the subscript t whenever possible without risk of confusion.

The main difference from small displacement case, which characterizes any problem in large displacements, concerns the need to constantly account for two sets of coordinates in two different configurations: the initial configuration Ω at the start of the motion and the final (or deformed) configuration Ω^φ (= "$\varphi(\Omega)$") at the end. We can write the corresponding position vectors for any particles in its initial and deformed configurations according to:

$$\mathbf{x} = x_i \mathbf{e}_i \; ; \; \mathbf{x}^\varphi = x_i^\varphi \mathbf{e}_i^\varphi \tag{4.3}$$

where $\mathbf{e}_i$ and $\mathbf{e}_i^\varphi$ unit base vectors chosen for initial and deformed configurations, respectively. By considering that any motion takes place in the Euclidean space, we can assume that the two sets of base vectors coincide:

$$\mathbf{e}_i^\varphi = \delta_{ij}\mathbf{e}_j \; , \; \delta_{ij} = \begin{cases} 1 \; ; \; i = j \\ 0 \; ; \; i \neq j \end{cases} \tag{4.4}$$

This observation is further exploited in computational phase. However, in the development of the theoretical formulation it is important for clarity to keep different notation for the base vectors in the initial versus the deformed configuration.

Within the framework of geometrically linear theory with small displacements and small displacement gradients, the two base vectors and two sets of coordinates in the initial and the deformed configurations are considered to coincide when computing the derivatives or the integrals; therefore, we can write:

[5] More precisely, we either impose a zero displacement where the supports are placed, or a known value of the displacement due to support settlement.

$$\text{if } \| \nabla \mathbf{d} \| \mapsto \| \nabla \mathbf{u} \| << 1 \Rightarrow \mathbf{x}^{\varphi} \approx \mathbf{x} + cst. \Rightarrow \begin{cases} \frac{\partial}{\partial x_i^{\varphi}}(\cdot) \approx \frac{\partial}{\partial x_i}(\cdot) \\ \int_{\Omega^{\varphi}}(\cdot) \approx \int_{\Omega}(\cdot) \end{cases} \tag{4.5}$$

We recall that the latter is the key result which allows to simplify the development of either strong or weak form of the equilibrium equations by simply computing the derivatives and integrals of the true stress acting in the deformed configuration with respect to the coordinates chosen in the initial configuration.

For the geometrically nonlinear theory with large displacements and large displacement gradients, this kind of simplification can no longer be justified. In other words, with large displacements and deformations, we are obliged to choose the configuration with which we are working in development of the theoretical formulation and numerical solution of a boundary value problem. In that sense, we can choose between the Lagrangian formulation considering that all the unknown variables are functions of coordinates x_i in the initial configuration, and the Eulerian formulation where all the variables depend upon the coordinates x_i^{φ} in the deformed configuration. In principle, the Eulerian formulation is well suitable for problems of fluid mechanics where the only configuration of interest is the current deformed configuration (for example, the problems of flooding of a city area) and where the constitutive behavior does not depend on the deformation trajectory (for Newtonian fluids, for example, the Cauchy stress is directly proportional to the spatial velocity gradient; see Duvaut [71]). The Lagrangian formulation is more suitable for solid mechanics, since it uses the configuration we should know the best – the initial configuration of the solid body. Moreover, this kind of formulation requires to take into account the complete deformation trajectory leading to a particular deformed configuration, which allows to define the corresponding evolution of the internal variables and the resulting value of stress for solid materials with inelastic behavior (for example, large strain plasticity and damage models).

We elaborate further on these ideas for 1D problem of an elastic truss-bar in the large displacement regime. In the initial configuration of the bar, denoted $\bar{\Omega} = [0, l]$, we suppose that the stress-free position of each particle is described by its position vector, with the only non-zero component x

$$x \in \bar{\Omega} \; ; \; \bar{\Omega} = [0, l] \tag{4.6}$$

We can describe the motion of the solid as the following transformation:

$$\varphi : \bar{\Omega} \times [0, T] \mapsto \mathbb{R} \tag{4.7}$$

where $t \in [0, T]$ is the pseudo-time loading parameter. For a fixed value of pseudo-time 't', we can describe the deformation of the solid body by specifying the new position of each particle:

$$x^{\varphi} = \varphi_t(x) \tag{4.8}$$

The one-dimensional case allows us to introduce the displacement field $d_t(x)$, and describe the motion according to:

$$x^{\varphi} := \varphi_t(x) = x + d_t(x) \tag{4.9}$$

The assembly of particles in the new position will constitute the deformed or current configuration of the solid body:

$$\bar{\Omega}_t^{\varphi} = [0, l_t^{\varphi}] \; ; \; l_t^{\varphi} = \varphi_t(l) \tag{4.10}$$

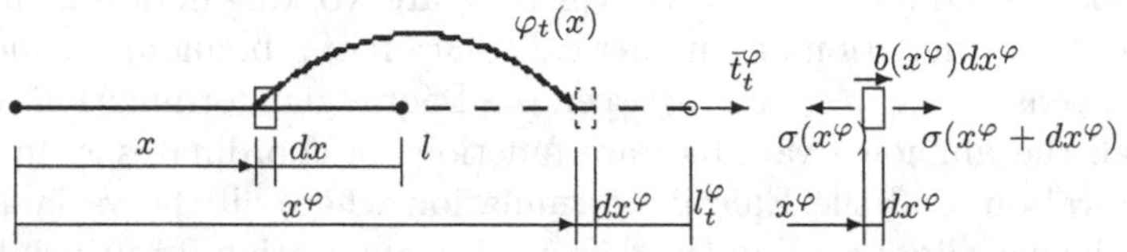

Fig. 4.2 Initial and deformed configurations of one-dimensional solid body in large displacement regime.

The imposed zero displacement on the Dirichlet boundary and the choice of the reference frame in Figure 4.2 imply that both the initial and deformed configurations include the origin of x-axis, with $\Gamma_u := \{0\}$:

$$d_t(0) = 0 \; (\varphi_t(0) = 0) \tag{4.11}$$

The Neumann boundary condition, corresponding to the imposed traction $\bar{t}_t^{\varphi}$ at the right end of the bar, can be written for the same case as:

$$\sigma_t^{\varphi}(l_t^{\varphi}) = \bar{t}_t^{\varphi} \tag{4.12}$$

where $\sigma_t^{\varphi}(\cdot)$ is the true or Cauchy stress. The boundary condition in (4.12) is written in spatial or Eulerian description, assuming that we are working with the deformed configuration and that all the variables are expressed as functions of coordinates x^{φ}. The Neumann boundary will change constantly for such a case according to: $\Gamma_{\sigma} = \{l_t^{\varphi} = \varphi_t(l)\}$. Choosing the material or Lagrangian description, considering that all the variables are expressed as the function of the coordinate x in the initial configuration, will allow us to fix the Neumann boundary with $\Gamma_{\sigma} = \{l\}$. However, this kind of choice will change the corresponding boundary condition, by requiring the appropriate representation of stress according to: $\sigma(\varphi(x)) = P(x)$.

Remark on the choice of formulation: Other than two basic theoretical formulation of mechanics at large displacements, Eulerian and Lagrangian, many other choices are possible. Namely, within the framework of the incremental/iterative analysis used for a nonlinear problem of this kind, we can move the reference frame in each increment or in each iteration and thus obtain so-called updated Lagrangian formulation (e.g. see Bathe [19]). This motion of the reference frame at each increment or iteration can also be performed by a rotation, in which case we obtain the co-rotational formulation (e.g. see Crisfield [60]). The main advantage of any such formulation is that moving frame would allow to exploit the results from the small strain theory pertinent to tangent operator or internal force vector, which could simplify their computations for nonlinear theory (ideally, by only using a simple transformation, such as the rotation). However, fairly vague hypothesis on small strains with large (or moderate) rotations which would allow this kind of simplification, as well as a large variety of the potential choices for the moving frame, makes any such formulation rather difficult to develop in a very rigorous manner. There are basically only a few exceptions to this rule, such as a 2-node truss-bar element in 1D, a 3-node CST element in 2D or a 4-node tetrahedral element in 3D. Each of these elements provides a constant approximation for the strain and rotation fields throughout the element domain and allows for an arbitrary placement of the moving reference frame separating large rotations from the small strains. Another type of theoretical formulation of interest is so-called arbitrary Eulerian–Lagrangian formulation (e.g. see Belytschko et al. [28]), which is suitable for the class of interaction problems where two interacting fields have very different nature of their motion properties (such as in fluid-structure interaction, with fluid motion which is handled by Eulerian and solid motion by Lagrangian formulation).

4.1.2 Deformation gradient

In the developments to follow we will favor the Lagrangian formulation, where the motion of a deformable solid body is described by specifying the motion of each particle. The latter can be represented by a point transformation (not a vector field) $\mathbf{x}^{\varphi} = \varphi(\mathbf{x})$; $\forall \mathbf{x} \in \Omega$. For any motion which is placed in the Euclidean space, we can also define the large displacement vector (see Figure 4.3) according to:

$$\mathbf{d}(\mathbf{x}) = \mathbf{x}^{\varphi} - \mathbf{x} \;\Leftrightarrow\; d_i(\mathbf{x}) = \varphi_j \delta_{ij} - x_i \tag{4.13}$$

If we would also like to define the total deformation field, we ought to describe not only the motion of a single particle but also the motion of its neighbors. For example, we can choose any neighboring particle placed at infinitesimal distance $d\mathbf{x}$ from the particle $\mathbf{x}$, in order to provide the local description of the deformation field. By denoting the corresponding displacement vector with $\mathbf{d}(\mathbf{x}+d\mathbf{x})$, we can describe the new position of the neighboring particle in the deformed configuration. The latter can be compared against an alternative

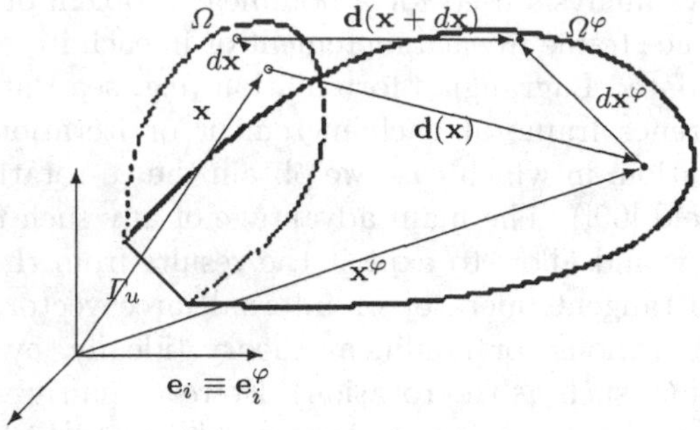

Fig. 4.3 Total displacement field.

way of describing the new position of the neighboring particle (see Figure 4.3); hence, we can write:

$$\mathbf{x}^\varphi + d\mathbf{x}^\varphi = \mathbf{x} + d\mathbf{x} + \mathbf{d}(\mathbf{x} + d\mathbf{x}) \implies d\mathbf{x}^\varphi = d\mathbf{x} + \mathbf{d}(\mathbf{x} + d\mathbf{x}) - \mathbf{d}(\mathbf{x}) \quad (4.14)$$

The last conclusion is drawn by appealing to the definition of the displacement vector in (4.13). By further exploiting the Taylor series formula, and truncating after the first-order term,[6] we can rewrite the last expression according to:

$$\begin{aligned} \mathbf{d}(\mathbf{x} + d\mathbf{x}) &= \mathbf{d}(\mathbf{x}) + \nabla\mathbf{d}(\mathbf{x})\; d\mathbf{x} + o(\| \,d\mathbf{x}\, \|) \\ \Leftrightarrow\; d_i(\mathbf{x} + d\mathbf{x}) &= d_i(\mathbf{x}) + \tfrac{\partial d_i}{\partial x_j}(\mathbf{x})\, dx_j + o(\| \,d\mathbf{x}\, \|) \end{aligned} \quad (4.15)$$

where $\nabla\mathbf{d}$ is the large displacement gradient. Comparison of the last two results readily shows that the infinitesimal position vector $d\mathbf{x}$ is transformed into its new position in the deformed configuration $d\mathbf{x}^\varphi$ by using the deformation gradient[7] $\mathbf{F}$:

$$d\mathbf{x}^\varphi = \underbrace{(\mathbf{I} + \nabla\mathbf{d})}_{\mathbf{F}}\, d\mathbf{x} \implies \boxed{d\mathbf{x}^\varphi = \mathbf{F} d\mathbf{x}} \;\; ; \; F_{ij} := \underbrace{\frac{\partial x_i}{\partial x_j}}_{\delta_{ij}} + \frac{\partial d_i}{\partial x_j} = \frac{\partial \varphi_i}{\partial x_j} \quad (4.16)$$

Since the deformation gradient $\mathbf{F}$ is a linear transformation (or a tensor) mapping a vector $d\mathbf{x}$ placed in the initial configuration[8] into a new vector $d\mathbf{x}^\varphi$ placed in the deformed configuration, we can conclude that $\mathbf{F}$ is a two-

[6] For this reason, this kind of formulation is also referred to as the *first gradient theory.*

[7] For a theoretical formulation of nonlinear solid mechanics set on a manifold (which allows a general choice of curvilinear coordinates), the displacement field or displacement gradient is not defined, but the deformation gradient $\mathbf{F}$ still is (e.g. see Marsden and Hughes [182], p. 59).

[8] More precisely, the infinitesimal vector dx is placed in the tangent space of the initial configuration, but in the Euclidean setting we do not have to distinguish between these two.

point tensor (e.g. see Marsden and Hughes [182] or Truesdell and Toupin [258]); we can thus write the corresponding component representation:

$$\boxed{\mathbf{F} = \frac{\partial \varphi_i}{\partial x_j} \mathbf{e}_i^{\varphi} \otimes \mathbf{e}_j} \tag{4.17}$$

By introducing the tensor notation for gradient operator $\boldsymbol{\nabla} = \frac{\partial}{\partial x_i}\mathbf{e}_i$, we can also write $\mathbf{F}$ according to:

$$\mathbf{F} = \boldsymbol{\varphi} \otimes \boldsymbol{\nabla} \; ; \; \boldsymbol{\varphi} = \varphi_i \mathbf{e}_i^{\varphi} \tag{4.18}$$

It is important to note that such a definition of the deformation gradient remains perfectly well-defined within the manifold framework. However, it is only in the Euclidean setting, where the displacement field $\mathbf{d}$ is defined, that we can also introduce the deformation gradient through the displacement gradient:

$$\nabla \mathbf{d} := \mathbf{d} \otimes \boldsymbol{\nabla} = \mathbf{F} - \mathbf{I} \; ; \; \nabla \mathbf{d} = \frac{\partial d_i}{\partial x_j} \mathbf{e}_i^{\varphi} \otimes \mathbf{e}_j \; ; \; \mathbf{I} := \mathbf{x} \otimes \boldsymbol{\nabla} = \delta_{ij} \mathbf{e}_i^{\varphi} \otimes \mathbf{e}_j \tag{4.19}$$

The components of tensors $\mathbf{F}$ and $\nabla \mathbf{d}$ can be written in matrix notation as:

$$\begin{aligned} \mathbf{F} := \nabla \boldsymbol{\varphi} \; &\mapsto \; \mathsf{F} = \begin{bmatrix} \frac{\partial \varphi_1}{\partial x_1} & \frac{\partial \varphi_1}{\partial x_2} & \frac{\partial \varphi_1}{\partial x_3} \\ \frac{\partial \varphi_2}{\partial x_1} & \frac{\partial \varphi_2}{\partial x_2} & \frac{\partial \varphi_2}{\partial x_3} \\ \frac{\partial \varphi_3}{\partial x_1} & \frac{\partial \varphi_3}{\partial x_2} & \frac{\partial \varphi_3}{\partial x_3} \end{bmatrix} \; ; \\ \nabla \mathbf{d} \; &\mapsto \; \mathsf{D} = \begin{bmatrix} \frac{\partial d_1}{\partial x_1} & \frac{\partial d_1}{\partial x_2} & \frac{\partial d_1}{\partial x_3} \\ \frac{\partial d_2}{\partial x_1} & \frac{\partial d_2}{\partial x_2} & \frac{\partial d_2}{\partial x_3} \\ \frac{\partial d_3}{\partial x_1} & \frac{\partial d_3}{\partial x_2} & \frac{\partial d_3}{\partial x_3} \end{bmatrix} \end{aligned} \tag{4.20}$$

It is easy to show that the deformation gradient not only controls the transformation of any infinitesimal vector from the initial to deformed configuration, but also the corresponding transformation of an infinitesimal surface element or an infinitesimal volume element (see Figure 4.4). Let dA be an infinitesimal surface element which can be constructed as the vector product

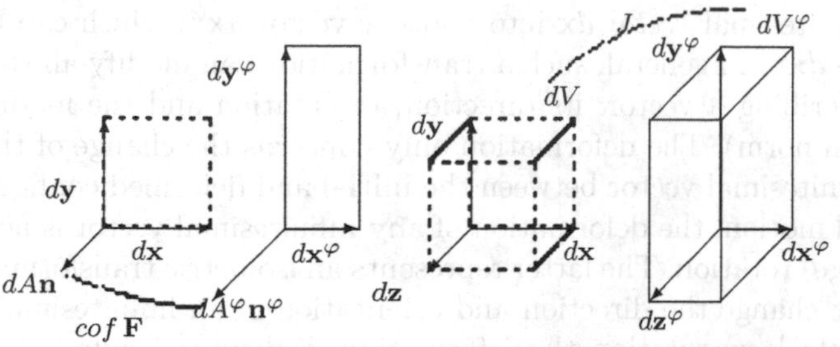

Fig. 4.4 Transformation of an infinitesimal surface and volume elements between initial and deformed configurations.

of two infinitesimal vectors $d\mathbf{x}$ and $d\mathbf{y}$, where $\mathbf{n} = (d\mathbf{x} \times d\mathbf{y}) / \parallel d\mathbf{x} \times d\mathbf{y} \parallel$ is the unit exterior normal. By using (4.16), we can define the new position of this surface element by using the cofactor of the deformation gradient $cof[\mathbf{F}]$:

$$\begin{aligned} dA^{\varphi}\mathbf{n}^{\varphi} &:= d\mathbf{x}^{\varphi} \times d\mathbf{y}^{\varphi} \\ &= (\mathbf{F}d\mathbf{x}) \times (\mathbf{F}d\mathbf{y}) \\ &= (det[\mathbf{F}]\mathbf{F}^{-T})\underbrace{(d\mathbf{x} \times d\mathbf{y})}_{dA\mathbf{n}} \\ &= dA(cof[\mathbf{F}])\mathbf{n} \end{aligned} \tag{4.21}$$

$$\Longrightarrow \quad \boxed{cof[\mathbf{F}] = det[\mathbf{F}]\mathbf{F}^{-T}} \quad ; \; \mathbf{F}^{-T} \equiv (\mathbf{F}^{-1})^T$$

The last result, yet referred to as the Nanson formula, remains valid in the framework of manifolds (e.g. see Marsden and Hughes [182] or Truesdell and Noll [257]). By exploiting the Nanson formula we can easily compute the change of an infinitesimal volume element between the initial and the deformed configurations, as the scalar product between the surface vector $dA\mathbf{n}$ with the infinitesimal vector $d\mathbf{z}$:

$$\begin{aligned} &dV^{\varphi} := d\mathbf{z}^{\varphi} \cdot dA\mathbf{n}^{\varphi} = \mathbf{F}d\mathbf{z} \cdot J\mathbf{F}^{-T}dA\mathbf{n} = J\,d\mathbf{z} \cdot dA\mathbf{n} = J\,dV \; ; \\ &\Longrightarrow \quad \boxed{J = det[\mathbf{F}]} \end{aligned} \tag{4.22}$$

The latter confirms that the change of any infinitesimal volume element is governed by the determinant of the deformation gradient. The same result can be obtained by using the standard formula for change of coordinates between the initial and the deformed configurations, from $\mathbf{x}$ to $\mathbf{x}^{\varphi} = \varphi(\mathbf{x})$, which allows us to write:

$$\begin{aligned} dV^{\varphi} &= (d\mathbf{x}^{\varphi} \times d\mathbf{y}^{\varphi}) \cdot d\mathbf{z}^{\varphi} \; ; \; \mathbf{x}^{\varphi} = \varphi(\mathbf{x}) \Rightarrow d\mathbf{x}^{\varphi} = \nabla\varphi\, d\mathbf{x} \\ &= (\mathbf{F}d\mathbf{x} \times \mathbf{F}d\mathbf{y}) \cdot \mathbf{F}d\mathbf{z} \\ &= \underbrace{det(\mathbf{F})}_{J}\underbrace{(d\mathbf{x} \times d\mathbf{y}) \cdot d\mathbf{z}}_{dV} \end{aligned} \tag{4.23}$$

Polar decomposition

As indicated by (4.16), the deformation gradient is a linear transformation of an infinitesimal *vector* $d\mathbf{x}$ into another vector $d\mathbf{x}^{\varphi}$, which can be written: $\mathbf{F} : d\mathbf{x} \mapsto d\mathbf{x}^{\varphi}$. In general, such a transformation can modify all three parameters describing a vector: its direction, orientation and the modulus (or its Euclidean norm). The deformation only concerns the change of the modulus of an infinitesimal vector between the initial and deformed configurations. In a general motion, the deformation of any infinitesimal vector is accompanied by a (large) rotation. The latter represents an isometric transformation which may only change the direction and orientation of an infinitesimal vector. In the case of a large rotation, the deformation gradient is denoted as $\mathbf{R}$. For such a special form of deformation gradient that leaves the norm of the transformed vector the same in the initial and deformed configurations, $\| d\mathbf{x}^{\varphi} \| = \| d\mathbf{x} \|$, we can show that:

$$
\begin{aligned}
d\mathbf{x} \cdot d\mathbf{x} &= d\mathbf{x}^{\varphi} \cdot d\mathbf{x}^{\varphi} \\
&= \mathbf{R}d\mathbf{x} \cdot \mathbf{R}d\mathbf{x} \\
&= d\mathbf{x} \cdot \underbrace{\mathbf{R}^T\mathbf{R}}_{\mathbf{I}} d\mathbf{x} \\
\Longrightarrow \quad & \mathbf{R}^T\mathbf{R} = \mathbf{I} \Leftrightarrow \mathbf{R}^{-1} = \mathbf{R}^T
\end{aligned}
\tag{4.24}
$$

We can thus conclude that the deformation gradient of an isometric transformation $\mathbf{R}$ has to be an orthogonal tensor. Therefore, for two successive large rotations, with $\mathbf{R}_1$ followed by $\mathbf{R}_2$, the resulting deformation gradient is (also) the orthogonal tensor that can be written as the multiplication of these two, respecting rigorously the order of rotations:

$$
\mathbf{F} \equiv \mathbf{R} = \mathbf{R}_2\mathbf{R}_1 \tag{4.25}
$$

We can also use the multiplicative decomposition of the deformation gradient in a more general case, where the corresponding transformation of an infinitesimal vector to its final position in the deformed configuration was produced not only by a rotation but also by a deformation. In other words, by using so-called polar decomposition, the deformation gradient can be written as a multiplicative split between an orthogonal tensor $\mathbf{R}$ of large rotation and a symmetric, positive-definite stretch tensor $\mathbf{U}$ that can provide the measure of large deformation. It is again very important to respect the order of transformations, with a large deformation followed by a large rotation:

$$
\boxed{\mathbf{F} = \mathbf{R}\mathbf{U}} \;\; ; \; \mathbf{R}^T = \mathbf{R}^{-1} \; ; \; \mathbf{U}^T = \mathbf{U} ; \; \| d\mathbf{x}^{\varphi} \| = \| \mathbf{U}\mathbf{x} \| \tag{4.26}
$$

Tensor $\mathbf{U}$ is the first large strain measure, yet referred to as the right stretch tensor.

The deformation gradient is a very important example of a two-point tensor, which transforms a vector in the initial configuration into another vector in the deformed configuration: $d\mathbf{x}^{\varphi} = \mathbf{F}d\mathbf{x}$. We will also consider the large rotation tensor $\mathbf{R}$ as a two-point tensor. This will further imply that the stretch tensor $\mathbf{U}$ represents the material deformation measure, producing the deformed vector that remains in the initial configuration:

$$
\begin{aligned}
& \mathbf{F} = \tfrac{\partial \varphi_i}{\partial x_j} \mathbf{e}_i^{\varphi} \otimes \mathbf{e}_j \;\&\; \mathbf{F} = \mathbf{R}\mathbf{U} \; ; \\
\Longrightarrow \quad & \boxed{\mathbf{R} = R_{ij}\mathbf{e}_i^{\varphi} \otimes \mathbf{e}_j} \; ; \; \boxed{\mathbf{U} = U_{ij}\mathbf{e}_i \otimes \mathbf{e}_j}
\end{aligned}
\tag{4.27}
$$

The rotation tensor interpretation as a two-point tensor, provides a consistent interpretation for an alternative form of the polar decomposition using the inverse order to (4.26), where a large rotation $\mathbf{R}$ is followed by a large deformation described by another stretch tensor $\mathbf{V}$. The latter is thus considered as the spatial deformation measure changing the norm of a vector in deformed configuration:

$$\boxed{\mathbf{F} = \mathbf{V}\mathbf{R}} \quad ; \Longrightarrow \quad \boxed{\mathbf{V} = V_{ij}\mathbf{e}_i^{\varphi} \otimes \mathbf{e}_j^{\varphi}} \tag{4.28}$$

In order to distinguish between the two stretch tensors, we refer to $\mathbf{U}$ as the right stretch tensor and to $\mathbf{V}$ as the left stretch tensor, with the terminology which implies their position within the polar decomposition formula. The corresponding graphic illustration of the polar decomposition is presented in Figure 4.5.

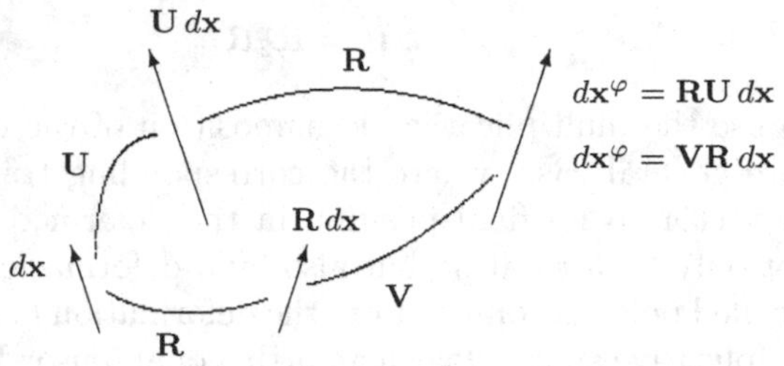

Fig. 4.5 Polar decomposition defining the left stretch tensor $\mathbf{V}$ and the right stretch tensor $\mathbf{U}$.

The direct computation of the polar decomposition can be given in 2D case, by exploiting the Cayley–Hamilton theorem that each symmetric tensor will satisfy its characteristic equation; see Table 4.1.

4.1.3 Large deformation measures

In large deformation regime, we can define other deformation measures[9] than the right or the left stretch tensors $\mathbf{U}$ and $\mathbf{V}$. For stress computations we do not need the rotations, but only the deformations. It is thus of particular interest for computational efficiency to use the deformation measures that can extract the deformation more directly, with no need to perform the polar decomposition of the deformation gradient, as for the case of the stretch tensors. One such measure is the Cauchy–Green right deformation tensor, which, thanks to the orthogonality of the rotation tensor, can be obtained with no need for polar decomposition:

[9] Theoretically, we can define an infinite number of them.

Table 4.1 Polar decomposition of deformation gradient $\mathbf{F}$ in 2D case.

Given: $\mathbf{F}$;	Find: $\mathbf{RU} = \mathbf{F}$, such that $\mathbf{R}^T = \mathbf{R}^{-1}$, $\mathbf{U} = \mathbf{U}^T$
eliminate rotation	$\mathbf{C} = \mathbf{F}^T\mathbf{F} = \mathbf{U}^T \underbrace{\mathbf{R}^T\mathbf{R}}_{I}\mathbf{U} = \mathbf{U}^2$
invariants of $\mathbf{C}$:	$I_c = trace[\mathbf{C}] = C_{11} + C_{22}$; $H_c = det[\mathbf{C}] = C_{11}C_{22} - C_{12}^2$
characteristic equation of $\mathbf{C}$:	$\lambda_C - I_C\,\lambda_C + H_C = 0$
by Cayley–Hamilton theorem:	$\mathbf{C}^2 - I_C\mathbf{C} + H_C\mathbf{I} = \mathbf{0}$
characteristic equation of $\mathbf{U}$:	$\lambda_U^2 - I_U\lambda_U + H_U = 0$
by Cayley–Hamilton theorem:	$\mathbf{U}^2 - I_U\mathbf{U} + H_U\mathbf{I} = \mathbf{0}$
since: $\mathbf{C} = \mathbf{U}^2 \Longrightarrow \underbrace{det\mathbf{C}}_{H_C} = (\underbrace{det\mathbf{U}}_{H_U})^2$	$\mathbf{C} - I_U\mathbf{U} + \sqrt{H_C}\mathbf{I} = \mathbf{0} \mid trace[\cdot]$
$\Rightarrow I_U{}^2 = I_C + 2\sqrt{H_c}$	$\Rightarrow I_U = \sqrt{I_C + 2\sqrt{H_C}}$
$\boxed{\mathbf{U} = \frac{1}{\sqrt{I_C+2\sqrt{H_C}}}(\mathbf{C} + \sqrt{H_c}\mathbf{I})}$	$\boxed{\mathbf{R} = \mathbf{F}\mathbf{U}^{-1}}$

$$\mathbf{C} := \mathbf{F}^T\mathbf{F} = \underbrace{\mathbf{U}^T}_{\mathbf{U}}\underbrace{\mathbf{R}^T\mathbf{R}}_{\mathbf{I}}\mathbf{U} = \mathbf{U}^2 \tag{4.29}$$

The last result shows that the right Cauchy–Green deformation tensor is the material deformation measure with $\mathbf{C} = C_{ij}\mathbf{e}_i \otimes \mathbf{e}_j$, which pertains to the initial configuration as the right stretch tensor $\mathbf{U}$. By the analogy with stretch tensors, we can also define the left Cauchy–Green deformation tensor as the spatial deformation measure:

$$\mathbf{B} := \mathbf{F}\mathbf{F}^T = \mathbf{V}\underbrace{\mathbf{R}\mathbf{R}^T}_{\mathbf{I}}\underbrace{\mathbf{V}^T}_{\mathbf{V}} = \mathbf{V}^2 = B_{ij}\mathbf{e}_i^{\varphi} \otimes \mathbf{e}_j^{\varphi} \tag{4.30}$$

A number of other large deformation measures can be derived from stretch tensors $\mathbf{U}$ and $\mathbf{V}$ and the Cauchy–Green strain tensors $\mathbf{C}$ and $\mathbf{B}$. The favorite choice for a number of works, which just mention large displacements problems, seems to be the Green–Lagrange deformation measure:

$$\boxed{\begin{aligned}\mathbf{E} &= \frac{1}{2}(\mathbf{F}^T\mathbf{F} - \mathbf{I}) = \frac{1}{2}(\mathbf{C} - \mathbf{I}) \\ &= \frac{1}{2}(\nabla\mathbf{d} + \nabla\mathbf{d}^T) + \tfrac{1}{2}\nabla\mathbf{d}^T\nabla\mathbf{d} \\ &= \frac{1}{2}(\frac{\partial d_i}{\partial x_j} + \frac{\partial d_j}{\partial x_i} + \frac{\partial d_k}{\partial x_i}\frac{\partial d_k}{\partial x_j})\mathbf{e}_i \otimes \mathbf{e}_j\end{aligned}} \tag{4.31}$$

Perhaps the most important reason for a frequent use of the Green–Lagrange deformation is its clear connection with the infinitesimal deformation measure for small strain regime, which is defined as the symmetric part of displacement gradient. Namely, it is easy to see that when the displacement gradient tends to zero, the quadratic term drops and the Green–Lagrange deformation measure reduces to:

$$\lim_{\nabla \mathbf{u} \mapsto \mathbf{0}} \mathbf{E} = \frac{1}{2}(\nabla \mathbf{u} + \nabla \mathbf{u}^T) =: \boldsymbol{\epsilon} \tag{4.32}$$

In the last expression, we used the following notation: $\| \nabla \mathbf{d} \| \mapsto \| \nabla \mathbf{u} \| << 1$. Therefore, the Green–Lagrange strain can (mainly) serve as a convenient choice for strain measure in the case where, despite the large displacements and rotations, the deformations remain small (for example, for plate and shell structures). In such a case, we can still exploit Hook's constitutive laws of linear elasticity (already used for linearized kinematics), which thus gives rise to so-called Saint-Venant–Kirchhoff material model.

It is also interesting to show the compatibility which exists between two different manners for separating the deformations from the rotations: the multiplicative decomposition of the deformation gradient through the polar factorization in (4.26) for large deformations and rotations versus the additive decomposition of the displacement gradient for the case of small deformations and rotations. In that respect, we first compute the consistent linearization of the right stretch tensor by exploiting its relation to the Green–Lagrange strain and the result in (4.32), which leads to:

$$\mathbf{U} = \sqrt{\mathbf{I} + 2\mathbf{E}} \implies \boxed{\lim_{\nabla \mathbf{u} \mapsto \mathbf{0}} \mathbf{U} = \mathbf{I} + \boldsymbol{\epsilon}} \tag{4.33}$$

With this result in hand, we can easily show that the consistent linearization of the inverse of the right stretch tensor $\mathbf{U}^{-1}$ will lead to:

$$\lim_{\nabla \mathbf{u} \mapsto \mathbf{0}} \mathbf{U}^{-1} = \mathbf{I} - \boldsymbol{\epsilon} \Rightarrow lin_{\nabla \mathbf{u} \mapsto \mathbf{0}}[\mathbf{U}\mathbf{U}^{-1}] = \mathbf{I} + \underbrace{\boldsymbol{\epsilon} - \boldsymbol{\epsilon}}_{=0} - \underbrace{\boldsymbol{\epsilon}^2}_{\approx 0} \tag{4.34}$$

The last result can further be used to obtain the consistent linearization of the orthogonal tensor of large rotations featuring the skew-symmetric tensor $\boldsymbol{\omega}$ representing the infinitesimal rotations:

$$\boxed{\begin{aligned} \lim_{\nabla \mathbf{u} \mapsto \mathbf{0}} \mathbf{R} &= (\mathbf{I} + \nabla \mathbf{u})(\mathbf{I} - \boldsymbol{\epsilon}) \\ &= \mathbf{I} + \underbrace{(\nabla \mathbf{u} - \boldsymbol{\epsilon})}_{\boldsymbol{\omega}} - \underbrace{\nabla \mathbf{u}\, \boldsymbol{\epsilon}}_{\approx 0} \\ &= \mathbf{I} + \boldsymbol{\omega} \end{aligned}} \tag{4.35}$$

Having established the last two results, we can finally construct the consistent linearization of the deformation gradient leading to:

$$\boxed{\begin{aligned}\lim_{\nabla\mathbf{u}\mapsto\mathbf{0}}\mathbf{F} &= \lim_{\nabla\mathbf{u}\mapsto\mathbf{0}}\mathbf{R}\mathbf{U}\\ &= (\mathbf{I}+\boldsymbol{\omega})(\mathbf{I}+\boldsymbol{\epsilon})\\ &= \mathbf{I}+\boldsymbol{\omega}+\boldsymbol{\epsilon}+\overbrace{\boldsymbol{\omega\epsilon}}^{\approx 0}\\ &= \mathbf{I}+\boldsymbol{\omega}+\boldsymbol{\epsilon}\end{aligned}} \tag{4.36}$$

We can say in conclusion that the polar factorization allows to master a non-commutative, multiplicative coupling between the large deformations and large rotations. This coupling weakens in a small displacement gradient case, where the rotations and the deformations are commutative and can simply be summed.

1D case: In closing this section we briefly examine one-dimensional case, where a number of notions we have introduced can be given a very clear illustration. In one-dimensional setting, the deformation gradient $\mathbf{F}$ takes a diagonal form:

$$d\mathbf{x}^{\varphi} = \mathbf{F}\,d\mathbf{x}\ ;\quad \mathbf{F} = \begin{bmatrix}\lambda(x) & 0 & 0\\ 0 & 1 & 0\\ 0 & 0 & 1\end{bmatrix} \tag{4.37}$$

The only component of deformation gradient for 1D case, that happens to be different from zero or one, is called stretch λ. With the corresponding choice of the reference frame, the stretch can be written:

$$dx^{\varphi} = \lambda_t(x)\,dx\ ;\quad \lambda_t(x) := \frac{d\varphi_t(x)}{dx} = 1 + \frac{\partial d_t(x)}{\partial x} \tag{4.38}$$

For a homogeneous strain field, the stretch is independent of x, and it can be computed as the ratio of the deformed and the initial length of the deformable body:

$$\lambda_t(x) \equiv \lambda_t = l_t^{\varphi}/l \tag{4.39}$$

It thus follows that no deformation is produced with the stretch equal to one, $\lambda_t = 1$. Moreover, with both initial and deformed length being positive, the stretch always remains positive. In other words, we cannot reduce the deformed length $dx^{\varphi} = \lambda_t(x)\,dx$ to zero, nor can we make it infinitely long:

$$0 < \lambda_t(x) < \infty\ ;\ \forall (x,t) \in [0,l] \times [0,T] \tag{4.40}$$

The stretch remains a useful deformation measure for 3D large deformations, even though it is not as easy to compute as in 1D case. Other deformation measures using the stretch can also be employed in 3D case. In fact, one can provide an infinite number of large deformation measures, which can all be expressed as a monotonically increasing function of the stretch $f(\lambda)$, verifying the following conditions:

$$f(\lambda) : (0,\infty) \mapsto \mathbb{R}\ ;\ f(1) = 0\ \&\ f'(1) = 1\ \&\ f'(\lambda) > 0\ ;\ \forall\lambda \tag{4.41}$$

The first of the conditions in (4.41) above will ensure that zero deformation is obtained for a unit stretch; the second condition allows us to recover from the proposed form the infinitesimal strain measure for the case of small displacement gradients; the final condition on monotonically increasing form of $f(\cdot)$ ensures that the infinite strain will be produced by infinite stretch. One such family of the functions $f(\cdot)$, which all verify the conditions imposed in (4.41), is proposed by Doyle and Ericksen [69] and also by Hill [104]:

$$f(\lambda) = \begin{cases} \frac{1}{m}(\lambda^m - 1) \; ; \; m \neq 0 \\ ln\lambda \; ; \; m = 0 \end{cases} \tag{4.42}$$

It is easy to see that the Green–Lagrange deformation is only a particular member of this family of large deformation measures, obtained for $m = 2$, which can be written in 1D case as:

$$E_t(x) = \frac{1}{2}\{[\lambda_t(x)]^2 - 1\} \tag{4.43}$$

One exceptional member of this family, which is a frequently used for experimental measurements, is obtained for $m = 0$ in the form of the natural or logarithmic large deformation measure:

$$\epsilon_t(x) = ln[\lambda_t(x)] \tag{4.44}$$

The vast diversity of large deformation measures may seem inconvenient for forcing us to define equally large number of the corresponding work-conjugate stress tensors, and identify the matching pair for each of them. However, such a large diversity of work-conjugate couples of stress and strain tensors in large deformations regime also provides an important advantage for easing the task of constructing the most appropriate constitutive model governing the material behavior for different deformation modes and various strain regimes, from very small to very large strains.

4.2 Equilibrium equations in large displacements

In geometrically nonlinear case where the displacements are no longer small enough to ignore the difference between the initial and final deformed configurations, the equilibrium equations are no longer represented in a unique manner. There exist a number of different possibilities for expressing equilibrium in the large deformation problems, each using a particular representation of the stress tensor. The most frequently used stress tensors are presented subsequently, and their relationship with the Cauchy or true stress is explained.

4.2.1 Strong form of equilibrium equations

If we wish to isolate an infinitesimal volume element from the equilibrated deformed configuration and maintain its equilibrium (see Figure 4.6), we ought to apply on this element the Cauchy stress; the latter is yet referred to as the true stress, since this is in reality the only stress that occurs in a deformed solid body undergoing large strains, displacements and rotations. The equilibrium equations for such an infinitesimal volume element can be written in the same format as the one presented in small displacement gradient case by using the coordinates $\mathbf{x}^{\varphi}$ in the deformed configuration:

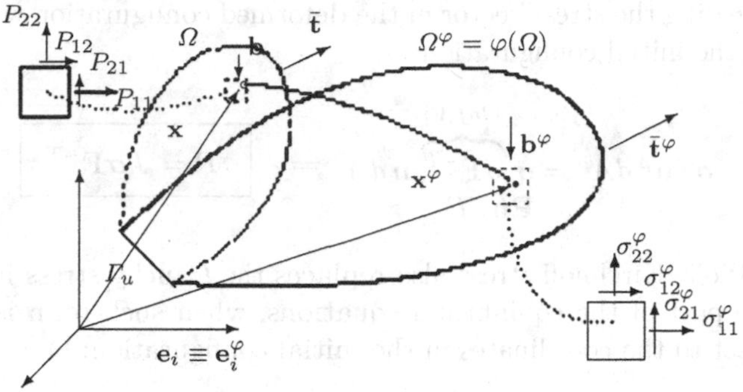

Fig. 4.6 Equilibrium of infinitesimal volume element in initial and deformed configurations.

$$\left.\begin{aligned} div^{\varphi}\boldsymbol{\sigma}^{\varphi} + \mathbf{b}^{\varphi} &= \frac{\partial \sigma^{\varphi}_{ij}}{\partial x^{\varphi}_{j}}\mathbf{e}^{\varphi}_{i} + b^{\varphi}_{i}\mathbf{e}^{\varphi}_{i} = \mathbf{0} \\ \boldsymbol{\sigma}^{\varphi,T} &= \boldsymbol{\sigma}^{\varphi} \Leftrightarrow \ \sigma^{\varphi}_{ij} = \sigma^{\varphi}_{ji} \end{aligned}\right\} \text{ in } \Omega^{\varphi} \quad (dV^{\varphi} = JdV) \tag{4.45}$$

However, contrary to small displacement case, we can soon realize the main disadvantage of such a form of equilibrium equations in the large displacement setting in that the deformed configuration and corresponding coordinates $\mathbf{x}^{\varphi}$, are known only once the problem is solved. Moreover, we cannot keep these coordinates fixed, since the deformed configuration keeps evolving during the overall large motion. For that reason, we seek to express the equilibrium equations in the known configuration that remains fixed for a deformable solid body, and the particular choice made herein is the one of the initial configuration. In other words, we will show how to find the external force description and the corresponding stress tensor that allow us to express equilibrium conditions with respect to the same infinitesimal volume element, but in its initial position (see Figure 4.6).

We begin with a simple coordinate transformation, which allows to express the external force contribution to the equilibrium equations in the initial configuration:

$$\left.\begin{array}{l}\mathbf{b}^{\varphi} dV^{\varphi} = \mathbf{b} dV \\ dV^{\varphi} = J\, dV\end{array}\right\} \implies \boxed{\mathbf{b}^{\varphi} J = \mathbf{b}} \tag{4.46}$$

We note in passing that the volume force field $\mathbf{b}$ defined with respect to the initial configuration are easy to quantify for homogeneous solid body, which is not the case for their counterpart $\mathbf{b}^{\varphi}$ in the deformed configuration. The corresponding transformation of the Cauchy stress contribution to equilibrium equations will give rise to the first Piola–Kirchhoff stress tensor $\mathbf{P}$. This stress tensor allows to express the Cauchy principle in the material description, connecting the stress vector in the deformed configuration with the unit normal in the initial configuration:

$$\begin{aligned} \boldsymbol{\sigma}^{\varphi} \mathbf{n}^{\varphi}\, dA^{\varphi} &= \boldsymbol{\sigma}\, \overbrace{J\mathbf{F}^{-T}}^{Cof[\mathbf{F}]}\, \mathbf{n}\, dA \implies \boxed{\mathbf{P} = J\boldsymbol{\sigma}\mathbf{F}^{-T}} \\ &= \mathbf{P}\mathbf{n}\, d\Gamma \end{aligned} \tag{4.47}$$

The first Piola–Kirchhoff stress also replaces the Cauchy stress in the stress divergence part of the equilibrium equations, when such term is computed with respect to the coordinates in the initial configuration:

$$\begin{aligned} J\, div^{\varphi} \boldsymbol{\sigma}^{\varphi} &= J\, \boldsymbol{\sigma}^{\varphi}\, \overbrace{\boldsymbol{\nabla}^{\varphi}}^{\mathbf{F}^{-T}\boldsymbol{\nabla}} \\ &= \underbrace{(J\, \overbrace{\boldsymbol{\sigma}}^{\boldsymbol{\sigma}^{\varphi}\circ\boldsymbol{\varphi}}\, \mathbf{F}^{-T})}_{P}\, \boldsymbol{\nabla} \\ &= \mathbf{P}\boldsymbol{\nabla} =: div\mathbf{P} \end{aligned} \tag{4.48}$$

The key result needed to obtain the last expression, $\boldsymbol{\nabla}^{\varphi} = \mathbf{F}^{-T}\boldsymbol{\nabla}$, can easily be confirmed by direct computation. Namely, for any function of the coordinates in the deformed configuration $h = h(x_1^{\varphi}, x_2^{\varphi}, x_3^{\varphi})$, we can also compute its derivative with respect to the coordinates in the initial configuration x_1, by using $x_i^{\varphi} = \varphi_i(x_1, x_2, x_3)$ and the chain rule to obtain:

$$\begin{aligned} \frac{\partial h(\varphi_i(x_j))}{\partial x_1} &= \frac{\partial h}{\partial x_1^{\varphi}} \frac{\partial \varphi_1}{\partial x_1} + \frac{\partial h}{\partial x_2^{\varphi}} \frac{\partial \varphi_2}{\partial x_1} + \frac{\partial h}{\partial x_3^{\varphi}} \frac{\partial \varphi_3}{\partial x_1} \\ &= \begin{bmatrix} \frac{\partial \varphi_1}{\partial x_1} & \frac{\partial \varphi_2}{\partial x_1} & \frac{\partial \varphi_3}{\partial x_1} \end{bmatrix} \begin{bmatrix} \frac{\partial h}{\partial x_1^{\varphi}} \\ \frac{\partial h}{\partial x_2^{\varphi}} \\ \frac{\partial h}{\partial x_3^{\varphi}} \end{bmatrix} \end{aligned} \tag{4.49}$$

In the same manner we can compute the partial derivatives $\frac{\partial h}{\partial x_2}$ and $\frac{\partial h}{\partial x_3}$, and write the final result in matrix notation:

$$\begin{bmatrix} \frac{\partial h}{\partial x_1} \\ \frac{\partial h}{\partial x_2} \\ \frac{\partial h}{\partial x_3} \end{bmatrix} = \begin{bmatrix} \frac{\partial \varphi_1}{\partial x_1} & \frac{\partial \varphi_2}{\partial x_1} & \frac{\partial \varphi_3}{\partial x_1} \\ \frac{\partial \varphi_1}{\partial x_2} & \frac{\partial \varphi_2}{\partial x_2} & \frac{\partial \varphi_3}{\partial x_2} \\ \frac{\partial \varphi_1}{\partial x_3} & \frac{\partial \varphi_2}{\partial x_3} & \frac{\partial \varphi_3}{\partial x_3} \end{bmatrix} \begin{bmatrix} \frac{\partial h}{\partial x_1^\varphi} \\ \frac{\partial h}{\partial x_2^\varphi} \\ \frac{\partial h}{\partial x_3^\varphi} \end{bmatrix} \tag{4.50}$$

By expressing this result in direct tensor notation, we can obtain:

$$\boldsymbol{\nabla}(h) = \mathbf{F}^T \boldsymbol{\nabla}^\varphi (h) \Rightarrow \boldsymbol{\nabla}^\varphi = \mathbf{F}^{-T}\boldsymbol{\nabla} \, ; \, \boldsymbol{\nabla}^\varphi = \frac{\partial}{\partial x_i^\varphi}\mathbf{e}_i^\varphi \, ; \, \boldsymbol{\nabla} = \frac{\partial}{\partial x_i}\mathbf{e}_i \tag{4.51}$$

By using the results in (4.46) and (4.48) above, we can rewrite the strong form of the equilibrium equations in the initial configuration:

$$J\, div^\varphi \boldsymbol{\sigma}^\varphi + J\,\mathbf{b}^\varphi = \mathbf{0} \;\Rightarrow\; \boxed{div\mathbf{P} + \mathbf{b} = \mathbf{0}} \tag{4.52}$$

One should not forget that the force equilibrium equations of this kind ought to be accompanied by the corresponding moment equilibrium equations, which confirm the symmetry of the Cauchy stress tensor and non-symmetry of the first Piola–Kirchhoff stress tensor. Namely, it follows from (4.47) that:

$$J\boldsymbol{\sigma}^{\varphi,T} = J\boldsymbol{\sigma}^\varphi \;\Rightarrow\; \boxed{\mathbf{F}\mathbf{P}^T = \mathbf{P}\mathbf{F}^T} \tag{4.53}$$

It is interesting to write the corresponding form of the equilibrium equations in 1D setting. In this case, the only component of the deformation gradient which can change is the stretch $\lambda_t(x)$, and any stress tensor has only one non-trivial component, with $\sigma^\varphi(x^\varphi)$ for Cauchy stress and $P(x)$ for the first Piola–Kirchhoff stress. It is easy to check from the corresponding 1D form of (4.47) that these two stress components will always have the same numerical value:

$$\mathbf{F} = \begin{bmatrix} \lambda & 0 & 0 \\ 0 & 1 & 0 \\ 0 & 0 & 1 \end{bmatrix} \; ; \; \mathbf{P}_t = J(\boldsymbol{\sigma}^\varphi \circ \boldsymbol{\varphi})\mathbf{F}^{-T} \;\Longrightarrow\; P(x) = \sigma^\varphi(x^\varphi) \tag{4.54}$$

We can obtain the same conclusion from the Cauchy principle, by exploiting 1D version of Nanson's formula, which leads to:

$$\begin{aligned} &\mathbf{n}_t^\varphi \circ \boldsymbol{\varphi}_t = J_t \mathbf{F}_t^{-T}\mathbf{n}_t \;\Longrightarrow\; n_t^\varphi(\varphi_t(x)) \equiv n_t(x) = 1 \\ &t_t := \sigma_t^\varphi(x^\varphi)\underbrace{n_t^\varphi(x^\varphi)}_{=1} = P_t(x)\underbrace{n_t(x)}_{=1} \;\Longrightarrow\; \sigma_t(x^\varphi)\circ\phi_t(x) = P_t(x) \end{aligned} \tag{4.55}$$

The last result and the chain rule application with

$$\frac{\partial P_t(x)}{\partial x} = \frac{\partial \sigma_t^\varphi(x^\varphi)}{\partial x^\varphi}\underbrace{\frac{\partial \varphi_t(x)}{\partial x}}_{\lambda_t(x)}$$

allow us to write the equilibrium equation of an infinitesimal segment in the deformed configuration, $dx^\varphi = \lambda\, dx$, according to:

$$\begin{aligned} 0 &= \lambda_t(x)[\frac{\partial \sigma_t^\varphi(\varphi_t(x))}{\partial x^\varphi} + \underbrace{b_t^\varphi(x^\varphi)}_{b_t(x)/\lambda_t(x)}\,] \\ &= \frac{\partial P_t(x)}{\partial x} + b_t(x) \end{aligned} \tag{4.56}$$

4.2.2 Weak form of equilibrium equations

The weak form of equilibrium equations in large displacements, or the principle of virtual work, can formally be written in the same manner as the one already presented for the small displacement gradient case. Namely, the virtual displacements still ought to be chosen as infinitesimal and kinematically admissible with respect to the Dirichlet boundary conditions, and we still imagine they are superposed on the deformed configuration. However, in large overall motion the deformed configuration can be significantly different from the initial configuration (see Figure 4.7), which prevents us from ignoring the difference between these two configurations and obliges us to parameterize the virtual displacement field by using the coordinates in the deformed configuration:

Weak form of elasticity at large displacements in spatial description (with respect Ω^φ:

Given: $b_i^\varphi : \Omega^\varphi \longrightarrow \mathbb{R}$, $\bar{t}_i^\varphi : {\Gamma_{\sigma_i}}^\varphi \longrightarrow \mathbb{R}$, $({\Gamma_{u_i}}^\varphi \equiv \Gamma_{u_i})$

Find: $\varphi_i \in \mathbb{V}_i$ such that $\forall\, w_i^\varphi \in \mathbb{V}_{i,0}$

$$\begin{aligned} 0 &= G(\boldsymbol{\varphi}; \mathbf{w}^\varphi) \\ &:= \int_{\Omega^\varphi} \hat{\boldsymbol{\epsilon}}^\varphi(\mathbf{w}^\varphi)\cdot \boldsymbol{\sigma}^\varphi dv^\varphi - \int_{\Omega^\varphi} \mathbf{w}^\varphi \cdot \mathbf{b}^\varphi dv^\varphi - \int_{\Gamma_\sigma^\varphi} \mathbf{w}^\varphi \cdot \bar{\mathbf{t}}^\varphi da^\varphi \\ &= \int_{\Omega^\varphi} \frac{\partial w_i^\varphi}{\partial x_j^\varphi} \sigma_{ij}^\varphi dv^\varphi - \int_{\Omega^\varphi} w_i^\varphi b_i^\varphi dv^\varphi - \sum_{i=1}^{n_{dm}} (\int_{\Gamma_{\sigma_i}^\varphi} w_i^\varphi \bar{t}_i^\varphi da^\varphi) \end{aligned} \tag{4.57}$$

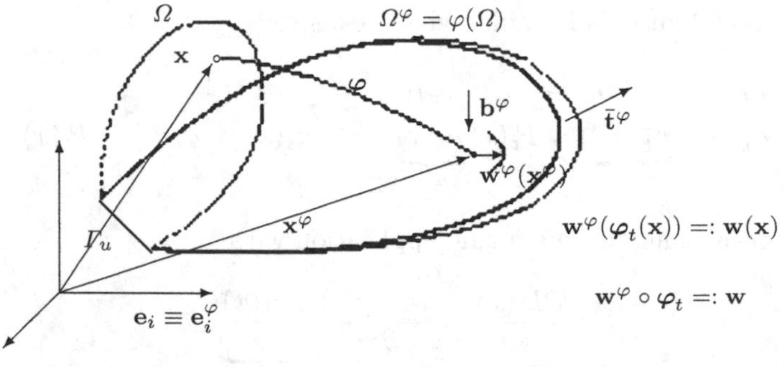

Fig. 4.7 Initial and final configurations and virtual displacement field.

with component form of virtual strain, stress and external load written explicitly as follows:

$$\hat{\boldsymbol{\epsilon}}^{\varphi}(\mathbf{w}^{\varphi}) = \frac{1}{2}(\frac{\partial w_i^{\varphi}}{\partial x_j^{\varphi}} + \frac{\partial w_j^{\varphi}}{\partial x_i^{\varphi}})\mathbf{e}_i^{\varphi} \otimes \mathbf{e}_j^{\varphi} \ , \ \boldsymbol{\sigma}^{\varphi} = \sigma_{ij}^{\varphi}\mathbf{e}_i^{\varphi} \otimes \mathbf{e}_j^{\varphi} \ ; \ \mathbf{b}^{\varphi} = b_i^{\varphi}\mathbf{e}_i^{\varphi} \ , \ \bar{\mathbf{t}}^{\varphi} = t_i \mathbf{e}_i^{\varphi}$$

We note that the virtual strains are defined as strain measures for linearized kinematics, which is in agreement with the hypothesis that the virtual displacements are infinitesimal. Moreover, the kinematically admissible virtual displacements take zero value on the boundary Γ_u^{φ}, regardless of the true value of the real displacements imposed on the same boundary. This implies that we only account for the virtual work of traction forces applied on Γ_σ^{φ}. The virtual work in (4.57) is expressed in terms of the Cauchy or true stress tensor, the only stress which acts in the deformed configuration. However, considering that the deformed configuration is unknown and that it keeps changing with motion, we should find a better basis for expressing the virtual work principle, with a particular preference towards the initial configuration. In this manner, we obtain a new representation of the true stress in terms of the second Piola–Kirchhoff stress tensor. The proper interpretation of the second Piola–Kirchhoff stress tensor can be obtained by rewriting the internal virtual work in terms of the corresponding integrals over the initial configuration. Namely, by using the change of coordinates between the deformed and the initial configurations, we can change the integration domain, and carry out the integral computation in the (known) initial configuration:

$$\begin{aligned} \textstyle\int_{\Omega^{\varphi}} \hat{\boldsymbol{\epsilon}}^{\varphi}(\mathbf{w}^{\varphi}) \cdot \boldsymbol{\sigma}^{\varphi}(\mathbf{x}^{\varphi})\, dV^{\varphi} &= \textstyle\int_{\Omega} \hat{\boldsymbol{\epsilon}}^{\varphi}(\hat{\mathbf{w}}(\boldsymbol{\varphi}(\mathbf{x}))) \cdot \underbrace{J(\mathbf{x})\boldsymbol{\sigma}^{\varphi}(\boldsymbol{\varphi}(\mathbf{x}))}_{\boldsymbol{\tau}(\mathbf{x})}\, dV \\ &= \textstyle\int_{\Omega} \hat{\boldsymbol{\epsilon}}(\mathbf{w}(\mathbf{x})) \cdot \boldsymbol{\tau}(\mathbf{x})\, dV \end{aligned} \tag{4.58}$$

Such a change of coordinates introduces a new stress tensor, referred to as the Kirchhoff stress $\boldsymbol{\tau}(\mathbf{x}) = J(\mathbf{x})\,\boldsymbol{\sigma}^{\varphi} \circ \boldsymbol{\varphi}(\mathbf{x})$. In (4.58) above, we could express the virtual displacements and virtual strains as the function of the coordinates in the initial configuration by exploiting the key result in (4.51):

$$\begin{aligned} \hat{\boldsymbol{\epsilon}}^{\varphi}(\mathbf{w}(\boldsymbol{\varphi}(\mathbf{x})) &= \tfrac{1}{2}(\tfrac{\partial w_i^{\varphi}}{\partial x_j^{\varphi}} + \tfrac{\partial w_j^{\varphi}}{\partial x_i^{\varphi}})\mathbf{e}_i^{\varphi} \otimes \mathbf{e}_j^{\varphi} \\ &= \tfrac{1}{2}(\mathbf{w}^{\varphi} \otimes \boldsymbol{\nabla}^{\varphi} + \boldsymbol{\nabla}^{\varphi} \otimes \mathbf{w}^{\varphi}) \\ &= \tfrac{1}{2}(\mathbf{w} \otimes \mathbf{F}^{-T}\boldsymbol{\nabla} + \mathbf{F}^{-T}\boldsymbol{\nabla} \otimes \mathbf{w}) \\ &= \tfrac{1}{2}((\mathbf{w} \otimes \boldsymbol{\nabla})\mathbf{F}^{-1} + \mathbf{F}^{-T}(\boldsymbol{\nabla} \otimes \mathbf{w})) \\ &= \tfrac{1}{2}(\nabla\mathbf{w}\mathbf{F}^{-1} + \mathbf{F}^{-T}\nabla\mathbf{w}) \end{aligned} \tag{4.59}$$

We can easily verify that the gradient of the virtual displacement is the directional derivative of the deformation gradient, which can be written:

$$\begin{aligned}\nabla \mathbf{w} &:= \frac{d}{d\varepsilon}\Big|_{\varepsilon=0} [\mathbf{F}_\varepsilon] \\ &= \frac{d}{d\varepsilon}\Big|_{\varepsilon=0} [\underbrace{(\boldsymbol{\varphi} + \varepsilon \mathbf{w})}_{\boldsymbol{\varphi}_\varepsilon} \otimes \boldsymbol{\nabla}] \\ &= \mathbf{w} \otimes \boldsymbol{\nabla}\end{aligned} \tag{4.60}$$

The same result can also be used to compute the directional derivative of the Green–Lagrange strain measure in the direction of the virtual displacement, and thus obtain:

$$\boxed{\boldsymbol{\Gamma} := \tfrac{d}{d\varepsilon}[\tfrac{1}{2}(\mathbf{F}_\varepsilon^T \mathbf{F}_\varepsilon - \mathbf{I})]|_{\varepsilon=0} = \frac{1}{2}(\mathbf{F}^T \nabla \mathbf{w} + \nabla \mathbf{w}^T \mathbf{F})} \tag{4.61}$$

We can also write this result in indicial notation leading to:

$$\begin{aligned}\hat{\boldsymbol{\Gamma}}(\boldsymbol{\varphi}; \mathbf{w}) &= \frac{1}{2}(\tfrac{\partial w_k}{\partial x_i}\tfrac{\partial \varphi_k}{\partial x_j} + \tfrac{\partial \varphi_k}{\partial x_i}\tfrac{\partial w_k}{\partial x_j})\mathbf{e}_i \otimes \mathbf{e}_j \\ &= \frac{1}{2}(\tfrac{\partial w_k}{\partial x_i}\tfrac{\partial (x_k + d_k)}{\partial x_j} + \tfrac{\partial (x_k + d_k)}{\partial x_i}\tfrac{\partial w_k}{\partial x_j})\mathbf{e}_i \otimes \mathbf{e}_j \\ &= \underbrace{\frac{1}{2}(\frac{\partial w_j}{\partial x_i} + \frac{\partial w_i}{\partial x_j} + \frac{\partial w_k}{\partial x_i}\frac{\partial d_k}{\partial x_j} + \frac{\partial d_k}{\partial x_i}\frac{\partial w_k}{\partial x_j})}_{\Gamma_{ij}}\, \mathbf{e}_i \otimes \mathbf{e}_j\end{aligned} \tag{4.62}$$

With this result in hand we can recast the internal virtual work in terms of the integral over the initial configuration which employs the virtual Green–Lagrange deformation. The latter can be shown to be work-conjugate to the second Piola–Kirchhoff stress. This kind of conclusion can be obtained by exploiting the auxiliary result concerning different manners for writing the scalar product of three second-order tensors $\mathbf{R}$, $\mathbf{S}$ and $\mathbf{T}$ according to:

$$\overbrace{\mathbf{R} \cdot (\mathbf{ST})}^{i} = \overbrace{(\mathbf{S}^T \mathbf{R}) \cdot \mathbf{T}}^{ii} = \overbrace{(\mathbf{R}\mathbf{T}^T) \cdot \mathbf{S}}^{iii} \tag{4.63}$$

This statement can be easily confirmed, by appealing to the definition of scalar product of two second order tensors $\mathbf{A} \cdot \mathbf{B} = A_{ij} B_{ij}$, as well as the definition of resulting transformation of a vector by successive applications of two second order tensors $\mathbf{AB} = A_{ik} B_{kj} \mathbf{e}_i \otimes \mathbf{e}_j$. This allows us to rewrite the last products in the indicial notation in three different ways, where we change the order of summation on dummy indices without changing the end result:

$$R_{ij}(S_{ik} T_{kj}) = (S_{ik} R_{ij}) T_{kj} = (R_{ij} T_{kj}) S_{ik} \tag{4.64}$$

This auxiliary result along with the results (4.59) and (4.61) allow us to rewrite the internal virtual work in (4.58) as follows:

$$
\begin{aligned}
&\int_{\Omega^\varphi} \hat{\boldsymbol{\epsilon}}^\varphi(\mathbf{w}^\varphi) \cdot \boldsymbol{\sigma}^\varphi \, dV^\varphi = \int_\Omega \hat{\boldsymbol{\epsilon}} \cdot \boldsymbol{\tau} dV \\
&= \int_\Omega \frac{1}{2} (\nabla \mathbf{w} \underbrace{\mathbf{F}^{-1}}_{i-iii} + \underbrace{\mathbf{F}^{-T}}_{i-ii} \nabla \mathbf{w}^T) \cdot \boldsymbol{\tau} dV \\
&= \int_\Omega \frac{1}{2} (\nabla \mathbf{w} \cdot \boldsymbol{\tau} \mathbf{F}^{-T} + \nabla \mathbf{w}^T \cdot \mathbf{F}^{-1} \boldsymbol{\tau}) dV \\
&= \int_\Omega \frac{1}{2} (\overbrace{\underbrace{\mathbf{F}^{-T}}_{i-ii} \mathbf{F}^T}^{I} \nabla \mathbf{w} \cdot \boldsymbol{\tau} \mathbf{F}^{-T} + \nabla \mathbf{w}^T \overbrace{\mathbf{F} \underbrace{\mathbf{F}^{-1}}_{i-iii}}^{I} \cdot \mathbf{F}^{-1} \boldsymbol{\tau}) dV \qquad (4.65) \\
&= \int_\Omega \frac{1}{2} (\mathbf{F}^T \nabla \mathbf{w} \cdot \mathbf{F}^{-1} \boldsymbol{\tau} \mathbf{F}^{-T} + \nabla \mathbf{w}^T \mathbf{F} \cdot \mathbf{F}^{-1} \boldsymbol{\tau} \mathbf{F}^{-T}) \, dV \\
&= \int_\Omega \frac{1}{2} (\mathbf{F}^T \nabla \mathbf{w} + \nabla \mathbf{w}^T \mathbf{F}) \cdot \mathbf{F}^{-1} \boldsymbol{\tau} \mathbf{F}^{-T} \, dV \\
&= \int_\Omega \boldsymbol{\Gamma} \cdot \mathbf{S} \, dV
\end{aligned}
$$

The main conclusion which can be drawn from this development concerns the explicit form of relationship between the second Piola–Kirchhoff stress and Cauchy stress tensor:

$$
\boxed{\mathbf{S} = \mathbf{F}^{-1} \boldsymbol{\tau} \mathbf{F}^{-T} = J \mathbf{F}^{-1} \boldsymbol{\sigma} \mathbf{F}^{-T}} \qquad (4.66)
$$

This relationship can also be written in the component form, stating that the components of the second Piola–Kirchhoff and the Cauchy stress are connected through:

$$
\begin{aligned}
S_{ij} \mathbf{e}_i \otimes \mathbf{e}_j &= J [\tfrac{\partial x_i}{\partial \varphi_k} \mathbf{e}_i \otimes \mathbf{e}^\varphi_k][\sigma_{lm} \mathbf{e}^\varphi_l \otimes \mathbf{e}^\varphi_m][\tfrac{\partial x_j}{\partial \varphi_n} \mathbf{e}^\varphi_n \otimes \mathbf{e}_j] \\
&= J \tfrac{\partial x_i}{\partial \varphi_l} \tfrac{\partial x_j}{\partial \varphi_m} \sigma_{lm} \mathbf{e}_i \otimes \mathbf{e}_j \qquad (4.67) \\
\Longrightarrow \quad & \boxed{S_{ij} = J \tfrac{\partial x_i}{\partial \varphi_l} \tfrac{\partial x_j}{\partial \varphi_m} \sigma_{lm}}
\end{aligned}
$$

With these results in hand, we can easily express the virtual work principle in the material description, making use of the Green–Lagrange strain which is work-conjugate to the second Piola–Kirchhoff stress:

$$
\begin{aligned}
0 &= G(\boldsymbol{\varphi}; \mathbf{w}) \\
&:= \int_\Omega \underbrace{\frac{1}{2} (\mathbf{F}^T \nabla \mathbf{w} + \nabla \mathbf{w}^T \mathbf{F})}_{\boldsymbol{\Gamma}} \cdot \mathbf{S} \, dV - \int_\Omega \mathbf{w} \cdot \mathbf{b} dV - \int_{\Gamma_h} \mathbf{w} \cdot \bar{\mathbf{t}} \, dA \qquad (4.68) \\
&= \int_\Omega \Gamma_{ij} S_{ij} dV - \int_\Omega w_i b_i dV - \sum_{i=1}^{n_{dm}} \int_{\Gamma_\sigma} w_i \bar{t}_i dA
\end{aligned}
$$

We also note that in both (4.58) and (4.68) we compute the internal virtual work by integrating over the initial configuration; however, each expression provides a different replacement of the Cauchy (or true) stress tensor, with either the Kirchhoff stress tensor in (4.58) or the second Piola–Kirchhoff stress tensor in (4.68). In conclusion, contrary to the case of linearized kinematics, for large displacement gradients problems there is no longer the unique choice of strain measure and conjugate stress for expressing the virtual work. We obtain instead a large variety of choices for strain measures and stress tensors.[10] We also conclude that it is possible to establish the clear link between seemingly unrelated choices of stress and strain tensors, and further enlarge the number of work-conjugate stress–strain couples for expressing the virtual work principle. For example, the first Piola–Kirchhoff stress tensor, which was initially introduced as the logical replacement of the Cauchy stress tensor for the strong form of equilibrium equations written in material description, can also be used to express the weak form of the same equations, or the virtual work principle. Namely, by taking into account the definitions of the two Piola–Kirchhoff stress tensors given in (4.47) and (4.66), respectively, we can easily conclude that:

$$\mathbf{S} = \mathbf{F}^{-1}\mathbf{P} \tag{4.69}$$

By further exploiting the auxiliary result in (4.63) on scalar product of three second order tensors, we can rewrite the internal virtual work in (4.68) according to:

$$\begin{aligned} \int_\Omega \mathbf{\Gamma} \cdot \mathbf{S}\, dV &= \int_\Omega \tfrac{1}{2}(\nabla \mathbf{w}^T \underbrace{\mathbf{F}}_{i-iii} + \underbrace{\mathbf{F}}_{i-ii} \nabla \mathbf{w}) \cdot \mathbf{S}\, dv \\ &= \int_\Omega \tfrac{1}{2}(\nabla \mathbf{w}^T \cdot \underbrace{\overbrace{\mathbf{S}}^{\equiv \mathbf{S}^T} \mathbf{F}^T}_{=\mathbf{P}^T} + \nabla \mathbf{w} \cdot \underbrace{\mathbf{F}\mathbf{S}}_{=\mathbf{P}})\, dv \\ &= \int_\Omega \nabla \mathbf{w} \cdot \mathbf{P}\, dv \end{aligned} \tag{4.70}$$

We can thus write not only the strong, but also the weak form of equilibrium equations by making use of the first Piola–Kirchhoff stress tensor:

$$\begin{aligned} 0 = G(\varphi; \mathbf{w}) &:= \int_\Omega \nabla \mathbf{w} \cdot \mathbf{P}\, dV - \int_\Omega \mathbf{w} \cdot \mathbf{b}\, dV - \int_{\Gamma_h} \mathbf{w} \cdot \bar{\mathbf{t}}\, dA \\ &= \int_\Omega \frac{\partial w_i}{\partial x_j} P_{ij} dV - \int_\Omega w_i b_i dV - \sum_{i=1}^{n_{dm}} \int_{\Gamma_\sigma} w_i \bar{t}_i dA \end{aligned} \tag{4.71}$$

In view of (4.60), the last result indicates that the first Piola–Kirchhoff stress tensor is work-conjugate to the deformation gradient.[11]

We finally provide a brief summary of all these results in 1D setting, which can further illustrate the relations between different work-conjugate stress–strain pairs for large deformation case. The 1D case allows to represent the

[10] We find, in fact, an infinite number of stress–strain conjugate couples in large deformation framework.

[11] We recall that the deformation gradient is not only measure of deformation but also of rotation, and we should thus find the way to eliminate any undesirable rotation contribution.

deformation gradient by the identity matrix, apart from a single diagonal component expressing the stretch as the ratio between the deformed and the initial length of an infinitesimal element, $\lambda = dx^{\varphi}/dx$. We start by expressing the internal virtual work by using the intrinsic form featuring the Cauchy stress:

$$G_{int}(\lambda; w^{\varphi}) := \int_{l_t^{\varphi}} \frac{dw^{\varphi}}{dx^{\varphi}} \sigma^{\varphi} \, dx^{\varphi} \tag{4.72}$$

and then rewrite this result in several alternative forms, which introduce different stresses as:

$$\begin{aligned} G_{int}(\lambda; w) &:= \int_l \underbrace{\frac{dw}{dx}}_{\frac{d}{d\epsilon}[F_\epsilon = \lambda_\epsilon]|_{\epsilon=0}} \sigma_P \, dx \\ &= \int_l \underbrace{\frac{dw}{dx}\frac{1}{\lambda}}_{\frac{d}{d\epsilon}[ln\lambda_\epsilon]|_{\epsilon=0}} \underbrace{\lambda\sigma}_{\tau} \, dx \\ &= \int_l \underbrace{\frac{dw}{dx}\lambda}_{\frac{d}{d\epsilon}[E_\epsilon = \frac{1}{2}(\lambda_\epsilon^2 - 1)]|_{\epsilon=0}} \underbrace{\frac{1}{\lambda}\sigma}_{S} \, dx \end{aligned} \tag{4.73}$$

With such a large variety of possible choices for stress and strain tensors available for large strain problem formulation, the question arises which one should we favor? Often an important criterion for choosing a particular stress–strain couple concerns the corresponding constitutive model formulation, which should provide the most reliable representation of a particular material behavior. These issues are studied in detail in the next section.

4.3 Linear elastic behavior in large displacements: Saint-Venant–Kirchhoff material model

All different possibilities for expressing the internal virtual work indicated in (4.73) are only different material representations of the same work, which is defined in terms of the Cauchy (or true) stress and infinitesimal deformation. In the case when the displacement gradient is small, all different possibilities for finite strain measure ought to reduce to this infinitesimal strain, and all different stresses will be the same as the Cauchy stress. We can therefore conclude that any such material description of an elastic constitutive law for large deformations should reduce to Hook's law for small deformation case. The constitutive model corresponding to Saint-Venant–Kirchhoff material takes this hypothesis one step further by postulating that Hook's law also remains valid for large deformations, defined in terms of the Green–Lagrange strain measure; this allows us to easily compute the corresponding values of the work-conjugate second Piola–Kirchhoff stress, as well as confirm the model capability to recover Hook's law for limit case of small deformations:

$$S = \mathsf{C}E \Longrightarrow \lim_{\lambda \mapsto 1, (du/dx)<<1} \{\frac{1}{\lambda}\sigma = \mathsf{C}[\frac{du}{dx} + \frac{1}{2}(\frac{du}{dx})^2]\} \Leftrightarrow \sigma = \mathsf{C} \underbrace{\epsilon}_{du/dx} \quad (4.74)$$

We can easily draw the same conclusion regarding Hook's law validity for the limit case of the small deformations, for a number of other large deformation measures. In other words, if the material behavior starts as linear elastic for the small deformation case, we should recover from different material models the same stress representation and the same elasticity tensor.
It is not possible to write the universal form of constitutive law in large deformation regime. For example, if we wish to express the Saint-Venant–Kirchhoff constitutive model, defined in (4.74) as Hook's law for Green–Lagrange strain and the second Piola–Kirchhoff stress, by using another work-conjugate couple of stress and strain tensors, it is no longer possible to keep the elasticity tensor with constant entries (e.g. see [21]). This can easily be shown for 1D case by expressing the Saint-Venant–Kirchhoff material law in terms of the first Piola–Kirchhoff stress and the stretch used as the conjugate large deformation measure, which allows us to obtain a constitutive law of nonlinear elasticity and the corresponding elastic tangent modulus:

$$\begin{aligned} & S = C\tfrac{1}{2}(\lambda^2 - 1) \ \& \ S = \lambda^{-1}P \Longrightarrow P = C\tfrac{1}{2}(\lambda^2-1)\lambda \\ & \Longrightarrow \tfrac{dP}{d\lambda} = \underbrace{C\frac{1}{2}(3\lambda^2 - 1)}_{C_{FF}} \end{aligned} \quad (4.75)$$

We can thus obtain the appropriate form of the constitutive relationship for any other particular choice of the large deformation measure that we would like to use, resulting with a wide variety of possible stress–strain relations to use for the same material model. The unique form of constitutive relation, which contains any such relation as a special case, can be written for a hyperelastic material model in terms of the strain energy potential. For the Saint-Venant–Kirchhoff material model, the strain energy for 1D case can be written:

$$\psi(\lambda) = \frac{1}{2}C[\frac{1}{2}(\lambda^2 - 1)]^2 \Longrightarrow \begin{cases} S = \frac{\partial \psi(\cdot)}{\partial(\frac{1}{2}(\lambda^2-1))} \\ P = \frac{\partial\psi(\cdot)}{\partial\lambda} \end{cases} \quad (4.76)$$

Turning now to 3D case, we can consider that the linear elastic behavior for small strain, even when accompanied by (very) large rotations, should again be described by Hook's law. Namely, it is quite clear that the Saint-Venant–Kirchhoff model reduces to Hook's law for the limit case of small deformation; namely, by using the corresponding linearization results in (4.31) and (4.66) for stress and strain tensors, we can easily obtain:

$$
\begin{aligned}
&\lim_{\nabla \mathbf{u} \mapsto \mathbf{0}} \underbrace{\frac{1}{2}(\nabla \mathbf{u} + \nabla \mathbf{u}^T + \nabla \mathbf{u}^T \nabla \mathbf{u})}_{\mathbf{E}} = \boldsymbol{\epsilon} := \tfrac{1}{2}(\nabla \mathbf{u} + \nabla \mathbf{u}^T) \\
&\lim_{\nabla \mathbf{u} \mapsto \mathbf{0}} \underbrace{J\mathbf{F}^{-1}\boldsymbol{\sigma}\mathbf{F}^{-T}}_{\mathbf{S}} = \boldsymbol{\sigma} \\
&\lim_{\nabla \mathbf{u} \mapsto \mathbf{0}} [\mathbf{S} = \mathcal{C}\mathbf{E}] \longrightarrow \boldsymbol{\sigma} = \mathcal{C}\boldsymbol{\epsilon}
\end{aligned}
\tag{4.77}
$$

The model of Saint-Venant–Kirchhoff employs the same stress-strain relationship in terms of Hook's law not only for small but also for large strain regime. In other words, the elasticity tensor with constant entries is used to connect the second Piola–Kirchhoff stress tensor and the Green–Lagrange deformation tensor:

$$
\begin{aligned}
\mathbf{S} &= \mathcal{C}\mathbf{E} \; ; \; \mathcal{C} = \lambda \mathbf{I} \otimes \mathbf{I} + 2\mu \mathcal{I} \\
&= \lambda \underbrace{(\mathbf{I} \cdot \mathbf{E})}_{tr(\mathbf{E})} \mathbf{I} + 2\mu \mathbf{E} \\
S_{ij} &= \mathcal{C}_{ijkl} E_{kl} \; ; \; \mathcal{C}_{ijkl} = \lambda \delta_{ij}\delta_{kl} + 2\mu \tfrac{1}{2}[\delta_{ik}\delta_{jl} + \delta_{il}\delta_{jk}] \\
&= \lambda E_{kk}\delta_{ij} + 2\mu E_{ij}
\end{aligned}
\tag{4.78}
$$

In the last expression, we use the standard interpretation of Lamé parameters λ and μ, which can also be replaced by Young's modulus and Poisson's ratio:

$$
\begin{aligned}
\lambda &= \tfrac{E\nu}{(1+\nu)(1-2\nu)} \; ; \; \mu = \tfrac{E}{2(1+\nu)} \\
E &= \tfrac{\mu(3\lambda+2\mu)}{\lambda+\mu} \; ; \; \nu = \tfrac{\lambda}{2(\lambda+\mu)}
\end{aligned}
\tag{4.79}
$$

Due to isotropy, the constitutive model of Saint-Venant–Kirchhoff can be defined through the corresponding relationship between the deviatoric and spherical tensor components:

$$
\begin{aligned}
dev[\mathbf{S}] &= 2\mu \, dev[\mathbf{E}] \;\; | \; dev[\mathbf{S}] = \mathbf{S} - \tfrac{1}{3}tr[\mathbf{S}]\mathbf{I} \\
\tfrac{1}{3}tr[\mathbf{S}] &= K tr[\mathbf{E}] \;\; | \; tr[\mathbf{S}] = \mathbf{S} \cdot \mathbf{I}
\end{aligned}
\tag{4.80}
$$

where $K = \lambda + \frac{2\mu}{3} = \frac{E}{3(1-2\nu)}$ is the bulk modulus. We can also write the inverse form of the Saint-Venant–Kirchhoff constitutive law, where the Green–Lagrange deformation tensor is expressed as a function of the second Piola–Kirchhoff stress tensor according to:

$$
\boxed{\mathbf{E} = \mathcal{C}^{-1}\mathbf{S} \; ; \; \mathcal{C}^{-1} = \tfrac{1}{2\mu}[\mathcal{I} - \tfrac{\lambda}{3\lambda+2\mu}\mathbf{I} \otimes \mathbf{I}]}
\tag{4.81}
$$

We note in passing that the inverse constitutive relationship is the only one which remains valid for the limit case of incompressible behavior with $\nu \mapsto 0.5$ (which implies $K \mapsto \infty$), where we can write:

$$
tr[\mathbf{E}] = 0 \; ; \; dev[\mathbf{E}] = \tfrac{1}{2\mu} dev[\mathbf{S}]
\tag{4.82}
$$

The Saint-Venant–Kirchhoff constitutive model is an example for hyperelastic material, with the corresponding quadratic form of the strain energy density in terms of the Green–Lagrange strains. Such a potential allows us to compute the second Piola–Kirchhoff stress tensor as its derivative with respect to the Green–Lagrange strain tensor:

$$W(\mathbf{E}) = \frac{1}{2}\mathbf{E}\cdot\boldsymbol{\mathcal{C}}\mathbf{E} \Rightarrow \mathbf{S} = \frac{\partial W}{\partial \mathbf{E}} = \boldsymbol{\mathcal{C}}\mathbf{E} \tag{4.83}$$

We further consider the case where the external loading applied to a solid body whose constitutive behavior is governed by the Saint-Venant–Kirchhoff model derives from a potential; we can take different potentials for volume force and surface traction:

$$\begin{aligned} \int_\Omega \mathbf{w}\cdot\mathbf{b}dV &= \frac{d}{d\varepsilon}\bigg|_{\varepsilon=0} \int_\Omega F(\boldsymbol{\varphi}_\varepsilon)dV \ ; \\ \int_{\Gamma^h} \mathbf{w}\cdot\bar{\mathbf{t}}dV &= \frac{d}{d\varepsilon}\bigg|_{\varepsilon=0} \int_{\Gamma_\sigma} H(\boldsymbol{\varphi}_\varepsilon)dA \end{aligned} \tag{4.84}$$

In this case, we can construct the total potential energy functional valid for the large strain regime as the corresponding integral of all different potentials:

$$\Pi(\boldsymbol{\varphi}) := \int_\Omega W(\mathbf{E})\,dV - \int_\Omega F(\boldsymbol{\varphi})\,dV - \int_{\Gamma_\sigma} H(\boldsymbol{\varphi})dA \tag{4.85}$$

The equilibrium state will then correspond to the first variation of such a functional:

$$0 = \frac{d}{d\alpha}\{\Pi(\boldsymbol{\varphi}_\alpha)\}\bigg|_{\alpha=0} = G(\boldsymbol{\varphi};\mathbf{w}) =: \int_\Omega \boldsymbol{\Gamma}\cdot\mathbf{S}\,dV - \int_\Omega \mathbf{w}\cdot\mathbf{b}\,dV - \int_{\Gamma_\sigma}\mathbf{w}\cdot\bar{\mathbf{t}}\,dA \tag{4.86}$$

The stability of this equilibrium state can be evaluated from the second variation of the total potential energy functional, requiring that the latter remains positive:

$$\min_{\boldsymbol{\varphi}}\{\Pi(\boldsymbol{\varphi})\} \ \Rightarrow \begin{cases} G(\boldsymbol{\varphi};\mathbf{w}) = 0 \\ \frac{d^2}{d\alpha^2}\{\Pi(\boldsymbol{\varphi}_\alpha)\} =: \int_\Omega \hat{\boldsymbol{\Gamma}}(\boldsymbol{\varphi},\mathbf{w})\cdot\boldsymbol{\mathcal{C}}\hat{\boldsymbol{\Gamma}}(\boldsymbol{\varphi},\mathbf{w})\,dV > 0 \end{cases} \tag{4.87}$$

This kind of requirement will preclude the geometric instability phenomena and restrict the kind of external loading which can be applied.

There still remains another difficulty for the Saint-Venant–Kirchhoff constitutive model, pertaining to material instability in large compressive deformation case, which precludes that a very large strain be accompanied by a very large value of the true or Cauchy stress. A clear illustration of this difficulty can be provided for a simple 1D case, where the strain energy can be written according to (4.76); we can thus easily compute the resulting values of the stress which accompany very large tensile and compressive strains, produced by an infinite and zero value of stretch, respectively:

$$\begin{aligned} \lambda \mapsto 1 \;\; &\Rightarrow P \mapsto 0 \;\; ; \psi \mapsto 0 \\ \lambda \mapsto \infty &\Rightarrow P \mapsto \infty \; ; \psi \mapsto \infty \\ \lambda \mapsto 0 \;\; &\Rightarrow P \mapsto 0 \;\; ; \psi \mapsto \mathsf{C}/8 \end{aligned} \tag{4.88}$$

In 1D case, where the first Piola–Kirchhoff stress P and the true stress σ share the same numerical value, the last result would imply that producing an infinite compressive deformation and reducing the deformed bar length to zero, the Saint-Venant–Kirchhoff material would require the zero value of the true stress:

$$\lambda \mapsto 0 \;\Leftrightarrow\; (u \mapsto -l) \Rightarrow \sigma \equiv P := C\frac{1}{2}\lambda(\lambda^2 - 1) \mapsto 0! \tag{4.89}$$

The last result represents a paradox that cannot be justified for any real material. One can explain what went wrong by computing the second variation of the strain energy, or the elastic tangent modulus for the present 1D framework, which quickly reveals that there is the minimum value of compressive stress with the corresponding stretch value beyond which the material instability will occur:

$$\frac{dP}{d\lambda} = \frac{C}{2}(3\lambda^2 - 1) = 0 \Rightarrow \lambda = \sqrt{1/3} \;\Rightarrow\; P_{min} = -\frac{1}{3\sqrt{3}}C \tag{4.90}$$

This result produced by the Saint-Venant–Kirchhoff model not to allow the stress decrease beyond the minimum value can clearly not be justified in elasticity.

An alternative point of view for explaining the Saint-Venant–Kirchhoff model deficiency in representing very large compressive strains can be provided through the loss of convexity of the strain energy. The latter occurs because of the presence of the inflection point corresponding to the value of stretch in (4.90), which in turn precludes any large value of compressive stress (see Figure 4.8).

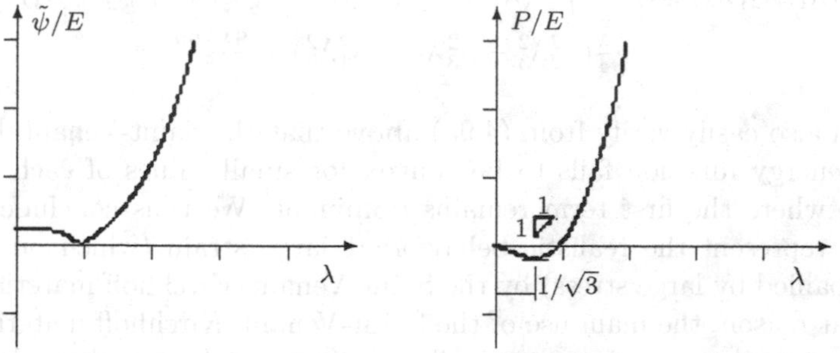

Fig. 4.8 Saint-Venant–Kirchhoff constitutive model: strain energy and the first Piola–Kirchhoff stress as function of stretch.

This kind of approach for establishing the well-posed form of the strain energy can be generalized to 3D case in terms of the polyconvexity conditions (see Ball [17]), which should provide the guarantees that the large strains remain accompanied by large stresses. In order to avoid a potential confusion with respect to the proper form of strain energy with respect to a very wide variety of available large deformation measures, the polyconvexity conditions can be stated with respect to the intrinsic measures of the large deformations; the latter includes the deformation gradient that controls the change of an infinitesimal vector (a line element), the cofactor of deformation gradient that controls the change of an infinitesimal surface element, and the determinant of deformation gradient that controls the change of an infinitesimal volume element. The polyconvexity conditions impose that the strain energy remains a convex function with respect to any of the intrinsic deformation measures, which can be written:

$$\begin{aligned} &\psi(\underbrace{\mathbf{F}, cof\mathbf{F}, det\mathbf{F}}_{\mathbf{A}}) \mapsto \text{convex function} \\ &\Longrightarrow \psi(\alpha \mathbf{A}_1 + (1-\alpha)\mathbf{A}_2) \leq \alpha\psi(\mathbf{A}_1) + (1-\alpha)\psi(\mathbf{A}_2) \end{aligned} \tag{4.91}$$

By following Ciarlet [50] we can show that the strain energy function for Saint-Venant–Kirchhoff constitutive model in 3D case does not verify the polyconvexity conditions; namely, the results in (4.31) and (4.83), will allow us to rewrite the Saint-Venant–Kirchhoff material strain energy as:

$$\tilde{W}(\mathbf{F}) := -\frac{3\lambda + 2\mu}{4} tr[\mathbf{C}] + \frac{\lambda + 2\mu}{8} tr[\mathbf{C}^2] + \frac{\lambda}{4} tr[cof[\mathbf{C}]] + \frac{9\lambda + 6\mu}{8} \ ; \tag{4.92}$$

where $\mathbf{C} = \mathbf{F}^T\mathbf{F}$. It is precisely the first terms with negative sign which precludes the polyconvexity of this strain energy and which can cause the difficulties in representing the true stress for very large compressive strains. The same result can be written in terms of the principal stretch values, directly revealing the potential problems with a finite value of strain energy for $\lambda_i \mapsto 0$:

$$\begin{aligned} \hat{W}(\lambda_1, \lambda_2, \lambda_3) := &-\tfrac{3\lambda+2\mu}{4}(\lambda_1^2 + \lambda_2^2 + \lambda_3^2) + \tfrac{\lambda+2\mu}{8}(\lambda_1^4 + \lambda_2^4 + \lambda_3^4) \\ &+\tfrac{\lambda}{4}(\lambda_2^2\lambda_3^2 + \lambda_3^2\lambda_1^2 + \lambda_1^2\lambda_2^2) + \tfrac{9\lambda+6\mu}{8} \end{aligned} \tag{4.93}$$

We can also easily verify from (4.93) above that the Saint-Venant–Kirchhoff strain energy function fails to be convex for small values of each principal stretch where the first term remains dominant. We thus conclude that we cannot represent the realistic behavior at large strain (which ought to be accompanied by large stress) by the Saint-Venant–Kirchhoff material model. For that reason, the main use of the Saint-Venant–Kirchhoff material model is limited to the case of geometrically nonlinear problems where despite the large displacements and large rotations, the strains remain small. In this class of problems we find the vast majority of geometrically nonlinear problems in structural mechanics, considering trusses, beams, plates, membranes and shells. For illustration of such structural mechanics models, we present subsequently the pertinent developments in the finite element implementation of the Saint-Venant–Kirchhoff model for a truss-bar and a plane membrane case, before turning to more general constitutive models capable of handling large deformation regime.

4.3.1 Weak form of Saint-Venant–Kirchhoff 3D elasticity model and its consistent linearization

We note that the theoretical formulation for both selected models, the truss-bar and the plane membrane, can be recovered as the special case of the boundary value problem for large displacement Saint-Venant–Kirchhoff linear elasticity in the general 3D setting, which can be written:
Weak form: Saint-Venant–Kirchhoff model for 3D large displacements elasticity
Given: b_i, $\bar{t}_i$
Find: $u_i \in S_i$ such that $\forall w_i \in V_i$

$$
\begin{aligned}
0 \quad &= G(\boldsymbol{\varphi}; \mathbf{w}) \\
&:= \int_{\Omega} \hat{\boldsymbol{\Gamma}}(\boldsymbol{\varphi}; \mathbf{w}) \cdot \mathbf{S} \, dv - \int_{\Omega} \mathbf{w} \cdot \mathbf{b} \, dV - \int_{\Gamma_\sigma} \mathbf{w} \cdot \bar{\mathbf{t}} \, dA \\
&:= \int_{\Omega} \Gamma_{ij} S_{ij} d\Omega - \int_{\Omega} w_i f_i d\Omega - \sum_{i=1}^{n_{sd}} \int_{\Gamma_{h_i}} w_i h_i d\Omega
\end{aligned}
$$

$$
\text{with:} \ \mathbf{S} = \boldsymbol{\mathcal{C}}\mathbf{E} \ ; \ \boldsymbol{\mathcal{C}} = \lambda \mathbf{I} \otimes \mathbf{I} + 2\mu \boldsymbol{\mathcal{I}} \ ; \ \mathbf{E} = \frac{1}{2}(\mathbf{F}^T\mathbf{F} - \mathbf{I})
$$

$$
\boldsymbol{\Gamma} = \frac{1}{2}\left(\frac{\partial w_k}{\partial x_i}\frac{\partial \varphi_k}{\partial x_j} + \frac{\partial \varphi_k}{\partial x_j}\frac{\partial w_k}{\partial x_i}\right)\mathbf{e}_i \otimes \mathbf{e}_j \ ; \ \mathbf{S} := S_{ij}\mathbf{e}_i \otimes \mathbf{e}_j \ ; \ \mathbf{b} = b_i \mathbf{e}_i \ ; \ \bar{\mathbf{t}} = \bar{t}_i \mathbf{e}_i \tag{4.94}
$$

where $\boldsymbol{\Gamma}$ is the virtual Green–Lagrange strain, $\mathbf{S}$ is the second Piola–Kirchhoff stress, $\boldsymbol{\mathcal{C}}$ is the elasticity tensor, whereas $\mathbf{b}$ and $\bar{\mathbf{t}}$ are external loads.
We will develop in this section the consistent linearization of the weak form of equilibrium equations in (4.94). When considering the use of the isoparametric finite element interpolations, the result of consistent linearization is directly applicable to constructing the tangent stiffness, which will be the main ingredient of Newton's iterative solution method applied to this highly nonlinear problem. Namely, the isoparametric finite elements provide the continuum consistent interpolations, which will allow the discretization and consistent linearization procedures to commute; we can thus compute the consistent tangent stiffness by simply discretizing the linearized problem.

We assume that the deformed configuration at time t, denoted as $\Omega^\varphi = \varphi_t(\Omega)$, is already computed, and that we ought to superpose onto this configuration the incremental field $\mathbf{u}^\varphi$, producing a new deformed configuration (see Figure 4.9) and the corresponding motion update φ_{t+dt}. We first assume that such an incremental displacement ought to be a small displacement. Moreover, by requiring the corresponding smoothness of motion, this small incremental displacement, which is a spatial field parameterized in the deformed configuration $\mathbf{u}^\varphi(\mathbf{x}^\varphi, t)$, can also be written in the material description; we can thus write:

$$\mathbf{u}(\mathbf{x}, t) = \hat{\mathbf{u}}^\varphi(\varphi_t(\mathbf{x}), t) \;\Leftrightarrow\; \mathbf{u}_t = \mathbf{u}_t^\varphi \circ \varphi_t \tag{4.95}$$

The material description allows us to easily compute the deformation gradient in the deformed configuration at time $t + dt$, which can be written:

$$\varphi_{t+dt}(\mathbf{x}) = \varphi_t(\mathbf{x}) + \mathbf{u}_t(\mathbf{x}) \implies \mathbf{F}_{t+dt}(\mathbf{x}) = \mathbf{F}_t(\mathbf{x}) + \nabla \mathbf{u}_t(\mathbf{x}) \tag{4.96}$$

We note in passing that the same result concerning the linearization of the deformation gradient can also be computed as the directional or Gâteaux derivative in the direction of the incremental displacement, according to:

$$\begin{aligned} Lin[\mathbf{F}_t] &:= \mathbf{F}_t + \frac{d}{d\varepsilon}[\underbrace{\varphi_{t,\varepsilon}}_{\varphi_t + \varepsilon\, \mathbf{u}_t} \otimes \boldsymbol{\nabla}]\Big|_{\varepsilon=0} \\ &= \mathbf{F}_t + \underbrace{\nabla \mathbf{u}_t}_{\mathbf{u}_t \otimes \boldsymbol{\nabla}} \end{aligned} \tag{4.97}$$

This equivalence of last two results is the consequence of the requirement on small incremental displacement superposed upon the configuration $\varphi_t(\Omega)$, which produce the update $\varphi_{t+dt}(\Omega)$. For the same reason, the directional derivative in the direction of incremental displacement can be used to perform the consistent linearization for other fields of particular interest in the updated configuration $\varphi_{t+dt}(\Omega)$; for example, the consistent linearization of the right Cauchy–Green strain tensor can be written:

$$\begin{aligned} Lin[\mathbf{C}_t] &:= \mathbf{C}_t + \frac{d}{d\varepsilon}[\mathbf{C}_{t,\varepsilon}]\Big|_{\varepsilon=0} \\ &= \mathbf{C}_t + \frac{d}{d\varepsilon}[\mathbf{F}_{t,\varepsilon}^T]\Big|_{\varepsilon=0} \mathbf{F}_t + \mathbf{F}_t^T \frac{d}{d\varepsilon}[\mathbf{F}_{t,\varepsilon}]\Big|_{\varepsilon=0} \\ &= \mathbf{C}_t + (\nabla \mathbf{u}_t^T \mathbf{F}_t + \mathbf{F}_t^T \nabla \mathbf{u}_t) \end{aligned} \tag{4.98}$$

By exploiting the results of this kind, we can easily obtain the linearized weak form of the equilibrium equations for large deformation linear elasticity with Saint-Venant–Kirchhoff material model:

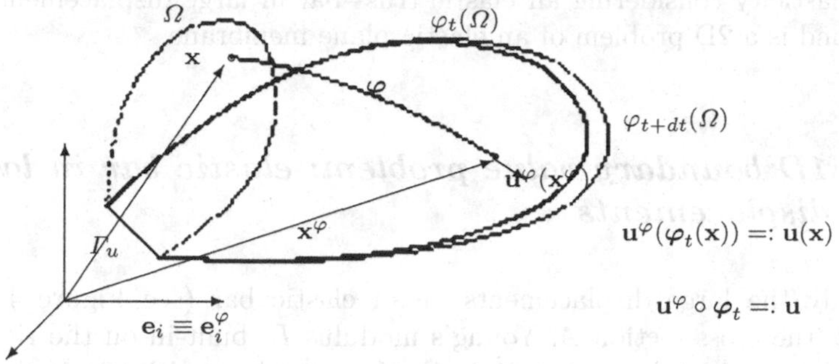

Fig. 4.9 Deformed configurations at time t and $t + dt$, as well as the incremental displacement field.

$$0 = Lin[G(\boldsymbol{\varphi};\mathbf{w})] := G(\boldsymbol{\varphi};\mathbf{w}) + \frac{d}{d\varepsilon}[G(\boldsymbol{\varphi}_\varepsilon;\mathbf{w})]\bigg|_{\varepsilon=0} \tag{4.99}$$

where the second term can be written explicitly as:

$$\begin{aligned}
&\tfrac{d}{d\varepsilon}[G(\boldsymbol{\varphi}_\varepsilon;\mathbf{w})]\bigg|_{\varepsilon=0} = \int_\Omega \tfrac{d}{d\varepsilon}[\hat{\Gamma}(\boldsymbol{\varphi}_{t,\varepsilon}\cdot \bar{\mathbf{S}}(\boldsymbol{\varphi}_t))]\bigg|_{\varepsilon=0} \\
&= \int_\Omega \tfrac{d}{d\varepsilon}[\tfrac{1}{2}(\nabla\mathbf{w}^T\mathbf{F}_\varepsilon \\
&+\mathbf{F}_\varepsilon^T\nabla\mathbf{w})]\bigg|_{\varepsilon=0}\cdot \mathbf{S}_t dV + \int_\Omega \boldsymbol{\Gamma}(\boldsymbol{\varphi}_t;\mathbf{w})\cdot \mathcal{C}\tfrac{d}{d\varepsilon}[\tfrac{1}{2}(\mathbf{F}_{t,\varepsilon}^T\mathbf{F}_{t,\varepsilon}-\mathbf{I})]\bigg|_{\varepsilon=0} dV \\
&= \int_\Omega \frac{1}{2}(\nabla\mathbf{w}^T\nabla\mathbf{u}+\nabla\mathbf{u}^T\nabla\mathbf{w})\cdot\mathbf{S}dV \\
&+\int_\Omega \tfrac{1}{2}(\nabla\mathbf{w}^T\mathbf{F}+\mathbf{F}^T\nabla\mathbf{w})\cdot\mathcal{C}\tfrac{1}{2}(\nabla\mathbf{u}^T\mathbf{F}+\mathbf{F}^T\nabla\mathbf{u})\,dV
\end{aligned} \tag{4.100}$$

We note that the final result is featuring the linearized Green–Lagrange deformation and its virtual counterpart, which can be written in component form according to:

$$\begin{aligned}
\tfrac{1}{2}(\nabla\mathbf{u}^T\mathbf{F}+\mathbf{F}^T\nabla\mathbf{u}) &= \frac{1}{2}\left(\frac{\partial u_k}{\partial x_i}\frac{\partial \varphi_k}{\partial x_j}+\frac{\partial \varphi_k}{\partial x_j}\frac{\partial u_k}{\partial x_i}\right)\mathbf{e}_i\otimes\mathbf{e}_j \\
\tfrac{1}{2}(\nabla\mathbf{w}^T\nabla\mathbf{u}+\nabla\mathbf{u}^T\nabla\mathbf{w}) &= \frac{1}{2}\left(\frac{\partial u_k}{\partial x_i}\frac{\partial w_k}{\partial x_j}+\frac{\partial w_k}{\partial x_j}\frac{\partial u_k}{\partial x_i}\right)\mathbf{e}_i\otimes\mathbf{e}_j
\end{aligned} \tag{4.101}$$

4.4 Numerical implementation of finite element method in large displacements elasticity

In order to illustrate all the main steps in constructing the finite element approximation of large displacement elasticity with Saint-Venant–Kirchhoff material model, we choose two model problems: the first is the simplest problem of 1D elasticity considering an elastic truss-bar in large displacements, and the second is a 2D problem of an elastic plane membrane.

4.4.1 1D boundary value problem: elastic bar in large displacements

We study the large displacements of an elastic bar (see Figure 4.10) of length l, the cross-section A, Young's modulus E, built-in on the right end and loaded on the left end by a traction force $\bar{t}_i$ along with a uniformly distributed loading b. Without loss of generality, we consider that the bar is straight and its motion is restricted to the plane x_1–x_2. By placing x_1 axis along the initial configuration of the bar and by taking into account that there remains only one non-zero stress component in the bar, we can considerably simplify the weak form of the boundary value problem for 3D case given in (4.94), where we finally obtain:

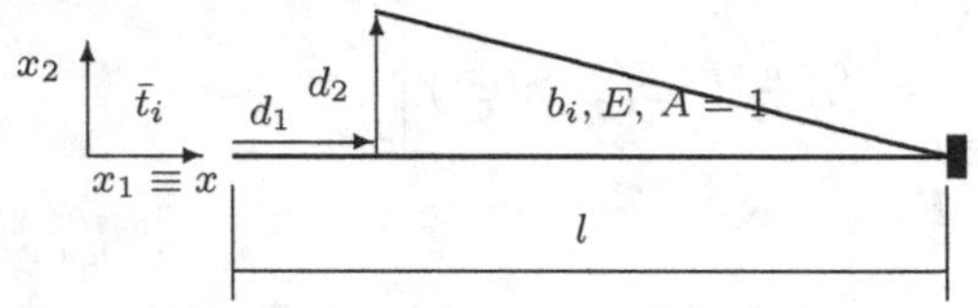

Fig. 4.10 Elastic truss-bar in large displacements: initial and deformed configurations for 1D model.

$$\begin{aligned} G(u_i; w_i) &:= \textstyle\int_0^l \Gamma_{11} S_{11} dx - \int_0^l w_i b_i dx - w_i(l)\bar{t}_i = 0 \\ \text{with:} \quad \Gamma_{11} &= \tfrac{d\varphi_1}{dx}\tfrac{dw_1}{dx} + \tfrac{d\varphi_2}{dx}\tfrac{dw_2}{dx} \\ S_{11} &= \underbrace{C_{1111}}_{E} E_{11} \; ; \; E_{11} = \frac{1}{2}[(\tfrac{d\varphi_1}{dx})^2 + (\tfrac{d\varphi_2}{dx})^2 - 1] \end{aligned} \tag{4.102}$$

where S_{11} is the second Piola–Kirchhoff stress, E_{11} is the Green–Lagrange deformation, and Γ_{11} is its variation.
This formulation is used as the starting point for constructing the discrete approximation of the solution by Galerkin's method. Such an approximation belongs to a finite dimensional space, which we choose as the appropriate subspace of the true, infinite dimensional solution space. The same kind of finite dimensional sub-space is chosen for constructing the discrete approximation of the weighting functions or virtual displacements, with $w^h \in \mathbb{V}^h_{i,0}$. Here, the subscript '0' indicates that the virtual displacements must remain equal to zero on the Dirichlet boundary, where the real displacement takes the

known value. The required regularity of this finite dimensional approximation space depends, in general, upon the choice of the strain measure and the constitutive law. For the choice made in (4.102), with the Green–Lagrange deformation and the Saint-Venant–Kirchhoff material model, we can admit the simplest discrete approximation for real and virtual displacement fields which are linear in each element (see Figure 4.11).

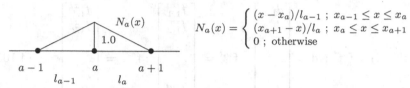

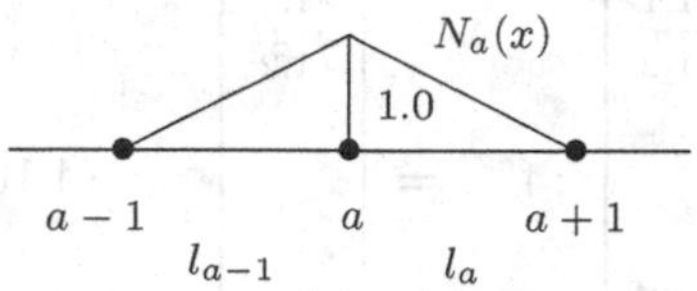

$$N_a(x) = \begin{cases} (x - x_a)/l_{a-1} \ ; \ x_{a-1} \le x \le x_a \\ (x_{a+1} - x)/l_a \ ; \ x_a \le x \le x_{a+1} \\ 0 \ ; \ \text{otherwise} \end{cases}$$

Fig. 4.11 Piece-wise linear discrete approximation of displacement field.

The linear approximation of the displacements implies that their derivatives remain constant in each element:

$$\begin{aligned} \frac{\partial \varphi_1^h}{\partial x} &= \sum_{a=1}^{n} \frac{dN_a(x)}{dx}(x_a + d_{a,1}) \\ \frac{\partial \varphi_2^h}{\partial x} &= \sum_{a=1}^{n} \frac{dN_a(x)}{dx} d_{a,2} \end{aligned} \quad \Longrightarrow \quad \frac{dN_a}{dx} = \begin{cases} \frac{1}{l_{a-1}} \ ; \ x_{a-1} \le x \le x_a \\ \frac{-1}{l_a} \ ; \ x_a \le x \le x_{a+1} \\ 0 \ ; \ \text{otherwise} \end{cases}$$

It thus follows that the discrete approximations of the Green–Lagrange strain and the second Piola–Kirchhoff stress will also be constant in each element. By introducing these approximations into the weak form in (4.102), we can obtain the final result in terms of the Galerkin equation expressing the principle of virtual work (or the weak form of equilibrium) within the framework of the chosen discrete approximation:

$$\sum_{a=1}^{n} \sum_{i=1}^{2} w_{a,i} r_{a,i} = 0 \tag{4.103}$$

$$r_{a,i} := \underbrace{\mathsf{e}_i^T \int_0^l \begin{bmatrix} \frac{dN_a}{dx} & 0 \\ 0 & \frac{dN_a}{dx} \end{bmatrix} \sum_{b=1}^{n} \begin{Bmatrix} \frac{dN_b}{dx}(x_b + d_{b,1}) \\ \frac{dN_b}{dx} d_{b,2} \end{Bmatrix} S_{11}^h dx}_{f_{a,i}^{int}(\mathsf{d})}$$

$$- \underbrace{\int_0^l N_a b_i dx - N_a(l) \bar{t}_i}_{f_{a,i}^{ext}}$$

$$S_{11} = E \tfrac{1}{2} [\textstyle\sum_{a=1}^{n} \sum_{b=1}^{n} \begin{Bmatrix} \frac{dN_a}{dx}(x_a + d_{a,1}) \\ \frac{dN_a}{dx} d_{a,2} \end{Bmatrix}^T \begin{Bmatrix} \frac{dN_b}{dx}(x_b + d_{b,1}) \\ \frac{dN_b}{dx} d_{b,2} \end{Bmatrix} - 1]$$

By considering arbitrary non-zero nodal values of virtual displacements $w_{a,i}$, where $a = 1, 2, ..., n$ and $i = 1, n_{dm}$, we can obtain from Galerkin's equation a set of nonlinear algebraic equations:

$$r_{a,i}(\mathbf{d}) = 0 \Longrightarrow f^{int}_{a,i}(\mathbf{d}) = f^{ext}_{a,i} \tag{4.104}$$

This set of equations governing the equilibrium of an elastic bar in large displacements can also be written in matrix notation according to:

$$\mathsf{f}^{int}(\mathsf{d}) = \mathsf{f}^{ext} \; ; \; \mathsf{d} = \begin{bmatrix} d_{1,1} \\ d_{1,2} \\ . \\ . \\ . \\ d_{n,1} \\ d_{n,2} \end{bmatrix} \; ; \; \mathsf{f}^{int} = \begin{bmatrix} f^{int}_{1,1} \\ f^{int}_{1,2} \\ . \\ . \\ . \\ f^{int}_{n,1} \\ f^{int}_{n,2} \end{bmatrix} \; ; \; \mathsf{f}^{ext} = \begin{bmatrix} f^{ext}_{1,1} \\ f^{ext}_{1,2} \\ . \\ . \\ . \\ f^{ext}_{n,1} \\ f^{ext}_{n,2} \end{bmatrix} \tag{4.105}$$

We show subsequently how to use the finite element method in order to construct the set of nonlinear equations in (4.105) more efficiently. To that end, we abandon the global point of view of Galerkin's method, in favor of the local point of view for each finite element. We can thus obtain an important benefit of processing in a unified manner all the finite elements of the same type, such as all truss-bar elements with the same number of nodes. For a 2-node bar element in large displacements (see Figure 4.12), the local description of this kind is defined in Table 4.2.

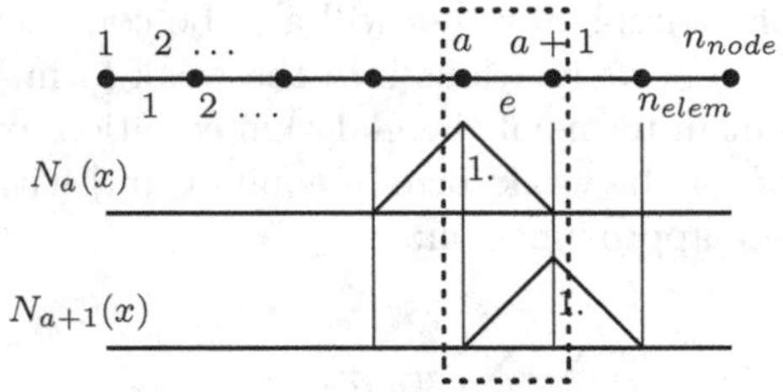

Fig. 4.12 Finite element model of a truss-bar in large displacements composed of 2-node truss-bar elements.

Table 4.2 Local description of a 2-node truss-bar element.

Ingredient	Global description	Local description
Domain	$[x_a, x_{a+1}]$	$[\xi_1, \xi_2]$
Nodes	$\{a, a+1\}$	$(1, 2)$
D.o.f.	$\{d_{a,1}, d_{a,2}, d_{a+1,1}, d_{a+1,2}\}$	$\{d^e_{1,1}, d^e_{1,2}, d^e_{2,1}, d^e_{2,2}\}$
Shape functions	$\{N_a(x), N_{a+1}(x)\}$	$\{N_1(\xi), N_2(\xi)\}$
Interpolations	$u^h_i(x) = N_a(x) d_{a,i} + N_{a+1}(x) d_{a+1,i}$	$u^h_i(\xi) = N^e_1(\xi) d^e_{1,i} + N^e_2(\xi) d^e_{2,i}$

The initial configuration of 2-node bar element is reconstructed from its parent element according to:

$$x(\xi)\Big|_{l^e} = \sum_{a=1}^{2} N_a^e(\xi) x_a;\; N_a^e = \frac{1}{2}(1 + \xi_a \xi),\; \xi_a = \begin{cases} -1\,, a = 1 \\ 1\,, a = 2 \end{cases} \tag{4.106}$$

For an isoparametric element, the same shape functions $N_a^e(\xi)$ are used for constructing the discrete approximation of the displacement field. The isoparametric elements will thus allow us to construct any deformed configuration very easily:

$$\begin{cases} \varphi_1^h(\xi)\Big|_{l^e} = \sum_{a=1}^{2} N_a^e(\xi)\,(x_a + d_{a,1}^e) \\ \varphi_2^h(\xi)\Big|_{l^e} = \sum_{a=1}^{2} N_a^e(\xi)\, d_{a,2}^e \end{cases} \tag{4.107}$$

In the Lagrangian formulation framework, we ought to construct the derivatives of the displacement components with respect to coordinate x in the initial configuration. For any isoparametric element the derivative computation ought to be carried out by exploiting the chain rule; for the chosen 2-node bar element this results with:

$$\begin{aligned} j(\xi)\Big|_{l^e} &= x_{,\xi} = \tfrac{x_2 - x_1}{2} = \tfrac{l^e}{2} \\ \Longrightarrow \quad \tfrac{\partial N_a}{\partial x} &= \tfrac{\partial N_a}{\partial \xi} \tfrac{1}{j(\xi)} = \tfrac{\xi_a}{2} \tfrac{2}{l^e} = \begin{cases} -1/l^e\,, a = 1 \\ \;\; 1/l^e\,, a = 2 \end{cases} \end{aligned} \tag{4.108}$$

We can thus easily confirm that the resulting discrete approximation of derivatives is constant within each element:

$$\frac{d\varphi_1^h}{dx} = \underbrace{\frac{\partial N_1}{\partial x}}_{-1/l^e}(x_1 + d_{1,1}) + \underbrace{\frac{\partial N_2}{\partial x}}_{1/l^e}(x_2 + d_{2,1})\,;\; \frac{d\varphi_2^h}{dx} = \underbrace{\frac{\partial N_1}{\partial x}}_{-1/l^e} d_{1,2} + \underbrace{\frac{\partial N_2}{\partial x}}_{1/l^e} d_{2,2} \tag{4.109}$$

By taking into account that $l^e = x_2^e - x_1^e$ and $0 = y_2^e - y_1^e$, we can rewrite the last result in matrix notation:

$$[(\frac{d\varphi_1}{dx})^2 + (\frac{d\varphi_2}{dx})^2]\Big|_{l^e} = \frac{2}{(l^e)^2} \mathsf{x}^{e\,T} \mathsf{H} \mathsf{d}^e + \frac{1}{(l^e)^2} \mathsf{d}^{e,T} \mathsf{H} \mathsf{d}^e + 1 \tag{4.110}$$

where we denote:

$$\mathsf{x}^e = \begin{bmatrix} x_1 \\ y_1 \\ x_2 \\ y_2 \end{bmatrix};\; \mathsf{d}^e = \begin{bmatrix} d_{1,1}^e \\ d_{1,2}^e \\ d_{2,1}^e \\ d_{2,2}^e \end{bmatrix};\; \mathsf{H}^e = \begin{bmatrix} 1 & 0 & -1 & 0 \\ 0 & 1 & 0 & -1 \\ -1 & 0 & 1 & 0 \\ 0 & -1 & 0 & 1 \end{bmatrix} \tag{4.111}$$

With these results in hand, we can write the corresponding discrete approximation of the Green–Lagrange strain for a 2-node bar element according to:

$$\boxed{E^h_{11}\Big|_{l^e} = \tfrac{1}{(l^e)^2}\mathsf{x}^{e\,T}\mathsf{H}^e\mathsf{d}^e + \tfrac{1}{2(l^e)^2}\mathsf{d}^{e\,T}\mathsf{H}^e\mathsf{d}^e} \tag{4.112}$$

For an elastic bar whose constitutive behavior is governed by the Saint-Venant–Kirchhoff material model, we can then obtain the corresponding finite element approximation of the second Piola–Kirchhoff stress by multiplying the last expression with Young's modulus:

$$\boxed{S^h_{11}\Big|_{l^e} = E(\tfrac{1}{(l^e)^2}\mathsf{x}^{e\,T}\mathsf{H}\mathsf{d}^e + \tfrac{1}{2(l^e)^2}\mathsf{d}^{e\,T}\mathsf{H}\mathsf{d}^e)} \tag{4.113}$$

The finite element approximation of the derivatives of virtual displacement field can be constructed in the same manner as the one for real displacement field in (4.110). This further allows us to obtain the corresponding finite element approximation of the virtual Green–Lagrange strain:

$$\boxed{\varGamma_{11} = \tfrac{1}{(l^e)^2}(\mathsf{x}^{e\,T} + \mathsf{d}^{e\,T})\mathsf{H}\mathsf{w}^e} \;\; ; \; \mathsf{w}^e = \begin{bmatrix} w^e_{1,1} \\ w^e_{1,2} \\ w^e_{2,1} \\ w^e_{2,2} \end{bmatrix} \tag{4.114}$$

It is interesting to note that the last result can also be obtained as the directional derivative of the Green–Lagrange strain approximation in (4.112) in the direction of virtual displacement.

With these results in hand, we can write a particular 2-node bar element contribution to the discrete approximation of the virtual work according to:

$$\begin{aligned}
\mathsf{w}^{e\,T}\mathsf{f}^{int,e}(\mathsf{d}^e) &= \textstyle\int_{l^e} \varGamma_{11} S_{11} dx \\
&= \mathsf{w}^{e\,T} \int_{l^e} \underbrace{\frac{1}{(l^e)^2}\mathsf{H}^T(\mathsf{x}^e + \mathsf{d}^e)S_{11}}_{const.}\, dx \\
&= \mathsf{w}^{e\,T}\mathsf{H}^T\frac{1}{l^e}(\mathsf{x}^e + \mathsf{d}^e)S_{11}\underbrace{\frac{1}{l^e}\int_{l^e} dx}_{=1} \\
\Rightarrow \quad & \boxed{\mathsf{f}^{int,e} = \mathsf{H}^T\frac{1}{l^e}(\mathsf{x}^e + \mathsf{d}^e)S^h_{11}}
\end{aligned} \tag{4.115}$$

where the second Piola–Kirchhoff stress approximation given in (4.113).

The same isoparametric finite element approximations are employed in constructing the linearized weak form, which is the basis of Newton's iterative solution procedure. For a 2-node bar element, this results with:

$$Lin[\mathsf{r}^e(\mathsf{d}^e)] = \mathsf{r}^e(\mathsf{d}^e) + \mathsf{K}^e(\mathsf{d}^e)\,\mathsf{u}^e \;; \quad \boxed{\mathsf{K}^e = \mathsf{K}^e_m + \mathsf{K}^e_g} \;;$$
$$\boxed{\mathsf{K}^e_m = \mathsf{H}^T \frac{1}{l^e}(\mathsf{x}^e + \mathsf{d}^e)\frac{E}{l^e}\frac{1}{l^e}(\mathsf{x}^{e\,T} + \mathsf{d}^{e\,T})\mathsf{H}\,;\; \mathsf{K}^e_g = \frac{S_{11}}{l^e}\mathsf{H}} \tag{4.116}$$

where K^e is the consistent tangent stiffness matrix for such element. We note that the tangent stiffness consists of two parts; the first K^e_m is referred to as the material stiffness and the second K^e_g is the geometric stiffness matrix which is characteristic of large displacement problems.

Even though the derivation of the element stiffness matrix and its residual force vector is carried out for a 2-node truss-bar element placed along x_1 axis of the global coordinate system, the final results will also apply to an arbitrary position of the element. Namely, any constant value of strain or stress, obtained from 2-node element approximation, will remain invariant under reference frame rotation by an arbitrary angle α; hence, for a general orientation of the 2-node truss-bar element, with $\cos\alpha = (x_2 - x_1)/l^e$ and $\sin\alpha = (y_2 - y_1)/l^e)$, we can still write:

$$\mathsf{f}^{int,e} = \frac{1}{l^e}\begin{bmatrix} -(x_2 - x_1) - (d_{2,1} - d_{1,1}) \\ -(y_2 - y_1) - (d_{2,2} - d_{1,2}) \\ (x_2 - x_1) + (d_{2,1} - d_{1,1}) \\ (y_2 - y_1) + (d_{2,2} - d_{1,2}) \end{bmatrix} S_{11} \tag{4.117}$$

Remark on higher order finite element interpolations: It is possible (e.g. see Ibrahimbegovic [112]) to generalize the finite element approximation presented in this section to a bar element with $n_{en} > 2$ nodes and a higher order interpolation; to that end, we employ the shape functions in terms of the Lagrange polynomials of order $n_{en} - 1$ (see Chapter 2). The elements of this kind can have an arbitrary curvilinear axis, parameterized by the arc-length s. The derivative computation of the element shape functions can be carried out by using the chain rule:

$$\begin{aligned}
&\frac{dN^e_a(\xi)}{ds} = \frac{1}{j(\xi)}\frac{dN^e_a(\xi)}{d\xi} \;;\; j(\xi) = \sqrt{(\tfrac{dx}{d\xi})^2 + (\tfrac{dy}{d\xi})^2} \;;\\
&\frac{dx}{d\xi} = \sum_{a=1}^{n_{en}} \frac{dN^e_a(\xi)}{d\xi} x^e_a \;;\; \frac{dy}{d\xi} = \sum_{a=1}^{n_{en}} \frac{dN^e_a(\xi)}{d\xi} y^e_a \;;\\
&\left.\frac{d\mathbf{x}(\xi)}{ds}\right|_{l^e} = \sum_{a=1}^{n_{en}} \frac{dN^e_a(\xi)}{ds}\mathsf{x}^e_a \;;\; \left.\frac{d\mathbf{u}(\xi)}{ds}\right|_{l^e} = \sum_{a=1}^{n_{en}} \frac{dN^e_a(\xi)}{ds}\mathsf{d}^e_a \;;\\
&\left.\frac{d\boldsymbol{\varphi}(\xi)}{ds}\right|_{l^e} = \sum_{a=1}^{n_{en}} \frac{dN^e_a(\xi)}{ds}(\mathsf{x}^e_a + \mathsf{d}^e_a) \;;
\end{aligned} \tag{4.118}$$

The only non-zero components of the second Piola–Kirchhoff stress and the internal force vector can then be computed (see [112]) according to:

$$\begin{aligned} S_{11}(\xi)\Big|_{l^e} &= E\tfrac{1}{2}[\tfrac{d\boldsymbol{\varphi}^T(\xi)}{ds}\tfrac{d\boldsymbol{\varphi}(\xi)}{ds} - 1] \\ \mathsf{f}^{int,e} &= [\mathsf{f}_a^{int,e}] \; ; \; \mathsf{f}_a^{int,e} = \textstyle\sum_{l=1}^{n_{en}-1} \tfrac{dN_a^e(\xi_l)}{ds}\tfrac{d\boldsymbol{\varphi}(\xi_l)}{ds} S_{11}(\xi_l) j(\xi_l) w_l \end{aligned} \tag{4.119}$$

We indicated in the last expression that the internal force vector for this higher order element is computed by numerical integration by using $n_{en} - 1$ Gauss quadrature points with corresponding choice of abscissas ξ_l and weights w_l. It is easy to verify that for $n_{en} = 2$, the last results allows us to recover the result in (4.117).

4.4.2 2D plane elastic membrane in large displacements

The second model problem used to illustrate the implementation details of the finite element method in large displacements is 2D plane elastic membrane with Saint-Venant–Kirchhoff constitutive law. The typical boundary value problem form this class can be written as follows:
Given: $b_i : \Omega \mapsto \mathbb{R}$; $\bar{t}_i : \Gamma_{\sigma_i} \mapsto \mathbb{R}$; $\bar{u}_i \,|_{\Gamma_{u_i}} = 0$; $i = 1, 2$
Find: $u_i \in \mathbb{V}_i$ such that $\forall w_i \in \mathbb{V}_{i,0}$

$$\begin{aligned} 0 = G(\boldsymbol{\varphi}; \mathbf{w}) &:= \textstyle\int_\Omega \hat{\boldsymbol{\Gamma}}(\boldsymbol{\varphi}; \mathbf{w}) \cdot \mathbf{S} dv - \int_\Omega \mathbf{w} \cdot \mathbf{b} dV - \int_{\Gamma_\sigma} \mathbf{w} \cdot \bar{\mathbf{t}} dA \\ \hat{\boldsymbol{\Gamma}}(\boldsymbol{\varphi}; \mathbf{w}) &= \frac{1}{2}(\nabla \mathbf{w}^T \mathbf{F} + \mathbf{F}^T \nabla \mathbf{w}) \; ; \; \mathbf{S} = \boldsymbol{\mathcal{C}} \mathbf{E} \; ; \; \mathbf{E} = \frac{1}{2}(\mathbf{F}^T \mathbf{F} - \mathbf{I}) \end{aligned} \tag{4.120}$$

where $\hat{\boldsymbol{\Gamma}}(\boldsymbol{\varphi}; \mathbf{w})$ is the virtual Green–Lagrange strain, $\mathbf{E}$ is the real Green–Lagrange strain and $\mathbf{S}$ related through the elasticity tensor $\boldsymbol{\mathcal{C}}$. The consistently linearized form of this problem, which provides the basis of Newton's iterative solution procedure, can then be obtained by computing the directional derivative of the last expression in the direction of the incremental displacement $\mathbf{u}$ resulting with:

$$\begin{aligned} \tfrac{d}{d\varepsilon}[G(\boldsymbol{\varphi}_\varepsilon; \mathbf{w})]\Big|_{\varepsilon=0} &:= \textstyle\int_\Omega \hat{\mathbf{H}}(\mathbf{w}, \mathbf{u}) \cdot \mathbf{S}\, dV + \int_\Omega \hat{\boldsymbol{\Gamma}}(\boldsymbol{\varphi}, \mathbf{w}) \cdot \boldsymbol{\mathcal{C}} \hat{\boldsymbol{\Gamma}}(\boldsymbol{\varphi}, \mathbf{u})\, dV \\ \hat{\boldsymbol{\Gamma}}(\boldsymbol{\varphi}, \mathbf{u}) &= \frac{1}{2}(\nabla \mathbf{u}^T \mathbf{F} + \mathbf{F}^T \nabla \mathbf{u}) \; ; \; \hat{\mathbf{H}} = \frac{1}{2}(\nabla \mathbf{w}^T \nabla \mathbf{u} + \nabla \mathbf{u}^T \nabla \mathbf{w}) \end{aligned} \tag{4.121}$$

with $\hat{\boldsymbol{\Gamma}}(\boldsymbol{\varphi}, \mathbf{u})$ and $\hat{\mathbf{H}}$ denoting, respectively, the increments of the real and virtual Green–Lagrange strains.
For constructing the discrete approximation of this problem by the finite element method, we employ the isoparametric interpolations. In particular, for a 4-node isoparametric element of this kind (see Figure 4.13), the discrete

approximation of the initial configuration and displacement field can easily be constructed according to:

$$\begin{bmatrix} x_1^h \\ x_2^h \end{bmatrix} =: \mathsf{x}^h \Big|_{\Omega^e} = \sum_{a=1}^{n_{en}} N_a(\xi,\eta)\mathsf{x}_a^e = \mathsf{N}^e\mathsf{x}^e \; ; \; \mathsf{x}^e = \begin{bmatrix} \mathsf{x}_1 \\ \vdots \\ \mathsf{x}_{en} \end{bmatrix}$$

$$\begin{bmatrix} u_1 \\ u_2 \end{bmatrix} =: \mathsf{u}^h \Big|_{\Omega^e} = \sum_{a=1}^{n_{en}} N_a(\xi,\eta)\mathsf{d}_a^e = \mathsf{N}^e\mathsf{d}^e \; ; \; \mathsf{d}^e = \begin{bmatrix} \mathsf{d}_1^e \\ \vdots \\ \mathsf{d}_{en}^e \end{bmatrix} \quad (4.122)$$

$$\mathsf{N}^e = [N_1\mathsf{I}_2, ..., N_{en}\mathsf{I}_2] \; ; \; N_a(\xi,\eta) = \tfrac{1}{4}(1+\xi_a\xi)(1+\eta_a\eta) \; ;$$

With the isoparametric finite element interpolations sharing the same shape functions for displacement field and element initial configuration, we can easily construct the corresponding discrete approximation of the deformed

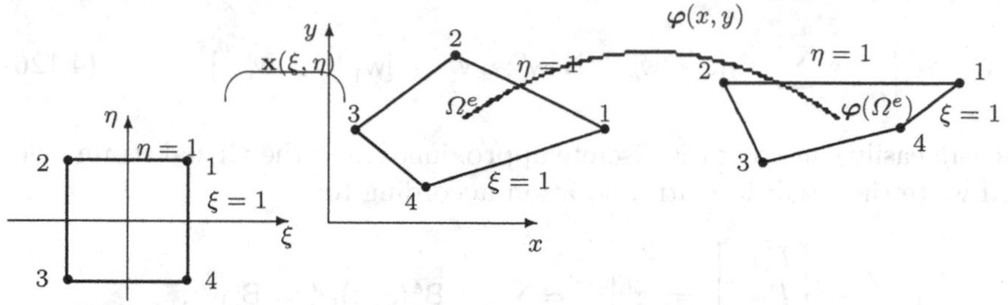

Fig. 4.13 Initial and deformed configurations, Ω^e and $\varphi(\Omega^e)$ for 4-node isoparametric element $Q4$ and its parent element.

configuration (see Figure 4.13), which employs the sum of the nodal values of two fields:

$$\varphi^h \Big|_{\Omega^e} = \sum_{a=1}^{n_{en}} N_a(\xi,\eta)(\mathsf{x}_a^e + \mathsf{d}_a^e) = \mathsf{N}^e(\mathsf{x}^e + \mathsf{d}^e) \quad (4.123)$$

We can also write the finite element interpolation of the Green–Lagrange strain, with standard ordering of strain tensor components within a vector:

$$\underbrace{\mathbf{E}}_{(2\times 2)} \Longrightarrow \begin{bmatrix} E_{11} \\ E_{22} \\ 2E_{12} \end{bmatrix} =: \mathsf{e}^h \Big|_{\Omega^e} = \begin{bmatrix} \frac{1}{2}\frac{\partial\varphi}{\partial x}^T\frac{\partial\varphi}{\partial x} - 1 \\ \frac{1}{2}\frac{\partial\varphi}{\partial y}^T\frac{\partial\varphi}{\partial y} - 1 \\ \frac{\partial\varphi}{\partial x}^T\frac{\partial\varphi}{\partial y} + \frac{\partial\varphi}{\partial y}^T\frac{\partial\varphi}{\partial x} \end{bmatrix} \; ;$$

$$\frac{\partial\varphi}{\partial x}\Big|_{\Omega^e} = \sum_{a=1}^{n_{en}} \frac{\partial N_a}{\partial x}(\xi,\eta)(\mathsf{x}_a^e + \mathsf{d}_a^e) \; ; \quad (4.124)$$

$$\frac{\partial\varphi}{\partial y}\Big|_{\Omega^e} = \sum_{a=1}^{n_{en}} \frac{\partial N_a}{\partial y}(\xi,\eta)(\mathsf{x}_a^e + \mathsf{d}_a^e)$$

The components of the work-conjugate second Piola-Kirchhoff stress tensor are also ordered within a vector, $\mathsf{s}^h = [S_{11}; S_{22}; S_{12}]^T$ before constructing their discrete approximation. The stress values can be obtained easily by multiplying the vector with strain components by the corresponding elasticity tensor components that are stored in a matrix; for Saint-Venant–Kirchhoff material, the latter takes the same form as in the case of linearized kinematics, and we can thus write:

$$\mathsf{s}^h = \mathsf{D}\mathsf{e}^e \Leftrightarrow \begin{bmatrix} S_{11} \\ S_{22} \\ S_{12} \end{bmatrix} = \begin{bmatrix} \bar{\lambda}+2\mu & \bar{\lambda} & 0 \\ \bar{\lambda} & \bar{\lambda}+2\mu & 0 \\ 0 & 0 & \mu \end{bmatrix} \begin{bmatrix} E_{11} \\ E_{22} \\ 2E_{12} \end{bmatrix} \; ; \; \bar{\lambda} = \frac{2\lambda\mu}{\lambda+2\mu} \tag{4.125}$$

where λ and μ are Lamé's parameters. For the isoparametric element of this kind, the finite element approximation of the virtual displacement field is constructed by using the same shape functions as the real displacement field leading to:

$$\mathsf{w}^h \Big|_{\Omega^e} = \sum_{a=1}^{n_{en}} N_a(\xi,\eta)\mathsf{w}_a^e = \mathsf{N}^e\mathsf{w}^e; \quad \mathsf{w}^e = [\mathsf{w}_1^{e\,T}, ..., \mathsf{w}_{en}^{e\,T}]^T \tag{4.126}$$

We can easily construct the discrete approximation of the virtual strain field, and write the result in matrix notation according to:

$$\underbrace{\boldsymbol{\Gamma}}_{(2\times2)} \longrightarrow \begin{bmatrix} \Gamma_{11} \\ \Gamma_{22} \\ 2\Gamma_{12} \end{bmatrix} =: \mathsf{g}^h \Big|_{\Omega^e} = \textstyle\sum_{a=1}^{n_{en}} \mathsf{B}_a^e(\xi,\eta)\mathsf{w}_a^e = \mathsf{B}^e\mathsf{w}^e \, ;$$
$$\mathsf{B}^e = [\mathsf{B}_1^e, ..., \mathsf{B}_{n_{en}}^e]; \; \mathsf{B}_a^e = \begin{bmatrix} \frac{\partial N_a}{\partial x}\frac{\partial \boldsymbol{\varphi}}{\partial x}^T \\ \frac{\partial N_a}{\partial y}\frac{\partial \boldsymbol{\varphi}}{\partial y}^T \\ \frac{\partial N_a}{\partial x}\frac{\partial \boldsymbol{\varphi}}{\partial y}^T + \frac{\partial N_a}{\partial y}\frac{\partial \boldsymbol{\varphi}}{\partial x}^T \end{bmatrix} \tag{4.127}$$

With these results in hand, we can now write the finite element approximation of the element internal force vector:

$$\int_\Omega \boldsymbol{\Gamma}\cdot\mathbf{S}\,dV = \mathsf{w}^T\mathsf{f}^{int,e} \; ; \; \mathsf{f}^{int,e} = [f_p^{int,e}]$$
$$f_p^{int,e} = \mathsf{e}_i^T \int_{\Omega^e} \mathsf{B}_a^T\mathsf{s}^h dV \; ; \; a = 1,\ldots,n_{en} \; ; \; i = 1,2 \; ; \; \mathsf{s}^h = [S_{11}; S_{22}; S_{12}]^T \tag{4.128}$$

The isoparametric interpolations are also chosen for the incremental displacement field, resulting with the corresponding discrete approximation that can be written:

$$\mathsf{u}^h \Big|_{\Omega^e} = \sum_{a=1}^{n_{en}} N_a(\xi,\eta)\mathsf{u}_a^e = \mathsf{N}^e\mathsf{u}^e; \quad \mathsf{u}^e = [\mathsf{u}_1^{e\,T}, ..., \mathsf{u}_{en}^{e\,T}]^T \tag{4.129}$$

The isoparametric elements, where the displacement value at each point is obtained as the linear combination of the nodal values, provide the continuum-consistent interpolations where the discretization and linearization commute. We can thus obtain the linearized form of the discrete approximation of the internal force vector from the discrete approximation of the linearized continuum problem in (4.121):

$$Lin[\mathsf{f}^{e,int} - \mathsf{f}^{e,ext}] := \mathsf{f}^{e,int} - \mathsf{f}^{e,ext} + \underbrace{\mathsf{K}^e}_{\partial f^{e,int}/\partial d^e} \mathsf{u}^e \tag{4.130}$$

The matrix K^e in the last expression denotes the element tangent stiffness, which consists of a material and of a geometric part:

$$\begin{aligned} &K^e_{pq} = \mathsf{e}_i^T \left[\int_{\Omega^e} \mathsf{B}_a^T \mathsf{D} \mathsf{B}_b dv + \int_{\Omega^e} \mathsf{G}_{ab} dv \right] \mathsf{e}_j \; ; \\ &p = n_{df}(a-1) + i \; ; \; q = n_{df}(b-1) + j \; ; \\ &\mathsf{G}_{ab} = S_{11} \frac{\partial N_a}{\partial x}\frac{\partial N_b}{\partial x} + S_{22} \frac{\partial N_a}{\partial y}\frac{\partial N_b}{\partial y} + S_{12} \left(\frac{\partial N_a}{\partial x}\frac{\partial N_b}{\partial y} + \frac{\partial N_a}{\partial y}\frac{\partial N_b}{\partial x} \right) \mathsf{I}_2 \end{aligned} \tag{4.131}$$

The geometric part of the tangent stiffness appears only in large displacement problems, and it depends directly on current stress values. This explains, in particular, how the pre-stressing can increase the stiffness of a deformable membrane in large displacement setting. The stress dependence of the geometric stiffness in this case concerns the second Piola–Kirchhoff stress, which is the consequence of the chosen material description of large displacement elasticity.
In the next section we show that the geometric stiffness can also be expressed as explicitly dependent upon the true or Cauchy stress, by using the spatial description of large displacement elasticity.

4.5 Spatial description of elasticity in large displacements

In this section we discuss the details of the finite element formulation of large displacements elasticity in spatial description, where the true stress and the strain field defined in the current configuration are used for writing the weak form of equilibrium equations. We can thus obtain a considerable gain in computational efficiency when computing the element arrays, the internal force vector and the element stiffness matrix, since they will have a sparse structure in spatial description, very much like the one used in linearized kinematic problems. More importantly, the spatial description is by far preferred choice in large displacements framework for providing the proper interpretation of the inelastic behavior by means of plasticity or damage criteria, since the physics of the inelastic behavior can be best expressed in terms of true or Cauchy stress. It is important to recall, however, that the inelastic behavior requires the spatial description (employing the stress acting in the deformed configuration) which accounts for the complete deformation trajectory right from the starting point in the initial configuration. Therefore, the spatial description of this kind is not quite the same as the standard Eulerian formulation, but rather more like an alternative manner of expressing the Lagrangian formulation by using the material description of the true or Cauchy stress in terms of the Kirchhoff stress.

We will illustrate herein two kinds of descriptions for a number of fields of direct interest for large displacement framework, such as velocity, acceleration and rates of strain and stress. We start by providing the Lagrangian description of the velocity vector for a particle in the deformed configuration a time 't', which can be written simply as the partial derivative of the motion '$\boldsymbol{\varphi}(\cdot)$' with respect to time:

$$\mathbf{v}(\mathbf{x},t) = \frac{\partial}{\partial t}\boldsymbol{\varphi}(\mathbf{x},t) \tag{4.132}$$

We note that the velocity vector defined above is placed in (the tangent plane of) the deformed configuration, but it is parameterized by the coordinates in the initial configuration; see Figure 4.14. In Eulerian description, the same velocity vector is parameterized with respect to the coordinates in the deformed configuration '$\mathbf{x}^{\varphi}$', and denoted with '$\mathbf{v}^{\varphi}(\mathbf{x}^{\varphi},t)$'. By assuming that the motion $\boldsymbol{\varphi}_t$ is sufficiently smooth to construct its inverse $\boldsymbol{\varphi}_t^{-1}$, we can easily establish the corresponding relationship between two different representations of the velocity vector:

$$\mathbf{v}(\mathbf{x},t) := \hat{\mathbf{v}}^{\varphi}(\underbrace{\boldsymbol{\varphi}_t(\mathbf{x})}_{\mathbf{x}^{\varphi}},t) \Leftrightarrow \mathbf{v}_t = \mathbf{v}_t^{\varphi} \circ \boldsymbol{\varphi}_t \implies \mathbf{v}^{\varphi} = \mathbf{v} \circ \boldsymbol{\varphi}^{-1} \tag{4.133}$$

We indicated in (4.133) above that the spatial description of the velocity vector is computed by the material derivative of the motion of a particle, where the particle remains the same and only coordinate representation will change. This important point can also be illustrated when computing the acceleration vector as the material derivative of the velocity vector. In material description, the acceleration vector is easy to obtain as the second partial derivative of the motion of a particle:

$$\begin{aligned} \mathbf{a}(\mathbf{x},t) &:= \tfrac{\partial}{\partial t}\mathbf{v}(\mathbf{x},t) \\ &= \tfrac{\partial^2}{\partial t^2}\boldsymbol{\varphi}(\mathbf{x},t) \end{aligned} \tag{4.134}$$

The computation of the material derivative in spatial description is more laborious, since we have to account for temporal evolution of a vector which remains attached to the same particle in the course of motion. We can carry out such a computation in three stages: first performing the pull-back or motion induced convective mapping to the initial configuration, followed by the second stage of computing partial time derivative of such a transformation result, and finally third stage with push-forward or convective mapping to

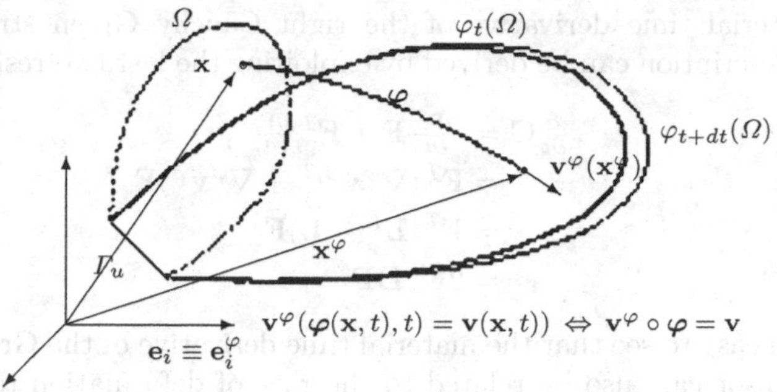

Fig. 4.14 Material and spatial descriptions of velocity vector.

the current configuration. For the material derivative of the velocity vector we thus obtain the following result:

$$\begin{aligned}\mathbf{a}^{\varphi}(\mathbf{x}^{\varphi},t) &:= \frac{\partial}{\partial t}\bigg|_{\mathbf{x}} \mathbf{v}^{\varphi}(\mathbf{x}^{\varphi},t) \\ &= [\tfrac{\partial}{\partial t}(\mathbf{v}^{\varphi}\circ\boldsymbol{\varphi})]\circ\boldsymbol{\varphi}^{-1} \\ &= \frac{\partial}{\partial t}\bigg|_{\mathbf{x}^{\varphi}} \mathbf{v}^{\varphi}(\mathbf{x}^{\varphi},t) + \underbrace{\mathbf{L}}_{\nabla^{\varphi}\mathbf{v}^{\varphi}}\ \underbrace{\mathbf{v}^{\varphi}}_{\partial\boldsymbol{\varphi}/\partial t}\end{aligned} \tag{4.135}$$

In the expression above, $\mathbf{L}$ is the spatial velocity gradient tensor that is defined as:

$$\mathbf{L} = \nabla^{\varphi}\mathbf{v}^{\varphi} \tag{4.136}$$

The last result will also allow us to write the rate of deformation in a compact form. To that end, we first compute the time derivative of the deformation gradient which provides an alternative interpretation of the spatial velocity gradient:

$$\begin{aligned}\tfrac{\partial}{\partial t}\mathbf{F} &:= \tfrac{\partial}{\partial t}\nabla\boldsymbol{\varphi} = \nabla(\tfrac{\partial\boldsymbol{\varphi}}{\partial t}) \equiv \nabla\mathbf{v} \\ &= (\mathbf{v}\otimes\boldsymbol{\nabla})(\mathbf{F}^{-1}\mathbf{F}) \\ &= (\mathbf{v}\otimes\mathbf{F}^{-T}\boldsymbol{\nabla})\mathbf{F} \qquad\qquad \Longrightarrow\ \mathbf{L} = \frac{\partial\mathbf{F}}{\partial t}\mathbf{F}^{-1} \\ &= (\mathbf{v}^{\varphi}\otimes\boldsymbol{\nabla}^{\varphi})\mathbf{F} \\ &= (\nabla^{\varphi}\mathbf{v}^{\varphi})\mathbf{F} \\ &= \mathbf{L}\mathbf{F}\end{aligned} \tag{4.137}$$

The rate of deformation tensor is then defined simply as the symmetric part of the velocity gradient:

$$\mathbf{D} = \frac{1}{2}(\mathbf{L} + \mathbf{L}^T) \tag{4.138}$$

The material time derivative of the right Cauchy–Green strain tensor in spatial description can be derived by exploiting the last two results to obtain:

$$\begin{aligned}\tfrac{\partial}{\partial t}\mathbf{C} &= \tfrac{\partial \mathbf{F}^T}{\partial t}\mathbf{F} + \mathbf{F}^T\tfrac{\partial \mathbf{F}}{\partial t}\\ &= \mathbf{F}^T(\nabla^{\varphi}\mathbf{v}^{\varphi,T} + \nabla^{\varphi}\mathbf{v}^{\varphi})\mathbf{F}\\ &= \mathbf{F}^T(\mathbf{L}^T + \mathbf{L})\mathbf{F}\\ &= 2\mathbf{F}^T\mathbf{D}\mathbf{F}\end{aligned} \tag{4.139}$$

It is then easy to see that the material time derivative of the Green–Lagrange strain tensor can also be related to the rate of deformation tensor through the deformation gradient:

$$\mathbf{E} = \frac{1}{2}(\mathbf{C} - \mathbf{I}) \Longrightarrow \frac{\partial \mathbf{E}}{\partial t} = \mathbf{F}^T\mathbf{D}\mathbf{F} \tag{4.140}$$

An interesting question to address concerns the strain tensor whose material derivative is equal to the rate of deformation tensor; it turns out, this is the Almansi strain tensor $\boldsymbol{\varepsilon}_A^{\varphi}$, which appears the natural replacement of the Green–Lagrange deformation in spatial description:

$$\boldsymbol{\varepsilon}_A^{\varphi} := \frac{1}{2}(\mathbf{I}^{\varphi} - \mathbf{B}^{-1}) = \mathbf{F}^{-T}(\mathbf{E}\circ\boldsymbol{\varphi}^{-1})\mathbf{F}^{-1} \tag{4.141}$$

The material time derivative of the Almansi strain tensor, which is a spatial field defined in the deformed configuration, ought again to be computed in three stages (pull-back, partial time derivative and push-forward), by exploiting the results in (4.137) to (4.141), to obtain:

$$\begin{aligned}\mathbf{D} &= \mathbf{F}^{-T}\{\tfrac{\partial}{\partial t}[\mathbf{F}^T\underbrace{\frac{1}{2}(\mathbf{I} - \mathbf{B}^{-1})}_{\epsilon_A}\mathbf{F}]\}\mathbf{F}^{-1}\\ &= \mathbf{F}^{-T}[\tfrac{\partial}{\partial t}\underbrace{\frac{1}{2}(\mathbf{C} - \mathbf{I})}_{E}]\mathbf{F}^{-1}\\ &= \tfrac{1}{2}(\mathbf{L} + \mathbf{L}^T)\end{aligned} \tag{4.142}$$

With the symmetric part of the spatial velocity gradient, which represents the strain rate, the remaining skew-symmetric part will represent the rate of large rotation or the spin tensor:

$$\mathbf{W} = \frac{1}{2}(\mathbf{L} - \mathbf{L}^T) \tag{4.143}$$

In the limit case of small displacement gradient theory (where $\| \nabla\mathbf{d} \| \mapsto \| \nabla\mathbf{u} \| \ll 1$), the last two results will consistently reduce to the standard form of infinitesimal strain rate and infinitesimal rotation rate, respectively:

$$\left.\begin{aligned}\lim_{\|\nabla u\|\mapsto 0}\mathbf{D} &= \tfrac{\partial \boldsymbol{\epsilon}}{\partial t}\\ \lim_{\|\nabla u\|\mapsto 0}\mathbf{W} &= \tfrac{\partial \boldsymbol{\omega}}{\partial t}\end{aligned}\right. \Rightarrow \lim_{\nabla u\mapsto 0}\mathbf{L} = \frac{\partial \mathbf{u}}{\partial t}\otimes\boldsymbol{\nabla} \tag{4.144}$$

The stress tensor which is energy-conjugate to the rate of deformation tensor is the Kirchhoff stress, since the result in (4.142) can be used to show that:

$$\underbrace{\mathbf{F}^{-1}\boldsymbol{\tau}\mathbf{F}^{-T}}_{\mathbf{S}} \cdot \frac{\partial}{\partial t}\mathbf{E} = \boldsymbol{\tau} \cdot \underbrace{\mathbf{F}^{-T}(\frac{\partial}{\partial t}\mathbf{E})\mathbf{F}^{-1}}_{\mathbf{D}} \tag{4.145}$$

This result can be confirmed by computing the material time derivative of the strain energy potential for the Saint-Venant–Kirchhoff material; the latter characterizes the constitutive behavior of a hyperelastic material, which is normally written in material description by using the conjugate pair of the second Piola–Kirchhoff stress and the Green–Lagrange strain. We can recast this result in spatial description, by using the Kirchhoff stress and the rate of deformation tensor:

$$\begin{aligned} \frac{\partial}{\partial t}\psi(\mathbf{E}) &= \underbrace{\frac{\partial \psi}{\partial \mathbf{E}}}_{\mathbf{S}} \cdot \frac{\partial \mathbf{E}}{\partial t} \\ &= \mathbf{F}\mathbf{S}\mathbf{F}^T \cdot \mathbf{F}^{-T}\frac{\partial \mathbf{E}}{\partial t}\mathbf{F}^{-1} \\ &= \boldsymbol{\tau} \cdot \mathbf{D} \end{aligned} \tag{4.146}$$

The consistent linearization of the Saint-Venant–Kirchhoff constitutive model in spatial description will require the proper definition of stress rate for Kirchhoff stress; the latter can be obtained as the corresponding transformation of the time derivative of the second Piola–Kirchhoff stress, leading to the notion of the Lie derivative of Kirchhoff stress tensor $L_v[\boldsymbol{\tau}]$:

$$\begin{aligned} &\boldsymbol{\tau} = \mathbf{F}\mathbf{S}\mathbf{F}^T \mid \frac{\partial}{\partial t} \Rightarrow \\ &\frac{\partial}{\partial t}\boldsymbol{\tau} = \underbrace{(\frac{\partial}{\partial t}\mathbf{F})\mathbf{F}^{-1}}_{\mathbf{L}}\underbrace{\mathbf{F}\mathbf{S}\mathbf{F}^T}_{\boldsymbol{\tau}} + \underbrace{\mathbf{F}\mathbf{S}\mathbf{F}^T}_{\boldsymbol{\tau}}\underbrace{\mathbf{F}^{-T}\frac{\partial}{\partial t}\mathbf{F}^T}_{\mathbf{L}^T} + \mathbf{F}(\frac{\partial}{\partial t}\mathbf{S})\mathbf{F}^T \Rightarrow \\ &L_v[\boldsymbol{\tau}] := \frac{\partial}{\partial t}\boldsymbol{\tau} - \mathbf{L}\boldsymbol{\tau} - \boldsymbol{\tau}\mathbf{L}^T = \mathbf{F}(\frac{\partial}{\partial t}\mathbf{S})\mathbf{F}^T \end{aligned} \tag{4.147}$$

The Lie derivative of the Kirchhoff stress $L_v[\boldsymbol{\tau}]$ is equal to the material time derivative of the Kirchhoff stress, which is a spatial field requiring again a three-stage-computation with pull-back or the convective transformation to the initial configuration, the partial time derivative computation and the push-forward or convected transformation of the end result to the deformed configuration; we thus obtain:

$$\begin{aligned} L_v[\boldsymbol{\tau}] &:= \{\mathbf{F}\frac{\partial}{\partial t}[\overbrace{\mathbf{F}^{-1}(\boldsymbol{\tau}\circ\boldsymbol{\varphi})\mathbf{F}^{-T}}^{\mathbf{S}}]\mathbf{F}^T\}\circ\boldsymbol{\varphi}^{-1} \\ &= \frac{\partial}{\partial t}\boldsymbol{\tau} - \underbrace{\mathbf{L}}_{-\mathbf{F}\frac{\partial}{\partial t}\mathbf{F}^{-1}}\boldsymbol{\tau} - \boldsymbol{\tau}\underbrace{\mathbf{L}^T}_{-(\frac{\partial}{\partial t}\mathbf{F}^{-T})\mathbf{F}^T} \end{aligned} \tag{4.148}$$

where we used the following auxiliary result:

$$\frac{\partial}{\partial t}[\mathbf{I} = \mathbf{F}\mathbf{F}^{-1}] \Rightarrow \mathbf{L} = -\mathbf{F}\frac{\partial}{\partial t}\mathbf{F}^{-1} \tag{4.149}$$

The Lie derivative of the Kirchhoff stress tensor, where the convective transformation is carried out by the deformation gradient, is yet called the Truesdell or Oldroyd stress rate. We can readily show that such a stress rate verifies the axiom of material indifference. First we show that the Kirchhoff stress tensor at the given deformed configuration also remains invariant under a superposed rigid body motion represented by a constant orthogonal tensor $\mathbf{Q} \in SO(3)$, since it transform according to:

$$\begin{aligned} \mathbf{F}^{srb} &= \mathbf{Q}\mathbf{F} \Rightarrow \tfrac{\partial}{\partial t}\mathbf{F}^{srb} = \mathbf{Q}\tfrac{\partial}{\partial t}\mathbf{F} + \boldsymbol{\Omega}\mathbf{Q}\mathbf{F} \ ; \\ \mathbf{L}^{srb} &= (\tfrac{\partial}{\partial t}\mathbf{F}^{srb})\mathbf{F}^{srb^{-1}} = \mathbf{Q}\mathbf{L}\mathbf{Q}^T + \boldsymbol{\Omega} \\ \boldsymbol{\tau}^{srb} &= \mathbf{Q}\boldsymbol{\tau}\mathbf{Q}^T \end{aligned} \tag{4.150}$$

where $\boldsymbol{\Omega} := (\frac{\partial}{\partial t}\mathbf{Q})\mathbf{Q}^T = -\boldsymbol{\Omega}^T$. The last result can then be exploited to confirm that the Truesdell or Oldroyd stress rate would also verify the invariance requirements (see [92]) in such a deformed configuration:

$$\begin{aligned} L_v[\boldsymbol{\tau}^{srb}] &= \mathbf{Q}(\tfrac{\partial}{\partial t}\boldsymbol{\tau})\mathbf{Q}^T - \mathbf{Q}\mathbf{L}\mathbf{Q}^T\mathbf{Q}\boldsymbol{\tau}\mathbf{Q}^T - \mathbf{Q}\boldsymbol{\tau}\mathbf{Q}^T\mathbf{Q}\mathbf{L}\mathbf{Q}^T \\ &= \mathbf{Q}L_v[\boldsymbol{\tau}]\mathbf{Q}^T \end{aligned} \tag{4.151}$$

Therefore, we conclude that any rate form of the constitutive law expressed in terms of the Truesdell or Oldroyd stress rate will ensure objectivity of material elastic response under a change of frame. For example, with these results in hand, we can easily compute the corresponding form of the elasticity tensor in the spatial description for the Saint-Venant–Kirchhoff material model. To that end, we can exploit the previous results on material and spatial descriptions of stress and strain tensors in (4.142) and (4.148), which allows us to write:

$$\frac{\partial \mathbf{S}}{\partial t} = \underbrace{\frac{\partial \mathbf{S}}{\partial \mathbf{E}}}_{\mathcal{C}}\frac{\partial \mathbf{E}}{\partial t} \Longrightarrow L_v[\boldsymbol{\tau}] = \frac{\partial \boldsymbol{\tau}}{\partial \boldsymbol{\epsilon}}\mathbf{D} \Leftrightarrow \underbrace{\frac{\partial \tau^{ab}}{\partial \varepsilon_{cd}}}_{\mathcal{C}^{\tau}_{abcd}} = F^a_i F^b_j F^c_k F^d_l \underbrace{\frac{\partial S^{ij}}{\partial E_{kl}}}_{\mathcal{C}_{ijkl}} \tag{4.152}$$

We note that the spatial representation of elasticity tensor for the Truesdell or Oldroyd stress rate can thus be obtained easily by a transformation from the standard form of elasticity tensor $\mathcal{C}$, with constant components in material description; the fact that such a transformation features only the deformation gradient, which is readily available at this stage of computation, is an important advantage of the Truesdell or Oldroyd stress rate for providing the spatial description of the constitutive response.

We note in passing that a number of other stress rates can be used instead, in fact infinitely many, theoretically. Each of those stress rates can be interpreted as an alternative form of the Lie derivative, where the deformation gradient is replaced by another operator in the expression (4.148) for the stress rate. One such example is the Green–McInnis–Naghdi stress rate of Kirchhoff stress tensor, where the pull-back and push-forward transformations are carried by the rotation tensor featuring in the polar decomposition formula $\mathbf{R} := \mathbf{F}\mathbf{U}^{-1}$; we can thus write:

$$
\begin{aligned}
L_w^{GMN}[\boldsymbol{\tau}] &= \{\mathbf{R}\tfrac{\partial}{\partial t}[\mathbf{R}^T(\boldsymbol{\tau}\circ\boldsymbol{\varphi})\mathbf{R}]\mathbf{R}^T\}\circ\boldsymbol{\varphi}^{-1} \\
&= \frac{\partial}{\partial t}\boldsymbol{\tau} - \boldsymbol{\tau}\boldsymbol{\Omega} + \boldsymbol{\tau}\boldsymbol{\Omega}
\end{aligned}
\tag{4.153}
$$

where $\boldsymbol{\Omega} = (\frac{\partial}{\partial t}\mathbf{R})\mathbf{R}^T$ is a skew-symmetric tensor. For a superposed rigid body motion, the Green–McInnis–Naghdi stress rate will yield the same result as the Oldroyd stress rate, but not in a more general case where the deformed configuration is produced with a non-zero strain field. It is more difficult to establish the compatibility between the Oldroyd stress rate and another popular choice, referred to as the Jaumann–Zaremba stress rate, where the corresponding pull-back and push-forward operator is the skew-symmetric part of the spatial velocity gradient $\mathbf{W} = \frac{1}{2}(\mathbf{L} - \mathbf{L}^T)$, resulting with:

$$
\overset{\nabla}{\boldsymbol{\tau}} = \frac{\partial}{\partial t}\boldsymbol{\tau} - \mathbf{W}\boldsymbol{\tau} + \boldsymbol{\tau}\mathbf{W}
\tag{4.154}
$$

The Jaumann–Zaremba stress rate of the Kirchhoff stress tensor was used as a frequent choice in the early attempts to construct the constitutive model of large deformation plasticity, featuring the additive decomposition of the rate of deformation tensor into elastic and plastic part. It has been since discovered that the plasticity model of this kind is unable to provide the correct representation of large plastic deformations produced by kinematic hardening (e.g. see [196]), or even large elastic deformations (e.g. see [145]), resulting for each case with large spurious oscillations, as illustrated in Figure 4.15 for a simple shear test. Another disadvantage of formulating the large deformation plasticity model in terms of the Jaumann–Zaremba stress rate pertains to a requirement to compute even the initial elastic response by using a time-integration scheme, which leads to serious difficulties in ensuring the computed result invariance, as illustrated in Kojic et Bathe [151]. For all those reasons, this kind of large deformation plasticity models are nowadays pretty much abandoned in favor of the plasticity model presented in the next section.

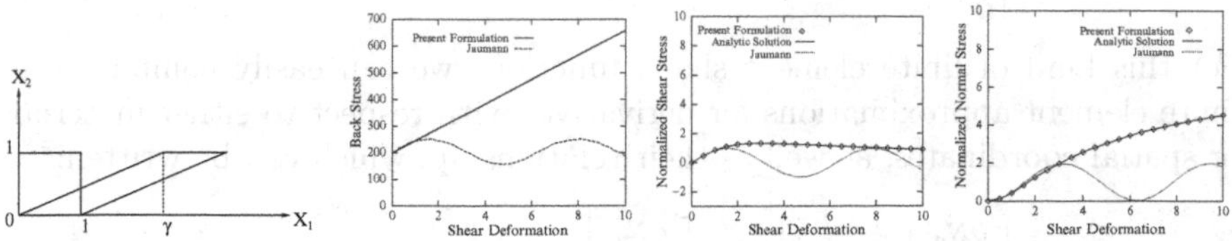

Fig. 4.15 Simple shear test: initial and deformed configurations, time history of kinematic hardening variable, shear stress and axial stress.

4.5.1 Finite element approximation of spatial description of elasticity in large displacements

In this section we will retrace the main steps in constructing the element residual vector and its stiffness matrix, when using the isoparametric finite element approximations and the spatial description of elasticity in large displacements. The starting point of this kind of formulation is the weak form of the equilibrium equations in spatial description, featuring the Kirchhoff stress tensor, which can easily be obtained from the material description through the corresponding transformation:

$$
\begin{aligned}
0 = G(\boldsymbol{\varphi};\mathbf{w}) &:= \int_{\Omega} \hat{\mathbf{\Gamma}}(\boldsymbol{\varphi},\mathbf{w}) \cdot \mathbf{S}\, dV \; - \; G^{ext}(\mathbf{w}) \\
&= \int_{\Omega} \tfrac{1}{2}(\mathbf{w} \otimes \underbrace{\mathbf{F}^{-T}\boldsymbol{\nabla}}_{\boldsymbol{\nabla}^{\varphi}} + \underbrace{\mathbf{F}^{-T}\boldsymbol{\nabla}}_{\boldsymbol{\nabla}^{\varphi}} \otimes \mathbf{w}) \cdot \mathbf{F}\mathbf{S}\mathbf{F}^{T}\, dV \; - \; G^{ext}(\mathbf{w}) \\
&= \int_{\Omega} \tfrac{1}{2}(\mathbf{w} \otimes \boldsymbol{\nabla}^{\varphi} + \boldsymbol{\nabla}^{\varphi} \otimes \mathbf{w}) \cdot (\boldsymbol{\tau} \circ \boldsymbol{\varphi})\, dV \; - \; G^{ext}(\mathbf{w})
\end{aligned}
\tag{4.155}
$$

It is important to note that the choice of the Kirchhoff stress tensor parameterized with the initial configuration coordinates also remains quite suitable for accommodating any inelastic constitutive model with internal variables, which requires to trace the complete motion evolution towards the current configuration.

This spatial description of the weak form can also be written in matrix notation, providing the basis for most efficient computations of element arrays. In particular, we choose a 4-node quadrilateral isoparametric element (see Figure 4.13), which employs the same shape functions for describing the initial and deformed configurations, as well as the virtual and incremental displacement fields:

$$
\begin{aligned}
&\mathsf{N}^e = [N_1 \mathsf{I}_2, \ldots, N_{en} \mathsf{I}_2] \\
&\mathsf{x}^h \Big|_{\Omega^e} \; \textstyle\sum_{a=1}^{n_{en}} N_a(\xi,\eta)\mathsf{x}_a^e = \mathsf{N}^e \mathsf{x}^e \; ; \\
&\mathsf{x}^{\varphi,h} \Big|_{\Omega^e} \; \textstyle\sum_{a=1}^{n_{en}} N_a(\xi,\eta)(\mathsf{x}_a^e + \mathsf{d}_a^e) = \mathsf{N}^e(\mathsf{x}^e + \mathsf{d}^e) \; ; \\
&\mathsf{w}^h \Big|_{\Omega^e} = \textstyle\sum_{a=1}^{n_{en}} N_a(\xi,\eta)\mathsf{w}_a^e = \mathsf{N}^e \mathsf{w}^e \; ; \\
&\mathsf{u}^h \Big|_{\Omega^e} = \textstyle\sum_{a=1}^{n_{en}} N_a(\xi,\eta)\mathsf{u}_a^e = \mathsf{N}^e \mathsf{u}^e
\end{aligned}
\tag{4.156}
$$

For this kind of finite element shape functions we can easily compute the finite element approximations for derivatives with respect to either material or spatial coordinates, as well as their relationship, which can be written:

$$
\boldsymbol{\nabla} N_a = \begin{bmatrix} \frac{\partial N_a}{\partial x_1} \\ \frac{\partial N_a}{\partial x_2} \end{bmatrix} \; ; \; \boldsymbol{\nabla}^{\varphi} N_a = \begin{bmatrix} \frac{\partial N_a}{\partial x_1^{\varphi}} \\ \frac{\partial N_a}{\partial x_2^{\varphi}} \end{bmatrix} \; ; \; \mathbf{F}^{-T}\boldsymbol{\nabla} N_a = \boldsymbol{\nabla}^{\varphi} N_a
\tag{4.157}
$$

If the finite element interpolations defined in (4.156) and (4.157) above are replaced in the weak form in spatial description given in (4.155), we can obtain the corresponding discrete approximation of equilibrium equations:

$$0 = \mathsf{f}^{int} - \mathsf{f}^{ext} \;\; ; \;\; \mathsf{f}^{int} = \mathop{\mathbb{A}}_{e=1}^{n_{elem}} \mathsf{f}^{int,e} \tag{4.158}$$

The explicit form of the element internal vector in spatial description can be written in terms of the Kirchhoff stress tensor components:

$$\mathsf{f}^{int,e} = [f_p^{int,e}] \, ; \; f_p^{int,e} = \mathsf{e}_i^T \int_{\Omega^e} \mathsf{B}_a^{\varphi,T} \mathsf{t}^h \, dV \; ;$$
$$\mathsf{B}_a^{\varphi} = \begin{bmatrix} \frac{\partial N_a}{\partial x_1^{\varphi}} & 0 \\ 0 & \frac{\partial N_a}{\partial x_2^{\varphi}} \\ \frac{\partial N_a}{\partial x_2^{\varphi}} & \frac{\partial N_a}{\partial x_1^{\varphi}} \end{bmatrix} ; \; \mathsf{t}^h = \begin{bmatrix} \tau_{11} \\ \tau_{22} \\ \tau_{12} \end{bmatrix} \tag{4.159}$$

It is interesting that the spatial description is characterized by the same sparse structure of the strain–displacement matrix B^{φ} as the one already defined for small strain case (as noted early on by [221] or [238]).
When Newton's iterative scheme is used for computing the solution to this kind of geometrically nonlinear problem, we ought to carry out the consistent linearization of the weak form in spatial description defined in (4.155), which results with:

$$Lin[G(\varphi; \mathbf{w})] := G(\varphi; \mathbf{w}) + \frac{d}{d\varepsilon}[G(\underbrace{\varphi_{\varepsilon}}_{\varphi + \varepsilon \mathbf{u}}, \mathbf{w})]\Big|_{\varepsilon=0} = 0 \tag{4.160}$$

where

$$\begin{aligned} \tfrac{d}{d\varepsilon}[G(\varphi_\varepsilon, \mathbf{w})]\Big|_{\varepsilon=0} &= \int_\Omega \tfrac{1}{2}(\mathbf{w} \otimes \nabla^{\varphi} + \nabla^{\varphi} \otimes \mathbf{w}) \cdot \tfrac{d}{d\varepsilon}[\boldsymbol{\tau}]\Big|_{\varepsilon=0} dV \\ &+ [\tfrac{1}{2}(\mathbf{w} \otimes \nabla^{\varphi} + \nabla^{\varphi} \otimes \mathbf{w})]\Big|_{\varepsilon=0} \cdot \boldsymbol{\tau} \, dV \\ &= \int_\Omega \{ \tfrac{1}{2}(\mathbf{w} \otimes \nabla^{\varphi} + \nabla^{\varphi} \otimes \mathbf{w}) \cdot \tfrac{\partial \boldsymbol{\tau}}{\partial \boldsymbol{\epsilon}} \tfrac{1}{2}(\mathbf{u} \otimes \nabla^{\varphi} + \nabla^{\varphi} \otimes \mathbf{u}) \, dV \\ &+ \boldsymbol{\tau} \cdot \tfrac{1}{2}[(\nabla^{\varphi} \otimes \mathbf{w})(\mathbf{u} \otimes \nabla^{\varphi}) + (\nabla^{\varphi} \otimes \mathbf{u})(\mathbf{w} \otimes \nabla^{\varphi})] \} \, dV \end{aligned} \tag{4.161}$$

With the continuum-consistent, isoparametric interpolations we can directly obtain from the last result the corresponding form of the consistent tangent matrix for each element; we note that such a matrix will again have the material and geometric terms, which can both be written in spatial description according to:

$$\mathsf{K}^e = [K_{pq}] \, ; \; K_{pq}^e = \mathsf{e}_i^T \left[\int_{\Omega^e} \mathsf{B}_a^{\varphi,T} \mathsf{D}^{\tau} \mathsf{B}_b^{\varphi} dV + \int_{\Omega^e} \mathsf{G}_{ab}^{\varphi} dV \right] \mathsf{e}_j \; ; \tag{4.162}$$

where:

$$\begin{aligned} &F_i^a F_j^b F_k^c F_l^d C_{ijkl} =: \mathcal{C}^{\tau}_{abcd} \mapsto \mathsf{D}^{\tau} \\ &\mathsf{G}^{\varphi}_{ab} = \tau_{11} \frac{\partial N_a}{\partial x_1^{\varphi}} \frac{\partial N_b}{\partial x_1^{\varphi}} + \tau_{22} \frac{\partial N_a}{\partial x_2^{\varphi}} \frac{\partial N_b}{\partial x_2^{\varphi}} + \tau_{12} \left(\frac{\partial N_a}{\partial x_1^{\varphi}} \frac{\partial N_b}{\partial x_2^{\varphi}} + \frac{\partial N_a}{\partial x_2^{\varphi}} \frac{\partial N_b}{\partial x_1^{\varphi}} \right) \mathsf{I}_2 \end{aligned} \tag{4.163}$$

4.6 Mixed variational formulation in large displacements and discrete approximations

The isoparametric finite elements are not always capable of providing the optimal discrete approximation for strain and stress fields in large displacements. A typical example of this kind concerns the presence of constraints in the strain field, such as quasi-incompressibility. In this case, we have to abandon the standard isoparametric interpolations in favor of either enhanced strain or assumed strain field, which allow, respectively, to add or to suppress some of the approximation terms arising from the standard isoparametric interpolations. The formulations of this kind can no longer be placed within the standard, displacement-type variational framework. We ought to construct a more general variational framework featuring independent strain and stress fields, as well as the corresponding finite element approximations that need not necessarily derive from the chosen displacement field.

4.6.1 Mixed Hu-Washizu variational principle in large displacements and method of incompatible modes

The computed response for large displacements can often be characterized by strong heterogeneities of the strain field and a highly anisotropic displacement gradients in a given deformed configuration. It is difficult, if not impossible for complex structures and non-proportional loading program, to provide a priori estimates of the dominant directions for large displacement gradients. This then leads to conclusion that the most reliable strategy for ensuring even quality of the discrete approximation is using the complete polynomial basis. We show here how to adapt the method of incompatible modes to large displacement setting in order to enrich the polynomial basis of a particular isoparametric element and construct a complete polynomial. We note that the method of incompatible modes can increase the order of interpolation with no need to increase the number of element nodes, providing this kind of element with more robustness with respect to distortion,[12] which is especially

[12] It is well known (e.g. see [19], [111] or [271]) that the higher order isoparametric elements with a large number of nodes, as for example 16-node or even 9-node quadrilateral

important for large displacement problems where initially regular elements can easily become distorted.
The enhancement provided by the method of incompatible modes is the most conveniently introduced at the level of strain field. To that end, we first assume an additive decomposition of the displacement gradient into a 'compatible' part $\nabla \mathbf{d}$, which is obtained by the standard computations from assumed isoparametric approximation of the displacement field, and 'incompatible' part $\tilde{\mathbf{D}}$, which is chosen appropriately[13] in order to enrich this field:

$$\bar{\mathbf{D}} = \nabla \mathbf{d} + \tilde{\mathbf{D}} \tag{4.164}$$

The additive decomposition of the displacement gradient will produce the corresponding multiplicative decomposition of the deformation gradient, which can be written:

$$\begin{aligned}\bar{\mathbf{F}} &= \mathbf{I} + \bar{\mathbf{D}} \\ &= \mathbf{I} + \nabla \mathbf{d} + \tilde{\mathbf{D}} \\ &= [\mathbf{I} + \underbrace{\tilde{\mathbf{D}}\mathbf{F}^{-1}}_{\mathbf{D}^{\varphi}}]\mathbf{F} \\ &= \tilde{\mathbf{F}}^{\varphi}\mathbf{F}\end{aligned} \tag{4.165}$$

where $\tilde{\mathbf{F}}^{\varphi}$ is the enhanced deformation gradient. In the spirit of incompatible mode method[14] in large displacements framework (see [122]), the enhanced displacement gradient can formally be constructed from the incompatible displacement field $\tilde{\mathbf{d}}$ by applying the same derivative computation routine as the one already used for compatible displacement field:

$$\tilde{\mathbf{D}} = \tilde{\mathbf{d}} \otimes \boldsymbol{\nabla} \implies \tilde{\mathbf{D}}^{\varphi} := (\mathbf{d} \otimes \boldsymbol{\nabla})\mathbf{F}^{-1} = \tilde{\mathbf{d}}^{\varphi} \otimes \boldsymbol{\nabla}^{\varphi} \tag{4.166}$$

With this kind of construction of the enhanced displacement gradient field, the mixed variational principle for incompatible mode method is reduced to:

$$\Pi(\boldsymbol{\varphi}, \bar{\mathbf{D}}, \mathbf{P}) := \int_{\Omega} \{W(\mathbf{I} + \bar{\mathbf{D}}) + \mathbf{P} \cdot [\nabla\boldsymbol{\varphi} - (\mathbf{I} + \bar{\mathbf{D}})]\}\, dV - \Pi_{ext}(\boldsymbol{\varphi}) \tag{4.167}$$

In the last expression, the first Piola–Kirchhoff stress tensor $\mathbf{P}$ plays the role of the Lagrange multiplier, which imposes in the weak sense the equivalence of the standard displacement gradient $(\nabla\boldsymbol{\varphi} - \mathbf{I})$ with the enhanced gradient field $\bar{\mathbf{D}}$. By computing the variations of this functional with respect to the displacement, strain and stress fields, we obtain the corresponding weak form

elements, can be very sensitive to element distortion, which can reduce drastically element performance.

[13] We can recover the polynomial interpolation basis of a 9-node element from the 4-node isoparametric element with quadratic incompatible modes, or yet the 16-node polynomial basis from 9-node isoparametric element with cubic incompatible modes.

[14] Alternative procedures are given in [237] and [245].

for this kind of problem featuring the equilibrium, the constitutive and the enhanced kinematics equations, which can be written respectively:

$$
\begin{aligned}
0 &:= \frac{d}{d\varepsilon}[\Pi(\varphi_\varepsilon, \bar{\mathbf{D}}, \mathbf{P})]\Big|_{\varepsilon=0} = G_\varphi(\varphi, \bar{\mathbf{D}}, \mathbf{P}; \mathbf{w}) := \int_\Omega \mathbf{P} \cdot \nabla \mathbf{w}\, dV - G_{ext}(\mathbf{w}) \\
0 &:= \frac{d}{d\varepsilon}[\Pi(\varphi, \bar{\mathbf{D}}_\varepsilon, \mathbf{P})]\Big|_{\varepsilon=0} = G_{\bar{D}}(\varphi, \bar{\mathbf{D}}, \mathbf{P}; \bar{\mathbf{H}}) := \int_\Omega (\tfrac{\partial W}{\partial \mathbf{F}} - \mathbf{P}) \cdot \bar{\mathbf{D}}\, dV \\
0 &:= \frac{d}{d\varepsilon}[\Pi(\varphi, \bar{\mathbf{D}}, \mathbf{P}_\varepsilon)]\Big|_{\varepsilon=0} = G_P(\varphi, \bar{\mathbf{D}}, \mathbf{P}; \mathbf{T}) := \int_\Omega (\nabla \mathbf{d} - \bar{\mathbf{D}}) \cdot \mathbf{T}\, dV
\end{aligned}
$$

For incompatible mode method, we further admit the additive decomposition of the displacement gradient into compatible and incompatible part, which can be written:

$$\bar{\mathbf{D}} = \nabla \mathbf{d} + \tilde{\mathbf{D}} \; ; \; \bar{\mathbf{H}} = \nabla \mathbf{w} + \tilde{\mathbf{H}} \tag{4.168}$$

We note that the enhancement concerns both real and virtual displacement fields, and that it provides enrichment of the compatible displacement gradient. By exploiting the last result, we can obtain a modified form of the mixed variational equations featuring the enhanced displacement gradient $\tilde{\mathbf{D}}$ among the independent variables:

$$
\begin{aligned}
0 &= G_\varphi(\varphi, \tilde{\mathbf{D}}, \mathbf{P}; \mathbf{w}) := \int_\Omega \mathbf{P} \cdot \nabla \mathbf{w}\, dV - G_{ext}(\mathbf{w}) \\
0 &= G_{\tilde{D}}(\varphi, \tilde{\mathbf{D}}, \mathbf{P}; \tilde{\mathbf{H}}) := \int_\Omega (\tfrac{\partial W}{\partial \mathbf{F}} - \mathbf{P}) \cdot \nabla \mathbf{w}\, dV \\
&\quad + \int_\Omega (\tfrac{\partial W}{\partial \mathbf{F}} - \mathbf{P}) \cdot \tilde{\mathbf{H}}\, dV \\
0 &= G_P(\varphi, \tilde{\mathbf{D}}, \mathbf{P}; \mathbf{T}) := \int_\Omega \tilde{\mathbf{D}} \cdot \mathbf{T}\, dV
\end{aligned}
\tag{4.169}
$$

In the final step of constructing the discrete approximation of these variational equations, we would like to provide the most convenient basis for implementation of the most general constitutive models where the stress is computed from the strains, as well as eventually from internal variables. This requires elimination of the assumed stress field from the discrete approximations, which can be done by imposing the following orthogonality condition:

$$\int_{\Omega^e} \tilde{\mathbf{D}}^h \cdot \mathbf{T}^h \, dV = 0 \; ; \; \int_{\Omega^e} \mathbf{P}^h \cdot \tilde{\mathbf{H}}^h \, dV = 0 \tag{4.170}$$

It is important to note that continuity is required neither from the stress nor enhanced displacement gradient field, which allows us to enforce the orthogonality condition independently in each element. The discrete approximation of mixed variational principle of this kind is thus reduced to:

$$
\begin{aligned}
0 &= G_\varphi(\varphi^h, \tilde{\mathbf{D}}^h, \cdot; \mathbf{w}^h) := \int_{\Omega^h} \tfrac{\partial W}{\partial \mathbf{F}^h} \cdot \nabla \mathbf{w}^h \, dV - G_{ext}(\mathbf{w}^h) \\
0 &= G_{\tilde{D}}(\varphi^h, \tilde{\mathbf{D}}^h, \cdot; \tilde{\mathbf{H}}^h) := \int_{\Omega^h} \tfrac{\partial W}{\partial \mathbf{F}^h} \cdot \nabla \mathbf{H}^h \, dV
\end{aligned}
\tag{4.171}
$$

The discrete approximation for displacement field is constructed by using the standard isoparametric interpolations, and the same choice is made for its variations and increments; we can thus write:

$$
\begin{aligned}
\mathsf{N}^e &= [N_1\mathsf{I}_2, ..., N_{en}\mathsf{I}_2] \\
\mathsf{x}^h\Big|_{\Omega^e} & \sum_{a=1}^{n_{en}} N_a(\xi,\eta)\mathsf{x}_a^e = \mathsf{N}^e\mathsf{x}^e \; ; \\
\mathsf{x}^{\varphi,h}\Big|_{\Omega^e} & \sum_{a=1}^{n_{en}} N_a(\xi,\eta)(\mathsf{x}_a^e + \mathsf{d}_a^e) = \mathsf{N}^e(\mathsf{x}^e + \mathsf{d}^e) \\
\mathsf{w}^h\Big|_{\Omega^e} & = \sum_{a=1}^{n_{en}} N_a(\xi,\eta)\mathsf{w}_a^e = \mathsf{N}^e\mathsf{w}^e \; ; \\
\mathsf{u}^h\Big|_{\Omega^e} & = \sum_{a=1}^{n_{en}} N_a(\xi,\eta)\mathsf{u}_a^e = \mathsf{N}^e\mathsf{u}^e
\end{aligned}
\tag{4.172}
$$

The method of incompatible modes (see [140]) will pick polynomials $M_b(\xi,\eta)$, which are of higher order than those already defined in the polynomial basis of the chosen isoparametric element. For example, in 2D case we can pick a quadrilateral 4-node isoparametric element with two incompatible modes $M_1(\xi) = 1 - \xi^2$ and $M_2(\eta) = 1 - \eta^2$, in order to complete the polynomial basis for such an element to a complete quadratic polynomial. Similarly, we can enrich a quadrilateral 9-node isoparametric element with two cubic terms $M_1(\xi) = \xi(1-\xi^2)$ and $M_2(\eta) = \eta(1-\eta^2)$ in order to provide a complete cubic polynomial for the enhanced element basis.
We note in passing that the choice of incompatible modes in large displacement setting remains the same as for the case of linearized kinematics, and that only change concerns the theoretical formulation.
The corresponding enhanced displacement gradient approximation is obtained from the standard gradient computation of the chosen incompatible displacement field, which can formally be written:

$$
\left.
\begin{aligned}
\tilde{\mathsf{d}}^h\Big|_{\Omega^e} &= \sum_{b=1}^{n_{im}} M_b(\xi,\eta)\boldsymbol{\alpha}_b^e \\
\tilde{\mathsf{w}}^h\Big|_{\Omega^e} &= \sum_{b=1}^{n_{im}} M_b(\xi,\eta)\boldsymbol{\beta}_b^e
\end{aligned}
\right.
\Longrightarrow
\begin{aligned}
\tilde{\mathbf{D}}^h\Big|_{\Omega^e} &= \sum_{b=1}^{n_{im}} \nabla M_b \otimes \boldsymbol{\alpha}_b^e \\
\tilde{\mathbf{H}}^h\Big|_{\Omega^e} &= \sum_{b=1}^{n_{im}} \nabla M_b \otimes \boldsymbol{\beta}_b^e
\end{aligned}
\tag{4.173}
$$

where n_{im} is the number of incompatible displacement modes. The enhanced displacement gradients ought to verify the orthogonality condition with respect to the chosen discrete approximation of the stress field defined in (4.170). It is important to note that any choice of the stress field discrete approximation must contain an element-wise constant stress field, which should ensure the convergence of the incompatible model method in the spirit of the patch test (e.g. see [19], [111], [271] or [253]). We can thus deduce the minimum convergence requirement imposing that the average value of the enhanced displacement gradient ought to be zero in each element:

$$
\int_{\Omega^e} \tilde{\mathbf{D}}^h \, dV = \mathbf{0} \; ; \; \int_{\Omega^e} \tilde{\mathbf{H}}^h \, dV = \mathbf{0}
\tag{4.174}
$$

We can easily construct the modified discrete approximation of the enhanced displacement gradient satisfying the patch-test condition in (4.174), which can be written:

$$\begin{aligned} &\tilde{\mathbf{D}}^h \Big|_{\Omega^e} = \textstyle\sum_{b=1}^{n_{im}} \nabla \hat{M}_b \otimes \boldsymbol{\alpha}_b^e \; ; \; \tilde{\mathbf{H}}^h \Big|_{\Omega^e} = \textstyle\sum_{b=1}^{n_{im}} \nabla \hat{M}_b \otimes \boldsymbol{\beta}_b^e \; ; \\ &\nabla \hat{M}_b = \nabla M_b - \tfrac{1}{\Omega^e} \textstyle\int_{\Omega^e} \nabla M_b \, dV \;\Rightarrow\; \textstyle\int_{\Omega^e} \nabla \hat{M}_b \, dV = \mathbf{0} \end{aligned} \tag{4.175}$$

This modified enhanced gradient approximation along with the isoparametric finite element interpolations of displacement field will allow us to eliminating the assumed stress field and reduce the number of variational equations; the remaining equations, featuring only the displacement and enhanced displacement gradient field as independent variables, can be written in terms of the second Piola–Kirchhoff stress approximation $\mathbf{S}^h$ that is computed from the chosen constitutive model according to:

$$\begin{aligned} 0 = G_\varphi(\boldsymbol{\varphi}^h, \tilde{\mathbf{D}}^h; \mathbf{w}) &:= \textstyle\int_{\Omega^h} \overbrace{\bar{\mathbf{F}}^{h,-1} \frac{\partial W}{\partial \bar{\mathbf{F}}^h}}^{\mathbf{S}^h} \cdot \tfrac{1}{2} (\nabla \mathbf{w}^{h,T} \bar{\mathbf{F}}^h + \bar{\mathbf{F}}^{h,T} \nabla \mathbf{w}^h) \, dV \\ &\quad - G_{ext}(\mathbf{w}^h) \\ 0 = G_{\tilde{D}}(\boldsymbol{\varphi}^h, \tilde{\mathbf{D}}^h; \tilde{\mathbf{H}}^h) &:= \textstyle\int_{\Omega^h} \underbrace{\bar{\mathbf{F}}^{h,-1} \frac{\partial W}{\partial \bar{\mathbf{F}}^h}}_{\mathbf{S}^h} \cdot \tfrac{1}{2} (\tilde{\mathbf{H}}^{h,T} \bar{\mathbf{F}}^h + \bar{\mathbf{F}}^{h,T} \tilde{\mathbf{H}}^h) \, dV \end{aligned} \tag{4.176}$$

By using the finite element interpolations for displacements and displacement gradients in (4.172) and (4.175), and imposing the independence of the virtual displacement and displacement gradient interpolation parameters, the weak form of equilibrium can be recast in terms of a set of nonlinear algebraic equations. The first part of this set are global equations, associated with the variations of the (compatible) displacement field, which ought to be computed by the finite element assembly procedure:

$$\begin{aligned} &0 = \mathop{\mathbb{A}}_{e=1}^{n_{elem}} \mathsf{w}^{e,T} (\mathsf{f}^{e,int}(\mathsf{d}^e, \boldsymbol{\alpha}^e) - \mathsf{f}^{e,ext}) \; ; \; \mathsf{f}^{e,int} = \textstyle\int_{\Omega^e} \mathsf{B}^{e,T} \mathsf{s} \, dV \; ; \\ &\mathsf{B}_a^e = \begin{bmatrix} \frac{\partial N_a}{\partial x} \frac{\partial \bar{\boldsymbol{\varphi}}}{\partial x}^T \\ \frac{\partial N_a}{\partial y} \frac{\partial \bar{\boldsymbol{\varphi}}}{\partial y}^T \\ \frac{\partial N_a}{\partial x} \frac{\partial \bar{\boldsymbol{\varphi}}}{\partial y}^T + \frac{\partial N_a}{\partial y} \frac{\partial \bar{\boldsymbol{\varphi}}}{\partial x}^T \end{bmatrix} \; ; \; \mathsf{s} = \begin{bmatrix} S_{11}^h(\mathsf{d}^e, \boldsymbol{\alpha}^e) \\ S_{22}^h(\mathsf{d}^e, \boldsymbol{\alpha}^e) \\ S_{12}^h(\mathsf{d}^e, \boldsymbol{\alpha}^e) \end{bmatrix} \end{aligned} \tag{4.177}$$

The second part concerns the variations with respect to enhanced gradient parameters, which amounts to a set of local equations defined independently in each particular element:

$$0 = \boldsymbol{\beta}^{e,T}\mathsf{h}^e(\mathsf{d}^e, \boldsymbol{\alpha}^e) \; e \in [1, n_{el}] \; ; \; \mathsf{h}^e = \int_\Omega \hat{\mathsf{G}}^{e,T}\mathsf{s}\, dV \; ;$$

$$\hat{\mathsf{G}}_b = \begin{bmatrix} \frac{\partial \hat{M}_b}{\partial x}\frac{\partial \bar{\varphi}}{\partial x}^T \\ \frac{\partial \hat{M}_b}{\partial y}\frac{\partial \bar{\varphi}}{\partial y}^T \\ \frac{\partial \hat{M}_b}{\partial x}\frac{\partial \bar{\varphi}}{\partial y}^T + \frac{\partial \hat{M}_b}{\partial y}\frac{\partial \bar{\varphi}}{\partial x}^T \end{bmatrix} \tag{4.178}$$

The solution to this highly nonlinear problem can be constructed by Newton's iterative method, which relies upon the consistent linearization of this system. The contribution of a particular element with incompatible modes towards the consistently linearized system can be written:

$$\begin{aligned} Lin[\mathsf{f}^{e,int} - \mathsf{f}^{e,ext}] &:= \mathsf{f}^{e,int} - \mathsf{f}^{e,ext} + \underbrace{\mathsf{K}^e}_{\partial f^{e,int}/\partial d^e} \mathsf{u}^e + \underbrace{\mathsf{F}^{e,T}}_{\partial f^{e,int}/\partial \alpha^e} \mathsf{a}^e \\ Lin[\mathsf{h}^e] &:= \mathsf{h}^e + \underbrace{\mathsf{F}^e}_{\partial h^e/\partial d^e} \mathsf{u}^e + \underbrace{\mathsf{H}^e}_{\partial h^e/\partial \alpha^e} \mathsf{a}^e \end{aligned} \tag{4.179}$$

where u and a are, respectively, the incremental nodal displacements and incremental element-wise enhanced displacement gradient parameters. The stiffness matrix K^e in the last expression takes exactly the same form as the one already defined for the corresponding isoparametric element (with no enhanced displacement gradients), which consists of the material and of the geometric parts:

$$\begin{aligned} K^e_{pq} &= \mathsf{e}_i^T \left[\int_{\Omega^e} \mathsf{B}_a^T \mathsf{D} \mathsf{B}_b dv + \int_{\Omega^e} \mathsf{G}_{ab} dv\right] \mathsf{e}_j \; ; \\ p &= n_{df}(a-1) + i \; ; \; q = n_{df}(b-1) + j \\ \mathsf{G}_{ab} &= S^h_{11}\frac{\partial N_a}{\partial x}\frac{\partial N_b}{\partial x} + S^h_{22}\frac{\partial N_a}{\partial y}\frac{\partial N_b}{\partial y} + S^h_{12}(\frac{\partial N_a}{\partial x}\frac{\partial N_b}{\partial y} + \frac{\partial N_a}{\partial y}\frac{\partial N_b}{\partial x})\mathsf{I}_2 \end{aligned} \tag{4.180}$$

Other element matrices that characterize only the element with enhanced displacement model, such as element matrix F^e, will also have a material and a geometric part. This can be written:

$$\begin{aligned} F^e_{pq} &= \mathsf{e}_i^T \left[\int_{\Omega^e} \hat{\mathsf{G}}_a^T \mathsf{D} \mathsf{B}_b dv + \int_{\Omega^e} \mathsf{L}_{ab} dv\right] \mathsf{e}_j \; ; \\ p &= n_{df}(a-1) + i \; ; \; q = n_{df}(b-1) + j \; ; \\ \mathsf{L}_{ab} &= S_{11}\frac{\partial \hat{M}_a}{\partial x}\frac{\partial N_b}{\partial x} + S_{22}\frac{\partial \hat{M}_a}{\partial y}\frac{\partial N_b}{\partial y} + S_{12}(\frac{\partial \hat{M}_a}{\partial x}\frac{\partial N_b}{\partial y} + \frac{\partial \hat{M}_a}{\partial y}\frac{\partial N_b}{\partial x})\mathsf{I}_2 \end{aligned} \tag{4.181}$$

and the same structure will characterize element matrix H^e:

$$\begin{aligned} H^e_{pq} &= \mathsf{e}_i^T \left[\int_{\Omega^e} \hat{\mathsf{G}}_a^T \mathsf{D} \hat{\mathsf{G}}_b dv + \int_{\Omega^e} \mathsf{M}_{ab} dv\right] \mathsf{e}_j \; ; \\ p &= n_{df}(a-1) + i \; ; \; q = n_{df}(b-1) + j \; ; \\ \mathsf{M}_{ab} &= S_{11}\frac{\partial \hat{M}_a}{\partial x}\frac{\partial \hat{M}_b}{\partial x} + S_{22}\frac{\partial \hat{M}_a}{\partial y}\frac{\partial \hat{M}_b}{\partial y} + S_{12}(\frac{\partial \hat{M}_a}{\partial x}\frac{\partial \hat{M}_b}{\partial y} + \frac{\partial \hat{M}_a}{\partial y}\frac{\partial \hat{M}_b}{\partial x})\mathsf{I}_2 \end{aligned} \tag{4.182}$$

We note in passing that all the element arrays can also be written in spatial description, where the second Piola-Kirchhoff stress is replaced by the Kirchhoff stress tensor $\boldsymbol{\tau} = \bar{\mathbf{F}}\mathbf{S}\bar{\mathbf{F}}^T$. The main advantage of such a description pertains in particular to a sparse structure of element strain–displacement matrices, which applies to both the compatible displacement field with:

$$\mathsf{f}^{e,int} = \int_{\Omega^e} \mathsf{B}^{\varphi,e,T}\mathsf{t}\, dV \ ; \ \mathsf{B}_a^{\varphi,e} = \begin{bmatrix} \frac{\partial N_a}{\partial x^\varphi} & 0 \\ 0 & \frac{\partial N_a}{\partial y^\varphi} \\ \frac{\partial N_a}{\partial y^\varphi} & \frac{\partial N_a}{\partial x^\varphi} \end{bmatrix} \ ; \ \mathsf{t} = \begin{bmatrix} \tau_{11} \\ \tau_{22} \\ \tau_{12} \end{bmatrix} \tag{4.183}$$

as well as to the incompatible displacement field where the element residual can be written:

$$\mathsf{h}^e = \int_{\Omega} \hat{\mathsf{G}}^{\varphi,e,T}\mathsf{t}\, dV \ ; \ \hat{\mathsf{G}}_b^{\varphi} = \begin{bmatrix} \frac{\partial \hat{M}_b}{\partial x^\varphi} & 0 \\ 0 & \frac{\partial \hat{M}_b}{\partial y^\varphi} \\ \frac{\partial \hat{M}_b}{\partial y^\varphi} & \frac{\partial \hat{M}_b}{\partial x^\varphi} \end{bmatrix} \tag{4.184}$$

The last two results are obtained from (4.176) by exploiting the relationship between the spatial and material descriptions of derivatives, which can be written in compact notation according to: $\boldsymbol{\nabla}^\varphi = \bar{\mathbf{F}}^{-T}\boldsymbol{\nabla}$, where $\boldsymbol{\nabla}^\varphi = [\frac{\partial}{\partial x_i^\varphi}]$ and $\boldsymbol{\nabla} = [\frac{\partial}{\partial x_i}]$. The same transformation will allow us to obtain the linearized form of the equilibrium equations in spatial description, where the key result needed pertains to the spatial description of tangent elasticity tensor, which can be written:

$$(L_v[\boldsymbol{\tau}])_{ab} = \underbrace{F_i^a F_j^b F_k^c F_l^d \frac{\partial S^{ij}}{\partial E_{kl}}}_{\mathcal{C}^\tau_{abcd}} \frac{1}{2}(\nabla\mathbf{u}^\varphi + \nabla\mathbf{u}^{\varphi,T})_{cd} \ ; \ \mathcal{C}^\tau_{abcd} \mapsto \mathsf{D}^\tau \tag{4.185}$$

With this result in hand, the spatial description of the stiffness matrix K^e can then be written according to:

$$\begin{aligned} &K^e_{pq} = \mathsf{e}_i^T \left[\textstyle\int_{\Omega^e} \mathsf{B}_a^{\varphi,T}\mathsf{D}^\tau\mathsf{B}_b^\varphi dv + \int_{\Omega^e} \mathsf{G}_{ab}^\varphi dv\right] \mathsf{e}_j \ ; \\ &p = n_{df}(a-1)+i \ ; \ q = n_{df}(b-1)+j \\ &\mathsf{G}^\varphi_{ab} = \tau_{11}\frac{\partial N_a}{\partial x^\varphi}\frac{\partial N_b}{\partial x^\varphi} + \tau_{22}\frac{\partial N_a}{\partial y^\varphi}\frac{\partial N_b}{\partial y^\varphi} + \tau_{12}(\frac{\partial N_a}{\partial x^\varphi}\frac{\partial N_b}{\partial y^\varphi} + \frac{\partial N_a}{\partial y^\varphi}\frac{\partial N_b}{\partial x^\varphi})\mathsf{I}_2 \end{aligned} \tag{4.186}$$

The same transformation can be carried out to provide the spatial description of element matrix F^e:

$$\begin{aligned} &F^e_{pq} = \mathsf{e}_i^T \left[\textstyle\int_{\Omega^e} \hat{\mathsf{G}}_a^{\varphi,T}\mathsf{D}^\tau\mathsf{B}_b^\varphi dv + \int_{\Omega^e} \mathsf{L}_{ab}^\varphi dv\right] \mathsf{e}_j \ ; \\ &p = n_{df}(a-1)+i \ ; \ q = n_{df}(b-1)+j \\ &\mathsf{L}^\varphi_{ab} = \tau_{11}\frac{\partial \hat{M}_a}{\partial x^\varphi}\frac{\partial N_b}{\partial x^\varphi} + \tau_{22}\frac{\partial \hat{M}_a}{\partial y^\varphi}\frac{\partial N_b}{\partial y^\varphi} + \tau_{12}(\frac{\partial \hat{M}_a}{\partial x^\varphi}\frac{\partial N_b}{\partial y^\varphi} + \frac{\partial \hat{M}_a}{\partial y^\varphi}\frac{\partial N_b}{\partial x^\varphi})\mathsf{I}_2 \end{aligned} \tag{4.187}$$

and also to furnish the spatial description of element matrix H^e:

$$H^e_{pq} = \mathsf{e}_i^T \left[\int_{\Omega^e} \hat{\mathsf{G}}_a^{\varphi,T} \mathsf{D}^\tau \hat{\mathsf{G}}_b^{\varphi} dv + \int_{\Omega^e} \mathsf{M}_{ab} dv \right] \mathsf{e}_j \; ;$$
$$p = n_{df}(a-1) + i \; ; \; q = n_{df}(b-1) + j \qquad (4.188)$$
$$\mathsf{M}^{\varphi}_{ab} = \tau_{11} \frac{\partial \hat{M}_a}{\partial x^\varphi} \frac{\partial \hat{M}_b}{\partial x^\varphi} + \tau_{22} \frac{\partial \hat{M}_a}{\partial y^\varphi} \frac{\partial \hat{M}_b}{\partial y^\varphi} + \tau_{12} \left(\frac{\partial \hat{M}_a}{\partial x^\varphi} \frac{\partial \hat{M}_b}{\partial y^\varphi} + \frac{\partial \hat{M}_a}{\partial y^\varphi} \frac{\partial \hat{M}_b}{\partial x^\varphi} \right) \mathsf{I}_2$$

In conclusion, the material or spatial descriptions of the linearized equilibrium equations for the incompatible mode method in large displacements take a fully equivalent form, which can be written at the level of a single element:

$$\begin{aligned} \mathsf{K}^e \mathsf{u}^e + \mathsf{F}^{e,T} \mathsf{a}^e &= \mathsf{f}^{ext,e} - \mathsf{f}^{int,e} \; ; \; \forall \mathsf{w}^e \\ \mathsf{F}^e \mathsf{u}^e + \mathsf{H}^e \mathsf{a}^e &= \mathsf{h}^e \; ; \; \forall \beta^e \end{aligned} \qquad (4.189)$$

The most convenient solution procedure for this kind of system, makes use of the operator split method (e.g., see Ibrahimbegovic and Kozar [130]). Namely, we will first iterate on the second equation, which is written for each element independently. We can thus find the corresponding value of the enhanced displacement gradient parameters $\bar{\boldsymbol{\alpha}}$ corresponding to a given value of element nodal displacement $\mathsf{d}^{e,(i)}$. For a particular value which will satisfy $\mathsf{h}^e(\mathsf{d}^{e,(i)}, \bar{\boldsymbol{\alpha}}) = \mathbf{0}$, we can then carry out the static condensation of the linearized form in (4.189), and thus recover the standard format of the element stiffness matrix that can directly be processed by the finite element assembly procedure:

$$\mathsf{a}^e = -\mathsf{H}^{e,-1} \mathsf{F}^e \mathsf{u}^e \implies \hat{\mathsf{K}}^e \mathsf{u}^e = \mathsf{f}^{ext,e} - \mathsf{f}^{int,e} \; ; \; \hat{\mathsf{K}}^e = \mathsf{K}^e - \mathsf{F}^{e,T} \mathsf{H}^{-1} \mathsf{F}^e \qquad (4.190)$$

We note that the assembly procedure is identical to the one carried out for isoparametric elements, which allows such an operator split solution procedure to fit easily within the standard computer code architecture. Having computed all element contributions, we will assemble and solve the set of global equations to provide nodal values of the incremental displacement $\mathsf{u}^{(i)}$ and carry out the corresponding displacements update $\mathsf{d}^{(i+1)} \longleftarrow \mathsf{d}^{(i)} + \mathsf{u}^{(i)}$; if the convergence check indicates the need, we can then restart the next iterative sweep of this operator split procedure by providing the next iterative guess of the nodal displacements for each element $\mathsf{d}^{e,(i+1)} = \mathsf{L}^e \mathsf{d}^{(i+1)}$.

4.6.2 Mixed Hu-Washizu variational principle in large displacements and assumed strain methods for quasi-incompressible behavior

The presence of quasi-incompressibility constraint, which requires that the volume change in any deformed configuration remains small, can lead to serious difficulties for standard isoparametric interpolations in providing a reliable representation of the corresponding deformation patterns. The resulting decrease in isoparametric element performance is often referred to as locking. In order to eliminate the locking phenomena, we can use the assumed strain method, where the discrete approximation for strain field is constructed independently from the chosen isoparametric interpolations of the displacements. The sound theoretical basis for constructing assumed strain interpolations cannot be provided by the standard displacement-type formulation. We will need instead the mixed Hu-Washizu variational principle considering the displacements and strains as independent fields. More precisely, in view of a quasi-incompressibility constraint on large displacements and strains, we consider that the volume change is controlled by an independent field $\Theta(\mathbf{x})$, which has no point-wise relationship with the part of deformation gradient $\mathbf{F}$ that controls the change of shape. We hope in this manner to be able to represent deformation patterns typical of quasi-incompressible material, with a large change of shape accompanied by a small volume change. With the independent field $\Theta(\mathbf{x})$ used for representing the complete volume change, the deformation gradient ought to be restricted to describing the change of shape only; thus, we construct a modified form of the deformation gradient corresponding to a volume-preserving deformation:

$$\tilde{\mathbf{F}} = J^{-1/3}\mathbf{F} \ ; \ , J = det\mathbf{F} \tag{4.191}$$

It is easy to check that the modified deformation gradient has a unit determinant, and thus it can only represent a volume-preserving deformation:

$$det[\tilde{\mathbf{F}}] := \underbrace{det[J^{-1/3}\mathbf{I}]}_{J^{-1}} \underbrace{det[\mathbf{F}]}_{J} = 1 \tag{4.192}$$

For mixed variational formulation development capable of representing a small change of volume accompanied by a large change of shape, we introduce the corresponding multiplicative decomposition of the deformation gradient:

$$\boxed{\bar{\mathbf{F}} = \Theta^{1/3}\tilde{\mathbf{F}}} \ ; \ \tilde{\mathbf{F}} = J^{-1/3}\mathbf{F} \ ; \ \mathbf{F} = \nabla\varphi \ ; \ J = det[\mathbf{F}] \tag{4.193}$$

We will present in detail an illustrative development of the mixed formulation of this kind for a hyperelastic material with strain energy density $W(\cdot)$. The latter is defined as a function of two variables: $\tilde{\mathbf{F}}$ and Θ, controlling the change of shape and change of volume, respectively. One can thus write:

$$W(\bar{\mathbf{C}}(\Theta, \varphi)) \ ; \ \bar{\mathbf{C}} = \bar{\mathbf{F}}^T\bar{\mathbf{F}} \ ; \ \bar{\mathbf{F}} = \Theta\tilde{\mathbf{F}} \tag{4.194}$$

By employing the assumed strain field interpolations, we do not have to change the stress computations, and the corresponding values of the stress tensor are again obtained in standard manner as the derivative of the strain energy potential with respect to the strain tensor; for example, the Kirchhoff stress tensor used for spatial description of finite deformation elasticity can be written:

$$\boldsymbol{\tau} = \bar{\mathbf{F}} \frac{2\partial W(\bar{\mathbf{C}})}{\partial \bar{\mathbf{C}}} \bar{\mathbf{F}}^T \tag{4.195}$$

With no essential restriction, we further consider an additive decomposition of the strain energy potential, with the first term denoted as $U(\Theta)$ controlling the change of volume, and the second term denoted as $\tilde{W}(\tilde{\mathbf{C}})$ which controls the change of shape:

$$W(\bar{\mathbf{C}}) = U(\Theta) + \tilde{W}(\tilde{\mathbf{C}}) \; ; \; \tilde{\mathbf{C}} = \tilde{\mathbf{F}}^T \tilde{\mathbf{F}} \tag{4.196}$$

This kind of strain energy potential further leads to the additive decomposition of the Kirchhoff stress tensor:

$$\begin{aligned} \boldsymbol{\tau} &:= \frac{2\,\partial W(\bar{\mathbf{C}})}{\partial \bar{\mathbf{C}}} \\ &= \frac{2\,dU}{d\Theta} \frac{\partial \Theta}{\partial \bar{\mathbf{C}}} + \frac{2\,\partial \tilde{W}}{\partial \tilde{\mathbf{C}}} \frac{\tilde{\mathbf{C}}}{\bar{\mathbf{C}}} \\ &= \frac{3\,dU}{d\Theta} \Theta \mathbf{I} + \tilde{\mathbf{F}} \frac{2\,\partial \tilde{W}}{\partial \tilde{\mathbf{C}}} \tilde{\mathbf{F}}^T \end{aligned} \tag{4.197}$$

where we exploited the following auxiliary results:

$$\bar{\mathbf{C}} = \Theta^{2/3} \tilde{\mathbf{C}} \Rightarrow \begin{aligned} \frac{\partial \Theta}{\partial \bar{\mathbf{C}}} &= \frac{3}{2} \Theta \bar{\mathbf{C}}^{-1} \\ \frac{\partial \tilde{\mathbf{C}}}{\partial \bar{\mathbf{C}}} &= \Theta^{-2/3} \end{aligned} \tag{4.198}$$

The quasi-incompressible behavior, which is characterized by a large value of bulk modulus proportional to the second derivative of the potential $U(\Theta)$, can now easily be accommodated within this kind of strain energy potential. Namely, we can introduce the mixed Hu-Washizu variational formulation of quasi-incompressible finite deformation elasticity featuring three independent fields, the displacement (or rather the motion) φ, the volume change Θ and the pressure field p; the corresponding total energy potential can then be written:

$$\Pi(\varphi, \Theta, p) := \int_{\Omega} \{W(\bar{\mathbf{C}}(\Theta^{1/3}, \varphi)) + p(J(\varphi) - \Theta)\}\, dV - \Pi_{ext}(\varphi) \tag{4.199}$$

where Π_{ext} is the potential energy of external forces. The main advantage of the variational formulation of this kind pertains to the fact that the equivalence between Θ and $J = det\mathbf{F}$ is not required point-wise, but only in the weak form sense. Therefore, the corresponding value of Θ for a very small change of volume in a particular deformed configuration for quasi-incompressible material, will not necessarily preclude a large change of shape in the same configuration, which is represented by the displacement field or rather the corresponding deformation gradient $\tilde{\mathbf{F}}$.

For further development of the variational equations for this mixed variational principle, which serves as the basis of finite element implementation, we need a couple of auxiliary results pertaining to the directional derivative computations; first, we can write:

$$\begin{aligned} \tfrac{d}{d\varepsilon}|_{\varepsilon=0}[\mathbf{F}_\varepsilon] &= \tfrac{d}{d\varepsilon}|_{\varepsilon=0}[(\mathbf{I}+\varepsilon\nabla^{\varphi}\mathbf{w}^{\varphi})\mathbf{F}] = \nabla^{\varphi}\mathbf{w}^{\varphi}\mathbf{F} \\ \tfrac{d}{d\varepsilon}|_{\varepsilon=0}[J_\varepsilon] &= J tr[\nabla^{\varphi}\mathbf{w}^{\varphi}] = J div^{\varphi}[\mathbf{w}^{\varphi}] \end{aligned} \tag{4.200}$$

which then allows us to compute:

$$\begin{aligned} \tfrac{d}{d\varepsilon}[\bar{\mathbf{F}}(\Theta,\varphi_t)]\Big|_{\varepsilon=0} &= \Theta^{1/3}\{J^{-1/3}\tfrac{d}{d\varepsilon}[\mathbf{F}_\varepsilon]\Big|_{\varepsilon=0} + \tfrac{d}{d\varepsilon}[J^{-1/3}]\Big|_{\varepsilon=0}\mathbf{F}\} \\ &= \Theta^{1/3}\{J^{-1/3}\nabla^{\varphi}\mathbf{F} - \tfrac{1}{3}J^{-4/3}\,Jtr[\nabla^{\varphi}]\mathbf{F} \\ &= (\nabla^{\varphi}\mathbf{w}^{\varphi} - \tfrac{1}{3}tr[\nabla^{\varphi}]\mathbf{I})\,\underbrace{\Theta^{1/3}J^{-1/3}\mathbf{F}}_{\bar{F}} \end{aligned} \tag{4.201}$$

By exploiting these results, we can easily obtain the first variational equation of the Hu-Washize mixed variational principle for quasi-incompressible finite elasticity, associated with the displacement variations ($\varphi_\varepsilon = \varphi + \varepsilon\mathbf{w}$):

$$\begin{aligned} 0 &= \tfrac{d}{d\varepsilon}[\Pi(\varphi_\varepsilon,\Theta,p)]\Big|_{\varepsilon=0} \\ &:= G_\varphi(\varphi,\Theta,p;\mathbf{w}) \\ &= \int_\Omega\{\tfrac{2\partial W}{\partial\mathbf{C}}\tfrac{1}{2}(\tfrac{d}{d\varepsilon}[\bar{\mathbf{F}}_\varepsilon^T]\Big|_{\varepsilon=0}\bar{\mathbf{F}} + \bar{\mathbf{F}}^T\tfrac{d}{d\varepsilon}[\bar{\mathbf{F}}_\varepsilon]\Big|_{\varepsilon=0}) + pJtr[\nabla^{\varphi}\mathbf{w}^{\varphi}]\}\,dV \\ &= \int_\Omega\{(\bar{\mathbf{F}}\tfrac{2\partial W}{\partial\mathbf{C}}\bar{\mathbf{F}}^T)\cdot dev[\nabla\mathbf{w}^{\varphi}] + Jpdiv[\mathbf{w}^{\varphi}]\}\,dV \end{aligned} \tag{4.202}$$

Second, the directional derivative of such a Hu-Washizu principle with respect to the independent volume-change variations $\Theta_\varepsilon = \Theta + \varepsilon\vartheta$ will lead to:

$$\begin{aligned} 0 &= \tfrac{d}{d\varepsilon}[\Pi(\varphi,\Theta_\varepsilon,p)]\Big|_{\varepsilon=0} := G_\Theta(\varphi,\Theta,p;\vartheta) \\ &= \int_\Omega \tfrac{2\partial W}{\partial\mathbf{C}}\cdot\tfrac{1}{2}(\tfrac{d}{d\varepsilon}[\Theta_\varepsilon^{1/3}]\Big|_{\varepsilon=0}\tilde{\mathbf{F}}^T\bar{\mathbf{F}} + \bar{\mathbf{F}}^T\tfrac{d}{d\varepsilon}[\Theta_\varepsilon^{1/3}]\Big|_{\varepsilon=0}\tilde{\mathbf{F}}) - p\tfrac{d}{d\varepsilon}[\Theta_\varepsilon]\Big|_{\varepsilon=0}dV \\ &= \int_\Omega\{\Theta^{-1}\tfrac{1}{3}tr[\bar{\mathbf{F}}\tfrac{2\partial W}{\partial\mathbf{C}}\bar{\mathbf{F}}^T] - p)\}\vartheta\,dV \end{aligned} \tag{4.203}$$

Finally, the directional derivative of the mixed Hu-Washizu principle for quasi-incompressible material with respect to the pressure field variation $p_\varepsilon = p + \varepsilon q$ will allow us to write:

$$0 = \frac{d}{d\varepsilon}[\Pi(\varphi,\Theta,p_\varepsilon)]\Big|_{\varepsilon=0} := G_p(\varphi,\Theta,p;q) := \int_\Omega q(J-\Theta)\,dV \tag{4.204}$$

It only remains now to make the most appropriate choice for the finite element approximations. In that sense we note that the mixed variational formulation of this kind requires the corresponding discrete approximations not only for the displacement field, but also for the independent volume change Θ and the pressure field p. It is important to note that no derivatives of the last two fields appear in the governing variational equations, and that their discrete approximations can be constructed in each element independently. In order to ensure the finite element accuracy of the mixed formulation, we ought to impose the most appropriate hierarchy between interpolation for the displacement field on one side against the pressure and volume change field interpolations on the other. For example, we can use the lowest order of approximation by combining a 4-node isoparametric displacement field, given in terms of linear polynomials, with an element-wise constant values for Θ and p. More precisely, the real, virtual and incremental displacement fields for each finite element are written in terms of the standard bilinear isoparametric approximations:

$$\begin{aligned}
&\mathsf{N}^e = [N_1 \mathsf{I}_2, ..., N_4 \mathsf{I}_2] \; ; \; N_a = \tfrac{1}{4}(1+\xi_a \xi)(1+\eta_a \eta) \\
&\mathsf{x}^h \Big|_{\Omega^e} \sum_{a=1}^4 N_a(\xi,\eta)\mathsf{x}_a^e = \mathsf{N}^e \mathsf{x}^e \; ; \\
&\mathsf{x}^{\varphi,h} \Big|_{\Omega^e} \sum_{a=1}^4 N_a(\xi,\eta)(\mathsf{x}_a^e + \mathsf{d}_a^e) = \mathsf{N}^e(\mathsf{x}^e + \mathsf{d}^e) \; ; \\
&\mathsf{w}^h \Big|_{\Omega^e} = \sum_{a=1}^4 N_a(\xi,\eta)\mathsf{w}_a^e = \mathsf{N}^e \mathsf{w}^e \; ; \\
&\mathsf{u}^h \Big|_{\Omega^e} = \sum_{a=1}^4 N_a(\xi,\eta)\mathsf{u}_a^e = \mathsf{N}^e \mathsf{u}^e
\end{aligned} \tag{4.205}$$

The constant value in each element is chosen for real and virtual pressure and volume change fields, which allows us to obtain from (4.204) the connection between Θ^h with volume change displacement field approximation according to:

$$\vartheta^h|_{\Omega^e} = \vartheta^e \; ; \; \Theta^h|_{\Omega^e} = \Theta^e = \tfrac{1}{\Omega^e} \int_{\Omega^e} J^h \, dV \tag{4.206}$$

The same kind of connection can be established between the independent pressure field and the stress tensor computed from constitutive equations in (4.203):

$$q^h|_{\Omega^e} = q^e \; ; \; p^h|_{\Omega^e} = p^e = \frac{1}{\Theta^e \Omega^e} \int_{\Omega^e} \frac{1}{3} tr[\bar{\mathbf{F}}^h \frac{2\,\partial W}{\partial \bar{\mathbf{C}}^h} \bar{\mathbf{F}}^{h,T}] \, dV \tag{4.207}$$

With these approximations in hand we can construct easily the corresponding discrete approximation for the equilibrium equations in (4.202), which can be written:

$$\begin{aligned}
&0 = \mathbb{A}_{e=1}^{n_{elem}} \mathsf{w}^{e,T}(\mathsf{f}^{e,int} - \mathsf{f}^{e,ext}) \; ; \; \mathsf{f}^{e,int} = [\mathsf{f}_a^{e,int}] \; ; \\
&\mathsf{f}_a^{e,int} = \int_{\Omega^e} \tilde{\mathsf{B}}_a^{\varphi,T} \mathsf{t} \, dV + p^e \int_{\Omega^e} J \mathsf{b}^{\varphi} \, dV
\end{aligned} \tag{4.208}$$

where the strain field approximation is defined as:

$$\tilde{\mathsf{B}}_a^\varphi = \begin{bmatrix} \frac{2}{3}\frac{\partial N_a}{\partial x^\varphi} & \frac{-1}{3}\frac{\partial N_a}{\partial y^\varphi} \\ \frac{-1}{3}\frac{\partial N_a}{\partial x^\varphi} & \frac{2}{3}\frac{\partial N_a}{\partial y^\varphi} \\ \frac{\partial N_a}{\partial y^\varphi} & \frac{\partial N_a}{\partial x^\varphi} \end{bmatrix} \; ; \; \mathsf{t} = \begin{bmatrix} \tau_{11} \\ \tau_{22} \\ \tau_{12} \end{bmatrix} \; ; \; \mathsf{b}_a^\varphi = \begin{bmatrix} \frac{\partial N_a}{\partial x^\varphi} \\ \frac{\partial N_a}{\partial y^\varphi} \end{bmatrix} \tag{4.209}$$

This kind of the strain field approximation can be placed within the framework of the assumed strain field method; first, we exploit the results in (4.206) and (4.207) in order to rewrite the pressure field approximation to the weak form of equilibrium equations according to:

$$p^e \int_{\Omega^e} J\mathsf{b}^\varphi \, dV = \int_{\Omega^e} \underbrace{\frac{1}{3} tr[\bar{\mathbf{F}} \frac{2\,\partial W}{\partial \bar{\mathbf{C}}} \bar{\mathbf{F}}^T]}_{\mathbf{I}\cdot\boldsymbol{\tau}} \, dV \; \underbrace{\frac{1}{\Omega^e \Theta^e} \int_{\Omega^e} J\mathsf{b}^\varphi \, dV}_{\bar{\mathsf{b}}^\varphi} \tag{4.210}$$

The last result allows us to obtain the corresponding $B - bar$ approximation equivalent to the small strain case:

$$\mathsf{f}_a^{e,int} = \int_{\Omega^e} \bar{\mathsf{B}}_a^{\varphi,T} \mathsf{t} \, dV \; ; \; \bar{\mathsf{B}}_a^\varphi = \tilde{\mathsf{B}}_a^\varphi + \begin{bmatrix} \frac{1}{3}\bar{\mathsf{b}}^{\varphi,T} \\ \frac{1}{3}\bar{\mathsf{b}}^{\varphi,T} \\ 0^T \end{bmatrix} \tag{4.211}$$

The mixed element performance is illustrated by a simple example, often referred to as Cook's cantilever, which considers a tapered beam built-in on one end and loaded with a transverse uniform load on another free-end. The numerical results for free-end vertical displacements (see Figure 4.16) confirm the superior performance of the mixed element approximation over the corresponding isoparametric element approximation, placing this kind of element at the same level of accuracy as the one obtained by the element with incompatible modes.

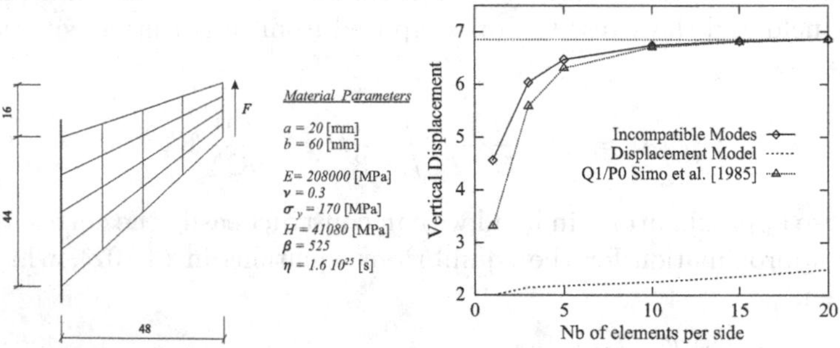

Fig. 4.16 Cook's cantilever: geometry, FE mesh and computed free-end displacement.

For this kind of mixed formulation, we can further increase the order of approximations. The optimal accuracy can thus be achieved in 2D case, by combining the quadratic polynomial approximation for displacements, provided by a 9-node isoparametric element, with a complete linear polynomial approximations for Θ and p. The element strain-displacement array for this higher order approximation can again be written in $B - bar$ format, by following this same procedure just described for 4-node isoparametric element; this is left as an exercise for the readers.

4.7 Constitutive models for large strains

In general, it is rather difficult to construct a constitutive model $\hat{\boldsymbol{\sigma}}(\cdot)$ which can represent in a reliable manner a wide variety of phenomena in the large strain regime. In fact, this is by far a more difficult task than the equivalent one we have already studied for small strain case. First difficulty for large strain constitutive model concerns the most appropriate measure of strain, since, contrary to the small strain case with a unique choice of the infinitesimal strain measure, there exists (theoretically) an infinite number of large strain measures and the corresponding work-conjugate stress tensors that provide the most suitable description of material behavior. First and foremost, one ought to choose between the strain measures defined in the initial configuration versus the strain measures defined in deformed configuration; the former grant the computational simplicity of using the known configuration, whereas the latter provide the solid basis for proper interpretation of the underlying physical phenomena in large strains in terms of Cauchy or true stress. For either choice of the material or spatial strain measure, we should first verify if it complies with the imposed restrictions that guarantee the response invariance in the large strain regime. This is elaborated upon in the next section.

4.7.1 Invariance restrictions on elastic response

The invariance restrictions on the elastic response discussed herein are typically concerned with an arbitrary choice of the reference frame. Any constitutive model ought to respect these invariance restrictions, which are thus more general than the polyconvexity conditions applicable only to hyperelastic materials.[15]

---

[15] The polyconvexity conditions impose that the strain energy of a hyperelastic material $\psi(\cdot)$ be defined as a convex function with respect to intrinsic deformation measures, the deformation gradient $\mathbf{F}$, its co-factor $cof[\mathbf{F}]$ and its determinant $det[\mathbf{F}]$, which in turn guarantees that very large strains will be accompanied by equally large stresses.

The first such restriction concerns the response invariance under the rigid body motion, defined either as the rigid body translation or the rigid body rotation, superposed upon a particular deformed configuration. The first invariance of the elastic response under superposed rigid body translation will require to exclude any direct dependence of the response with respect to the motion $\boldsymbol{\varphi}$, and to allow only dependence upon the motion derivatives or the deformation gradient $\mathbf{F} = \nabla\boldsymbol{\varphi}$. With respect to the invariance requirement under translation, the elastic response can therefore be written in terms of the Cauchy stress $\boldsymbol{\sigma}$ as a function of the deformation gradient $\mathbf{F}$:

$$\hat{\boldsymbol{\sigma}} : (\mathbf{x}, \mathbf{F}) \in \bar{\Omega} \times \mathbb{M}_+ \mapsto \boldsymbol{\sigma} \in \mathbb{S} \tag{4.212}$$

How should we interpret this expression? We recall that the Cauchy stress is the true stress acting in a given deformed configuration that provides the most reliable physical basis for interpretation of a particular constitutive behavior. However, the preference for Lagrangian formulation in solid mechanics will require that we express the Cauchy stress as a function of coordinates in the initial configuration and employ the material description of response:

$$\boldsymbol{\sigma}^{\varphi}(\mathbf{x}^{\varphi}) = \hat{\boldsymbol{\sigma}}(\mathbf{x}, \nabla\boldsymbol{\varphi}(\mathbf{x})) \ , \ \forall \ \mathbf{x}^{\varphi} = \boldsymbol{\varphi}(\mathbf{x}) \in \Omega^{\varphi} \tag{4.213}$$

The second aspect of this invariance restriction, often referred to as the axiom of material indifference, imposes that the material description of the elastic response remains unaffected by the rigid body rotation superposed upon a particular deformed configuration. As shown in Figure 4.17, any such rotated deformed configuration can be produced by a constant orthogonal tensor $\mathbf{Q}^+$.

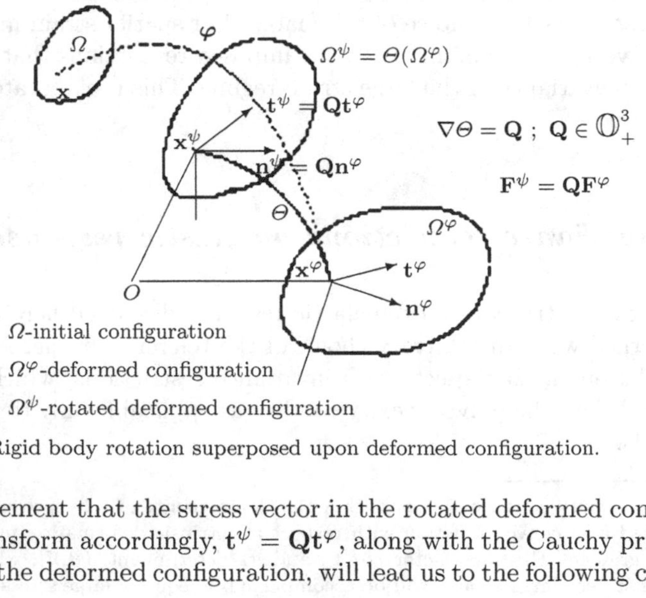

Fig. 4.17 Rigid body rotation superposed upon deformed configuration.

The requirement that the stress vector in the rotated deformed configuration should transform accordingly, $\mathbf{t}^{\psi} = \mathbf{Q}\mathbf{t}^{\varphi}$, along with the Cauchy principle expressed in the deformed configuration, will lead us to the following conclusion:

$$\mathbf{t}^{\psi} = \begin{cases} \boldsymbol{\sigma}^{\psi}(\mathbf{x}^{\psi})\mathbf{n}^{\psi} \\ \quad \overbrace{\mathbf{Q}^T\mathbf{n}^{\psi}} \quad \Rightarrow \hat{\boldsymbol{\sigma}}(\mathbf{x}, \mathbf{QF}) = \mathbf{Q}\,\hat{\boldsymbol{\sigma}}(\mathbf{x}, \mathbf{F})\,\mathbf{Q}^T \\ \mathbf{Q}\boldsymbol{\sigma}^{\varphi}(\mathbf{x}^{\varphi})\;\mathbf{n}^{\varphi} \end{cases} \tag{4.214}$$

The last result remains true for any choice of the proper orthogonal tensor $\mathbf{Q}$ representing the superposed rotation. A particular choice of this rotation tensor that coincides with rotational part of the deformation gradient $\mathbf{R}$, defined from the polar decomposition theorem with $\mathbf{F} = \mathbf{RU}$, will allow us to simplify further the last result and to conclude that the Cauchy stress can be expressed as a function of the stretch tensor:

$$\begin{aligned} \hat{\boldsymbol{\sigma}}(\mathbf{x}, \underbrace{\mathbf{F}}_{\mathbf{RU}}) &= \mathbf{R}\,\hat{\boldsymbol{\sigma}}(\mathbf{x}, \mathbf{U})\,\mathbf{R}^T \\ &= \mathbf{F}\mathbf{U}^{-1}\hat{\boldsymbol{\sigma}}(\mathbf{U})\mathbf{U}^{-T}\mathbf{F}^T \end{aligned} \tag{4.215}$$

The material description of the last result allows us to provide a very compact form of the elastic response satisfying the axiom of material indifference in terms of the second Piola–Kirchhoff stress tensor $\mathbf{S}$, which is a natural replacement of the Cauchy stress $\boldsymbol{\sigma}$ in the material description of the weak form of equilibrium equations through $\mathbf{S} = J\mathbf{F}^{-1}\boldsymbol{\sigma}\mathbf{F}^{-T}$; we thus obtain:

$$\begin{aligned} \hat{\mathbf{S}}(\mathbf{F}) &= (det\mathbf{F})\mathbf{F}^{-1}\;\hat{\boldsymbol{\sigma}}(\mathbf{x}, \mathbf{F})\,\mathbf{F}^{-T} \\ &= (det\mathbf{U})\mathbf{U}^{-1}\hat{\boldsymbol{\sigma}}(\mathbf{U})\mathbf{U}^{-T} \\ &= \tilde{\mathbf{S}}(\mathbf{C}) \end{aligned} \tag{4.216}$$

where we used that $\mathbf{U} = \mathbf{C}^{1/2}$. The last result is certainly the most compact form of the elastic response verifying the requirements of material indifference, which states that the second Piola–Kirchhoff stress can be directly expressed as a function of the right Cauchy–Green large strain tensor. This also implies that the material indifference of elastic response for any hyperelastic material will impose that the strain energy density should be defined as a function of only $\mathbf{C}$ (not rotational part), and that the corresponding Piola–Kirchhoff stress value should be obtained according to:

$$\psi(\mathbf{x}, \mathbf{F}) = \tilde{\psi}(\mathbf{x}, \mathbf{C}) \;\Rightarrow\; \mathbf{S} := \frac{\partial\tilde{\psi}(\cdot)}{\partial\mathbf{C}}\frac{\partial\mathbf{C}}{\partial\mathbf{E}} = 2\frac{\partial\tilde{\psi}(\cdot)}{\partial\mathbf{C}} \tag{4.217}$$

The second kind of invariance we can impose to restrict the admissible form of elastic response in material description concerns the case of homogeneous and isotropic material. The homogeneity of the elastic response is a physically based restriction which implies that the response will be the same for any chosen particle. It is clear that such a restriction is very unlikely to hold in any deformed configuration produced under heterogeneous stress field, since different particles could be in very different strain regimes and thus have very different current response. Therefore, the homogeneity of the elastic response can be verified easily only for the initial configuration with no residual stress or strain; for the material description of homogeneous material response, the chosen particle plays no role and can be dropped from the final response function:

$$\begin{aligned} \hat{\boldsymbol{\sigma}}(\mathbf{x},\mathbf{F}) &= \tilde{\boldsymbol{\sigma}}(\mathbf{F}) \\ \hat{\mathbf{S}}(\mathbf{x},\mathbf{F}) &= \tilde{\mathbf{S}}(\mathbf{F}) \end{aligned} \tag{4.218}$$

The second part of this important restriction on elastic response, which pertains to the initial configuration, is the response isotropy. This kind of restriction applies to materials with no preferential directions, where the elastic response at the level of a single particle remains exactly the same in all directions. Therefore, any change of reference frame by a rigid body rotation superposed upon the initial configuration (represented by an orthogonal tensor $\mathbf{Q}$; see Figure 4.18), must leave the elastic response invariant; we can thus write:

$$\hat{\boldsymbol{\sigma}}(\mathbf{F}) = \hat{\boldsymbol{\sigma}}(\mathbf{F}\mathbf{Q}^T) \; ; \; \mathbf{Q}\mathbf{Q}^T = \mathbf{I} \tag{4.219}$$

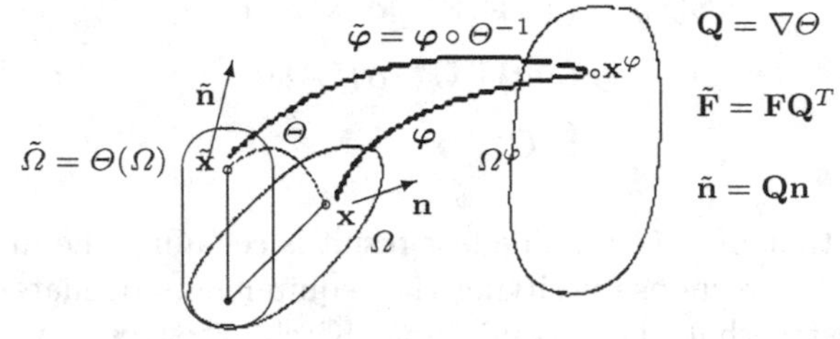

Fig. 4.18 Reference frame change under superposed rigid body rotation in the initial configuration.

The Cauchy principle written in material description, expressing the stress vector with respect to initial configurations obtained either before or after rotation $\mathbf{Q}$, will allow us conclude that:

$$\mathbf{t} = \begin{cases} \tilde{\mathbf{P}}\tilde{\mathbf{n}} = \tilde{\mathbf{F}}\tilde{\mathbf{S}}\tilde{\mathbf{n}} \\ \mathbf{P}\mathbf{n} = \overbrace{\mathbf{F}}^{\tilde{\mathbf{F}}\mathbf{Q}}\,\mathbf{S}\overbrace{\mathbf{n}}^{\mathbf{Q}^T\tilde{\mathbf{n}}} \end{cases} \Rightarrow \hat{\mathbf{S}}(\mathbf{F}\mathbf{Q}^T) = \mathbf{Q}\hat{\mathbf{S}}(\mathbf{F})\mathbf{Q}^T \tag{4.220}$$

For a special choice of the reference frame rotation that is equal to the rotation tensor $\mathbf{R}$ obtained from the polar decomposition of the deformation gradient with $\mathbf{F} = \mathbf{VR}$, we can further express the isotropic elastic response in terms of the stretch tensor $\mathbf{V}$ according to:

$$\begin{aligned}\hat{\mathbf{S}}(\mathbf{F}) &= \mathbf{R}^T \hat{\mathbf{S}}(\overbrace{\mathbf{F}\mathbf{R}^T}^{\mathbf{V}})\mathbf{R} \\ &= \mathbf{F}^{-1}\mathbf{V}\,\hat{\mathbf{S}}(\mathbf{V})\,\mathbf{V}^T\mathbf{F}^{-T}\end{aligned} \tag{4.221}$$

The last result can be rewritten in spatial description in terms of the true or Cauchy stress tensor $\boldsymbol{\sigma} = (1/J)\mathbf{FSF}^T$, $J = det[\mathbf{F}]$, leading to:

$$\begin{aligned}\hat{\boldsymbol{\sigma}}(\mathbf{F}) &= \tfrac{1}{det[\mathbf{F}]}\mathbf{F}\mathbf{F}^{-1}\mathbf{V}\hat{\mathbf{S}}(\mathbf{V})\mathbf{V}^T\mathbf{F}^{-T}\mathbf{F}^T \\ &= \tfrac{1}{det[\mathbf{V}]}\mathbf{V}\hat{\mathbf{S}}(\mathbf{V})\mathbf{V}^T \\ &= \hat{\boldsymbol{\sigma}}(\mathbf{B})\end{aligned} \tag{4.222}$$

where we used that $\mathbf{V} = \mathbf{B}^{1/2}$. We have shown that the most convenient manner for expressing the elastic response of an isotropic material in terms of the Cauchy stress tensor as a function of the left Cauchy–Green tensor. In summary, the material indifference leads to preferred form of the elastic response written in terms of the right Cauchy–Green strain tensor, whereas the isotropy of elastic response is preferably described in terms of the left Cauchy–Green strain tensor. The logical question then is how to best describe the elastic response that satisfies both of these invariance requirements, the isotropy and the material indifference. The answer is simple to provide if we recall that two Cauchy–Green strain tensors share the same principal invariants $i_{jB} \equiv i_{jC}\,;\, j = 1, 2, 3$, which allows us to write:

$$\begin{aligned}i_{1C} &:= tr[\underbrace{\mathbf{F}^T\mathbf{F}}_{\mathbf{C}}] = F_{ij}F_{ij} \equiv tr[\underbrace{\mathbf{F}\mathbf{F}^T}_{\mathbf{B}}] =: i_{1B} \\ i_{2C} &:= \tfrac{1}{2}((tr[\mathbf{C}])^2 - tr[\mathbf{C}^2]) \equiv \tfrac{1}{2}((tr[\mathbf{B}])^2 - tr[\mathbf{B}^2]) =: i_{2B} \\ i_{3C} &:= det[\mathbf{F}^T\mathbf{F}] = (det[\mathbf{F}])^2 \equiv det[\mathbf{F}\mathbf{F}^T] =: i_{3B}\end{aligned} \tag{4.223}$$

Such an equivalence of principal invariants for left and right Cauchy–Green strain tensors is not surprising, considering that the principal invariants of a tensor are the coefficients of the characteristic equation for computing its eigenvalues; for two Cauchy–Green strain tensors, their eigenvalues can be expressed in terms of the principal stretches $\lambda_i, i = 1, 2, 3$ according to:

$$\begin{aligned}0 &:= (\lambda_i^2)^3 - i_{1C}(\lambda_i^2)^2 + i_{2C}(\lambda_i^2) - i_{3C} \\ &= det[\mathbf{C} - \lambda_i^2\mathbf{I}] \\ &= det[\mathbf{F}^T]det[\lambda_i^{-2}\mathbf{I} - \mathbf{B}^{-1}]det[\mathbf{F}]\lambda_i^2\end{aligned} \tag{4.224}$$

The last two results are of special interest for a hyperelastic material. In this case, the simplest manner to describe an invariant response, verifying both the requirements of material indifference and of isotropy is in terms of the strain energy potential written as a function of three invariants of the Cauchy–Green strain tensors:

$$\boxed{\psi(i_{1C}, i_{2C}, i_{3C})} \tag{4.225}$$

One example of this kind of elastic response representation is given for the case of incompressible material (incompressibility implies that $i_{3C} = 1$), which is known as the Mooney–Rivlin material model [189]; the corresponding strain energy potential can be written:

$$W(i_{1C}, i_{2C}) := \alpha_1(i_{1C} - 3) + \alpha_2(i_{2C} - 3) \; ; \tag{4.226}$$

where α_1 and α_2 are the material parameters. Another useful example from the same family is so-called neo-Hookean material model with the strain energy defined by:

$$W(i_{1C}) := \frac{1}{2}\mu(i_{1C} - 3) \tag{4.227}$$

For the subsequent stress tensor computation for such a response, the following auxiliary result concerning the partial derivatives of the principal invariants proves to be very useful:

$$\begin{array}{l} i_{1C} = tr[\mathbf{C}] \;\Rightarrow\; \frac{\partial i_{1C}}{\partial \mathbf{C}} = \mathbf{I} \\ i_{2C} = \frac{1}{2}((tr[\mathbf{C}])^2 - tr[\mathbf{C}^2]) \;\Rightarrow\; \frac{\partial i_{2C}}{\partial \mathbf{C}} = i_{1C}\mathbf{I} - \mathbf{C} \\ i_{3C} = det[\mathbf{C}] \;\Rightarrow\; \frac{\partial i_{3C}}{\partial \mathbf{C}} = i_{3C}\mathbf{C}^{-1} \end{array} \tag{4.228}$$

With these results in hand, we can provide from (4.217) an explicit representation of the second Piola–Kirchhoff stress tensor in terms of the right Cauchy–Green strain tensor and its invariants:

$$\boxed{\mathbf{S} := \frac{2\partial\psi(\cdot)}{\partial \mathbf{C}} = 2(\frac{\partial\psi}{\partial i_{1C}} + \frac{\partial\psi}{\partial i_{2C}} i_{1C})\mathbf{I} - 2\frac{\partial\psi}{\partial i_{2C}}\mathbf{C} + \frac{\partial\psi}{\partial i_{3C}}\mathbf{C}^{-1}} \tag{4.229}$$

An alternative manner to describe an isotropic hyperelastic response is by means of the Kirchhoff stress tensor representation in terms of the left Cauchy–Green strain tensor and its invariants:

$$\boxed{\boldsymbol{\tau} := \mathbf{F}\mathbf{S}\mathbf{F}^T = \frac{\partial\psi}{\partial i_{3B}}\mathbf{I} + 2(\frac{\partial\psi}{\partial i_{1B}} + \frac{\partial\psi}{\partial i_{2B}} i_{1B})\mathbf{B} - 2\frac{\partial\psi}{\partial i_{2B}}\mathbf{B}^2} \tag{4.230}$$

4.7.2 Constitutive laws for large deformations in terms of principal stretches

A very elegant alternative for the construction of an elastic constitutive response that satisfies the material indifference and isotropy restrictions, relies upon the strain energy potential defined in terms of the principal stretches (see Hill [105] or Ogden [209]). The principal stretches are already used in (4.224) above for defining the eigenvalues of the Cauchy–Green deformation tensors. That result is in agreement with the basic definition of the principal stretches as the solutions to the characteristic equation, or the eigenvalues, of the stretch tensors **U** and **V**. The latter derives from the standard eigenvalue problem for computing the principal values of the stretch tensor, which can be written either in material description (by using **U**) or in the spatial description (by using **V**):

$$\begin{aligned} (\mathbf{U} - \lambda_i \mathbf{I})\mathbf{n}_i &= \mathbf{0} \\ (\mathbf{V} - \lambda_i \mathbf{I}^{\varphi})\mathbf{m}_i &= \mathbf{0} \end{aligned} \tag{4.231}$$

We note that the computed eigenvalues (or the principal stretches $\lambda_i\,;\, i = 1,2,3$) remain the same in either material or spatial description, but the corresponding eigenvectors $\mathbf{n}_i$ and $\mathbf{m}_i$ do not. The computed solutions to these eigenvalue problems can be exploited to obtain the spectral decomposition of the deformation gradient, rotation tensor and both stretch tensors:

$$\boxed{\begin{aligned} \mathbf{F} &= \textstyle\sum_{i=1}^{3} \lambda_i \mathbf{m}_i \otimes \mathbf{n}_i \\ \mathbf{R} &= \textstyle\sum_{i=1}^{3} \mathbf{m}_i \otimes \mathbf{n}_i \\ \mathbf{U} &= \textstyle\sum_{i=1}^{3} \lambda_i \mathbf{n}_i \otimes \mathbf{n}_i \\ \mathbf{V} &= \textstyle\sum_{i=1}^{3} \lambda_i \mathbf{m}_i \otimes \mathbf{m}_i \end{aligned}} \tag{4.232}$$

The last result holds for the principal vectors that form the ortho-normal principal frames, both in material and spatial descriptions:

$$\begin{aligned} &\mathbf{n}_i \cdot \mathbf{I}\mathbf{n}_j = \mathbf{m}_i \cdot \mathbf{I}^{\varphi}\mathbf{m}_j = \delta_{ij} := \begin{cases} 1\;;\; i = j \\ 0\;;\; i \neq j \end{cases}\;; \\ &\mathbf{I} = \delta_{ij}\mathbf{e}_i \otimes \mathbf{e}_j\;;\; \mathbf{I}^{\varphi} = \delta_{ij}^{\varphi}\mathbf{e}_i^{\varphi} \otimes \mathbf{e}_j^{\varphi} \end{aligned} \tag{4.233}$$

Such a choice of the principal frame allows us to easily obtain the spectral decomposition of the transpose of the deformation gradient:

$$\mathbf{n}_i = \lambda_i^{-1}\mathbf{F}^T\mathbf{m}_i \;\Rightarrow\; \mathbf{F}^T = \sum_{i=1}^{3} \lambda_i \mathbf{n}_i \otimes \mathbf{m}_i \tag{4.234}$$

The last result and the one in (4.232) will also allow us to obtain the spectral decomposition for both Cauchy–Green strain tensors:

$$\begin{aligned} \mathbf{C} &:= \mathbf{F}^T\mathbf{F} = \textstyle\sum_{i=1}^3 \lambda_i^2 \mathbf{n}_i \otimes \mathbf{n}_i \; ; \\ \mathbf{B} &:= \mathbf{F}\mathbf{F}^T = \textstyle\sum_{i=1}^3 \lambda_i^2 \mathbf{m}_i \otimes \mathbf{m}_i \end{aligned} \tag{4.235}$$

The last result is in agreement with the relationship between the stretch tensors and the Cauchy–Green strain tensors, where $\mathbf{C} = \mathbf{U}^2$ and $\mathbf{B} = \mathbf{V}^2$. With these results in hand, we can easily express the principal invariants of both Cauchy–Green strain tensors defined in (4.223) in terms of the principal stretches:

$$\begin{aligned} i_{1C} &= \lambda_1^2 + \lambda_2^2 + \lambda_3^2 \\ i_{2C} &= \lambda_1^2\lambda_2^2 + \lambda_2^2\lambda_3^2 + \lambda_3^2\lambda_1^2 \\ i_{3C} &= \lambda_1^2\lambda_2^2\lambda_3^2 \end{aligned} \tag{4.236}$$

Any isotropic hyperelastic response that obeys the principle of material indifference can thus be expressed in terms of the strain energy potential defined as a function of the principal stretches. For example, we can write such a strain energy potential for Mooney–Rivlin material model:

$$\psi(\lambda_1, \lambda_2, \lambda_3) = \alpha_1(\lambda_1^2 + \lambda_2^2 + \lambda_3^2 - 3) + \alpha_2(\lambda_1^2\lambda_2^2 + \lambda_2^2\lambda_3^2 + \lambda_3^2\lambda_1^2 - 3) \tag{4.237}$$

Similarly, the neo-Hookean material model can be defined through the strain energy dependent upon the principal stretches, which is written:

$$\psi(\lambda_1, \lambda_2, \lambda_3) = \frac{1}{2}\mu(\lambda_1^2 + \lambda_2^2 + \lambda_3^2 - 3) \tag{4.238}$$

We note that both material models are used for representing incompressible behavior, where the principal stretches must satisfy the incompressibility constraint, with:

$$\lambda_1\lambda_2\lambda_3 = 1 \;\Rightarrow\; J = 1 \tag{4.239}$$

We also note that both models are used for describing the elastic material behavior, where the unit value of each principal stretch is accompanied by the zero stress value:

$$\forall i\,;\; \lambda_i, \mapsto 1 \;\Rightarrow\; \psi(\lambda_i) \mapsto 0 \tag{4.240}$$

The main advantage of any hyperelastic constitutive model with strain energy potential defined in terms of the principal stretches concerns the simple manner to check the polyconvexity conditions; the latter, enforcing that the large stresses should accompany very large values strains, will simply imply that:

$$\begin{aligned} \psi(\lambda_i) &\mapsto \infty \,\text{if}\, \{\lambda_1, \lambda_2, \lambda_3\} \mapsto \infty \text{ (in tension)} \\ \psi(\lambda_i) &\longrightarrow \infty \,\text{if}\, \{\lambda_1, \lambda_2, \lambda_3\} \longrightarrow 0^+ \text{ (in compression)} \end{aligned} \tag{4.241}$$

In other words, for a hyperelastic constitutive model defined in terms of principal stretches, the polyconvexity conditions will require the strain energy convexity with respect to each principal stretch. We will further discuss in detail one such constitutive model verifying the polyconvexity conditions, given in terms of so-called Ogden's material. The latter is capable of representing very large deformations of rubberlike materials where the principal stretches may be an order of magnitude larger than one. The rubberlike material is considered incompressible, and thus the internal energy of Ogden's material ought to be chosen to accommodate the incompressibility constraint, with $\lambda_1\lambda_2\lambda_3 = 1$.

We further discuss the details of the finite element implementation of Ogden's constitutive model, starting with special cases. In 1D case, we can exploit the incompressibility constraint to reduce the number of principal stretches to compute to one, by using that $\lambda_1 \equiv \lambda \Rightarrow \lambda_2 = \lambda_3 = 1/\sqrt{\lambda}$. We can thus write the strain energy of Ogden's material for 1D case according to:

$$\psi(\lambda) = \sum_r \frac{\mu_r}{\alpha_r}(\lambda^{\alpha_r} + 2\lambda^{-0.5\alpha_r} - 3) \; ; \; \sum_r \alpha_r \mu_r = 2\mu \tag{4.242}$$

We indicated above that the material parameters α_r and μ_r must obey the condition $\sum_r \alpha_r \mu_r = 2\mu$, which ensures that we recover 1D version of Hooke's law for an incompressible material. It is easy to check that such a constitutive model verifies the polyconvexity condition:

$$\begin{aligned} \lambda \mapsto \infty &\Rightarrow \psi(\lambda) \mapsto +\infty \\ \lambda \mapsto 0^+ &\Rightarrow \psi(\lambda) \mapsto +\infty \end{aligned} \tag{4.243}$$

and that it also provides the accompanying zero value of strain energy for unit value of stretch:

$$\lambda \mapsto 1 \Rightarrow \psi(\lambda) \mapsto 0 \tag{4.244}$$

The strain energy defined in terms of principal stretch will still allow the straightforward computation of any chosen stress. For example, by exploiting the relationship between the stretch λ and the Green–Lagrange deformation in 1D case: $E = \frac{1}{2}(\lambda^2 - 1)$, we can compute the second Piola–Kirchhoff stress, from the corresponding 1D strain energy potential for Ogden's material model:

$$\begin{aligned} S &= \frac{\partial \psi}{\partial E} \\ &= \frac{\partial \psi}{\partial \lambda} \underbrace{\frac{\partial \lambda}{\partial E}}_{1/\lambda} \\ &= \frac{1}{\lambda}\frac{\partial \psi}{\partial \lambda} \\ &= \textstyle\sum_r \mu_r [\lambda^{(\alpha_r - 2)} - \lambda^{(-0.5\alpha_r - 2)}] \end{aligned} \tag{4.245}$$

In the same manner, we can also compute the elastic tangent modulus as the derivative of the second Piola–Kirchhoff stress with respect to the Green–Lagrange strain, which results with:

$$\begin{aligned} \mathcal{C} &= \frac{\partial S}{\partial E} \\ &= \frac{1}{\lambda}\frac{\partial S}{\partial \lambda} \\ &= \frac{1}{\lambda^2}\frac{\partial \psi}{\partial \lambda^2} \\ &= \textstyle\sum_r \mu_r[(\alpha_r - 2)\lambda^{(\alpha_r - 4)} + (0.5\alpha_r + 2)\lambda^{(-0.5\alpha_r - 4)}] \end{aligned} \tag{4.246}$$

The last results for the second Piola–Kirchhoff stress and tangent modulus can directly be inserted into (4.115) and (4.116) respectively, providing the internal force vector and tangent stiffness matrix of a truss-bar element in large displacements made of rubberlike Ogden's material.
We now turn to finite element implementation of Ogden's material model in either 2D or 3D case. The Valanis hypothesis can now be exploited to write the internal energy as the sum of the functions that each depends only on one principal stretch:

$$\psi(\lambda_1, \lambda_2, \lambda_3) = \tilde{\psi}(\lambda_1) + \tilde{\psi}(\lambda_2) + \tilde{\psi}(\lambda_3) \tag{4.247}$$

The restriction on zero stress at unit stretch, the polyconvexity conditions and incompressibility constraint will further restrict the admissible form of any such function:

$$\tilde{\psi}(1) = 0 \; ; \; \tilde{\psi}(0) = \tilde{\psi}(\infty) \mapsto \infty \; ; \; \tilde{\psi}'(1) + \tilde{\psi}''(1) = 2\mu \tag{4.248}$$

For a general 3D case, Ogden's material model [209] provides one example of this kind very suitable for describing the incompressible behavior of rubberlike materials, where the strain energy can be written:

$$\psi(\lambda_i) = \sum_r \frac{\mu_r}{\alpha_r}(\lambda_1^{\alpha_r} + \lambda_2^{\alpha_r} + \lambda_3^{\alpha_r} - 3) \; ; \; \sum_r \mu_r \alpha_r = 2\mu \tag{4.249}$$

The chosen values of material parameters α_r and μ_r should verify the incompressibility constraint, with $\sum_r \mu_r \alpha_r = 2\mu$.
Two special 2D cases follow from this general result valid in 3D. The first is the plane strain case, which is obtained for a fixed, unit value of the third principal stretch $\lambda_3 = 1$. The strain energy for the plane strain 2D case can then be written:

$$\lambda_3 = 1 \Longrightarrow \psi(\lambda_1, \lambda_2) = \sum_r \frac{\mu_r}{\alpha_r}(\lambda_1^{\alpha_r} + \lambda_2^{\alpha_r} - 2) \tag{4.250}$$

where the incompressibility constraint now reads $\lambda_1 \lambda_2 = 1$. The second special case is 2D plane stress, where the incompressibility constraint is used to eliminate the third stretch with $\lambda_3 = 1/\lambda_1\lambda_2$. The strain energy for plane stress 2D case can thus be written:

$$\psi(\lambda_1, \lambda_2) = \sum_r \frac{\mu_r}{\alpha_r}[\lambda_1^{\alpha_r} + \lambda_2^{\alpha_r} + (\lambda_1\lambda_2)^{-\alpha_r} - 3] \tag{4.251}$$

In computations of stress and tangent elasticity tensor components of Ogden's material model, an important role is played by the auxiliary result developed subsequently pertaining to derivatives of the principal values of a tensor. This result can be obtained by applying the Gâteaux derivative formalism to the corresponding eigenvalue problem statement, leading to:

$$\begin{aligned}
\mathbf{0} &= \{\frac{d}{d\varepsilon}\Big|_{\varepsilon=0} [\overbrace{\mathbf{C}_\varepsilon}^{\mathbf{C}+\varepsilon\, d\mathbf{C}} - \lambda_{i,\varepsilon}^2 \mathbf{I}]\mathbf{n}_{i,\varepsilon}\} \cdot \mathbf{n}_i \\
&= \{[\underbrace{\frac{\partial \mathbf{C}}{\partial \mathbf{C}}}_{\mathcal{I}} d\mathbf{C} - 2\lambda_i \left(\tfrac{\partial \lambda_i}{\partial \mathbf{C}} \cdot d\mathbf{C}\right)]\mathbf{n}_i + (\mathbf{C} - \lambda_i^2 \mathbf{I}) d\mathbf{n}_i = \mathbf{0}\} \cdot \mathbf{n}_i \\
&= (d\mathbf{C}\,\mathbf{n}_i) \cdot \mathbf{n}_i - 2\lambda_i \left(\tfrac{\partial \lambda_i}{\partial \mathbf{C}} \cdot d\mathbf{C}\right) \underbrace{\mathbf{n}_i \cdot \mathbf{n}_i}_{\mathbf{I}} + d\mathbf{n}_i \cdot \underbrace{(\mathbf{C} - \lambda_i^2 \mathbf{I})\mathbf{n}_i}_{=0} \\
&= (\mathbf{n}_i \otimes \mathbf{n}_i) \cdot d\mathbf{C} - 2\lambda_i \tfrac{\partial \lambda_i}{\partial \mathbf{C}} \cdot d\mathbf{C} \\
\Rightarrow &\quad \boxed{\tfrac{\partial \lambda_i}{\partial \mathbf{C}} = \tfrac{1}{2\lambda_i} \mathbf{n}_i \otimes \mathbf{n}_i}
\end{aligned} \tag{4.252}$$

With this result in hand, we can carry out a simple chain rule computation of derivatives, which provides the second Piola–Kirchhoff stress from the strain energy potential written in terms of principal stretches:

$$\begin{aligned}
\mathbf{S} &= \tfrac{\partial \psi(\lambda_1,\lambda_2)}{\partial \mathbf{E}} \\
&= 2\tfrac{\partial \psi(\lambda_1,\lambda_2)}{\partial \mathbf{C}} \\
&= 2\textstyle\sum_{i=1}^{2} \tfrac{\partial \psi(\lambda_1,\lambda_2)}{\partial \lambda_i} \tfrac{\partial \lambda_i}{\partial \mathbf{C}} \\
&= \textstyle\sum_{i=1}^{2} \tfrac{1}{\lambda_i} \tfrac{\partial \psi(\lambda_1,\lambda_2)}{\partial \lambda_i} \mathbf{n}_i \otimes \mathbf{n}_i \\
\Rightarrow &\quad \boxed{\mathbf{S} = \textstyle\sum_{i=1}^{2} s_i \mathbf{n}_i \otimes \mathbf{n}_i \;;\; s_i = \tfrac{1}{\lambda_i} \tfrac{\partial \psi(\lambda_1,\lambda_2)}{\partial \lambda_i}}
\end{aligned} \tag{4.253}$$

Needless to say, for any of the special cases one should select the corresponding form of the strain energy potential for Ogden's material model; for example, for plane strain 2D case we will have:

$$s_i = \lambda_i^{-1} \frac{\partial W(\lambda_j)}{\partial \lambda_i} = \lambda_i^{-2} \sum_r \mu_r [\lambda_i^{\alpha_r} - (\lambda_1\lambda_2)^{-\alpha_r}] \tag{4.254}$$

Going one step further in the directional derivative computation, we can obtain the elastic tangent modulus, which can be written:

$$\begin{aligned}\mathcal{C} &= \frac{\partial \mathbf{S}}{\partial \mathbf{E}} \\ &= 2\frac{\partial \mathbf{S}}{\partial \mathbf{C}} \\ &= \underbrace{2\sum_{i=1}^{2} \frac{\partial s_i}{\partial \mathbf{C}} \mathbf{n}_i \otimes \mathbf{n}_i}_{\mathcal{C}_{mat}} + \underbrace{2\sum_{i=1}^{2} s_i \frac{\partial}{\partial \mathbf{C}} (\mathbf{n}_i \otimes \mathbf{n}_i)}_{\mathcal{C}_{geo}}\end{aligned} \tag{4.255}$$

By exploiting the auxiliary result in (4.252), we can provide the closed form expression for the material part of the tangent elasticity tensor in terms of its reduced form in principal axes D_{ij}:

$$\begin{aligned}\mathcal{C}_{mat} &:= \sum_{i=1}^{2} 2\frac{\partial s_i}{\partial \mathbf{C}} \mathbf{n}_i \otimes \mathbf{n}_i \\ &= 2\sum_{i=1}^{2} \sum_{j=1}^{2} \frac{\partial s_i}{\partial \lambda_j} \frac{\partial \lambda_j}{\partial \mathbf{C}} \mathbf{n}_i \otimes \mathbf{n}_i \\ &= \sum_{i=1}^{2} \sum_{j=1}^{2} \underbrace{\left(\frac{1}{\lambda_j} \frac{\partial s_i}{\partial \lambda_j} \right)}_{D_{ij}} [\mathbf{n}_i \otimes \mathbf{n}_i] \otimes [\mathbf{n}_j \otimes \mathbf{n}_j]\end{aligned} \tag{4.256}$$

We note that the principal axes components of the reduced tangent elastic modulus are computed in accordance with a particular form of the strain energy potential; for example, in plane stress 2D case for Ogden's material model, we will have:

$$\boxed{D_{ij} := \frac{1}{\lambda_j \lambda_j} \frac{\partial^2 \psi}{\partial \lambda_j \partial \lambda_j} = \lambda_i^{-2} \lambda_j^{-2} \{ \sum_r \mu_r \alpha_r [\delta_{ij} \lambda_j^{\alpha_r} + (\lambda_1 \lambda_2)^{-\alpha_r}] \}} \tag{4.257}$$

The explicit form of the geometric part of the tangent elasticity tensor can also be obtained, but with somewhat more laborious computations that require a systematic application of the auxiliary result in (4.252). Namely, by exploiting the spectral decompositions of the right Cauchy–Green strain tensor and unit tensor, we can provide the corresponding representation for rank-one tensor products of eigenvectors:

$$\begin{aligned}&\mathbf{C} = \sum_{i=1}^{2} \lambda_i^2 \mathbf{n}_i \otimes \mathbf{n}_i \; ; \; \mathbf{I} = \sum_{i=1}^{2} \mathbf{n}_i \otimes \mathbf{n}_i \\ &\Rightarrow \begin{array}{l} \mathbf{C} - \lambda_2^2 \mathbf{I} = (\lambda_1^2 - \lambda_2^2) \mathbf{n}_1 \otimes \mathbf{n}_1 \\ \mathbf{C} - \lambda_1^2 \mathbf{I} = -(\lambda_1^2 - \lambda_2^2) \mathbf{n}_2 \otimes \mathbf{n}_2 \end{array} \Rightarrow \begin{array}{l} \mathbf{n}_1 \otimes \mathbf{n}_1 = \frac{1}{\lambda_1^2 - \lambda_2^2} (\mathbf{C} - \lambda_2^2 \mathbf{I}) \\ \mathbf{n}_2 \otimes \mathbf{n}_2 = \frac{1}{\lambda_1^2 - \lambda_2^2} (\mathbf{C} - \lambda_1^2 \mathbf{I}) \end{array}\end{aligned} \tag{4.258}$$

The result is valid only for distinct eigenvalues, $\lambda_1^2 \neq \lambda_2^2$, since otherwise $\mathbf{C} = \lambda^2 \mathbf{I}$. This kind of representation of the rank-one tensor products of eigenvectors allows us to reduce the directional derivative computation to the systematic application of the auxiliary result in (4.252); for the first eigenvector, we can thus write:

$$
\begin{aligned}
\frac{\partial}{\partial \mathbf{C}}(\mathbf{n}_1 \otimes \mathbf{n}_1) &= \frac{\partial}{\partial \mathbf{C}}\left[\frac{1}{\lambda_1^2-\lambda_2^2}(\mathbf{C}-\lambda_2^2\mathbf{I})\right] \\
&= \frac{1}{\lambda_1^2-\lambda_2^2}[\boldsymbol{\mathcal{I}} - \underbrace{\mathbf{I}}_{(\mathbf{n}_1\otimes\mathbf{n}_1+\mathbf{n}_2\otimes\mathbf{n}_2)} \otimes(2\lambda_2\tfrac{1}{2\lambda_2}\mathbf{n}_2\otimes\mathbf{n}_2)] - \frac{1}{(\lambda_1^2-\lambda_2^2)^2} \\
&\underbrace{(\mathbf{C}-\lambda_2^2\mathbf{I})}_{(\lambda_1^2-\lambda_2^2)\mathbf{n}_1\otimes\mathbf{n}_1} \otimes\left(2\lambda_1\tfrac{1}{2\lambda_1}\mathbf{n}_1\otimes\mathbf{n}_1 - 2\lambda_2\tfrac{1}{2\lambda_2}\mathbf{n}_2\otimes\mathbf{n}_2\right) \\
&= \frac{1}{\lambda_1^2-\lambda_2^2}[\boldsymbol{\mathcal{I}} - (\mathbf{n}_1\otimes\mathbf{n}_1)\otimes(\mathbf{n}_2\otimes\mathbf{n}_2) - (\mathbf{n}_2\otimes\mathbf{n}_2)\otimes(\mathbf{n}_2\otimes\mathbf{n}_2) \\
&-(\mathbf{n}_1\otimes\mathbf{n}_1)\otimes(\mathbf{n}_1\otimes\mathbf{n}_1) + (\mathbf{n}_1\otimes\mathbf{n}_1)\otimes(\mathbf{n}_2\otimes\mathbf{n}_2)] \\
&= \frac{1}{\lambda_1^2-\lambda_2^2}[\boldsymbol{\mathcal{I}} - (\mathbf{n}_1\otimes\mathbf{n}_1)\otimes(\mathbf{n}_1\otimes\mathbf{n}_1) - (\mathbf{n}_2\otimes\mathbf{n}_2)\otimes(\mathbf{n}_2\otimes\mathbf{n}_2)]
\end{aligned}
\tag{4.259}
$$

In the same manner, we can also compute the directional derivative of the second eigenvector:

$$
\begin{aligned}
\frac{\partial}{\partial \mathbf{C}}(\mathbf{n}_2 \otimes \mathbf{n}_2) &= \frac{\partial}{\partial \mathbf{C}}\left[-\frac{1}{\lambda_1^2-\lambda_2^2}(\mathbf{C}-\lambda_1^2\mathbf{I})\right] \\
&= -\frac{1}{\lambda_1^2-\lambda_2^2}[\boldsymbol{\mathcal{I}} - \underbrace{\mathbf{I}}_{(\mathbf{n}_1\otimes\mathbf{n}_1+\mathbf{n}_2\otimes\mathbf{n}_2)} \otimes(2\lambda_1\tfrac{1}{2\lambda_1}\mathbf{n}_1\otimes\mathbf{n}_1)] - \frac{-1}{(\lambda_1^2-\lambda_2^2)^2} \\
&\underbrace{(\mathbf{C}-\lambda_2^2\mathbf{I})}_{-(\lambda_1^2-\lambda_2^2)\mathbf{n}_2\otimes\mathbf{n}_2} \otimes\left(2\lambda_1\tfrac{1}{2\lambda_1}\mathbf{n}_1\otimes\mathbf{n}_1 - 2\lambda_2\tfrac{1}{2\lambda_2}\mathbf{n}_2\otimes\mathbf{n}_2\right) \\
&= -\frac{1}{\lambda_1^2-\lambda_2^2}[\boldsymbol{\mathcal{I}} - (\mathbf{n}_1\otimes\mathbf{n}_1)\otimes(\mathbf{n}_1\otimes\mathbf{n}_1) - (\mathbf{n}_2\otimes\mathbf{n}_2)\otimes(\mathbf{n}_1\otimes\mathbf{n}_1) \\
&-(\mathbf{n}_2\otimes\mathbf{n}_2)\otimes(\mathbf{n}_1\otimes\mathbf{n}_1) + (\mathbf{n}_2\otimes\mathbf{n}_2)\otimes(\mathbf{n}_2\otimes\mathbf{n}_2)] \\
&= -\frac{1}{\lambda_1^2-\lambda_2^2}[\boldsymbol{\mathcal{I}} - (\mathbf{n}_1\otimes\mathbf{n}_1)\otimes(\mathbf{n}_1\otimes\mathbf{n}_1) - (\mathbf{n}_2\otimes\mathbf{n}_2)\otimes(\mathbf{n}_2\otimes\mathbf{n}_2)]
\end{aligned}
\tag{4.260}
$$

With these results in hand, we can write the explicit form of the tangent elasticity tensor, by using the direct tensor notation:

$$
\boxed{
\begin{aligned}
\boldsymbol{\mathcal{C}} &:= 2\frac{\partial \mathbf{S}}{\partial \mathbf{C}} \\
&= \textstyle\sum_{i=1}^{2}\sum_{j=1}^{2} D_{ij}(\mathbf{n}_i\otimes\mathbf{n}_i)\otimes(\mathbf{n}_j\otimes\mathbf{n}_j) \\
&+2\frac{s_1-s_2}{\lambda_1^2-\lambda_2^2}[\boldsymbol{\mathcal{I}} - (\mathbf{n}_1\otimes\mathbf{n}_1)\otimes(\mathbf{n}_1\otimes\mathbf{n}_1) - (\mathbf{n}_2\otimes\mathbf{n}_2)\otimes(\mathbf{n}_2\otimes\mathbf{n}_2)]
\end{aligned}
}
\tag{4.261}
$$

In the finite element implementation, all the result we obtained for Ogden's material model are further recast in matrix notation. First, we choose the coordinate representation for eigenvectors in 2D case in terms of a single parameter, the angle α between the first principal direction and x_1 axis; thus we can write:

$$\mathbf{n}_1 \mapsto \begin{bmatrix} cos\alpha \\ sin\alpha \end{bmatrix} \equiv \begin{bmatrix} c \\ s \end{bmatrix} \quad ; \quad \mathbf{n}_2 \mapsto \begin{bmatrix} -sin\alpha \\ cos\alpha \end{bmatrix} = \begin{bmatrix} -s \\ c \end{bmatrix} \tag{4.262}$$

With the standard convention for matrix notation, we then choose to order the second Piola–Kirchhoff stress tensor components in a vector $\mathbf{S} \mapsto \mathsf{s}^T = [S_{11}, S_{22}, S_{12}]$; this allows us to recast the spectral decomposition result given in (4.253) for the second Piola–Kirchhoff stress tensor in terms of:

$$\mathbf{S} = \sum_{i=1}^{2} s_i \mathbf{n}_i \otimes \mathbf{n}_i \;\mapsto\; \underbrace{\mathsf{s}^h}_{(3\times1)} = \underbrace{\mathsf{T}}_{(3\times2)} \underbrace{\hat{\mathsf{s}}_p^h}_{(2\times1)}$$

$$\Leftrightarrow \underbrace{\begin{bmatrix} S_{11} \\ S_{22} \\ S_{12} \end{bmatrix}}_{\mathsf{s}^h} = \underbrace{\begin{bmatrix} c^2 & s^2 \\ s^2 & c^2 \\ sc & -sc \end{bmatrix}}_{\mathsf{T}} \underbrace{\begin{bmatrix} s_1 \\ s_2 \end{bmatrix}}_{\hat{\mathsf{s}}_p^h} \tag{4.263}$$

The computed stress tensor components can further be placed in the corresponding slots in (4.128) defining the internal force vector of an isoparametric element for Ogden's material for 2D case.
Furthermore, we can also write the tangent elasticity tensor in matrix form, connecting the second Piola–Kirchhoff stress components with the Green–Lagrange strain tensor components, packed in a vector $\mathbf{E} \mapsto \mathsf{e}^T = [E_{11}, E_{22}, E_{12}]$; for 2D case with Ogden's material, we thus obtain:

$$\mathcal{C} = \frac{\partial \mathbf{S}}{\partial \mathbf{E}} \longrightarrow \underbrace{\mathsf{C}}_{(3\times3)} = \mathsf{T} \underbrace{\frac{\partial \hat{\mathsf{s}}_p}{\partial \hat{\mathsf{e}}_p}}_{D_{ij}} \mathsf{T}^T + \frac{s_1 - s_2}{\lambda_1^2 - \lambda_2^2} \mathsf{g}\,\mathsf{g}^T \; ; \; \mathsf{g} = \begin{bmatrix} -sin\,2\alpha \\ sin\,2\alpha \\ cos\,2\alpha \end{bmatrix} \tag{4.264}$$

This compact form of the geometric part of tangent elasticity tensor in matrix notation was obtained thanks to the following auxiliary result:

$$[\mathcal{I} - (\mathbf{n}_1 \otimes \mathbf{n}_1) \otimes (\mathbf{n}_1 \otimes \mathbf{n}_1) - (\mathbf{n}_2 \otimes \mathbf{n}_2) \otimes (\mathbf{n}_2 \otimes \mathbf{n}_2)] \mapsto$$

$$\begin{bmatrix} 1 & & \\ & 1 & \\ & & 1/2 \end{bmatrix} - \begin{bmatrix} c^2 \\ s^2 \\ sc \end{bmatrix} \begin{bmatrix} c^2 \\ s^2 \\ sc \end{bmatrix}^T - \begin{bmatrix} s^2 \\ c^2 \\ -sc \end{bmatrix} \begin{bmatrix} s^2 \\ c^2 \\ -sc \end{bmatrix}^T =: \frac{1}{2} \mathsf{g}\,\mathsf{g}^T \tag{4.265}$$

The matrix representation of the elasticity tensor or this kind can further be used in (4.131) in order to obtain the consistent tangent matrix of an isoparametric element for 2D Ogden's material model. We have thus defined all the ingredients needed for the efficient solution of a boundary value problem describing large displacements and deformations of rubberlike materials in 2D case. The results of an illustrative problem of this kind are given in Figure 4.19, which shows the deformed configuration of a rubber sheet obtained in a simple tension test under very large stretch. The deformed configuration is computed with a finite element mesh of 16×6 isoparametric 4-node membrane elements. We can observe the characteristic reduction of the sheet width produced by stretching, as the result of enforcing the incompressibility constraint. We note in passing that the solution of the same problem for plane strain case, would require the finite element approximations capable of dealing with the incompressibility constraint.

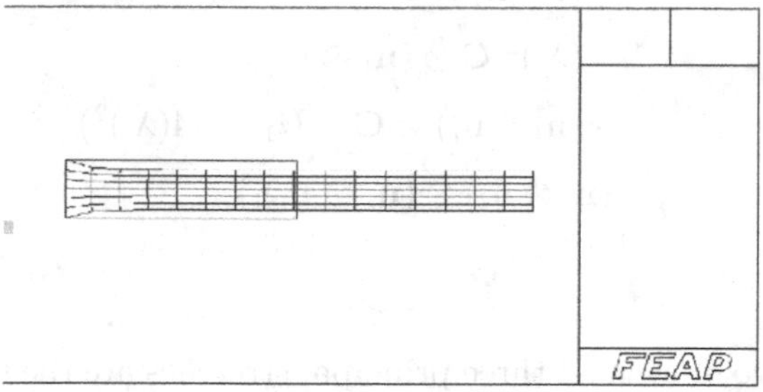

Fig. 4.19 Deformed configuration of rubber sheet in a simple traction test, computed with 16×6 membrane isoparametric 4-node elements.

In closing this section we briefly comment on how to obtain the stress tensor and tangent elasticity tensor for Ogden's material in general 3D case. Needless to say, such a computation is more laborious than the one just described for 2D case, and requires the mixed finite element interpolations capable of dealing with the incompressibility constraint.
First, by exploiting the spectral decomposition of the right Cauchy–Green strain tensor, we can obtain the corresponding representation of the principal vectors:

$$\begin{aligned} \mathbf{n}_i \otimes \mathbf{n}_i &= \frac{\lambda_i^2}{D_i}[\mathbf{C} - (i_{1C} - \lambda_i^2)\mathbf{I} + i_{3C}\lambda_i^{-2}\mathbf{C}^{-1}] \\ D_i &= (\lambda_i^2 - \lambda_j^2)(\lambda_i^2 - \lambda_k^2) \end{aligned} \tag{4.266}$$

where $i_{iC}; i = 1,2,3$ are the principal invariants of strain tensor $\mathbf{C}$ defined in (4.223). Having computed the principal directions, we can easily complete the computation of the second Piola–Kirchhoff stress tensor in terms of its principal values, which are obtained from the corresponding internal energy potential:

$$\mathbf{S} = \sum_{i=1}^{3} \lambda_i \frac{\partial \psi}{\partial \lambda_i} \mathbf{n}_i \otimes \mathbf{n}_i \tag{4.267}$$

We can also provide the closed form result for the tangent elasticity tensor, which consists of the material and geometric part:

$$\mathcal{C} = \mathcal{C}_{mate} + \mathcal{C}_{geom} \tag{4.268}$$

The material part of elasticity tensor can be written:

$$\mathcal{C}^{ep}_{mate,n+1} = \sum_{i=1}^{3}\sum_{j=1}^{3} \frac{\partial^2 \psi}{\partial \lambda_i \partial \lambda_j} (\mathbf{n}_i \otimes \mathbf{n}_i) \otimes (\mathbf{n}_j \otimes \mathbf{n}_j) \tag{4.269}$$

whereas the geometric part is defined as:

$$\begin{aligned} \mathcal{C}_{geom} = \tfrac{2\tau_i}{D_i} \{ & \mathcal{I}_C - \mathbf{C} \otimes \mathbf{C} - i_{3C}(\lambda_i)^{-2} [\mathcal{I} \\ & -(\mathbf{I} - \mathbf{n}_i \otimes \mathbf{n}_i) \otimes (\mathbf{I} - \mathbf{n}_i \otimes \mathbf{n}_i)] \\ & +(\lambda_i)^2 [\mathbf{C} \otimes (\mathbf{n}_i \otimes \mathbf{n}_i) \\ & +(\mathbf{n}_i \otimes \mathbf{n}_i) \otimes \mathbf{C} + (i_{1C} - 4(\lambda_i)^2) \\ & (\mathbf{n}_i \otimes \mathbf{n}_i) \otimes (\mathbf{n}_i \otimes \mathbf{n}_i)]\} \end{aligned} \tag{4.270}$$

Remarks:

1. In the case two or even all three principal stretches are the same, we ought to modify the spectral decomposition result for $\mathbf{C}$ accordingly:

$$\mathbf{C} = \begin{cases} \sum_{i=1}^{3} \lambda_i^2 \mathbf{n}_i \otimes \mathbf{n}_i \ ; \ \lambda_1 \neq \lambda_2 \neq \lambda_3 \\ \lambda_1^2 \mathbf{I} + (\lambda_3^2 - \lambda_1^2) \mathbf{n}_3 \otimes \mathbf{n}_3 \ ; \ \lambda_1 = \lambda_2 \neq \lambda_3 \\ \lambda_1^2 \mathbf{I} \ ; \ \lambda_1 = \lambda_2 = \lambda_3 \end{cases} \tag{4.271}$$

2. The principal direction framework can conveniently be exploited in order to obtain the generalized deformation measures as first proposed by Doyle and Ericksen [69]; all such strain measure can be written in a unified manner according to:

$$\begin{aligned} \mathbf{E}^m &:= \tfrac{1}{m}(\mathbf{C}^{m/2} - \mathbf{I}) \\ &= \begin{cases} \sum_{i=1}^{3} \frac{1}{m}(\lambda_i^m - 1) \mathbf{n}_i \otimes \mathbf{n}_i \ ; \ \lambda_1 \neq \lambda_2 \neq \lambda_3 \\ \frac{1}{m}(\lambda_1^2 - 1)\mathbf{I} + \frac{1}{m}(\lambda_3^2 - \lambda_1^2) \mathbf{n}_3 \otimes \mathbf{n}_3 \ ; \ \lambda_1 = \lambda_2 \neq \lambda_3 \\ \frac{1}{m}(\lambda_1^2 - 1)\mathbf{I} \ ; \ \lambda_1 = \lambda_2 = \lambda_3 \end{cases} \end{aligned} \tag{4.272}$$

We note that the Green–Lagrange strain measure corresponds to $m = 2$.

4.8 Plasticity and viscoplasticity for large deformations

In this section we present a detailed development of a representative constitutive model capable of dealing with very large elastic and inelastic deformations. For that purpose, we choose the large deformation plasticity model first proposed by Lee [165] and Mandel [177], which is based on the multiplicative decomposition of the deformation gradient into elastic and plastic parts. We also discuss the corresponding modifications of such model that must be introduced to provide the corresponding viscoplasticity model capable of representing very large deformations. It is important to note that all these developments are presented in the standard Euclidean setting, rather than a more general framework of differential manifolds (very much favored by the initial works on the subject: Marsden and Hughes [182], Moran et al. [190] or Ibrahimbegovic [113], Simo [235]). The main reason for this choice is in the mathematical complexity of the manifold setting, which can often be a hinderance from clear understanding of the physical basis of the plasticity model. Moreover, the vast majority of practical applications can be studied in the Euclidean setting, with only rare exceptions (such as shells, see Ibrahimbegovic [114]) where the true benefits of the manifold setting are readily apparent.

The main objectives of the developments presented in this section can be stated as follows:

(i) Construct a theoretical formulation of large deformation plasticity, based upon the multiplicative decomposition of the deformation gradient, which allows us to describe large elastic and large plastic deformations, in either material or spatial description

(ii) Develop several refinements of such a plasticity model by taking into account the strain rate effects with viscoplasticity model extension, as well the isotropic and kinematic hardening phenomena in large deformation with generalized Prager-Ziegler hardening model [268]

(iii) Present the operator split solution procedure for the time-discretized evolution problem in large strain plasticity, with the corresponding computation of admissible values of stress in principal axes framework, as proposed initially by Hill [105]

(iv) Obtain the explicit form of the internal force vector and the tangent stiffness matrix for an isoparametric element for large strain plasticity and viscoplasticity models, which can directly be implemented in a finite element computer code

4.8.1 Multiplicative decomposition of deformation gradient

The usual hypothesis on additive decomposition of total deformation into elastic and plastic component, which can be justified for small deformation

case (e.g. see Lubliner [172]), is no longer applicable for the case of large deformations. The first apparent reason is in multiplicity of choices for large deformation measures along with the nonlinear relationships among them, which prevent that the additive decomposition for one large strain measure would carry over to all the others. The second reason is in the multiplicative split between strain and rotation through the polar decomposition of the deformation gradient applicable to large elastic strains, which is unlikely to remain compatible with the additive decomposition of the total strain.
These two reasons are quite sufficient justification to abandon any formulation of large strain plasticity using an additive decomposition of the total strain, and turn towards more pertinent formulation of large strain plasticity that is based on the multiplicative decomposition of the deformation gradient. More precisely, by following the proposal of Lee [165] and Mandel [177], the deformation gradient $\mathbf{F}$ for a particular deformed configuration produced by large elastoplastic deformations is written as a multiplicative split between the elastic part $\mathbf{F}^e$, which is produce by elastic deformation only, and the plastic part $\mathbf{F}^p$, which is produced by plastic deformation. Thus, we can write:

$$\boxed{\mathbf{F} = \mathbf{F}^e \mathbf{F}^p} \tag{4.273}$$

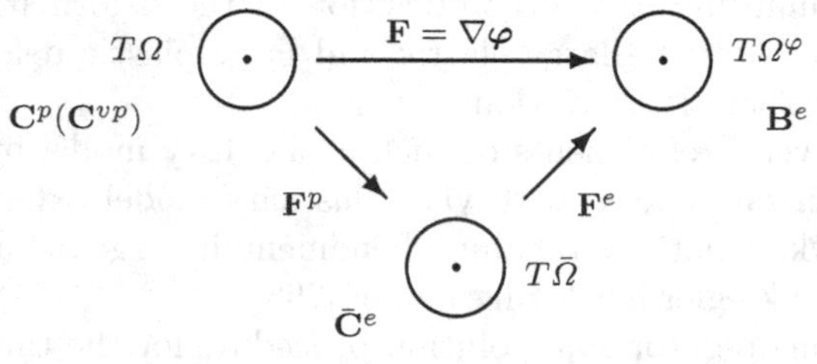

Fig. 4.20 Multiplicative decomposition of deformation gradient introducing an intermediate configuration.

In Figure 4.20, we illustrate the corresponding local interpretation of the multiplicative decomposition of deformation gradient introducing the intermediate configuration placed in-between the initial and deformed configurations. In particular, the intermediate configuration can be produced by the elastic unloading (corresponding to mapping $\mathbf{F}^{e-1}$) from the deformed configuration that will reduce the stress to zero. For that reason, the intermediate configuration is yet referred to as the stress-free configuration. We note in passing that such elastic unloading is carried out locally (for each particle

independently), which implies that the intermediate configuration need not necessarily represent a collection of compatible neighborhoods.
A very sound justification of the key hypothesis on multiplicative decomposition of deformation gradient under large elastoplastic deformations can be provided by micro-mechanics of mono-crystals with a single plastic slip system, first proposed by G.I. Taylor (e.g. see Asaro [13]). As illustrated in Figure 4.21, we can identify the plastic slip for that case with a dislocation motion which will not affect the chosen orthogonal frame. This allows us to write plastic deformation $\mathbf{F}^p = \mathbf{I} + \gamma \mathbf{m} \otimes \mathbf{n}$, which is followed by the elastic deformation $\mathbf{F}^e$ introducing the rotation and deformation of crystal lattice.

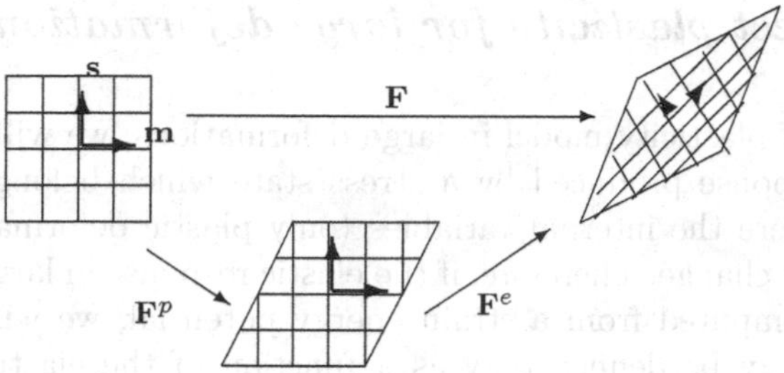

Fig. 4.21 Multiplicative decomposition of deformation gradient for elastoplastic deformation of mono-crystal with a single plastic slip system.

The multiplicative decomposition of deformation gradient can be given a very clear interpretation 1D case, where it leads to a multiplicative decomposition of total stretch into elastic and plastic part. The latter can further be related to length of a 1D bar for a homogeneous stress field. Namely, let us consider an elastoplastic bar of initial length l_0, built-in on one end and pulled by an axial traction force applied at the free end. Such a 1D problem is statically determinate and we can easily compute that the first Piola–Kirchhoff stress will remain constant along the bar and equal to the applied traction force divided by nominal section. By assuming homogeneous material, we can further conclude that the deformation is also constant along the bar and expressed as the deformed length divided by initial length of the bar, if the stretch is chosen for large strain measure. By increasing the applied traction and producing the stress beyond the elasticity limit, we will obtain the homogeneous elastoplastic deformation and the corresponding deformed length of the bar denoted as l_t. Because of the presence of plastic deformation, the unloading of the bar at this stage will result with an intermediate configuration with length l_p placed somewhere in-between the initial length l_0 and final length l_t. In any such case, it is easy to verify that the multiplicative decomposition of the deformation gradient is reduced to the multiplicative decomposition of the total stretch $\lambda = l_t/l_0$ into elastic part $\lambda^e = l_t/l_p$ and plastic part $\lambda^p = l_p/l_0$. Namely, we can write:

$$\begin{bmatrix} l_t/l_0 & 0 & 0 \\ 0 & 1 & 0 \\ 0 & 0 & 1 \end{bmatrix} =: \mathbf{F} = \mathbf{F}^e\mathbf{F}^p \Leftrightarrow \underbrace{(l_t/l_0)}_{\lambda} = \underbrace{(l_t/l_p)}_{\lambda^e}\underbrace{(l_p/l_0)}_{\lambda^p} \tag{4.274}$$

4.8.2 Perfect plasticity for large deformations

For the perfect plasticity model in large deformations, we will first consider the elastic response produced by a stress state which belongs to the elastic domain where the internal variables (only plastic deformations, for this model) will not change. Therefore, if the elastic response in large deformation plasticity is computed from a strain energy potential, we will require that the strain energy be defined only as a function of the elastic deformation gradient $\mathbf{F}^e$.

The admissible form of such an elastic response in large deformation plasticity and pertinent invariance restrictions to be imposed were the subject of much controversy ever since the pioneering works of Lee [165] and Mandel [177]. In particular, the proper role of large rotation in defining the intermediate configuration has long been disputed (e.g. see discussions in Casey and Naghdi [43], Dashner [63], Lee [166], Naghdi [194], or Nemat-Nasser [200]). In order to avoid this debate, we consider herein only the case of isotropic elastic response, which is applicable to a large number of practical problems dealing with metals and alloys.[16] The isotropy of the elastic response implies that any rigid body motion, represented by an orthogonal tensor $\mathbf{Q}$ superposed upon the intermediate configuration, should leave the elastic response and computed stress value invariant. This will then impose the left Cauchy–Green deformation tensor as the natural choice for large elastic strain measure:

[16] The isotropy of elastic response is applicable to vast majority of metals with zero residual plastic deformation; in fact, what is often represented by anisotropic plasticity models, could be more appropriately considered as isotropic elastoplastic material with residual stress field and initial plastic deformation. The case in point concerns Hill's anisotropic plasticity criterion [103] used for sheet metal forming, which is a simplified representation of the initial stress and plastic deformation state of metal sheet produced by lamination procedure. One could alternatively tackle this problem by using the isotropic plasticity model, which first considers lamination to produce the final deformed configuration with the corresponding stress and plastic deformation state, and then carry on with the sheet metal forming by starting from this final lamination state.

$$\begin{aligned}&\mathbf{F}^{esrb} \mapsto \mathbf{F}^{e}\mathbf{Q}^{T} \; ; \; (\mathbf{Q}^{T} = \mathbf{Q}^{-1}) \\ &\Rightarrow \psi(\mathbf{B}^{esrb}) = \psi(\mathbf{F}^{e} \underbrace{\mathbf{Q}^{T}\mathbf{Q}}_{\mathbf{I}} \mathbf{F}^{e\,T} \equiv \psi(\mathbf{B}^{e}) \; ; \; \mathbf{B}^{e} = \mathbf{F}^{e}\mathbf{F}^{e,T}\end{aligned} \tag{4.275}$$

The choice of the left Cauchy–Green elastic strain implies that the rigid body rotation superposed upon the intermediate configuration plays no role in defining the isotropic elastic response, which eliminates any further controversy. In the case of zero plastic deformation, with $\mathbf{F}^{p} = \mathbf{I}$ and $\mathbf{F}^{e} = \mathbf{F}$, the deformation measure in (4.275) is equivalent to Almansi's strain measure.
The axiom of material indifference, requiring the response invariance under the rigid body motion superposed upon the deformed configuration, will place further restrictions on admissible form of the isotropic elastic response in (4.275). Enforcing the material indifference of elastic response for a large rotation $\mathbf{Q}$ superposed upon the deformed configuration, will impose that the internal energy potential remains dependant only upon the invariants of the left Cauchy–Green elastic deformation tensor. Following the developments presented in the previous section, instead of strain tensor invariants, we will construct the properly invariant form of elastic response in terms of the principal elastic stretches:

$$\mathbf{F}^{e,srb} \mapsto \mathbf{Q}\mathbf{F}^{e} \; ; \; (\mathbf{Q}^{T} = \mathbf{Q}^{-1}) \Rightarrow \psi(\lambda_i^{e}) \tag{4.276}$$

The principal elastic stretches λ_i^{e} in (4.276) above are the solutions to the following eigenvalue problem:

$$[\mathbf{B}^{e} - (\lambda_i^{e})^2 \mathbf{I}^{\varphi}]\mathbf{m}_i = \mathbf{0} \; ; \; \mathbf{m}_i \cdot \mathbf{m}_j = \delta_{ij} \; ; \; \delta_{ij} = \begin{cases} 1 \; ; \; i = j \\ 0 \; ; \; i \neq j \end{cases} \tag{4.277}$$

The last result can be exploited for providing the spectral decomposition of the left Cauchy–Green elastic deformation tensor in terms of elastic principal stretches:

$$\boxed{\mathbf{B}^{e} = \textstyle\sum_{i=1}^{3} (\lambda_i^{e})^2 \mathbf{m}_i \otimes \mathbf{m}_i} \tag{4.278}$$

We note in passing that, even for 3D case, the elastic principal stretches are the same as the total principal stretches in the absence of the plastic deformation.
One advantage of the principal axes formulation of large deformation plasticity, as already recognized by Hill [105], concerns the resulting simplicity where tensor calculus is to a large extent replaced by scalar computations. Another important advantage of this kind of approach is in its capability to easily provide the material description of the large deformation plasticity

problem (see [113]). Namely, it is sufficient to that end to simply rewrite the eigenvalue problem in (4.277) in material description, which results with:

$$\begin{aligned} \mathbf{0} &= \mathbf{F}^{-1}\left\{[\mathbf{B}^e - (\lambda_i^e)^2\mathbf{I}^\varphi]\mathbf{F}^{-T}\mathbf{F}^T\mathbf{m}_i\right\} \\ &= [(\underbrace{\mathbf{F}^{p\,T}\mathbf{F}^p}_{\mathbf{C}^p})^{-1} - (\lambda_i^e)^2(\underbrace{\mathbf{F}^T\mathbf{F}}_{\mathbf{C}})^{-1}]\underbrace{\mathbf{F}^T\mathbf{m}_i}_{\mathbf{n}_i} \\ &= \boxed{[\mathbf{G}^p - (\lambda_i^e)^2\mathbf{C}^{-1}]\mathbf{n}_i} \quad ; \ \mathbf{n}_i \cdot \mathbf{C}^{-1}\mathbf{n}_j = \delta_{ij} \end{aligned} \tag{4.279}$$

We can see from (4.279) that the only difference between the spatial and the material descriptions concerns the principal directions, whereas the principal values or elastic principal stretches remain the same in either description. This result is a direct confirmation the invariance of the principal axes formulation. Furthermore, the same result implies that all of the following descriptions of the strain energy potential are equivalent and interchangeable:

$$\psi(\lambda_i^e) = \psi(\mathbf{B}^e) = \psi(\mathbf{C}, \mathbf{G}^p) \tag{4.280}$$

The formulation of large strain plasticity model of this kind will also require the invariant form of the plasticity criterion. In analogy with the previous result for strain energy potential, we can construct various forms of this plasticity criterion, which are all equivalent to the one specified in terms of the principal values of Kirchhoff stress tensor τ_i:

$$\phi(\tau_i) = \phi(\boldsymbol{\tau}) = \phi(\mathbf{S}, \mathbf{C}) \tag{4.281}$$

In (4.279) above, we introduced the tensor $\mathbf{G}^p$, equal to the inverse of the right Cauchy–Green plastic deformation tensor, which appears as the natural choice of internal variable for large strain plasticity in material description:

$$\boxed{\mathbf{G}^p \equiv (\mathbf{C}^p)^{-1} := (\mathbf{F}^{p\,T}\mathbf{F}^p)^{-1} = \mathbf{F}^{-1}\mathbf{B}^e\mathbf{F}^{-T}} \tag{4.282}$$

We will choose subsequently the material description for remaining developments of other model ingredients. Thus, we can write the second principle of thermodynamics in material description according to:

$$\begin{aligned} 0 \leq D &:= \mathbf{S} \cdot \tfrac{1}{2}\tfrac{\partial \mathbf{C}}{\partial t} - \tfrac{\partial}{\partial t}\hat{\psi}(\mathbf{C}, \mathbf{G}^p) \\ &= (\mathbf{S} - 2\tfrac{\partial \hat{\psi}}{\partial \mathbf{C}}) \cdot \tfrac{1}{2}\tfrac{\partial \mathbf{C}}{\partial t} - 2\tfrac{\partial \hat{\psi}}{\partial \mathbf{G}^p} \cdot \tfrac{1}{2}\tfrac{\partial \mathbf{G}^p}{\partial t} \end{aligned} \tag{4.283}$$

With respect to evolution of internal variable in (4.282) above, we can distinguish between two possibilities:

(i) *Elastic processus* which does not admit any change of the plastic deformation with $\frac{\partial \mathbf{G}^p}{\partial t} = \mathbf{0}$, leading to zero plastic dissipation. In this case, we can obtain from the second principle the constitutive equations for the second Piola–Kirchhoff stress, which can be written in the form of a hyperelastic potential with a frozen value of plastic deformation:

$$\mathcal{D} = 0 \implies \mathbf{S} = 2\frac{\partial \hat{\psi}}{\partial \mathbf{C}}\bigg|_{G^p} = 2\sum_{i=1}^{3} \frac{\partial \psi}{\partial \lambda_i^e} \frac{\partial \lambda_i^e}{\partial \mathbf{C}} \tag{4.284}$$

It is easy to show that the constitutive equations of this kind can further be simplified if placed within the principal axes framework. Namely, by computing the directional derivative of the eigenvalue problem in (4.279) with respect to right Cauchy–Green strain tensor $\mathbf{C}$, we can obtain:

$$\begin{aligned} \mathbf{0} = \{-2\lambda_i^e(\tfrac{\partial \lambda_i^e}{\partial \mathbf{C}} \cdot d\mathbf{C})\mathbf{C}^{-1}\mathbf{n}_i - (\lambda_i^e)^2 \tfrac{\partial \mathbf{C}^{-1}}{\partial \mathbf{C}} \mathbf{n}_i \\ + \left[\mathbf{G}^p - (\lambda_i^e)^2 \mathbf{C}^{-1}\right] d\mathbf{n}_i\} \cdot \mathbf{n}_i \\ \Rightarrow \tfrac{\partial \lambda_i^e}{\partial \mathbf{C}} = \tfrac{\lambda_i^e}{2} \mathbf{C}^{-1}\mathbf{n}_i \otimes \mathbf{C}^{-1}\mathbf{n}_i \end{aligned} \tag{4.285}$$

where we have used the following auxiliary result:

$$\mathbf{I} = \mathbf{C}^{-1}\mathbf{C} \Rightarrow \frac{\partial \mathbf{C}^{-1}}{\partial \mathbf{C}} \cdot d\mathbf{C} = \mathbf{C}^{-1} d\mathbf{C}\mathbf{C}^{-1} \tag{4.286}$$

Furthermore, the isotropy of the elastic response implies that the second Piola–Kirchhoff stress tensor shares the same eigenvectors with this strain tensor; we can thus write the corresponding eigenvalue problem in material description:

$$(\mathbf{S} - \tau_i \mathbf{C}^{-1})\mathbf{n}_i = \mathbf{0} \Rightarrow \mathbf{S} = \sum_{i=1}^{3} \tau_i \mathbf{C}^{-1}\mathbf{n}_i \otimes \mathbf{C}^{-1}\mathbf{n}_i \; ; \; \tau_i = \frac{\partial \psi}{\partial \lambda_i^e} \lambda_i^e \tag{4.287}$$

The last result that follows from (4.284) and (4.285), reduces all the remaining stress computations to scalar fields. The latter is the most important advantage of the principal axes formulation for large strain plasticity.

(ii) *Plastic processus* that introduces further evolution of the plastic deformation, is triggered when the plastic criterion is satisfied with zero value of yield function (recall that a positive value of yield function is not admissible, whereas negative values correspond the elastic processus). The corresponding value of the plastic dissipation can then be defined from the second principle of thermodynamics, by assuming that the constitutive equations in (4.287) above remain valid for plastic processus; we can thus write:

$$0 < D^p := -\mathbf{C}\mathbf{S}\mathbf{G}^{p\,-1} \cdot \frac{1}{2} \frac{\partial \mathbf{G}^p}{\partial t} \equiv -\mathbf{S} \cdot \frac{1}{2} \mathbf{C} \frac{\partial \mathbf{G}^p}{\partial t} \mathbf{G}^{p\,-1} \tag{4.288}$$

In order to write such a compact form of the plastic dissipation, we exploited the auxiliary result regarding the directional derivative of (4.279) with respect to $\mathbf{G}^p$:

$$\begin{aligned} \mathbf{0} = \{d\mathbf{G}^p\mathbf{n}_i - 2\lambda_i^e(\tfrac{\partial \lambda_i^e}{\partial \mathbf{G}^p} \cdot d\mathbf{G}^p)\mathbf{C}^{-1}\mathbf{n}_i + [\mathbf{G}^p - (\lambda_i^e)^2\mathbf{C}^{-1}]d\mathbf{n}_i\} \cdot \mathbf{n}_i \\ \implies \tfrac{\partial \lambda_i^e}{\partial \mathbf{G}^p} = \tfrac{1}{2\lambda_i^e} \mathbf{n}_i \otimes \mathbf{n}_i \end{aligned} \tag{4.289}$$

along with the following key result (see [113]) that allows us to establish the connection between two directional derivatives in (4.285) and (4.289):

$$\frac{\partial \lambda_i^e}{\partial \mathbf{C}} = \mathbf{C}^{-1} \frac{\partial \lambda_i^e}{\partial \mathbf{G}^p} \mathbf{G}^p \Longrightarrow \underbrace{\frac{2\partial \psi}{\partial \mathbf{C}}}_{\mathbf{S}} = \mathbf{C}^{-1} \frac{2\partial \psi}{\partial \mathbf{G}^p} \mathbf{G}^p \tag{4.290}$$

The remaining model ingredients can be obtained by appealing to the principle of maximum plastic dissipation in (4.288) above. Namely, among all stress fields which can be considered admissible with respect to the chosen plasticity criterion, we will choose the stress that will maximize the plastic dissipation; the principle of maximum plastic dissipation can be presented in terms of a constrained minimization problem, which can be recast as unconstrained minimization by using the Lagrange multiplier method:

$$L^p(\mathbf{S}, \frac{\partial \gamma}{\partial t})\bigg|_{\mathbf{C}} = -D^p(\mathbf{S}, \bar{\mathbf{C}}) + \frac{\partial \gamma}{\partial t} \phi(\mathbf{S}, \bar{\mathbf{C}})) \longrightarrow min \tag{4.291}$$

The corresponding Kuhn–Tucker optimality conditions for this kind of minimization problem will produce the evolution equation for plastic deformation:

$$0 = \frac{\partial L^p}{\partial \mathbf{S}} := \frac{1}{2} \mathbf{C} \dot{\mathbf{G}}^p \mathbf{G}^{p\,-1} + \dot{\gamma} \frac{\partial \hat{\phi}}{\partial \mathbf{S}} \Longrightarrow \dot{\mathbf{G}}^p = -2\dot{\gamma} \mathbf{C}^{-1} \frac{\partial \hat{\phi}}{\partial \mathbf{S}} \mathbf{G}^p \tag{4.292}$$

accompanied by the loading/unloading conditions:

$$\phi(\mathbf{S}, \bar{\mathbf{C}}) \leq 0 \; ; \; \dot{\gamma} \geq 0 \; ; \; \dot{\gamma}\, \phi = 0 \tag{4.293}$$

Remark on viscoplasticity: We can easily generalize the present development to account for eventual response sensitivity with respect to the rate of deformation, and obtain the viscoplasticity model placed within the chosen principal axes framework. For viscoplasticity model, we will not prohibit any positive value of the yield function, but rather consider that precisely those values will trigger the further evolution of the viscoplastic deformation. The viscoplasticity model of this kind can be obtained from the penalty formulation of the constrained minimization problem resulting from the principle of maximum plastic dissipation:

$$\begin{aligned} & L^{vp}(\mathbf{S}) = -D^{vp}(\mathbf{S}) + P(\hat{\phi}(\mathbf{S}, \bar{\mathbf{C}})) \longrightarrow min \\ & 0 < D^{vp} := -\mathbf{C}\mathbf{S}\mathbf{G}^{vp\,-1} \cdot \tfrac{1}{2} \dot{\mathbf{G}}^{vp} \equiv -\mathbf{S} \cdot \tfrac{1}{2} \mathbf{C} \dot{\mathbf{G}}^{vp} \mathbf{G}^{vp\,-1} \end{aligned} \tag{4.294}$$

The plastic admissibility constraint on stress field is relaxed, and replaced by the chosen penalty functional $P(\cdot)$ of Perzyna [222]:

$$P(\hat{\phi}(\cdot)) = \begin{cases} \frac{1}{2\eta} [\hat{\phi}(\cdot)]^2 \; ; \hat{\phi} > 0 \\ \quad 0 \qquad\quad ; \hat{\phi} \leq 0 \end{cases} \tag{4.295}$$

where η is the viscosity coefficient. The gradient of this penalty function can be written:

$$P'(\hat{\phi}(\cdot)) = \frac{1}{\eta} < \hat{\phi}(\cdot) >:= \begin{cases} \frac{1}{\eta}\hat{\phi}(\cdot) \; ; \hat{\phi} > 0 \\ 0 \quad ; \hat{\phi} \leq 0 \end{cases} \tag{4.296}$$

We can further impose the corresponding optimality conditions, which will provide the evolution equation for viscoplastic deformation:

$$\begin{aligned} 0 = \frac{\partial L^{vp}}{\partial \mathbf{S}} &:= \frac{1}{2}\mathbf{C}\dot{\mathbf{G}}^{vp}\mathbf{G}^{vp\,-1} + \frac{1}{\eta} < \hat{\phi} > \frac{\partial \hat{\phi}}{\partial \mathbf{S}} \\ \Rightarrow \dot{\mathbf{G}}^{vp} &= -2\dot{\gamma}\mathbf{C}^{-1}\frac{\partial \hat{\phi}}{\partial \mathbf{S}}\mathbf{G}^{vp} \; ; \; \dot{\gamma} = \frac{<\hat{\phi}>}{\eta} \end{aligned} \tag{4.297}$$

For the case of vanishing viscosity coefficient $\eta \mapsto 0$, the proposed viscoplasticity model will reduce to the perfect plasticity model studied previously with the corresponding yield criterion $\hat{\phi} = 0$, which is imposed by a finite value of $\dot{\gamma}$ in (4.297). Moreover, for any value of $\dot{\gamma}$, the evolution equation for viscoplastic deformation can formally be written as the evolution equation of plastic deformation in (4.292); for that reason, the computational procedure to be presented in the remainder of this section for plasticity model is directly applicable to viscoplasticity.
The evolution equation for the plastic deformation can further be simplified if written in the principal axes framework. This can be done by exploiting the auxiliary result on computing the directional derivative of eigenvalue problem in (4.287) with respect to the second Piola–Kirchhoff stress:

$$\mathbf{0} = d\mathbf{S}\mathbf{n}_i - (\frac{\partial \tau_i}{\partial \mathbf{S}} \cdot d\mathbf{S})\mathbf{C}^{-1}\mathbf{n}_i + [\mathbf{S} - \tau_i\mathbf{C}^{-1}]d\mathbf{n}_i \mid \cdot \mathbf{n}_i \implies \frac{\partial \tau_i}{\partial \mathbf{S}} = \mathbf{n}_i \otimes \mathbf{n}_i \tag{4.298}$$

For the invariant form of the plasticity criterion in (4.281) expressed in terms of principal values of the stress tensor τ_i, we can further write:

$$\boxed{\dot{\mathbf{G}}^p = -2\dot{\gamma}[\textstyle\sum_{i=1}^{3} \frac{\partial \phi}{\partial \tau_i}\mathbf{C}^{-1}(\mathbf{n}_i \otimes \mathbf{n}_i)\mathbf{G}^p]} \tag{4.299}$$

We can obtain the exact analytic solution of such an evolution equation by making use of the exponential mapping formula (e.g. see Hirsch and Smale [107]); for example, direct application of the exponential mapping for computing evolution of a tensor $\mathbf{X}$, whose is governed by tensor $\mathbf{A}$, will result with:

$$\dot{\mathbf{X}}(t) = \mathbf{A}\mathbf{X}(t) \; ; \; \mathbf{X}(0) = \mathbf{I} \Rightarrow \mathbf{X}(t) = exp[\mathbf{A}t] := \sum_{n=0}^{\infty} [\mathbf{I} + \frac{1}{n}(t\mathbf{A})]^n \tag{4.300}$$

We can obtain the closed form solution for the exponential mapping computation of the evolution of plastic deformation tensor in (4.299), by taking into account the eigenvectors orthogonality with respect to **C**, which can be written:

$$\begin{aligned} \mathbf{G}^p(t) &= \sum_{i=1}^{3} exp[-2\dot{\gamma}t\tfrac{\partial\phi}{\partial\tau_i}\mathbf{C}^{-1}(\mathbf{n}_i\otimes\mathbf{n}_i)] \\ &= \sum_{i=1}^{3} exp[-2\dot{\gamma}t\tfrac{\partial\phi}{\partial\tau_i}]\mathbf{C}^{-1}(\mathbf{n}_i\otimes\mathbf{n}_i) \end{aligned} \tag{4.301}$$

It follows from the last result that large plastic deformation remains incompressible, for any pressure insensitive yield criterion imposing that:

$$0 = \frac{\partial\phi}{\partial p} := 3\sum_{i=1}^{3}\frac{\partial\phi}{\partial\tau_i}\;;\; p = (\tau_1+\tau_2+\tau_3)/3 \tag{4.302}$$

Namely, by making use of the following identity (e.g. see Gurtin [96]):

$$det(exp[\mathbf{A}t]) = exp(tr[\mathbf{A}t]) \tag{4.303}$$

we can easily show that the determinant of the plastic deformation tensor in (4.301) above remains equal to one:

$$\begin{aligned} det(\mathbf{G}^p(t)) &= exp\{tr[-2\tfrac{<\phi>}{\eta}t\sum_{i=1}^{3}(\tfrac{\partial\phi}{\partial\tau_i}\mathbf{C}^{-1}\mathbf{n}_i\otimes\mathbf{n}_i)]\} \\ &= exp\{-2\tfrac{<\phi>}{\eta}t\sum_{i=1}^{3}\underbrace{\frac{\partial\phi}{\partial\tau_i}}_{=0}\} \\ &= 1 \end{aligned} \tag{4.304}$$

We can finally conclude that there cannot be any change in volume brought by the plastic deformation, since the last result implies that:

$$\left.\begin{aligned} det\mathbf{G}^p(t) &= 1 \\ det\mathbf{G}^p &= (det\mathbf{F}^p)^2 \end{aligned}\right\} \Rightarrow \boxed{det\mathbf{F}^p(t) = 1} \tag{4.305}$$

Any material obeying such a plasticity criterion will remain plastically incompressible. One example of the material with plastic incompressibility is defined by the von Mises plasticity criterion, where $\phi(\hat{\tau}_i) = \| \sum_{i=1}^{3}\hat{\tau}_i^2 \| - \sqrt{2/3}\sigma_y$, $\hat{\tau}_i = \tau_i - \frac{1}{3}\sum_{k=1}^{3}\tau_k)$. For the ultimate load computations, where the plastic deformation is much larger than the elastic deformation, the constitutive behavior of this kind of material can be considered as quasi-incompressible. As shown in the previous chapter, the quasi-incompressible material would require mixed finite element approximation to ensure the optimal solution accuracy.

4.8.3 Isotropic and kinematic hardening in large deformation plasticity

For increasing the predictive capabilities of the plasticity model to account for hardening effects, we ought to include additional internal variables; the first one is a scalar variable ζ representing the isotropic hardening, and the second is the tensor variable $\boldsymbol{\Xi}$ representing the kinematic hardening. The Helmholtz free energy for large deformation plasticity model with hardening can then be written:

$$\boxed{\tilde{\psi}(\mathbf{C}, \mathbf{G}^p, \boldsymbol{\Xi}, \zeta)} \tag{4.306}$$

where $\boldsymbol{\Xi}$ is the material description of the kinematic hardening tensor variable. We will limit our development to the kinematic hardening model with a generalization of the Prager-Ziegler kinematic hardening model to large deformation plasticity. We thus consider that $\boldsymbol{\Xi}$ is a strain tensor (that we refer to as back strain, in analogy to back stress), which is proportional to the plastic deformation:[17]

$$\boldsymbol{\Xi} = c\,\mathbf{G}^p \tag{4.307}$$

where 'c' is a given constant. The last result directly implies that both strain tensors $\boldsymbol{\Xi}$ and $\mathbf{G}^p$ share the same principal vectors. We can thus define the corresponding eigenvalue problem:

$$[\boldsymbol{\Xi} - (z_i)^2 \mathbf{C}^{-1}]\mathbf{n}_i = \mathbf{0} \tag{4.308}$$

where z_i are the principal values of the back strain tensor. All the subsequent developments can easily be placed within the proposed principal axes framework. First, we can provide the corresponding invariant form of the free energy in terms of principal values of strain tensors and isotropic hardening variable:

$$\psi(\lambda_i^e, z_i, \zeta) \tag{4.309}$$

We can further compute the back stress tensor $\boldsymbol{\Upsilon}$ sharing the same principal vectors as those in (4.308), which allows us to set the corresponding eigenvalue problem:

$$(\boldsymbol{\Upsilon} - \upsilon_i \mathbf{C}^{-1})\mathbf{n}_i = \mathbf{0} \tag{4.310}$$

where υ_i are the principal values of the back stress. The last result would allow us to recast the plasticity criterions for large deformation plasticity model with hardening in an invariant form written in terms of the principal values of stress tensors and stress-like hardening variable for isotropic hardening:

$$\tilde{\phi}(\mathbf{C}, \mathbf{S}, \boldsymbol{\Upsilon}, q) = \phi(\tau_i, \upsilon_i, q) \tag{4.311}$$

[17] For developments with a more general nonlinear kinematic hardening for large strain plasticity, we refer to [226].

Having defined the main ingredients of such a plasticity model with hardening, all the remaining ingredients are provided in the same manner as already described for perfect plasticity model. First, we define the constitutive equations for stress and the plastic dissipation by appealing to the second principle of thermodynamics:

$$
\begin{aligned}
0 \leq D &:= \mathbf{S} \cdot \tfrac{1}{2}\tfrac{\partial \mathbf{C}}{\partial t} - \tfrac{\partial}{\partial t}\tilde{\psi}(\mathbf{C}, \mathbf{G}^p, \boldsymbol{\Xi}, \zeta) \\
&= (\mathbf{S} - 2\tfrac{\partial \tilde{\psi}}{\partial \mathbf{C}}) \cdot \tfrac{1}{2}\tfrac{\partial \mathbf{C}}{\partial t} - 2\tfrac{\partial \tilde{\psi}}{\partial \mathbf{G}^p} \cdot \tfrac{1}{2}\tfrac{\partial \mathbf{G}^p}{\partial t} - 2\tfrac{\partial \tilde{\psi}}{\partial \boldsymbol{\Xi}} \cdot \tfrac{1}{2}\tfrac{\partial \boldsymbol{\Xi}}{\partial t} - \tfrac{\partial \tilde{\psi}}{\partial \zeta} \cdot \tfrac{\partial \zeta}{\partial t}
\end{aligned}
\tag{4.312}
$$

We can deal with two different cases:
(i) Elastic processus which implies that all the internal variables remain fixed and the plastic dissipation reduces to zero; the last result then provides the corresponding set of constitutive equations for computing the stress tensor and stress-like hardening variables:

$$
\frac{\partial \mathbf{G}^p}{\partial t} = \mathbf{0}; \; (\frac{\partial \mathbf{F}^p}{\partial t} = \mathbf{0}); \; \frac{\partial \boldsymbol{\Xi}}{\partial t} = \mathbf{0}; \; \frac{\partial \zeta}{\partial t} = 0 \; \Rightarrow \; \mathbf{S} = 2\frac{\partial \tilde{\psi}}{\partial \mathbf{C}}; \; \boldsymbol{\Upsilon} = 2\frac{\partial \tilde{\psi}}{\partial \boldsymbol{\Xi}}; \; q = -\frac{\partial \tilde{\psi}}{\partial \zeta}
\tag{4.313}
$$

The main novelty with respect to perfect plasticity is in that the same free energy potential also provides the constitutive equations for the back stress tensor, which can be written in principal axes space as:

$$
\boxed{
\begin{aligned}
\boldsymbol{\Upsilon} &= \textstyle\sum_{i=1}^{3} 2\frac{\partial \psi}{\partial z_i}\frac{\partial z_i}{\partial \mathbf{C}} \\
&= \textstyle\sum_{i=1}^{3} \frac{\partial \psi}{\partial z_i} z_i \mathbf{C}^{-1}\mathbf{n}_i \otimes \mathbf{C}^{-1}\mathbf{n}_i \quad ; \quad \upsilon_i = \frac{\partial \psi}{\partial z_i} z_i
\end{aligned}
}
\tag{4.314}
$$

Moreover, by the analogy with the key result in (4.290), we can conclude that:

$$
\frac{\partial z_i}{\partial \mathbf{C}} = \mathbf{C}^{-1}\frac{\partial z_i}{\partial \boldsymbol{\Xi}}\boldsymbol{\Xi}
\tag{4.315}
$$

(ii) Plastic processus which produces the evolution of internal variables and the corresponding plastic dissipation defined with:

$$
0 < D^p := -\mathbf{S} \cdot \frac{1}{2}\mathbf{C}\frac{\partial \mathbf{G}^p}{\partial t}\mathbf{G}^{p-1} - \boldsymbol{\Upsilon} \cdot \frac{1}{2}\mathbf{C}\frac{\partial \boldsymbol{\Xi}}{\partial t}\boldsymbol{\Xi}^{-1} + q\frac{\partial \zeta}{\partial t}
\tag{4.316}
$$

We note that the internal variable evolution is produced only by stress and back stress which verify the plasticity criterion in (4.311). By appealing to the principle of maximum plastic dissipation, we can find the corresponding form of evolution equations as the solution to the following constrained minimization principle:

$$\min_{\forall(S,\Upsilon,q)} \max_{\dot{\gamma}} L^p \Big|_{\mathbf{C}} \; ; \; L^p(\mathbf{S}, \boldsymbol{\Upsilon}, q, \dot{\gamma}) = -D^p(\mathbf{S}, \boldsymbol{\Upsilon}, q) + \dot{\gamma}\phi(\mathbf{S}, \boldsymbol{\Upsilon}, q) \Rightarrow$$

$$0 = \frac{\partial L^p}{\partial \mathbf{S}} := \tfrac{1}{2}\mathbf{C}\dot{\mathbf{G}}^p \mathbf{G}^{p\,-1} + \dot{\gamma}\frac{\partial \tilde{\phi}}{\partial \mathbf{S}} \Rightarrow \boxed{\dot{\mathbf{G}}^p = -2\dot{\gamma}\mathbf{C}^{-1}\frac{\partial \tilde{\phi}}{\partial \mathbf{S}}\mathbf{G}^p}$$

$$0 = \frac{\partial L^p}{\partial \boldsymbol{\Upsilon}} := \tfrac{1}{2}\mathbf{C}\dot{\boldsymbol{\Xi}}\boldsymbol{\Xi}^{-1} + \dot{\gamma}\frac{\partial \tilde{\phi}}{\partial \boldsymbol{\Upsilon}} \Rightarrow \boxed{\dot{\boldsymbol{\Xi}} = -2\dot{\gamma}\mathbf{C}^{-1}\frac{\partial \tilde{\phi}}{\partial \boldsymbol{\Upsilon}}\boldsymbol{\Xi}}$$

$$0 = \frac{\partial L^p}{\partial q} := -\zeta + \dot{\gamma}\frac{\partial \tilde{\phi}}{\partial q} \Rightarrow \boxed{\dot{\zeta} = \dot{\gamma}\frac{\partial \tilde{\phi}}{\partial q}} \tag{4.317}$$

The difference between the elastic and plastic processes can again be indicated in terms of the loading–unloading conditions, which can be written:

$$\frac{\partial \gamma}{\partial t} \geq 0 \, ; \phi \leq 0 \, ; \, \frac{\partial \gamma}{\partial t}\phi = 0 \tag{4.318}$$

By making use of the invariant form of the yield criterion in terms of principal values of stress, along with the directional derivative with respect to stress tensor in (4.298), we can write the principal axes representation of the back strain evolution equation:

$$\dot{\boldsymbol{\Xi}} = -2\dot{\gamma}\Big(\sum_{i=1}^{3} \frac{\partial \phi}{\partial v_i}\mathbf{C}^{-1}\mathbf{n}_i \otimes \mathbf{n}_i\Big)\boldsymbol{\Xi} \tag{4.319}$$

By exploiting again the exponential mapping and the orthogonality of eigenvectors with respect to $\mathbf{C}$, we can obtain the closed form solution for the last equation and specify the corresponding evolution of the back strain tensor according to:

$$\boldsymbol{\Xi}(t) = exp\Big[\sum_{i=1}^{3} -2\frac{<\phi>}{\eta}t\frac{\partial \phi}{\partial v_i}\Big]\mathbf{C}^{-1}(\mathbf{n}_i \otimes \mathbf{n}_i)) \tag{4.320}$$

With this result in hand, it is easy to verify that the back strain tensor must remain deviatoric for any yield criterion with $\sum_i \partial\phi/\partial v_i = 0$

$$\begin{aligned} det\boldsymbol{\Xi}(t) &= exp\{tr[-2\tfrac{<\phi>}{\eta}t\textstyle\sum_{i=1}^{3}(\frac{\partial \phi}{\partial v_i}\mathbf{C}^{-1}\mathbf{n}_i \otimes \mathbf{n}_i)]\} \\ &= exp\{-2\tfrac{<\phi>}{\eta}\underbrace{\sum_{i=1}^{3}\frac{\partial \phi}{\partial v_i}}_{=0}\} \\ &= 1 \end{aligned} \tag{4.321}$$

One example of this kind is the von Mises plasticity criterion with Prager–Ziegler kinematic hardening and isotropic hardening, where the yield function can be written:

$$\phi(\tau_i, \upsilon_i, q) := [\sum_{i=1}^{3}(\tilde{\tau}_i + \upsilon_i)^2]^{1/2} - \sqrt{\frac{2}{3}}(\sigma_y - q) \; ; \; \tilde{\tau}_i = \tau_i - \frac{1}{3}\sum_{k=1}^{3}\tau_k \quad (4.322)$$

4.8.4 Spatial description of large deformation plasticity

In this section we will recast all the governing equations of the proposed large deformation plasticity model from material to spatial description. The key role in this development is played by the eigenvalue problem in (4.279), which can be used to show that the Kirchhoff stress tensor $\boldsymbol{\tau}$ (used in spatial description) shares the same eigenvalues τ_i with the second Piola–Kirchhoff stress tensor $\mathbf{S} = \mathbf{F}^{-1}\boldsymbol{\tau}\mathbf{F}^{-T}$ (used in material description). Only the principal vectors of these two stress tensors will be different, with the corresponding results for spatial and material descriptions. With respect to result in (4.279), we can establish the relationship between these two sets of eigenvectors according to $\mathbf{m}_i = \mathbf{F}^{-T}\mathbf{n}_i$; the latter allows us to write:

$$\begin{aligned} \mathbf{0} &= \mathbf{F}(\mathbf{S} - \tau_i \mathbf{C}^{-1})\mathbf{n}_i \\ &= [\underbrace{\mathbf{F}\mathbf{S}\mathbf{F}^T}_{\boldsymbol{\tau}} - \tau_i \mathbf{F}(\mathbf{F}^T\mathbf{F})^{-1}\mathbf{F}^T]\underbrace{\mathbf{F}^{-T}\mathbf{n}_i}_{\mathbf{m}_i} \\ &= [\boldsymbol{\tau} - \tau_i \mathbf{I}^{\varphi}]\mathbf{m}_i \end{aligned} \quad (4.323)$$

We can also provide the Kirchhoff stress tensor representation in terms of its spectral decomposition, by using the principal values τ_i. These values are computed as the corresponding partial derivatives of the invariant form of strain energy potential expressed as a function of the elastic principal stretches λ_i^e; we thus obtain:

$$\begin{aligned} \boldsymbol{\tau} := \mathbf{F}\mathbf{S}\mathbf{F}^T &= \textstyle\sum_{i=1}^{3} \frac{\partial \psi}{\partial \lambda_i^e}\lambda_i^e \mathbf{F}(\mathbf{F}^T\mathbf{F})^{-1}\mathbf{n}_i \otimes \mathbf{F}(\mathbf{F}^T\mathbf{F})^{-1}\mathbf{n}_i \\ &= \textstyle\sum_{i=1}^{3} \frac{\partial \psi}{\partial \lambda_i^e}\lambda_i^e \mathbf{m}_i \otimes \mathbf{m}_i \\ &= \textstyle\sum_{i=1}^{3} \tau_i \mathbf{m}_i \otimes \mathbf{m}_i \; ; \; \Rightarrow \; \tau_i = \frac{\partial \psi}{\partial \lambda_i^e}\lambda_i^e \end{aligned} \quad (4.324)$$

We note in passing that the Kirchhoff stress is a very good replacement of the true or Cauchy stress for von Mises plasticity model, if we remain limited to small elastic deformations. Namely, with $J^p = 1$ and $J^e \approx 1$, these two stresses take similar values $\sigma_{ij} \approx \tau_{ij}$, and they can be used interchangeably.

We can further extend this spatial description to the large strain plasticity model with Prager–Ziegler kinematic hardening. In particular, we can thus obtain the spatial representation of the back strain tensor $\mathbf{Z} = \mathbf{F}\boldsymbol{\Xi}\mathbf{F}^T$ that shares the same eigenvalues z_i, which can be shown from:

$$\begin{aligned} \mathbf{0} &= \mathbf{F}(\boldsymbol{\Xi} - z_i^2\mathbf{C}^{-1})\mathbf{n}_i \\ &= [\underbrace{\mathbf{F}\boldsymbol{\Xi}\mathbf{F}^T}_{\mathbf{Z}} - z_i^2\mathbf{F}(\mathbf{F}^T\mathbf{F})^{-1}\mathbf{F}^T]\underbrace{\mathbf{F}^{-T}\mathbf{n}_i}_{\mathbf{m}_i} \\ &= [\mathbf{Z} - z_i^2\mathbf{I}^{\varphi}]\mathbf{m}_i \end{aligned} \tag{4.325}$$

The same kind of spatial description can be provided for the back stress tensor, $\boldsymbol{\upsilon} = \mathbf{F}\boldsymbol{\Upsilon}\mathbf{F}^T$. Namely, the most convenient representation is constructed by the spectral decomposition, employing the principal values υ_i that are computed from the strain energy potential in terms of z_i:

$$\begin{aligned} \boldsymbol{\upsilon} := \mathbf{F}\boldsymbol{\Upsilon}\mathbf{F}^T &= \textstyle\sum_{i=1}^3 \frac{\partial\psi}{\partial z_i} z_i \mathbf{F}(\mathbf{F}^T\mathbf{F})^{-1}\mathbf{n}_i \otimes \mathbf{F}(\mathbf{F}^T\mathbf{F})^{-1}\mathbf{n}_i \\ &= \textstyle\sum_{i=1}^3 \upsilon_i \mathbf{m}_i \otimes \mathbf{m}_i \; ; \; \upsilon_i = \frac{\partial\psi}{\partial z_i} z_i \end{aligned} \tag{4.326}$$

Finally, we can also recast the internal variable evolution equation in spatial description, by making use of the result in (4.282); we find in this manner that the most convenient choice of the internal variable in spatial representation is the left Cauchy–Green elastic deformation tensor:

$$\begin{aligned} \tfrac{\partial}{\partial t}\mathbf{B}^e &= \tfrac{\partial}{\partial t}[\mathbf{F}\mathbf{G}^p\mathbf{F}^T] \\ &= \tfrac{\partial \mathbf{F}}{\partial t}\mathbf{F}^{-1}\mathbf{F}\mathbf{G}^p\mathbf{F}^T + \mathbf{F}\mathbf{G}^p\mathbf{F}^T\mathbf{F}^{-T}\tfrac{\partial \mathbf{F}^T}{\partial t} + \mathbf{F}\dot{\mathbf{G}}^p\mathbf{F}^T \\ &= \mathbf{L}\mathbf{B}^e + \mathbf{B}^e\mathbf{L}^T + \mathbf{F}\tfrac{\partial \mathbf{G}^p}{\partial t}\mathbf{F}^T \end{aligned} \tag{4.327}$$

This evolution equation can also be expressed in a more compact form by appealing to the Lie derivative formalism and making use of the principal axis representation of evolution equation in (4.299), which leads to:

$$\begin{aligned} L_v[\mathbf{B}^e] := \tfrac{\partial \mathbf{B}^e}{\partial t} - \mathbf{L}\mathbf{B}^e - \mathbf{B}^e\mathbf{L}^T &= \mathbf{F}\tfrac{\partial \mathbf{G}^p}{\partial t}\mathbf{F}^T \\ &= -2\dot{\gamma}\textstyle\sum_{i=1}^3 \frac{\partial\phi}{\partial\tau_i}\mathbf{F}\mathbf{C}^{-1}\mathbf{n}_i \otimes \mathbf{n}_i\mathbf{G}^p\mathbf{F}^T \\ &= -2\dot{\gamma}\textstyle\sum_{i=1}^3 \frac{\partial\phi}{\partial\tau_i}\underbrace{\mathbf{F}(\mathbf{F}^T\mathbf{F})^{-1}\mathbf{F}^T}_{\mathbf{I}}\,\mathbf{m}_i \otimes \mathbf{m}_i\underbrace{\mathbf{F}\mathbf{G}^p\mathbf{F}^T}_{\mathbf{B}^e} \\ &= -2\dot{\gamma}(\textstyle\sum_{i=1}^3 \frac{\partial\phi}{\partial\tau_i}\mathbf{m}_i \otimes \mathbf{m}_i)\mathbf{B}^e \end{aligned} \tag{4.328}$$

In the same manner we can obtain the spatial description of the evolution equation for the back strain, which can also be written by using the Lie derivative formalism according to:

$$L_v[\mathbf{Z}] := \frac{\partial \mathbf{Z}}{\partial t} - \mathbf{L}\mathbf{Z} - \mathbf{Z}\mathbf{L}^T = -2\frac{<\phi>}{\eta}(\sum_{i=1}^3 \frac{\partial\phi}{\partial \upsilon_i}\mathbf{m}_i \otimes \mathbf{m}_i)\mathbf{Z} \tag{4.329}$$

4.8.5 Numerical implementation of large deformation plasticity

In this section we present the computational procedure providing the numerical values of the state variable for a large deformation plasticity model with isotropic and kinematic hardening. In spatial description, this reduces to computing the left Cauchy–Green elastic deformation tensor $\mathbf{B}^e$, the back strain tensor $\mathbf{Z}$ for kinematic hardening, the scalar variable ζ for isotropic hardening and the corresponding value of position vector $\boldsymbol{\varphi}$; the latter is in fact needed for the computation of the corresponding value of deformation gradient:

$$\mathbf{F} = [\frac{\partial \boldsymbol{\varphi}}{\partial x_1}; \frac{\partial \boldsymbol{\varphi}}{\partial x_2}; \frac{\partial \boldsymbol{\varphi}}{\partial x_3}] \tag{4.330}$$

Theoretically, the state variable values ought to be computed at each point (or for each particle); practically however, the finite element method will reduce this requirement significantly. For example, the position vector is obtained by using the finite element interpolation of the displacement field from nodal values for each element:

$$\boldsymbol{\varphi}(\boldsymbol{\xi}_l, t)\bigg|_{\Omega^e} = \sum_{a=1}^{n_{en}} N_a^e(\boldsymbol{\xi}_l)(\mathbf{x}_a + \mathbf{d}_a^e(t)) \tag{4.331}$$

where $\boldsymbol{\xi}_l$; $l = 1, 2, \ldots, n_{in}$ are the chosen points of numerical integration (e.g. Gauss quadrature). Moreover, the finite element method will also allow us to reduce the internal variable computations only to integration points. Therefore, the numerical integration leads to very significant computational savings. The same applies to the element internal force vector and its tangent stiffness matrix computations, requiring the stress and tangent modulus values only at quadrature points.

Similar computational saving will characterize the time evolution computations, by providing the values only at the chosen instants within the time interval of interest. Typically, the state variable evolution is computed by using a single step time-integration scheme, which allows us to further reduce the complexity of any such evolution problem and to construct the solution over one step at the time; for a typical step, we can thus write:

Central problem of large deformation plasticity ($\forall \boldsymbol{\xi}_l$):

Given: $\boldsymbol{\varphi}_n = \boldsymbol{\varphi}(\boldsymbol{\xi}_l, t_n)$, $\mathbf{B}_n^e = \mathbf{B}^e(\boldsymbol{\xi}_l, t_n)$, $\mathbf{Z}_n = \mathbf{Z}(\boldsymbol{\xi}_l, t_n)$, $\zeta_n = \zeta(\boldsymbol{\xi}_l, t_n)$,
$h = t_{n+1} - t_n > 0$

Find: $\boldsymbol{\varphi}_{n+1}$, $\mathbf{B}_{n+1}^e$, $\mathbf{Z}_{n+1}$, ζ_{n+1}

also find: $\boldsymbol{\tau}_{n+1}$, $\boldsymbol{v}_{n+1}$, q_{n+1}

such that the equilibrium equations are satisfied:

$$0 = G(\varphi_{n+1}, \mathbf{B}^e_{n+1}, \mathbf{Z}_{n+1}, \zeta_{n+1}, \varphi) := \underbrace{\int_{\Omega} \boldsymbol{\tau}_{n+1} \cdot \frac{1}{2}(\nabla^{\varphi}\mathbf{w}^{\varphi} + \nabla^{\varphi}\mathbf{w}^{\varphi,T})\, dV}_{\sum_{e=1}^{n_{elem}} \sum_{l=1}^{n_{in}} \boldsymbol{\tau}(\xi_l, t_{n+1}) \cdot \frac{1}{2}(\nabla^{\varphi}\mathbf{w}^{\varphi} + \nabla^{\varphi}\mathbf{w}^{\varphi,T})|_{\xi_l}\, j(\xi_l) w_l} - G_{ext_{n+1}} \tag{4.332}$$

and the stress values at all Gauss quadrature points remain plastically admissible:

$$[\phi(\boldsymbol{\tau}_{n+1}, \boldsymbol{v}_{n+1}, q_{n+1}) \leq 0 \ ; \ \dot{\gamma} \geq 0 \ ; \ \dot{\gamma}\, \phi(\boldsymbol{\tau}_{n+1}, \boldsymbol{v}_{n+1}, q_{n+1}) = 0]_{\xi_l} \tag{4.333}$$

The operator split solution method applied to the central problem in (4.332) above, plays a very important role in large deformation regime for further reducing the problem complexity. Namely, in the first, global phase of computations, we will deal only with geometric nonlinearities in providing the new deformed configuration and carrying out the convective transport of internal variables computed at the previous time step. In the second, local phase of operator split computation, we will keep this deformed configuration fixed and seek the final values of the internal variables which guarantee the plastic admissibility of stress. Therefore, geometric nonlinearities are limited to global phase of such an operator split procedure, whereas the local phase deals with material nonlinearities only. Such an approach will allow us to exploit the internal variable computational procedures that are developed for small strain problems within the large strain regime.

A more detailed account of the operator split computational procedure for this large strain plasticity model is given next. Let us suppose we are given the best iterative value[18] $\varphi^{(i)}_{n+1}$ for the solution of equilibrium equations in (4.332). We can then readily complete the global phase of computations by performing the convective transport of internal variables. The latter is carried out under hypothesis that the current step remains elastic with $\dot{\gamma}^{trial} = 0$, which leads to:

$$\begin{aligned} &\frac{\partial \mathbf{B}^e}{\partial t} - \mathbf{L}\mathbf{B}^e - \mathbf{B}^e\mathbf{L}^T = \mathbf{0} \ ; \ \mathbf{B}^e(0) = \mathbf{B}^e_n \\ &\frac{\partial \mathbf{Z}}{\partial t} - \mathbf{L}\mathbf{Z} - \mathbf{Z}\mathbf{L}^T = 0 \ ; \ \mathbf{Z}(0) = \mathbf{Z}_n \\ &\frac{\partial \zeta}{\partial t} = 0 \ ; \ \zeta(0) = \zeta_n \end{aligned} \tag{4.334}$$

It is readily apparent that the exact solution of the last equation in (4.334) above can be written:

$$\zeta^{trial}_{n+1} = \zeta_n \tag{4.335}$$

[18] In each new time step the first iterative value is chosen as the corresponding converged value from the previous step.

It is less apparent that we can also find the exact solution of first two equations in (4.334); the first among them is equal to the time derivative of (4.282) computed while keeping the plastic deformation fixed:

$$\frac{\partial}{\partial t}\mathbf{B}^e = \frac{\partial}{\partial t}[\mathbf{F}\mathbf{G}^p\mathbf{F}^T]\bigg|_{\bar{\mathbf{G}}^p} \quad ; \quad \mathbf{B}^e(0) = \mathbf{B}^e_n := \mathbf{F}_n\mathbf{G}^p_n\mathbf{F}^T_n \tag{4.336}$$

The exact solution of the last equation can thus easily be computed by using the convective transport of the left Cauchy–Green elastic deformation tensor $\mathbf{B}^e_n$, carried out with $\tilde{\mathbf{F}}_{n+1} = \mathbf{F}_{n+1}\mathbf{F}^{-1}_n$, which results with:

$$\begin{aligned} \mathbf{B}^{e\,,\,trial}_{n+1} &= \mathbf{F}_{n+1}\mathbf{G}^p_n\mathbf{F}^T_{\;n+1} \\ &= \tilde{\mathbf{F}}_{n+1}\mathbf{B}^e_n\tilde{\mathbf{F}}^T_{n+1} \end{aligned} \tag{4.337}$$

The computed result is denoted as the trial value of the left Cauchy–Green elastic strain tensor $\mathbf{B}^{e\,,\,trial}_{n+1}$, which is in accordance with the elastic trial step of the operator split solution method. Having computed the result in (4.337) above, the trial values of elastic stretches $\lambda^{e,\,trial}_{n+1}$ can easily be computed as the solution to the corresponding eigenvalue problem:

$$[\mathbf{B}^{e\,,\,trial}_{n+1} - (\lambda^{e\,,\,trial}_{i\,,\,n+1})^2\mathbf{I}^{\varphi}]\mathbf{m}^{trial}_{i\,,\,n+1} = 0 \tag{4.338}$$

The same procedure can also be used to obtain the exact solution of the second equation in (4.334) for convective transport of back strain tensor, and provide the corresponding trial value of principal strains:

$$\mathbf{Z}^{trial}_{n+1} = \bar{\mathbf{F}}_{n+1}\mathbf{Z}_n\bar{\mathbf{F}}^T_{\;n} \Rightarrow [\mathbf{Z}^{trial}_{n+1} - z^{trial}_{i,n+1}\mathbf{I}^{\varphi}]\mathbf{m}^{trial}_{i\,,\,n+1} = 0 \tag{4.339}$$

The computational efficiency for the subsequent local phase of the operator split method is ensured by keeping the subsequent computations in the principal axes framework. Namely, this allows us to only deal with the corresponding principal values of tensors. In particular, for the given trial principal values of elastic stretch and back deformation tensor, we can compute from the invariant form of the strain energy potential the corresponding principal values of the second Piola–Kirchhoff stress and the back stress:

$$\tau^{trial}_{i\,,\,n+1} = \lambda^{trial}_{n+1}\frac{\partial\psi(\lambda^{e,\,trial}_{i,\,n+1},\cdot)}{\partial\lambda^e_i}\;;\; \upsilon^{trial}_{i,\,n+1} = z^{trial}_{n+1}\frac{\partial\psi(z^{trial}_{i,\,n+1},\cdot)}{\partial z_i}\;;\; q^{trial}_{n+1} = -\frac{\partial\psi(\zeta^{trial}_{n+1})}{\partial\zeta} \tag{4.340}$$

We can thus provide the corresponding trial value at time t_{n+1} for the invariant form of the yield function in principal axes; this allows us verify if the step indeed remains elastic, if the yield function takes a negative value:

$$\phi(\tau^{trial}_{i,n+1}, \upsilon^{trial}_{i,n+1}, q^{trial}_{n+1}) \leq 0 \tag{4.341}$$

In the opposite case, a positive trial value of yield function $\phi(\tau_{i\,,\,n+1}^{trial}, \upsilon_{i\,,\,n+1}^{trial}, q_{n+1}^{trial}) > 0$ indicates that the trial state cannot be considered as plastically admissible. This confirms that the step is not elastic, and that we must compute the evolution of internal variables and find the corresponding (positive) value of plastic multiplier γ_{n+1}. We note again that the local computation of this kind will take place in the fixed deformed configuration (with no transport terms needed), if we start from the corresponding trial values provided by the global phase of the operator split method:

$$\begin{aligned} &\tfrac{\partial}{\partial t}\mathbf{B}^e = -2\tfrac{\partial\gamma}{\partial t}\left(\textstyle\sum_{i=1}^3 \tfrac{\partial\phi}{\partial\tau_i}\mathbf{m}_i\otimes\mathbf{m}_i\right)\mathbf{B}^e \quad ; \quad \mathbf{B}^e(0) = \mathbf{B}_{n+1}^{e\,,\,trial} \\ &\tfrac{\partial}{\partial t}\mathbf{Z} = -2\tfrac{\partial\gamma}{\partial t}\left(\textstyle\sum_{i=1}^3 \tfrac{\partial\phi}{\partial\upsilon_i}\mathbf{m}_i\otimes\mathbf{m}_i\right)\mathbf{Z} \quad ; \quad \mathbf{Z}(0) = \mathbf{Z}_{n+1}^{trial} \\ &\tfrac{\partial\zeta}{\partial t} = \tfrac{\partial\gamma}{\partial t}\tfrac{\partial\phi}{\partial q} \quad ; \quad \zeta(0) = \zeta_{n+1}^{trial} \end{aligned} \tag{4.342}$$

For the local phase of the operator split method we can provide an approximate solution computed by an implicit time-integration scheme. For example, the first-order, backward Euler scheme can be used to obtain the solution of the last evolution equations in (4.342), leading to:

$$\zeta_{n+1} = \zeta_{n+1}^{trial} + \gamma_{n+1}\frac{\partial\phi}{\partial q}\bigg|_{n+1} \tag{4.343}$$

The first two evolution equations in (4.342) are integrated by another first-order, implicit scheme, which is based upon a discrete approximation to the exponential mapping (see [107]):

$$\begin{aligned} \mathbf{B}_{n+1}^e &= \left[\textstyle\sum_{i=1}^3 exp(-2\gamma_{n+1}\tfrac{\partial\phi}{\partial\tau_i})\mathbf{m}_{i\,,\,n+1}\otimes\mathbf{m}_{i\,,\,n+1}\right]\mathbf{B}_{n+1}^{e\,,\,trial} \\ \mathbf{Z}_{n+1} &= \left[\textstyle\sum_{i=1}^3 exp(-2\gamma_{n+1}\tfrac{\partial\phi}{\partial\upsilon_i})\mathbf{m}_{i\,,\,n+1}\otimes\mathbf{m}_{i\,,\,n+1}\right]\mathbf{Z}_{n+1}^{trial} \end{aligned} \tag{4.344}$$

An efficient implementation of the exponential mapping exploits the principal axes framework, which requires the spectral decomposition for all strain tensors:

$$\begin{aligned} \mathbf{B}_{n+1}^e &= \textstyle\sum_{i=1}^3(\lambda_{i\,,\,n+1}^e)^2\,\mathbf{m}_{i\,,\,n+1}\otimes\mathbf{m}_{i\,,\,n+1}\ ; \\ \mathbf{B}_{n+1}^{e\,,\,trial} &= \textstyle\sum_{i=1}^3(\lambda_{i\,,\,n+1}^{e\,,\,trial})^2\,\mathbf{m}_{i\,,\,n+1}^{trial}\otimes\mathbf{m}_{i\,,\,n+1}^{trial}\ ; \\ \mathbf{Z}_{n+1} &= \textstyle\sum_{i=1}^3(z_{i\,,\,n+1})^2\,\mathbf{m}_{i\,,\,n+1}\otimes\mathbf{m}_{i\,,\,n+1}\ ; \\ \mathbf{Z}_{n+1}^{trial} &= \textstyle\sum_{i=1}^3(z_{i\,,\,n+1}^{trial})^2\,\mathbf{m}_{i\,,\,n+1}^{trial}\otimes\mathbf{m}_{i\,,\,n+1}^{trial} \end{aligned} \tag{4.345}$$

The last two results imply that the principal direction of trial elastic stretch are the same as those for final elastic stretch values; moreover, they are also shared by the back deformation tensor:

$$\boxed{\mathbf{m}_{i,n+1}\otimes\mathbf{m}_{i,n+1} \equiv \mathbf{m}_{i,n+1}^{trial}\otimes\mathbf{m}_{i,n+1}^{trial}} \tag{4.346}$$

This is the key result which allows us to significantly increase the computational efficiency in the second, local phase of computations; namely, with the last result in hand, we cast recast the internal variables updates in (4.344) in terms of the multiplicative updates of elastic stretches and principal values of back strain tensor:

$$\boxed{(\lambda^e_{i\,,n+1})^2 = (\lambda^{e\,,trial}_{i\,,n+1})^2\, exp[-2\gamma_{n+1}\frac{\partial\phi}{\partial\tau_i}\Big|_{n+1}] \;;\; i=1,2,3}$$
$$\boxed{(z_{i\,,n+1})^2 = (z^{trial}_{i\,,n+1})^2\, exp[-2\gamma_{n+1}\frac{\partial\phi}{\partial v_i}\Big|_{n+1}] \;;\; i=1,2,3} \quad (4.347)$$

Furthermore, by using the logarithmic strain measures that are computed from the corresponding values of elastic principal stretch and the principal values of back strain tensor:

$$\begin{aligned} \epsilon^e_{i\,,n+1} &= ln(\lambda^e_{i\,,n+1}) \;;\; \epsilon^{e\,,trial}_{i\,,n+1} = ln(\lambda^{e\,,trial}_{i\,,n+1}) \\ \kappa_{i\,,n+1} &= ln(z_{i\,,n+1}) \;;\; \kappa^{trial}_{i\,,n+1} = ln(z^{trial}_{i\,,n+1}) \end{aligned} \quad (4.348)$$

we can replace the multiplicative updates of elastic stretch in (4.347) by the corresponding additive updates:

$$\begin{aligned} \epsilon^e_{i\,,n+1} &= \epsilon^{e\,,trial}_{i\,,n+1} - \gamma_{n+1}\frac{\partial\phi}{\partial\tau_i}\Big|_{n+1} \\ \kappa_{i\,,n+1} &= \kappa^{trial}_{i\,,n+1} - \gamma_{n+1}\frac{\partial\phi}{\partial v_i}\Big|_{n+1} \;;\; i=1,2,3 \end{aligned} \quad (4.349)$$

The internal variables updates of this kind are formally equivalent to the those used for implicit time integration scheme for small strain plasticity. This result allows us to directly apply to large strain regime the standard solution algorithms for plastic strain computations in small strain case, and carry out the computations of elastic principal values. For the given invariant form of the yield function, we can thus compute the corresponding principal values of plastically admissible stress as the solution of the following set of nonlinear algebraic equations:

$$\mathbf{r}_{n+1} := \left\{ \begin{array}{l} \epsilon^e_{i,\,n+1} - \epsilon^{e\,,trial}_{i,\,n+1} + \gamma_{n+1}\partial_{\tau_i}\phi_{n+1} \\ \phi_{n+1} \\ \kappa_{i,\,n+1} - \kappa^{trial}_{i,\,n+1} + \gamma_{n+1}\partial_{v_i}\phi_{n+1} \\ -\zeta_{n+1} + \zeta_n + \gamma_{n+1}\partial_q\phi_{n+1} \end{array} \right\} = \mathbf{0} \;;\; i=1,2,3 \quad (4.350)$$

with $\epsilon^e_{i,n+1}$, $\kappa_{i,n+1}$, ζ_{n+1} and γ_{n+1} as unknowns. It might be possible to simplify this task considerably by reducing it to a single algebraic equation for plastic multiplier computation. The first condition to fulfill for this kind of reduction concerns a quadratic form of the yield function in terms of principal values of Kirchhoff stress and back stress tensors; one such example is given by the invariant form of von Mises plasticity criterion, which can be written:

$$\phi(\tilde{\tau}_i, \upsilon_i, q) := [(\tilde{\tau}_1 + \upsilon_1)^2 + (\tilde{\tau}_2 + \upsilon_2)^2 + (\tilde{\tau}_3 + \upsilon_3)^2]^{1/2} - \sqrt{\frac{2}{3}}(\tau_y - q) \quad (4.351)$$

where $\tilde{\tau}_i = \tau_i - \frac{1}{3}(\sum_{k=1}^{3} \tau_k)$. This yield function will impose the incompressibility constraint upon the plastic deformation (with $J^p = 1$) and the back stress tensor. Furthermore, for the case where the elastic deformations remain very small (with $J = J^e \approx 1$), the difference between the Kirchhoff stress resulting from this computation and the true Cauchy stress will also be small $\tau_i \approx \sigma_i$. Therefore, the proposed criterion can be quite reliable in representing the true physical phenomena of large deformation plasticity.
In order to complete the large deformation plasticity model description, we will pick a simple example of the strain energy potential defined as a quadratic form in terms of logarithmic strain measure for elastic principal stretches and principal values of back strain tensor:

$$\begin{aligned}\psi(J^e, \tilde{\lambda}^e_i, z_i, \zeta) := \tfrac{1}{2}K(lnJ^e)^2 + \tfrac{1}{2}2\mu[(ln\tilde{\lambda}^e_1)^2 + (ln\tilde{\lambda}^e_2)^2 + (ln\tilde{\lambda}^e_3)^2] \\ + \tfrac{1}{2}\tfrac{2H}{3}[(lnz_1)^2 + (lnz_2)^2 + (lnz_3)^2] + \Xi(\zeta)\end{aligned} \quad (4.352)$$

where $\Xi(\cdot)$ is the corresponding potential for isotropic hardening, whereas $\tilde{\lambda}^e_i$ are defined according to:

$$\tilde{\lambda}^e_i = (J^e)^{-1/3}\lambda^e_i \,;\; i = 1,2,3 \,;\; J^e = \lambda^e_1\lambda^e_2\lambda^e_3 \quad (4.353)$$

The set of nonlinear algebraic equations in (4.350) to be solved for general plasticity model can now be reduced to a single scalar equation, with the plastic multiplier γ_{n+1} as the only unknown. If the hardening potential $\Xi(\zeta)$ corresponds to saturation hardening, such an equation will be perfectly suitable for Newton's iterative procedure, producing at each iterative sweep the following update:

$$\{2\mu + \frac{2}{3}[\Xi''(\zeta^{(k)}_{n+1}) + H] + \}\Delta\gamma^{(k)}_{n+1} = \phi^{(k)}_{n+1} \Rightarrow \gamma^{(k+1)}_{n+1} = \gamma^{(k)}_{n+1} + \Delta\gamma^{(k)}_{n+1} \quad (4.354)$$

The final converged solution of this equation for the plastic multiplier $\bar{\gamma}_{n+1}$, which ensures the plastically admissibility of stress field with $\phi(\bar{\gamma}_{n+1}) = 0$, is used to carry out the final updates of internal variables as defined as

(4.350). The computed principal values of internal variables can then be used to recover all the components of strain tensors:

$$\begin{aligned} \mathbf{B}^e_{n+1} &= \textstyle\sum_{i=1}^3 [exp(\epsilon^e_{i\,,\,n+1})]^2\, \mathbf{m}_{i\,,\,n+1} \otimes \mathbf{m}_{i\,,\,n+1} \\ \mathbf{Z}_{n+1} &= \textstyle\sum_{i=1}^3 [exp(\kappa_{i\,,\,n+1})]^2\, \mathbf{m}_{i\,,\,n+1} \otimes \mathbf{m}_{i\,,\,n+1} \end{aligned} \tag{4.355}$$

The von Mises plasticity model ensures that the final principal directions are identical to their trial values, and that the final principal vector also verifies the following identity:

$$\begin{aligned} \mathbf{m}_{i,n+1} \otimes \mathbf{m}_{i,n+1} &= \tfrac{1}{d_{i,n+1}} [\mathbf{B}^e - (\lambda^e_{j,n+1})^2 \mathbf{I}^\varphi][\mathbf{B}^e - (\lambda^e_{k,n+1})^2 \mathbf{I}^\varphi] \; ; \\ d_{i,n+1} &= ((\lambda^e_{i,n+1})^2 - (\lambda^e_{j,n+1})^2)((\lambda^e_{i,n+1})^2 - (\lambda^e_{k,n+1})^2) \end{aligned} \tag{4.356}$$

Therefore, the computed principal values of elastic stretch will also allow us to provide the principal values of the plastically admissible Kirchhoff stress according to:

$$\begin{bmatrix} \tau_{1,n+1} \\ \tau_{2,n+1} \\ \tau_{3,n+1} \end{bmatrix} = \begin{bmatrix} (K + \frac{4\mu}{3}) & (K + \frac{2\mu}{3}) & (K + \frac{2\mu}{3}) \\ (K + \frac{2\mu}{3}) & (K + \frac{4\mu}{3}) & *K + \frac{2\mu}{3}) \\ (K + \frac{2\mu}{3}) & (K + \frac{2\mu}{3}) & (K + \frac{4\mu}{3}) \end{bmatrix} \begin{bmatrix} \epsilon^e_{1,n+1} \\ \epsilon^e_{2,n+1} \\ \epsilon^e_{3,n+1} \end{bmatrix} \tag{4.357}$$

Once the principal values are known, it is easy to recover the Kirchhoff stress tensor physical components by making use of the spectral decomposition result:

$$\boldsymbol{\tau}_{n+1} = \sum_{i=1}^{3} \tau_{i,n+1} \mathbf{m}_{i,n+1} \otimes \mathbf{m}_{i,n+1} \; ; \tag{4.358}$$

With the computed value of the Kirchhoff stress, we can easily obtain the internal force vector by accounting for the corresponding contribution of each particular element to the global set of equilibrium equations; for 2D plane strain case, the internal force vector of a typical element can be written:

$$\begin{aligned} &\mathsf{f}^{int,e} = [f_p^{int,e}] \; ; \; f_p^{int,e} = \mathsf{e}_i^T \int_{\Omega^e} \bar{\mathsf{B}}_a^{\varphi,T} \mathsf{t}_{n+1}\, dV \; ; \\ &\bar{\mathsf{B}}_a^{\varphi} = \tilde{\mathsf{B}}_a^{\varphi} + \begin{bmatrix} \frac{1}{3}\bar{\mathsf{b}}^{\varphi} \\ \frac{1}{3}\bar{\mathsf{b}}^{\varphi} \\ \frac{1}{3}\bar{\mathsf{b}}^{\varphi} \\ 0^T \end{bmatrix} ; \; \mathsf{t}^h = \begin{bmatrix} \tau_{11,n+1} \\ \tau_{22,n+1} \\ \tau_{33,n+1} \\ \tau_{12,n+1} \end{bmatrix} \end{aligned} \tag{4.359}$$

where we indicated that $B - bar$ strain interpolation should be used (rather than the standard isoparametric interpolations), in order to accommodate the quasi-incompressibility constraint.

If the equilibrium equations are not satisfied within the prescribed tolerance, we seek a new improved value of displacements in the deformed configuration $\varphi_{n+1}^{(i+1)}$; the latter can be obtained by using the iterative update with the incremental displacement $\mathbf{u}_{n+1}^{(i)}$, which is computed from the linearized set of equilibrium equations:

$$
\begin{aligned}
&Lin[G(\varphi_{n+1}^{(i)};\mathbf{w}) = G(\varphi_{n+1}^{(i)};\mathbf{w} + \tfrac{d}{d\varepsilon}[G(\overbrace{\varphi_{n+1,\varepsilon}^{(i)}}^{\varphi_{n+1}^{(i)}+\varepsilon \mathbf{u}_{n+1}^{(i)}};\mathbf{w})]\Big|_{\varepsilon=0} \\
&\tfrac{d}{d\varepsilon}[G(\varphi_{n+1,\varepsilon}^{(i)};\mathbf{w})]\Big|_{\varepsilon=0} = \int_\Omega \tfrac{1}{2}(\nabla^\varphi \mathbf{u}^T \nabla^\varphi \mathbf{w} + \nabla^\varphi \mathbf{w}^T \nabla^\varphi \mathbf{u}) \cdot \boldsymbol{\tau}\, dV \\
&+ \int_\Omega \tfrac{1}{2}(\nabla^\varphi \mathbf{w} + \nabla^\varphi \mathbf{w}^T) \cdot \mathcal{C}_{n+1}^{ep} \tfrac{1}{2}(\nabla^\varphi \mathbf{u} + \nabla^\varphi \mathbf{u}^T)\, dV
\end{aligned}
\tag{4.360}
$$

The main new ingredient needed for constructing this linearized equation is the consistent tangent elastoplastic modulus. The latter is computed by accounting for contributions of all three elastic principal stretches. This applies even to 2D plane strain case, where the total principal stretch will remain constant $\lambda_3 = 1$, but not the corresponding elastic principal stretch $\lambda_3^e \neq 1$ (unless there is no plastic deformation). The consistent tangent elastoplastic modulus consists of two parts (e.g. see [124]), with the first referred to as the material and the second as the geometric component:

$$
\mathcal{C}_{n+1}^{ep} = \mathcal{C}_{mate,n+1}^{ep} + \mathcal{C}_{geom,n+1}^{ep} \tag{4.361}
$$

The material part of tangent elastoplastic tensor will depend upon the elastoplastic moduli in principal directions:

$$
\mathcal{C}_{mate,n+1}^{ep} = \sum_{i=1}^{3}\sum_{j=1}^{3} \frac{\partial \tau_{i,n+1}}{\partial \epsilon_{j,n+1}} (\mathbf{m}_{i,n+1} \otimes \mathbf{m}_{i,n+1}) \otimes (\mathbf{m}_{j,n+1} \otimes \mathbf{m}_{j,n+1}) \tag{4.362}
$$

whereas the geometric part will be directly proportional to the computed, plastically admissible principal values of the Kirchhoff stress

$$
\begin{aligned}
\mathcal{C}_{geom,n+1}^{ep} = \tfrac{2\tau_i}{d_{i,n+1}} \{ & \mathcal{I}_{B^e} - \mathbf{B}_{n+1}^e \otimes \mathbf{B}_{n+1}^e - i_{3B^e}(\lambda_{i,n+1}^e)^{-2}[\mathcal{I} \\
& -(\mathbf{I} - \mathbf{m}_{i,n+1} \otimes \mathbf{m}_{i,n+1}) \otimes (\mathbf{I} - \mathbf{m}_{i,n+1} \otimes \mathbf{m}_{i,n+1})] \\
& +(\lambda_{i,n+1}^e)^2[\mathbf{B}_{n+1}^e \otimes (\mathbf{m}_{i,n+1} \otimes \mathbf{m}_{i,n+1}) \\
& +(\mathbf{m}_{i,n+1} \otimes \mathbf{m}_{i,n+1}) \otimes \mathbf{B}_{n+1}^e + (i_{1B^e} - 4(\lambda_{i,n+1}^e)^2) \\
& (\mathbf{m}_{i,n+1} \otimes \mathbf{m}_{i,n+1}) \otimes (\mathbf{m}_{i,n+1} \otimes \mathbf{m}_{i,n+1})]\}
\end{aligned}
\tag{4.363}
$$

We provide an illustrative result to show the capabilities of this kind of large strain plasticity and viscoplasticity models. Namely, as shown in Figure 4.22, the plasticity model correctly predicts the necking phenomena related to the peak resistance of the cylindrical specimen in simple tension test, which is accompanied by the appearance of the shear bands inclined at 45° with respect to the loading direction. A number of successive deformed shapes are presented in Figure 4.23 in order to illustrate very large displacements and rotations, which this 3D plasticity model is capable to handle.

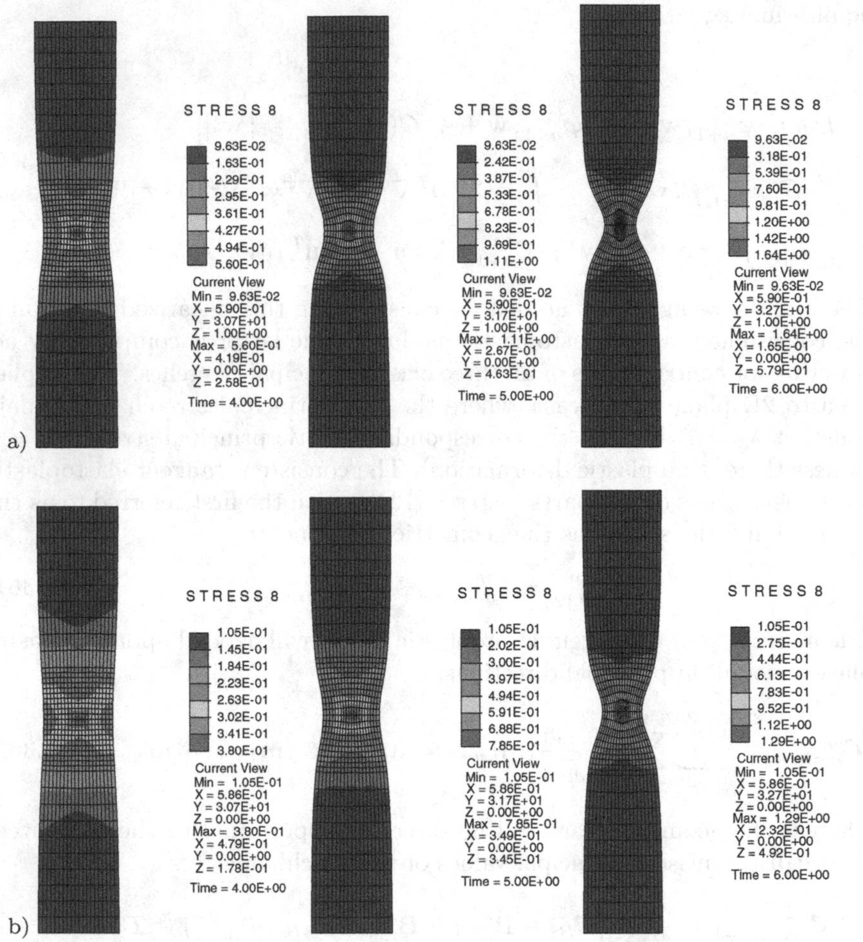

Fig. 4.22 Contours of hardening variable computed for: **(a)** plasticity model and **(b)** viscoplasticity model, corresponding to imposed displacement values $\bar{u} = 4$, 5 or 6 [mm].

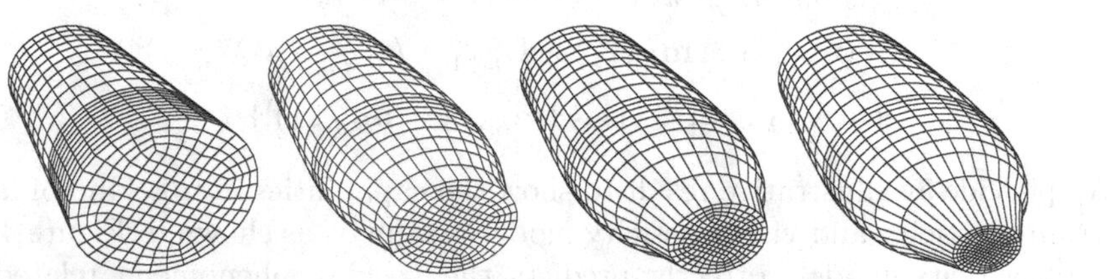

Fig. 4.23 Necking phenomena and successive deformed shapes at (t = 0, 6, 7, 8) computed by large strain 3D plasticity model.

Chapter 5
Changing boundary conditions: contact problems

After detailed discussions of the three main sources of nonlinearity in a boundary value problem in mechanics, pertaining to kinematics, equilibrium and constitutive equations, we turn in this chapter to the studies of the last potential source of nonlinearity that concerns the changing boundary conditions. The problems of this kind are yet referred to as contact problems. The boundary conditions for contact problem are neither those of Dirichlet with imposed value of displacement, nor of Neumann with imposed traction, but rather the boundary conditions where we cannot impose the exact values other than the limitation (or constraint) that the displacements and/or traction should obey. One example of this kind is the unilateral contact, corresponding to a deformable body placed next to a rigid obstacle that cannot be penetrated and thus prevents the free deformation of the body under an arbitrary loading. We can also have a bilateral contact with two or more deformable solid bodies that enter in contact, or yet a self-contact where a body can undergo very large displacements and deformations which bring different parts of the body in contact with each other. All such contact problems are of great practical interest, and are discussed in this chapter.

The outline of this chapter is as follows. We start with the simplest possible problem of unilateral contact in 1D setting, which allows us nonetheless to illustrate all the basic solution methods for the class of boundary value problems with a contact constraint on the configuration space. Namely, we discuss how to apply to such a 1D contact problem the Lagrange multiplier method for handling unilateral contact constraint, then the penalty method and finally the augmented Lagrangian method. We show than all these methods for solving the 1D unilateral contact problems also apply to bilateral contact and auto-contact in 2D and 3D cases, but with an additional difficulty which appears related to the search for the contact zone after possibly very large displacements and deformations of the bodies in contact. An efficient search algorithm for the contact zone, referred to as 'slide line', is then presented in order to illustrate the corresponding procedure for 2D case. We also discuss another very important issue for 2D and 3D contact problems related to the finite element discretization of the contact boundary and the corresponding discrete approximation of the contact pressure. We finish this chapter with a discussion of several possible extensions of the basic frictionless contact model, such as accounting for the frictional phenomena in contact, the contact surface asperities deformation etc.

5.1 Unilateral 1D contact problem

For a three-dimensional contact problem, and especially for the non-smooth body boundaries, the search for contact zone can very easily become a dominant phase of solution procedure in terms of computational cost and complexity, and completely overshadow the mechanics of contact. In order to avoid such risk and to be able to first focus upon the mechanics of contact, we start with the simplest possible 1D setting for unilateral contact problem, where an elastic bar in extension reaches a rigid obstacle. The contact zone search is such a case will remain trivial and the contact will reduce to a single point. We note in passing that the mechanics of 1D contact remains very much the same as in 2D and 3D cases, and that all the basic solution methods for dealing with contact constraint carry over to 2D and 3D case with no major modification.

In particular, we present for such a 1D unilateral contact problem three different methods for accommodating the presence of contact constraint in the configuration space: the method of Lagrange multipliers, the penalty method and the augmented Lagrangian method. The Lagrange multiplier method is very much the same as the one used for solving the optimization problem in the presence of a constraint (or a limitation) which is encountered previously in dealing with the principle of maximum plastic dissipation. However, contrary to the constrained minimization problem in plasticity where the constraint is imposed locally for any particular point,[1] the constraint imposed in contact problems is global in that it affects simultaneously all the points on the contact boundary. Because of this kind of difference in constraint application (local versus global), the Lagrange multiplier method used for contact problems is somewhat more elaborate than the Lagrange multiplier method used in plasticity. Namely, the changing boundary conditions tend to complicate implementation of the Lagrange multiplier method, and even prevent the iterative procedure from converging. We thus often use the penalty method instead to increase the robustness of the solution procedure for contact problem, although the penalty method implies that we only settle for an approximation of the contact constraint. It is only the augmented Lagrangian

[1] In plasticity, we check independently for each numerical integration point in the domain that the stress components at that point obeys the plastic admissibility constraint, thus placing the corresponding constraint on the displacement field producing the stress which verifies the chosen yield condition.

method which can both enhance the computational robustness and enforce the contact constraint exactly.
These three basic methods for solution of unilateral 1D contact problem are also the currently dominant group of methods for solving the contact problems of practical interest, such as bilateral contact and self-contact in 2D or 3D setting. The pertinent modifications which are needed for such an extension are discussed in the second part of this chapter.

5.1.1 Strong form of 1D elasticity in presence of unilateral contact constraint

In this section, we revisit the theoretical formulation of the boundary value problem in 1D elasticity in large displacements, for the case when the body in motion can enter in contact with a rigid obstacle, which is referred to as the unilateral contact problem. The 1D elasticity in large displacements is chosen mostly for convenience of having all the model ingredients defined in the previous chapter. It is important to emphasize that all the solution methods for unilateral contact problem of this kind will remain applicable for any other boundary value problem among those we have already studied (e.g. with the main source of nonlinearity pertinent to the inelastic constitutive behavior).
The strong form of the boundary value problem we consider will regroup the equations expressing the equilibrium, the elastic constitutive behavior and the large displacement kinematics, as defined originally in (4.102), (4.75) and (4.43), respectively. We consider that the initial length of the elastic bar is equal to l, and we position the reference frame such that the initial configuration corresponds to an interval of x-axis:

$$\bar{\Omega} = [0, l] \tag{5.1}$$

The left end of the bar is held fixed; hence, for any deformed configuration we can write:

$$\varphi_t(0) = 0 \tag{5.2}$$

A horizontal traction $\bar{t}$ is the only load applied at free end of the bar. For a unit cross-section of the bar, the statically determined problem of this kind will allow us to conclude that:

$$\sigma_t(l_t) \equiv P_t(l) = \bar{t}_t \tag{5.3}$$

where $l_t = \varphi_t(l)$ is the deformed bar length. In order to introduce a unilateral contact constraint in this problem, we consider that a rigid obstacle is placed at distance $\bar{x} > l$ with respect to the origin (see Figure 5.1). The initial distance between the right end of the bar and the rigid obstacle is given as:

$$g_0 := \hat{g}(l) = l - \bar{x} \tag{5.4}$$

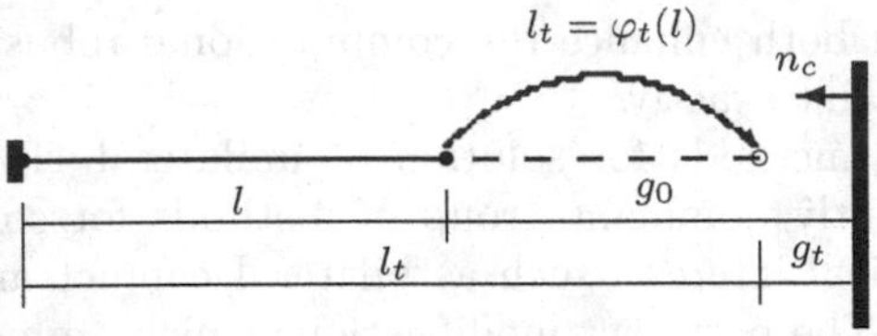

Fig. 5.1 Unilateral 1D contact problem: initial and deformed configurations of elastic bar in large displacements.

In the deformed configuration at time 't', the current distance to the obstacle is defined with:

$$g_t := \hat{g}(l_t) = \underbrace{\varphi_t(l)}_{l_t} - \bar{x} \tag{5.5}$$

The bar cannot penetrate the obstacle. With the chosen sign convention, this implies that the distance between the bar and obstacle is always negative or zero:

$$g_t \leq 0 \tag{5.6}$$

The equality sign in the last equations implies that the elastic bar has entered in contact with the rigid obstacle. At that stage, according to the action–reaction principle, the action of the bar produces equal and opposite reaction of the obstacle. With the sign convention for such a reaction force defined by the unit exterior normal to the obstacle, we can consider that the reaction can be either positive (for the case of contact) or zero (if there is no contact of the bar with obstacle); we can thus write:

$$p_t \geq 0 \tag{5.7}$$

The values of contact gap in (5.6) and of contact reaction in (5.7) cannot be equal to zero simultaneously. There is always only one of them which is zero; this observation allows us to write the contact constraint in terms of Kuhn-Tucker condition:

$$p_t \, g_t = 0 \tag{5.8}$$

The unilateral contact conditions between a deformable solid and a rigid obstacle in (5.6), (5.7) and (5.8) are yet referred to as the Signorini conditions. *Remark on subdifferential*: Following terminology of [Clarke [52], Ekeland and Temam [73], Moreau [191], or Oden [205]], we can define the indicator function (for example, see [73], p. 8):

$$\Psi_{\mathbb{R}^-}(g_t) = \begin{cases} 0 \; ; & g_t \leq 0 \\ +\infty \; ; & g_t > 0 \end{cases} \tag{5.9}$$

With the indicator function being non-differentiable in the classical sense, the only proper interpretation of its derivative can be made in terms of subdifferential, which is defined through the tools of convex analysis:

$$\partial\Psi_{\mathbb{R}^-}(g_t) = \{p_t | 0 \geq p_t[g_t^* - g_t] \ ; \forall g_t^* \in \mathbb{R}^-\} \tag{5.10}$$

Introducing the indicator function and its subdifferential allows us to furnish an elegant description of Signorini conditions in terms of a "contact law", connecting the contact pressure with the contact gap through the Moreau subdifferential:

$$p_t = \partial\Psi_{\mathbb{R}^-}(g_t) \tag{5.11}$$

This kind of contact law accepts multiple values of the contact pressure (or obstacle reaction) in the case of contact with $g_t = 0$, which can be handled through judicious application of the convex analysis tools to carry out the computations (e.g. see Moreau [191]). We do not elaborate any further on this particular approach, since it is not easily applicable to more complex contact laws. In anticipating future increase of refined contact laws based on micro-mechanics (some of the refinements are discussed at the end of this chapter), we will give preference to the kind of computational methods which can handle efficiently any model complexity since they rely on the standard Gâteaux derivatives rather than subdifferentials.

5.1.2 *Weak form of unilateral 1D contact problem and its finite element solution*

In this section we present the weak form of the unilateral 1D contact problem, which is the starting point in constructing the corresponding finite element approximation. We discuss three different methods for constructing the solution of contact problem, which takes into account the contact constraint: Lagrange multipliers method, penalty method and augmented Lagrangian method. The chosen model problem is 1D elasticity in large displacements, which is studied in detail in the previous chapter.

5.1.2.1 Lagrange multipliers method for contact

The main idea of the Lagrange multiplier method in application to the contact problem on hands is to enlarge the configuration space with respect to nodal displacements only by including the contact pressure p_t as the Lagrange multiplier independent from displacements. We thus eliminate the need to use the non-differentiable contact law in (5.11). However, the presence of a new variable in terms of contact pressure requires an additional equation as a supplement to the set of equilibrium equations. The corresponding modified form of the principle of minimum of total potential energy provides certainly the best way for picturing the role of this extra equation. Namely, we can take into account the presence of the contact constraint in (5.8), and transform the minimization of the total potential energy of the elastic bar into a constrained minimization problem:

$$\min_{\hat{g}(\varphi_t(l))\leq 0} \Pi(\varphi_t) \longrightarrow \max_{p_t\geq 0}\min_{\forall\varphi_t}[L^c] \; ; \; L^c(\varphi_t, p_t) := \pi(\varphi_t) + p_t\hat{g}(\varphi_t(l)) \quad (5.12)$$

The constrained minimization problem in (5.12) is conceptually the same as the one we encountered with the principle of maximum plastic dissipation. Therefore, we can again follow the same procedure as for plasticity in order to obtain the corresponding form of the Kuhn-Tucker optimality conditions for present contact problem (e.g. see Luenberger [175], p. 314); the latter can be written:

$$0 = \frac{d}{ds}[L^c(\varphi_t(x) + s\,w(x), p_t)]|_{s=0} := G(\varphi_t(x); w(x)) + p_t w(l) \quad (5.13)$$
$$0 = \frac{d}{ds}[L^c(\varphi_t(x), p_t + s\,q)]\bigg|_{s=0} := q\hat{g}(\varphi_t(x)) \; ; \; p_t \geq 0 \; ; \; \hat{g}(\varphi_t(l)) \leq 0$$

where q is the contact pressure variation, whereas $G(\varphi_t(x); w(x))$ is the weak form of 1D large strain elasticity problem with no contact. We can thus find two different cases: the first where the contact is not activated with zero contact pressure $p(t) = 0$ and the second case of active contact with persistent contact pressure $p(t) > 0$. The first case reduces to the standard boundary value problem in 1D large deformation elasticity, which can be solved by the finite element method as we have already presented in the previous chapter. However, for the second case of active contact, the standard finite element discretization will produce a set of $n_{eq} + 1$ algebraic nonlinear equations, where first n_{eq} unknowns are the nodal displacements $\mathbf{d}(t)$ and the last unknown is the contact pressure $p(t)$: this set of equations can be written:

$$\begin{bmatrix} \hat{\mathsf{f}}^{int}(\mathsf{d}(t)) - \mathsf{f}^{ext}(t) + p(t)\mathsf{1}_{n_c} \\ \hat{g}(\mathsf{1}_{n_c} \cdot \mathsf{d}(t)) \end{bmatrix} = 0 \quad (5.14)$$

The vector $\mathbf{1}_{n_c}$ in the expression above will have all the components equal to zero, except from a unit value of the component $n_c \in [1, n_{eq}]$ corresponding to the node in contact with rigid obstacle. For example, with node numbers increasing from left to right end of the bar, we will have $n_c = n_{eq}$ and only the last one among all equilibrium equations will be modified by the presence of contact pressure. The resulting set of equilibrium equations along with active contact constraint can be solved by using the incremental-iterative procedure.

At a typical iteration (i), we can write the consistently linearized form of the system in (5.14), which reads:

$$\begin{bmatrix} \mathsf{K}_{n+1}^{(i)} & \mathbf{1}_{n_c} \\ \mathbf{1}_{n_c}^T & 0 \end{bmatrix} \begin{bmatrix} \mathsf{u}_{n+1}^{(i)} \\ \Delta p_{n+1}^{(i)} \end{bmatrix} = \begin{bmatrix} \mathsf{f}_{n+1}^{ext} - \hat{\mathsf{f}}^{int}(\mathsf{d}_{n+1}^{(i)}) - p_{n+1}^{(i)} \mathbf{1}_{n_c} \\ -\hat{g}(\mathbf{1}_{n_c} \cdot \mathsf{d}_{n+1}^{(i)}) \end{bmatrix} \tag{5.15}$$

The computed solution of the linearized system in (5.15) above is unique, if the tangent stiffness matrix $\mathsf{K}_{n+1}^{(i)}$ is positive definite. The iterative values of nodal displacements and contact pressure are updated after each solve according to:

$$\mathsf{d}_{n+1}^{(i+1)} = \mathsf{d}_{n+1}^{(i)} + \mathsf{u}_{n+1}^{(i)} \; ; \; p_{n+1}^{(i+1)} = p_{n+1}^{(i)} + \Delta p_{n+1}^{(i)} \tag{5.16}$$

The main disadvantage of the Lagrange multiplier procedure applied to contact problems concerns a potential increase of computational cost, which stems from breaking the sparse structures of the tangent stiffness matrix with the presence of contact constraint. We can avoid this cost increase by using the node numbering placing the nodes in contact and Lagrange multipliers at the end of the set of linearized equilibrium equations to be solved. However, the benefits of such ordering cannot be guaranteed for the case of large contact sliding where the nodes in contact can change at each iteration. In fact, the problems with large contact sliding are notorious for the lack of robustness of Newton's iterative scheme, and convergence difficulties imposed by a rapid change of contact regime (from active to inactive and back) for each node. Two alternative methods are thus proposed in order to avoid these difficulties in dealing with contact problems, the penalty method and the augmented Lagrangian method. Both of them are presented next for the chosen model problem of 1D elasticity in large displacements.

Remark on duality: We give a brief comment upon a very important notion of duality, and illustrate how it applies to the contact problem on hands. For that purpose, we consider the case with active contact and denote with p_t the solution to the minimization problem in (5.12) for the contact pressure. Moreover, we consider that the minimization problem defined in (5.12) remains well defined for any value of contact pressure p_t^* that belongs to the close neighborhood of the true solution p_t, which further allows us to introduce the dual function:

$$\phi(p_t^*) = \min_{\varphi_t(x)} [\Pi(\varphi_t(x)) + p_t^* \hat{g}(\varphi_t(l))] \tag{5.17}$$

From the definition of the dual function in (5.17) above, we can easily see that its gradient (or rather the corresponding Gâteaux derivative in the direction of virtual displacement) is equal to the gap function for testing impenetrability condition:

$$\frac{d\phi(p_t^*)}{dp_t^*} := \overbrace{\{\frac{d}{ds}[\Pi(\varphi_t(x)+s\,w(x))+p_t^*\hat{g}(\varphi_t(x)+s\,w(x))]|_{s=0}}^{G_{contact}(\cdot)=0}\frac{d\tilde{\varphi}(p_t^*)}{dp_t^*}$$
$$+\underbrace{\frac{d}{dp_t^*}[p_t^*\hat{g}(\varphi_t(x))]}_{\hat{g}(\varphi_t(x))=0}$$

Hence, for the case of active contact, we conclude that the first derivative of the dual function is equal to zero. The second derivative of the dual function can further be obtained by a successive application of the chain rule resulting with:

$$\frac{d^2\phi(p_t^*)}{(dp_t^*)^2} = \underbrace{\frac{d}{ds}[\hat{g}(\varphi_{t,s}(x))]\Big|_{s=0}}_{w(l)}\frac{d\tilde{\varphi}_t(p_t^*)}{dp_t^*} \tag{5.18}$$

The last term in the equation above can be obtained from the Gâteaux derivative of the weak form for active contact with respect to p^*, which leads to:

$$\overbrace{\frac{d}{ds}[G(\varphi_{t,s}(x),w(x))]\Big|_{s=0}}^{\frac{d^2}{ds^2}[\Pi(\varphi_{t,s}(x))]\big|_{s=0}}\frac{d\tilde{\varphi}_t}{dp_t^*}+\frac{d}{dp_t^*}[p_t^* w(l)]=0 \tag{5.19}$$
$$\Longrightarrow \frac{d\tilde{\varphi}_t}{dp_t^*} = -\{\frac{d^2}{ds^2}[\Pi(\varphi_{t,s}(x))]\Big|_{s=0}\}^{-1}w(l)$$

We can restate the last result within the discrete approximation constructed by using the finite element method, in terms of the tangent stiffness matrix $\mathbf{K}_{n+1}$; by hypothesis that the tangent stiffness is a positive definite matrix, implying that for any node in contact $\mathbf{1}_{n_c}^T\mathbf{K}_{n+1}^{-1}\mathbf{1}_{n_c}>0$, we can conclude that:

$$\frac{d^2\phi(p_{n+1})}{d(p_{n+1}^*)^2} = -\mathbf{1}_{n_c}^T\mathbf{K}_{n+1}^{-1}\mathbf{1}_{n_c}<0 \tag{5.20}$$

By assuming the sufficient smoothness of the dual function, we can use the computed values of its first and second derivatives at the solution point p_t in order to provide the Taylor formula estimate for the dual function value corresponding to any other value of contact pressure p_t^* in a close neigborhood of the exact solution:

$$\phi(p_t^*)-\phi(p_t)\approx\underbrace{\frac{d\phi(p_t)}{dp_t}}_{g_t=0}(p_t^*-p_t)+\frac{1}{2}\underbrace{\frac{d^2\phi(p_t)}{d(p_t^*)^2}}_{<0}\underbrace{(p_t^*-p_t)^2}_{>0}<0 \tag{5.21}$$

The last results indicates that the exact solution for contact pressure corresponds to the maximum of the dual function:

$$p_t = arg\{\max_{\forall p_t^*} \phi(p_t^*)\} \tag{5.22}$$

This observation and the principle of minimum potential energy corresponding to the exact solution of displacement confirm jointly the standard min–max or the saddle point interpretation of the contact problem in (5.12).

5.1.2.2 Penalty method

If we want to avoid the presence of the Lagrange multiplier (as an extra unknown with respect to nodal displacements), and thus keep the standard format of equilibrium equations even in the presence of contact, we can turn to the penalty method. The best possibility in that sense would be to employ the exact penalty method (e.g. see Luenberger [175]), where the indicator function is used to produce a modified form of the total energy potential for large strain 1D elasticity accounting for eventual presence of the contact constraint:

$$\min_{\hat{g}(\varphi_t(l))\leq 0} \Pi(\varphi_t(x)) \implies \min_{\forall \varphi_t}[\Pi(\varphi_t(x)) + \Psi_{\mathbb{R}^-}(\hat{g}(\varphi_t(l)))] \tag{5.23}$$

where $\Psi_{\mathbb{R}^-}(\cdot)$ is already defined in (5.9). We note that all displacement values are considered admissible from the standpoint of such a functional, and not only those which are in agreement with the contact constraint. However, the presence of indicator function puts a very high (or rather an infinite) penalty to those displacements which lead to obstacle penetration by the bar, which is supposed to discourage any such displacement and to achieve practically the same goal as the Lagrange multiplier method. The main disadvantage of the exact penalty method is the obligation to use the subdifferentials. We can eliminate this disadvantage and be able to use again the standard tools of the Gâteaux derivative computations, by accepting an approximation of the indicator function which remains everywhere differentiable:

$$\Psi_{\mathbb{R}^-}(g_t) \approx P(g_t) \tag{5.24}$$

This kind of approximation will still put rather high (but no longer infinite) penalty on the displacements leading to the obstacle penetration, and the positive values of distance function $g_t > 0$ will also be considered admissible. For the standard penalty method, we typically choose a simple quadratic form of the indicator function approximation, which can be written:

$$P(g_t) = \begin{cases} \frac{1}{2}k(g_t)^2 \ ; \ g_t \geq 0 \\ 0 \ ; \ g_t < 0 \end{cases} \tag{5.25}$$

where k is the penalty parameter. We will show shortly that the quadratic penalty function of this kind is equivalent to replacing the rigid obstacle by an elastic spring, with elasticity coefficient equal to the penalty parameter. We can easily accept other, more elaborate penalty functions provided that each potential candidate will also increase the penalty for larger penetrations. For example, a micro-mechanically based analysis of the contact surface asperities can provide a clear physical interpretation for the penalty function giving the power law for contact (e.g. see Ibrahimbegovic and Wilson [141]).
The penalty method allows us to formulate the contact problem as the unconstrained minimization of the total potential energy functional augmented by a penalty term. This leads to the kind of formulation where the displacement values remain the only unknowns, with no additional unknowns as for the Lagrange multiplier procedure; we can thus write the penalty formulation of the contact problem according to:

$$\begin{aligned} 0 &= \tfrac{d}{ds}[\Pi(\varphi_t(x) + s\,w(x)) + P(\hat{g}(\varphi_t(l) + s\,w(l)))]|_{s=0} \\ &:= G(\varphi_t(x), w(x)) + k < \hat{g}(\varphi_t(l)) > w(l) \end{aligned} \tag{5.26}$$

where:

$$< \hat{g}(\varphi_t(l)) > := \begin{cases} g(\varphi_t(l))\ ;\ g(\varphi_t(l)) \geq 0 \\ 0\ ;\ g(\varphi_t(l)) < 0 \end{cases} \tag{5.27}$$

It is also possible to interpret the penalty method as a method providing an approximation for contact pressure (or Lagrange multiplier) which can be written as a function of displacement:

$$\tilde{p}_t := k < \hat{g}(\varphi_t(l)) > = \begin{cases} k\,g(\varphi_t(l))\ ;\ g(\varphi_t(l)) \geq 0 \\ 0\ ;\ g(\varphi_t(l)) < 0 \end{cases} \tag{5.28}$$

This kind of expression for contact pressure computation allows us to write the penalty-method-based weak form of contact problem in (5.26) in the same manner as the one in (5.13) obtained with the Lagrange multiplier method, but with no need for supplementary condition on contact pressure:

$$\hat{\mathsf{f}}^{int}(\mathsf{d}(t)) - \mathsf{f}^{ext}(t) + 1_{n_c}\tilde{p}(\mathsf{d}(t)) = 0 \tag{5.29}$$

The linearized form of the system in (5.29) above for the case of active contact will feature the rank-one update of the tangent stiffness matrix for the same problem without contact, which can be written:

$$[\mathsf{K}^{(i)}_{n+1} + k 1_{n_c} 1^T_{n_c}]\mathsf{u}^{(i)}_{n+1} = \mathsf{f}^{ext}_{n+1} - \hat{\mathsf{f}}^{int}(\mathsf{d}^{(i)}_{n+1}) - 1_{n_c}\tilde{p}^{(i)}_{n+1} \tag{5.30}$$

This kind of modification of the stiffness matrix, which is the result of a particular form of vector 1_{n_c} with only one non-zero component, will modify the tangent stiffness matrix by placing the chosen penalty parameter at the main diagonal in the slot reserved for the node in contact. We can further appeal to the Gerschagorin theorem (e.g. see Dahlquist and Bjorck [62], p. 209) to find out that the increase of the maximum eigenvalue of such modified stiffness matrix will be proportional to the penalty parameter, k. It thus follows that choosing a very large value of the penalty parameter, which is desirable in order to ensure good approximation of impenetrability condition, will lead to poor conditioning[2] of the modified tangent stiffness matrix in the presence of contact. A successful remedy is then to bound the penalty parameter value with respect to the largest eigenvalue (or simply, with respect to the largest tangent stiffness component among those placed on the main diagonal), even though this will reduce the quality of impenetrability condition representation (see [75]). If such an approximation of impenetrability condition is unacceptable, but we still want to have a computationally efficient approach, we can abandon the penalty in favor or the augmented Lagrangian method.

5.1.2.3 Augmented Lagrangian method

The augmented Lagrangian method exploits a judicious combination of the penalty and Lagrange multiplier method, which is capable of keeping only the advantageous features of either of its predecessors (e.g. see [89]). Namely, this kind of method will again use a penalty term, but with the value of penalty coefficient k_0 sufficiently small in order to avoid the bad conditioning of the modified tangent stiffness in the presence of contact. Furthermore, the augmented Lagrangian method also employs the Lagrange multipliers for representing the contact pressure, with the goal of computing the corresponding values of Lagrange multipliers that enforce the impenetrability condition in contact. However, the presence of the penalty term will make the iterative solution procedure (much) more robust than the corresponding one we have described for the Lagrange multiplier method. More precisely, the augmented Lagrangian method is employed within an operator split solution procedure, which allows us to fix the current value of Lagrange multiplier (or the contact pressure) and recover the same simple problem as already solved with penalty method with only unknowns as the nodal displacements. For each converged value of displacement, the iterative update of the contact pressure is carried out in the second part of this operator split procedure, until the impenetrability condition is fully enforced (within the prescribed tolerance). Perhaps the simplest manner to introduce the augmented Lagrangian method is through the extension of the penalty method, presented in the previous section. To that end, we consider the case where the contact is already initiated and the contact pressure takes a non-zero, yet unknown value. Given the best (iterative) guess for contact pressure, $p^{(k)}$, the contact problem can be

---

[2] Recall that the condition number of the stiffness matrix K, $cond(K) = \lambda_{max}/\lambda_{min}$, allows us to quantify the effect that a perturbation on load vector or stiffness matrix components will have on computed displacements.

formulated as the constrained minimisation problem of the corresponding Lagrangian which can be written:

$$\min_{\hat{g}(\varphi_t(l))\leq 0} [\Pi(\varphi_t(x)) + p_t^{(k)} \hat{g}(\varphi_t(l))] \; ; \; p_t^{(k)} > 0 \tag{5.31}$$

We note in passing that for the correct choice of contact pressure with $p_t^{(k)} \equiv p_t$, we will also have $p_t \hat{g}(\varphi_t(l)) = 0$, and the last problem of constrained minimisation will become identical to the original contact problem in (5.12) and thus lead to the same result. By appealing now to penalty method, we can recast the constrained minimisation in (5.31) above in a format without constraint, which we can write:

$$\min_{\forall \varphi_t} [\Pi(\varphi_t(x)) + p_t^{(k)} \hat{g}(\varphi_t(l)) + \frac{1}{2} k_0 [\hat{g}(\varphi_t(l))]^2 \tag{5.32}$$

In the last expression we choose the penalty parameter k_0 with a sufficiently small value that does not produce a poor conditioning of the tangent stiffness matrix. We note again that, even with such a (sufficiently) small value of the penalty coefficient, this kind of formulation can still enforce the impenetrability condition with the right choice of the contact pressure or Lagrange multiplier.

On the other hand, if the chosen iterative value of the contact pressure $p_t^{(k)}$ does not allow to ensure the impenetrability, with $\hat{g}(\phi_t(l))\big|_{p^{(k)}} \neq 0$, the minimisation problem in (5.32) can be used in order to provide a new, improved value of the contact pressure; the latter is done by using the same pressure update as for the penalty method in (5.28) resulting with:

$$p_t^{(k+1)} = p_t^{(k)} + k_0 \hat{g}(\varphi_t(l))\big|_{p_t^{(k)}} \tag{5.33}$$

With this new iterative value of contact pressure, we can restart, if needed, a new iterative cycle of computations to also obtain the corresponding displacement values.

An alternative derivation of the augmented Lagrangian method, which can shed some light on the best manner for contact pressure update, is provided by appealing to duality. By considering again the case of active contact with $g_t = 0$, we can modify the original contact problem of constrained minimisation with a penalty term; it is important to note that such a penalty term should not change the exact solution of the original contact problem, and that we can still write:

$$\min_{\hat{g}(\varphi_t(l))=0} [\Pi(\varphi_t(x)) + \frac{1}{2} k_0 [\hat{g}(\varphi_t(l))]^2 \tag{5.34}$$

For the augmented Lagrangian in (5.34) above, we can define the dual function in the same manner as in (5.17) applied to the Lagrange multiplier method, resulting with:

$$\phi(p_t) = \min_{\varphi_t}[\Pi(\varphi_t(x)) + p_t\hat{g}(\varphi_t(l)) + \frac{1}{2}k_0[\hat{g}(\varphi_t(l))]^2] \quad (5.35)$$

Moreover, following the same procedure as the one presented previously for Lagrange multiplier method, we can obtain the Hessian of this dual function, as well as its discrete approximation:

$$\frac{d^2}{ds^2}[\phi(p_{n+1} + s\, q_{n+1})]|_{s=0} = -1^T_{n_c}[\mathsf{K}_{n+1} + k_0 1_{n_c} 1^T_{n_c}]^{-1} 1_{n_c} \quad (5.36)$$

By taking into account that the gradient of this dual function will reduce to the impenetrability condition, we can then obtain the fully consistent update procedure for contact pressure, which can be written:

$$p_t^{(k+1)} = p_t^{(k)} + \{1/1^T_{n_c}[\mathsf{K}_{n+1} + k_0 1_{n_c} 1^T_{n_c}]^{-1} 1_{n_c}\}\, \hat{g}(\varphi_t(l))\Big|_{p^{(k)}} \quad (5.37)$$

If we further assume that the penalty coefficient k_0 is sufficiently large with respect to the components of the stiffness matrix of the deformable solid in contact, we can approximate by k_0^{-1} the Hessian component for the contact node, which allows us to recover the simple update formula for contact pressure proposed in (5.33).

The simplified update formula for contact pressure ought to be further modified to account for changing contact conditions, in the case where the contact does not persist. To that end, we introduce an auxiliary variable $z_t \geq 0$ that allows to replace the original minimisation problem with inequality constraint defined in (5.12) by the corresponding constrained minimisation problem with equality constraint; the latter will provide the steady manner for computing the contact pressure under changing contact conditions, which can be written:

$$\min_{\hat{g}(\varphi_t(x))+z_t=0} \Pi(\varphi_t(x)) \; ; \; z_t \geq 0 \quad (5.38)$$

We can easily obtain the dual function for this kind of contact problem formulation to replace the one in (5.35) according to:

$$\phi(p_t) = \min_{\varphi_t, z_t \geq 0}[\Pi(\varphi_t(x)) + p_t(\hat{g}(\varphi_t(l)) + z_t) + \frac{1}{2}k_0(\hat{g}(\varphi_t(l)) + z_t)^2 \quad (5.39)$$

We can further obtain the analytic result for the minimisation of this dual function with respect to the auxiliary variable z_t, leading to the following result:

$$p_t + k_0(\hat{g}(\varphi_t(l)) + z_t) = 0 \quad (5.40)$$

By considering that the auxiliary variable must take a non-negative value, $z_t \geq 0$, the last result will allow us to specify how the latter can be computed; we thus write:

$$z_t = \begin{cases} -\hat{g}(\varphi_t(l)) - \frac{1}{k_0}p_t \; ; \text{ if } -\hat{g}(\varphi_t(l)) - \frac{1}{k_0}p_t \geq 0 \\ 0 \; ; \text{ if } -\hat{g}(\varphi_t(l)) - \frac{1}{k_0}p_t < 0 \end{cases} \quad (5.41)$$

If the last result is introduced into (5.39) it will allow us to obtain the final form of the augmented Lagrangian functional, which is applicable to the unilateral contact problem with impenetrability condition specified in terms of inequality constraint:

$$\min_{\varphi}\{\Pi(\varphi_t(x)) + \frac{1}{2k_0}[< p_t + k_0\hat{g}(\varphi_t(l)) >^2 -(p_t)^2] \qquad (5.42)$$

where $< \cdot >$ denotes again the Macauley bracket which filters only non-negative values of the argument. The augmented Lagrangian method provides not only the solid basis for theoretical formulation of a contact problem, but also for its numerical solution. Namely, the computational advantage of the augmented Lagrangian method becomes obvious when used within the framework of the operator split procedure, where we can separate the global computation of all the nodal values of displacements from the local (limited to contact zone) computation of the contact pressure. The first phase of such an operator split procedure corresponds to finding the solution of the minimisation problem in (5.42) at a fixed value of contact pressure, $p_t^{(k)}$. This is global phase of computations providing the solution to the following equilibrium problem:

$$0 = G(\varphi_t(x), w(x)) + < k_0\hat{g}(\varphi_t(l)) + p_t^{(k)} > w(l) \qquad (5.43)$$

where $G(\varphi_t(x), w(x))$ is the weak form of equilibrium equations for deformable body (here, represented with 1D model of finite elasticity) without contact. By choosing the isoparametric finite element interpolations, and a particular value of pseudo-time t_{n+1}, we can reduce the last expression to a set of nonlinear algebraic equations, which can be written:

$$\hat{\mathsf{f}}^{int}(\mathbf{d}_{n+1}) - \mathsf{f}^{ext}_{n+1} + 1_{n_c} < k_0\hat{g}(1_{n_c}\cdot \mathsf{d}_{n+1}) + p^{(k)}_{n+1} >= 0 \qquad (5.44)$$

This set of nonlinear algebraic equations is solved by Newton's iterative procedure. At each iteration '(i)', we will solve a particular linearized form of the system:

$$\begin{aligned}[\mathsf{K}^{(i)}_{n+1} + k_0 1_{n_c} 1^T_{n_c}]\mathsf{u}^{(i)}_{n+1} &= \mathsf{f}^{ext}_{n+1} - \hat{\mathsf{f}}^{int}(\mathsf{d}^{(i)}_{n+1}) \\ &\quad -1_{n_c} < k_0\hat{g}(1_{n_c}\cdot \mathsf{d}^{(i)}_{n+1}) + p^{(k)}_{n+1} > \end{aligned} \qquad (5.45)$$

which allows us to recover a new iterative value of nodal displacements, obtained through the corresponding update:

$$\mathsf{d}^{(i+1)}_{n+1} = \mathsf{d}^{(i)}_{n+1} + \mathsf{u}^{(i)}_{n+1} \qquad (5.46)$$

This iterative procedure typically requires several iteration to converge. It is important to recall that in all those iteration the contact pressure value $p^{(k)}_{n+1}$ is kept constant.

With the converged values of nodal displacements, denoted as $\bar{\mathbf{d}}_{n+1}$, we start the local phase of the operator split procedure for computing the contact pressure. More precisely, we obtain the contact pressure update by using the appropriate modification of the result in (5.37) to account for the inequality constraint:

$$p_{n+1}^{(k+1)} = < p_{n+1}^{(k)} + k_0 \hat{g}(\mathbf{1}_{n_c} \cdot \bar{\mathbf{d}}_{n+1}) > \quad (5.47)$$

The computation of contact pressure is carried out for each contact element independently, which corresponds to the local phase of the operator split procedure. The iterative procedure in the local phase should stop when the correct value of contact pressure is reached, enforcing exactly (within a given tolerance) the impenetrability condition. It is important to note that the iterative procedure in local phase is done with a single iteration at the time, followed by a number of global iterations on equilibrium equations with the updated value of contact pressure.

The augmented Lagrangian method can thus keep the best features of previously presented methods, with no increase of the number of equations to solve in the global phase (for the given iterative value of contact pressure), as well as with exact enforcement of the impenetrability condition once we converged with local iterations towards the exact contact pressure.

In closing this section, we would like to briefly comment upon the important difference in applying the operator split solution procedure to contact as opposed to the previously described operator split for plasticity problems. In the later case, the global phase of solving the equilibrium equations is advanced one iteration at time, followed by a number of iterations in the local phase at each Gauss point for computing the plastic multiplier (the Lagrange multiplier for plasticity) and the corresponding values of internal variables enforcing the admissibility of stress field in the sense of plasticity criterion. In other words, the local phase of the operator split procedure for plasticity is carried out in the inner loop. The inverse order is followed in the operator split procedure applied to contact. Namely, one local iteration for computing contact pressure (the Lagrange multiplier corresponding to contact) is followed by a number of global iterations on the equilibrium equations for computing the nodal values of displacements. Hence, what is the local computation for contact is carried out in the outer loop of the operator split procedure. This difference in the order of application between the global and local phase of the operator split procedure corresponds to the quite different nature of the constraint in plasticity versus the one in contact.

5.2 Contact problems in 2D and 3D

The solution methods which we presented for simple contact problem in 1D case will apply with practically no modification to 2D and 3D contact problems. The main novelty in 2D and 3D case that increases the contact problem complexity in large displacement setting, pertains to the search of the contact zone that can change during the motion (see Figure 5.2).

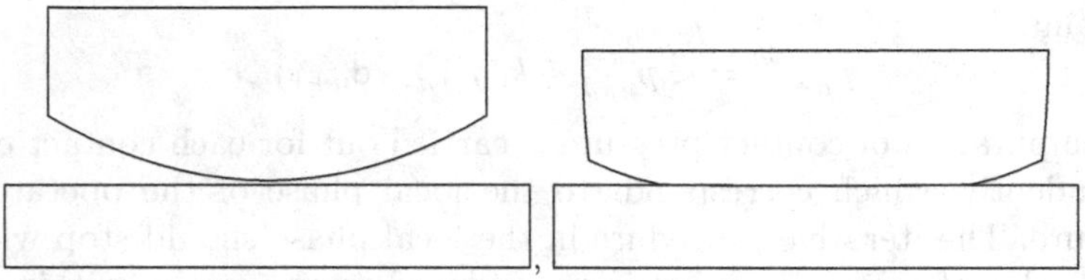

i) Contact initiation. ii) Developed contact zone.

Fig. 5.2 Detection of contact zone for 2D case.

Therefore, our first goal in the subsequent developments is to show how to identify the contact zone or the boundary Γ_c, which is reduced to a line in 2D or to a surface in 3D contact problems. We elaborate only upon 2D case presenting the slide-line algorithm as an efficient procedure for describing the evolution of the contact zone; the latter leads to the corresponding propagations of the contact boundary Γ_c at the expense of the Neumann boundary Γ_σ. The slide line algorithm presented herein targets the case where the boundary Γ_σ is represented with straight segments, which would typically follow from a 2D finite element mesh constructed with 4-node isoparametric elements. The rigid obstacle boundary is also represented with piece-wise straight segments (see Figure 5.3).

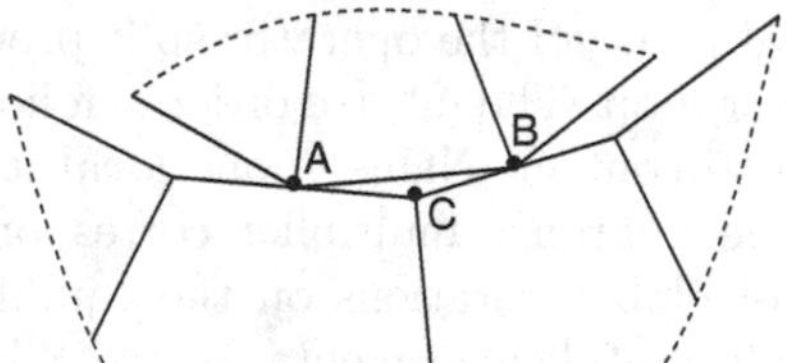

Fig. 5.3 Finite element representation of the rigid obstacle and deformable solid boundary; note that the nodes on the rigid obstacle boundary are not in contact.

We consider that the penetration of the rigid obstacle by the deformable body which enters in contact is inadmissible (except for the penalty method), and that it has to be eliminated by the corresponding iterative procedure to provide the final displacement in agreement with contact constraint. The admissibility is reestablished if the contact boundary Γ_c for such a case is reduced to a single point $\bar{\mathbf{y}}$, which we find as the orthogonal projection of the point φ

onto the rigid body obstacle boundary. We assume that $\forall \varphi$ we can provide the unique point $\bar{\mathbf{y}}$ as the shortest distance to the obstacle according to:

$$\| \varphi - \bar{\mathbf{y}} \| = \min_{\forall \mathbf{y} \in \Gamma_{rigide}} \| \varphi - \mathbf{y} \| \tag{5.48}$$

In practical computations, we can define the measure of penetration by the gap function, whose value is computed as the oriented distance:

$$\hat{g}(\varphi) := -\mathbf{n}_k \cdot (\varphi - \mathbf{y}_k) \tag{5.49}$$

where $\mathbf{n}_k$ is the unit exterior normal to rigid obstacle at point $\bar{\mathbf{y}}$, and $\mathbf{y}_k$ is the closest nodal point to $\bar{\mathbf{y}}$. An admissible equilibrium state in 2D unilateral contact is thus defined with non-positive value of the gap function:

$$\hat{g}(\varphi) \leq 0 \tag{5.50}$$

More precisely, the zero value of the gap function indicates the contact between the solid deformable body and the rigid obstacle (at point $\bar{\mathbf{y}}$), whereas a negative value of the gap function indicates that there is no contact and that the body and the obstacle are separated. The positive values of gap imply the obstacle penetration, which is considered inadmissible, except for the penalty method.

In large displacement setting with large sliding in the contact zone, the boundary nodes which enter in contact with rigid obstacle can change in each increment or even in each iteration. This can often lead to convergence difficulties without an efficient contact zone search algorithm. The latter can be provided for 2D case in terms of the slide-line algorithm, which can define the closest point $\bar{\mathbf{y}}$ on the rigid obstacle boundary for any point φ placed on the Neumann boundary Γ_σ^φ of a particular deformed configuration. Although all the points are concerned theoretically, practically we check only the finite element nodes that are likely to enter in contact with the rigid obstacle. Therefore, we can identify two different cases, pertaining to a convex or a concave rigid obstacle form (see Figure 5.4, for an illustration obtained with a piece-wise linear approximation of the obstacle boundary). For the former case of convex obstacle, we can use the gap computation as defined in (5.49) above when the point φ is placed either in sub-domain A_{k-1} or in sub-domain A_{k+1}. We can define the closest projection to be in node $\mathbf{y}_k$ if the point φ is placed in sub-domain A_k. In summary, we can write the gap function for the convex obstacle case according to:

$$\hat{g}(\varphi) = \begin{cases} -\mathbf{n}_{k-1} \cdot (\varphi - \mathbf{y}_{k-1}) \; ; \; \forall \varphi \in A_{k-1} \\ -\mathbf{n}_k \cdot (\varphi - \mathbf{y}_k) \; ; \; \forall \varphi \in A_{k+1} \\ \| \varphi - \mathbf{y}_k \| \; ; \; \forall \varphi \in A_k \end{cases} \tag{5.51}$$

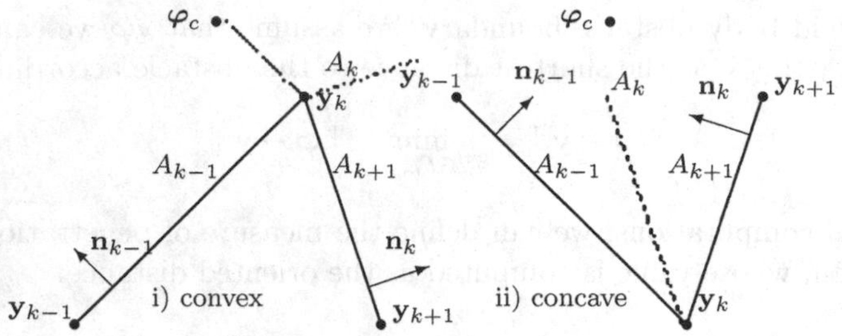

Fig. 5.4 Finite element piece-wise linear approximation of rigid obstacle and deformable solid boundaries: convex and concave cases.

For the concave obstacle case, the sub-domain A_k degenerates into a line. There thus remain only two possibilities to define the closest point projection and the corresponding value of the gap function:

$$\hat{g}(\boldsymbol{\varphi}) = \begin{cases} \min(g_1, g_2) \; ; \; g_1 \neq g_2 \, ; \, g_1 = -\mathbf{n}_{k-1} \cdot (\boldsymbol{\varphi} - \mathbf{y}_{k-1}) \, , \\ \qquad\qquad\qquad\qquad\qquad g_2 = -\mathbf{n}_k \cdot (\boldsymbol{\varphi} - \mathbf{y}_k) \\ \parallel \boldsymbol{\varphi} - \mathbf{y}_k \parallel \; ; \; g_1 = g_2 \end{cases} \tag{5.52}$$

Having defined the gap function value for each case of practical interest, we further turn to defining the contact term contribution to the weak form of equilibrium equations. Recall that we find no contact term in the strong form of the equilibrium equations, but rather an additional constraint that contact will place upon the displacement field along the contact boundary Γ_c; for example, the strong form of equilibrium equations corresponding to 2D or 3D elasticity problem with contact can be written:

$$\begin{aligned} & div[\hat{\mathbf{P}}(\boldsymbol{\varphi})] + \mathbf{b} = \mathbf{0} && \text{in } \Omega \\ & \mathbf{Pn} = \bar{\mathbf{t}} && \text{on } \Gamma_\sigma \\ & \mathbf{u} = \bar{\mathbf{u}} (= \mathbf{0}) && \text{on } \Gamma_u \\ & \hat{g}(\boldsymbol{\varphi}) \leq 0 \, ; p_c := -\mathbf{n} \cdot \mathbf{Pn} \geq 0 \, ; \, p_c \hat{g}(\cdot) = 0 \text{ on } \Gamma_c \end{aligned} \tag{5.53}$$

The contact term contribution to equilibrium equations only appears in the weak form of such equations. In analogy with the contact problem in 1D case, we can provide three different versions for writing the weak form, by using either the Lagrange multiplier method, the penalty method or the augmented Lagrangian method. We can thus write:

Weak form of 2D/3D elasticity in large displacements in presence of contact (on Γ_c)

Given: $\mathbf{b}; \bar{\mathbf{t}}; \bar{\mathbf{u}} = \mathbf{0}$

Find: $\boldsymbol{\varphi}$, such that $\forall \mathbf{w}$

(with: $G(\varphi;\mathbf{w}) := \int_\Omega (\nabla \mathbf{w} \cdot \mathbf{P} - \mathbf{w} \cdot \mathbf{b})\, dV - \int_{\Gamma_\sigma} \mathbf{w} \cdot \bar{\mathbf{t}}\, dA$)
which verifies (i), (ii) or (iii), with
(i) The Lagrange multipliers method, imposing exactly the impenetrability constraint, with $p_t \geq 0$; $\hat{g}(\varphi(\varphi)) \leq 0$; $p_t g(\varphi) = 0$,

$$\begin{aligned} 0 &= G_{contact}(\varphi, p_t; \mathbf{w}) := G(\varphi;\mathbf{w}) + \int_{\Gamma_c} p_t (\mathbf{w} \cdot \mathbf{n})\, dA \\ 0 &= G_{contact}(\varphi, p_t; q) := \int_{\Gamma_c} q \hat{g}(\varphi)\, dA \end{aligned} \tag{5.54}$$

where q is the contact pressure variation
(ii) The penalty method relaxing the impenetrability constraint and using the corresponding approximation of contact pressure $p_t = kg(\varphi)$)

$$0 = G_{contact}(\varphi;\mathbf{w}) := G(\varphi;\mathbf{w}) + \int_{\Gamma_c} k\hat{g}(\varphi(\mathbf{x}_c))(\mathbf{w} \cdot \mathbf{n})\, dA \tag{5.55}$$

(iii) The augmented Lagrangian method using a judicious combination of penalty and Lagrange multiplier methods, which inherits only the advantages of both:

$$\begin{aligned} &0 = G_{contact}(\varphi;\mathbf{w}) := G(\varphi;\mathbf{w}) + \int_{\Gamma_c} < p^{(k)} + k_0 \hat{g}(\varphi(\mathbf{x}_c)) > (\mathbf{n} \cdot \mathbf{w})\, dA \\ &p^{(k+1)} =< p^{(k)} + k_0 g(\varphi(\mathbf{x}_c)) > \end{aligned} \tag{5.56}$$

Having defined the contact surface by using the *slide line* algorithm and having made the final choice of the solution method among three we presented, we can further carry on with the finite element discretization to obtain as the final result a set of nonlinear algebraic equations. The latter is the same as the corresponding result for 1D contact problem. Hence, the subsequent solution procedure will simply follow in the steps already traced for 1D case.

5.2.1 Contact between two deformable bodies in 2D case

In this section we will generalize the contact problem by considering that the obstacle is not a rigid body, but another solid deformable body. In other words, we will study a bilateral contact problem. We present two different approaches to constructing the solution for this kind of problem: the classical "master–slave" approach and the more modern approach where a special contact element, yet referred to as 'mortar element', is placed between two deformable bodies in contact.
When solving a bilateral contact problem with the master–slave approach, we consider that one of the solid bodies in contact is the "master" which dictates the contact conditions in the sense of computing the direction of the normal and providing the value of the gap function. The other solid body remains the

"slave", which follows the motion of the master in agreement with the contact constraint. If one of the solid bodies in bilateral contact is (much) stiffer than the other, it is naturally designated as the master. However, very often the difference in stiffness of two solid bodies in contact is not so pronounced,[3] and the choice of master body is not so clear. In such a case we use two-pass algorithm, where each of two solid deformable bodies gets to play the role of master and the results of two analyses are finally combined.

We study here in more detail this kind of master–slave approach for 2D case. Having selected the solid deformable body which plays the role of a master (with respect to the other that is considered as slave), we can apply the slide-line algorithm for contact zone search in a very much the same manner as already described for unilateral contact, but with the master body replacing the rigid obstacle. The corresponding value of the gap function can be obtained by (5.51) or (5.52) for convex or concave form of master body, respectively. The only difference with respect to unilateral contact concerns the exterior normal computation, which no longer remains constant but changes along with the motion of the master body. The simplest discrete approximation for master–slave approach is based on a 3-node contact element (see Wriggers and Simo [266]), which consists of a straight segment of the master body boundary placed between two nodes '1' and '2', and a single node 's' of the slave body. This kind of approximation allows us to define the gap function according to:

$$\left.\begin{aligned} \mathbf{m}^m &= (\boldsymbol{\varphi}_2 - \boldsymbol{\varphi}_1)/\parallel \boldsymbol{\varphi}_2 - \boldsymbol{\varphi}_1 \parallel \\ \mathbf{n}^m &= \mathbf{e}_3 \times \mathbf{m}^m \end{aligned}\right\} \tag{5.57}$$
$$\Longrightarrow \hat{g}(\boldsymbol{\varphi}_s, \boldsymbol{\varphi}_1, \boldsymbol{\varphi}_2) = -(\boldsymbol{\varphi}_s - \boldsymbol{\varphi}_1) \cdot \mathbf{n}^m$$

With the final position of these 3 nodes defined in a particular deformed configuration with $\boldsymbol{\varphi}_1 = \mathbf{x}_1 + \mathbf{d}_1$, $\boldsymbol{\varphi}_2 = \mathbf{x}_2 + \mathbf{d}_2$ and $\boldsymbol{\varphi}_s = \mathbf{x}_s + \mathbf{d}_s$, along with their variations $\mathbf{w}_1$, $\mathbf{w}_2$ and $\mathbf{w}_s$, we can easily compute the variation of the tangent vector to the contact surface:

$$\begin{aligned} \frac{d}{dt}[\mathbf{m}^m_t]\bigg|_{t=0} &= \frac{d}{dt}[\hat{\mathbf{m}}^m(\overbrace{\boldsymbol{\varphi}_{1,t}}^{\boldsymbol{\varphi}_1 + t\mathbf{w}_1}, \overbrace{\boldsymbol{\varphi}_{2,t}}^{\boldsymbol{\varphi}_2 + t\mathbf{w}_2})]\bigg|_{t=0} \\ &= [-\frac{1}{\|\boldsymbol{\varphi}_2 - \boldsymbol{\varphi}_1\|^2}\frac{d}{dt}(\boldsymbol{\varphi}_{2,t} - \boldsymbol{\varphi}_{1,t}) \cdot (\boldsymbol{\varphi}_2 - \boldsymbol{\varphi}_1) \\ &\quad + \frac{1}{\|\boldsymbol{\varphi}_2 - \boldsymbol{\varphi}_1\|}\frac{d}{dt}(\boldsymbol{\varphi}_{2,t} - \boldsymbol{\varphi}_{1,t})]\bigg|_{t=0} \\ &= \frac{1}{\|\boldsymbol{\varphi}_2 - \boldsymbol{\varphi}_1\|}[\mathbf{I} - \mathbf{m}^m \otimes \mathbf{m}^m](\mathbf{w}_2 - \mathbf{w}_1) \\ &= \frac{1}{\|\boldsymbol{\varphi}_2 - \boldsymbol{\varphi}_1\|}(\mathbf{n}^m \otimes \mathbf{n}^m)(\mathbf{w}_2 - \mathbf{w}_1) \end{aligned} \tag{5.58}$$

[3] The same is true when we have to deal with the problem of self-contact where two parts of the same solid deformable body enter in contact, which can also be placed within the bilateral contact framework.

as well as the variation of the exterior normal:

$$\left.\frac{d}{dt}[\mathbf{n}_t^m]\right|_{t=0} = \mathbf{e}_3 \times \left.\frac{d}{dt}[\mathbf{m}_t^m]\right|_{t=0}$$
$$= \frac{1}{\|\boldsymbol{\varphi}_2-\boldsymbol{\varphi}_1\|} \overbrace{(\mathbf{e}_3 \times \mathbf{n}^m)}^{-\mathbf{m}^m}[\mathbf{n}^m \cdot (\mathbf{w}_2 - \mathbf{w}_1)] \tag{5.59}$$
$$= -\frac{1}{\|\boldsymbol{\varphi}_2-\boldsymbol{\varphi}_1\|}(\mathbf{m}^m \otimes \mathbf{n}^m)(\mathbf{w}_2 - \mathbf{w}_1)$$

With these results in hand, we can easily obtain the corresponding variation of the gap function, which can be written:

$$\left.\frac{d}{dt}[\hat{g}(\boldsymbol{\varphi}_t)]\right|_{t=0} = \left.\frac{d}{dt}[(\boldsymbol{\varphi}_{s,t} - \boldsymbol{\varphi}_{1,t}) \cdot \mathbf{n}_t^m]\right|_{t=0}$$
$$= (\mathbf{w}_s - \mathbf{w}_1) \cdot \mathbf{n}^m - \frac{\boldsymbol{\varphi}_s-\boldsymbol{\varphi}_1}{\|\boldsymbol{\varphi}_s-\boldsymbol{\varphi}_1\|} \cdot [\mathbf{m}^m \otimes \mathbf{n}^m](\mathbf{w}_2 - \mathbf{w}_1) \tag{5.60}$$
$$= (\mathbf{w}_s - \mathbf{w}_1) \cdot \mathbf{n}^m - \bar{\xi}\mathbf{n}^m \cdot (\mathbf{w}_2 - \mathbf{w}_1)$$
$$= \mathbf{w}_s - (1-\bar{\xi})\mathbf{w}_1 - \bar{\xi}\mathbf{w}_2$$

where $\bar{\xi} = \frac{\boldsymbol{\varphi}_s-\boldsymbol{\varphi}_1}{\|\boldsymbol{\varphi}_s-\boldsymbol{\varphi}_1\|} \cdot \mathbf{m}^m$. The corresponding contribution of such a contact element to the discrete approximation of equilibrium equations can then be written:

$$0 = G^e_{contact}(\boldsymbol{\varphi}; \mathbf{w}) := \int_{\Gamma_c} \left.\frac{d}{dt}[\hat{g}(\boldsymbol{\varphi}_t)]\right|_{t=0} p_c \, d\Gamma = \mathsf{w}^e \cdot p_c \hat{\mathsf{f}}_c^{e,int}(\mathbf{d})\,;$$
$$\mathsf{w}^e = \begin{bmatrix} \mathsf{w}_s \\ \mathsf{w}_1 \\ \mathsf{w}_2 \end{bmatrix}\,;\ \hat{\mathsf{f}}_c^{e,int} := \begin{bmatrix} \mathbf{n}^m \cdot \mathsf{w}_s \\ -(1-\bar{\xi})\mathbf{n}^m \cdot \mathsf{w}_1 \\ -\bar{\xi}\mathbf{n}^m \cdot \mathsf{w}_2 \end{bmatrix} \tag{5.61}$$

where p_c is computed by using one of the solution methods for contact defined in (5.54) to (5.56). The consistent linearization of the equilibrium equations can then be carried out by exploiting the auxiliary result in (5.60), but with incremental displacement u replacing the variation; we thus obtain the following result:

$$Lin\left[\begin{bmatrix} \mathsf{w}^e \cdot p_c \hat{\mathsf{f}}_c^{e,int} \\ +q_c \hat{g} \end{bmatrix}\right] = \begin{bmatrix} \mathsf{w}^e \\ q_c \end{bmatrix} \cdot \left\{ \begin{bmatrix} \hat{\mathsf{f}}_c^{e,int} \\ \hat{g} \end{bmatrix} + \begin{bmatrix} \mathsf{K}_c^e & \hat{\mathsf{f}}_c^{e,int,T} \\ \hat{\mathsf{f}}_c^{e,int} & 0 \end{bmatrix} \begin{bmatrix} \mathsf{u}^e \\ \Delta p_c \end{bmatrix} \right\}$$

with:

$$\mathsf{w}^e \cdot \mathsf{K}_c^e \mathsf{u}^e = -\frac{p_c}{\|\boldsymbol{\varphi}_2-\boldsymbol{\varphi}_1\|}\{(\mathsf{u}_s - (1-\bar{\xi})\mathsf{u}_1 - \bar{\xi}\mathsf{u}_2) \cdot [\mathbf{m}^m \otimes \mathbf{n}^m]$$
$$(\mathsf{w}_2 - \mathsf{w}_1) + (\mathsf{w}_s - (1-\bar{\xi})\mathsf{w}_1 - \bar{\xi}\mathsf{w}_2) \cdot [\mathbf{n}^m \otimes \mathbf{m}^m](\mathsf{u}_2 - \mathsf{u}_1) \tag{5.62}$$
$$+ \frac{g_s}{\|\boldsymbol{\varphi}_s-\boldsymbol{\varphi}_1\|}(\mathsf{u}_2 - \mathsf{u}_1) \cdot [\mathbf{n}^m \otimes \mathbf{n}^m](\mathsf{w}_2 - \mathsf{w}_1)\}$$

We can note that the tangent stiffness matrix of a contact element obtained for the bilateral contact takes a considerably more elaborate form than the same matrix for unilateral contact, since the normal to contact surface in large displacements no longer remains fixed. The finite element assembly will then allow us to take into account and store appropriately the contributions of all contact elements towards the tangent stiffness matrix in (5.62) and the internal force vector in (5.61).
In order to complete the description of the proposed procedure for bilateral contact problems, we also briefly discuss the corresponding modifications for the case of non-smooth contact surface. This type of modification is needed in the case where we use irregular representation of the contact boundary in terms of the piece-wise linear approximation (see Figure 5.5), where in passing from one contact element to another we obtain a jump in the normal direction to contact surface. Moreover, in non-smooth transition from one contact element to another, the gap function variation can be difficult to compute:

$$\hat{g}(\boldsymbol{\varphi}_s, \boldsymbol{\varphi}_k) = \| \boldsymbol{\varphi}_s - \boldsymbol{\varphi}_k \| \Rightarrow \frac{d}{dt}[\hat{g}_t]\bigg|_{t=0} = \frac{\boldsymbol{\varphi}_s - \boldsymbol{\varphi}_k}{\| \boldsymbol{\varphi}_s - \boldsymbol{\varphi}_k \|} \cdot (\mathbf{w}_s - \mathbf{w}_k) \qquad (5.63)$$

We present subsequently two possibilities for dealing with this kind of difficulty by using a modified contact formulations. The first modified formulation of the contact problem considers a non-smooth representation of the contact boundary and defines the gap function in the form of multi-surface criteria, quite similar to those discussed for plasticity model. More precisely, at the point of passing from one contact element to another, both the gap function and the contact pressure are represented as the appropriate linear combinations of the corresponding values in two neighboring elements; for example, for the augmented Lagrangian method we can write the gap and contact pressure, respectively:

$$g_i^{(k)} := -\mathbf{n}_i^{m\,(k)} \cdot (\boldsymbol{\varphi}_s^{(k)} - \boldsymbol{\varphi}_i^{(k)}) \; ; \; < p_{ci}^{(k)} + k_0 g_i^{(k)} > \; ; \; i = 1, 2 \qquad (5.64)$$

The computational procedure remains the same as described previously, except for this modified definition of the gap function for the node where we pass from one element to another, where both contact elements attached to the node will contribute. The contact reaction force at the particular node is then expressed as a vector with two components, defined from the contact pressure values for both contact elements attached to this node.
The second modified formulation, which provides an efficient approach for dealing with a bilateral contact problem with irregular contact surface, concerns replacement of such a surface with the corresponding smooth shape (see Figure 5.5). One replacement of this kind that ensures a smooth contact boundary representation can be provided by using the Hermite polynomials. In order to construct this kind of representation, we start by computing at each node of the contact boundary the common reference frame shared

by two neighboring contact elements; to that end, we first find the tangent vectors in each element, $\mathbf{t}_1$ and $\mathbf{t}_2$:

$$\mathbf{t}_1 = \boldsymbol{\varphi}_k - \boldsymbol{\varphi}_{k-1} \; ; \; \mathbf{t}_2 = \boldsymbol{\varphi}_{k+1} - \boldsymbol{\varphi}_k \tag{5.65}$$

where the first contact element is placed between nodes $k-1$ and k, whereas the second is placed between the nodes k and $k+1$. We can compute at the particular node k the radius of curvature $\boldsymbol{\varphi}_k - \mathbf{c}_k$ and the curvature κ_k according to:

$$\begin{aligned}
\mathbf{c}_k &= \boldsymbol{\varphi}_k + \tfrac{1}{2}[(\mathbf{t}_1 \cdot \mathbf{t}_1)\mathbf{t}_2 + (\mathbf{t}_2 \cdot \mathbf{t}_2)\mathbf{t}_1] \times \tfrac{\mathbf{t}_2 \times \mathbf{t}_1}{\|\mathbf{t}_2 \times \mathbf{t}_1\|} \\
\mathbf{n}_k &= \pm \tfrac{\mathbf{c}_k - \boldsymbol{\varphi}_k}{\|\mathbf{c}_k - \boldsymbol{\varphi}_k\|} \; ; \; \mathbf{m}_k = \mathbf{e}_3 \times \mathbf{n}_k \\
\kappa_k &= \tfrac{1}{\mathbf{n}_k \cdot (\mathbf{c}_k - \boldsymbol{\varphi}_k)}
\end{aligned} \tag{5.66}$$

Having completed this kind of parameter computations for all nodes, we can easily complete the construction of the corresponding smooth representation of contact boundary. For example, for a particular contact element placed between nodes k and $k+1$, we can write:

$$\phi(s) = \underbrace{\phi_{k+1} - \phi_k}_{=0} + \phi'_k H_1(s) + \phi'_{k+1} H_2(s) + \phi''_k H_3(s) + \phi''_{k+1} H_4(s) \tag{5.67}$$

where $H_i(s)\,;\, i = 1, 2, 3, 4$ are the Hermite polynomials defining the smooth contact boundary in terms of arc-length coordinate s, running from 0 to final length $l = \int_k^{k+1} s\, ds$:

$$\begin{aligned}
H_1(s) &= \tfrac{1}{2l^4}(l/2 - s)^3(l/2 + s)(5l + 6s) \; ; \\
H_2(s) &= \tfrac{1}{2l^4}(l/2 - s)(l/2 + s)^3(5l - 6s) \; ; \\
H_3(s) &= \tfrac{1}{2l^3}(l/2 + s)^2(l/2 - s)^3 \; ; \\
H_4(s) &= \tfrac{1}{2l^3}(l/2 + s)^3(l/2 - s)^2
\end{aligned} \tag{5.68}$$

The parameters for the smooth contact boundary representation of this kind are computed in the local reference frames placed at each contact node:

$$\begin{aligned}
\phi'_k &:= \tan\alpha_k = -\frac{\mathbf{m}_k \times (\boldsymbol{\varphi}_{k+1} - \boldsymbol{\varphi}_k) \cdot \mathbf{e}_3}{\mathbf{m}_k \cdot (\boldsymbol{\varphi}_{k+1} - \boldsymbol{\varphi}_k)} \; ; \\
\phi'_{k+1} &:= \tan\alpha_{k+1} = \frac{\mathbf{m}_{k+1} \times (\boldsymbol{\varphi}_{k+1} - \boldsymbol{\varphi}_k) \cdot \mathbf{e}_3}{\mathbf{m}_{k+1} \cdot (\boldsymbol{\varphi}_{k+1} - \boldsymbol{\varphi}_k)} \; ; \\
\phi''_k &= \kappa_k [1 + (\phi'_k)^2]^{3/2} \; ; \; \phi''_{k+1} = \kappa_{k+1} [1 + (\phi'_{k+1})^2]^{3/2}
\end{aligned} \tag{5.69}$$

This kind of representation of contact boundary is sufficiently smooth to ensure the continuity of the second derivative when passing from one segment to another. Therefore, we can count on the quadratic convergence rate of Newton's iterative method when applied to this class of contact problems.

5.2.2 Mortar element method for contact

The mortar element method (e.g. see Bernardi et al. [33]) is a successful alternative for solving the bilateral contact problems, which provides a very

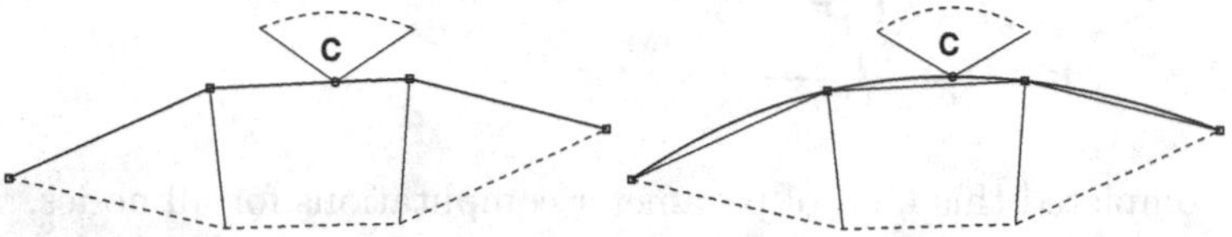

Fig. 5.5 Contact boundary: irregular representation and corresponding smooth representation.

important advantage of ensuring an improved accuracy for contact pressure computations. The main new idea advanced with mortar method is to avoid the issue of selecting the master body in bilateral contact by introducing an intermediate surface between two solid bodies in contact, which not necessarily coincides with any of the boundaries of the bodies in contact. The intermediate surface, denoted by Γ_c, then provides the unique basis to formulate bilateral contact problem and to carry out the corresponding integrals computations. In this manner, the mortar element method provides the intrinsic representation of the contact pressure, contrary to the master–slave contact element approach where we must accept in general two different boundaries in contact.

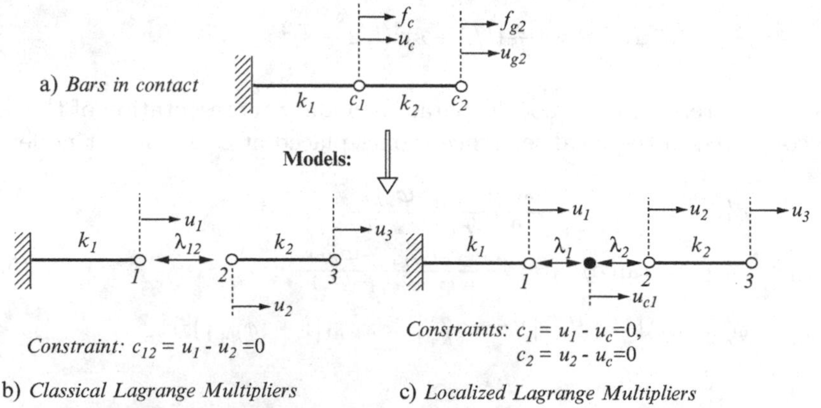

Fig. 5.6 Contact between two 1D bars: intermediate point between two bars in contact, classical Lagrange multipliers and localized Lagrange multipliers.

For clarity, we first start by presenting the mortar method in the simplest possible 1D case, where two elastic bars enter in contact during the motion constrained to horizontal axis; see Figure 5.6. In such a case, the intermediate surface is reduced to a single point denoted with $\Gamma_c = \{c\}$. We will assign to contact point 'c' an independent (incremental) displacement 'u_c', which allows us to specify its final position in deformed configuration 'φ_c'. The independence of motion of node c allows us to define the contact pressure in the intrinsic manner, with no need to privilege one body or another. More precisely, we will introduce the localized Lagrange multipliers (e.g. see Park et al. [217]), which here correspond to contact forces $p_c^{(1)} = \lambda_1$ and $p_c^{(2)} = \lambda_2$ applied on node c by body one and two, respectively. The discrete approximation of the weak form of equilibrium equations for this kind of formulation of bilateral contact problem can thus be written:

$$\begin{aligned} 0 &= G(\boldsymbol{\varphi}^{(1)}, p_c^{(1)}; \mathbf{w}^{(1)}) := \mathbf{w}^{(1)^T} (\mathbf{f}^{int,(1)} - \mathbf{f}^{ext,(1)} + p_c^{(1)} \mathbf{1}_{n_c}^{(1)}) \; ; \\ 0 &= G(\boldsymbol{\varphi}^{(2)}, p_c^{(2)}; \mathbf{w}^{(2)}) := \mathbf{w}^{(2)^T} (\mathbf{f}^{int,(2)} - \mathbf{f}^{ext,(2)} + p_c^{(2)} \mathbf{1}_{n_c}^{(2)}) \; ; \\ 0 &= w_c (p_c^{(1)} + p_c^{(2)}) \end{aligned} \tag{5.70}$$

Among these three equations, the first two are the equilibrium equations for each body, whereas the last one is the equilibrium equation for the intermediate contact point. We note that the last equation is the only one featuring both localized Lagrange multipliers. Other than that, there is no coupling of the localized Lagrange multiplier, and each remains restricted to the corresponding equilibrium equation of only one body. In other words, the localized Lagrange multiplier for mortar element applies to only one body in contact (and the intermediate point c), which explains the name *localized* Lagrange multiplier. This is in sharp contrast with the classical definition of Lagrange multipliers in bilateral contact, which is conjugate to relative displacement between two bodies in contact and thus couples the equilibrium equations of both bodies. Moreover, the definition of each localized Lagrange multiplier will remain unique even in the case where more than two solid bodies enter in contact (see Park et al. [217]), contrary to the classical Lagrange multipliers for the same case that are no longer uniquely defined nor linearly independent.

The contact problem formulation with localized Lagrange multipliers and the resulting structure of equilibrium equations allow a straightforward application of the operator split solution procedure. We start by first solving in parallel the corresponding equilibrium equation for each body in contact, for the given value of the localized Lagrange multiplier. We then carry on with the solution of the equilibrium equations for the nodes placed at the intermediate contact boundary Γ_c, which provides the bases for computing the localized Lagrange multipliers updates.

We study subsequently a bilateral contact problem in 2D case. The main difficulty with respect to 1D case we already discussed pertains to ensuring that the choice of the intermediate surface Γ_c between two solid bodies in contact remains unique. The latter is solved by Papadopoulous and Taylor [215] for 2D case of bilateral contact by using the contact boundary representation based on 3-node contact elements; see Figure 5.7.

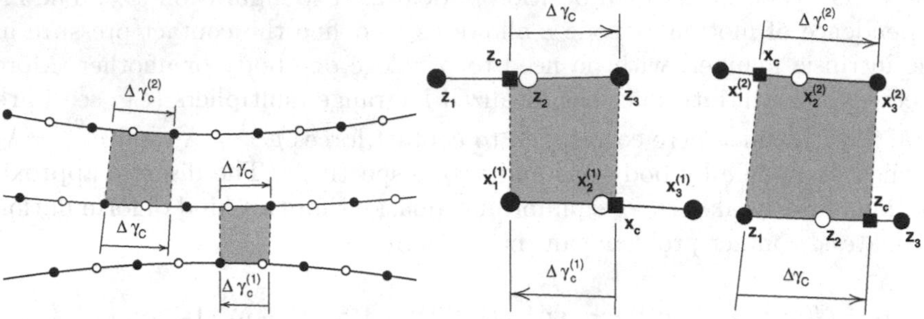

Fig. 5.7 Representation of contact boundary for solid bodies in contact by using 3-node contact elements.

The proposed approach is defined as follows: we construct the orthogonal projection for the corner nodes of each contact element which are placed on the boundaries of both bodies in contact, $\partial\Omega^{(1)}$ and $\partial\Omega^{(2)}$. The chosen order of accounting for successive boundary nodes should allow us to define the segments with 4-nodes, where each segment only concerns one contact element, placed either on $\partial\Omega^{(1)}$ or on $\partial\Omega^{(2)}$. We can easily show that the number of segments remains the same regardless of the order of passing from node to node along boundaries $\partial\Omega^{(\alpha)}$; $\alpha = 1, 2$.
Having chosen the segments, we further define the corresponding parameterization of the intermediate contact boundary Γ_c, such that each segment pertains to the corresponding parent element with value of natural coordinate $-1 \le \xi \le 1$. This kind of isoparametric parameterization is constructed by using 3-node elements, with the central node always placed in the center of each element with $z_2 = \frac{1}{2}(z_1 + z_2)$, which allows us to simplify the corresponding computations related to contact term by using the standard numerical integration (see Figure 5.8).
The weak form of the equilibrium equations with mortar elements used for contact can thus be written:

$$
\begin{aligned}
0 &= G_c(\boldsymbol{\varphi}^{(\alpha)}, \boldsymbol{\varphi}_c, p_c^{(\alpha)}; \mathbf{w}^{\alpha}, \mathbf{w}_c, q_c^{(\alpha)}) \\
&= G(\boldsymbol{\varphi}^{(\alpha)}; \mathbf{w}^{(\alpha)}) + \textstyle\int_{\Gamma_c^h} \{p_c^{(\alpha)}(\mathbf{w}^{(\alpha)} - \mathbf{w}_c) \cdot \mathbf{n}_c\} ds \qquad (5.71) \\
0 &= \textstyle\int_{\Gamma_c^h} \{q_c^{(\alpha)}(\boldsymbol{\varphi}^{(\alpha)} - \boldsymbol{\varphi}_c) \cdot \mathbf{n}_c\} ds \ ; \ (\alpha) = 1, 2
\end{aligned}
$$

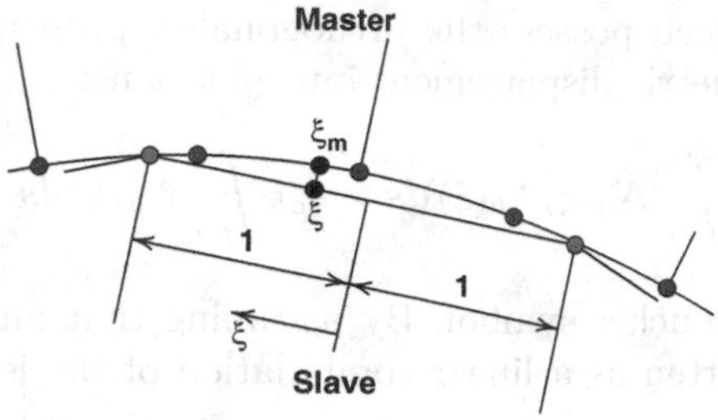

Fig. 5.8 Numerical integration with mortar element after projection on intermediate contact surface.

where $G(\varphi^{(\alpha)}; \mathbf{w}^{(\alpha)})$ is the corresponding part of the weak form of equilibrium equations for each body without contact, whereas Γ_c^h is the corresponding mortar element representation of the intermediate contact surface. It is important to note that all the numerical integration for contact is carried out at the contact boundary

$$\int_{\Gamma_c} f(\xi) d\xi = \int_{\Gamma_{m,s}} \tilde{f}(\tilde{\xi}) d\tilde{\xi} \tag{5.72}$$

where $\tilde{\xi}$ is the projection of the chosen parameterization along $\partial\Omega^{(\alpha)}$ onto Γ_c. The discrete approximation of the equilibrium equations is obtained by using the isoparametric interpolation of the position vector:

$$\varphi(\xi)\Big|_{\Gamma_c^h} = \sum_{a=1}^{3} N_a(\xi)\varphi_a \tag{5.73}$$

The localized Lagrange multipliers or contact pressure field is now discretized by making the appropriate choice of special shape functions $\hat{N}_a$:

$$p_c^{(\alpha)}\Big|_{\Gamma_c^h} = \sum_{a=1}^{3} \hat{N}_a(\xi) p_{c,a}^{(i)} \tag{5.74}$$

In principle, the shape functions for contact pressure discrete approximation are not necessarily the same as those used for displacements, and moreover their choice is not unique. For example, we can choose the contact pressure shape function in the form of Dirac function $\hat{N}_a(\mathbf{x}) = \delta(\mathbf{x}_a - \mathbf{x}_c)$, where $\mathbf{x}_c$ is the contact point, which will allow us to recover from the present mortar element framework the master–slave contact elements. The choice of contact pressure shape functions which are the same as the isoparametric shape functions, with $\hat{N}_a = N_a$, would produce the standard mortar elements. Perhaps the most interesting is the choice of dual shape functions for contact pressure (see Figure 5.9), which possess the orthogonality property with respect to the chosen isoparametric displacement interpolations:

$$\int_{\Gamma_c^h} \hat{N}_a(\xi) N_b(\xi)\, ds = \delta_{ab} \int_{\Gamma_c^h} N_a(\xi)\, ds \tag{5.75}$$

where δ_{ab} is the Kroncker symbol. By assuming that any such dual shape function can be written as a linear combination of the isoparametric shape functions:

$$\widehat{N}_a = \alpha_1 \bar{N}_1 + \alpha_2 \bar{N}_2$$

we can obtain from the orthogonality condition the corresponding values of coefficients α_i given as:

$$\frac{h}{6}\begin{bmatrix} 2\,1 \\ 1\,2 \end{bmatrix} \begin{bmatrix} \alpha_1 \\ \alpha_2 \end{bmatrix} = \frac{h}{2}\begin{bmatrix} 1 \\ 0 \end{bmatrix} \Longrightarrow \alpha_1 = 2\ ;\ \alpha_2 = -1$$

In the same manner we can also define the dual functions corresponding to the isoparametric shape functions using quadratic polynomials, with the results illustrated in Figure 5.9.

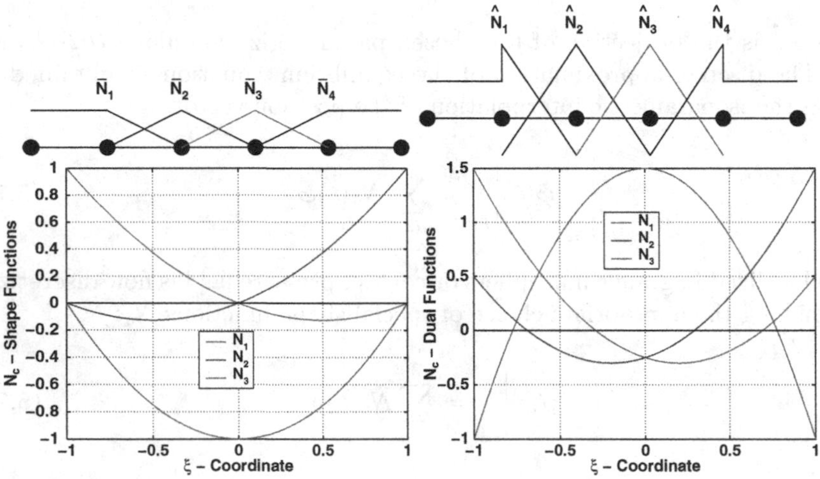

Fig. 5.9 Mortar elements, isoparametric shape functions for displacements and dual shape functions for contact pressure, with either linear or quadratic polynomials.

With the choice of the dual functions for the contact pressure interpolations, the corresponding contribution of any mortar element to the weak form of equilibrium equations can be written in a diagonal form:

$$\sum_a \sum_b p_{c,a} \int_{-1}^{1} \hat{N}_a(\xi)[N_b(\tilde{\xi})\varphi_b^{(\alpha)} - N_b(\xi)\varphi_{c,b}]\, j(\xi) d\xi\ ; \tag{5.76}$$

We can conclude that the choice of dual shape functions will result in a diagonal form of the mortar element contribution, which will allow us to increase the computational efficiency.

5.2.3 Numerical examples of contact problems

In this section we present the results of several numerical simulations, which will illustrate very good performance of the presented mortar elements.

5.2.3.1 Patch test for contact problem

The most basic condition which should be verified by any contact element in order to ensure the convergence of the finite element method, is its capability to exactly represent the constant stress field or, in other words, to pass the patch test. One version of the patch test for contact problems is proposed by Taylor and Papadopoulous [251] (see Figure 5.10), where the same uniformly distributed loading is applied on both solid bodies in contact, which requires

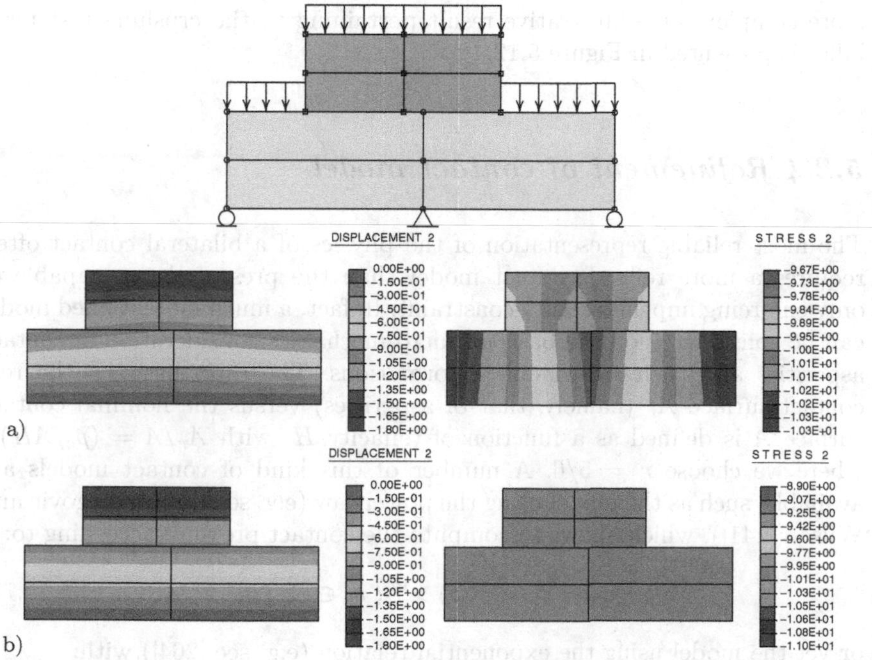

Fig. 5.10 Patch test for contact enforcing correct representation of a constant stress state: mesh and boundary conditions, vertical displacement and stress computed with **(a)** master–slave contact element and **(b)** mortar contact element.

that the stress remains constant everywhere. The main difficulty in the chosen test is introduced by different positions of boundary nodes of two bodies, and the fact that the nodes in contact do not coincide. The results presented in Figure 5.10 clearly show that the mortar elements can pass this patch test, contrary to the master–slave contact element.

5.2.3.2 Hertz 2D contact problem

We also present the results of numerical analysis for a 2D version of the Hertz contact problem considering two spheres in contact. The computations are carried out by standard master–slave approach and by morter contact elements. We can see in Figure 5.11 that the mortar elements can represent much better the stress transfer between two bodies in contact than the standard master–slave elements.

5.2.3.3 Crushing test for a billet

The same method based on mortar contact elements can be applied to 3D contact problems. The main phases of development remain the same as in 2D, but the contact zone search as well as the implementation details are more complex. One illustrative result pertaining to the crushing test for a billet is presented in Figure 5.12.

5.2.4 Refinement of contact model

The most reliable representation of the physics of a bilateral contact often requires a more refined contact model than the presented one capable of only enforcing impenetrability constraint. In fact, a much more refined model can be constructed on the basis of micro-mechanics consideration of contact asperities and their irreversible deformations. The ratio between the real contact surface A_r (namely, that of asperities) versus the nominal contact surface A is defined as a function of tenacity H, with $A_r/A = (p_c/AH)^r$, where we choose $r = 5/6$. A number of this kind of contact models are available, such as the model using the power-law (e.g. see Ibrahimbegovic and Wilson [141]), which allows to compute the contact pressure according to:

$$p_c = c_\nu < \hat{g}(\varphi) >^m ;\ m \in [2, 3.3] \tag{5.77}$$

or yet the model using the exponential relation (e.g. see [264]) with:

$$p_c = c_1\, exp[-c_2 \hat{g}(\varphi)] \tag{5.78}$$

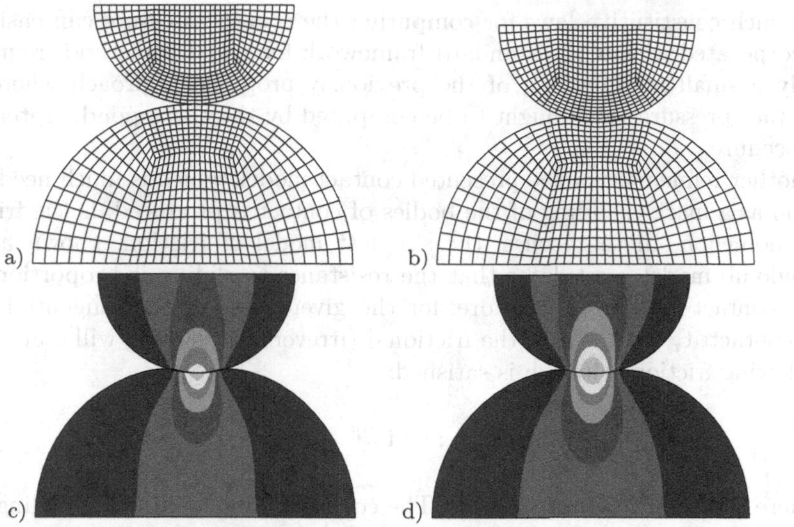

Fig. 5.11 Hertz contact problem: **(a)** initial configuration, **(b)** deformed configuration, **(c)** vertical stress computed by master–slave contact elements, **(d)** vertical stress computed by mortar elements.

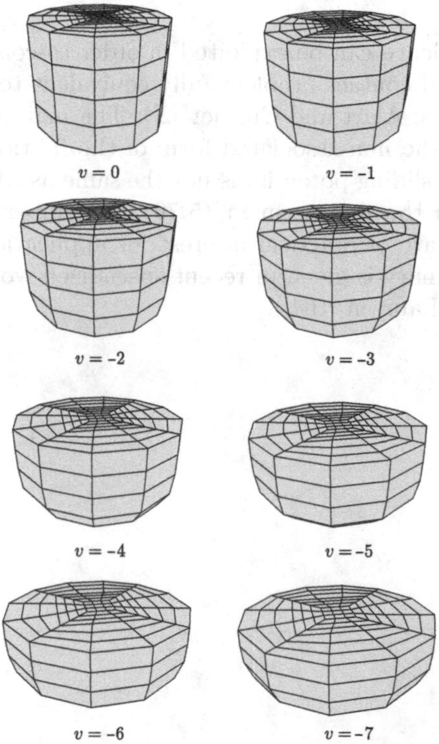

Fig. 5.12 Crushing test for a billet: successive deformed configurations.

All such constitutive laws for computing the contact pressure can easily be incorporated within the standard framework for penalty method, requiring only a small modification of the previously proposed approach where the contact pressure value ought to be computed by the corresponding iterative procedure.

Another refinement of the presented contact model, which is much needed as soon as a marked sliding of the bodies of contact is produced, is the friction phenomenon. In that sense, the simplest model of friction, known as the Coulomb model, postulates that the resistance to sliding is proportional to the contact pressure. Therefore, for the given value of the tangential force in contact, $\mathbf{t}_\tau = \mathbf{t}^\nu - p_c \mathbf{n}$, the frictional (irreversible) sliding will start if the following friction criterion is satisfied:

$$\phi(p_c, \mathbf{t}_\tau) := \| \mathbf{t}_\tau \| - \mu |p_c| = 0 \tag{5.79}$$

where μ is the friction coefficient. The corresponding loading–unloading conditions can be presented in Kuhn-Tucker form, in very much the same manner as already done for plasticity:

$$\phi(\cdot) \leq 0\,;\; \dot{\gamma} \geq 0\,;\; \dot{\gamma}\phi(\cdot) = 0 \tag{5.80}$$

This kind of equivalence can be exploited in order to construct the solution method for frictional contact problem fully equivalent to the one presented for plasticity (e.g. see Alart and Curnier [2]). The main difference from the plasticity concerns the non-associated form of the frictional contact model, where the frictional sliding potential is not the same as the friction criterion, but one only keeps the first term in (5.79). A number of other potential refinements, which are of practical interest for applications, are possible; a very extensive summary is given in recent specialized works on contact (see Wriggers [264] and Laursen [164]).

Chapter 6
Dynamics and time-integration schemes

Until now, we have considered different sources of nonlinearities in a boundary value problem under a quasi-static loading. A tacit hypothesis in any such problem is that the external loading increases slowly enough so that the inertia effects can be neglected.
Starting with this chapter, we will consider a nonlinear problem in dynamics, or yet referred to as the initial-boundary value problem, where the external loading increases rapidly forcing us to take into account the corresponding inertia effects. In a nonlinear initial-boundary value problem, the pseudo-time parameter describing a particular loading program is replaced by real time, which we will still denote with 't'. Moreover, the equilibrium equations for statics are replaced by the equations of motion for dynamics. Discrete finite element approximation of equations of motion in dynamics, no longer results with a set of algebraic equations, but rather with a set of *ordinary differential equations* in time. We will thus explore in this chapter the general method for solving such set of equations by using the time-integration schemes, either explicit or implicit, replacing respectively the incremental analysis and incremental/iterative strategy used for nonlinear problems in statics.
The main goal of this chapter is to provide a detailed presentation of the main ingredients of any initial-boundary value problem in mechanics, as well as the novelties brought by dynamics, both in the theoretical formulation and in numerical solution by the finite element method. A special care is dedicated to trying to improve to quality of the numerical solution provided by a time-integration scheme, by enforcing in the discrete approximation the conservation of some important motion properties, such as the linear and angular momenta or the total energy. These modifications turn to be very important for time-integration schemes applied to a long interval in time, where they proved successful in controlling the truncation error and providing the good accuracy of displacements. If we also want to ensure the results accuracy for displacement derivatives (or strains) and for stress, further modifications of the time-integration schemes are needed, capable of introducing the numerical dissipation of high-frequency modes, which are not well resolved by a coarse finite element mesh.

A. Ibrahimbegovic, *Nonlinear Solid Mechanics: Theoretical Formulations and Finite Element Solution Methods*, Solid Mechanics and its Applications 160,

In the second part of this chapter, we revisit the problem of time-integration of evolution problem in plasticity in the framework of dynamics, as well as the contact problem in dynamics with a special attention to impact phenomena.

6.1 Initial boundary value problem

6.1.1 Strong form of elastodynamics

For the sake of illustration of the finite element method applied to nonlinear dynamics we choose the model problem of 1D elasticity in large displacements. This kind of choice is made mostly for convenience of being the first example studied in two previous chapters. It is important to note that the main solution steps for the chosen model problem will remain the same for a number of other problems in nonlinear dynamics.
We will choose the reference frame in which the initial configuration of 1D solid corresponds to a closed interval $\bar{\Omega} = [0, l]$. Each particle of this configuration is identified by its position vector:

$$x \in \bar{\Omega} \equiv [0, l] \tag{6.1}$$

We apply the dynamic loads in terms of time-dependent distributed loading $b(x, t)$, producing the motion of the body that can be described by the trajectory of each particle from its initial position x to the current position at time t, denoted as $\varphi(x, t)$. Alternatively, the motion can also be described by the displacement field specifying the displacement value along x for each particle:

$$x^{\varphi} = x + d(x, t) =: \varphi(x, t) \equiv \varphi_t(x) \tag{6.2}$$

The deformed configuration is defined as the assembly of all particles in their current position:

$$x^{\varphi} \in \bar{\Omega}_t^{\varphi} = [0, l_t^{\varphi}] \tag{6.3}$$

Velocity and acceleration of motion can be written in either spatial or material description, by using x^{φ} or x as the independent variable, respectively. The point transformation in (6.2) allows us to easily switch from one description to other, as already shown previously for static equilibrium equations. In this chapter we carry on with this development for dynamics, in order to show how to represent inertia effects in either spatial or material description.
The velocity in *material description* can simply be computed as the partial derivative of motion in (6.2) with respect to time while keeping the particle material position fixed; this results with:

$$v(x, t) = \left.\frac{\partial \varphi(x, t)}{\partial t}\right|_{\bar{x}} \tag{6.4}$$

It is important to understand that the velocity vector in (6.4) above acts in the current configuration, but that it is parameterized in material description with respect to the particle position in the initial configuration through the point transformation $x^{\varphi} = \varphi_t(\bar{x})$. By assuming that the motion is sufficiently regular to define the inverse of this transformation, φ_t^{-1}, the velocity vector can also be expressed in spatial description:

$$\hat{v}(\varphi_t^{-1}(x^{\varphi}), t) =: v^{\varphi}(x^{\varphi}, t) \equiv \left.\frac{\partial \varphi(\varphi_t^{-1}(x^{\varphi}), t)}{\partial t}\right|_{\bar{x}^{\varphi}} \tag{6.5}$$

We can also elaborate upon the relationship between spatial and material descriptions of the particle acceleration. Namely, in material description the latter can be expressed simply as the second partial derivative of the motion with respect to time:

$$a(x,t) = \left.\frac{\partial v(x,t)}{\partial t}\right|_{x} \equiv \left.\frac{\partial^2 \phi(x,t)}{\partial t^2}\right|_{x} \tag{6.6}$$

On the other hand, the spatial description of the acceleration vector will result with a so-called convective term:

$$a^{\varphi}(x^{\varphi}, t) = \frac{\partial v^{\varphi}(x^{\varphi}, t)}{\partial t} + \frac{\partial v^{\varphi}(x^{\varphi}, t)}{\partial x^{\varphi}} v^{\varphi}(x^{\varphi}, t) \tag{6.7}$$

Mass conservation principle postulates that in any deformed configuration of a deformable solid body, Ω^{φ}, the total mass of the body will remain constant. Therefore, the mass density $\rho^{\varphi}(x^{\varphi}, t)$ will change in each particular deformed configuration Ω^{φ}, but the total mass will not:

$$m(\Omega^{\varphi}) = \int_{\Omega^{\varphi}} \rho^{\varphi}(x^{\varphi}, t)\, dx^{\varphi} \tag{6.8}$$

With $\rho(x)$ denoting the mass density in the initial configuration, the principle of mass conservation will further allow us to write:

$$\int_{\Omega^{\varphi}} \rho^{\varphi}(x^{\varphi}, t)\, dx^{\varphi} = \int_{\Omega} \rho(x)\, dx \tag{6.9}$$

By considering that the mass conservation principle also applies to an arbitrary subdomain, we can obtain the local form of this mass conservation principle:

$$\rho^{\varphi}(\underbrace{x^{\varphi}}_{\varphi_t(x)}, t) = \rho(x)/\lambda(x,t) \; ; \; \lambda = dx^{\varphi}/dx \tag{6.10}$$

We note in passing that the last result derived for 1D case under consideration, can easily be generalized to 3D case by properly accounting for the change of infinitesimal volume through the determinant of the deformation gradient:

$$\rho^{\varphi}\, dV^{\varphi} = \rho\, dV \implies \boxed{\rho^{\varphi} = \rho / J} \; ; \; J = det[\mathbf{F}] \tag{6.11}$$

With these results in hand, we can provide the spatial and material descriptions of the local form of equations of motion. The former, which applies to the current position of the particle 'x^{φ}' in a given deformed configuration, reads:

$$\rho^{\varphi}(x^{\varphi},t)a^{\varphi}(x^{\varphi},t) = \frac{\partial \sigma^{\varphi}(x^{\varphi},t)}{\partial x^{\varphi}} + b^{\varphi}(x^{\varphi}) \; ; \text{ in } dx^{\varphi} \tag{6.12}$$

The driving force of motion in spatial description is defined by the expression equivalent to the statics equilibrium equations written in term of the Cauchy or true stress, as defined previously in (4.45).
We have already shown that the static equilibrium equations can also be written in material description by using the first Piola–Kirchhoff stress $P(x,t)$ and the distributed loading $b(x,t)$ per unit volume of the initial configuration. We can apply this kind of transformation to the present problem in dynamics by adding the inertia term in material description and the principle of mass conservation defined in (6.6) and (6.10), respectively. The end result is the material description of the equations of motion, which can be written:

$$\rho(x)a(x,t) = \frac{\partial P(x,t)}{\partial x} + b(x,t) \; ; \text{ in } dx \tag{6.13}$$

It is important to note that the material description results with the linear inertia term, which follows from using the fixed reference frame where $a(x,t) = \frac{\partial^2 \varphi(x,t)}{\partial t^2}$. The linear form of inertia terms in material description of equations of motion also applies to 3D case, which is written by using the direct tensor notation:

$$\rho(\mathbf{x})\mathbf{a}(\mathbf{x},t) = div\mathbf{P}(\mathbf{x},t) + \mathbf{b}(\mathbf{x},t) \; ; \; \forall \mathbf{x} \in \Omega \,\&\, \forall t \in [0,T] \tag{6.14}$$

6.1.2 Weak form of equations of motion

A possible starting point in the development of the weak form of the equations of motion is provided by the d'Alembert principle. The latter postulates that the snap-shot of motion taken at time 't' can be described formally with the equilibrium equations. However, unlike the statics problem, these equilibrium equations should also include an extra external load in terms of the inertia force; the latter, denoted as $f_{inertia} = -\rho(x^{\varphi},t)a(x^{\varphi},t)$, is proportional to the mass and directed opposite to acceleration. We can thus write:

$$0 = -\rho^{\varphi}(x^{\varphi},t)a^{\varphi}(x^{\varphi},t) + \frac{\partial \sigma^{\varphi}(x^{\varphi},t)}{\partial x^{\varphi}} + b^{\varphi}(x^{\varphi},t) \tag{6.15}$$

This point of view of d'Alembert provides an important conceptual advantage in allowing us to reduce a new problem of describing the motion in dynamics to a familiar problem of equilibrium in statics. The latter can then be treated by the principle of virtual work, here called the d'Alembert principle, stating that the work of internal forces on chosen virtual displacement should be equal to the virtual work of external forces, including the inertia force among them. In this approach, the time is kept fixed at the chosen value $\bar{t}$ corresponding to a particular deformed configuration; hence, the virtual displacement field is independent of time, and can be denoted: $w^{\bar{\varphi}} = \hat{w}^{\varphi}(\varphi(x, \bar{t})$. Following in the steps already explained for problems in statics, we will multiply the equations of motion by the chosen virtual displacement (or weighting function), integrate over the domain and exploit the integration by parts to provide the final weak form of the 1D problem in elastodynamics:

$$\begin{aligned} 0 &= G_{dyn}(\varphi_t; w^{\bar{\varphi}}) \\ &:= \int_{\Omega^{\varphi}} w^{\bar{\varphi}}[\rho^{\varphi}(x^{\varphi},t)\, a^{\varphi}(x^{\varphi},t) - \tfrac{\partial \sigma^{\varphi}(x^{\varphi},t)}{\partial x^{\varphi}} - b^{\varphi}(x^{\varphi},t)]\, dx^{\varphi} \\ &= \int_{\Omega^{\varphi}} [w^{\bar{\varphi}} \rho^{\varphi}(x^{\varphi},t) a^{\varphi}(x^{\varphi},t) + \tfrac{dw^{\bar{\varphi}}}{dx^{\varphi}} \sigma^{\varphi}(x^{\varphi},t) \\ &\qquad -w^{\bar{\varphi}} b^{\varphi}(x^{\varphi},t)]\, dx^{\varphi} - w^{\bar{\varphi}}(l_t)\bar{t}^{\varphi}(t) \end{aligned} \tag{6.16}$$

By exploiting result in (6.10), along with the material parameterization of the virtual displacement field $w(x) = \hat{w}^{\bar{\varphi}}(\varphi(x, \bar{t}))$, the weak form of equations of motion can be recast in material description:

$$\begin{aligned} 0 &= G_{dyn}(\varphi_t(x); w(x)) \\ &:= \int_{\Omega} [w(x)\rho(x)a(x,t)\, dx \\ &\quad \underbrace{+ \frac{dw(x)}{dx} P(x,t) - w(x)b(x,t)]\, dx - w(l)\bar{t}(t)}_{G_{stat}} \end{aligned} \tag{6.17}$$

where $G_{stat}(\cdot)$ is the corresponding result previously derived for 1D elastostatics. We can obtain the same kind of result for 3D case, and write it by using the direct tensor notation:

$$0 = G_{dyn}(\boldsymbol{\varphi}_t(\mathbf{x}); \mathbf{w}(\mathbf{x})) := \int_{\Omega} \mathbf{w}(\mathbf{x}) \cdot \rho(\mathbf{x})\mathbf{a}(\mathbf{x},t)\, dV + G_{stat}(\boldsymbol{\varphi}_t(\mathbf{x}); \mathbf{w}(\mathbf{x})) \tag{6.18}$$

Another approach to obtain the weak form of equations of motion is based upon the variational formulation, which applies to hyperelastic materials. Here, the strain energy density $\psi(\varphi)$ and external loads jointly define the potential energy functional $\Pi(\varphi_t(x))$, which can be written:

$$\Pi(\varphi) = \int_{\Omega} \psi(\varphi)\, dx - \Pi_{ext}(\varphi) \tag{6.19}$$

We recall that the first variation of such a functional is equivalent to the weak form of equilibrium equations in statics, and moreover, a positive value of the second variation indicates that the equilibrium state remains stable:

$$\begin{aligned} 0 &= G_{stat}(\varphi_t(x), w(x)) := \tfrac{d}{ds}[\Pi(\varphi_{t,s}(x))]|_{s=0} \; ; \\ 0 &< \tfrac{d^2}{ds^2}[\Pi(\varphi_{t,s}(x))]|_{s=0} \end{aligned} \tag{6.20}$$

It is possible to generalize this result to dynamics, by appealing to the Hamilton variational principle employing the total energy functional, which consists of both potential and kinetic energy:

$$H(\varphi(x,t), v(x,t)) := T(v(x,t)) + \Pi(\varphi(x,t)) \tag{6.21}$$

In the material description, the kinetic energy can be written as a quadratic form in velocities:

$$T(v(x,t)) = \int_\Omega \frac{1}{2}\rho(x)[v(x,t)]^2 \, dx \tag{6.22}$$

We can appeal to the principle of least action, to postulate that in the dynamic motion of a hyperelastic body over a time interval $[t_1, t_2]$, the zero value of the first variation of total energy (or yet referred to Hamiltonian) will produce the equations of motion, whereas a positive value of the second variation will confirm the motion stability:

$$\begin{aligned} 0 &= G_{dyn} := \tfrac{d}{ds}[\textstyle\int_{t_1}^{t_2} H(\varphi_{t,s}(x))\, dt]|_{s=0} \; ; \\ 0 &< \tfrac{d^2}{ds^2}[\textstyle\int_{t_1}^{t_2} H(\varphi_{t,s}(x))\, dt]|_{s=0} \end{aligned} \tag{6.23}$$

It is easy to confirm the last result, by using the directional derivative computation. In fact, the only new term with respect to statics concerns the directional derivative of the kinetic energy in (6.22), which can be written:

$$\begin{aligned} &\tfrac{d}{ds}[\textstyle\int_{t_1}^{t_2} T(\frac{\partial(\varphi_{t,s}(x))}{\partial t})]|_{s=0}\, dt = \int_{t_1}^{t_2} \frac{\partial T(v)}{\partial v}\frac{\partial^2 \varphi_{t,s}}{\partial s \partial t}\, dt \\ &= [\frac{\partial T(v)}{\partial v} \underbrace{\frac{\partial \varphi_{t,s}}{\partial s}}|_{s=0}]_{t_1}^{t_2} - \int_{t_1}^{t_2} \frac{\partial}{\partial t}[\frac{\partial T(v)}{\partial v}] w \, dt \\ &= [\frac{\partial T(v(t_2))}{\partial v} \underbrace{w(t_2)}_{0} - \frac{\partial T(v(t_1))}{\partial v} \underbrace{w(t_1)}_{=0}] - \int_{t_1}^{t_2}\int_\Omega w\, \rho v(x,t)\, dx\, dt \end{aligned} \tag{6.24}$$

The last result is obtained by imposing a supplementary admissibility condition on zero value of the variations at both limits of time interval, with $w(t_1) = 0$ and $w(t_2) = 0$, which allows us to recover the inertia term of equations of motion from the first variation of the kinetic energy. The last result along with the variation of the potential energy producing $G_{stat} = \frac{d}{ds}[\Pi]|_{s=0}$, will confirm that the first variation of the Hamiltonian corresponds to the equations of motion (6.17):

$$0 = \frac{d}{ds}[\int_{t_1}^{t_2} H(\varphi_{t,s}(x))\, dt]|_{s=0} \equiv G_{dyn} \tag{6.25}$$

In the absence of external load with $\Pi^{ext}(\varphi(x,t)) = 0$, the Hamilton principle implies that the total energy of the motion remains conserved:

$$0 = \frac{d}{ds}[\int_{t_1}^{t_2} H(\varphi_{t,s}(x))\, dt]|_{s=0} \implies H(t) = cst. \tag{6.26}$$

This is an important observation which is further exploited in the design of time-integration schemes.

6.1.3 Finite element approximation for mass matrix

The weak form of the equations of motion in (6.17) above is used as the starting point for constructing the discrete approximation based upon the finite element method. Considering that the discrete approximation of the static part $G_{stat}(\cdot)$ in (6.17) is already discussed,[1] we only elaborate upon the finite element approximation of the inertia term, which allows us to define the mass matrix.

We will use the isoparametric finite element interpolations for a truss-bar element with n_{en} nodes, based upon the shape functions defined in terms of the natural coordinate, $N_a(\xi)$ $a = 1, 2 \ldots n_{en}$. The initial configuration of the element is thus represented with:

$$x(\xi)\Big|_{\Omega^e} = \sum_{a=1}^{n_{en}} N_a(\xi) x_a \tag{6.27}$$

The finite element approximation of the virtual displacement field can be constructed with the isoparametric finite elements, by exploiting the same shape functions and fixed (in time) nodal values w_a:

$$w^h(\xi)\Big|_{\Omega^e} = \sum_{a=1}^{n_{en}} N_a(\xi) w_a \tag{6.28}$$

The real displacement field, which will express for each particle the distance between its initial and current positions, depends in dynamics on both space and time. With the separation of variables approach, we can represent the

[1] We have already studied in Chapters 3, 4 and 5 the quasi-static problems, where the pseudo-time parameter 't' was introduced to describe a particular loading program, along with their finite element implementation. There is no modification in dynamics to any of the finite element procedures we described, other than considering parameter 't' as the real time.

displacement field as the product of functions that only depend on space and functions that only depend on time; the former are chosen again as the isoparametric finite element shape functions and the latter will be chosen as the time-dependent nodal values of displacements. For example, for incremental displacement field we can write:

$$u^h(\xi,t)\Big|_{\Omega^e} = \sum_{a=1}^{n_{en}} N_a(\xi)\, u_a(t) \tag{6.29}$$

where $u_a(t)$ are the nodal valued of the incremental displacement. Summing up all the contributions of this kind during the incremental sequence until given time t, we obtain the total displacement nodal value in the current configuration, $d_a(t)$. The main advantage of using the isoparametric finite element interpolations pertains to reconstructing the current configuration, which will reduce to a simple addition of the nodal values:

$$\varphi^h(\xi,t)\Big|_{\Omega^e} = \sum_{a=1}^{n_{en}} N_a(\xi)(x_a + d_a(t)) \tag{6.30}$$

By computing the time derivative of the last expression, we can easily obtain the discrete approximation of the velocity field in material description:

$$v^h(\xi,t)\Big|_{\Omega^e} = \sum_{a=1}^{n_{en}} N_a(\xi)\dot{d}_a(t) \tag{6.31}$$

where $\dot{d}_a(t) = \frac{d}{dt} d_a(t)$ are the corresponding nodal values. The subsequent time derivative of this expression will produce the finite element approximation of the acceleration field in the material description which can be written:

$$a^h(\xi,t)\Big|_{\Omega^e} = \sum_{a=1}^{n_{en}} N_a(\xi)\ddot{d}_a(t) \tag{6.32}$$

where $\ddot{d}_a(t)$ are nodal values of acceleration. By introducing these finite element approximations into the weak form of equations of motion in (6.17), we obtain again the corresponding Galerkin equation. For arbitrary nodal values of virtual displacement, we finally obtain the space-discretized form of the equations of motion which is described by a set of *ordinary differential equations in time*, which is written:

$$\mathsf{M}\ddot{\mathsf{d}}(t) + \hat{\mathsf{f}}^{int}(\mathsf{d}(t)) = \mathsf{f}^{ext}(t) \tag{6.33}$$

Such a discretization procedure with respect to space coordinates only is often referred to as *semi-discretization*. Despite the conceptual difference between real time and pseudo-time 't' used for describing a particular loading program, we can use in dynamics exactly the same finite element procedure for constructing internal and external load vectors $\mathsf{f}^{int}(\mathsf{d}(t))$ and $\mathsf{f}^{ext}(t)$, as already explained for quasi-statics problems. Therefore, the main novelty in dynamics concerns the computation of inertia term and the mass matrix, M. The finite element method can be employed for mass matrix computation, to take into account different element contributions through the finite element assembly procedure:

$$\mathsf{M} = \mathbb{A}_{e=1}^{n_{elem}} \mathsf{M}^e \tag{6.34}$$

where M^e is an element mass matrix. For a truss-bar element, each component of the element mass matrix can be obtained according to:

$$\begin{aligned} \mathsf{M}^e = [M^e_{ab}] \ ; \ M^e_{ab} &= \int_{\Omega^e} N_a(\xi)\rho(x)N_b(\xi)\, dx \\ &= \sum_{l=1}^{n_{in}} N_a(xi_l)\hat{\rho}(x(\xi_l))N_b(\xi_l)j(\xi_l)w_l \end{aligned} \tag{6.35}$$

We have indicated in the last expression that the corresponding integrals for mass matrix components are again computed by using the numerical integration. Such a numerical integration will employ a higher order quadrature rule than the one used for internal force and tangent stiffness matrix computations, since the shape functions rather than their derivatives appear in the integrand for mass matrix computation. Thus, for a 2-node truss-bar element, two-point Gauss quadrature is needed to correctly compute the mass matrix components with:

$$\mathsf{M}^e = \frac{A\rho h^e}{6} \begin{bmatrix} 2 & 1 \\ 1 & 2 \end{bmatrix} \tag{6.36}$$

The mass matrix in (6.36) is referred to as *consistent*, since it is computed with the same shape functions for displacements and accelerations. We have previously shown that an explicit time-integration scheme can much gain in efficiency when using a *diagonal* form of the mass matrix. Among different possibilities for constructing a diagonal mass matrix (e.g. see Zienkiewicz and Taylor [271]), one of the simplest will transform the consistent mass matrix by summing up all the components in each row into a single component placed on the main diagonal. For consistent mass for 2-node truss-bar element in (6.36) above, such a diagonal mass matrix can be written:

$$\tilde{\mathsf{M}}^e = \frac{A\rho h^e}{2} \begin{bmatrix} 1 & 0 \\ 0 & 1 \end{bmatrix} \tag{6.37}$$

The diagonal form of the mass matrix provides intuitively logical results, where half of the total mass is placed at each of two element nodes.

In 2D and 3D cases, the mass matrix of an isoparametric element can be written:

$$\begin{aligned} \mathsf{M}^e = [M^e_{ab}]\ ;\ M^e_{ab} &= \int_{\Omega^e} N_a(\mathbf{x})\rho(\mathbf{x})N_b(\mathbf{x})\, dV \\ &= \sum_{l=1}^{n_{in}} N_a(\boldsymbol{\xi}_l)\rho(\mathbf{x}(\boldsymbol{\xi}_l))N_b(\boldsymbol{\xi}_l)\, j(\boldsymbol{\xi}_l) w_l \end{aligned} \tag{6.38}$$

where the total number of Gauss quadrature points n_{in} is typically superior to the one used for the stiffness matrix computation of the same element. The diagonal form of this matrix can again be obtained by the row-sum technique as for 1D case.
The main advantage of the inertia frame for large motion is a mass matrix with constant entries, and a linear form of the inertia term in the equations of motion. The main source of nonlinearity remains pertinent to the internal force vector, which can again be computed with the same procedure as the one used in statics.

6.2 Time-integration schemes

The final product of semi-discretization procedure carried out by the finite element method in nonlinear dynamics is a system of nonlinear *ordinary differential equations* in time. Any such nonlinear dynamics problem is quite different from the nonlinear equilibrium problem in statics, which is governed by a set of nonlinear algebraic equations parameterized by pseudo-time. Despite this difference, in constructing the solution to a nonlinear dynamics problem by using the time-integration schemes, we will find again all the main ingredients of incremental/iterative solution procedure for quasi-static problems. On the other hand, the traditional methods for solving linear dynamics problems, such as modal superposition method[2] (e.g. see Clough and Penzien [53] or Gradin and Rixen [84]), are not of much interest for nonlinear dynamics problems, since the free-vibration modes would change with each incremental/iterative modification of the tangent stiffness matrix.
We present subsequently the main steps in solving the nonlinear dynamics problems by using the time-integration schemes. To that end, the time-interval of interest $[0, T]$ is subdivided in the chosen number of time steps,

[2] For a free vibration problem in linear dynamics, with d_0 and v_0 as the initial displacement and velocity, the solution is obtained with mode superposition $\mathsf{d}_{free}(t) = \sum_{i=1}^{n} \boldsymbol{\phi}_i(\boldsymbol{\phi}_i^T \mathsf{M}\mathsf{d}_0 \cos\omega_i t + (1/\omega_i)\boldsymbol{\phi}_i^T\mathsf{M}\mathsf{v}_0 \sin\omega_i t)$, where $\boldsymbol{\phi}_i$ and ω_i are natural modes of free vibrations and natural frequencies; the latter can be computed as the solution to the eigenvalue problem $[\mathsf{K} - \omega_i^2\mathsf{M}]\boldsymbol{\phi}_i = 0$. The mode superposition method can also be employed to construct the solution to a forced vibration problem in linear dynamics, where the forcing term is supplied by Duhamel's integral with $\mathsf{d}_{forced}(t) = \mathsf{d}_{free}(t) + \sum_{i=1}^{m} \boldsymbol{\phi}_i(\int_0^t (\boldsymbol{\phi}_i^T \mathsf{M}\mathsf{f}(s)/\omega_i) \sin\omega_i(t-s)\, ds)$.

which will specify the time instants where the selected time-integration scheme should deliver the solution:

$$0 < t_1 < t_2 < \ldots < t_n < t_{n+1} < \ldots < T \tag{6.39}$$

This is equivalent to the incremental analysis for quasi-statics problems; however, contrary to statics, the complete solution requires not only the nodal values of displacements, but also of velocities and accelerations. For example, the solution at time t_n is defined with:

$$\mathsf{d}_n = \mathsf{d}(t_n) \; ; \; \mathsf{v}_n = \dot{\mathsf{d}}(t_n) \; ; \; \mathsf{a}_n = \ddot{\mathsf{d}}(t_n) \tag{6.40}$$

In order to simplify the computer code architecture, we employ in general one-step time-integration schemes that construct the solution over a single step at the time. For a typical time step starting at time t_n, the time-integration scheme should deliver the nodal values of displacements, velocities and acceleration at time t_{n+1}, which verify the equations of motion; this can formally be stated in terms of:

Central problem in nonlinear dynamics for $[t_n, t_{n+1}]$

$$\begin{aligned} &\text{Given:} && \mathsf{d}_n, \mathsf{v}_n, \mathsf{a}_n \\ &\text{Find:} && \mathsf{d}_{n+1}, \mathsf{v}_{n+1}, \mathsf{a}_{n+1} \\ &\text{such that:} && \mathsf{M}\mathsf{a}_{n+1} + \hat{\mathsf{f}}^{int}(\mathsf{d}_{n+1}) = \mathsf{f}^{ext}_{n+1} \end{aligned} \tag{6.41}$$

One possibility to solve the central problem in dynamics is by exploiting the one-step time-integration schemes which were introduced previously for first-order differential equations. In fact, we can always recast a set of second-order differential equations that characterize the nonlinear dynamics in terms of the set of the first-order equations containing twice as many equations. We will switch this way to so-called state space form of the system, where both displacements and velocities are considered as independent variables:

$$\dot{\mathsf{z}}(t) = \mathsf{h}(\mathsf{z}(t)) \; ; \; \mathsf{z} = \begin{bmatrix} \mathsf{d} \\ \mathsf{v} \end{bmatrix} \; ; \; \mathsf{h}(\mathsf{z}) = \begin{bmatrix} \mathsf{v} \\ \mathsf{M}^{-1}(\mathsf{f}^{ext} - \hat{\mathsf{f}}^{int}(\mathsf{d})) \end{bmatrix} \tag{6.42}$$

We could thus directly apply the time-integration schemes developed previously for integrating first-order systems, such as the heat transfer equation or internal variable evolution equations. However, this would not lead to the most efficient implementation. The computational efficiency can be improved significantly with the time-integration schemes applicable directly to the original form of the second-order differential equations. Several time-integration schemes of this kind, both explicit and implicit, are presented subsequently.

6.2.1 Central difference (explicit) scheme

The central difference scheme is based upon the *second order approximation* of the differential equations of motion, with a truncated series development employing only the first derivative of displacements and velocities in (6.42). This kind of approximation can be constructed in a systematic manner by using the central difference operator (e.g., see Dahlquist and Bjorck [62], p. 353); we thus obtain the corresponding approximation of the first differential equation in (6.42), which can be written:

$$\dot{\mathsf{d}}(t) = \mathsf{v}(t) \implies \frac{\mathsf{d}_{n+1} - \mathsf{d}_{n-1}}{2\,h} = \mathsf{v}_n \tag{6.43}$$

The last result confirms the explicit form of the central difference approximation, since in computing the displacement d_{n+1} we only need the displacement and the velocity in two previous steps. However, for simplicity of computer code architecture, we prefer the numerical implementation of the central difference scheme where the only values needed for computing the displacement at time t_{n+1} are those from the previous time t_n. In order to ensure such a format of the central difference scheme, we use the corresponding approximations of the velocity at the mid-point of each time step, which allow us to write:

$$\left.\begin{aligned} \tfrac{1}{h}(\mathsf{d}_{n+1} - \mathsf{d}_n) &= \mathsf{v}_{n+1/2} \\ \tfrac{1}{h}(\mathsf{d}_n - \mathsf{d}_{n-1}) &= \mathsf{v}_{n-1/2} \end{aligned}\right\} \Rightarrow \frac{1}{2\,h}(\mathsf{d}_{n+1} - \mathsf{d}_{n-1}) = \frac{1}{2}(\mathsf{v}_{n+1/2} + \mathsf{v}_{n-1/2}) \tag{6.44}$$

The comparison of last two results reveals an alternative possibility to write the corresponding approximation to velocity at time t_n according to:

$$\mathsf{v}_n = \frac{1}{2}(\mathsf{v}_{n+1/2} + \mathsf{v}_{n-1/2}) \tag{6.45}$$

The same kind of central difference approximation based upon the mid-point values can also be constructed for acceleration vector:

$$\mathsf{a}(t) = \dot{\mathsf{v}}(t) \implies \mathsf{a}_n = \frac{\mathsf{v}_{n+1/2} - \mathsf{v}_{n-1/2}}{h} \tag{6.46}$$

Summing up the last two results, we obtain the corresponding approximation for the velocity vector in the middle of the interval of interest, which can be written:

$$\mathsf{v}_{n+1/2} = \mathsf{v}_n + \frac{h}{2}\mathsf{a}_n \tag{6.47}$$

By replacing the last result into (6.44), we can obtain the final form of the central difference approximation for the displacement vector at time t_{n+1}:

$$\mathsf{d}_{n+1} = \mathsf{d}_n + h\mathsf{v}_n + \frac{h^2}{2}\mathsf{a}_n \tag{6.48}$$

This kind of approximation for displacement is fully explicit in that it only employs the known values of displacements, velocities and accelerations at time t_n.

The final form of the central difference approximation of velocity at time t_{n+1} can be obtained as a linear combination of the results in (6.45) and (6.47) above, as well as the equivalent result for subsequent time step,[3] which results with:

$$\mathsf{v}_{n+1} = \mathsf{v}_n + \frac{h}{2}(\mathsf{a}_n + \mathsf{a}_{n+1}) \tag{6.49}$$

In the last expression, the acceleration vectors a_n and a_{n+1} are computed directly from the equations of motion at times t_n and t_{n+1}, respectively. The most efficient implementation of the central difference scheme is presented in Table 6.1.

Table 6.1 Central difference scheme in nonlinear dynamics.

Initialize: $\mathsf{d}_0, \mathsf{v}_0 \longrightarrow \mathsf{a}_0 = \mathsf{M}^{-1}(\mathsf{f}_0^{ext} - \hat{\mathsf{f}}^{int}(\mathsf{d}_0))$

Compute for each step: $n = 0, 1, 2, ...$ (given: $\mathsf{d}_n, \mathsf{v}_n, \mathsf{a}_n$ and h)

$$\begin{aligned} &\mathsf{d}_{n+1} = \mathsf{d}_n + h\mathsf{v}_n + \tfrac{h^2}{2}\mathsf{a}_n \\ &\mathsf{M}\mathsf{a}_{n+1} + \hat{\mathsf{f}}^{int}(\mathsf{d}_{n+1}) = \mathsf{f}_{n+1}^{ext} \implies \mathsf{a}_{n+1} = \mathsf{M}^{-1}(\mathsf{f}_{n+1}^{ext} - \hat{\mathsf{f}}^{int}(\mathsf{d}_{n+1})) \\ &\mathsf{v}_{n+1} = \mathsf{v}_n + \tfrac{h}{2}(\mathsf{a}_n + \mathsf{a}_{n+1}) \end{aligned} \tag{6.50}$$

Next: $n \longleftarrow n+1$

The most costly phase in the proposed implementation of the central difference scheme clearly pertains to the solution of set of algebraic equations for computing the acceleration vector a_{n+1}. However, this cost can be reduced considerably by using a diagonal form of the mass matrix, which allows to obtain the solution with the number of operations equal only to number of equations 'n'. With this kind of computational efficiency, we can easily accept a very small time steps that is often required to meet the conditional stability of the explicit scheme. For example, for 1D hyperelastic bar (with free energy density $\psi(\lambda)$) and 2-node finite element approximations (with a typical element length l^e), the conditional stability of the central difference scheme requires (e.g. see [208]) the time step 'h' no larger than:

$$h \leq \frac{l^e}{\sqrt{6}c_{max}} \; ; \; c_{max} = \max_{\forall \lambda > 0}[\frac{\tilde{C}(\lambda)}{\rho}]^{1/2} \; ; \; \tilde{C}(\lambda) = \frac{d^2\psi(\lambda)}{d\lambda^2} \tag{6.51}$$

Interestingly enough, by using a diagonal form of the mass matrix, the largest acceptable time step which guarantees the stability of central difference scheme increases to:

$$h \leq \frac{l^e}{\sqrt{2}c_{max}} \tag{6.52}$$

[3] We can also write that $\mathsf{v}_{n+1} = \frac{1}{2}(\mathsf{v}_{n+3/2} + \mathsf{v}_{n+1/2})$ and $\mathsf{v}_{n+3/2} = \mathsf{v}_{n+1} + \frac{h}{2}\mathsf{a}_{n+1}$.

However, for a number of cases of practical interest, the restriction placed by the conditional stability of the explicit central difference scheme can be too severe, forcing us to take a time step which is too small with respect to the required result accuracy. For that reason, the central difference scheme is mostly used in applications to dynamics phenomena of short duration, such as impact problems, explosions or wave propagation with very short duration; in any such problem, the computed response would have a significant contribution of high-frequency modes and would require very small time steps, which can easily be handled with central difference scheme.

Remarks:

1. For nonlinear inelastic behavior in dynamics, the central difference scheme provides a very interesting alternative to fully implicit schemes. Namely, we can solve the equations of motion by the explicit central difference scheme, combined with an implicit scheme solution of a small set of the evolutions equations for internal variables at each numerical integration point. This kind of explicit–implicit approach to nonlinear dynamics problems with inelastic behavior can be implemented within the framework of the operator split procedure. We thus obtain the highest computational efficiency in solving a large set of equations of motion that provides the best value of displacements $\mathsf{d}_{n+1}^{(i)}$ as well as the corresponding value of the total deformation field $\epsilon_{n+1}^{(i)}$, along with results reliability for the local computations of internal variables and admissible value of stress that is carried out by an implicit scheme (e.g. the backward Euler scheme). The latter should be implemented in the manner which guarantees the convergence of the local computation providing the admissibility of the stress in the sense of the chosen plasticity or damage criteria. It is important to note that only stress values are needed for this kind of strategy, with no need to compute the tangent moduli.

2. We have implemented this kind of strategy for a coupled damage-plasticity constitutive model suitable for representing the behavior of concrete under impact (see [100]). This model employs a judicious combination of damage model of Mazars (see [186]), for representing the concrete cracking with the threshold related to the principal elastic strains, and Gurson-like (see [95] or [199]) viscoplasticity model, capable of representing the concrete hardening behavior under compressive stress. A graphic illustration of the elastic domain defined by the proposed criterion is represented in Figure 6.1. We note a considerable complexity of the constitutive model of this kind with a fairly long list of internal variables (such as the damage variable d_{n+1}, viscoplastic strains ϵ_{n+1}^{vp}, the porosity f^*, as well as the hardening variables; see [100]), which would represent quite a significant challenge for guaranteeing the convergence of the fully implicit schemes. However, the explicit computations of equations of motion, accompanied by only implicit computations of internal variables does not have any such difficulty. A couple of illustrative results which concern the impact computation on a concrete slab, both for the cases with and without perforation, are presented in Figure 6.2.

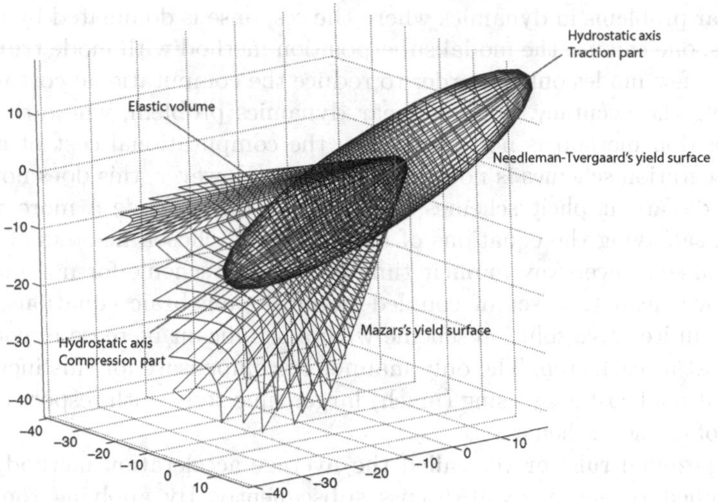

Fig. 6.1 Elastic domain for coupled damage-plasticity model of concrete under impact (see [100]).

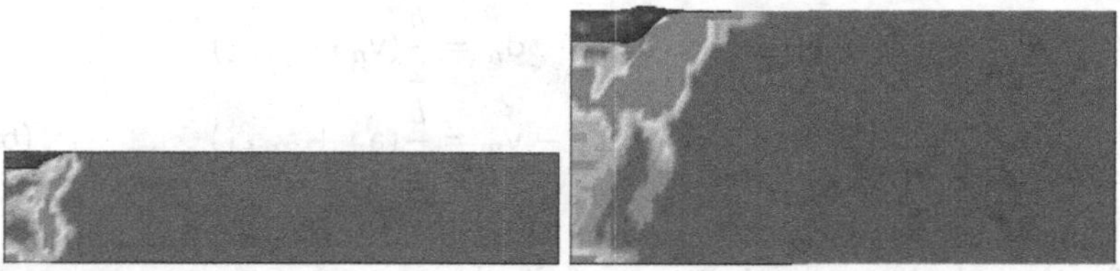

Fig. 6.2 Impact on a reinforced concrete slab: contours of damage variable for the cases with and without perforation.

6.2.2 Trapezoidal rule or average acceleration (implicit) scheme

For a large number of problems in low velocity dynamics, such as the problems typical of earthquake engineering with a significant damping, the dynamic response is to a large extent governed by a fairly few, low-frequency modes. One finds that the contribution of high-frequency modes for such problems is not very pronounced, and that they are quickly damped out. Integrating the equations of motion for one such problem in dynamics by using an explicit scheme, with a very small time step that must be selected to guarantee the stability, would imply a significant, yet unnecessary increase in computational cost. We thus have to turn to implicit schemes, in order to allow for larger time steps.

For linear problems in dynamics where the response is dominated by low frequencies, one can use the modal superposition method with mode truncation keeping a few modes only, in order to reduce the computational cost without sacrificing the accuracy. For nonlinear dynamics problem, where the mode superposition method is not applicable, the computational cost of implicit time-integration schemes is not easy to reduce. However, this does not mean we will discard implicit schemes, for they we can provide a more reliable solution satisfying the equations of motion for all chosen instants of the incremental sequence. Any implicit time-integration scheme for nonlinear dynamics will lead to a set of coupled nonlinear algebraic equations, which requires an iterative solution scheme with a significant increase in computational cost at each step. The only manner to compensate for this increase of computational cost is by using (much) larger time steps with respect to those typical of explicit schemes.

The trapezoidal rule, or yet called the average acceleration method, is the first implicit scheme we will discuss subsequently. By applying the trapezoidal rule (e.g., see Dahlquist and Bjorck [62], p. 347) to the equations of motion, we obtain the second order approximation to evolution equations for displacement and velocity, which can be written:

$$
\begin{aligned}
\dot{\mathsf{d}}(t) = \mathsf{v}(t) \implies \mathsf{d}_{n+1} - \mathsf{d}_n &= \frac{h}{2}(\mathsf{v}_n + \mathsf{v}_{n+1}) \\
\dot{\mathsf{v}}(t) = \mathsf{a}(t) \implies \mathsf{v}_{n+1} - \mathsf{v}_n &= \frac{h}{2}(\mathsf{a}_n + \mathsf{a}_{n+1})
\end{aligned} \tag{6.53}
$$

Rewriting the result in $(6.53)_1$ we can obtain the corresponding approximation for the velocity vector in terms of displacement increment:

$$
\mathsf{v}_{n+1} = -\mathsf{v}_n + \frac{2}{h}(\mathsf{d}_{n+1} - \mathsf{d}_n) \tag{6.54}
$$

With this result in hand, we can then obtain from $(6.53)_2$ the same kind of approximation of the acceleration vector at time t_{n+1} which reads:

$$
\mathsf{a}_{n+1} = -\mathsf{a}_n - \frac{4}{h}\mathsf{v}_n + \frac{4}{h^2}(\mathsf{d}_{n+1} - \mathsf{d}_n) \tag{6.55}
$$

Note that both of these approximations are implicit in the sense that they depend upon the displacement value at time t_{n+1}. Combining these velocity and acceleration approximations with the equations of motion written at time t_{n+1}, we can obtain the most suitable implementation of the trapezoidal rule as presented in Table 6.2.

In terms of computational cost, the most demanding phase in the proposed implementation of the trapezoidal rule pertains to iterative solution of the set of nonlinear algebraic equations with displacements at time t_{n+1} as

Table 6.2 Trapezoidal rule or average acceleration scheme for nonlinear dynamics.

Initialize: $\mathsf{d}_0, \mathsf{v}_0 \longrightarrow \mathsf{M}^{-1}(\mathsf{f}^{ext} - \hat{\mathsf{f}}^{int}(\mathsf{d}_0))$

At each step: $n = 0, 1, 2, ...$ (Given: $\mathsf{d}_n, \mathsf{v}_n, \mathsf{a}_n$ and h)

Find: $\mathsf{d}_{n+1}, \mathsf{v}_{n+1}, \mathsf{a}_{n+1}$

Iterate: $(i) = 1, 2, ...$

$$\mathsf{a}_{n+1}^{(i)} = \hat{\mathsf{a}}(\mathsf{d}_{n+1}^{(i)}) := \frac{4}{h^2}(\mathsf{d}_{n+1}^{(i)} - \mathsf{d}_n) - \frac{4}{h}\mathsf{v}_n - \mathsf{a}_n$$
$$\mathsf{M}\hat{\mathsf{a}}(\mathsf{d}_{n+1}^{(i)}) + \hat{\mathsf{f}}^{int}(\mathsf{d}_{n+1}^{(i)}) = \mathsf{f}_{n+1}^{ext} \longrightarrow \mathsf{d}_{n+1}^{(i+1)}$$
$$\mathsf{v}_{n+1}^{(i)} = \frac{2}{h}(\mathsf{d}_{n+1}^{(i)} - \mathsf{d}_n) - \mathsf{a}_n$$

IF $\| \mathsf{f}_{n+1}^{ext} - \mathsf{M}\mathsf{a}_{n+1}^{(i)} - \hat{\mathsf{f}}^{int}(\mathsf{d}_{n+1}^{(i)}) \|$ THEN $(i) \to (i+1) \longrightarrow \bar{\mathsf{d}}_{n+1}$

ELSE $n \to n+1$

unknowns. If Newton's method is employed for this computation, we have to solve at each iteration the consistently linearized form of the system, which can be written:

$$[\frac{4}{h^2}\mathsf{M} + \mathsf{K}_{n+1}^{(i)}]\mathsf{u}_{n+1}^{(i)} = \mathsf{f}_{n+1}^{ext} - \hat{\mathsf{f}}^{int}(\mathsf{d}_{n+1}^{(i)}) - \mathsf{M}\hat{\mathsf{a}}(\mathsf{d}_{n+1}^{(i)}) \qquad (6.56)$$

The matrix $\mathsf{K}_{n+1} = \frac{\partial \hat{\mathsf{f}}^{int}(\mathsf{d}_{n+1})}{\partial \mathsf{d}_{n+1}}$ above is the tangent stiffness already defined for nonlinear problems of static equilibrium. The tangent operator in dynamics is yet referred to as the effective tangent stiffness, which includes both the true tangent stiffness for statics and the corresponding contribution from the mass matrix. The presence of the mass matrix (with full rank) in the effective tangent stiffness has the regularizing effect, which ensures the correct rank. We can thus always obtain the unique solution for incremental displacement, and carry out the displacement update according to:

$$\mathsf{d}_{n+1}^{(i+1)} = \mathsf{d}_{n+1}^{(i)} + \mathsf{u}_{n+1}^{(i)} \qquad (6.57)$$

This iterative procedure continues until the chosen convergence tolerance is reached, providing the final displacement value $\bar{\mathsf{d}}_{n+1}$ for that particular step. For linear dynamics problems, where the internal force can be written as $\hat{\mathsf{f}}^{int}(\mathsf{d}_{n+1}) = \mathsf{K}\mathsf{d}_{n+1}$, the trapezoidal rule guarantees the unconditional stability of computations regardless of the time step size. Moreover, the trapezoidal rule ensures the energy conservation in linear dynamics problems, for the free vibration phase of motion in the absence of external loading. Namely, with the time derivative of total energy identical to the equations of motion, it is easy to conclude that:

$$\frac{d}{dt}[E(t)] = \underbrace{\dot{\mathsf{d}}(t) \cdot (\mathsf{M}\ddot{\mathsf{d}}(t) + \mathsf{K}\mathsf{d}(t))}_{G_{dyn}(\varphi;\mathsf{v})=0} \implies E(t) = cst. \qquad (6.58)$$

The discrete approximation of free vibration problem in linear dynamics provided by the trapezoidal rule will also ensure the conservation of the total energy with:

$$
\begin{aligned}
0 &= \frac{\mathsf{d}_{n+1}-\mathsf{d}_n}{h}\cdot \mathsf{M}\frac{\mathsf{a}_n+\mathsf{a}_{n+1}}{2} + \frac{\mathsf{d}_{n+1}-\mathsf{d}_n}{h}\cdot \mathsf{K}\frac{\mathsf{d}_n+\mathsf{d}_{n+1}}{2}\\
&= \frac{1}{2}(\mathsf{v}_{n+1}+\mathsf{v}_n)\cdot \mathsf{M}(\mathsf{v}_{n+1}-\mathsf{v}_n) + \frac{1}{2}(\mathsf{d}_{n+1}-\mathsf{d}_n)\cdot \mathsf{K}(\mathsf{d}_{n+1}-\mathsf{d}_n)\\
&= (\frac{1}{2}\mathsf{v}_{n+1}\cdot \mathsf{M}\mathsf{v}_{n+1} + \frac{1}{2}\mathsf{d}_{n+1}\cdot \mathsf{K}\mathsf{d}_{n+1}) - (\frac{1}{2}\mathsf{v}_n\cdot \mathsf{M}\mathsf{v}_n + \frac{1}{2}\mathsf{d}_n\cdot \mathsf{K}\mathsf{d}_n)\\
&= E_{n+1}-E_n
\end{aligned}
\tag{6.59}
$$

The result of this kind is an additional confirmation of the unconditional stability of the trapezoidal rule method for linear dynamics problems. The unconditional stability of the same scheme in nonlinear dynamics is guaranteed only (see Belytschko and Schoeberk [30]) in the case where the discrete approximation of the internal energy verifies the following condition:[4]

$$
\psi_{n+1} := \psi_n + \mathsf{u}_{n+1}\cdot(\hat{\mathsf{f}}^{int}(\mathsf{d}_{n+1}) + \hat{\mathsf{f}}^{int}(\mathsf{d}_n))/2 \geq 0 \tag{6.60}
$$

Such a condition is violated (e.g. see Hughes [109]) for nonlinear dynamics problems with softening behavior.[5] Moreover, the trapezoidal rule can no longer ensure the energy conservation for nonlinear dynamics problems, which is one of the main motivations for study of other implicit time-integration schemes, such as mid-point rule.

Remark on Newmark family of algorithms: The central difference scheme and the trapezoidal rule are the most frequently employed members of the Newmark family of algorithms (e.g. see [19], [111], or [270]). Any particular member of this family can be chosen by specifying the values for the Newmark parameters β and γ, in order to obtain the corresponding approximations for velocity and acceleration:

$$
\begin{aligned}
\mathsf{d}_{n+1} &= \mathsf{d}_n + h\mathsf{v}_n + h^2[(\tfrac{1}{2}-\beta)\mathsf{a}_n + \beta\mathsf{a}_{n+1}]\\
\mathsf{v}_{n+1} &= \mathsf{v}_n + h[(1-\gamma)\mathsf{a}_n + \gamma\mathsf{a}_{n+1}]\\
\mathsf{M}\mathsf{a}_{n+1} &+ \hat{\mathsf{f}}^{int}(\mathsf{d}_{n+1}) = \mathsf{f}^{ext}
\end{aligned}
\tag{6.61}
$$

Hence, by choosing $\beta = 0$ and $\gamma = 1/2$ we recover the central difference scheme, whereas for $\beta = 1/4$ and $\gamma = 1/2$ the general expression will reduce to the trapezoidal rule approximation. We can thus easily provide a general

[4] We note in passing that the inertia term in Lagrangian formulation of nonlinear dynamics still remains linear, hence the stability condition for nonlinear dynamics only concerns the internal energy.

[5] The softening behavior implies the stress decrease with increasing strain; several methods for dealing with softening behavior problems are discussed in Chapter 8.

computer code employing both time-integration schemes and leaving to users to make the most suitable choice between explicit and implicit schemes with the appropriate values of Newmark parameters.

6.2.3 Mid-point (implicit) scheme and its modifications for energy conservation and energy dissipation

Another implicit scheme of particular interest for low-frequency nonlinear dynamics is the mid-point rule. The scheme of this kind is especially interesting in a modified version enforcing the total energy conservation (in the absence of external forces). This will in general be sufficient to ensure the satisfying performance of the scheme in computations over very long time interval, where the accumulation of round-off errors can seriously affect the quality of the computed results. Furthermore, in this manner it is possible (see Simo [236]) to prove the unconditional stability of the mid-point scheme for all the problems in nonlinear dynamics where the constitutive behavior derives from a potential (for example, for a nonlinear hyperelasticity). For any such problem, as indicated in Table 6.3, the computational procedure using the mid-point scheme would require a number of iterations in each time step.

Table 6.3 Mid-point scheme in nonlinear dynamics.

Initialize: $\mathsf{d}_0, \mathsf{v}_0$
For each step $n = 1, 2, \ldots$:
Given: $\mathsf{d}_n, \mathsf{v}_n$ and h
Compute: $\mathsf{d}_{n+1}, \mathsf{v}_{n+1}$ such that
Iterate: $(i) = 1, 2, \ldots$

$$\frac{\mathsf{d}_{n+1}^{(i)} - \mathsf{d}_n}{h} = \mathsf{v}_{n+1/2}^{(i)}$$
$$\mathsf{M}\frac{\mathsf{v}_{n+1}^{(i)} - \mathsf{v}_n}{h} + \hat{\mathsf{f}}^{int}(\mathsf{d}_{n+1/2}^{(i)}) = \mathsf{f}_{n+1/2}^{ext}$$

Next iteration: $(i) \longleftarrow (i+1)$
Next time step: $n \longleftarrow n+1$

The mid-point scheme will verify the equations of motion in the middle of any given step, such as $t_{n+1/2}$. This property also ensures the correct contribution of kinetic energy towards the total energy conservation by mid-point rule. Namely, by selecting the weighting function as $\mathsf{w} = (\mathsf{v}_{n+1} - \mathsf{v}_n)$, the mid-point scheme representation of the inertia term of the weak form of the equations of motion can be written:

$$\begin{aligned} \int_\Omega \rho \mathsf{v}_{n+1/2} \cdot (\mathsf{v}_{n+1} - \mathsf{v}_n)\, dV &= \tfrac{1}{2}(\mathsf{v}_{n+1} \cdot \mathsf{M}\mathsf{v}_{n+1} - \mathsf{v}_n \cdot \mathsf{M}\mathsf{v}_n) \\ &= T_{n+1} - T_n \end{aligned} \tag{6.62}$$

This corresponds to the kinetic energy increase, with T_n and T_{n+1} as the corresponding values of kinetic energy at time t_n and t_{n+1}, respectively.
In order to achieve the same result for internal energy term, we must use the algorithmic constitutive equations. Namely, the correct increment value of the internal energy contribution towards the total energy conservation is computed by the mid-point scheme only for linear dynamics problems, where the internal energy can be written as a quadratic form in displacements. Moreover, the mid-point rule leads to the same result as the one provided by the trapezoidal rule for linear dynamics:

$$\hat{\mathsf{f}}^{int}(\mathsf{d}_{n+1/2}) := \mathsf{K}\ \overbrace{\mathsf{d}_{n+1/2}}^{\frac{1}{2}(\mathsf{d}_n+\mathsf{d}_{n+1})} = \frac{1}{2}[\hat{\mathsf{f}}^{int}(\mathsf{d}_n) + \hat{\mathsf{f}}^{int}(\mathsf{d}_{n+1})] \tag{6.63}$$

None of these findings remains true for nonlinear dynamics. Namely, the internal energy increment is no longer computed correctly by the mid-point rule approximation in nonlinear dynamics problems, where the internal force is a nonlinear function of displacement:

$$\hat{\mathsf{f}}^{int}(\mathsf{d}_{n+1/2}) \neq \frac{1}{2}[\hat{\mathsf{f}}^{int}(\mathsf{d}_n) + \hat{\mathsf{f}}^{int}(\mathsf{d}_{n+1})] \tag{6.64}$$

For example, for a hyperelastic material behavior at large strains, the internal force vector computed with the mid-point scheme can be expressed as a nonlinear function of the corresponding value of the second Piola–Kirchhoff stress tensor:

$$\begin{aligned} &\mathsf{w}^e \cdot \mathsf{f}^{int,e}_{n+1/2} = \textstyle\int_{\Omega^e} \frac{1}{2}(\mathbf{F}^T_{n+1/2}\nabla\mathbf{w} + \nabla\mathbf{w}^T\mathbf{F}_{n+1/2}) \cdot \mathbf{S}_{n+1/2}\, dV\ ; \\ &\mathbf{S}_{n+1/2} = 2\frac{\partial\psi(\mathbf{C}_{n+1/2})}{\partial\mathbf{C}} \end{aligned} \tag{6.65}$$

We can enforce the energy conservation by the mid-point scheme only if the algorithmic constitutive equations are used to compute the modified internal force:

$$\mathsf{f}^{int}_{cons} \cdot (\mathsf{d}_{n+1} - \mathsf{d}_n) = W_{n+1} - W_n \tag{6.66}$$

For the case of hyperelastic behavior at large strain, the latter will lead to the algorithmic value of stress, denoted as $\mathbf{S}_{cons}$, which is not the same as the one computed in (6.65); therefore, each finite element contribution to the algorithmic internal force can be obtained from:

$$\mathsf{w}^e \cdot \mathsf{f}^{int,e}_{cons} = \int_{\Omega^e} \frac{1}{2}(\mathbf{F}^T_{n+1/2}\nabla\mathbf{w} + \nabla\mathbf{w}^T\mathbf{F}_{n+1/2}) \cdot \mathbf{S}_{cons}\, dV \tag{6.67}$$

We can finally choose to construct the given value of stress ensuring the energy conservation for a hyperelastic material by using the algorithmic constitutive equations:

$$\begin{aligned} \mathbf{S}_{cons} = 2\frac{\partial\psi(\mathbf{C}_{n+1/2})}{\partial\mathbf{C}} &+ \frac{1}{\|\mathbf{C}_{n+1}-\mathbf{C}_n\|^2}[\psi(\mathbf{C}_{n+1}) - \psi(\mathbf{C}_n) \\ &- \frac{\partial\psi(\mathbf{C}_{n+1/2})}{\partial\mathbf{C}} \cdot (\mathbf{C}_{n+1} - \mathbf{C}_n)](\mathbf{C}_{n+1} - \mathbf{C}_n) \end{aligned} \tag{6.68}$$

By selecting the virtual displacement equal to incremental displacement, $\mathrm{w} = (\varphi_{n+1} - \varphi_n)$, and by taking into account the symmetry of the algorithmic stress tensor $\mathbf{S}_{cons} = \mathbf{S}^T_{cons}$, the result in (6.67) can be replaced with:

$$
\begin{aligned}
(\varphi_{n+1} - \varphi_n) \cdot \mathbf{f}^{int,e}_{cons} &= \int_{\Omega^e} \mathbf{S}_{cons} \cdot \overbrace{\mathbf{F}^T_{n+1/2}}^{(\mathbf{F}_n + \mathbf{F}_{n+1})/2} (\mathbf{F}_{n+1} - \mathbf{F}_n)\, dV \\
&= \int_{\Omega^e} \tfrac{1}{2}\mathbf{S}_{cons} \cdot (\mathbf{C}_{n+1} - \mathbf{C}_n)\, dV \\
&= \int_{\Omega^e} \psi(\mathbf{C}_{n+1})\, dV - \int_{\Omega^e} \psi(\mathbf{C}_n)\, dV \\
&= W(\mathbf{C}_{n+1}) - W(\mathbf{C}_n)
\end{aligned}
\tag{6.69}
$$

We can thus conclude that the algorithmic constitutive equations will indeed enforce the energy conservation with the mid-point scheme computation in a free-vibration phase.

For the Saint-Venant–Kirchhoff material model, with a quadratic form of the internal energy in term of the Green–Lagrange strain tensor, the algorithmic constitutive equation ensuring the energy conservation by the mid-point rule can be simplified further, with the end result which corresponds to the trapezoidal rule computation:

$$
\mathbf{S}_{cons} = \mathcal{C}\frac{1}{2}(\mathbf{C}_n + \mathbf{C}_{n+1}) \tag{6.70}
$$

An important advantage of the algorithmic constitutive equations is in their ability to eliminate any undesirable contribution from rotations towards the stress computation by the mid-point rule, thus enforcing the continuum consistent result for large time steps. The advantage is lost with the standard implementation of the mid-point scheme where the right Cauchy–Green strain for constitutive equations is computed from the mid-point configuration $\varphi_{n+1/2} = \frac{1}{2}(\varphi_n + \varphi_{n+1})$, which produces an artificial coupling between the deformations and the rotations according to:

$$
\begin{aligned}
\hat{\mathbf{C}}(\varphi_{n+1/2}) &= \frac{1}{2}(\mathbf{F}_n + \mathbf{F}_{n+1})^T \frac{1}{2}(\overbrace{\mathbf{F}_n}^{\mathbf{R}_n\mathbf{U}_n} + \overbrace{\mathbf{F}_{n+1}}^{\mathbf{R}_{n+1}\mathbf{U}_{n+1}}) \\
&= \frac{1}{4}(\mathbf{C}_n + \mathbf{C}_{n+1} + \mathbf{U}_n\mathbf{R}^T_n\mathbf{R}_{n+1}\mathbf{U}_{n+1} + \mathbf{U}_{n+1}\mathbf{R}^T_{n+1}\mathbf{R}_n\mathbf{U}_n)
\end{aligned}
\tag{6.71}
$$

If the time step goes to zero, the algorithmic constitutive equations for the mid-point scheme will approach the consistent form used for continuum model, and the difference remains of higher order:

$$
\mathbf{S}_{cons} = \underbrace{2\frac{\partial \psi(\mathbf{C}_{n+1/2})}{\partial \mathbf{C}}}_{S_{n+1/2}} + O(h^2) \tag{6.72}
$$

This implies that the proposed modification of the mid-point scheme with the algorithmic constitutive equation will not affect the order of convergence with respect to the basic mid-point scheme.
Yet another modification of the mid-point scheme discussed herein concerns providing the capability to dissipate the energy of high frequency modes (see [133]). Such a feature is of strong practical interest for stress computation accuracy, which can often be plagued by unresolved high frequency modes.[6] The proposed cures from linear dynamics based upon the modified Newmark method [101] have not always proved successful for nonlinear dynamics problems (e.g. see [24] or [132]), where the high frequency modes would (constantly) change in the course of motion with each modification of the tangent stiffness matrix. One way for constructing a modification of the mid-point scheme capable of numerical dissipation of high-frequency modes (with no need to compute them exactly), is to start from the algorithmic constitutive equations designed for energy conservation with mid-point scheme, and add the following dissipative term:

$$\mathbf{S}_{diss} = \mathbf{S}_{cons} + 2D_{\psi}(\mathbf{C}_{n+1} - \mathbf{C}_n) / \parallel \mathbf{C}_{n+1} - \mathbf{C}_n \parallel \tag{6.73}$$

where D_{ψ} is the chosen numerical dissipation coefficient. With the choice of such a modification where D_{ψ} is multiplied by the difference of the right Cauchy–Green strains at time t_{n+1} and t_n, the main change will come to high-frequency modes (with the period equal or smaller than the chosen time step), without much affecting the low-frequency modes (with the period much larger than the chosen time step) where the strain difference within the time step $\mathbf{C}_{n+1} - \mathbf{C}_n$ is likely to remain negligible. We can set the dissipation coefficient value in the manner which will not reduce the order of the mid-point scheme:

$$\begin{aligned} D_{\psi} &= 4\alpha[\tfrac{1}{2}(\psi(\mathbf{C}_{n+1}) + \psi(\mathbf{C}_n)) - \psi(\mathbf{C}_{n+1/2})] \; ; \\ &\Longrightarrow \lim_{\parallel \mathbf{C}_{n+1} - \mathbf{C}_n \parallel \mapsto 0} \frac{D_{\psi}}{\parallel \mathbf{C}_{n+1} - \mathbf{C}_n \parallel} \mapsto 0 \end{aligned} \tag{6.74}$$

The numerical dissipation is thus controlled by parameter α. By direct computation which follows the steps for energy conserving modification in (6.69), it is easy to check that such a modified algorithmic constitutive equation will dissipate the energy, and that the amount of dissipation will be directly proportional to D_{ψ}:

$$(\boldsymbol{\varphi}_{n+1} - \boldsymbol{\varphi}_n) \cdot \mathbf{f}^{int}_{diss} = \int_{\Omega} (\psi_{n+1} - \psi_n) \, dV + \int_{\Omega} D_{\psi} \, dV \tag{6.75}$$

---

[6] Namely, it is well known (e.g. see [249]) that a coarse finite element mesh cannot represent high frequency modes with sufficient accuracy, and that the best strategy is to introduce numerical dissipation (e.g. see [111]), which will quickly damp out the contribution of these modes.

In order to also ensure the dissipation of the kinetic energy, we further introduce the corresponding modification of the velocity computations with the mid-point scheme proportional to parameter β; for a typical node 'a', the latter results with:

$$\frac{\varphi_{a,n+1}-\varphi_{a,n}}{h}=\mathbf{v}_{n+1/2}+\beta\frac{\|\mathbf{v}_{a,n+1}\|-\|\mathbf{v}_{a,n}\|}{\|\mathbf{v}_{a,n+1}\|+\|\mathbf{v}_{a,n}\|}\mathbf{v}_{a,n+1/2} \tag{6.76}$$

With this result in hand and a diagonal form of the mass matrix, we can further express the inertia term contribution to the equations of motion:

$$\begin{aligned}
I &= (\varphi_{a,n+1}-\varphi_{a,n})\cdot(\mathbf{f}^{iner}_{cons}+\mathbf{f}^{iner}_{diss})\\
&= \textstyle\sum_{a=1}^{n}(\varphi_{a,n+1}-\varphi_{n,a})\cdot\mathbf{M}_a(\mathbf{v}_{a,n+1}-\mathbf{v}_{n,a})\\
&= \textstyle\sum_{a=1}^{n}\mathbf{v}_{a,n+1/2}\cdot\mathbf{M}_a(\mathbf{v}_{a,n+1}-\mathbf{v}_{a,n})\\
&\quad+\textstyle\sum_{a=1}^{n}\beta\frac{\|\mathbf{v}_{a,n+1}\|-\|\mathbf{v}_{a,n}\|}{\|\mathbf{v}_{a,n+1}\|+\|\mathbf{v}_{a,n}\|}\mathbf{v}_{n+1/2}\cdot(\mathbf{v}_{a,n+1}-\mathbf{v}_{a,n})\\
&= (1/2)\textstyle\sum_{a=1}^{n}\mathbf{v}_{a,n+1}\cdot\mathbf{M}_a\mathbf{v}_{a,n+1}-(1/2)\sum_{a=1}^{n}\mathbf{v}_{a,n}\cdot\mathbf{M}_a\mathbf{v}_{a,n}\\
&\quad+(1/2)\beta\textstyle\sum_{a=1}^{n}\mathbf{M}_a(\|\mathbf{v}_{a,n+1}\|^2-\|\mathbf{v}_{a,n}\|^2)\\
&= T_{n+1}-T_n+D_T\ ;\ D_T=(1/2)\beta\int_\Omega\rho(\|\mathbf{v}_{n+1}\|^2-\|\mathbf{v}_n\|^2)\,dV
\end{aligned} \tag{6.77}$$

The dissipation of inertia term contribution will mostly affect the high-frequency modes, where we are more likely to find a large difference in velocities at the beginning and at the end of the time step; moreover, the amount of dissipation will be proportional to the chosen value of β parameter.

6.2.3.1 Numerical example: Flexible beam on a guiding rod

We illustrate the performance of mid-point rule with energy conserving or energy dissipating algorithmic constitutive equations with a multibody dynamics system that consists of a flexible beam attached to a massive cylinder which is allowed to slide vertically along and rotate about the guiding rod. The beam has elastic constitutive behavior at large displacements described by Saint-Venant–Kirchhoff model, with Young's modulus $E = 10{,}000$, Poisson's ratio $\nu = 0.25$ and mass density $\rho = 0.02$. The finite element model of the beam consists of $4 \times 4 \times 16 = 256$ 3D hexahedral solid elements, each with 8 nodes (see Figure 6.3). The beam is attached to a massive cylinder, with external and internal radii equal to 8 and 4, respectively. A very large value of cylinder mass of 2,000 is chosen in order to stabilize the cylinder in sliding; moreover, the sliding rod radius is chosen to be equal to 3.9, in order to avoid frictional phenomena during the cylinder sliding.
The loading program consists of two different phases (see Figure 6.3). In the first phase, we apply a vertical force in the center of the beam increasing

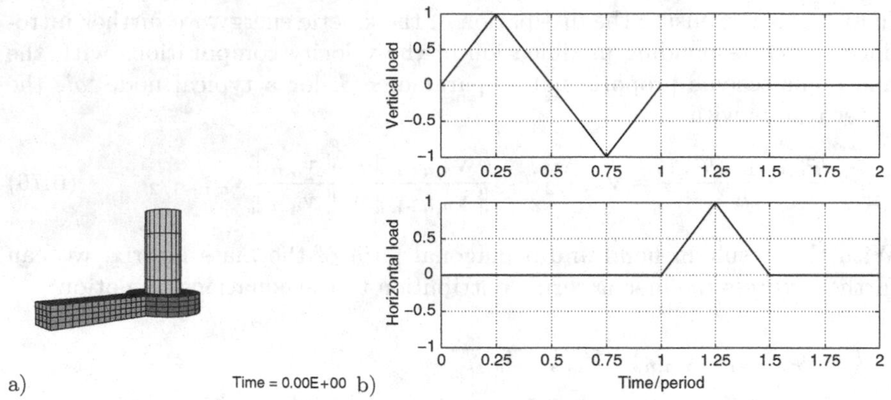

Fig. 6.3 Flexible beam on a guiding rod: **(a)** finite element mesh, **(b)** loading program.

linearly to a peak-value of 3,500 at time 0.25 T, and then decreasing to −3,500 at 0.75 T, to finally come to zero value at time 1.0 T; the chosen value of parameter T = 0.874 corresponds to the first mode of the flexible beam, if rigidly attached (or with built-in end). In the second loading phase, we apply a horizontal force in the center of the beam, starting at 1.0 T, which increases to a peak-value at time 1.25 T and goes back to zero at 1.5 T.
The time-integration of the equations of motion of this system is carried out for first 4 s, by using the mid-point scheme with energy conservation with the time step $h = 0.001$. The successive deformed shapes obtained by this computations are presented in Figure 6.4. In Figure 6.5, we show that the mid-point scheme with algorithmic constitutive equations can indeed ensure the energy conservation in the free-vibration phase, which starts after the second phase of chosen loading program. We have also illustrated on the same figure that the energy-conserving modification is not capable (in general) to ensure the sufficient accuracy of the stress computations, which is often plagued by the high-frequency noise:

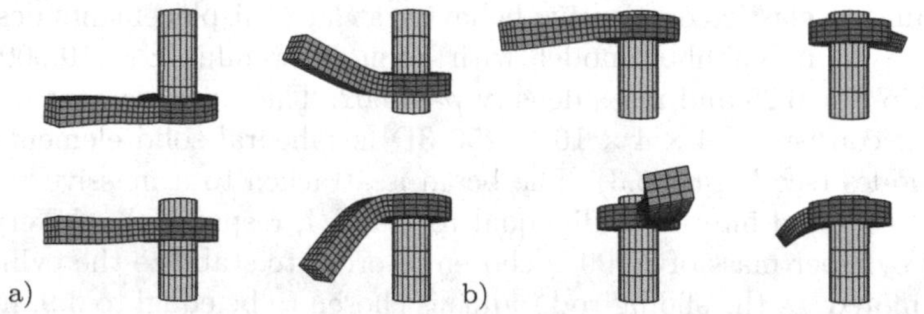

Fig. 6.4 Flexible beam on a guiding rod: successive deformed shapes at each 0.2 s. for: **(a)** first phase **(b)** second phase of loading program.

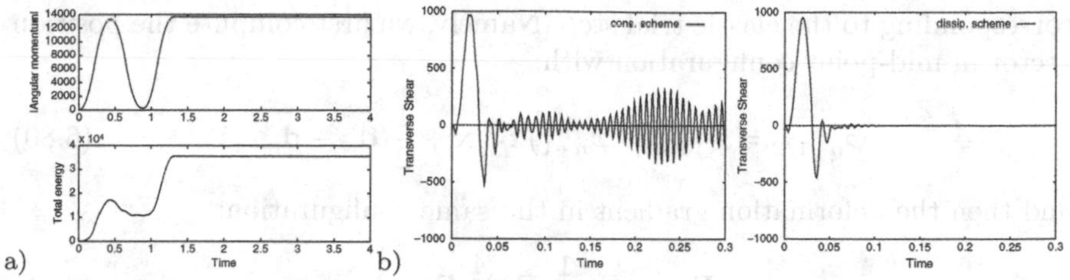

Fig. 6.5 Flexible beam on a guiding rod: time-evolution of **(a)** total energy and angular momentum **(b)** shear force close to support.

6.3 Mid-point (implicit) scheme for finite deformation plasticity

In this section we will revisit the finite deformation plasticity problem within the context of dynamics. The mid-point rule will be used in order to carry out the time-integration for such a problem in each time step. The outline of the numerical implementation is given in Table 6.4.

Table 6.4 Mid-point scheme for finite deformation plasticity in dynamics.

$$\text{Given: } \mathbf{d}_n,\ \mathbf{C}^p_n,\ \zeta_n,\ \mathbf{v}_n$$
$$\text{Find: } \mathbf{d}_{n+1},\ \mathbf{C}^p_{n+1},\ \zeta_{n+1},\ \mathbf{v}_{n+1}, \text{ such that:}$$

$$\begin{aligned} &\frac{\mathbf{d}_{n+1}-\mathbf{d}_n}{h} = \mathbf{v}_{n+1/2} \\ &\mathbf{M}\frac{\mathbf{v}_{n+1}-\mathbf{v}_n}{h} + \hat{\mathbf{f}}^{int}(\mathbf{d}_{n+1/2}, \mathbf{C}^p_{n+1/2}, \zeta_{n+1/2}) = \mathbf{f}^{ext}_{n+1/2} \end{aligned} \tag{6.78}$$

The internal force vector is computed with the plastically admissible value of stress:

$$\begin{aligned} \mathrm{w}^e \cdot \hat{\mathbf{f}}^{int,e}(\mathbf{d}_{n+1/2}, \mathbf{C}^p_{n+1/2}, \zeta_{n+1/2}) &= \int_{\Omega^e} \tfrac{1}{2}(\mathbf{F}^T_{n+1/2}\nabla\mathbf{w} \\ &+\nabla\mathbf{w}^T\mathbf{F}_{n+1/2}) \cdot \overbrace{\mathbf{F}^{-1}_{n+1/2}\boldsymbol{\tau}_{n+1/2}\mathbf{F}_{n+1/2}}^{\mathbf{S}(\mathbf{d}_{n+1/2},\mathbf{C}^p_{n+1/2},\zeta_{n+1/2})}\, dV \\ \dot{\gamma}_{n+1/2} \geq 0\ ;\ \phi(\boldsymbol{\tau}_{n+1/2}, q_{n+1/2})\ &;\ \dot{\gamma}_{n+1/2}\phi(\cdot|_{n+1/2}) = 0 \end{aligned} \tag{6.79}$$

We will elaborate subsequently only upon the local step of the operator-split procedure for plasticity, which concerns the computation of plastically admissible stress for mid-point scheme, as well as the corresponding values of internal variables. This computation is carried out in two stages. In the first one, we will compute the mid-step configuration and the value of stress corresponding to the elastic trial step. Namely, we first compute the position vector in mid-point configuration with:

$$\boldsymbol{\varphi}_{n+1/2} = \frac{1}{2}(\boldsymbol{\varphi}_n + \boldsymbol{\varphi}_{n+1}) := \mathbf{x} + \frac{1}{2}(\mathbf{d}_n + \mathbf{d}_{n+1}) \tag{6.80}$$

and then the deformation gradient in the same configuration:

$$\mathbf{F}_{n+1/2} = \frac{1}{2}(\mathbf{F}_n + \mathbf{F}_{n+1}) \tag{6.81}$$

The elastic trial value of left Cauchy–Green strain tensor can be computed by the corresponding transformation to this mid-step configuration with no change in plastic deformation from the previous step, resulting with:

$$\mathbf{b}^{e,trial}_{n+1/2} = \mathbf{F}_{n+1/2}\mathbf{G}^p_n\mathbf{F}^T_{n+1/2} \; ; \; \mathbf{G}^p_n = [\mathbf{C}^p_n]^{-1} \tag{6.82}$$

We can also obtain the last result with a convective transport of the elastic left Cauchy–Green strain tensor at time t_n, which is equal to $\mathbf{b}^e_n = \mathbf{F}_n\mathbf{G}^p_n\mathbf{F}^T_n$s:

$$\mathbf{b}^{e,trial}_{n+1/2} = \mathbf{f}_{n+1/2}\mathbf{b}^e_n\mathbf{f}^T_{n+1/2} \; ; \; \mathbf{f}_{n+1/2} = \mathbf{F}_{n+1/2}\mathbf{F}^{-1}_n \tag{6.83}$$

We can then compute the corresponding trial values of elastic principal stretches $\lambda^{e,trial}_{i,n+1/2}$ and the principal directions $\mathbf{m}^{trial}_{n+1/2}$ as the solution of the eigenvalue problem featuring $\mathbf{b}^{e,trial}_{n+1/2}$:

$$[\mathbf{b}^{e,trial}_{n+1/2} - (\lambda^{e,trial}_{i,n+1/2})^2]\mathbf{m}^{trial}_{i,n+1/2} = \mathbf{0} \; ; \; \mathbf{m}^{trial}_{i,n+1/2} \cdot \mathbf{m}^{trial}_{j,n+1/2} = \delta_{ij} \tag{6.84}$$

where δ_{ij} is the Kronecker symbol. The corresponding trial value of principal Kirchhoff stress $\tau^{trial}_{i,n+1/2}$ and hardening variable $q^{trial}_{n+1/2}$ at mid-step, are then computed from the invariant form of the strain energy potential:

$$\tau^{trial}_{i,n+1/2} = \lambda^{e,trial}_{i,n+1/2}\frac{\partial\psi(\lambda^{e,trial}_{i,n+1/2},\zeta_n)}{\partial\lambda^e} \; ; \; q^{trial}_{n+1/2} = -\frac{\partial\psi(\lambda^{e,trial}_{i,n+1/2},\zeta_n)}{\partial\zeta} \tag{6.85}$$

It is now easy to compute the trial value of the yield function by using the invariant form featuring the principal values:

$$\phi^{trial}_{n+1/2} := \phi(\tau^{trial}_{i,n+1/2}, q^{trial}_{n+1/2}) \tag{6.86}$$

If this trial value of the yield function is negative $\phi^{trial}_{n+1/2} \leq 0$, the trial elastic state is accepted as final, and the internal variables will not change with respect to the previous step.
In the opposite case with $\phi^{trial}_{n+1/2} > 0$, the internal variable values should be updated in accordance with the mid-point scheme:

$$\begin{aligned} \mathbf{b}^e_{n+1/2} &= exp[-2\gamma_{n+1/2}\frac{\partial\phi(\boldsymbol{\tau}_{n+1/2},q_{n+1/2})}{\partial\boldsymbol{\tau}}]\mathbf{b}^{trial}_{n+1/2} \\ \zeta_{n+1/2} &= \zeta_n + \gamma_{n+1/2}\frac{\partial\phi(\boldsymbol{\tau}_{n+1/2},q_{n+1/2})}{\partial q} \end{aligned} \tag{6.87}$$

The plastic multiplier at mid-step $\gamma_{n+1/2}$ is either computed from the plastic consistency condition at that time for the plasticity model with:

$$\gamma_{n+1/2} \geq 0 \; ; \; \phi(\boldsymbol{\tau}_{n+1/2}, q_{n+1/2}) \leq 0 \; ; \; \gamma_{n+1/2}\phi(\boldsymbol{\tau}_{n+1/2}, q_{n+1/2}) = 0 \quad (6.88)$$

or from the corresponding value of the stress for viscoplasticity model:

$$\gamma_{n+1/2} = \frac{h}{2\eta} < \phi(\boldsymbol{\tau}_{n+1/2}, q_{n+1/2}) > \quad (6.89)$$

It is important to note that in either case the validity of the plasticity criterion is imposed at the mid-point value of the time step. It is also important to note that the corresponding computation of the evolution of the elastic left Cauchy–Green strain tensor is carried out in the principal axes space; thus, we ought to write more precisely that:

$$\begin{aligned} &\epsilon^e_{i,n+1/2} = \epsilon^{e,trial}_{i,n+1/2} - \gamma_{n+1/2}\frac{\partial\phi(\tau_{i,n+1/2}, q_{i,n+1/2})}{\partial\tau_i} \; ; \; \epsilon^{e,trial}_{i,n+1/2} = ln\lambda^{e,trial}_{i,n+1/2} \\ &\mathbf{b}^e_{n+1/2} = \textstyle\sum_{i=1}^3 (\lambda^e_{i,n+1/2})^2 \mathbf{m}^{trial}_{n+1/2} \otimes \mathbf{m}^{trial}_{n+1/2} \; ; \; \lambda^e_{i,n+1/2} = exp[\epsilon^e_{i,n+1/2}] \end{aligned} \quad (6.90)$$

With these results in hand, it is easy to obtain the plastically admissible mid-point value of the stress tensor, by exploiting the computed values of internal variables, the invariant form of the strain energy potential and the spectral decomposition result:

$$\boldsymbol{\tau}_{n+1/2} = \sum_{i=1}^{3} \lambda^e_{i,n+1/2} \frac{\partial\psi(\lambda^e_{i,n+1/2}, \zeta_{n+1/2})}{\partial\lambda^e_i} \mathbf{m}^{trial}_{i,n+1/2} \otimes \mathbf{m}^{trial}_{i,n+1/2} \quad (6.91)$$

We note finally that the internal variable values are also needed at the end of the step at time t_{n+1} in order to provide the corresponding elastic trial value for the next step. Such values can easily be obtained starting from the computed admissible values at mid-point state at $t_{n+1/2}$, which leads to:

$$\begin{aligned} \mathbf{b}^e_{n+1} &= exp[-2\gamma_{n+1}\frac{\partial\phi(\boldsymbol{\tau}_{n+1}, q_{n+1})}{\partial\boldsymbol{\tau}}]\mathbf{b}^{e,trial}_{n+1} \; ; \\ \mathbf{b}^{e,trial}_{n+1} &= \tilde{\mathbf{f}}_{n+1}\mathbf{b}^e_{n+1/2}\tilde{\mathbf{f}}^T_{n+1} \; ; \; \tilde{\mathbf{f}}_{n+1} = \mathbf{F}_{n+1}\mathbf{F}^{-T}_{n+1/2} \; ; \\ \zeta_{n+1} &= \zeta_{n+1/2} + \gamma_{n+1}\frac{\partial\phi(\boldsymbol{\tau}_{n+1}, q_{n+1})}{\partial q} \end{aligned} \quad (6.92)$$

where the plastic multiplier is defined for either plasticity model:

$$\gamma_{n+1} \geq 0 \; ; \; \phi_{n+1} \leq 0 \; ; \; \gamma_{n+1}\phi_{n+1} = 0 \quad (6.93)$$

or viscoplasticity model:

$$\gamma_{n+1} = \frac{h}{2\eta} < \phi_{n+1} > \quad (6.94)$$

This computation of internal variables at the end of the step does not represent a significant increase in computational cost of the mid-point scheme. It ought to be done only once within each step, after the operator split solution procedure has converged and delivered the admissible values of displacements and all internal variables at mid-point value.

6.4 Contact problem and time-integration schemes

6.4.1 Mid-point (implicit) scheme for contact problem in dynamics

The mid-point scheme can also successfully be applied to contact problems in dynamics, where we seek to better control the high-frequency content of motion. We consider a model problem of bilateral contact between two solid deformable bodes occupying the domains Ω^α ; $\alpha = 1, 2$. The weak form of equations of motion can be written:

$$\begin{aligned} &\sum_{\alpha=1}^{2}[\int_{\Omega^\alpha} \rho^\alpha \mathbf{v}^\alpha \cdot \mathbf{w}^\alpha \, dV + \int_{\Omega^\alpha} \mathbf{P}^\alpha \cdot \nabla \mathbf{w}^\alpha \, dV] \\ &= \sum_{\alpha=1}^{2}[\int_{\Omega^\alpha} \mathbf{b}^\alpha \cdot \mathbf{w}^\alpha \, dV + \int_{\Gamma_\sigma^\alpha} \bar{\mathbf{t}}^\alpha \cdot \mathbf{w}^\alpha \, dA] + \int_{\Gamma_c^\alpha} \mathbf{t} \cdot (\mathbf{w}^1 - \mathbf{w}^2(\bar{\mathbf{y}})) \, dA \end{aligned} \tag{6.95}$$

In order to guarantee the solution uniqueness, we impose the following restrictions upon the boundaries of two solid bodies in contact:

$$\Gamma^\alpha = \Gamma_u^\alpha \bigcup \Gamma_\sigma^\alpha \bigcup \Gamma_c^\alpha \; ; \; \Gamma_u^\alpha \bigcap \Gamma_\sigma^\alpha \bigcap \Gamma_c^\alpha = \emptyset \tag{6.96}$$

We consider here the large displacements of solid bodies in contact, but with the constitutive behavior of each body which remains elastic. However, the presented developments will also apply to other constitutive models for the bodies in contact (for example, for inelastic behavior), with the only modification pertaining to the corresponding computations of stresses $\mathbf{P}^\alpha$ in accordance with the chosen constitutive model. The contact is characterized by the last term of the equations of motion in (6.95). We will consider frictionless contact case, where the contact force is oriented in the direction of unit exterior normal $\mathbf{n}$ and imposes the pressure p between the bodies in contact; we can thus write:

$$\mathbf{t} = p \, \mathbf{n}(\bar{\mathbf{y}}) \; ; \; p \geq 0 \tag{6.97}$$

We indicated in the last expression that the unit exterior normal vector is constructed at the obstacle point $\bar{\mathbf{y}}$, which is defined as the closest projection to $\mathbf{x}$ according to:

$$\bar{\mathbf{y}}(\mathbf{x}) = arg \min_{\mathbf{y}\in\Gamma^2} \| \boldsymbol{\varphi}^1(\mathbf{x}) - \boldsymbol{\varphi}^2(\mathbf{y}) \| \tag{6.98}$$

The gap function for contact can then be defined with:

$$g(\mathbf{x}) = \mathbf{n} \cdot [\boldsymbol{\varphi}^1(\mathbf{x}) - \boldsymbol{\varphi}^2(\bar{\mathbf{y}}(\mathbf{x}))] \geq 0 \tag{6.99}$$

Since the gap function measures the distance along the unit exterior normal, we can also write:

$$\boldsymbol{\varphi}^1(\mathbf{x}) - \boldsymbol{\varphi}^2(\bar{\mathbf{y}}(\mathbf{x})) = g(\mathbf{x})\, \mathbf{n}(\bar{\mathbf{y}}(\mathbf{x})) \tag{6.100}$$

With these results in hand, we can rewrite the contact term contribution to the weak form of equations of motion according to:

$$\int_{\Gamma_c^\alpha} \mathbf{t} \cdot (\mathbf{w}^1 - \mathbf{w}^2(\bar{\mathbf{y}}))\, dA = \int_{\Gamma^\alpha_{contact}} p\, \mathbf{n} \cdot (\mathbf{w}^1 - \mathbf{w}^2(\bar{\mathbf{y}}))\, dA \tag{6.101}$$

The impenetrability condition will then impose that the product between the contact pressure and gap between the bodies must always remain zero, since either the gap or the contact pressure is zero:

$$p \geq 0 \; ; \; g(\mathbf{x}) \geq 0 \; ; \; p\, g(\mathbf{x}) = 0 \tag{6.102}$$

In the case of persisting contact, the consistency condition will require that not only the gap function but also its time-derivative has to remain equal to zero; we can thus obtain:

$$0 = \dot{g} = \mathbf{n} \cdot (\mathbf{v}^1 - \mathbf{v}^2(\bar{\mathbf{y}})) + g \underbrace{\dot{\mathbf{n}} \cdot \mathbf{n}}_{=0} \tag{6.103}$$

were we exploited the result in (6.100) and the orthogonality between the unit exterior normal and its time-derivative. We can also provide the local form of consistency condition for dynamic contact in the following format:

$$p\, \dot{g} = 0 \tag{6.104}$$

We can show that it is precisely the consistency condition enforcement which allows us to impose the total energy conservation for a dynamic contact problem between two hyperelastic solid bodies. More precisely, by choosing the velocity at time t as the weighting function, the weak form of the equations of motion in the absence of external forces can be written:

$$\underbrace{\sum_{\alpha=1}^{2} [\int_{\Omega^\alpha} \rho^\alpha \mathbf{v}^\alpha \cdot \dot{\mathbf{v}}^\alpha \, dV}_{dT/dt} + \underbrace{\int_{\Omega^\alpha} \frac{\partial \psi^\alpha}{\partial \mathbf{F}^\alpha} \cdot \nabla \mathbf{v}^\alpha \, dV]}_{dW/dt} = \underbrace{\int_{\Gamma_c^\alpha} p \dot{g} \, d\Gamma}_{=0} \tag{6.105}$$

We can thus conclude that the total energy remains constant during such a motion in dynamics, even in the presence of contact, as long as the contact consistency condition is imposed. We further show that the mid-point scheme can be modified accordingly in order to enforce the total energy conservation in the discrete approximation setting.
By using the mid-point scheme for dynamic contact, the central problem over a typical time step is described as given in Table 6.5.

Table 6.5 Mid-point scheme for bilateral contact problem in dynamics.

Given: $\mathbf{d}_n^\alpha$, $\mathbf{v}_n^\alpha$ and $h = t_{n+1} - t_n$
Find: $\mathbf{d}_{n+1}^\alpha$, $\mathbf{v}_{n+1}$,
such that $\forall \mathbf{w}^\alpha$:

$$\begin{aligned} &\tfrac{1}{h}(\mathbf{d}_{n+1} - \mathbf{d}_n) = \mathbf{v}_{n+1/2} \\ &0 = G_{dyn}(\boldsymbol{\varphi}_{n+1/2}^\alpha; \mathbf{w}^\alpha) + G_{contact}(\boldsymbol{\varphi}_{n+1/2}^\alpha; \mathbf{w}^\alpha) \end{aligned} \tag{6.106}$$

with:

$$\begin{aligned} &G_{dyn}(\boldsymbol{\varphi}_{n+1/2}^\alpha; \mathbf{w}^\alpha) = \textstyle\sum_{\alpha=1}^2 [\int_{\Omega^\alpha} \rho^\alpha \tfrac{1}{h}(\mathbf{v}_{n+1}^\alpha - \mathbf{v}_n^\alpha) \cdot \mathbf{w}^\alpha \, dV \\ &\quad + \int_{\Omega^\alpha} \mathbf{F}_{n+1/2}^\alpha \mathbf{S}^\alpha \cdot \nabla \mathbf{w}^\alpha \, dV] \\ &G_{contact}(\boldsymbol{\varphi}_{n+1/2}^\alpha; \mathbf{w}^\alpha) := \int_{\Gamma_c^\alpha} \mathbf{t} \cdot [\mathbf{w}^2(\bar{\mathbf{y}}) - \mathbf{w}^1] \, dA \\ &\quad = \int_{\Gamma_c^\alpha} p_c \mathbf{n}_{n+1/2} \cdot [\mathbf{w}^2(\bar{\mathbf{y}}) - \mathbf{w}^1] \, dA \end{aligned} \tag{6.107}$$

In order to ensure the energy conservation by mid-point scheme for dynamic contact problem of this kind, we ought to not only use the algorithmic constitutive equations for computing the corresponding value of second Piola–Kirchhoff stress $\mathbf{S}_{cons} = \mathcal{C}\frac{1}{2}(\mathbf{C}_n + \mathbf{C}_{n+1})$, but also modify the contact term contribution accordingly. Namely, for a particular choice of the virtual displacement given as $\mathbf{w}^\alpha = \boldsymbol{\varphi}_{n+1}^\alpha - \boldsymbol{\varphi}_n^\alpha$, we can show that the mid-point scheme representation of the contact term of the weak form of equations of motion can be written:

$$\begin{aligned} E_{c,n+1} - E_{c,n} &= G_{contact}(\boldsymbol{\varphi}_{n+1}^\alpha; \mathbf{w}^\alpha) \\ &:= \int_{\Gamma_c^\alpha} p_c \mathbf{n}_{n+1/2} \cdot (\mathbf{v}_{n+1/2}^1 - \mathbf{v}_{n+1/2}^2(\bar{\mathbf{y}})) \, dA \\ &= \int_{\Gamma_c^\alpha} p_c \mathbf{n}_{n+1/2} \cdot [(\boldsymbol{\varphi}_{n+1}^1 - \boldsymbol{\varphi}_{n+1}^2(\bar{\mathbf{y}})) \\ &\quad -(\boldsymbol{\varphi}_n^1 - \boldsymbol{\varphi}_n^2(\bar{\mathbf{y}}))] \, dA \\ &= \int_{\Gamma_c^\alpha} p_c (g_{n+1}^d - g_n^d) \, dA \end{aligned} \tag{6.108}$$

where we introduced the corresponding dynamic gap furnished by mid-point scheme:

$$g_{n+1}^d = g_n^d + \mathbf{n}_{n+1/2} \cdot [(\boldsymbol{\varphi}_{n+1}^1 - \boldsymbol{\varphi}_{n+1}^2(\bar{\mathbf{y}})) - (\boldsymbol{\varphi}_n^1 - \boldsymbol{\varphi}_n^2(\bar{\mathbf{y}}))] \tag{6.109}$$

We further use the penalty method in order to obtain the corresponding value of contact pressure for the given value of dynamic gap, and moreover provide the modified algorithmic form of such a relationship with:

$$p_{c,cons} = k\frac{1}{2}(g^d_{n+1} + g^d_n) \tag{6.110}$$

where k is the penalty parameter. For the case under consideration where the relationship between the contact pressure and displacements is nonlinear, the proposed algorithmic constitutive relation for contact pressure is not the same as the one used by the standard mid-point approximation, which can be written as: $p_{n+1/2} = k\, g_{n+1/2}$. However, the algorithmic constitutive equations is precisely the one which will enforce the energy conservation, since it allows to rewrite (6.108) according to:

$$\begin{aligned} E_{c,n+1} - E_{c,n} &= \tfrac{1}{2}k(g^d_{n+1}g^d_{n+1} - g^d_n g^d_n) \\ &= W(g^d_{n+1}) - W(g^d_n) \end{aligned} \tag{6.111}$$

where $W(g^d_{n+1})$ is the elastic potential for computing the contact pressure. If we also want to compute the accurate time-history of contact pressure, we should further modify the proposed algorithmic constitutive relation for contact pressure, by adding the corresponding dissipation of high-frequency modes. With penalty method, this would lead to the following constitutive relation for the contact pressure:

$$p_{c,diss} = p_{c,cons} + \alpha k(g^d_{n+1} - g^d_n) \tag{6.112}$$

We can show that parameter α will control the energy dissipation during contact, since this modified algorithmic equation for contact pressure leads to:

$$E_{c,n+1} - E_{c,n} = W(g^d_{n+1}) - W(g^d_n) + \alpha k(g^d_{n+1} - g^d_n)^2 \tag{6.113}$$

It is also easy to check that for any low-frequency mode with the period much larger than the chosen time step, the change of dynamic gap and thus the energy dissipation will quite likely remain insignificant. The same kind of scheme can also be applied to frictional contact problems (see [8]).

6.4.2 Central difference (explicit) scheme and impact problem

The dynamic contact of very short duration, or impact, belongs to the class of problems where the high-frequency content should neither be neglected nor eliminated, and the discussion from the previous section no longer pertains. For this reason and also for the short duration of loading phase, the small time steps and explicit time-integration schemes are indeed the most suitable for impact. The main difficulty one has to deal with in the presence of impact concerns to the solution discontinuities with respect to the time evolution (e.g. see [252]). More precisely, the chosen time instants do not necessarily coincide with the time of impact or time of separation, and the computed solution is not necessarily regular. This is illustrated by an example considering the impact of an elastic bar, launched against a rigid obstacle by an initial velocity (see Figure 6.6). By denoting with t_c the time when the bar impacts the obstacle, and with t_r the time when it separates, we can obtain the corresponding result for time-evolution of the contact pressure, or Lagrange multiplier, as presented in Figure 6.6.

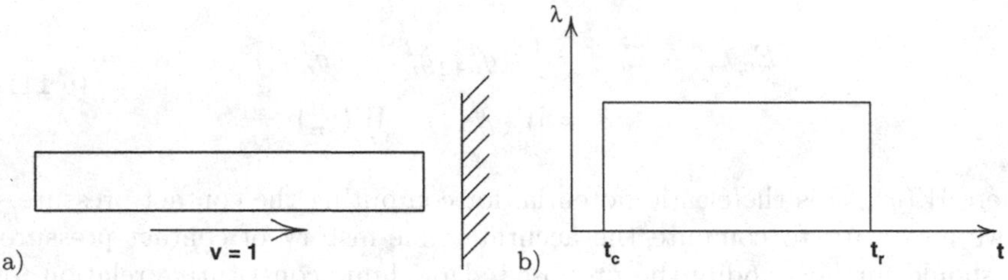

Fig. 6.6 Impact of an elastic bar against rigid obstacle: **(a)** problem definition, **(b)** time-evolution of contact pressure, with t_c time of contact and t_r time of separation.

At any moment in time before the impact, we can compute the motion of the bar by using the standard central difference scheme:

$$\begin{aligned} \mathsf{d}_{n+1} &= \mathsf{d}_n + h\mathsf{v}_n + \tfrac{1}{2}h^2\mathsf{a}_n \\ \mathsf{a}_{n+1} &= \mathsf{M}^{-1}(\mathsf{f}_{n+1} - \mathsf{K}\mathsf{d}_{n+1}) \\ \mathsf{v}_{n+1} &= \mathsf{v}_n + \tfrac{1}{2}h(\mathsf{a}_n + \mathsf{a}_{n+1}) \end{aligned} \tag{6.114}$$

where $\mathsf{d}_{n+1}, \mathsf{v}_{n+1}$ and a_{n+1} are, respectively, the nodal values of displacement, velocities and acceleration at time t_{n+1}.
Once the bar enters in contact with the rigid obstacle, we also have to impose the impenetrability condition upon computed displacements and velocities, which can be accomplished by the following constraint:

$$\Pi_{contact}(\varphi_t, p_t) = \Pi(\varphi_t) + p_t \overbrace{[\underbrace{\varphi^c(t)}_{x^c+d^c(t)} - \underbrace{\varphi^o(t)}_{x^o=cst.}]}^{g_t} \tag{6.115}$$

where p_t is the contact pressure and g_t is the gap function. The last term is written in Kuhn-Tucker form, indicating that we ought to have either the contact pressure or the gap to obstacle equal to zero. We would like here to impose the same constraint within the context of time-discretization upon the results computed by the central difference scheme. The idea is relatively simple; we note that the central difference scheme imposes the influence of the acceleration a_{n+1} upon the displacement at subsequent time t_{n+2} according to:

$$\mathsf{d}_{n+2} = \mathsf{d}_{n+1} + h\mathsf{v}_{n+1} + \frac{1}{2}h^2\mathsf{a}_{n+1} \tag{6.116}$$

Hence, we will have to take the last result into account along with the constraint placed upon the displacement by contact, which finally leads to a modified form of the contact terms where the values at time t_{n+1} and t_{n+2} are both affected:

$$\begin{aligned} &\Pi_{contact}(\varphi_{n+1}, \varphi^c_{n+1}, p_{n+1}) = \Pi(\varphi_{n+1}) + p_{n+1} g_{n+2} \\ &:= \Pi(\varphi_{n+1}) + p_{n+1}(x^c + d^c_{n+2} - x^o) \end{aligned} \tag{6.117}$$

The proposed modification of the central difference scheme, will thus require implicit-like computations but only for the nodes in contact.
We further illustrate the performance of the proposed method with an example of elastic bar impacting a rigid obstacle. The chosen properties for the elastic bar are: Young's modulus $E = 1$, cross-section $A = 1$ and mass density $\rho = 1$; furthermore, we choose initial velocity $v = 1$, time of impact $t_c = 2.8$ and time of separation $t_r = 22.8$. The numerical results obtained with the proposed modification of the central difference scheme are presented in Figure 6.7. The computations are performed with two different meshes using two time steps: first with a mesh of 10 elements and time step $h = 1/2$ and second with a mesh of 100 elements and the time step $h = 1/20$.

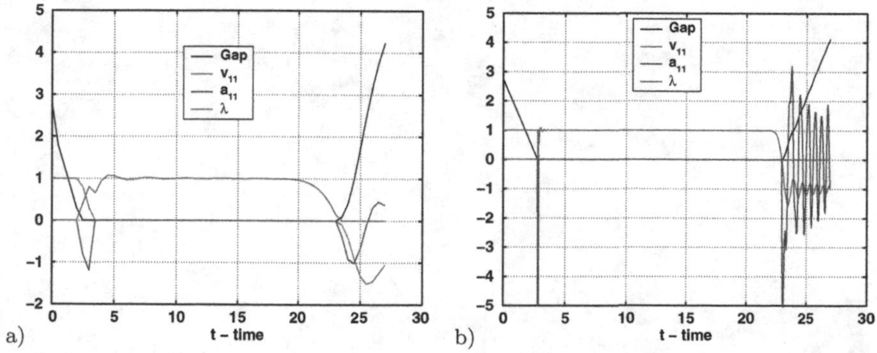

Fig. 6.7 Impact of an elastic bar on obstacle: time evolution of displacement, velocity and acceleration computed with **(a)** 10 elements and **(b)** 100 elements.

We can see that the computed displacement values are excellent with respect to contact impenetrability constraint satisfaction. However, the values of velocity and acceleration remain rather polluted with the numerical noise. Moreover, the quality of result for contact pressure is slightly reduced before the separation time.

Chapter 7
Thermodynamics and solution methods for coupled problems

In this chapter we study one of the most important examples of multi-physics problems, the coupling between mechanics and heat transfer that is naturally placed within the thermodynamics framework. All the model problems selected herein for illustration of thermodynamical coupling are non-stationary. We start by revisiting the initial-boundary value problem in elastodynamics for both small and large strain case. We then turn to inelastic problems, with plasticity model first, and then finally to contact. For each of the chosen model problems, we will underline the changes which are brought by taking into account the thermodynamical coupling, such as the temperature dependence of mechanical properties. More importantly, we will present for each problem how to construct the corresponding operator split solution method, which combines the finite element solution procedure already presented for non-stationary heat transfer on one hand with the corresponding solution method for mechanics problem on the other.
We note that all the problems discussed in this chapter rely crucially upon the thermodynamics framework, not only for development of their theoretical formulation but also for the most suitable numerical implementation and the most robust solution method; the case in point is the unconditional stability of the operator split solution method, achieved by employing adiabatic phase of computation, contrary to only conditional stability brought by isothermal phase. We also note that a number of other multi-physics problems share the same structure of governing equations with thermodynamics, such as hydro-mechanical coupling for porous media or the coupling between mechanics and physical chemistry for material durability problems. Therefore, we expect that the developments presented in this chapter will be directly applicable to a number of other multi-physics problems of practical interest.
The outline of the chapter is as follows. In the first section we present the theoretical formulation of thermodynamics applied to elastic behavior in small and in large strains. In the same section, we will present the finite element solution method for this kind of problems, based upon different operator

split procedures. The extension of these methods to the case of plasticity is considered in the next section. The last section considers the thermodynamical coupling for contact problems.

7.1 Thermodynamics of reversible processes

In this section, we deal with the evolution problems for both mechanical and thermal behavior of deformable solids, as well as their coupling. We will limit ourselves to phenomenological models[1] in thermodynamics, and thus the developments to follow are suitable for theoretical formulations in continuum mechanics or porous media.[2]

7.1.1 Thermodynamical coupling in 1D elasticity

In this section, we study the elastic behavior in dynamics corresponding to the motion from initial, stress-free configuration. The elasticity in dynamics implies that the complete removal of the external loading would allow us to start the free-vibration phase, where the total energy of any elastic system (with no damping) remains constant; the free vibrations of such an elastic system allow for the free exchange between the kinetic and potential energy.[3] In thermodynamics, the same principle of energy conservation will remain valid only for hyperelastic behavior at a constant temperature, and it is known as the first principle of thermodynamics on energy conservation; however, there is no free exchange between mechanical and thermal energies for an evolution problem with non-homogeneous temperature field, with the exchange restricted by the second principle in thermodynamics.

[1] Multi-scale models are considered only in the last chapter of the book.

[2] The porous media framework can be used as an extension of continuum mechanics for studying the multi-physics problems of thermo-hydro-mechanical coupling or coupling between mechanics and physical chemistry, [169].

[3] The exchange of kinetic and potential energy can easily be shown with a simple example of a mechanical system composed of an assembly of a concentrated mass m and an elastic spring with stiffness k; namely, if the motion starts from the initial displacement u_0, we obtain the solution $u(t) = u_0 \cos(\omega t)$ for displacement and $\dot{u}(t) = -u_0\omega \sin(\omega t)$ for velocity, which allows us to construct the evolution of the potential energy $\Pi(t) := \frac{1}{2}k[u(t)]^2 = \frac{1}{2}ku_0^2 \cos^2(\omega t)$ and the kinetic energy $K(t) := \frac{1}{2}m[\dot{u}(t)]^2 = \frac{1}{2}m\omega^2 u_0^2 \sin^2(\omega t)$. By taking into account that the natural frequency of this system is $\omega = \sqrt{k/m}$, we can easily confirm that the total energy remains constant $E(t) := K(t) + \Pi(t) = \frac{1}{2}ku_0^2$ in free vibration phase, and that there is free exchange in the course of motion between the potential and kinetic energy with each reaching its maximum value of $\frac{1}{2}ku_0^2$ at one point in time. The same proof can be supplied for a complex elastic systems with n degrees of freedom by using the modal superposition method.

In order to illustrate the first principle of thermodynamics, we consider the segment $x_1 - x_2$ of an elastic bar as an isolated system, with no heat exchange with the rest, which should keep constant the total energy $E(t)$; the latter is constructed as the sum of elastic potential energy $\Pi(t)$ and kinetic energy $K(t)$, which can be written:

$$E(t) := \Pi(t) + K(t) = \int_{x_1}^{x_2} e(\frac{\partial u}{\partial x}, s)\, dx + \frac{1}{2}\int_{x_1}^{x_2} \rho(\frac{\partial u}{\partial t})^2\, dx \tag{7.1}$$

where $e(\cdot)$ is the internal energy density and ρ is the mass density. The change of total energy of the system can be brought about by two sources. The first is the mechanical power of the external loading:

$$P = \int_{x_1}^{x_2} b\frac{\partial u}{\partial t}\, dx + [\sigma(x_2)\frac{\partial u(x_2)}{\partial t} - \sigma(x_1)\frac{\partial u(x_1)}{\partial t}] \tag{7.2}$$

and the second is the heating of the system:

$$Q = \int_{x_1}^{x_2} r\, dx - [q(x_2) - q(x_1)] \tag{7.3}$$

where b is the distributed loading, r is the distributed heat source, $\sigma(x_i)n(x_i)$ (with $n(x_i) = 1$ or $n(x_i) = -1$) are driving tractions at the end of the segment and $q(x_i)$ is the heat flux.

The first principle of thermodynamics postulates that in the absence of external loading and heating the total energy remains constant, and, more generally, that the rate of change of the total energy will be proportional to given sources according to:

$$\frac{\partial E}{\partial t} = P + Q \tag{7.4}$$

For the chosen segment of the elastic bar at small strain the first principle of thermodynamics can be written explicitly according to:

$$\begin{aligned} \tfrac{\partial}{\partial t}[\textstyle\int_{x_1}^{x_2}\left(e + \tfrac{1}{2}\rho\left(\tfrac{\partial u}{\partial t}\right)^2\right) dx] = \textstyle\int_{x_1}^{x_2} b\tfrac{\partial u}{\partial t}\, dx + \left[\sigma\tfrac{\partial u}{\partial t}\right]_{x_1}^{x_2} \\ + \textstyle\int_{x_1}^{x_2} r\, dx + [q]_{x_1}^{x_2} \end{aligned} \tag{7.5}$$

In the limit case with $x_2 \mapsto x_1 = x$, we can obtain the local (or strong) form of the first principle of thermodynamics:

$$\begin{aligned} \tfrac{\partial}{\partial t}e &= -\rho\tfrac{\partial^2 u}{\partial t^2}\tfrac{\partial u}{\partial t} + b\tfrac{\partial u}{\partial t} + \tfrac{\partial}{\partial x}\left(\sigma\tfrac{\partial u}{\partial t}\right) + r - \tfrac{\partial q}{\partial x} \\ &= \tfrac{\partial u}{\partial t}\underbrace{[-\rho\frac{\partial^2 u}{\partial t^2} + \frac{\partial \sigma}{\partial x} + b]}_{=0} + \sigma\tfrac{\partial}{\partial t}(\tfrac{\partial u}{\partial x}) + r - \tfrac{\partial q}{\partial x} \end{aligned} \tag{7.6}$$

An alternative result for the strong form of the first principle in thermodynamics for elastic bar in small strains can be obtained by taking into account the equations and motion $\rho\frac{\partial^2 u}{\partial t^2} - \left(\frac{\partial \sigma}{\partial x} + b\right) = 0$ and small strain kinematics $\epsilon = \frac{\partial u}{\partial x}$, which allows us to recast the last result according to:

$$\frac{\partial e}{\partial t} = \sigma \frac{\partial \epsilon}{\partial t} + r - \frac{\partial q}{\partial x} \tag{7.7}$$

From this general form, which is valid for thermoelasticity, we can obtain two special cases. The first concerns the absence of external loading and constant deformation field ($\frac{\partial \epsilon}{\partial t} = 0$), with the strong form of the first principle of thermodynamics reducing to standard heat equation:

$$\frac{\partial e}{\partial t} := \underbrace{\frac{\partial e}{\partial \theta}}_{c} \frac{\partial \theta}{\partial t} = r - \underbrace{\frac{\partial q}{\partial \frac{\partial \theta}{\partial x}}}_{-k} \frac{\partial^2 \theta}{\partial x^2} \tag{7.8}$$

where c is the heat capacity, and k is heat conductivity coefficient. The second special case is obtained from the first principle of thermodynamics in (7.7) as an adiabatic process at fixed entropy, $s = cst. \implies r - \frac{\partial q}{\partial x} = 0$, where we recover the standard result on hyperelastic constitutive model:

$$s = cst. \Leftrightarrow \frac{\partial s}{\partial t} = 0 \implies r - \frac{\partial q}{\partial x} = 0 \implies \frac{\partial e}{\partial t} = \sigma \frac{\partial \epsilon}{\partial t} \Rightarrow \sigma = \frac{\partial e}{\partial \epsilon}|_s \tag{7.9}$$

An alternative derivation of this special result can be made in terms of Helmholtz energy potential, which is defined by Legendre transformation from internal energy, $\psi(\epsilon, \theta) = e(\epsilon, s) - \theta s$; we will now keep the temperature fixed to obtain will hyperelastic constitutive equations:

$$\sigma := \underbrace{\left.\frac{\partial e(\epsilon, s)}{\partial \epsilon}\right|_s}_{adiabatic} = \underbrace{\left.\frac{\partial \psi(\epsilon, \theta)}{\partial \epsilon}\right|_\theta}_{isothermal} \tag{7.10}$$

We thus conclude that different thermodynamics potentials will reduce to hyperelasticity: either adiabatic case with fixed entropy, or isothermal case with fixed temperature.
The first principle of thermodynamics will also imply the zero dissipation only for a homogeneous temperature field. For a non-homogeneous temperature field (where the temperature is no longer constant within the domain), we will also appeal to the second principle of thermodynamics, in order to properly account for the temperature evolution in time and the resulting non-zero dissipation value. For the case of thermoelasticity, the second principle will not provide any new information, but rather confirm validity of Fourier's law for heat transfer.

With an infinitesimal increase in entropy at a constant temperature proportional to the corresponding increase in heating and inversely proportional to the temperature, we can define the total entropy[4] of the segment:

$$dS = \frac{dQ}{\theta} \; ; \; S = \int_{x_1}^{x_2} s \, dx \; ; \; Q = \int_{x_1}^{x_2} r \, dx - [q]_{x_1}^{x_2} \tag{7.11}$$

In homogeneous temperature field (with $\frac{\partial \theta}{\partial x} = 0$), we can further obtain the local form of the last result, which is identical to the first principle of thermodynamics:

$$\frac{\partial s}{\partial t} = \frac{r}{\theta} - \frac{\partial}{\partial x}\left(\frac{q}{\theta}\right)\bigg|_{\frac{\partial \theta}{\partial x}=0} \Leftrightarrow \underbrace{\overbrace{\frac{\partial e}{\partial \epsilon}}^{\sigma} \frac{\partial \epsilon}{\partial t} + \overbrace{\frac{\partial e}{\partial s}}^{\theta} \frac{\partial s}{\partial t}}_{\frac{\partial e}{\partial t}} = \sigma \frac{\partial \epsilon}{\partial t} + r - \frac{\partial q}{\partial x} \tag{7.12}$$

For a non-homogeneous temperature field (with $\frac{\partial \theta}{\partial x} \neq 0$), the second principle of thermodynamics will ensure that the heat flow is directed from warmer to colder part, and thus the entropy increase always remains superior to the heat source:

$$dS \geq \frac{dQ}{\theta} \Leftrightarrow \int_{x_1}^{x_2} \frac{\partial s}{\partial t} \geq \int_{x_1}^{x_2} \frac{r}{\theta} \, dx - \frac{q}{\theta}\bigg|_{x_1}^{x_2} \tag{7.13}$$

For the limit case with $x_2 \rightarrow x_1 = x$, we can obtain the local form of the second principle, specifying the positive value of the dissipation by conduction D_{cond}:

$$0 \leq D_{cond} =: [\theta \frac{\partial s}{\partial t} - (r - \frac{\partial q}{\partial x})] - \frac{1}{\theta} q \frac{\partial \theta}{\partial x} \tag{7.14}$$

The last result also remains in agreement with zero dissipation value and the first principle in (7.12) for a homogeneous temperature field (with $\frac{\partial \theta}{\partial x} = 0$). For a non-homogeneous temperature field, the first principle and the last result will jointly imply that the dissipation by conduction should remain positive, and provide the confirmation of Fourier's law for heat transfer (with a positive value of conductivity coefficient):

$$D_{cond} := -\frac{1}{\theta} q \frac{\partial \theta}{\partial x} > 0 \; ; \; q = -k \frac{\partial \theta}{\partial x} \; ; \; k > 0 \; ; \; \theta > 0 \tag{7.15}$$

The proposed framework of thermodynamics and the results obtained for thermoelasticity carry over to the geometrically nonlinear case (see [119]); we consider herein only the model of 1D thermoelasticity for illustration. We find that all the results presented in (7.1) to (7.14) remain valid, if rewritten in the corresponding format for large deformation setting. For example, by

[4] Recall that the entropy is extensive variable, which is proportional to the total mass (e.g. see Ericksen [74]).

using the material description of 1D finite deformation elasticity, we can state the local form of the first principle of thermodynamics according to:

$$
\begin{aligned}
\tfrac{\partial}{\partial t} e(\lambda, s) &= -\rho \tfrac{\partial^2 \varphi}{\partial t^2} \tfrac{\partial \varphi}{\partial t} + \tfrac{\partial}{\partial x}(P \tfrac{\partial \varphi}{\partial t}) + \tfrac{\partial \varphi}{\partial t} b + r - \tfrac{\partial q_0}{\partial x} \\
&= \tfrac{\partial \varphi}{\partial t} \underbrace{[-\rho \frac{\partial^2 \varphi}{\partial t^2} + \frac{\partial P}{\partial x} + b]}_{=0} + P \tfrac{\partial}{\partial t} \tfrac{\partial \varphi}{\partial x} + r - \tfrac{\partial q_0}{\partial x} \\
&= P \tfrac{\partial}{\partial t} \lambda + r - \tfrac{\partial q_0}{\partial x}
\end{aligned}
\tag{7.16}
$$

where P is the first Piola–Kirchhoff stress, $\lambda = 1 + \frac{\partial d}{\partial x}$ is the stretch and q_0 is the heat flux expressed with respect to the initial configuration. We can also express the last result by using the Helmholtz free-energy potential and thus obtain the constitutive equations for stress and entropy, as well as the reduced form of the first principle:

$$
\begin{aligned}
&(\tfrac{\partial \psi}{\partial \lambda} - P) \tfrac{\partial \lambda}{\partial t} + (\tfrac{\partial \psi}{\partial \theta} + s) \tfrac{\partial \theta}{\partial t} + \theta \tfrac{\partial s}{\partial t} = r - \tfrac{\partial q_0}{\partial x} \\
&\Longrightarrow P = \tfrac{\partial \psi}{\partial \lambda} \; ; \; s = -\tfrac{\partial \psi}{\partial \theta} \; ; \; \theta \tfrac{\partial s}{\partial t} = r - \tfrac{\partial q_0}{\partial x}
\end{aligned}
\tag{7.17}
$$

Furthermore, by appealing to the second principle of thermodynamics at large strains, we can again confirm the validity of Fourier's law for heat transfer, which is written for the heat flux in material description:

$$
\begin{aligned}
0 \leq D_{cond} &:= r - \theta \tfrac{\partial}{\partial x}(q_0/\theta) \\
&= -\underbrace{[\theta \frac{\partial s}{\partial t} - (r - \frac{\partial q_0}{\partial x}]}_{=0} - \tfrac{1}{\theta} q_0 \tfrac{\partial \theta}{\partial x} \quad \Longrightarrow q_0 = -k \frac{\partial \theta}{\partial x}
\end{aligned}
\tag{7.18}
$$

We can also use the spatial description of mechanical and thermal response for thermoelasticity in geometrically nonlinear setting, which employs the objects placed in the current configuration but parameterized with respect to the initial configuration. For example, the mechanical response can be defined in terms of the Kirchhoff stress computation:

$$
\tau := \lambda P = \lambda \frac{\partial \psi}{\partial \lambda} \tag{7.19}
$$

and thermal response can be expressed in terms of heat flux:

$$
q := \lambda q_0 = -k \lambda \frac{\partial \theta}{\partial x} \tag{7.20}
$$

This kind of choice for stress tensor and heat flux would allow us to write the weak form of governing equations for 1D thermoelasticity according to:

$$
\begin{aligned}
0 = G_M(\varphi, \theta; w) &:= \textstyle\int_l \{ w (\rho \frac{\partial^2 \varphi}{\partial t} - b) + \frac{1}{\lambda} \frac{dw}{dx} \tau \} \, dx - w(l) \bar{t} \\
0 = G_T(\varphi, \theta; \vartheta) &:= \textstyle\int_l \{ \vartheta (\theta \frac{\partial s}{\partial t} - r) - \frac{1}{\lambda} \frac{\partial \vartheta}{\partial x} q \} \, dx - \vartheta(l) \bar{h}
\end{aligned}
\tag{7.21}
$$

These equations provide the starting point for constructing the finite element approximations. The derivations of these equations can also be carried out in a more general framework of manifolds (e.g. see [119]).

7.1.2 Thermodynamics coupling in 3D elasticity and constitutive relations

The main steps in development of the governing equations of thermodynamics for 3D case remain very much the same as those traced for 1D case. The end result of applying the finite element method to solving 3D problems also remains the same, apart the need to use the vectors and tensors for corresponding representations of displacements $\mathbf{u} = u_i \mathbf{e}_i$, strains $\boldsymbol{\epsilon} = \epsilon_{ij} \mathbf{e}_i \otimes \mathbf{e}_j$ and stresses $\boldsymbol{\sigma} = \sigma_{ij} \mathbf{e}_i \otimes \mathbf{e}_j$. This difference does not concern the entropy nor temperature, which remain the scalar fields even in 3D case. We can thus directly write the corresponding local form of the first principle of thermodynamics:

$$
\begin{aligned}
r - \underbrace{\nabla \cdot \mathbf{q}}_{div[\mathbf{q}]} &= -\boldsymbol{\sigma} \cdot \tfrac{\partial \boldsymbol{\epsilon}}{\partial t} + \tfrac{\partial e(\boldsymbol{\epsilon}, s)}{\partial t} \\
&= -\boldsymbol{\sigma} \cdot \tfrac{\partial \boldsymbol{\epsilon}}{\partial t} + \tfrac{\partial}{\partial t}[\psi(\boldsymbol{\epsilon}, \theta) + s\theta] \\
&= (\tfrac{\partial \psi}{\partial \boldsymbol{\epsilon}} - \boldsymbol{\sigma}) \cdot \tfrac{\partial \boldsymbol{\epsilon}}{\partial t} + (\tfrac{\partial \psi}{\partial \theta} + s) \tfrac{\partial \theta}{\partial t} + \theta \tfrac{\partial s}{\partial t}
\end{aligned}
\tag{7.22}
$$

The independent variations of the last expression with respect to the strain and temperature will provide the constitutive equations for stress and entropy, respectively:

$$
\boldsymbol{\sigma} = \frac{\partial \psi}{\partial \boldsymbol{\epsilon}} \; ; \; s = -\frac{\partial \psi}{\partial \theta} \tag{7.23}
$$

along with the reduced form of the first principle of thermodynamics:

$$
\theta \frac{\partial s}{\partial t} = r - div[\mathbf{q}] \tag{7.24}
$$

By taking into account that two constitutive equations in (7.23) above derive from the same free energy potential, we can easily confirm the Maxwell relation as the equality of mixed partial derivatives of such a potential:

$$
\boldsymbol{\beta} := -\frac{\partial \boldsymbol{\sigma}}{\partial \theta} = -\frac{\partial^2 \psi}{\partial \theta \partial \boldsymbol{\epsilon}} \equiv -\frac{\partial^2 \psi}{\partial \boldsymbol{\epsilon} \partial \theta} = \frac{\partial s}{\partial \boldsymbol{\epsilon}} \tag{7.25}
$$

We denote as $\boldsymbol{\beta}$ the second mixed derivative of the free energy potential, and further refer to it as the thermal stress. The second derivative of the free-energy potential with respect to the temperature is used to provide an alternative definition of the heat capacity coefficient $c := \frac{\partial e}{\partial \theta}$; namely, by appealing to the Legendre transform, we can obtain:

$$c := \frac{\partial}{\partial \theta} \underbrace{(\psi(\epsilon, \theta) + \theta s)}_{e} = \underbrace{\frac{\partial \psi}{\partial \theta}}_{-s} + s + \theta \underbrace{\frac{\partial s}{\partial \theta}}_{-\frac{\partial^2 \psi}{\partial \theta^2}} = -\theta \frac{\partial^2 \psi}{\partial \theta^2} \tag{7.26}$$

Finally, the second derivative of the free energy potential with respect to the strain tensor will define the (tangent) elasticity tensor:

$$\boldsymbol{\mathcal{C}} := \frac{\partial \boldsymbol{\sigma}}{\partial \boldsymbol{\epsilon}} = \frac{\partial^2 \psi}{\partial \boldsymbol{\epsilon} \partial \boldsymbol{\epsilon}} \tag{7.27}$$

For nonlinear thermoelasticity, the elasticity tensor and heat capacity coefficient remain (nonlinear) functions of state variables, the strain tensor $\boldsymbol{\epsilon}$ and temperature θ:

$$\hat{c}(\boldsymbol{\epsilon}, \theta) \ ; \ \hat{\boldsymbol{\mathcal{C}}}(\boldsymbol{\epsilon}, \theta) \tag{7.28}$$

However, both will remain constant values for the case of linear thermoelasticity, where the free energy potential can be written as a quadratic form in state variables:

$$\psi(\boldsymbol{\epsilon}, \theta) = \psi_0 - s_0(\theta - \theta_0) + \underbrace{\boldsymbol{\sigma}_0}_{=0} \cdot \boldsymbol{\epsilon} - \frac{1}{2\theta_0} c_0 (\theta - \theta_0)^2 - \boldsymbol{\beta}_0 \cdot \boldsymbol{\epsilon}(\theta - \theta_0) + \frac{1}{2}(\boldsymbol{\mathcal{C}}_0 \boldsymbol{\epsilon}) \cdot \boldsymbol{\epsilon} \tag{7.29}$$

where:

$$\begin{array}{c} \psi_0 = \psi(\mathbf{0}, \theta_0) \ ; \ \boldsymbol{\sigma}_0 := \frac{\partial \psi}{\partial \boldsymbol{\epsilon}}|_0 = 0 \ ; \ s_0 = -\frac{\partial \psi}{\partial \theta}|_0 \\ c_0 = -\theta_0 \frac{\partial^2 \psi}{\partial \theta^2}|_0 \ ; \ \boldsymbol{\beta}_0 = \frac{\partial^2 \psi}{\partial \theta \partial \boldsymbol{\epsilon}}|_0 \ ; \ \boldsymbol{\mathcal{C}}_0 = \frac{\partial^2 \psi}{\partial \boldsymbol{\epsilon} \partial \boldsymbol{\epsilon}} \end{array} \tag{7.30}$$

Without loss of generality, we will also choose zero value of the strain in the initial configuration, but not of absolute temperature:

$$\boldsymbol{\epsilon}_0 = \mathbf{0} \ ; \ \theta_0 \neq 0 \tag{7.31}$$

For linear thermoelasticity this choice implies that the stress tensor consists of the strain-dependent mechanical part and the thermal stress proportional to temperature increase:

$$\boldsymbol{\sigma} := \frac{\partial \psi}{\partial \boldsymbol{\epsilon}} = \boldsymbol{\mathcal{C}}_0 \boldsymbol{\epsilon} - (\theta - \theta_0) \boldsymbol{\beta}_0 \tag{7.32}$$

For the same model, we can compute the entropy as the sum of its initial value, structural heating contribution and the term proportional to temperature increase with the heat capacity as the proportionality coefficient:

$$s := -\frac{\partial \psi}{\partial \theta} = s_0 + \boldsymbol{\beta}_0 \cdot \boldsymbol{\epsilon} + \frac{1}{\theta_0} c_0 (\theta - \theta_0) \tag{7.33}$$

For the case of isotropy, where the same response applies to each direction, we can obtain a simplified form of the governing equations for thermoelasticity with only four parameters; two of them (such as the Lamé parameters λ and μ) are needed to define the mechanical response:

$$\mathcal{C}_0 = \lambda \mathbf{I} \otimes \mathbf{I} + 2\mu \mathcal{I} \Leftrightarrow \mathcal{C}_{ijkl} = \lambda \delta_{ij}\delta_{kl} + 2\mu \frac{1}{2}(\delta_{ik}\delta_{jl} + \delta_{il}\delta_{jk}) \tag{7.34}$$

Two remaining parameters will define the thermal response, with the heat capacity c_0 and the heat conductivity coefficient k, defining the ratio between the temperature gradient and heat flux according to Fourier's law:

$$\mathbf{q} = -k\,\boldsymbol{\nabla}\theta \tag{7.35}$$

For the case of isotropic thermoelasticity the thermal stress $\boldsymbol{\beta}_0$ can be written: $\boldsymbol{\beta}_0 = \beta\mathbf{I}$; hence, the constitutive equations for stress and entropy will take a simplified form:

$$\boldsymbol{\sigma} = 2\mu\boldsymbol{\epsilon} + (\lambda tr[\boldsymbol{\epsilon}] - \beta(\theta - \theta_0)\mathbf{I} \; ; \; s = s_0 + \frac{1}{\theta_0}c_0(\theta - \theta_0) + \beta tr[\boldsymbol{\epsilon}] \tag{7.36}$$

The last result indicates that the temperature change has the same effect as the spherical part of strain tensor, producing only a volume change for isotropic material. An alternative manner for quantifying the thermomechanical coupling for isotropic case is provided in terms of the thermal dilatation coefficient, which can be defined as $\alpha = \beta/(3\lambda + 2\mu)$; the last result follows from:

$$\begin{aligned} tr[\boldsymbol{\sigma}] &= (3\lambda + 2\mu)tr[\boldsymbol{\epsilon}] - 3\beta(\theta - \theta_0) \\ &\Longrightarrow tr[\boldsymbol{\epsilon}] = \tfrac{1}{3\lambda+2\mu}tr[\boldsymbol{\sigma}] + 3\underbrace{\frac{\beta}{3\lambda + 2\mu}}_{\alpha}(\theta - \theta_0) \\ \boldsymbol{\epsilon} &= \tfrac{1}{2\mu}\boldsymbol{\sigma} - \tfrac{\lambda}{2\mu(3\lambda+2\mu)}tr[\boldsymbol{\sigma}]\mathbf{I} + \alpha(\theta - \theta_0)\mathbf{I} \end{aligned} \tag{7.37}$$

The same kind of development can be carried out in geometrically nonlinear setting, but the end result in no longer unique. For example, the thermodynamics coupling in large strain elasticity can be described in two alternative manners (see [119]), the first which employs the additive decomposition of stress rate and the second employing the multiplicative decomposition of deformation gradient. The former can be used for the spatial description of thermoelasticity, with an additive split of the stress rate of the Kirchhoff stress $\boldsymbol{\tau}$ into mechanical and thermal parts:

$$L_v[\boldsymbol{\tau}(\mathbf{B}, \theta)] := \mathbf{F}\frac{\partial}{\partial t}[\underbrace{\mathbf{F}^{-1}\boldsymbol{\tau}\mathbf{F}^{-T}}_{\mathbf{S}}]\mathbf{F}^T = \mathcal{C}^\tau\,\mathbf{D} + \underbrace{\frac{\partial \boldsymbol{\tau}}{\partial \theta}}_{\boldsymbol{\beta}}\frac{\partial \theta}{\partial t} \tag{7.38}$$

In the last expression, $L_v[\boldsymbol{\tau}]$ is the stress rate expressed in terms of the Lie derivative of Kirchhoff stress, $\mathcal{C}^\tau$ is the spatial elasticity tensor in large deformation, $\mathbf{D} = \frac{1}{2}(\dot{\mathbf{F}}\mathbf{F}^{-1} + \mathbf{F}^{-T}\dot{\mathbf{F}}^T)$ is the rate of deformation tensor and β is the thermal stress. The second possibility for describing the large strain thermoelasticity can be used for the material description; here we can introduce the multiplicative decomposition of the thermomechanics deformation gradient $\tilde{\mathbf{F}}$ into mechanical part $\mathbf{F}$ and thermal part $\mathbf{F}_\theta$, with the latter concerned only with the temperature-dependent volume change:

$$\tilde{\mathbf{F}} = \mathbf{F}_\theta \mathbf{F} \; ; \; \mathbf{F}_\theta = (1 + \alpha(\theta - \theta_0))\mathbf{I} \tag{7.39}$$

We can provide a clear illustration of the last result for a homogeneous deformation case in 1D setting, where the multiplicative decomposition of deformation gradient can be written in terms of mechanical stretch λ_t and thermal stretch λ_θ:

$$\frac{l_\theta}{l_t}\frac{l_t}{l} \Leftrightarrow \lambda_\theta \lambda_t \tag{7.40}$$

In the last expression, l_t is the (mechanically) deformed length of a bar with initial length l, to which we further apply a homogeneous temperature increase leading to final length l_θ. It is easy to verify that the multiplicative decomposition of this kind will imply the corresponding additive decomposition of the time derivative of logarithmic strain:

$$\frac{1}{\lambda_\theta \lambda_t}\frac{\partial}{\partial t}(\lambda_\theta \lambda_t) = \frac{1}{\lambda_\theta}\frac{\partial \lambda_\theta}{\partial t} + \frac{1}{\lambda_t}\frac{\partial \lambda_t}{\partial t} \tag{7.41}$$

which allows us to establish the equivalence of these two descriptions of thermodynamical coupling in large strain elasticity.

Another special case of thermodynamical coupling which merits our special attention is the adiabatic process, with no external heat source. According to the reduced form of the first principle in (7.24), the adiabatic process will enforce the constant value of entropy (with $\dot{s} = 0$), which will result with:

$$s := s_0 + \frac{1}{\theta_0}c(\theta - \theta_0) + \beta_0 tr[\epsilon] = s_0 (= cst.) \tag{7.42}$$

Therefore, all the temperature increase between the initial and current configurations for the adiabatic case is entirely due to the structural heating, and can be expressed as:

$$\theta - \theta_0 = -\frac{\theta_0}{c}\beta_0 tr[\epsilon] \; ; \; tr[\epsilon] = \mathbf{I} \cdot \epsilon \tag{7.43}$$

By exploiting this result for thermoelasticity, we can again formally express the stress tensor in terms of strains by replacing the elasticity tensor with the adiabatic tensor $\mathcal{C}_{ad}$:

$$\begin{aligned} \boldsymbol{\sigma} &= \mathcal{C}_0 \epsilon - (\theta - \theta_0)\beta_0 \mathbf{I} \\ &= \mathcal{C}_0 + \tfrac{\theta_0}{c}\beta_0(\mathbf{I} \cdot \epsilon)\beta_0 \mathbf{I} \\ &= \underbrace{[\mathcal{C}_0 + \frac{\theta_0}{c}\beta_0 \mathbf{I} \otimes \beta_0 \mathbf{I}]}_{\mathcal{C}_{ad}} \epsilon \end{aligned} \tag{7.44}$$

7.2 Initial-boundary value problem in thermoelasticity and operator split solution method

In this section we derive the strong and the weak form of an initial-boundary value problem of thermodynamical coupling for 3D elasticity, further referred to as 3D thermoelasticity. We also present the solution method for such problems, obtained by using the finite element method and the time-integration schemes.

7.2.1 Weak form of initial-boundary value problem in 3D elasticity and its discrete approximation

In order to simplify notation, we will subsequently omit subscript "0", further denoting $\boldsymbol{\mathcal{C}}_0 \mapsto \boldsymbol{\mathcal{C}}$, $\boldsymbol{\beta}_0 \mapsto \boldsymbol{\beta}$, $(\theta - \theta_0) \mapsto \theta$. The free energy potential of 3D linear thermoelasticity can thus be stated according to:

$$\psi(\epsilon, \theta) = \frac{1}{2}\epsilon \cdot \boldsymbol{\mathcal{C}}\epsilon - \theta\beta\mathbf{I} \cdot \epsilon + \frac{1}{2}c\theta^2 \tag{7.45}$$

This kind of potential allows us to obtain all the constitutive equations for dependent variables, along with the reduced form of the first principle of thermodynamics for 3D thermoelasticity:

$$\begin{aligned} &\boldsymbol{\sigma} := \tfrac{\partial \psi}{\partial \epsilon} = \boldsymbol{\mathcal{C}}\epsilon \; ; \; s := -\tfrac{\partial \psi}{\partial \theta} = c\theta - \beta\mathbf{I} \cdot \epsilon \; ; \; \theta\dot{s} = r - \boldsymbol{\nabla} \cdot \mathbf{q} \\ &\beta\mathbf{I} = \tfrac{\partial s}{\partial \epsilon} \; ; \; c = \theta\tfrac{\partial s}{\partial \theta} = -\theta\tfrac{\partial^2 \psi}{\partial \theta^2} \; ; \; \boldsymbol{\mathcal{C}} = \tfrac{\partial \boldsymbol{\sigma}}{\partial \epsilon} = \tfrac{\partial^2 \psi}{\partial \epsilon \partial \epsilon} \end{aligned} \tag{7.46}$$

By appealing to Fourier's law for isotropic case where the heat flux is directly proportional to the temperature gradient, $\mathbf{q} = -k\boldsymbol{\nabla}\theta$, we can write an alternative form of the first principle of thermodynamics for 3D linear thermoelasticity, which looks very much alike the standard equation for non-stationary heat transfer, apart an additional source term proportional to the structural heating:

$$\begin{aligned} r + \boldsymbol{\nabla} \cdot \mathbf{q} &= \theta\tfrac{\partial s}{\partial t} \\ &= \underbrace{-\theta\frac{\partial s}{\partial \theta}}_{c}\tfrac{\partial \theta}{\partial t} + \theta \underbrace{\frac{\partial s}{\partial \epsilon}}_{\beta\mathbf{1}} \cdot \tfrac{\partial \epsilon}{\partial t} \\ \Rightarrow \quad & \boxed{c\tfrac{\partial \theta}{\partial t} = div[k\boldsymbol{\nabla}\theta] + \boldsymbol{\sigma}_\theta\tfrac{\partial \epsilon}{\partial t} + r} \; ; \; \boldsymbol{\sigma}_\theta = -\beta\theta\mathbf{I} \end{aligned} \tag{7.47}$$

To this equations we add the standard form of the equations of motion in order to fully describe the thermodynamical coupling, with the only novelty which concerns the dependence of stress tensor on both deformation $\boldsymbol{\epsilon}$ (or displacement $\mathbf{u}$) and temperature θ; hence we can write:

$$\boxed{\rho \frac{\partial^2}{\partial t^2}\mathbf{u} = div[\boldsymbol{\sigma}] + \mathbf{b}} \; ; \; \boldsymbol{\sigma} = \boldsymbol{\mathcal{C}}\boldsymbol{\epsilon} + \boldsymbol{\sigma}_\theta \; ; \; \boldsymbol{\sigma}_\theta = -\beta\theta\mathbf{I} \tag{7.48}$$

The equations of motion can also be written as a set of first-order equation, by including the velocity $\mathbf{v}$ among the state variables:

$$\begin{aligned} \frac{\partial}{\partial t}\mathbf{u} &= \mathbf{v} \\ \rho\frac{\partial}{\partial t}\mathbf{v} &= div[\boldsymbol{\mathcal{C}}\boldsymbol{\epsilon} - \beta\theta\mathbf{I}] + \mathbf{b} \end{aligned} \tag{7.49}$$

The description of the strong form of the initial-boundary value problem of 3D linear thermoelasticity is complete with the set of equations in (7.47) and (7.49), accompanied by the initial conditions:

$$\theta(\mathbf{x},0) = \theta_0(\mathbf{x}) \; ; \; \mathbf{u}(\mathbf{x},0) = \mathbf{u}_0(\mathbf{x}) \; ; \; \mathbf{v}(\mathbf{x},0) = \mathbf{v}_0(\mathbf{x}) \; ; \; \forall \mathbf{x} \in \Omega \tag{7.50}$$

as well as the chosen set of boundary conditions; for example, the homogeneous Dirichlet boundary conditions:

$$\theta(\mathbf{x},t) = 0 \; ; \; \mathbf{u}(\mathbf{x},t) = \mathbf{0} \; ; \; \forall \mathbf{x} \in \partial\Omega \; ; \; \forall t \in [0,T] \tag{7.51}$$

will provide the sufficient guarantee for the solution uniqueness in linear thermoelasticity; this could be shown from verification of two key conditions: (i) The evolution problem of thermodynamic coupling in 3D linear thermoelasticity for homogeneous Dirichlet boundary condition obeys the following inequality:

$$\frac{d}{dt}L(\mathbf{z}) = -\int_\Omega \boldsymbol{\nabla}\theta \cdot k\boldsymbol{\nabla}\theta \, dV < 0 \tag{7.52}$$

where $\mathbf{z}$ is a vector that gathers all the state variables:

$$\mathbf{z} = \begin{bmatrix} \mathbf{u} \in H_0^1(\Omega)^{n_{dim}} \\ \mathbf{v} \in L_2(\Omega)^{n_{dim}} \\ \theta \in L_2(\Omega) \end{bmatrix} \tag{7.53}$$

and $L(\mathbf{z})$ is a linear form in state variables that is defined with (see [61]):

$$L(\mathbf{z}) = \frac{1}{2}\int_\Omega \left[\nabla\mathbf{u} \cdot \boldsymbol{\mathcal{C}}\nabla\mathbf{u} + \rho\mathbf{v}\cdot\mathbf{v} + c\theta^2\right] dV \tag{7.54}$$

(ii) The state variables evolution is a contraction, which ensures that any eventual difference from the initial state will remain bounded in time:

$$\|\mathbf{z}_t^{(2)} - \mathbf{z}_t^{(1)}\| < \|\mathbf{z}_0^{(2)} - \mathbf{z}_0^{(1)}\| \tag{7.55}$$

Therefore, the uniqueness of solution $\mathbf{z}_t$ is guaranteed with respect to a unique choice of the initial condition $\mathbf{z}_0$. The proof is by contradiction. Namely, let $\mathbf{z}_t^{(1)}$ and $\mathbf{z}_t^{(2)}$ be two different solutions corresponding to the same initial condition $\mathbf{z}_0$. The difference between these two solution, further denoted as $\mathbf{z}_t = \mathbf{z}_t^{(2)} - \mathbf{z}_t^{(1)}$ will verify the evolution problem defined above with the homogeneous Dirichlet boundary conditions. It is then easy to show that such a difference between two solutions can be computed according to:

$$\begin{aligned}
\tfrac{d}{dt}L(\mathbf{z}) &= \int_\Omega \left[\nabla\dot{\mathbf{u}}\cdot\boldsymbol{\mathcal{C}}\nabla\mathbf{u} + \rho\dot{\mathbf{v}}\cdot\mathbf{v} + c\dot{\theta}\theta\right] dV \\
&= \int_\Omega \Big[\nabla\mathbf{v}\cdot\boldsymbol{\mathcal{C}}\nabla\mathbf{u} + \rho\tfrac{1}{\rho}div\left[\boldsymbol{\mathcal{C}}\nabla\mathbf{u} - \beta\theta\mathbf{I}\right]\cdot\mathbf{v} \\
&\quad + c\left(\tfrac{1}{c}div\left[k\boldsymbol{\nabla}\theta\right] - \tfrac{1}{c}\beta\mathbf{I}\cdot\nabla\dot{\mathbf{u}}\right)\theta\Big]\, dV \\
&= \int_\Omega [\nabla\mathbf{v}\cdot\boldsymbol{\mathcal{C}}\nabla\mathbf{u} - \nabla\mathbf{v}\cdot\boldsymbol{\mathcal{C}}\nabla\mathbf{u} + \theta\beta\mathbf{I}\cdot\nabla\mathbf{v} \\
&\quad - \theta\beta\mathbf{I}\cdot\nabla\mathbf{v} - \nabla\theta k\nabla\theta]\, dV \\
&= -\int_\Omega \nabla\theta\cdot k\nabla\theta\, dV < 0
\end{aligned} \tag{7.56}$$

which confirms that the linear form remains bounded in time $L[\mathbf{z}_t] \leq L[\mathbf{z}_0]$. This result and the Korn inequality (e.g. see Duvaut and Lions [72]), which allows to bound the function by its derivatives for homogeneous Dirichlet boundary conditions, will further imply that the difference between two solutions has to remain bounded:

$$L[\mathbf{z}_t] \leq L[\mathbf{z}_0] \Rightarrow \parallel \mathbf{z}_t^{(2)} - \mathbf{z}_t^{(1)} \parallel \leq \parallel \mathbf{z}_0^{(2)} - \mathbf{z}_0^{(1)} \parallel \tag{7.57}$$

Therefore, we conclude that there cannot be any difference between two solutions, if they both start from the same initial condition. In other words, with $\mathbf{z}_0^{(2)} = \mathbf{z}_0^{(1)}$ we can write:

$$\|\mathbf{z}_t^{(2)} - \mathbf{z}_t^{(1)}\| \leq \|\mathbf{z}_0 - \mathbf{z}_0\| \;\Rightarrow\; \mathbf{z}_t^{(1)} = \mathbf{z}_t^{(2)} \tag{7.58}$$

For constructing the finite element approximation for the initial-boundary value problem in thermoelasticity, we will make use of the weak form which can be written:

Weak form of 3D linear thermoelasticity

Given: $\mathbf{b}$, r, as well as $\overline{\mathbf{u}}\big|_{\Gamma_u} = \mathbf{0}$, $\overline{\theta}\big|_{\Gamma_\theta} = 0$

Find: $\mathbf{u}$, θ such that $\forall\mathbf{w} \in H_0^1(\Omega)^{n_{dim}}$, $\forall\vartheta \in L_2(\Omega)$ that verify:

(i) Weak form of energy conservation:

$$\int_\Omega \vartheta c\frac{\partial\theta}{\partial t}\, dV = -\int_\Omega \boldsymbol{\nabla}\vartheta\cdot k\boldsymbol{\nabla}\theta\, dV - \int_\Omega \vartheta\beta\mathbf{I}\cdot\frac{\partial\nabla^s\mathbf{u}}{\partial t}\, dV + \int_\Omega \vartheta r\, dV \tag{7.59}$$

(ii) Weak form of equations of motion:

$$\int_{\Omega} \mathbf{w} \cdot \rho \frac{\partial^2 \mathbf{u}}{\partial t^2} \, dV = - \int_{\Omega} \nabla^s \mathbf{w} \cdot \boldsymbol{\mathcal{C}} \cdot \nabla^s \mathbf{u} \, dV + \int_{\Omega} \beta \mathbf{I} \cdot \nabla^s \mathbf{w} \theta \, dV + \int_{\Omega} \mathbf{b} \cdot \mathbf{w} \, dV \tag{7.60}$$

We will look for the solution by using the separation of variables between the space and time, and the finite element method to carry out the space discretization. In particular, we can choose the isoparametric interpolations for real and virtual displacement fields:

$$\begin{aligned} \mathsf{u}|_{\Omega^e} &= \textstyle\sum_{a=1}^{n_{en}} N_a(\xi) \mathsf{d}_a^u(t) \Rightarrow \frac{\partial \mathsf{u}}{\partial t}\Big|_{\Omega^e} = \sum_{a=1}^{n_{en}} N_a(\xi) \dot{\mathsf{d}}_a^u(t)\,; \\ \mathsf{w}|_{\Omega^e} &= \textstyle\sum_{a=1}^{n_{en}} N_a(\xi) \mathsf{w}_a \end{aligned} \tag{7.61}$$

along with the same kind of approximations for the temperature field and its variation:

$$\begin{aligned} \theta|_{\Omega^e} &= \textstyle\Sigma_{a=1}^{n_{en}} N_a(\xi) d_a^\theta(t) \Rightarrow \frac{\partial \theta}{\partial t}\Big|_{\Omega^e} = \sum_{a=1}^{n_{en}} N_a(\xi) \dot{d}_a^\theta(t)\,; \\ \vartheta|_{\Omega^e} &= \textstyle\sum_{a=1}^{n_{en}} N_a(\xi) \vartheta_a \end{aligned} \tag{7.62}$$

In (7.61) and (7.62) above, we indicated that the chosen finite element interpolations allow us to easily construct the finite element approximation for time derivatives of the displacement and temperature fields. We have also indicated that the displacement and temperature variations are independent on time, since they are related to a particular deformed configuration. Moreover, by considering that all the nodal values of displacement and temperature variations can be chosen arbitrarily, the discrete approximation of the weak form of 3D linear thermoelasticity reduces to a set of ordinary differential equations in time, which will specify the evolution of nodal values of displacements and temperatures:

$$\begin{aligned} \mathsf{M}^\theta \dot{\mathsf{d}}^\theta + \mathsf{K}^\theta \mathsf{d}^\theta + \mathsf{F}^\theta \dot{\mathsf{d}}^u &= \mathsf{f}^\theta \\ \mathsf{M}^u \ddot{\mathsf{d}}^u - \mathsf{F}^u \mathsf{d}^\theta + \mathsf{K}^u \mathsf{d}^u &= \mathsf{f}^u \end{aligned} \tag{7.63}$$

We note the lack of symmetry of this system,[5] where the matrices F^θ and F^u, which are the transpose of each other, have the opposite signs. For isoparametric finite element interpolations, the explicit form of each matrix in the

[5] In fact, the lack of symmetry of thermodynamical coupling will also characterize the discrete approximation of the problem, which follows from the different types of variables concerned by coupling, with displacement time-derivative (or velocity) on one side versus temperature on the other.

system above can easily be obtained; for thermal component of the coupled system, we can write:

$$\begin{cases} \mathsf{M}^{\theta} = \mathbb{A}_{e=1}^{n_{elem}} \mathsf{M}^{\theta,e} \; ; \; \mathsf{M}^{\theta,e} = \left[\mathsf{M}_{ab}^{\theta,e}\right] \; ; \; \mathsf{M}_{ab}^{\theta,e} = \int_{\Omega^e} N_a c \mathbf{I} N_b \, dV \\ \mathsf{K}^{\theta} = \mathbb{A}_{e=1}^{n_{elem}} \mathsf{K}^{\theta,e} \; ; \; \mathsf{K}^{\theta,e} = \left[\mathsf{K}_{ab}^{\theta,e}\right] \; ; \; \mathsf{K}_{ab}^{\theta,e} = \int_{\Omega^e} \mathsf{b}_a^T k \mathsf{b}_b \, dV \\ \mathsf{F}^{\theta} = \mathbb{A}_{e=1}^{n_{elem}} \mathsf{F}^{\theta,e} \; ; \; \mathsf{F}^{\theta,e} = \left[\mathsf{F}_{ab}^{\theta,e}\right] \; ; \; \mathsf{F}_{ab}^{\theta,e} = \int_{\Omega^e} N_a \beta \mathsf{b}_b^T \, dV \\ \mathsf{f}^{\theta} = \mathbb{A}_{e=1}^{n_{elem}} \mathsf{f}^{\theta,e} \; ; \; \mathsf{f}^{\theta,e} = \left[f_a^{\theta,e}\right] \; ; \; f_a^{\theta,e} = \int_{\Omega^e} N_a r \, dV \end{cases} \tag{7.64}$$

whereas for the mechanical part we will have:

$$\begin{cases} \mathsf{M}^{u} = \mathbb{A}_{e=1}^{n_{elem}} \mathsf{M}^{u,e} \; ; \; \mathsf{M}^{u,e} = [\mathsf{M}_{ab}^{u,e}] \; ; \; \mathsf{M}_{ab}^{u,e} = \int_{\Omega^e} N_a \rho \mathsf{I} N_b \, dV \\ \mathsf{K}^{u} = \mathbb{A}_{e=1}^{n_{elem}} \mathsf{K}^{u,e} \; ; \; \mathsf{K}^{u,e} = [\mathsf{K}_{ab}^{u}] \; ; \; \mathsf{K}_{ab}^{u,e} = \int_{\Omega^e} \mathsf{B}_a^T \mathsf{C} \mathsf{B}_b \, dV \\ \mathsf{F}^{u} = \mathbb{A}_{e=1}^{n_{elem}} \mathsf{F}^{u,e} \; ; \; \mathsf{F}^{u,e} = [\mathsf{F}_{ab}^{u,e}] \; ; \; \mathsf{F}_{ab}^{u,e} = \int_{\Omega^e} \mathsf{b}_a \beta N_b \, dV \\ \mathsf{f}^{u} = \mathbb{A}_{e=1}^{n_{elem}} \mathsf{f}^{u,e} \; ; \; \mathsf{f}^{u,e} = [\mathsf{f}_a^{u}] \; ; \; \mathsf{f}_a^{u} = \int_{\Omega^e} N_a \mathsf{b} \, dV \end{cases} \tag{7.65}$$

In (7.65) and (7.64) above, we used the standard isoparametric finite element interpolations to construct the corresponding approximation for the spherical part of strain tensor governing the volume change:

$$\hat{\mathsf{e}}(\mathsf{u}^h)|_{\Omega^e} = \textstyle\sum_{a=1}^{n_{en}} \mathsf{B}_a \mathsf{d}_a^u \; ; \; \mathsf{B}_a = \begin{bmatrix} \frac{\partial N_a}{\partial x_1} & 0 & 0 \\ 0 & \frac{\partial N_a}{\partial x_2} & 0 \\ 0 & 0 & \frac{\partial N_a}{\partial x_3} \\ \frac{\partial N_a}{\partial x_2} & \frac{\partial N_a}{\partial x_1} & 0 \\ 0 & \frac{\partial N_a}{\partial x_3} & \frac{\partial N_a}{\partial x_2} \\ \frac{\partial N_a}{\partial x_3} & 0 & \frac{\partial N_a}{\partial x_1} \end{bmatrix} \; ;$$

$$sph[\epsilon^h]|_{\Omega^e} \mapsto \theta^h|_{\Omega^e} = \textstyle\sum_{a=1}^{n_{en}} \mathsf{b}_a^T \mathsf{d}_a^u \; ; \; \mathsf{b}_a = \begin{bmatrix} \frac{\partial N_a}{\partial x_1} \\ \frac{\partial N_a}{\partial x_2} \\ \frac{\partial N_a}{\partial x_3} \end{bmatrix} \tag{7.66}$$

However, it is of interest for practical problems to increase the order of this approximation to be able to match the order of corresponding isoparametric element approximation for temperature field dependent volume increase. This goal can be accomplished within the proposed isoparametric interpolation framework by appealing to the method of incompatible modes for enhancing the volume change representation (e.g. see [88]); we can then write the corresponding enhanced strain approximation for each element Ω^e:

$$\hat{\mathsf{e}}(\mathsf{u}^h)|_{\Omega^e} = \textstyle\sum_{a=1}^{n_{en}} \mathsf{B}_a \mathsf{d}_a^u + \sum_{b=1}^{n_{im}} \hat{\mathsf{G}}_b \boldsymbol{\alpha}_b^u \; ; \; \hat{\mathsf{G}}_a = \begin{bmatrix} \frac{\partial \hat{M}_a}{\partial x_1} & 0 & 0 \\ 0 & \frac{\partial \hat{M}_a}{\partial x_2} & 0 \\ 0 & 0 & \frac{\partial \hat{M}_a}{\partial x_3} \\ \frac{\partial \hat{M}_a}{\partial x_2} & \frac{\partial \hat{M}_a}{\partial x_1} & 0 \\ 0 & \frac{\partial \hat{M}_a}{\partial x_3} & \frac{\partial \hat{M}_a}{\partial x_2} \\ \frac{\partial \hat{M}_a}{\partial x_3} & 0 & \frac{\partial \hat{M}_a}{\partial x_1} \end{bmatrix} \tag{7.67}$$

$$sph[\hat{\mathsf{e}}(\mathsf{u}^h)] \mapsto \theta^h|_{\Omega^e} = \textstyle\sum_{a=1}^{n_{en}} \mathsf{b}_a^T \mathsf{d}_a^u + \sum_{b=1}^{n_{im}} \hat{\mathsf{g}}_b \boldsymbol{\alpha}_b^u ; \; \hat{\mathsf{g}}_a = \begin{bmatrix} \frac{\partial \hat{M}_a}{\partial x_1} \\ \frac{\partial \hat{M}_a}{\partial x_2} \\ \frac{\partial \hat{M}_a}{\partial x_3} \end{bmatrix}$$

In the last equation, $M_b(\xi)$ is the chosen incompatible mode, which should increase the order of approximation of the volume change representation due to mechanical strains; for example, we use the incompatible modes $M_1 = 1 - \xi^2$ and $M_2 = 1 - \eta^2$ for $Q4$ isoparametric element, or the modes $M_1 = \xi(1-\xi^2)$ and $M_2 = \eta(1-\eta^2)$ for $Q9$ element. We have also indicated in (7.67) above that each of these modes ought to be orthogonalized with respect to a constant stress field (in each finite element) in order to satisfy the patch test:

$$\frac{\partial \hat{M}_a}{\partial x_i} = \frac{\partial M_a}{\partial x_i} - \frac{1}{\Omega^e} \int_{\Omega^e} \frac{\partial \hat{M}_a}{\partial x_i} \, dV \; ; \; i = 1, 2, 3 \tag{7.68}$$

It is important to note that any such modification of incompatible modes will not change the global set of equations in (7.65) and (7.64), except for a new form of the stiffness matrix and the structural heating term:

$$\begin{aligned} &\mathsf{M}^\theta \dot{\mathsf{d}}^\theta + \mathsf{K}^\theta \mathsf{d}^\theta + \mathsf{F}^\theta \dot{\mathsf{d}}^u + \mathsf{P}^\theta \dot{\boldsymbol{\alpha}}^u = \mathsf{f}^\theta \\ &\mathsf{M}^u \ddot{\mathsf{d}}^u - \mathsf{F}^u \mathsf{d}^\theta + \mathsf{K}^u \mathsf{d}^u + \mathsf{F}^\alpha \boldsymbol{\alpha}^u = \mathsf{f}^u \\ &\mathsf{F}^{\alpha,e,T} \mathsf{d}^{u,e} + \mathsf{H}^{\alpha,e} \boldsymbol{\alpha}^{u,e} = 0 \; ; \; \forall e \in [1, n_{el}] \end{aligned} \tag{7.69}$$

where the explicit form of each new matrix is recorded here:

$$\begin{aligned} &\mathsf{P}^\theta = \mathop{\mathbb{A}}_{e=1}^{n_{elem}} \mathsf{P}^{\theta,e} \; ; \; \mathsf{P}^{\theta,e} = [\mathsf{P}_{ab}^{\theta,e}] \; ; \; \mathsf{P}_{ab}^{\theta,e} = \textstyle\int_{\Omega^e} N_a \beta \hat{\mathsf{g}}_b^T \, dV \\ &\mathsf{F}^\alpha = \mathop{\mathbb{A}}_{e=1}^{n_{elem}} \mathsf{F}^{\alpha,e} \; ; \; \mathsf{F}^{\alpha,e} = [\mathsf{F}_{ab}^{\alpha,e}] \; ; \; \mathsf{F}_{ab}^{\alpha,e} = \textstyle\int_{\Omega^e} = \mathsf{B}_a^T \mathbf{C} \hat{\mathsf{G}}_b \, dV \\ &\mathsf{H}^{\alpha,e} = [\mathsf{H}_{ab}^{\alpha,e}] \; ; \; \mathsf{H}_{ab}^{\alpha,e} = \textstyle\int_{\Omega^e} \hat{\mathsf{G}}_a^T \mathbf{C} \hat{\mathsf{G}}_b \, dV \end{aligned} \tag{7.70}$$

We will thus obtain a supplementary group of equations in (7.69); however, those equation are algebraic and thus can be used to express the incompatible mode parameters $\boldsymbol{\alpha}^u$ in terms of the nodal displacement values d^u for each element:

$$\boldsymbol{\alpha}^{u,e} = -\mathsf{H}^{e,-1}\mathsf{F}^{\alpha,e,T}\mathsf{d}^e \; ; \; \forall e \in [1, n_{el}] \tag{7.71}$$

By exploiting this relation, it is possible to reduce the system and recover the same number of global equations to solve as in the case defined by (7.63) where we used the standard isoparametric interpolations without incompatible modes:

$$\begin{aligned} \mathsf{M}^\theta\dot{\mathsf{d}}^\theta + \mathsf{K}^\theta\mathsf{d}^\theta + \hat{\mathsf{F}}^\theta\dot{\mathsf{d}}^u &= \mathsf{f}^\theta \\ \mathsf{M}^u\ddot{\mathsf{d}}^u - \mathsf{F}^u\mathsf{d}^\theta + \hat{\mathsf{K}}^u\mathsf{d}^u &= \mathsf{f}^u \end{aligned} \tag{7.72}$$

The explicit form of modified matrices is written below:

$$\begin{aligned} &\hat{\mathsf{K}}^u = \mathop{\mathbb{A}}_{e=1}^{n_{elem}} \hat{\mathsf{K}}^{u,e} \; ; \hat{\mathsf{K}}^{u,e} = \mathsf{K}^{u,e} - \hat{\mathsf{F}}^{\alpha,e}\hat{\mathsf{H}}^{\alpha,e,-1}\hat{\mathsf{F}}^{\alpha,e,T} \\ &\hat{\mathsf{F}}^{\theta,e} = \mathsf{F}^{\theta,e} - \mathsf{P}^{\theta,e}\hat{\mathsf{H}}^{\alpha,e,-1}\hat{\mathsf{F}}^{\alpha,e,T} \end{aligned} \tag{7.73}$$

We conclude that the final set of differential equations for coupled problem in thermoelasticity contains only the nodal displacements and nodal temperatures as unknowns. The numerical solution to such a set of equations is obtained by one-step time-integration schemes, which allows us to reduce the task on hands to:

Central problem in 3D linear thermoelasticity - $t \in [t_n, t_{n+1}]$

Given: $\mathsf{d}^\theta_n = \mathsf{d}^\theta(t_n)$; $\mathsf{v}^\theta_n = \dot{\mathsf{d}}^\theta(t_n)$; $\mathsf{d}^u_n = \mathsf{d}^u(t_n)$; $\mathsf{v}^u_n = \dot{\mathsf{d}}^u(t_n)$; $\mathsf{a}^u_n = \ddot{\mathsf{d}}^u(t_n)$; $h = t_{n+1} - t_n$

Find: d^θ_{n+1} ; v^θ_{n+1} ; d^u_{n+1} ; v^u_{n+1} ; a^u_{n+1}

such that:

$$\begin{aligned} \mathsf{M}^\theta\mathsf{v}^\theta_{n+1} + \mathsf{K}^\theta\mathsf{d}^\theta_{n+1} + \mathsf{F}^\theta\mathsf{v}^u_{n+1} &= \mathsf{f}^\theta_{n+1} \\ \mathsf{M}^u\mathsf{a}^u_{n+1} - \mathsf{F}^u\mathsf{d}^\theta_{n+1} + \mathsf{K}^u\mathsf{d}^u_{n+1} &= \mathsf{f}^u_{n+1} \end{aligned} \tag{7.74}$$

For constructing the solution to this problem we can exploit the time-integration schemes proposed for previously studied problems in dynamics and non-stationary heat transfer. In particular, we can use the α-scheme for thermal part of the system:

$$\begin{cases} \widetilde{\mathsf{d}}^\theta_{n+1} := \mathsf{d}^\theta_n + (1-\alpha)h\mathsf{v}^\theta_n \\ \frac{1}{h\alpha}\left(\mathsf{M}^\theta + h\alpha\mathsf{K}^\theta\right)\mathsf{d}^\theta_{n+1} + \frac{\gamma}{\beta h}\mathsf{F}^\theta(\mathsf{d}^u_{n+1} - \mathsf{d}^u_n) = \mathsf{f}^\theta_{n+1} + \frac{1}{h\alpha}\mathsf{M}^\theta\widetilde{\mathsf{d}}^\theta_{n+1} \\ \qquad\qquad\qquad\qquad\qquad\qquad\qquad\qquad - \mathsf{F}^\theta\widetilde{\mathsf{v}}^u_{n+1} \\ \mathsf{v}^\theta_{n+1} = (\mathsf{d}^\theta_{n+1} - \widetilde{\mathsf{d}}^\theta_{n+1})/(h\alpha) \end{cases} \tag{7.75}$$

along with the Newmark time-integration scheme for the mechanical part:

$$\begin{cases} \mathsf{d}^u_{n+1} = \mathsf{d}^u_n + \mathsf{u}_{n+1} \\ \widetilde{\mathsf{v}}^u_{n+1} = \frac{\beta-\gamma}{\beta}\mathsf{v}^u_n + \frac{(\beta-0.5\gamma)h}{\beta}\mathsf{a}^u_n \\ \widetilde{\mathsf{a}}^u_{n+1} = -\frac{1}{\beta h}\mathsf{v}^u_n - \frac{0.5-\beta}{\beta}\mathsf{a}^u_n \\ -\mathsf{F}^u\mathsf{d}^\theta_{n+1} + \left(\frac{1}{\beta h^2}\mathsf{M}^u + \mathsf{K}^u\right)\mathsf{u}_{n+1} = \mathsf{f}^u_{n+1} - \mathsf{M}^u\widetilde{\mathsf{a}}^u_{n+1} - \mathsf{K}^u\mathsf{d}^u_n \\ \mathsf{v}^u_{n+1} = \frac{\gamma}{\beta h}\mathsf{u}_{n+1} + \widetilde{\mathsf{v}}^u_{n+1} \\ \mathsf{a}^u_{n+1} = \frac{1}{\beta h^2}\mathsf{u}_{n+1} + \widetilde{\mathsf{a}}^u_{n+1} \end{cases} \tag{7.76}$$

We should choose the values for scheme parameters as $\alpha = 1/2$ for thermal part and $\beta = 1/4$ and $\gamma = 1/2$ for mechanical part, in order to provide the second order accuracy to discrete approximation of each sub-problem. In either case we recover an implicit scheme, with the most costly computational phase related to the solution of the system of linear algebraic equations with nodal values of displacements and temperatures at time t_{n+1} as unknowns:

$$\begin{bmatrix} (\frac{2}{h}\mathsf{M}^\theta + \mathsf{K}^\theta) & \frac{2}{h}\mathsf{F}^\theta \\ -\mathsf{F}^u & (\frac{4}{h^2}\mathsf{M}^u + \mathsf{K}^u) \end{bmatrix} \begin{bmatrix} \mathsf{d}^\theta_{n+1} \\ \mathsf{d}^u_{n+1} \end{bmatrix} = \begin{bmatrix} \mathsf{f}^\theta_{n+1} + \frac{2}{h}\mathsf{M}^\theta\widetilde{\mathsf{d}}^\theta_{n+1} + \mathsf{F}^\theta\widetilde{\mathsf{v}}^u_{n+1} \\ \mathsf{f}^u_{n+1} - \frac{2}{h}\mathsf{F}^u\mathsf{d}^u_n + \mathsf{M}^u\widetilde{\mathsf{a}}^u_{n+1} \end{bmatrix} \tag{7.77}$$

This set of algebraic equations is non-symmetric, which implies that the cost of the direct solution will be doubled with respect to the symmetric case. Even more important reason than non-symmetry against the direct solution of the coupled problem is in eventually large difference in time-evolution scales of the mechanical versus thermal component, which are decided by different parameters of the coupled system and driven by independent source terms. Therefore, we are very much interested in operator split solution scheme for the coupled problem in thermodynamics, where the solution of mechanical component and thermal component are not sought simultaneously.

7.2.2 Operator split solution method for 3D thermoelasticity

In this section we discuss two different methods based upon the operator split procedure for thermodynamical coupling in 3D thermoelasticity, as well as the details of their numerical implementation. We would like to elaborate upon a general solution of the set of algebraic equations in (7.77) (which includes the special choice obtained for α time-integration scheme for thermal part and Newmark's scheme for mechanical part of the problem). For that reason, we start the development to follow from the set of differential equations (written in first order form) governing the evolution problem in 3D linear thermoelasticity:

$$\dot{\mathbf{z}}(t) = \mathbf{A}\mathbf{z}(t) \ ; \ \mathbf{z}(0) = \mathbf{z}_0 \ ;$$
$$\mathbf{z} = \begin{bmatrix} \mathbf{u} \\ \mathbf{v} \\ \theta \end{bmatrix} \ ; \ \mathbf{A}\mathbf{z} = \begin{bmatrix} \mathbf{v} \\ \frac{1}{\rho} div[\boldsymbol{\mathcal{C}}\nabla^s \mathbf{u} - \theta\beta\mathbf{I}] \\ \frac{1}{c} div[k\boldsymbol{\nabla}\theta] - \frac{1}{c}\theta\beta\mathbf{I}\cdot\nabla^s\dot{\mathbf{u}} \end{bmatrix} \tag{7.78}$$

Ideally, we would compute the exact solution to this problem by using the exponential mapping formula:

$$\mathbf{z}(t) = exp[t\mathbf{A}]\mathbf{z}_0 \ ; \ exp[\mathbf{A}] = \lim_{n\mapsto\infty} (\mathbf{I} + \frac{1}{n}\mathbf{A})^n \tag{7.79}$$

However, the closed form result is not available for an arbitrary form of the operator $\mathbf{A}$, and we employ instead the following discrete approximation:

$$\begin{aligned} \mathbf{z}_{n+1} &= exp[h\mathbf{A}]\mathbf{z}_n \ ; \ h = t_{n+1} - t_n \\ &\approx [\mathbf{I} + h\mathbf{A}]\mathbf{z}_n \end{aligned} \Rightarrow \ (\mathbf{z}_{n+1} - \mathbf{z}_n)/h = \mathbf{A}\mathbf{z}_n \tag{7.80}$$

The last result provides only the first-order approximation, equivalent to the explicit Euler scheme, which is used only to illustrate the proposed operator split methodology.
In applying the operator split method to the evolution problem of thermodynamical coupling (see [267]), we assume that the operator $\mathbf{A}$ is split additively in two parts, $\mathbf{A} = \mathbf{A}_1 + \mathbf{A}_2$, and that the evolution problem for each of them can be solved independently by using the same kind of explicit Euler approximation as the one employed above for the coupled problem:

$$\begin{aligned} \mathbf{z}_{n+1} &= exp[h(\mathbf{A}_1 + \mathbf{A}_2)]\mathbf{z}_n \\ &\approx [\mathbf{I} + h\mathbf{A}_1] \underbrace{[\mathbf{I} + h\mathbf{A}_2]\mathbf{z}_n}_{intermediate\ solution} \end{aligned} \tag{7.81}$$

The last result indicates that the intermediate values obtained by the first solution phase are immediately exploited for the computations in the second phase. The order of approximation remains independent upon the order of application between two operators $\mathbf{A}_1$ and $\mathbf{A}_2$; however, the final value computed at the end is affected by the order of application of two operators $\mathbf{A}_1$ and $\mathbf{A}_2$, except if they share the same eigenvectors.

The same reasoning applies to the implicit time-integration schemes, as shown next for the coupled problem in 3D linear thermoelasticity. For solution of this problem, where the operator $\mathbf{A}$ is defined in (7.78), we can choose between two possible operator split implementations: isothermal and adiabatic, which are presented subsequently.
(i) *Isothermal split method* will first solve the mechanical problem at fixed value of temperature, and then the thermal component with fixed values of displacements and velocities. The corresponding operators $\mathbf{A}_1$ and $\mathbf{A}_2$ for isothermal split are defined as follows:

$$\dot{\mathbf{z}}(t) = (\mathbf{A}_1 + \mathbf{A}_2)\mathbf{z}(t) \; ; \; \mathbf{z} = \begin{bmatrix} \mathbf{u} \\ \mathbf{v} \\ \theta \end{bmatrix} ;$$

$$\mathbf{A}_1\mathbf{z} = \begin{bmatrix} \mathbf{v} \\ \frac{1}{\rho} div[\boldsymbol{\mathcal{C}}\nabla^s\mathbf{u} - \theta\beta\mathbf{I}] \\ 0 \end{bmatrix} \; ; \; \mathbf{A}_2\mathbf{z} = \begin{bmatrix} \mathbf{0} \\ \mathbf{0} \\ \frac{1}{c} div[k\boldsymbol{\nabla}\theta] - \frac{1}{c}\theta\beta\mathbf{I}\cdot\nabla^s\dot{\mathbf{u}} \end{bmatrix} \tag{7.82}$$

We will apply the implicit mid-point scheme to each of two sub-problems. The mechanical component will be the first to be integrated with the fixed temperature value from the previous step; we can present either the weak form of this problem

$$\begin{aligned} &\int_\Omega \boldsymbol{v} \cdot \tfrac{1}{h}(\mathbf{u}_{n+1} - \mathbf{u}_n)\, dV = \int_\Omega \boldsymbol{v} \cdot \mathbf{v}_{n+\frac{1}{2}} \}\, dV \\ &\int_\Omega \mathbf{w} \cdot \tfrac{\rho}{h}(\mathbf{v}_{n+1} - \mathbf{v}_n)\, dV - \int_\Omega [\nabla^s\mathbf{w} \cdot (\boldsymbol{\mathcal{C}}\nabla^s\mathbf{u}_{n+\frac{1}{2}} - \beta\theta_n\mathbf{I})]\, dV \\ &\qquad = \int_\Omega \mathbf{w} \cdot \mathbf{b}_{n+\frac{1}{2}}\, dV \end{aligned} \tag{7.83}$$

or its discrete approximation obtained by the finite element method, which can be written:

$$\mathsf{M}^u\mathsf{a}^u_{n+\frac{1}{2}} + \mathsf{K}^u\mathsf{d}^u_{n+\frac{1}{2}} - \mathsf{F}^u\mathsf{d}^\theta_n = \mathsf{f}^\theta_{n+\frac{1}{2}} \tag{7.84}$$

Having obtained the corresponding value of velocity from the first phase of computations, it is further employed in the second phase for computing the structural heating term; we can thus write the weak form of the thermal sub-problem to be solved:

$$\begin{aligned} &\int_\Omega \{\vartheta \tfrac{c}{h}(\theta_{n+1} - \theta_n) + \int_\Omega \boldsymbol{\nabla}\vartheta \cdot k\boldsymbol{\nabla}^s\theta_{n+\frac{1}{2}} + \int_\Omega \theta\beta\mathbf{I}\cdot\nabla^s\mathbf{v}_{n+\frac{1}{2}}\}\, dV \\ &\qquad = \int_\Omega \vartheta r_{n+\frac{1}{2}}\, dV \end{aligned} \tag{7.85}$$

along with the corresponding result of the finite element approximation of this problem which is reduced to a set of algebraic equations:

$$\mathsf{M}^{\theta}\mathsf{v}^{\theta}_{n+\frac{1}{2}} + \mathsf{K}^{\theta}\mathsf{d}^{\theta}_{n+\frac{1}{2}} + \mathsf{F}^{\theta}\mathsf{v}^{u}_{n+\frac{1}{2}} = \mathsf{f}^{\theta}_{n+\frac{1}{2}} \tag{7.86}$$

Despite unconditional stability of the mid-point rule for each sub-problem, the isothermal operator split method provides only a conditional stability for coupled problem of thermodynamics in the context of linear thermoelasticity. For example, for 1D case, with the finite element solution constructed with 2-node truss-bar elements and a diagonal form of the mass matrix, the critical value of time step can be written:

$$h \leq \frac{2l^e}{\delta\sqrt{E/\rho}} \; ; \; \delta = \frac{\beta}{\sqrt{Ec}} \tag{7.87}$$

where l^e is the element length. This lack of unconditional stability of isothermal operator split solution method is an important motivation for giving the preference to another implementation capable of providing the stability of computations.

(ii) *Adiabatic operator split* method starts with solving dynamics problem first while keeping the entropy fixed (which does not preclude the temperature evolution), followed by solving the heat transfer problem with fixed displacement and velocity and the resulting temperature. We can formally write the governing equations for each sub-problem with the following choice of the operator $\mathbf{A}_1$ and $\mathbf{A}_2$:

$$\dot{\mathbf{z}}(t) = (\mathbf{A}_1 + \mathbf{A}_2)\mathbf{z}(t) \; ;$$

$$\mathbf{z} = \begin{bmatrix} \mathbf{u} \\ \mathbf{v} \\ \theta \end{bmatrix} ; \; \mathbf{A}_1\mathbf{z} = \begin{bmatrix} \mathbf{v} \\ \frac{1}{\rho}div[\boldsymbol{\mathcal{C}}\nabla^s\mathbf{u} - \theta\beta\mathbf{I}] \\ -\frac{1}{c}\theta\beta\mathbf{I}\cdot\nabla^s\dot{\mathbf{u}} \end{bmatrix} ; \; \mathbf{A}_2\mathbf{z} = \begin{bmatrix} \mathbf{0} \\ \mathbf{0} \\ \frac{1}{c}div[k\boldsymbol{\nabla}\theta] \end{bmatrix} \tag{7.88}$$

We note in particular that the second phase of computation is reduced to standard non-stationary heat transfer problem. In the first phase of computation, the entropy is kept fixed, since according to (7.88) it holds

$$0 = \underbrace{c\dot{\theta} + \beta\theta\mathbf{I}\cdot\nabla^s\mathbf{v}}_{\theta\dot{s}} \Rightarrow s = cst. \tag{7.89}$$

We can note that the constraint on keeping the entropy fixed will imply the corresponding temperature evolution, which can be computed directly from the last equation. Within the context of finite element method implementation, the computation of this kind can be carried out locally, at each Gauss numerical integration point, resulting with the corresponding intermediate value of temperature $\tilde{\theta}$ defined at that point:

$$\theta\dot{s} = 0 \Leftrightarrow \dot{\theta} = -\frac{\beta\theta}{c}\mathbf{I}\cdot\nabla^s\mathbf{v} \Longrightarrow \tilde{\theta} = \theta_n - \frac{\beta}{c}\theta_n\mathbf{I}\cdot(\nabla^s\mathbf{u} - \nabla\mathbf{u}_n) \tag{7.90}$$

By exploiting this result, we can carry out the computation in the first phase of adiabatic split in standard manner, except for the adiabatic tensor $\mathcal{C}_{ad} = \mathcal{C} + \frac{1}{c}\beta\mathbf{I} \otimes \beta\mathbf{I}$ replacing the elasticity tensor:

$$
\begin{aligned}
&\int_\Omega \boldsymbol{v} \cdot \tfrac{1}{h}(\mathbf{u}_{n+1} - \mathbf{u}_n)\, dV = \int_\Omega \boldsymbol{v} \cdot \mathbf{v}_{n+\frac{1}{2}}\, dV \\
&\int_\Omega \mathbf{w} \cdot \tfrac{\rho}{h}(\mathbf{v}_{n+1} - \mathbf{v}_n)\, dV + \int_\Omega \nabla^s \mathbf{w} \cdot \mathcal{C}_{ad} \nabla^s \mathbf{u}_{n+\frac{1}{2}}\, dV \\
&\qquad = \int_\Omega \mathbf{w} \cdot \mathbf{b}_{n+\frac{1}{2}}\, dV + \int_\Omega \nabla^s \mathbf{w} \cdot (\theta_n \tfrac{\beta}{c}\mathbf{I} \cdot \nabla^s \mathbf{u}_n)\beta\mathbf{I}\, dV
\end{aligned}
\tag{7.91}
$$

All the modifications of the standard problem in dynamics, that ought to be introduced in the first phase of computations with the adiabatic operator split method, are fairly modest, and they can be handled at the level of a single element. The same applies to the second stage of adiabatic operator split procedure, where one finds in fact a non-stationary heat transfer problem, which can be formulated and solved in standard manner:

$$
\int_\Omega \frac{c}{h}(\theta_{n+1} - \tilde{\theta})\, dV + \int_\Omega \boldsymbol{\nabla}\vartheta \cdot k \boldsymbol{\nabla}\theta_{n+\frac{1}{2}}\, dV = \int_\Omega \vartheta r_{n+\frac{1}{2}}\, dV \tag{7.92}
$$

Remark: In fact one can show (see [147]) the instability of the isothermal split for linear thermoelasticity with any kind of time integration scheme, and even the exact integration for each component of a set of governing differential–algebraic equations. This can be done by exploiting the analogy between the stability analysis of the operator split procedure and convergence condition of Gauss–Seidel iterations presented in [11]; this results with the following stability condition for 1D thermoelasticity:

$$
\delta^2 := \frac{3}{4}\frac{\beta^2\theta_0}{Ec} < 1 \tag{7.93}
$$

where β is the thermal stress, θ_0 the reference temperature, E Young's modulus and c the heat capacity coefficient. We can also show (see [147]) that the adiabatic split can be interpreted as the corresponding preconditioning of the Gauss–Seidel scheme, which enforces that the stability condition will always remain verified:

$$
\frac{1}{4}\frac{\delta^2}{1+\delta^2} < 1
$$

7.2.3 Numerical examples in thermoelasticity

7.2.3.1 Thermoelastic coupling for 1D thermoelasticity

The first example seeks to validate the stability criterion presented in the previous remark. To that end, we consider a simple one-dimensional thermoelastic truss-bar, set in motion by imposed displacement time variation at the left end (see Figure 7.1). The finite element model of the bar consists of 100 elements. The chosen thermomechanical properties of the bar are given in the table below.

Young's modulus	E	$200\,000\ N \cdot mm^{-2}$
Area	A	$1\ mm^2$
Mass density	ρ	$7.80 \cdot 10^{-9}\ N \cdot s^2 \cdot mm^{-4}$
Specific heat capacity	c	$1,2\ N \cdot mm^{-2} \cdot K^{-1}$
Reference temperature	θ_0	$1\ K$
Conductivity	k	$0.15\ N \cdot s^{-1} \cdot K^{-1}$
Expansion coefficient	α	$1.5 \cdot 10^{-5}\ K^{-1}$
Thermal stress	$\beta = E\alpha$	$3.00\ N \cdot mm^{-2} \cdot K^{-1}$

Thermoelastic bar: mechanical and thermal parameters

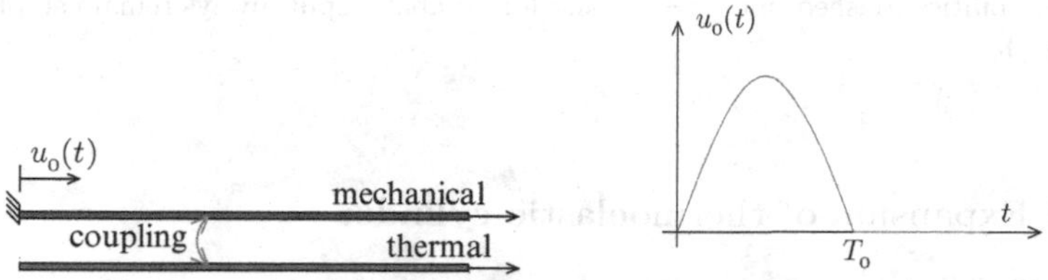

Fig. 7.1 Boundary value problem in 1D thermoelasticity and imposed loading.

The evolution process of thermoelastic coupling develops as follows: first the displacement imposed at one end of the bar generates a compressive wave, which further leads to a local increase in temperature (very much like when compressing a perfect gas); the wave-reflection on the other end of the bar then generates a traction wave, with the corresponding decrease in temperature. We note that the variations in temperature remain essentially local, since the time scale of non stationary heat transfer is very large compared to the time scale of the mechanical wave propagation.

The results of this example allow us to test the instability criterion for operator split solution procedure for thermomechanical coupling. Namely, for the chosen set of thermomechanical properties, we obtain $\delta^2 < 1$ and the isothermal split indeed remains stable. However, by increasing the thermal stress to 100β, we get the value of stability criterion close to 1, and indeed find the marked instability developing with isothermal split; see Figure 7.2. The result obtained with adiabatic split, however, does not show any instability.

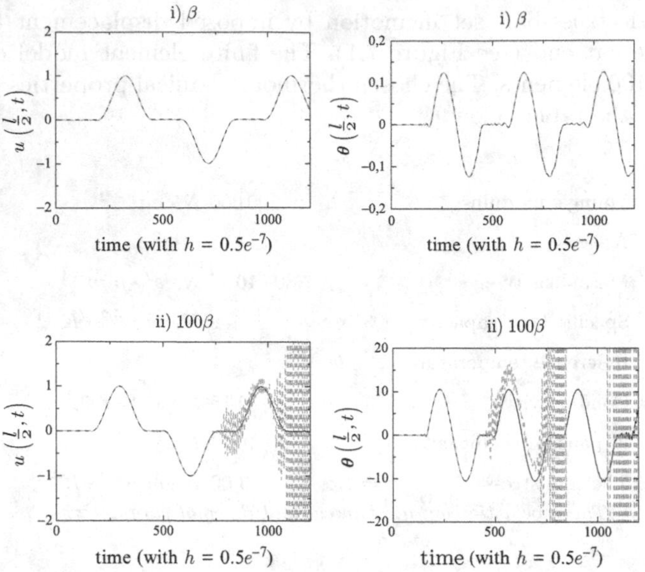

Fig. 7.2 Computed displacement $u\left(\frac{l}{2},t\right)$ and temperature $\theta\left(\frac{l}{2},t\right)$ for isothermal split shown instabilities (dashed line [- - -]) and for adiabatic split always remain stable (solid line [——]).

7.2.3.2 Expansion of thermoelastic cylinder

We present an example concerning the expansion of a thermoelastic thick cylinder under internal pressure (see Figure 7.3). This example illustrates the instability of computations which occur with isothermal operator split method for the case when the thermal stress are important, as well as the capability of the adiabatic operator split method to eliminate this instability. The thick cylinder has internal radius $a = 5$ mm, external radius $b = 15$ mm and linear thermoelastic constitutive behavior. The chosen material parameters are given in the table below:

bulk modulus	κ	$164206\, N/mm^2$
shear modulus	μ	$80139.8\, N/mm^2$
mass density	ρ_0	$7.8\, 10^{-9}\, Ns^2/mm^4$
conductivity coefficient	k	$45.0\, N/JK$
heat capacity	c	$0.46\, 10^9 Nm^2/s^2K$

Thermolastic cylinder: mechanical and thermal parameters

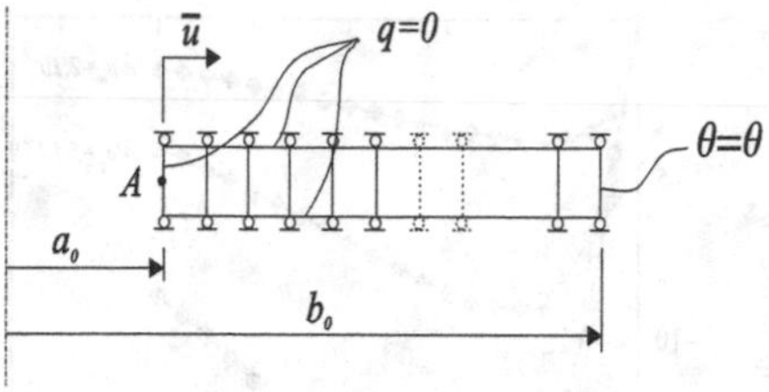

Fig. 7.3 Thermolastic cylinder: axisymmetric model, geometric properties and boundary conditions.

We construct an axisymmetric finite element model for a unit slice of the cylinder, by using ten finite elements $Q4$. For the axisymmetric model, we consider the cylinder to be very long, with neither displacements nor temperature flux along the axis of the cylinder, and with the boundary conditions with zero displacement and heat flux chosen accordingly (see Figure 7.3). The zero value of heat flux is also imposed on the internal boundary, whereas the imposed value of temperature $\theta_0 = 293$ is imposed on the external boundary of the cylinder. The computation is performed under displacement control imposed on internal boundary, and the pressure is obtained as the corresponding reaction value. The zero traction is imposed on the external cylinder boundary and the corresponding displacement are left free. The cylinder expansion imposes very large displacements and rotations, with the internal radius initial value being finally tripled, with $a = 3a_0 = 15$ mm.
The first analysis is performed for the case of weak thermomechanical coupling, for the heat expansion coefficient $\alpha = 10^{-5}$. In order to illustrate the sensitivity of heat diffusion process with respect to time, we have considered three different values of rate of increase of imposed loading with: $\dot{a}/a_0 = 10^{-1}$, $\dot{a}/a_0 = 5 \times 10^{-2}$ and $\dot{a}/a_0 = 2 \times 10^{-2}$ $[1/s]$. The computed evolution of the temperature at the internal boundary of the cylinder with respect to the imposed displacement is presented in Figure 7.4, both for the case of simultaneous solution procedure (the reference value) and the isothermal operator split method. We can see that the latter remains stable, and that two methods give practically the same results. We can also see that the temperature at the internal boundary reaches higher values for the faster cylinder expansion, which is quite logical since the heat diffusion effect does not have enough time to act in order to reduce the peak temperature.
The second analysis is performed for the case of strong thermomechanical coupling, where the heat expansion coefficient is increased 10 times to $\alpha = 10^{-4}$. The computation is carried out for two different values of time step, with $h = 0.1s$ and $h = 1s$, and the computed results for temperature evolution at the internal boundary are presented in Figure 7.5. We can now observe the loss of stability for isothermal operator split method, which becomes more pronounced with the increase of rate of expansion; in fact, the highest rate of expansion does not even allow us to obtain the converged results.

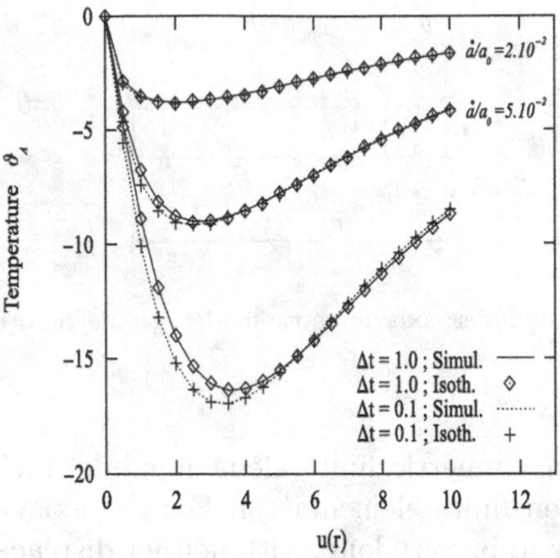

Fig. 7.4 Thermoelastic cylinder: internal boundary temperature evolution with respect to imposed displacement for weak coupling ($\alpha = 10^{-5}$).

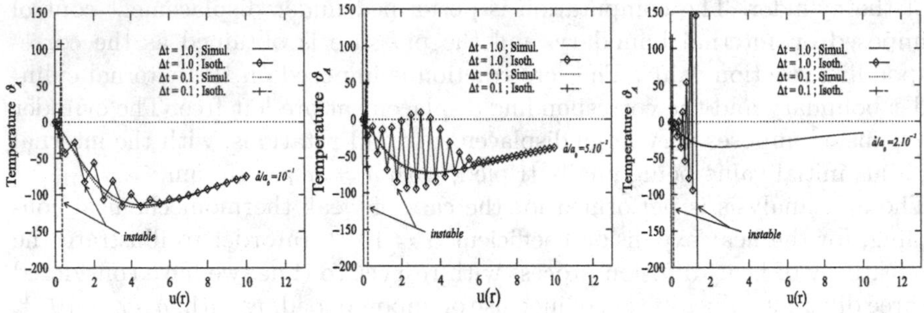

Fig. 7.5 Thermoelastic cylinder: internal boundary temperature evolution with respect to imposed displacement for strong coupling ($\alpha = 10^{-4}$).

7.3 Thermodynamics of irreversible processes

In this section we will generalize the class of problems of thermodynamical coupling including those dealing with irreversible processes. We choose in particular two model problems, such as the case of inelastic constitutive behavior described with plasticity model and frictional contact, each coupled with heat transfer problem and resulting temperature field. The main goal of this section is to present all the ingredients for these representative models of thermodynamical coupling, and to discuss the corresponding operator split solution methods.

7.3.1 Thermodynamics coupling for 1D plasticity

In order to introduce the main ideas pertaining to theoretical developments of a complex model for thermodynamical coupling in large strain plasticity, we start with the simplest choice of 1D setting, or elastoplastic bar. The deformed configuration of such a bar, stretched along x-axis, can be defined by specifying the position vector for each particle with:

$$\varphi(x,t) = x + d(x,t) \tag{7.94}$$

where x and t are independent variables of space and time and $d(x,t)$ is the displacement along the bar axis. For any value of x, we can define the velocity $v(x,t)$ and the stretch $\lambda(x,t)$ according to:

$$v(x,t) = \frac{\partial \varphi(x,t)}{\partial t} \; ; \; \lambda(x,t) = \frac{\partial \varphi(x,t)}{\partial x} \tag{7.95}$$

Equations of motion describing the trajectory of a particle between its initial and deformed position can be written:

$$\rho \frac{\partial v(x,t)}{\partial t} = \frac{\partial P(x,t)}{\partial x} + b(x,t) \tag{7.96}$$

where ρ is the bar mass density, P is the first Piola–Kirchhoff stress and b distributed loading per unit length of initial configuration. We assume that the stress P can be obtained from the internal energy potential; for thermo-plasticity with hardening, such a potential can be written:

$$e(\lambda^e, \zeta, s^e) \tag{7.97}$$

where λ^e is the elastic stretch, ζ is internal variable for isotropic hardening, and s^e is elastic entropy. For a homogeneous strain field, the elastic stretch can be defined as the ratio between the deformed length of the bar l^φ and the residual length l^p obtained upon complete unloading, which will not be the same as the initial length l in the presence of plastic deformation; we can thus write:

$$\lambda = \lambda^e \lambda^p \;\Leftrightarrow\; \frac{l^\varphi}{l} = \frac{l^\varphi}{l^p}\frac{l^p}{l} \tag{7.98}$$

With the choice of logarithmic strain, this multiplicative decomposition of stretch is transformed into the additive decomposition of total deformation into elastic and plastic component, which is formally equivalent to the standard result in the small strain setting:

$$\epsilon = \epsilon^e + \epsilon^p \; ; \; \epsilon = ln\lambda \; ; \; \epsilon^e = ln\lambda^e \; ; \; \epsilon^p = ln\lambda^p \tag{7.99}$$

The same kind of additive decomposition can be used for entropy, which is an extensive variable (e.g. see Ericksen [74]), which defines the corresponding elastic and plastic components:

$$s = s^e + s^p \tag{7.100}$$

The elastic components of strain and entropy are produced with an elastic motion of crystals, whereas their plastic components are associated with dislocation motion. For an elastic processes, both plastic entropy and plastic deformation will remain frozen, and the internal energy potential can thus be written:

$$e(\epsilon^e, \zeta, s^e) \tag{7.101}$$

With this result in hand, we can establish the local form of the first principle of thermodynamics in the context of thermodynamical coupling for 1D thermoplasticity. We first consider a segment of the bar between x_1 and x_2, for which we can write:

$$\frac{d}{dt}\int_{x_1}^{x_2} (e + \frac{1}{2}\rho v^2)\, dx = \int_{x_1}^{x_2} vb\, dx + [Pv]_{x_1}^{x_2} + \int_{x_1}^{x_2} r\, dx - [q_0]_{x_1}^{x_2} \tag{7.102}$$

In the limit, with $x_2 \mapsto x_1 \equiv x$, we can then obtain the local form of the first principle:

$$\underbrace{\dot{e}(\lambda^e, \zeta, s^e)}_{\frac{\partial e}{\partial \epsilon^e}\dot{\epsilon}^e + \frac{\partial e}{\partial \zeta}\dot{\zeta} + \frac{\partial e}{\partial s^e}\dot{s}^e} + \underbrace{(\rho\dot{v} - \frac{\partial P}{\partial x} - b)\, v}_{=0} = \underbrace{P\dot{\lambda}}_{\tau(\dot{\epsilon}^e + \dot{\epsilon}^p)} + r - \frac{\partial q_0}{\partial x} \tag{7.103}$$

We thus have two cases to consider: an elastic processes first, where the values of internal variables (plastic entropy and plastic strain) does not change; the last result will simplify to:

$$(\frac{\partial e}{\partial \epsilon^e} - \tau)\dot{\epsilon}^e + \frac{\partial e}{\partial s^e}\dot{s}^e = r - \frac{\partial q_0}{\partial x} \tag{7.104}$$

In order to allow an independent evolutions of deformation with respect to entropy, the following result must hold:

$$\tau = \frac{\partial e}{\partial \epsilon^e} \tag{7.105}$$

By assuming that the temperature derives from the same potential, we can also conclude that:

$$\theta = \frac{\partial e}{\partial s^e} \tag{7.106}$$

The last result will allow to simplify (7.104) and make it again equal to the reduced form of the first principle:

$$\theta \dot{s}^e = r - \frac{\partial q_0}{\partial x} \tag{7.107}$$

Instead of entropy we can choose the temperature as the state variable, since its evolution is easier to control. To that end, we can apply the Legendre transformation switching the roles between entropy and temperature and replacing the internal energy potential with the Helmholtz free-energy:

$$\psi(\epsilon^e, \zeta, \theta) = e(\epsilon^e, \zeta, s^e) - s^e\theta \tag{7.108}$$

The first principle of thermodynamics can then be rewritten:

$$(\frac{\partial \psi}{\partial \epsilon^e} - \tau)\dot{\epsilon}^e + (\frac{\partial \psi}{\partial \theta} + s^e)\dot{\theta} + \theta \dot{s}^e = r - \frac{\partial q_0}{\partial x} \tag{7.109}$$

This further allows us to confirm the reduced form of the first principle in (7.107) by expressing the stress and the elastic entropy with the following constitutive equations:

$$\tau = \frac{\partial \psi}{\partial \epsilon^e} \; ; s^e = -\frac{\partial \psi}{\partial \theta} \tag{7.110}$$

We also postulate that all these constitutive relations remain applicable to a plastic processes, but the reduced form of the first principle on energy conservation no longer applies unless we account for the plastic dissipation:

$$\underbrace{\frac{\partial e}{\partial \epsilon^e}\dot{\epsilon}^e + \frac{\partial e}{\partial \zeta}\dot{\zeta} + \frac{\partial e}{\partial s^e}\dot{s}^e}_{\dot{e}} - \tau\dot{\epsilon}^e + q\dot{\zeta} - \theta)\dot{s}^e = \underbrace{\tau\dot{\epsilon}^p + q\dot{\zeta} + \theta\dot{s}^p}_{D^p} - \theta\dot{s}^e + r - \frac{\partial q_0}{\partial x} \tag{7.111}$$

The validity of this result ought to further be approved by thermodynamics. Namely, contrary to the temporal evolution of free-vibration processes in dynamics, where one can count with the free exchange between kinetic and potential energies, the same spontaneous exchange is not produced between mechanical and thermal energy. It is precisely the role of the second principle of thermodynamics to limit this kind of exchange. In essence, the second principle of thermodynamics postulates that the dissipation is never negative; for example, for the bar segment between x_1 and x_2, we can obtain:

$$\theta \frac{d}{dt} \int_{x_1}^{x_2} s \, dx \geq \int_{x_1}^{x_2} r \, dx - [q_0]_{x_1}^{x_2} \tag{7.112}$$

In the limit case with $x_2 \mapsto x_1 \equiv x$, we can obtain the local form of dissipation:

$$\begin{aligned} 0 \leq D^p &= \theta \dot{s} - (r - \frac{\partial q_0}{\partial x}) \\ &= \theta \underbrace{\dot{s}}_{\dot{s}^e + \dot{s}^p} - \frac{\partial}{\partial t} e(\epsilon^e, \zeta, s^e) + \tau \dot{\epsilon} \end{aligned}$$

$$= \underbrace{(\tau - \frac{\partial \psi}{\partial \epsilon^e})}_{=0}\dot{\epsilon}^e + \underbrace{(s^e + \frac{\partial \psi}{\partial \theta})}_{=0}\dot{\theta} + \theta \dot{s}^p + \tau \dot{\epsilon}^p + \underbrace{(-\frac{\partial \psi}{\partial \zeta})}_{=q}\dot{\zeta} \tag{7.113}$$

$$= \underbrace{\theta \dot{s}^p}_{D^p_{ther}} + \underbrace{\tau \dot{\epsilon}^p + q\dot{\zeta}}_{D^p_{meca}}$$

For thermodynamical coupling in large strain plasticity, we can write the additive decomposition of the total dissipation into mechanical part $D^p_{meca} = \tau\dot{\epsilon}^p + q\dot{\zeta}$ and thermal part $D^p_{ther} = \theta\dot{s}^p$. We note in passing that the same kind of additive split also holds in small deformation setting.

With total dissipation in hand, we can apply the principle of maximum plastic dissipation in order to derive the corresponding evolution equations for internal variables. In that sense, we seek among all plastically admissible values of stress the one which will maximize the plastic dissipation; this can be written in terms of the constrained minimization problem, or rather its unconstrained counterpart presented as min–max problem featuring the appropriate Lagrangian functional:

$$\max_{\phi(\tau,q,\theta)=0} D^p(\tau, q, \theta) \Leftrightarrow \min_{\forall(\tau,q,\theta)} \max_{\dot{\gamma}>0} L^p(\tau, q, \theta, \dot{\gamma}) \; ; \tag{7.114}$$
$$L^p(\cdot) = -D^p(\cdot) + \dot{\gamma}\phi(\cdot)$$

The Kuhn-Tucker optimality conditions corresponding to such a Lagrangian can readily be written:

$$\begin{aligned} 0 = \tfrac{\partial L^p}{\partial \tau} &:= -\dot{\epsilon}^p + \dot{\gamma}\tfrac{\partial \phi}{\partial \tau} \implies \dot{\epsilon}^p = \dot{\gamma}\tfrac{\partial \phi}{\partial \tau} \\ 0 = \tfrac{\partial L^p}{\partial q} &:= -\dot{\zeta} + \dot{\gamma}\tfrac{\partial \phi}{\partial q} \implies \dot{\zeta} = \dot{\gamma}\tfrac{\partial \phi}{\partial q} \\ 0 = \tfrac{\partial L^p}{\partial \theta} &:= -\dot{s}^p + \dot{\gamma}\tfrac{\partial \phi}{\partial \theta} \implies \dot{s}^p = \dot{\gamma}\tfrac{\partial \phi}{\partial \theta} \end{aligned} \tag{7.115}$$

including the standard loading/unloading conditions for plasticity:

$$\dot{\gamma} \geq 0 \; ; \; \phi \leq 0 \; ; \; \dot{\gamma}\phi = 0 \tag{7.116}$$

The same evolution equations are applicable to thermodynamical coupling in viscoplasticity, except that the plastic multiplier is no longer computed from loading/unloading conditions but from the following constitutive relation:

$$\dot{\gamma} = \frac{1}{\eta} < \phi > \tag{7.117}$$

where η is the viscosity coefficient and $< \cdot >$ is the Macauley bracket.

We further write explicitly the evolution equations for a 1D finite plasticity model, with the simplest form of the temperature-dependent yield criterion:

$$0 \geq \phi(\tau, q, \theta) := |\tau| - (\sigma_y(\theta) - q) \tag{7.118}$$

In this case, we can easily show that the mechanical part of plastic dissipation can further be simplified to:

$$\begin{aligned} D^p_{meca} &= \tau\dot{\epsilon}^p + q\dot{\zeta} \\ &= \dot{\gamma}\underbrace{(\overbrace{\tau sign(\tau)}^{|\tau|} + q - \sigma_y(\theta))}_{\phi} + \dot{\gamma}\sigma_y(\theta) \\ &= \dot{\gamma}\sigma_y(\theta) \end{aligned} \tag{7.119}$$

The experimental results for metallic materials show that the elastic domain and the yield stress value are reduced with increasing temperature; by postulating the linear dependence on temperature increase, we can write for the yield stress:

$$\sigma_y(\theta) = \sigma_y(\theta_0)[1 - \omega(\theta - \theta_0)] \ ; \ \omega > 0 \tag{7.120}$$

where ω is the chosen material parameter specifying the yield stress sensitivity to temperature change. For simplicity, the same kind of linear temperature dependence is assumed for hardening variable that controls the evolution of the plasticity threshold due to hardening:

$$q = -K(\theta)\,\zeta \ ; \ K(\theta) = K(\theta_0)[1 - \omega(\theta - \theta_0)] \tag{7.121}$$

where $K(\theta)$ is the corresponding value of the temperature-dependent hardening modulus.
For the chosen model of 1D thermoplasticity, we can also write the following free-energy potential:

$$\psi(\epsilon^e, \zeta, \theta) = \underbrace{\frac{1}{2}\epsilon^e E \epsilon^e}_{\bar{\psi}(\epsilon^e)} + \underbrace{\frac{1}{2}\zeta K \zeta}_{\Xi(\zeta)} + \underbrace{c[(\theta - \theta_0) - \theta ln\frac{\theta}{\theta_0}]}_{T(\theta)} + \underbrace{\alpha E(\theta - \theta_0)\epsilon^e}_{M(\epsilon^e,\theta)} \tag{7.122}$$

With such a choice of potential, we can obtain the corresponding constitutive equations for the Kirchhoff stress including the thermal stress component:

$$\tau := \frac{\partial \psi}{\partial \epsilon^e} = \underbrace{E\epsilon^e}_{\frac{\partial \bar{\psi}}{\partial \epsilon^e}} \underbrace{-\alpha E(\theta - \theta_0)}_{\frac{\partial M}{\partial \epsilon^e}} \tag{7.123}$$

The same potential also provides the entropy for the elastic processes according to:

$$s := -\frac{\partial \psi}{\partial \theta} = \underbrace{-c ln(\theta/\theta_0)}_{\frac{\partial T}{\partial \theta}} \underbrace{-\alpha E \epsilon^e}_{\frac{\partial M}{\partial \theta}} \tag{7.124}$$

With these results in hand, we can easily obtain the weak form of the coupled problem in thermoplasticity; for the mechanical part, we can write:

$$\begin{aligned} 0 = G_M(\varphi, \theta, \epsilon^p, \zeta, s^p; w) &:= \int_l (w\rho\ddot{\varphi} + \tfrac{1}{\lambda}\tfrac{dw}{dx}\tau \; - \; wb)\, dx - w(l)\bar{t} \; ; \\ \phi(\hat{\tau}(\lambda, \theta, \epsilon^p), \hat{q}(\theta, \zeta)) &\leq 0 \; ; \; \dot{\gamma} \geq 0 \; ; \; \dot{\gamma}\phi = 0 \end{aligned} \tag{7.125}$$

where the Kirchhoff stress $\hat{\tau}(\cdot)$ and heat flux $\hat{q}(\cdot)$ ought to be computed from constitutive equations in (7.123) and (7.121), respectively. The weak form of the thermal part of this coupled problem can be written:

$$\begin{aligned} 0 = G_T(\varphi, \theta, \epsilon^p, \zeta, s^p; \vartheta) &:= \int_l \vartheta(\theta\dot{s}^e \underbrace{-\tau\dot{\epsilon}^p - q\dot{\zeta}}_{-D^p_{meca}} - r + \tfrac{\partial q_0}{\partial x}\, dx \; ; \\ \phi(\hat{\tau}(\lambda, \theta, \epsilon^p), \hat{q}(\theta, \zeta)) &\leq 0 \; ; \; \dot{\gamma} \geq 0 \; ; \; \dot{\gamma}\phi = 0 \end{aligned} \tag{7.126}$$

where D^p_{meca} is the plastic dissipation, which contributes a supplementary term to the heat source. It is important to note that the weak form equations in (7.125) and (7.126) hold for the corresponding values of internal variables that provide the plastically admissible stress state.

7.3.2 Thermodynamics coupling in 3D plasticity

In this section we seek to generalize the presented 1D thermoplasticity model to 3D case. To that end, instead of the product of elastic and plastic stretch in (7.98), for 3D case we employ a multiplicative decomposition of the deformation gradient into elastic and plastic part:

$$\mathbf{F} = \mathbf{F}^e\mathbf{F}^p \; ; \; \forall \mathbf{x} \in \Omega \tag{7.127}$$

The strong and weak forms of the motion equations for 3D case can be written:

$$\rho\dot{\mathbf{v}} = \underbrace{\boldsymbol{\tau}\boldsymbol{\nabla}^{\varphi}}_{\mathbf{P}\boldsymbol{\nabla}} + \mathbf{b} \; ; \; \text{in } \Omega \tag{7.128}$$

and

$$\int_{\Omega} \mathbf{w} \cdot \rho\dot{\mathbf{v}}\, dV = \int_{\Omega} (-\underbrace{\nabla^{\varphi}\mathbf{w} \cdot \boldsymbol{\tau}}_{\nabla\mathbf{w}\cdot\mathbf{P}} + \mathbf{w} \cdot \mathbf{b})\, dV + \int_{\Gamma_\sigma} \mathbf{w} \cdot \bar{\mathbf{t}}\, dA \tag{7.129}$$

where $\mathbf{P}$ and $\boldsymbol{\tau}$ denote, respectively, the first Piola–Kirchhoff and Kirchhoff stress tensors, $\mathbf{b}$ is the volume force and $\bar{\mathbf{t}}$ is the boundary traction.
We will keep for 3D case the additive decomposition of entropy into elastic and plastic part, as postulated in (7.100). The strong and weak forms for the

thermal part of the coupled problem in 3D thermoplasticity, presenting the energy balance result, can be written:

$$\theta \dot{s} = -\mathbf{q}_0 \cdot \nabla^{\varphi} + r + D^p \Leftrightarrow \theta \dot{s}^e = -\mathbf{q}_0 \cdot \nabla^{\varphi} + r + D^p_{meca} \tag{7.130}$$

and

$$\int_{\Omega} \vartheta \theta \dot{s}^e \, dV - \int_{\Omega} \nabla^{\varphi} \vartheta \cdot \mathbf{q}_0 \, dV = \int_{\Omega} \vartheta (D^p_{meca} + r) \, dV + \int_{\Gamma_q} \vartheta \mathbf{q} \cdot \boldsymbol{\nu} \, dA \tag{7.131}$$

where D^p_{meca} is the plastic dissipation produced by the mechanical part. It is possible to rewrite the strong form of the thermal component of 3D thermoplasticity to show explicitly the contribution of the mechanical component to the source terms in energy balance equation. More precisely, by using the additive decomposition of entropy in (7.100) along with the constitutive equation for its elastic part, we can write:

$$\begin{aligned} \theta \dot{s} &= \theta \dot{s}^p + \theta \dot{s}^e \\ &= \theta \dot{s}^p \underbrace{-\theta \frac{\partial^2 \psi}{\partial \theta^2}}_{c} \dot{\theta} - \theta \tfrac{\partial}{\partial \theta} [\mathbf{F}^e \tfrac{2 \partial \psi}{\partial \mathbf{C}^e} \mathbf{F}^{e,T} \cdot \mathbf{D}^e] \\ &= \theta \dot{s}^p + c \dot{\theta} - \theta \tfrac{\partial}{\partial \theta} [\boldsymbol{\tau} \cdot \mathbf{D}^e] \end{aligned} \tag{7.132}$$

In (7.132) above, the second term on the right hand side depends upon the heat capacity $c = -\theta \frac{\partial^2 \psi}{\partial \theta^2}$ and the last term is the structural heating expressed in terms of the Kirchhoff stress and elastic rate of deformation tensor, $\mathbf{D}^e = \frac{1}{2}(\mathbf{L}^e + \mathbf{L}^{e,T})$, with $\mathbf{L}^e = \dot{\mathbf{F}}^e \mathbf{F}^{e,-1}$. With this result in hand, we can restate the energy balance equation in (7.130) according to:

$$c \dot{\theta} = -\mathbf{q}_0 \cdot \nabla^{\varphi} + r + D^p_{meca} + \theta \frac{\partial}{\partial \theta} (\boldsymbol{\tau} \cdot \mathbf{D}^e) \tag{7.133}$$

This alternative form for energy balance shows clearly the mechanical dissipation and structural heating to be two additional heat sources. In this case (and number of other similar cases with inelastic behavior) the mechanical dissipation is by far the dominant heat source, which allows in general to drop the structural heating term altogether.
We will further exploit the principal axes framework, which was already used for 3D finite plasticity model, and carry on with the developments suitable for present case of 3D thermoplasticity in large strains. All the ingredients of such a 3D model can be obtained as the corresponding generalization of the results already presented for 1D case. For example, the governing potential of Helmholtz free-energy can be written in terms of elastic strains, temperature and hardening variable:

$$\psi(\mathbf{C}^e, \theta, \zeta) = \bar{\psi}(\mathbf{C}^e) + \Xi(\zeta) + T(\theta) - \underbrace{(\theta - \theta_0) G(J^e)}_{M(J^e, \theta)} \tag{7.134}$$

In (7.134) above, $J^e = det\mathbf{C}^e$ represents the elastic volume change, which is the only one affected by the thermomechanical coupling for this kind of model of 3D thermoplasticity. The thermal part of the potential for 3D case in (7.134) is the same as for 1D case, since the temperature is the scalar field; we can thus write:

$$T(\theta) = c[(\theta - \theta_0) - \theta ln(\theta/\theta_0)] \tag{7.135}$$

With such choice of potential, the constitutive relation can be used to compute the elastic part of entropy according to:

$$s^e = c\, ln(\theta/\theta_0) + G(J^e) \tag{7.136}$$

From this expression we can obtain the temperature evolution in terms of elastic strain and entropy:

$$\theta = \theta_0 exp\{[s^e - G(J^e)]/c\} =: \hat{\theta}(\mathbf{C}^e, s^e) \tag{7.137}$$

With this result in hand, we can also define the internal energy potential through the Legendre transformation:

$$e(\mathbf{C}^e, s^e, \zeta) = \psi(\mathbf{C}^e, \zeta, \hat{\theta}(s^e)) + s^e \hat{\theta}(\mathbf{C}^e, s^e) \tag{7.138}$$

The invariant form of any of these potentials can easily be provided by exploiting the principal axes framework. First, we ought to compute the principal values of the elastic left Cauchy–Green strain tensor, or elastic principal stretches, as the solution to the following eigenvalue problem:

$$[\mathbf{B}^e - (\lambda_i^e)^2 \mathbf{I}]\mathbf{m}_i = \mathbf{0} \tag{7.139}$$

We can then use this result to obtain the invariant form of the free-energy according to:

$$\psi(\lambda_i^e, \zeta, \theta) \tag{7.140}$$

A simple choice for the mechanical part of the free-energy potential of this kind is written as the quadratic form in natural strains computed from elastic principal stretches:

$$\bar{\psi}(\lambda_i^e) = \frac{1}{2}\lambda(ln\lambda_1^e + ln\lambda_2^e + ln\lambda_3^e)^2 + \mu[(ln\lambda_1^e)^2 + (ln\lambda_2^e)^2 + (ln\lambda_3^e)^2] \tag{7.141}$$

where λ and μ are two Lame's parameters. The main term of the free-energy potential which controls the thermomechanical coupling can also be adapted to this kind of strain measure leading to:

$$M(J^e, \theta) = -3\alpha(\lambda + \frac{2}{3}\mu)(\theta - \theta_0)(ln\lambda_1^e + ln\lambda_2^e + ln\lambda_3^e)/(\lambda_1^e\lambda_2^e\lambda_3^e) \tag{7.142}$$

From the chosen free-energy potential of thermoplasticity we can obtain the principal values of the Kirchhoff stress; with such a result and principal vectors of the elastic left Cauchy–Green strain tensor in (7.139), we can also provide[6] the spectral decomposition representation of the Kirchhoff stress tensor, and thus obtain all of its physical components:

$$\begin{aligned} &\boldsymbol{\tau} = \textstyle\sum_{i=1}^{3} \tau_i \mathbf{m}_i \otimes \mathbf{m}_i + p_\theta \mathbf{I} \\ &\tau_i := \lambda_i^e \frac{\partial \bar{\psi}}{\partial \lambda_i^e} \; ; \; p_\theta = -3\alpha(\lambda + \tfrac{2}{3}\mu)(\theta - \theta_0)\frac{1-(ln\lambda_1^e + ln\lambda_2^e + ln\lambda_3^e)}{\lambda_1^e \lambda_2^e \lambda_3^e} \end{aligned} \tag{7.143}$$

The principal axes framework also provides an invariant form of the yield criterion in terms of the principal values of Kirchhoff stress; for example, we can generalize the von Mises plasticity criterion with isotropic hardening to thermodynamical framework, where we can write:

$$\begin{aligned} 0 \geq \phi(\tau_i, q, \theta) &:= \sqrt{\tfrac{2}{3}(\tau_1^2 + \tau_2^2 + \tau_3^2 - \tau_1\tau_2 - \tau_2\tau_3 - \tau_3\tau_1)} \\ &-\sqrt{\tfrac{2}{3}}(\sigma_y(\theta) - \hat{q}(\zeta, \theta)) \end{aligned} \tag{7.144}$$

For thermoplasticity, we ought to admit the temperature-dependence of all parameters; for example, with a simple linear dependence upon the temperature of limit of elasticity and saturation hardening, we can obtain the consistent modification of plasticity threshold for any temperature change:

$$\begin{aligned} &\hat{q}(\zeta, \theta) := -\{[\sigma_\infty(\theta) - \sigma_y(\theta)][1 - exp[-\beta\zeta)] + K(\theta)\zeta\} \\ &\sigma_\infty(\theta) = \sigma_\infty(\theta_0)[1 - \omega(\theta - \theta_0)] \\ &\sigma_y(\theta) = \sigma_y(\theta_0)[1 - \omega(\theta - \theta_0)] \\ &K(\theta) = K(\theta_0)[1 - \omega(\theta - \theta_0)] \end{aligned} \tag{7.145}$$

With the specified choice of the yield function, we can further obtain the explicit form of the evolution equations for internal variables by appealing to the principle of maximum plastic dissipation. We thus obtain the evolution equation for plastic entropy:

$$\begin{aligned} \dot{s}^p := \dot{\gamma}\frac{\partial \phi}{\partial \theta} &= \dot{\gamma}\sqrt{\frac{2}{3}}\frac{\partial}{\partial \theta}[\sigma_y(\theta) - \hat{q}(\zeta, \theta)] \\ &= \dot{\gamma}\sqrt{\tfrac{2}{3}}\omega[\sigma_y(\theta_0) - \hat{q}(\zeta, \theta_0)] \end{aligned} \tag{7.146}$$

the evolution equation of the isotropic hardening variable:

$$\dot{\zeta} := \dot{\gamma}\frac{\partial \phi}{\partial q} = \sqrt{\frac{2}{3}}\dot{\gamma} \tag{7.147}$$

---

[6] Due to isotropy, the Kirchhoff stress tensor shares the same principal vectors with the elastic left Cauchy–Green strain tensor.

and the evolution equation for the elastic left Cauchy–Green strain tensor:

$$\dot{\mathbf{B}}^e - \mathbf{L}\mathbf{B}^e - \mathbf{B}^e\mathbf{L}^T = -2\dot{\gamma}(\sum_{i=1}^{3} \frac{\partial \phi}{\partial \tau_i} \mathbf{m}_i \otimes \mathbf{m}_i)\, \mathbf{B}^e \tag{7.148}$$

7.3.3 Operator split solution method for 3D thermoplasticity

Having defined all the ingredients of the coupled model of 3D thermoplasticity, we briefly review in this section the corresponding solution procedure based upon the finite element method and an implicit time-integration scheme (for more details see [9]). In order to provide the optimal discrete approximation by finite elements, we can use the incompatible mode method capable of balancing the order of polynomials used for strain field representation with the temperature field representation. In such a solution procedure of the coupled thermoplasticity problem, combining the finite element approximation and time-integration scheme, we will employ the operator split method separating the mechanical and thermal sub-problems with either isothermal or adiabatic split. Each of sub-problems is further split into a global phase for computing either the nodal displacements or temperatures at chosen instant of time and a local problem for computing the internal variables at Gauss quadrature points at the same moment in time. It is important to clarify that the internal variable evolution can be triggered both in mechanical and in thermal phase of the operator split computations. The main steps for operator split computation remain the same both for isothermal and adiabatic split, in that we seek the nodal values of temperature in the thermal phase and the nodal displacements and velocities in the mechanical phase. The temperature will also evolve in mechanical phase for the adiabatic split, but such temperature evolution remains only local computation, carried out at each Gauss point; this can be written:

$$\begin{array}{ll}
1 : \text{Mecanical phase} + \text{adiabatic loading} & 2 : \text{Heat conduction} \\
\dot{\boldsymbol{\varphi}} = \mathbf{v} & \dot{\boldsymbol{\varphi}} = \mathbf{0} \\
\rho\dot{\mathbf{v}} = \boldsymbol{\tau}\boldsymbol{\nabla}^{\varphi} + \mathbf{b} & \dot{\mathbf{v}} = \mathbf{0} \\
\theta\dot{s}^e = 0 & \theta\dot{s}^e = D^p_{meca} - \mathbf{q}_0 \cdot \boldsymbol{\nabla}^{\varphi} + r \\
\theta\dot{s}^p = 0 & \dot{s}^p = \dot{\gamma}\frac{\partial\phi}{\partial\theta}
\end{array} \tag{7.149}$$

It is important to note that each of two phases of the operator split method can also contain the corresponding evolution of the internal variables computed from:[7]

[7] The first of these evolution equations is equivalent to the one in (7.148).

$$\mathbf{L}^p = \dot{\gamma}\mathbf{C}^{e,-1}\frac{\partial \phi}{\partial \mathbf{S}} \; ; \; \dot{\zeta} = \dot{\gamma}\frac{\partial \phi}{\partial q} \tag{7.150}$$

The plastic multiplier in these equations is picked up in agreement with the plastic admissibility of stress, in order to satisfy the corresponding loading/unloading conditions: $\dot{\gamma} \geq 0 \,; \phi \leq 0 \,; \dot{\gamma}\phi = 0$. We note that the yield criterion activation and resulting internal variable computations can be triggered either by a modification of the displacement field or by modification of the temperature field, since either strain or temperature change can affect the stress value and the plasticity threshold. By using the one-step implicit time-integration schemes, the thermodynamical coupling in finite strain 3D thermoplasticity can be reduced to:

Central problem of 3D thermoplasticity

Given: state variables at time t_n, $\{\varphi_n, \theta_n, \mathbf{B}^e_n, \zeta_n, s^p_n\}$

Find: the corresponding values at t_{n+1}, resulting with plastically admissible stress which will verify the equations of motion and energy balance equation:

$$\begin{aligned} 0 = G_M(\varphi_{n+1}, \theta_{n+1}, \mathbf{B}^e_{n+1}, \zeta_{n+1}, s^p_{n+1}; \mathbf{w}) := \textstyle\int_\Omega (\mathbf{w} \cdot \rho \ddot{\varphi}_{n+1} \\ + \tfrac{1}{2}(\nabla^\varphi \mathbf{w} + \nabla^\varphi \mathbf{w}^T) \cdot \boldsymbol{\tau}_{n+1} - \mathbf{w} \cdot \mathbf{b}_{n+1})\, dV - \textstyle\int_{\Gamma_\sigma} \mathbf{w} \cdot \bar{\mathbf{t}}_{n+1}\, dA \,; \\ \phi(\hat{\boldsymbol{\tau}}(\lambda^e_{i,n+1}, \theta_{n+1}), \hat{q}(\theta_{n+1}, \zeta_{n+1})) \leq 0 \; ; \; \dot{\gamma}_{n+1} \geq 0 \; ; \; \dot{\gamma}_{n+1}\phi_{n+1} = 0 \end{aligned} \tag{7.151}$$

$$\begin{aligned} 0 = G_T(\varphi_{n+1}, \theta_{n+1}, \mathbf{B}^e_{n+1}, \zeta_{n+1}, s^p_{n+1}; \vartheta) := \textstyle\int_\Omega \vartheta(\theta_{n+1}\dot{s}^e_{n+1} \\ \underbrace{-\boldsymbol{\tau}_{n+1} \cdot \dot{\mathbf{D}}^p_{n+1} - q_{n+1}\dot{\zeta}_{n+1}}_{-D^p_{meca,n+1}} - r_{n+1} + div[\mathbf{q}_{0,n+1}]\, dV \,; \\ \phi(\hat{\boldsymbol{\tau}}(\lambda^e_{i,n+1}, \theta_{n+1}), \hat{q}(\theta_{n+1}, \zeta_{n+1})) \leq 0 \; ; \; \dot{\gamma}_{n+1} \geq 0 \; ; \; \dot{\gamma}_{n+1}\phi_{n+1} = 0 \end{aligned} \tag{7.152}$$

The main steps of the operator split solution method for thermodynamical coupling in large strain 3D thermoplasticity can be stated as follows:

1. We start by the mechanical phase of the coupled problem, with the best iterative guess for the displacement field $\mathbf{d}^{(i)}_{n+1}$, which allows us to compute the corresponding values of state variables for elastic trial step; namely, first we can write the trial elastic value of the left Cauchy–Green strain tensor:

$$\mathbf{B}^{e,trial}_{n+1} = \tilde{\mathbf{f}}_{n+1}\,\mathbf{B}^e_n\,\tilde{\mathbf{f}}^T_{n+1} \; ; \; \mathbf{f}_{n+1} = \mathbf{F}_{n+1}\mathbf{F}^{-1}_n \tag{7.153}$$

2. We can then compute the spectral decomposition of this strain tensor to obtain the trial values of elastic principal stretches:

$$\begin{aligned} [\mathbf{B}^{e,trial}_{n+1} - (\lambda^{e,trial}_{i,n+1})^2\mathbf{I}]\mathbf{m}^{trial}_{i,n+1} = \mathbf{0} \\ \Rightarrow \mathbf{B}^{e,trial}_{n+1} = \textstyle\sum_{i=1}^3 (\lambda^{e,trial}_{i,n+1})^2 \mathbf{m}^{trial}_{i,n+1} \otimes \mathbf{m}^{trial}_{i,n+1} \end{aligned}$$

$$\begin{aligned}
\mathbf{m}_{i,n+1}^{trial} \otimes \mathbf{m}_{i,n+1}^{trial} &= \frac{1}{(\lambda_{j,n+1}^{e,trial})^2-(\lambda_{i,n+1}^{e,trial})^2}[\mathbf{B}_{n+1}^{e,trial} - (\lambda_{j,n+1}^{e,trial})^2\mathbf{I}] \\
&\otimes \frac{1}{(\lambda_{k,n+1}^{e,trial})^2-(\lambda_{i,n+1}^{e,trial})^2}[\mathbf{B}_{n+1}^{e,trial} - (\lambda_{k,n+1}^{e,trial})^2\mathbf{I}] \; ; \\
i = 1,2,3 \; ; \; j &= 1+mod(3,i) \; ; \; k = 1+mod(3,j)
\end{aligned} \tag{7.154}$$

3. The corresponding trial values of the Kirchhoff stress tensor can be obtained from the invariant form of the free-energy potential, written in terms of elastic principal stretches:

$$\begin{aligned}
\tau_{i,n+1}^{trial} &= \lambda_{i,n+1}^{e,trial}\frac{\partial\psi}{\partial\lambda_{i,n+1}^{e,trial}} + p_{\theta,n+1}\mathbf{I} \\
p_{\theta,n+1} &= -3\alpha(\lambda+\tfrac{2}{3}\mu)(\theta_{n+1}-\theta_n)\frac{1-(ln\lambda_{1,n+1}^{e,trial}+ln\lambda_{2,n+1}^{e,trial}+ln\lambda_{3,n+1}^{e,trial})}{\lambda_{1,n+1}^{e,trial}\lambda_{2,n+1}^{e,trial}\lambda_{3,n+1}^{e,trial}} \\
\lambda_{i,n+1}^{e,trial}\frac{\partial\psi}{\partial\lambda_{i,n+1}^{e,trial}} &= (\lambda+2\mu)ln\lambda_{i,n+1}^{e,trial} + \lambda(ln\lambda_{i,n+1}^{e,trial}+ln\lambda_{i,n+1}^{e,trial}) \\
\boldsymbol{\tau}_{n+1}^{trial} &= \textstyle\sum_{i=1}^{3}\tau_{i,n+1}^{trial}\mathbf{m}_{i,n+1}^{trial}\otimes\mathbf{m}_{i,n+1}^{trial}
\end{aligned} \tag{7.155}$$

4. The intermediate temperature at fixed entropy can be obtained according to:

$$\tilde{\theta}_n = \theta_n exp\{[s_n^e + 3\alpha(\lambda+\frac{2}{3}\mu)\frac{(ln\lambda_{1,n+1}^{e,trial}+ln\lambda_{2,n+1}^{e,trial}+ln\lambda_{3,n+1}^{e,trial})}{\lambda_{1,n+1}^{e,trial}\lambda_{2,n+1}^{e,trial}\lambda_{3,n+1}^{e,trial}}]/c\} \tag{7.156}$$

5. Which will then allow us to check the trial value of the yield criterion:

$$\begin{aligned}
\phi_{n+1}^{trial} &= \sqrt{\tfrac{2}{3}}[(\tau_{1,n+1}^{trial})^2 + (\tau_{2,n+1}^{trial})^2 + (\tau_{3,n+1}^{trial})^2 - \tau_{1,n+1}^{trial}\tau_{2,n+1}^{trial} \\
&-\tau_{2,n+1}^{trial}\tau_{3,n+1}^{trial} - \tau_{3,n+1}^{trial}\tau_{1,n+1}^{trial}] - \sqrt{\tfrac{2}{3}}[\sigma_y(\tilde{\theta}_n) - q_n]
\end{aligned} \tag{7.157}$$

IF ($\phi_{n+1}^{trial} \leq 0$) THEN

the step remains elastic, and the trial values can be accepted as final, resulting with:

$$\boldsymbol{\tau}_{n+1} = \boldsymbol{\tau}_{n+1}^{trial} \; ; \; \boldsymbol{C}_{n+1}^{ep} = \boldsymbol{C} \tag{7.158}$$

ELSE

the step is plastic, and we ought to correct the trial values:

6. Carry out the return mapping algorithm computations in principal axes, in order to obtain the final values of internal variables and plastically admissible stress:

$$\hat{\phi}(\gamma_{n+1}) = 0 \;\Rightarrow\; \bar{\gamma}_{n+1}$$

$$\zeta_{n+1} = \zeta_n + \sqrt{\tfrac{2}{3}}\bar{\gamma}_{n+1} \;\Rightarrow\; q_{n+1} = q_n - K\zeta_{n+1}$$

$$\epsilon_{i,n+1}^e = \epsilon_{i,n+1}^{e,trial} - \bar{\gamma}_{n+1}\frac{\partial\phi}{\partial\tau_{i,n+1}} \;\Rightarrow\; \lambda_{i,n+1}^e = exp[-\bar{\gamma}_{n+1}\frac{\partial\phi}{\partial\tau_{i,n+1}}]\lambda_{i,n+1}^{e,trial}$$

$$\tau_{i,n+1} = (\lambda + 2\mu) ln\lambda^e_{i,n+1} + \lambda(ln\lambda^e_{j,n+1} + ln\lambda^e_{k,n+1})$$
$$p_{\theta,n+1} = -3\alpha(\lambda + \tfrac{2}{3}\mu)(\tilde{\theta}_n - \theta_n)\frac{1-(ln\lambda^e_{1,n+1}+ln\lambda^e_{2,n+1}+ln\lambda^e_{3,n+1})}{\lambda^e_{1,n+1}\lambda^e_{2,n+1}\lambda^e_{3,n+1}} \qquad (7.159)$$
$$\boldsymbol{\tau}_{n+1} = \textstyle\sum_{i=1}^{3} \tau_{i,n+1}\mathbf{m}^{trial}_{i,n+1} \otimes \mathbf{m}^{trial}_{i,n+1} + p_{\theta,n+1}\mathbf{I}$$

7. Compute the consistent tangent elastoplasticity tensor:

$$\begin{aligned}\boldsymbol{\mathcal{C}}^{ep}_{n+1} &= \textstyle\sum_{i=1}^{3}\sum_{j=1}^{3} \underbrace{\frac{\partial \tau_{i,n+1}}{\partial \epsilon^e_{j,n+1}}}_{\hat{\mathcal{C}}^{ep}_{ij}} (\mathbf{m}^{trial}_{i,n+1} \otimes \mathbf{m}^{trial}_{i,n+1}) \\ &\otimes(\mathbf{m}^{trial}_{j,n+1} \otimes \mathbf{m}^{trial}_{j,n+1}) + \textstyle\sum_{i=1}^{3} 2\tau_{i,n+1}\boldsymbol{\mathcal{G}}^{trial}_{i,n+1}\end{aligned} \qquad (7.160)$$

where:

$$\begin{aligned}\boldsymbol{\mathcal{G}}^{trial}_{i,n+1} &= \tfrac{1}{d_i}\{\boldsymbol{\mathcal{I}}_{B^e} - \mathbf{B}^{e,trial}_{n+1} \otimes \mathbf{B}^{e,trial}_{n+1} - I_3(\lambda^e_{i,n+1})^{-2} \\ &[\boldsymbol{\mathcal{I}} - (\mathbf{I} - \mathbf{m}^{e,trial}_{i,n+1} \otimes \mathbf{m}^{e,trial}_{i,n+1}) \otimes (\mathbf{I} - \mathbf{m}^{e,trial}_{i,n+1} \otimes \mathbf{m}^{e,trial}_{i,n+1})] \\ &+(\lambda^{e,trial}_{i,n+1})^2[\mathbf{B}^e_{n+1} \otimes (\mathbf{m}^{e,trial}_{i,n+1} \otimes \mathbf{m}^{e,trial}_{i,n+1}) + (\mathbf{m}^{e,trial}_{i,n+1} \otimes \mathbf{m}^{e,trial}_{i,n+1}) \otimes \mathbf{B}^e_{n+1}]\} \\ &+(I_1 - 4(\lambda^{e,trial}_{i,n+1})^2)(\mathbf{m}^{e,trial}_{i,n+1} \otimes \mathbf{m}^{e,trial}_{i,n+1}) \otimes (\mathbf{m}^{e,trial}_{i,n+1} \otimes \mathbf{m}^{e,trial}_{i,n+1}) \;;\end{aligned}$$

.

$$\boldsymbol{\mathcal{I}}_{B^e} = \tfrac{1}{2}(B^e_{ik}B^e_{jl} + B^e_{il}B^e_{jk})\mathbf{e}_i \otimes \mathbf{e}_j \otimes \mathbf{e}_k \otimes \mathbf{e}_l \;;$$
$$\boldsymbol{\mathcal{I}} = \tfrac{1}{2}(\delta_{ik}\delta_{jl} + \delta_{il}\delta_{jk})\mathbf{e}_i \otimes \mathbf{e}_j \otimes \mathbf{e}_k \otimes \mathbf{e}_l \;;$$
$$d_i = [(\lambda^{e,trial}_{i,n+1})^2 - (\lambda^{e,trial}_{j,n+1})^2][(\lambda^{e,trial}_{i,n+1})^2 - (\lambda^{e,trial}_{k,n+1})^2] \;;$$
$$I_1 = tr[\mathbf{B}^e_{n+1}] \;;\; I_3 = (J^e_{n+1})^2$$

compute the adiabatic tangent tensor:

$$\boldsymbol{\mathcal{C}}^{ep}_{ad,n+1} = \boldsymbol{\mathcal{C}}^{ep}_{n+1} + \frac{1}{c}\beta\mathbf{I} \otimes \beta\mathbf{I} \qquad (7.161)$$

8. The end of mechanical phase–compute the new iterative value of displacement:

$$\textstyle\sum_a \mathsf{w}^u_a(\sum_b \mathsf{K}^u_{ab}\mathsf{u}^u_b = \mathsf{f}^{u,ext}_a - \mathsf{f}^{u,int}_a)$$
$$\mathsf{w}_a \cdot \mathsf{f}^{u,int}_a = \textstyle\int_\Omega \tfrac{1}{2}(\nabla^\varphi \mathsf{w} + \nabla^\varphi \mathbf{w}^T) \cdot \boldsymbol{\tau}\, dV$$

$$\mathsf{w}_a^u \cdot \mathsf{K}_{ab}^u \mathsf{u}_b^u = \int_\Omega \tfrac{1}{2}(\nabla^\varphi \mathbf{w} + \nabla^\varphi \mathbf{w}^T) \cdot \boldsymbol{\mathcal{C}}_{ad,n+1}^{ep} \tfrac{1}{2}(\nabla^\varphi \mathbf{u} + \nabla^\varphi \mathbf{u}^T)\, dV$$
$$+ \int_\Omega \tfrac{1}{2}[(\nabla^\varphi \mathbf{w} + \nabla^\varphi \mathbf{w}^T)\nabla^\varphi \mathbf{u} + (\nabla^\varphi \mathbf{w} + \nabla^\varphi \mathbf{w}^T)\nabla^\varphi \mathbf{u}^T] \cdot \boldsymbol{\tau}\, dV \tag{7.162}$$

.

$$\Rightarrow \mathsf{d}_{n+1}^u \longleftarrow \mathsf{d}_{n+1}^u + \mathsf{u}_{n+1}^u$$

9. Carry out the temperature computation in thermal phase:

$$\begin{aligned}
&\textstyle\sum_a \mathsf{w}_a^\theta (\sum_b \mathsf{K}_{ab}^\theta\, \mathsf{u}_b^\theta = \mathsf{f}_{a,n+1}^{\theta,ext} - \mathsf{f}_{a,n+1}^{\theta,int}) \\
&\boldsymbol{\vartheta}_{a,n+1} \mathsf{f}_{a,n+1}^{\theta,int} = \int_\Omega (\vartheta D_{meca,n+1}^p + \boldsymbol{\nabla}^\varphi \vartheta \cdot \mathbf{q}_{0,n+1})\, dV \\
&\textstyle\sum_a \sum_b \mathsf{w}_a^\theta \mathsf{K}_{ab,n+1}^{theta} \mathsf{u}_{b,n+1}^{theta} = \int_\Omega [\vartheta (\frac{\partial s_{n+1}^e}{\partial \theta} + \frac{\partial D_{meca,n+1}^p}{\partial \theta})\theta \\
&\quad - \boldsymbol{\nabla}^\varphi \vartheta \cdot k \boldsymbol{\nabla}^\varphi \theta_{n+1}]\, dV
\end{aligned} \tag{7.163}$$

.

$$\Rightarrow \mathsf{d}_{n+1}^\theta \longleftarrow \mathsf{d}_{n+1}^\theta + \mathsf{u}_{n+1}^\theta$$

10. Check plasticity criterion for a change of temperature:

$$\begin{aligned}
\phi_{n+1} = &\sqrt{\tfrac{2}{3}}[(\tau_{1,n+1})^2 + (\tau_{2,n+1})^2 + (\tau_{3,n+1})^2 - \tau_{1,n+1}^{trial}\tau_{2,n+1} \\
&- \tau_{2,n+1}\tau_{3,n+1} - \tau_{3,n+1}\tau_{1,n+1}] - \sqrt{\tfrac{2}{3}}[\sigma_y(\tilde{\theta}_n) - q_n]
\end{aligned} \tag{7.164}$$

IF $(\phi_{n+1} > 0)$

11. Carry out the return mapping algorithm computations to obtain the new values of internal variables and plastically admissible stress values:

$$\begin{aligned}
&\hat{\phi}(\gamma_{n+1}) = 0 \implies \bar{\gamma}_{n+1} \\
&\zeta_{n+1} = \zeta_n + \sqrt{\tfrac{2}{3}}\bar{\gamma}_{n+1} \;\Rightarrow\; q_{n+1} = q_n - K\zeta_{n+1} \\
&\epsilon_{i,n+1}^e = \epsilon_{i,n+1}^{e,trial} - \bar{\gamma}_{n+1}\frac{\partial \phi}{\partial \tau_{i,n+1}} \;\Rightarrow\; \lambda_{i,n+1}^e = exp[-\bar{\gamma}_{n+1}\frac{\partial \phi}{\partial \tau_{i,n+1}}]\lambda_{i,n+1}^{e,trial} \\
&\tau_{i,n+1} = (\lambda + 2\mu) ln\lambda_{i,n+1}^e + \lambda(ln\lambda_{j,n+1}^e + ln\lambda_{k,n+1}^e) \\
&p_{\theta,n+1} = -3\alpha(\lambda + \tfrac{2}{3}\mu)(\tilde{\theta}_n - \theta_n)\frac{1-(ln\lambda_{1,n+1}^e + ln\lambda_{2,n+1}^e + ln\lambda_{3,n+1}^e)}{\lambda_{1,n+1}^e \lambda_{2,n+1}^e \lambda_{3,n+1}^e} \\
&\boldsymbol{\tau}_{n+1} = \textstyle\sum_{i=1}^3 \tau_{i,n+1}\mathbf{m}_{i,n+1}^{trial} \otimes \mathbf{m}_{i,n+1}^{trial} + p_{\theta,n+1}\mathbf{I}
\end{aligned} \tag{7.165}$$

7.3.4 Numerical example: thermodynamics coupling in 3D plasticity

7.3.4.1 Thermoplastic cylinder under internal pressure

This illustrative example of thermodynamical coupling in large strain thermoplasticity was first proposed in [5]. It considers a thick cylinder with internal radius $a_0 = 100$ [mm] and external radius $b_0 = 200$ [mm], submitted to internal pressure. We consider a finite element model constructed with a structured mesh of ten elements $Q4$, and impose the same boundary conditions as for the previously studied thermoelastic case shown in Figure 7.3. The constitutive behavior of the cylinder is described with 3D large strain thermoplasticity model, where the temperature increase reduces the value of yield stress. The chosen mechanical and thermal properties of the cylinder are given in table below:

Bulk modulus	$\kappa = 58333$	N/mm^2
Shear modulus	$\mu = 26926$	N/mm^2
Mass density	$\rho = 2.7 \times 10^{-9}$	Ns^2/mm^4
Yield stress	$\sigma_y = 70$	N/mm^2
Hardening modulus	$K_{iso} = 210$	N/mm^2
Conductivity coef.	$k = 150$	N/sK
Heat capacity coef.	$c = 0.9 \times 10^9$	mm^2/s^2K
Coef. of thermal expansion	$\alpha = 23.8 \times 10^{-6}$	K^{-1}
Coef. of thermal softening	$\omega = 3 \times 10^{-4}$	K^{-1}

Thermoplastic cylinder under internal pressure: mechanical and thermal properties

The computation is carried out under displacement control, until reaching the deformed configuration where the internal radius of the cylinder is expanded to $a = 2.3\,a_0 = 230$ mm. In the final deformed configuration, a large part of the cylinder becomes plastic, and the plastic dissipation provides much more important than heat source from structural heating. The computations are performed by using two different time-integration schemes: the backward Euler scheme with 100 time steps and the trapezoidal rule with ten time steps only. The results obtained for time-evolution of pressure (versus imposed displacements) and temperature, presented in Figure 7.6a, are in a very good agreement with the reference results given in [5]. Different rates for load increase are considered, all placed in-between two limit values of time for reaching the final deformed configuration (i.e. between $t \to 0$ and $t \to \infty$); as illustrated in Figure 7.6b, the temperature profile is quite sensitive to the chosen loading rate, with a slower load increase that leaves more time for heat diffusion to produce a more homogeneous temperature profile. We also note that the temperature profile can be connected to plastic dissipation.

7.3.4.2 Necking of a steel specimen

In this example, we present the results for numerical simulations of simple tension test on a 2D and 3D models of steel specimen. The computations are carried out under displacement control and go well beyond the peak response

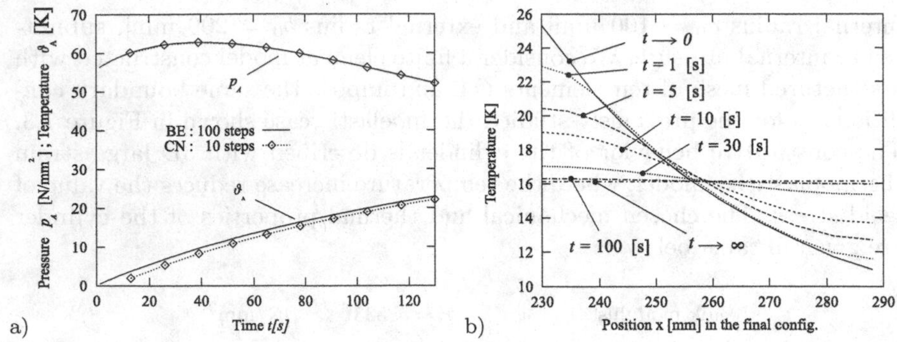

Fig. 7.6 Thermoplastic cylinder: **(a)** time-history of pressure and temperature at inner radius **(b)** temperature space-distribution in final deformed configuration as a function of rate of load increase.

into softening regime. The latter requires a viscoplasticity model (see next chapter) in order to ensure the response objectivity with respect to the chosen mesh. We choose the value for viscosity coefficient $\eta = 50$. The chosen values for other material parameters are given in table below:

Bulk modulus	$\kappa = 164206$	N/mm^2
Shear modulus	$\mu = 801938$	N/mm^2
Mass density	$\rho = 7.8 \times 10^{-9}$	Ns^2/mm^4
Yield stress	$\sigma_y = 450$	N/mm^2
Saturation limit	$\sigma_\infty = 715$	N/mm^2
Hardening modulus	$K_{iso} = 129.24$	N/mm^2
Coef. of saturation (exp.)	$\beta = 16.93$	
Coef. of conductivity	$k = 45$	N/sK
Heat capacity coef.	$c = 0.46 \times 10^9$	mm^2/s^{2}K
Coef. of thermal expansion	$\alpha = 1 \times 10^{-5}$	K^{-1}
Coef. of thermal softening	$\omega = 0.002$	K^{-1}

Necking of specimen in tension test: mechanical and thermal properties

The specimen length is $l = 53.334$ [mm], its width $b = 12.826$ [mm]. The analysis is carried out for 2D plane strain model with unit thickness. The finite element model considers only a quarter of the specimen (see Figure 7.7), represented with 200 $Q4$ isoparametric elements with incompatible modes. The computation is carried out under displacement control until reaching the final value of imposed displacement $\bar{u} = 8$ mm, with the chosen rate of load increase $\dot{u} = 1$ mm/s. In the final deformed configuration we obtain

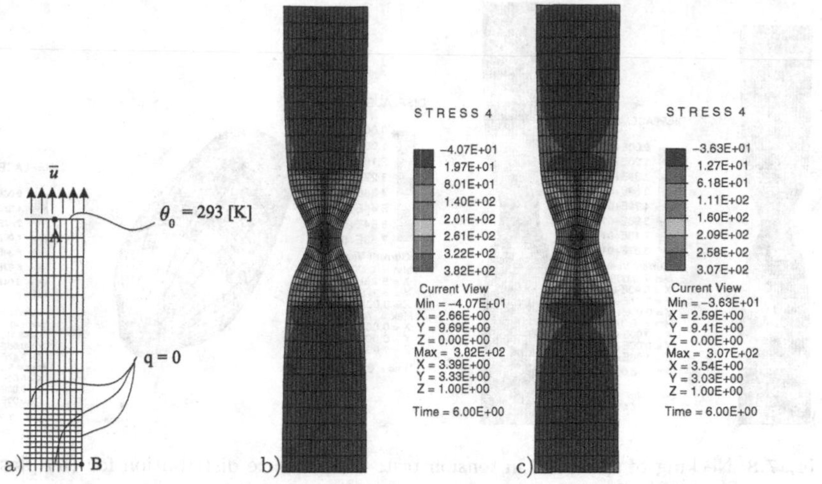

Fig. 7.7 Necking of specimen in tension test: **(a)** mesh and boundary conditions **(b)** shear stress in plasticity **(c)** shear stress in viscoplasticity.

very much visible shear bands inclined with respect to the loading direction as typical of metallic materials, similar results are obtained with either plasticity or viscoplasticity model with von Mises temperature dependent yield criterion. It is important to note that the present case of thermodynamical coupling makes it unnecessary to introduce any (geometric or material) perturbation in order to trigger a single shear band creation and related necking phenomena; in fact, the thermal softening for metallic materials and non-homogeneous temperature filed are sufficient to generate the same final result, with the peak value of temperature that occurs in the center of specimen (see Figure 7.8). We also show in Figure 7.8 that the same kind of observation can be made for computations performed with a 3D model. Those results are obtained for a cylindrical specimen of length $l = 53.334$ [mm], radius $r = 6.413$ [mm], and the same mechanical and thermal properties as in 2D case; the finite element model is constructed for a 1/8 of the specimen with a total of 960 8-node solid elements with incompatible modes.

7.4 Thermomechanical coupling in contact

In this section, we will outline the main steps for constructing the solution method for thermomechanical coupling for contact problems (see [265], [64] or [163] for details). We start by a simple contact problem in small displacement setting, where the contact surface is not modified, in any significant manner, when passing from initial to deformed configuration, which allows to focus

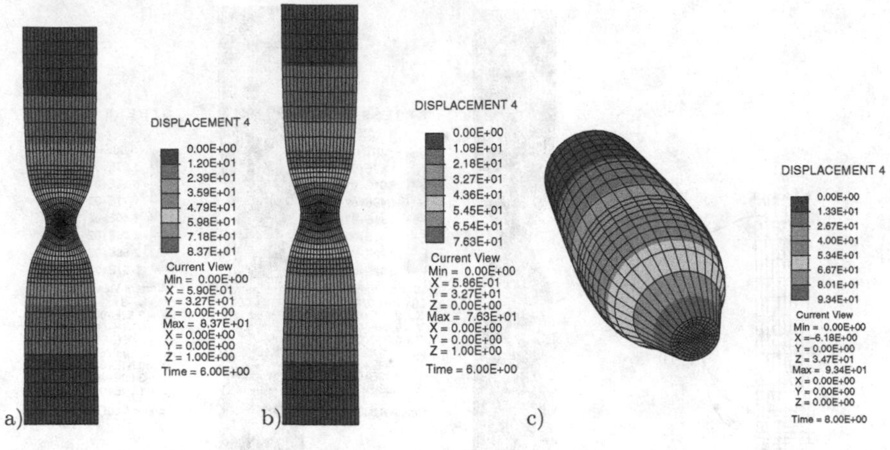

Fig. 7.8 Necking of specimen in tension test: temperature distribution for **(a)** plasticity **(b)** viscoplasticity **(c)** model 3D.

mostly upon the thermomechanical aspects of contact. The most general case with large frictional sliding is only visited briefly at the end of this section. Within the small displacement framework, we consider that 2 nodes in contact in the initial configuration will remain in contact in the deformed configuration. We will consider the frictionless contact case, where the only behavior to define concerns the normal interface. For a thermodynamical contact problem, we define not only contact gap in displacement but also in temperature:

$$g_u = u_2 - u_1 \; ; \; g_\theta = \theta_2 - \theta_1 \tag{7.166}$$

All the values of mechanical and thermal coefficient can be expressed per unit area of contact surface; the mechanical response is described in terms of its compliance:

$$g_u = 1.363\, \delta [-ln(5.589 \frac{p}{H_c}]^{1/2} \tag{7.167}$$

where p is the contact pressure, δ is the standard variation of asperity profiles with respect to the contact surface and H_c is the toughness coefficient of the contact surface. The power law for contact can be used to account for the increased resistance to crushing of asperities (e.g. see [141]).
The conductivity coefficient of asperities is also defined as power law, according to:

$$k_c = \frac{125 k \tilde{m}}{\delta} [\frac{p}{c_1} (1.6177 \frac{10^6 \delta}{\tilde{m}})^{-c_2}]^{0.95/(1+0.0711 c_2)} \tag{7.168}$$

where k is the conductivity coefficient of bodies in contact, $\tilde{m}$ is the mean slope of asperities, and c_1 and c_2 the parameters which can be obtained from micro-hardness tests. If the space between the asperities is filled-in with gas, the conductivity coefficient ought to be modified, and the same holds for heat capacity coefficient. We can also easily account for heat radiation effect between two surfaces in contact, which remains proportional to the difference of temperatures (to the power four), with the coefficient of proportionality equal to Stefan–Boltzmann constant.

The same kind of contact model can be implemented in a more general framework, considering bilateral contact of solids undergoing large deformations, displacements and frictional sliding on the contact boundary. The motion of each of two such bodies in contact can be described by position vector $\varphi^{(\alpha)}$; the gap function g_c is defined with the orthogonal projection of each body onto the contact surface Γ_c, such that $g_c(\mathbf{x}, t) \leq 0$. The first principle of thermodynamics for two bodies in the presence of contact can be written:

$$
\begin{aligned}
\dot{E} &= P + Q \\
\dot{E} &= \int_{\Omega^{(1)} \bigcup \Omega^{(2)}} \tfrac{d}{dt}[\dot{E}^{(\alpha)} + \tfrac{1}{2}\rho^{(\alpha)} \mathbf{v}^{(\alpha)} \cdot \mathbf{v}^{(\alpha)}]\, dV + \int_{\Gamma_c} \dot{e}_c \, dA \\
P &= \int_{\Omega^{(1)} \bigcup \Omega^{(2)}} \mathbf{b}^{(\alpha)} \cdot \mathbf{v}^{(\alpha)}\, dV + \int_{\Gamma_\sigma^{(1)} \bigcup \Gamma_\sigma^{(2)}} \bar{\mathbf{t}}^{(\alpha)} \cdot \mathbf{v}^{(\alpha)}\, dA \\
&\quad - \int_{\Gamma_c} \mathbf{t} \cdot (\varphi^{(2)} - \varphi^{(1)})\, dA \\
Q &= \int_{\Omega^{(1)} \bigcup \Omega^{(2)}} r^{(\alpha)}\, dV - \int_{\Gamma_q^{(1)} \bigcup \Gamma_q^{(2)}} \mathbf{q} \cdot \mathbf{n}^{(\alpha)}\, dA + \int_{\Gamma_c} (q_c^{(1)} + q_c^{(2)})\, dA
\end{aligned}
$$

where e_c is the internal energy density per unit area of contact surface, $\mathbf{t}$ is the stress vector acting on the contact surface and $q_c^{(\alpha)}$ is the heat flux in the contact interface. The latter is by convention considered positive, if $q_c^{(\alpha)}$ will result in the heat flow from the body towards the contact interface. By considering the limit case with shrinking contact surface, we can also obtain the local form of the energy balance equation:

$$
\dot{e}_c = -\mathbf{t} \cdot (\dot{\varphi}^{(1)} - \dot{\varphi}^{(2)}) + q_c^{(1)} + q_c^{(2)} \tag{7.169}
$$

The second principle of thermodynamics will further restrict the thermomechanical exchange in the presence of contact; we can either provide the global form of the second principle:

$$
\begin{aligned}
\dot{S} &\geq - \int_{\Gamma_q^{(1)} \bigcup \Gamma_q^{(2)}} \frac{\mathbf{q} \cdot \mathbf{n}^{(\alpha)}}{\theta}\, dA + \int_{\Gamma_c} \left(\frac{q_c^{(1)}}{\theta^{(1)}} + \frac{q_c^{(2)}}{\theta^{(2)}}\right) dA \\
S &= \int_{\Omega^{(1)} \bigcup \Omega^{(2)}} s^{(\alpha)}\, dV + \int_{\Gamma_c} s_c \, dA
\end{aligned}
\tag{7.170}
$$

or the corresponding local form applicable to the contact surface:

$$
\dot{s}_c \geq \frac{q_c^{(1)}}{\theta^{(1)}} + \frac{q_c^{(2)}}{\theta^{(2)}} \tag{7.171}
$$

where $\theta^{(1)}$ and $\theta^{(2)}$ are the temperatures on the surface of the body (1) and body (2), respectively, when we approach the body surface from inside. The components of the Piola stress vector, $\mathbf{t} = \mathbf{P}\mathbf{n}$, can be represented in the convective frame with base vectors $\boldsymbol{\tau}_\alpha$ produced by the motion of bodies in contact, which allows us to construct the normal to the surface of contact:

$$\boldsymbol{\tau}_\alpha = \mathbf{F}\mathbf{t}_\alpha \; ; \; \boldsymbol{\nu} = \frac{\boldsymbol{\tau}_1 \times \boldsymbol{\tau}_2}{\| \boldsymbol{\tau}_1 \times \boldsymbol{\tau}_2 \|} \tag{7.172}$$

In the chosen reference frame, we can present the stress vector in terms of its components:

$$\mathbf{t} = t_N \boldsymbol{\nu} - t_{T_\alpha} \boldsymbol{\tau}_\alpha \tag{7.173}$$

We note that t_N is the contact pressure which will be work-conjugate to penetration g_u, whereas the stress components t_{T_α} are conjugate to corresponding tangential sliding components along the contact surface, which can be computed as:

$$g_{T_\alpha} = \int_{t_c}^{t} \dot{\zeta}_\alpha \, dt \tag{7.174}$$

We will asume the additive decomposition of tangential sliding into elastic and irreversible (or plastic) part according to:

$$g_{T_\alpha} = g^e_{T_\alpha} + g^p_{T_\alpha} \tag{7.175}$$

We will also assume that the internal energy potential only depends upon the elastic part of the tangential component, leading to: $e(g, g^e_{T_\alpha}, s^e)$. We can further replace the internal energy potential with the free energy potential by appealing to the Legendre transform, which can be written:

$$\psi(g, g_{T_\alpha}, \theta_c) = e(g, g_{T_\alpha}, s^e) - s^e \theta_c \tag{7.176}$$

where θ_c is the temperature on the contact surface, which is in general different from the temperatures $\theta^{(1)}$ or $\theta^{(2)}$ for two bodies in contact. From the second principle of thermodynamics we can obtain the corresponding expression for dissipation:

$$\begin{aligned} 0 \leq D := (t_N - \tfrac{\partial \psi}{\partial g})\dot{g} - (s^e + \tfrac{\partial \psi}{\partial \theta_c})\dot{\theta}_c + (t_{T_\alpha} - \tfrac{\partial \psi}{\partial g_{T_\alpha}})\dot{g}_{T_\alpha} + s^p \dot{\theta}_c \\ + t_{T_\alpha}\dot{g}^p_{T_\alpha} + \tfrac{q_c^{(1)}}{\theta^{(1)}}(\theta^{(1)} - \theta_c) + \tfrac{q_c^{(2)}}{\theta^{(2)}}(\theta^{(2)} - \theta_c) \end{aligned} \tag{7.177}$$

The last result allows us to define the constitutive relations for the elastic case with no frictional sliding:

$$t_N = \frac{\partial \psi}{\partial g} \; ; \; t_{T_\alpha} - \frac{\partial \psi}{\partial g_{T_\alpha}} \; ; \; s^e = -\frac{\partial \psi}{\partial \theta_c} \tag{7.178}$$

For example, a simple choice with a quadratic form of free energy potential for thermomechanical coupling in contact, which can be written:

$$\psi(g, g_{T_\alpha}, \theta_c) := \frac{1}{2} E_N g^2 + \frac{1}{2} g_{T_\alpha} C_{\alpha\beta} g_{T_\beta} - \frac{c_c}{2\theta_0}(\theta_c - \theta_0)^2 \tag{7.179}$$

will allow us to obtain the simple linear relations for normal and tangent contact force components with respect to penetration and sliding, respectively:

$$t_N = E_N g \ ; \ t_{T_\alpha} = C_{\alpha\beta} g_{T_\beta} \tag{7.180}$$

as well as the constitutive relation for the elastic entropy:

$$s^e = c_c(\theta - \theta_0) \tag{7.181}$$

By assuming that the constitutive relations remain applicable to the irreversible process with frictional sliding, we can obtain from the second principle the corresponding expression for local dissipation according to:

$$0 < D^f := \underbrace{s^p \dot{\theta}_c}_{D^f_{ther}} + \underbrace{t_{T_\alpha} \dot{g}^p_{T_\alpha}}_{D^f_{meca}} + \underbrace{\frac{q_c^{(1)}}{\theta^{(1)}}(\theta^{(1)} - \theta_c) + \frac{q^{(2)}}{\theta^{(2)}}(\theta^{(2)} - \theta_c)}_{D^f_{cont}} \tag{7.182}$$

The first term in (7.182) is the thermal dissipation, the second the mechanical dissipation of frictional sliding and the last term is the dissipation due to heat conduction process. Each of these terms should remain positive or zero. The requirement of a positive value of dissipation by conduction places the restriction on heat exchange in a heat transfer problem with a non-homogeneous temperature field. The first two dissipation terms, on the other hand, will allow us to obtain the evolution equations for internal variables in thermomechanical frictional contact. This can be done by appealing to the principle of maximum plastic dissipation, where we consider all admissible values of temperatures and contact forces in the sense of the frictional contact criterion:

$$0 \geq \phi(t_N, t_{T_\alpha}, \theta_c) := (t_{T_\alpha} G_{\alpha\beta} t_{T_\beta})^{1,2} - \mu(\theta_c)\, t_N \tag{7.183}$$

with $\mu(\theta_c)$ as the temperature dependent friction coefficient for coupled thermomechanics problems. Thus, we will pick the value of temperature that will maximize the thermal dissipation:

$$\begin{aligned} &\max_{\phi(\cdot)=0} D^{cf}_{ther} \ \Leftrightarrow \ \max_{\dot{\gamma}} \min_{\theta} L^{cf}_{ther}(\cdot, \theta_c, \dot{\gamma}) \ ; \\ &L^{cf}_{ther}(\cdot) = -D^{df}_{ther}(\cdot) + \dot{\gamma}\phi(\cdot) \\ &0 = \frac{\partial L^{cf}_{ther}}{\partial \theta_c} \implies \dot{s}^p := \dot{\gamma}\frac{\partial \phi}{\partial \theta_c} = -\dot{\gamma}\frac{\partial \mu(\theta_c)}{\partial \theta_c} \end{aligned} \tag{7.184}$$

The same principle will select the tangential contact force which will maximize the mechanical dissipation for fixed value of contact pressure:

$$\begin{aligned} &\max_{\phi(\cdot)=0} D^{cf}_{meca} \ \Leftrightarrow \ \max_{\dot{\gamma}} \min_{t_{T_\alpha}} L^{cf}_{meca}(\cdot, t_{T_\alpha}, \dot{\gamma}) \ ; \\ &L^{cf}_{meca} = -D^{df}_{meca}(\cdot) + \dot{\gamma}\phi(\cdot) \\ &0 = \frac{\partial L^{cf}_{meca}}{\partial t_{T_\alpha}} \implies \dot{g}^p_{T_\alpha} = \dot{\gamma}\frac{\partial \phi}{\partial t_{T_\alpha}} \end{aligned} \tag{7.185}$$

These evolution equations for internal variables ought to be accompanied by the loading/unloading conditions which can be written:

$$\dot{\gamma} \geq 0 \; ; \; \phi(\cdot) \leq 0 \; ; \; \dot{\gamma}\phi(\cdot) = 0 \tag{7.186}$$

By exploiting the constitutive equations and the Legendre transform, we can rewrite the first principle of thermodynamics in a reduced form:

$$\theta_c \dot{s}^e = D^{cf}_{meca} + q^{(1)}_c + q^{(2)}_c \tag{7.187}$$

If the heat flux on each surface in contact is proportional to temperature, with the coefficient of proportionality dependent upon the contact pressure, we can rewrite the last expression according to:

$$q^{(1)}_c = \gamma^{(1)}_{c(t_N)}\theta^{(1)} \; ; \; q^{(2)}_c = \gamma_{c(t_N)}\theta^{(2)} \tag{7.188}$$

We could thus write the energy balance in the modified form:

$$c_0 \dot{\theta}_c = D^{cf}_{meca} + \gamma^{(1)}_{c(t_N)}\theta^{(1)} + \gamma_{c(t_N)}\theta^{(2)} \tag{7.189}$$

The numerical solution of thermodynamical coupling in contact is sought by an operator split procedure, based on adiabatic split. Thus, in the first phase we will keep the entropy fixed with:

$$\theta_c \dot{s}^e = 0 \; ; \; \theta_c \dot{s}^p = 0 \tag{7.190}$$

whereas the second phase will reduce to the corresponding heat transfer problem, with additional source related to mechanical dissipation:

$$\theta_c \dot{s}^e = D^{df}_{meca} + q^{(1)}_c + q^{(2)}_c \; ; \; \theta_c \dot{s}^p = \dot{\gamma}\theta_c \frac{\partial \phi}{\partial \theta_c} \tag{7.191}$$

The details of this numerical solution procedure are equivalent to the corresponding one already explained for plasticity, and are thus omitted.

Chapter 8
Geometric and material instabilities

If all the equilibrium states in the course of a particular loading program remain stable, the solution of a nonlinear problem in solid mechanics should not be very complicated to find. Namely, by means of incremental analysis and iterative method of Newton, the solution task for a nonlinear problem is reduced to solving the corresponding linear problems defined through incremental/iterative procedure. Each linear problem of this kind provides the best linear approximation of the true nonlinear problem for the current values of state variables in a given increment and/or iteration. Unfortunately, not all equilibrium states in nonlinear mechanics will always remain stable. Therefore, the main goal of this chapter is to present different causes of instable equilibrium states and present the solution methods for nonlinear mechanics problems in presence of instability.

The instability phenomena imply in general that a small perturbation of loading (mechanical, thermal etc.) can lead to a disproportional amplification in computed response (displacements, deformations and stresses). The instability phenomena can be present in large displacement or large deformation setting of geometrically nonlinear problems, for the case when the geometric part of the stiffness matrix will modify the material part of the stiffness matrix in the way which renders the tangent stiffness matrix singular. This kind of instability is often called geometric instability. However, the instability phenomena can also be present in small displacements and small strains, when the material behavior is characterized by softening. The latter is the kind of constitutive behavior which typically precedes the fracture, where the stress decreases even with increasing values of strain. The instability phenomena of this kind are referred to as material instability. Both material and geometric instability occur together in a number of problems in nonlinear mechanics related to engineering applications. It is thus important to learn how to detect the presence of instabilities and how to solve the problems of this kind, which is the main objective of this chapter.

The outline of the chapter is as follows: we first recall the linear instability problem or buckling of Euler, where displacements and deformations remain small before arriving to critical equilibrium point where geometric instability

A. Ibrahimbegovic, *Nonlinear Solid Mechanics: Theoretical Formulations and Finite Element Solution Methods*, Solid Mechanics and its Applications 160,

problem is produced. We then develop the finite element method solution procedure for buckling problems, which allows to define the instability detection criterion and the buckling load computation. We show that the same detection criterion applies to nonlinear instability problems, where the displacements, rotations and strains can be quite large before arriving at the critical equilibrium state. We also present the method for solving the problems in nonlinear mechanics in presence of instability. First, we discuss the arc-length solution procedure, which allows us to trace the complete force–displacement diagram in presence of instability (including post-buckling response), with no need to precisely locate the critical equilibrium points. We then present the method for direct computation of instability points by Newton's scheme, for the case where precise detection of those points happens to be important.

In the second part of this chapter, we discuss the material instabilities or yet called localization problems. First, we show that the origin of the phenomena of localization is traced to softening constitutive behavior of materials. We recall several methods proposed to deal with the localization phenomena, which are commonly referred to as localization limiters. We give a more detailed presentation of two simple methods among available limiters of localization, which can be used with any standard finite element code (with no need for code modification), employing only the mesh-dependent material parameters. The first of these methods uses the mesh-dependent value of softening modulus to ensure the desired plastic dissipation, whereas the second employs the viscoplastic regularization to achieve the same goal with the time-step–dependent viscosity parameter. We then present the localization limiter that has currently become the favorite choice, which provides a proper reinterpretation of the strain field in the case of softening, by introducing either displacement or deformation discontinuities within the standard finite element approximations. We also present how to enhance such a localization limiter for better representation of the failure state of massive structure, where we find the interaction of two kinds of dissipative mechanisms: the volumetric dissipation in so-called "fracture process zone" and the dissipation at displacement discontinuity. This kind of model is adapted to both plasticity and damage constitutive models.

8.1 Geometric instabilities

8.1.1 Buckling, nonlinear instability and detection criteria

8.1.1.1 Linear instability or buckling

Euler's buckling: is the classical example of (linear) geometric instability, where we consider the stability of a beam submitted to the compressive axial force P and displaced by a perturbation with a small transverse displacement $v(x)$ (see Figure 8.1). We can judge the stability of such an equilibrium state with respect to the perturbation effects; namely, if the beam returns to the original configuration (before perturbation), the equilibrium state is stable, whereas, on the other hand, if the displacement increases in a disproportional manner with respect to the initial perturbation, the equilibrium state is unstable. Somewhere in-between these two is placed so-called critical equilibrium state, which marks the transition from stable to unstable equilibrium states (or vice versa). For the critical equilibrium state, the beam will remain in equilibrium under the critical value of axial force P_{cr}, in the perturbed configuration produced with transverse displacement $v(x)$. Therefore, we can establish the equilibrium equation in this deformed configuration according to:

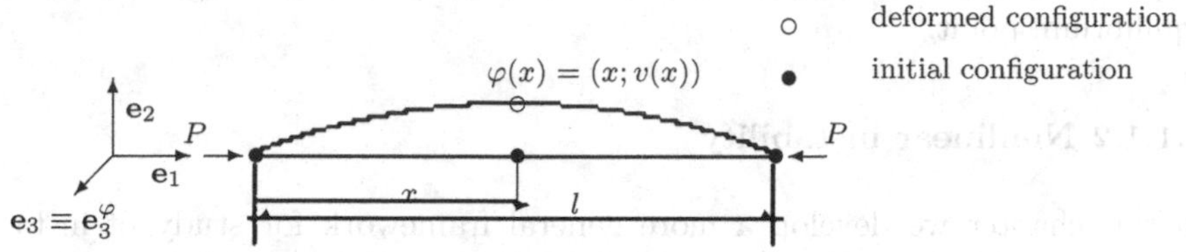

Fig. 8.1 Euler's buckling: initial and deformed configuration of beam under critical force in the deformed configuration produced by perturbation.

$$M(x) = -P_{cr}v(x) \tag{8.1}$$

where $M(x)$ is the bending moment in the beam.

The last equation is the only nonlinear equation (with equilibrium established in the deformed configuration), which we use in this problem. Given that the displacement $v(x)$ is small, we will use linear kinematics equation to compute the curvature of the beam according to: $\kappa(x) = \frac{d^2v(x)}{dx^2}$. Moreover, we will consider linear elastic constitutive equation in terms of Hook's law: $\sigma(x, y) = E\epsilon(x, y)$, along with the classical hypothesis for Euler's beam on plane sections that remain plane leading to a linear variation of strain through the beam thickness $\epsilon(x, y) = y\kappa(x)$; the last two equations jointly lead to the constitutive equation in terms of moment–curvature relation, which can be written:

$$M(x) = EI\kappa(x) \equiv EI\frac{d^2v(x)}{dx^2} \; ; \; M(x) = \int_A y\sigma(x, y)\, dA \; ; \; I = \int_A y^2\, dA$$

where E is Young's modulus and I is the moment of inertia of the beam. By exploiting this equation we can express the equilibrium equation directly in terms of displacement field, whose analytic solution provides the well-known result for Euler buckling load (see Figure 8.1), which can be written:

$$\left.\begin{array}{l} 0 = EI\frac{d^4v(x)}{dx^4} + P\frac{d^2v(x)}{dx^2}\ ; \\ v(0) = v(l) = 0\ ; \\ EI\frac{d^2v(0)}{dx^2} = EI\frac{d^2v(l)}{dx^2} = 0; \end{array}\right\} \Longrightarrow \left\{\begin{array}{l} v_{cr}(x) = \pm \sin\frac{\pi x}{l} \\ P = \pi^2 EI/l^2 \end{array}\right. \qquad (8.2)$$

The critical equilibrium point of this kind is referred to as bifurcation, with a typical symmetry breaking post-bifurcation response. In order to obtain the result for Euler buckling, we combined a linear kinematics equation and linear constitutive equation with a nonlinear equilibrium equation, with the latter being established in the deformed configuration. These are typical ingredients of what we refer to as the linear instability problem, where the displacements, strains and stresses remain small before reaching the critical equilibrium point.

8.1.1.2 Nonlinear instability

In this chapter we develop a more general framework for study of instability problems than the one used for solving the Euler buckling problem, where we assume that the displacements, rotations and strains can be (very) large before arriving to the critical equilibrium point, and moreover we admit nonlinear inelastic constitutive behavior. We will call those the problems of nonlinear instability. The increase of complexity of nonlinear instability problems is such, that it is practically impossible to obtain the analytic solution (of the strong form) for any such problem, comparable to the one we have constructed for linear instability problem of Euler buckling. Therefore, we can only obtain the corresponding solution of the weak form by using the finite element method.

We will first address the nonlinear instability problems where displacement can be large when arriving at the critical equilibrium point, but the constitutive behavior remains elastic. We will introduce this class of problems by means of a simple example, considering a shallow truss composed of two truss-bar elements with 2-nodes, which is loaded by a vertical force at the apex (see Figure 8.2).

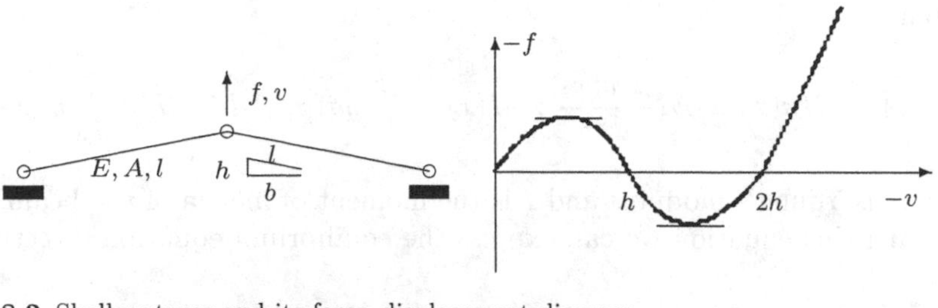

Fig. 8.2 Shallow truss and its force–displacement diagram.

By assuming that the elastic constitutive behavior of each bar is described by Saint-Venant–Kirchhoff material model, and by taking into account the symmetry, we can write the weak form of the equilibrium equations according to:

$$G^{ext} = G^{int} \Leftrightarrow wf = w\, \underbrace{2AS_{11}\frac{1}{l}(\overbrace{y_2 - y_1}^{h} + v)}_{f^{int}} \; ; \; S_{11} = E\tfrac{1}{l^2}(hv + \tfrac{1}{2}v^2) \tag{8.3}$$
$$\Longrightarrow f = \tfrac{2EA}{l^3}(h+v)(hv + \tfrac{1}{2}v^2)$$

In the last expression, we took into account that the second Piola–Kirchhoff stress is constant in each bar, and that it can be written explicitly in terms of vertical displacement v. It is easy to see that maintaining equilibrium will require zero value of vertical external force for the displacement values $v = -h$ and $v = -2h$. These two equilibrium states correspond, respectively, to the truss equilibrium state where both bars are horizontal and the equilibrium state where bars have been moved through to the opposite side so that the deformed length of each bar becomes the same as in the initial configuration, which implies zero internal force. However, even if they share the same (zero) value of external force, these two equilibrium states are not the same: the first is unstable, since any small force increase will lead to very large increase of displacement, whereas the second is stable with a small force increase accompanied by a proportionally small displacement increase. It thus follows that in-between these two states we will eventually find a critical equilibrium state, which is the first to bring disproportionately large displacement increase accompanying a small force increase; this condition can be written in the inverse form, which allows us to conclude that the zero value of tangent stiffness for this truss indicates a critical equilibrium state:

$$\frac{dv}{df} \mapsto \infty \;\&\; f = f^{int} \Longrightarrow \frac{df^{int}}{dv} =: K = 0 \tag{8.4}$$

In fact, in this example, we can compute two critical equilibrium states, which correspond to:

$$0 = K := \frac{df^{int}}{dv} = \frac{2EA}{l^3}(h+v)^2 + \frac{2EA}{l^3}(hv + \frac{1}{2}v^2) \Longrightarrow \tag{8.5}$$

$$\begin{aligned} v_{cr_1} &= -h(1 - \sqrt{3}/3) \Longrightarrow f_{cr_1} = -\tfrac{2EAh^3}{l^3}\tfrac{\sqrt{3}}{9} \\ v_{cr_2} &= -h(1 + \sqrt{3}/3) \Longrightarrow f_{cr_2} = \tfrac{2EAh^3}{l^3}\tfrac{\sqrt{3}}{9} \end{aligned} \tag{8.6}$$

The first critical state concerns passing from stable to unstable equilibrium states, whereas the second critical state marks return to stable equilibrium states.

8.1.1.3 Detection criterion for critical equilibrium state

On the basis of discussion in the previous section, we can provide the first general criterion for detecting a critical equilibrium state in problems of practical interest with a large number of equilibrium equations. Namely, we could check when the tangent stiffness matrix becomes singular, with its determinant taking zero value at the critical state of equilibrium:

$$\det[\hat{\mathsf{K}}(\mathsf{d}_{cr})] = 0 \; ; \; \hat{\mathsf{K}}(\mathsf{d}_{cr}) = \frac{\partial \mathsf{f}^{int}(\mathsf{d}_{cr})}{\partial \mathsf{d}} \tag{8.7}$$

Although theoretically correct, the criterion for detection of equilibrium state in terms of zero determinant is not practical to use, for two reasons. First, the cost of computing the determinant of $n \times n$ matrix is prohibitively high (of the order of $n!$) and second, a rapid increase in determinant values in the neighborhood of the critical point can lead to significant convergence difficulties of Newton's iterative method for computing the corresponding displacement d_{cr} resulting with the zero value of determinant. We are thus prompted to seek yet other detection criteria for critical equilibrium state and the more suitable choice for computation of critical equilibrium points. Detection criterion for instability based upon variation of total potential energy is an alternative criterion for detecting a critical equilibrium state. The critical point is reached for the equilibrium state where the second variation of the total potential energy becomes zero. For a hyperelastic constitutive model, it is easy to establish direct connection of this energy criterion with the one previously proposed based upon singularity of the tangent stiffness matrix; namely, for a geometrically nonlinear problem with the total potential energy $\Pi(\varphi)$, we can obtain the first and the second variations and establish their connection with the equilibrium equation and the tangent stiffness matrix, respectively:

$$\begin{aligned} D_{\mathsf{w}}\Pi(\varphi) &:= \tfrac{d}{d\varepsilon}[\Pi(\varphi_{\varepsilon}]_{\varepsilon=0} = G(\varphi;\mathsf{w}) = \mathsf{w}^T(\hat{\mathsf{f}}^{int}(\mathsf{d}) - \mathsf{f}^{ext}) = 0 \\ D_{\mathsf{w}}[D_{\mathsf{w}}\Pi(\varphi)] &:= \tfrac{d^2}{d\varepsilon^2}[\Pi(\varphi_{\varepsilon})]_{\varepsilon=0} = \tfrac{d}{d\varepsilon}[G(\varphi_{\varepsilon};\mathsf{w})]_{\varepsilon=0} = \mathsf{w}^T\mathsf{K}\mathsf{w} \end{aligned} \tag{8.8}$$

With these results in hand, we can express the total potential energy modification in the neighborhood of the chosen equilibrium state φ, for any new equilibrium state which is produced by a small, kinematically admissible perturbation w. To that end, we can use the Taylor series representation to write:

$$\begin{aligned} \Pi(\varphi + \mathsf{w}) &\approx \Pi(\varphi) + D_{\mathsf{w}}\Pi(\varphi) + D_{\mathsf{w}}[D_{\mathsf{w}}\Pi(\varphi)] \\ &\approx \Pi(\varphi) + \underbrace{G(\varphi;\mathsf{w})}_{=0} + \mathsf{w}^T\mathsf{K}\mathsf{w} \end{aligned} \tag{8.9}$$

With the first variation which is equal to zero (since it represents equilibrium equation of an equilibrium state), the difference in total potential energy of these two adjacent equilibrium states will be controlled by the second variation. We can thus conclude that the given equilibrium state is stable, if any other equilibrium state in its neighborhood, which is produced by a kinematically admissible perturbation, will impose an increase in energy. For such a case, removing the perturbation will allow the structure to go back to the original equilibrium state before perturbation. We can also conclude that any stable equilibrium state will have the positive definite tangent stiffness matrix, which is written:

$$\begin{aligned} &\Pi(\varphi + \mathsf{w}) > \Pi(\varphi)\,;\ \forall \mathsf{w} \in \mathbb{V}_0 \implies D_w[D_w\Pi(\varphi] := \mathsf{w}^T\mathsf{K}\mathsf{w} > 0 \\ &\Leftrightarrow \mathsf{K} \text{ positive definite} \end{aligned} \tag{8.10}$$

The equilibrium state is considered as unstable, if there exists a small, kinematically admissible perturbation that will reduce the total potential energy. The tangent stiffness matrix of the unstable equilibrium state is negative definite:

$$\begin{aligned} &\Pi(\varphi + \mathsf{w}) < \Pi(\varphi)\,;\ \exists \mathsf{w} \in \mathbb{V}_0 \implies D_w[D_w\Pi(\varphi] := \mathsf{w}^T\mathsf{K}\mathsf{w} < 0 \\ &\Leftrightarrow \mathsf{K} \text{ negative definite} \end{aligned} \tag{8.11}$$

The principle of minimum potential energy only applies to stable equilibrium states, which guarantees the return to the original equilibrium state following any perturbation. The same principle does not apply to unstable equilibrium states, where a small perturbation will produce the state with smaller energy which would no longer allow to recover the original equilibrium state after perturbation removal. In fact, the unstable equilibrium state perturbation is very likely not to remain limited to small displacements, but rather to lead to large displacements and strains with the great risk of subsequent structural failure. Needless to say, the risk of unstable, or even critical equilibrium states, should not in general be acceptable.
The critical equilibrium state indicates passing from stable to unstable states (or vice versa), with the second variation of its total potential energy equal to zero. The tangent stiffness matrix of the critical equilibrium state is a singular matrix, for which we can write:

$$\begin{aligned} &\Pi(\varphi + \mathsf{w}) = \Pi(\varphi)\,;\ \forall \mathsf{w} \in \mathbb{V}_0 \implies D_w[D_w\Pi(\varphi] := \mathsf{w}^T\mathsf{K}\mathsf{w} = 0 \\ &\Leftrightarrow \mathsf{K} \text{ singular} \end{aligned} \tag{8.12}$$

A graphical illustration of these results deduced from the detection criterion based upon the total potential energy variations is given in Figure 8.3, for the case where the potential energy potential is defined from gravity field.

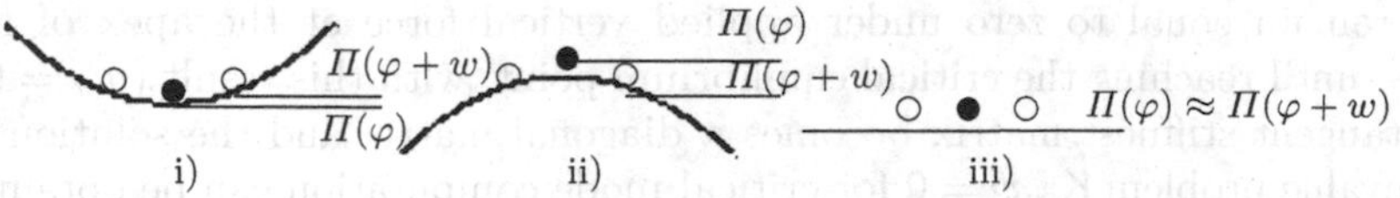

Fig. 8.3 Total potential energy of equilibrium state: (i) stable - $\Pi(\varphi) \mapsto min.$, (ii) unstable – $\Pi(\varphi) \mapsto max.$, (iii) critical $D_w[D_w\Pi(\varphi_{cr})] = 0$.

Detection criterion based on zero eigenvalue is another detection criterion for instability, which can be used to verify if the tangent stiffness is a singular matrix. Namely, by choosing for virtual displacement vector in (8.12) the instability mode ψ with zero eigenvalue, we can further write:

$$\mathsf{w} = \psi \implies 0 = [\mathsf{K} - \underbrace{\lambda}_{=0} \mathsf{I}]\psi = \mathsf{K}\psi \tag{8.13}$$

The main advantage of such a criterion is in providing not only the indication on the critical equilibrium state, but also on the kind of perturbation which reveals instability, in terms of the corresponding eigenvector ψ.
We will further illustrate the main advantage of this detection criterion, which can indicate the type of instability mode by computing the eigenvector of the tangent stiffness at the critical equilibrium point associated with zero eigenvalue. For that reason, we will go back to the example of a simple truss structure, which consists of two bars with linear hyperelastic constitutive behavior described by the Saint–Venant–Kirchoff model. We allow this time for any form of the initial configuration, for either shallow or deep truss, with the corresponding values of b and h (see Figure 8.2). We use again for each bar a 2-node isoparametric finite element, which allows us to construct the tangent stiffness matrix by the finite element assembly procedure:
Element 1:

$$\mathsf{K}^{(1)} := \mathsf{K}_m^{(1)} + \mathsf{K}_g^{(1)} = \begin{pmatrix} b+u \\ h+v \end{pmatrix} \tfrac{AE}{l^3} \begin{pmatrix} b+u \\ h+v \end{pmatrix}^T + \tfrac{S^{(1)}A}{l} \begin{bmatrix} 1 & 0 \\ 0 & 1 \end{bmatrix} ;$$
$$S^{(1)} = \tfrac{E}{l^2}(bu + hv + \tfrac{1}{2}u^2 + \tfrac{1}{2}v^2)$$

Element 2:

$$\mathsf{K}^{(2)} := \mathsf{K}_m^{(2)} + \mathsf{K}_g^{(2)} = \begin{pmatrix} -b+u \\ h+v \end{pmatrix} \tfrac{AE}{l^3} \begin{pmatrix} -b+u \\ h+v \end{pmatrix}^T + \tfrac{S^{(2)}A}{l} \begin{bmatrix} 1 & 0 \\ 0 & 1 \end{bmatrix} ;$$
$$S^{(2)} = \tfrac{E}{l^2}(-bu + hv + \tfrac{1}{2}u^2 + \tfrac{1}{2}v^2)$$

which results with:

$$\mathsf{K} = \mathsf{K}^{(1)} + \mathsf{K}^{(2)} := \frac{2AE}{l^3} \begin{bmatrix} u^2 + b^2 & u(h+v) \\ u(h+v) & (h+v)^2 \end{bmatrix} + \frac{2AE}{l^3}(hv + \frac{1}{2}u^2 + \frac{1}{2}v^2) \begin{bmatrix} 1 & 0 \\ 0 & 1 \end{bmatrix}$$

The symmetry of the structure imposes that the horizontal displacement will remain equal to zero under applied vertical force at the apex of the truss, until reaching the critical equilibrium point; with this result ($u_{cr} = 0$), the tangent stiffness matrix becomes a diagonal matrix and the solution to eigenvalue problem $\mathsf{K}_{cr}\psi = 0$ for critical mode computation can be obtained in closed form. We can have two cases: the first, were the perturbation of the equilibrium configuration brought by instability mode will push down the truss further in the same direction $\psi^T = (0,1)$; we thus recover the previously presented result on nonlinear instability of this truss with very large value of vertical displacement v_{cr} at the critical equilibrium point:

$$0 = \mathsf{K}_{cr}\psi := \frac{2AE}{l^3}\begin{bmatrix} b^2 + (hv_{cr} + \frac{1}{2}v_{cr}^2) & 0 \\ 0 & (h+v_{cr})^2 + (hv + \frac{1}{2}v_{cr}^2) \end{bmatrix} \begin{bmatrix} 0 \\ 1 \end{bmatrix}$$

$$\Longrightarrow \begin{array}{l} u_{cr} = 0 \\ v_{cr} = -h(1 \pm \sqrt{3}/3) \end{array}$$

This kind of critical point provides the maximum load level which can be carried by the truss when both bars are compressed, and it is referred to as the limit load point.

In the second case, the perturbation is produced by a horizontal displacement with instability mode $\psi^T = (1,0)$, leading to the bifurcation point, where lateral motion, either to the left or to the right, breaks the symmetry of the structure:

$$0 = \mathsf{K}_{cr}\psi := \frac{2AE}{l^3}\begin{bmatrix} b^2 + (hv_{cr} + \frac{1}{2}v_{cr}^2) & 0 \\ 0 & (h+v_{cr})^2 + (hv_{cr} + \frac{1}{2}v_{cr}^2) \end{bmatrix} \begin{bmatrix} 1 \\ 0 \end{bmatrix}$$

$$\Longrightarrow \begin{array}{l} u_{cr} = 0 \\ v_{cr} = -h \pm \sqrt{h^2 - 2b^2} \end{array}$$

We note that the shallow truss with $h^2 - 2b^2 < 0$, cannot have the bifurcation point. The latter will be typical, however, of a deep truss with $h \gg b$ (which guarantees that $h^2 - 2b^2 > 0$), where the last expression will provide the solution for vertical displacement v_{cr} at bifurcation point, leading to phenomena very much equivalent to the Euler buckling. In conclusion, the last detection criterion for nonlinear instability, based upon the computation of the eigenvectors of tangent stiffness matrix revealing the type of instability, remain applicable to the linear instability problems and provide an equivalent result to Euler's buckling solution. Moreover, the same conclusion remains valid for much more complex structures [137] than the simple truss used herein for illustration.

8.1.2 Solution methods for boundary value problem in presence of instabilities

In this section we discuss two solution methods for the boundary value problem in presence of geometric instabilities. We note in passing that the same methods are applicable to material instability problems, which are discussed later in this chapter. The first method will allow us to completely trace the force–displacement diagram for the given loading program, which could either enforce loading or unloading in order to maintain equilibrium in the presence of critical limit points. This method, referred to as arc-length, does not seek to detect the critical equilibrium points, but just to allow the continuation in the post-critical regime. The second method presented in this section provides the direct computation of the critical equilibrium points, and it thus provides the complementary information to the one given by the arc-length method. Two methods are often used together, especially when the direct computation of the critical points is carried out by Newton's iterative procedure where we need the best possible initial guess to ensure convergence. By using these two methods combined, we can trace the complete force–displacement diagram, detect the critical points, identifies their true nature between limit and bifurcation point, and explore any post-bifurcation path.

8.1.2.1 Arc-length method

The starting point in our presentation of the arc-length method is the finite element based discrete approximation of a boundary value problem in nonlinear mechanics with the eventual presence of instabilities. The main novelty with respect to the nonlinear mechanics problems solved previously (by using incremental/iterative solution methods) is the need to ensure the continuation in response computation after passing through a critical equilibrium point, which should allow within a particular loading program to either increase or reduce the level of external loading. For that reason, the external load vector is written as the product between a time–invariant vector f_0, indicating only the loading direction, and the time-dependent loading parameter $\lambda(t)$, specifying the current level of loading with:

$$\mathsf{f}^{ext}(t) = \lambda(t)\mathsf{f}_0^{ext} \tag{8.14}$$

The set of nonlinear equilibrium equations in this case obtained as the end result of the finite element discretization can be written:

$$\mathsf{r}(\mathsf{d}, \lambda) := \hat{\mathsf{f}}^{int}(\mathsf{d}(t)) - \lambda(t)\mathsf{f}_0^{ext} = 0 \tag{8.15}$$

where $\lambda(t)$ is an additional unknown, whose value can either increase or decrease during the given loading program. We thus obtain a set of 'n' nonlinear algebraic equations with '$n+1$' unknowns. The arc-length method will then seek to provide an additional equation in order to establish a well-posed problem, where the number of unknowns is equal to number of equations. The most general form of such an additional equation, which ought to be solved together with (8.15), can be written:

$$g(\mathsf{d}(t), \lambda(t)) = 0 \tag{8.16}$$

The main role of the additional equation is to stabilize the behavior (e.g. see Ibrahimbegovic et al. [129]) of the original structure in presence of instability, so that we can carry on with the computation for the complete loading program. A number of different proposals to achieve this stabilization have been made (see Batoz and Dhatt [22], Crisfield [59], Riks [228] etc.). For example, for a problem with a limit point specifying the maximum force bearing capacity, we can simply use an additional equation imposing the displacement increments; in other words, even though the original problem will lead to a loss of stability under imposed force upon reaching the limit point followed subsequently by failure (or a gradual loss of resistance in post-peak regime), we can add the displacement increment control which would stabilize the post-peak response in the course of failure and allow to carry out the computation of complete failure trajectory. Many other choices for defining the additional equation in (8.16) will be acceptable, especially those which account for the control of dominant failure mechanisms; the case in point, is the additional equation which will control crack-mouth opening displacement in a brittle failure problem corresponding to three-point bending test, or control of plastic sliding in shear bands in a traction test indicating the ductile failure with necking of a metallic speciment, which will both lead to material instability phenomena.

With the lack of a sound physical basis for additional equation when dealing with instability of a more complex structure, we can choose the additional equation for arc-length method that controls both the increments of displacements and increments of force, which can be written:

$$g(\mathsf{d}(t), \lambda(t)) := \sqrt{\| \dot{\mathsf{d}} \|^2 + |\dot{\lambda}|^2} - \dot{s} = 0 \tag{8.17}$$

where $\dot{s}$ is the chosen value which controls the rate of increase of the arc-length parameter along the deformation trajectory. This kind of additional equation can also be employed within the framework of incremental/iterative analysis, where the problem to be solved reduces to a set of '$n+1$' nonlinear algebraic equations with d_{n+1} and λ_{n+1} as unknowns:

$$\begin{aligned} \mathsf{r}(\mathsf{d}_{n+1}, \lambda_{n+1}) &:= \hat{\mathsf{f}}^{int}(\mathsf{d}_{n+1}) - \lambda_{n+1}\mathsf{f}_0^{ext} = 0 \\ g(\mathsf{d}_{n+1}, \lambda_{n+1}) &:= \sqrt{\| (\mathsf{d}_{n+1} - \mathsf{d}_n) \|^2 + (\lambda_{n+1} - \lambda_n)^2} - \Delta s_{n+1} = 0 \end{aligned} \tag{8.18}$$

where Δs_{n+1} will specify the chosen increment of arc-length (used for running the incremental analysis instead of the time-step). In the higher-dimensional vector space of displacement components d_{n+1} and load parameter λ_{n+1}, the additional equation can be interpreted as a hyper-circle, with the parameter Δs_{n+1} as its radius. The computation in each increment will look for the intersection of this kind of hyper-circle centered at the equilibrium point at time t_n with the point of force–displacement diagram corresponding to equilibrium state at time t_{n+1}. In this manner, we can also converge with this computations to an equilibrium state at time t_{n+1} which is unstable; moreover, we can decrease, if needed, the circle radius Δs_{n+1} in order to ensure convergence and reduce the number of iterations in each step while dealing with a complex instability problem with snap-back (which can be handled with neither imposed force nor imposed displacement increment). In each iteration (i) of the arc-length computations we carry out the consistent linearization of the system in (8.18), solve for unknown incremental displacements $\Delta \mathsf{d}_{n+1}^{(i)}$ and increment in load parameter $\Delta\lambda_{n+1}^{(i)}$ and perform the corresponding updates:

$$(i) = 1, 2, \ldots$$

$$\begin{bmatrix} \mathsf{K} & -\mathsf{f}_0^{ext} \\ \frac{\partial g}{\partial \mathsf{d}} & \frac{\partial g}{\partial \lambda} \end{bmatrix}_{n+1}^{(i)} \begin{bmatrix} \Delta \mathsf{d}_{n+1}^{(i)} \\ \Delta \lambda_{n+1}^{(i)} \end{bmatrix} = \begin{bmatrix} \lambda_{n+1}^{(i)} \mathsf{f}_0^{ext} - \hat{\mathsf{f}}^{int}(\mathsf{d}_{n+1}^{(i)}) \\ -g(\mathsf{d}_{n+1}^{(i)}, \lambda_{n+1}^{(i)}) \end{bmatrix} \tag{8.19}$$

$$\mathsf{d}_{n+1}^{(i+1)} = \mathsf{d}_{n+1}^{(i)} + \Delta \mathsf{d}_{n+1}^{(i)} \; ; \; \lambda_{n+1}^{(i+1)} = \lambda_{n+1}^{(i)} + \Delta\lambda_{n+1}^{(i)}$$

We carry on with this computation until the convergence is reached within a specified tolerance. Only then can we check if the critical point has been located within the current step, in the case where the determinant sign would change with respect to the one at the beginning of the step:

$$det[\mathsf{K}_{n+1})] = -det[\mathsf{K}_n] \implies \exists t_{cr} \in [t_n, t_{n+1}] \; ; \; det[\mathsf{K}_{cr}] = 0 \tag{8.20}$$

It is important to note that the fulfillment of the condition in (8.20) above only indicates the presence of instability between the times t_n and t_{n+1}, without detecting precisely when, or providing more information on the type of instability, between the limit or bifurcation point. If we would like to obtain all the detailed information of this kind, including the exact time t_{cr} where the critical equilibrium point occurs, we can use the method presented in the next section.

8.1.2.2 Direct computation of critical equilibrium points

The proposed procedure for direct computation of critical points seeks to detect the critical equilibrium states for nonlinear instability phenomena and converge the iterative procedure directly to this state, with no need to visit any other equilibrium state. The final result of this computation leads to the displacement vector and critical load parameter value for the critical equilibrium state, accompanied by the supplementary information which can reveal if we have detected a limit or a bifurcation point. The direct computation is carried out by Newton's iterative procedure; therefore, it is very important to have a very good initial guess in order to ensure the convergence of this computation. For that reason, we rely upon the arc-length method in order to get closer to the critical point and to provide the best iterative guess for initial values which will be placed within the radius of convergence of Newton's method. Practically, as soon as the criterion in (8.20) for change-of-sign of the determinant of tangent stiffness indicates the presence of the critical point within the current step, we will take the displacement and load parameter from the beginning of that step for the initial values of our direct computation. Thus, besides the equilibrium equations for critical state, the direct computation needs an additional condition, which can be written in two different manners:

$$\frac{d^2}{d\varepsilon^2}[\Pi(\varphi_{cr,\varepsilon})]\bigg|_{\varepsilon=0} = 0 \Leftrightarrow \begin{cases} \det[\mathsf{K}_{cr}] = 0 \\ \mathsf{K}_{cr}\boldsymbol{\psi} = 0 \end{cases} \tag{8.21}$$

where $\mathsf{K}_{cr} = \partial \hat{\mathsf{f}}^{int}(\mathsf{d}_{cr})/\partial \mathsf{d}$ is the tangent stiffness matrix at the critical equilibrium state. The final choice we will make depends upon what kind of information about the critical state we would like to provide. By choosing the first condition, we can obtain the formulation where we only compute the critical values of displacements and load parameter, by solving the following augmented system:

$$\begin{bmatrix} \hat{\mathsf{f}}^{int}(\mathsf{d}_{cr}) - \lambda_{cr}\mathsf{f}_0 \\ det[\mathsf{K}_{cr}] \end{bmatrix} = 0 \mapsto (\mathsf{d}_{cr}, \lambda_{cr}) \tag{8.22}$$

This is, in general, a well-posed problem with $n+1$ nonlinear algebraic equations with the same number of unknowns, including the displacement at the critical state and the corresponding load parameter $(\mathsf{d}_{cr}, \lambda_{cr})$. By using Newton's iterative procedure for solving this problem, at each iteration we perform the consistent linearization of these equations, obtain the increments for critical displacements and load parameters and carry out the corresponding updates:

$$\begin{aligned} &(i) = 1, 2, \ldots \\ &\begin{bmatrix} \mathsf{K} & \mathsf{f}_0^{ext} \\ tr[\mathsf{K}^{-1}\frac{\partial \mathsf{K}}{\partial \mathsf{d}}] & 0 \end{bmatrix}_{cr}^{(i)} \begin{bmatrix} \Delta \mathsf{d}_{cr}^{(i)} \\ \Delta \lambda_{cr}^{(i)} \end{bmatrix} = \begin{bmatrix} \lambda_{cr}^{(i)}\mathsf{f}_0 - \mathsf{f}^{int}(\mathsf{d}_{cr}^{(i)}) \\ -1 \end{bmatrix} \\ &\mathsf{d}_{cr}^{(i+1)} = \mathsf{d}_{cr}^{(i)} + \Delta \mathsf{d}_{cr}^{(i)} \;;\; \lambda_{cr}^{(i+1)} = \lambda_{cr}^{(i)} + \Delta \lambda_{cr}^{(i)} \end{aligned} \tag{8.23}$$

The iterative procedure will continue until the convergence is reached within the specified tolerance. One disadvantage of this method concerns the lack of information about the computed equilibrium point. In order to obtain this kind of information we need an alternative formulation of the extended system, where the critical mode $\boldsymbol{\psi}$ which renders the tangent stiffness matrix singular is also computed; this kind of formulation leads to a set of '$2n+1$' nonlinear algebraic equations with the same number of unknowns:

$$\begin{bmatrix} \hat{\mathsf{f}}^{int}(\mathsf{d}_{cr}) - \lambda_{cr}\mathsf{f}_0 \\ \mathsf{K}_{cr}\,\boldsymbol{\psi} \\ l(\boldsymbol{\psi}) \end{bmatrix} = 0 \;\mapsto\; (\mathsf{d}_{cr}, \lambda_{cr}, \boldsymbol{\psi}) \tag{8.24}$$

The last of equations in (8.24) above represents a complementary condition which is chosen in agreement with the kind of instability we would like to detect, between a limit and a bifurcation point. The precise form of the complementary condition to impose can be obtained from the linearized form of equilibrium equations at the critical point, by scalar-multiplication with the critical mode $\boldsymbol{\psi}$ and by taking into account the symmetry of the tangent stiffness matrix to obtain the following result:

$$\begin{aligned} 0 &= \boldsymbol{\psi}^T \{\overbrace{\hat{\mathsf{f}}^{int}(\mathsf{d}_{cr}) - \lambda_{cr}\mathsf{f}_0}^{=0} + \mathsf{K}_{cr}\,\Delta\mathsf{d} - \mathsf{f}_0^{ext}\Delta\lambda\} \\ &= \Delta\mathsf{d}^T \underbrace{\mathsf{K}_{cr}\boldsymbol{\psi}}_{=0} - \boldsymbol{\psi}^T\mathsf{f}_0^{ext}\Delta\lambda \end{aligned} \tag{8.25}$$

The last result can be verified for a critical equilibrium state pertaining to a limit point where the external load has reached the maximum value with no further increase, which implies $\Delta\lambda = 0$. In such a case, one complementary condition that can be used in (8.24) will serve to normalize the computed critical mode with:

$$l(\boldsymbol{\psi}) := \parallel \boldsymbol{\psi} \parallel -1 = 0 \tag{8.26}$$

However, the result in (8.25) above can also be verified for the case where the external load can still increase, but the critical mode will push the structure in the direction that is orthogonal to the applied external load direction resulting with $\boldsymbol{\psi}^T\mathsf{f}_0^{ext} = 0$; we will thus find a bifurcation point, if the complementary equation in (8.24) is written:

$$l(\boldsymbol{\psi}) := \boldsymbol{\psi}^T\mathsf{f}_0^{ext} = 0 \tag{8.27}$$

The direct computation of the critical point with the corresponding information on the type of critical mode, can also be carried out by Newton's iterative procedure, where at each iteration we obtain the consistently linearized form of our system, compute the increments in nodal displacements, load parameter and critical mode, and carry out the corresponding updates:

$$(i) = 1, 2, \ldots$$

$$\begin{bmatrix} \mathsf{K} & 0 & -\mathsf{f}_0 \\ \partial(\mathsf{K}\boldsymbol{\psi})/\partial\mathsf{d} & \mathsf{K} & \partial(\mathsf{K}\boldsymbol{\psi})/\partial\lambda \\ 0 & \partial l/\partial\boldsymbol{\psi} & 0 \end{bmatrix}_{cr}^{(i)} \begin{bmatrix} \Delta\mathsf{d}_{cr}^{(i)} \\ \Delta\boldsymbol{\psi}^{(i)} \\ \Delta\lambda_{cr}^{(i)} \end{bmatrix} = \begin{bmatrix} \lambda_{cr}^{(i)}\mathsf{f}_0 - \mathsf{f}^{int}(\mathsf{d}_{cr}^{(i)}) \\ -\mathsf{K}_{cr}^{(i)}\,\boldsymbol{\psi}^{(i)} \\ -l(\boldsymbol{\psi}^{(i)}) \end{bmatrix}$$

$$\mathsf{d}_{cr}^{(i+1)} = \mathsf{d}_{cr}^{(i)} + \Delta\mathsf{d}_{cr}^{(i)} \;;\; \lambda_{cr}^{(i+1)}) = \lambda_{cr}^{(i)} + \Delta\lambda_{cr}^{(i)} \;;\; \boldsymbol{\psi}_{cr}^{(i+1)} = \boldsymbol{\psi}_{cr}^{(i)} + \Delta\boldsymbol{\psi}_{cr}^{(i)}$$

The iterations are stopped when the specified tolerance is reached. The main difficulty in applying this solution procedure is related to the choice of initial values for unknowns in the iterative procedure. Namely, we can again use the displacements and load parameter value provided by the arc-length solution procedure for the beginning of the step where the presence of the critical point was noticed (by the change of the sign of determinant of the tangent stiffness matrix), but there is no indication, in general, what to use for the first iterative guess of the critical bifurcation mode $\psi^{(1)}$; a poor guess of this kind can significantly slow down or even prevent the iterative procedure from converging.

After passing through a critical bifurcation point, we can explore different post-bifurcation responses, where the computation along each path can be started with a small perturbation of the critical equilibrium state in the direction of the computed critical mode with: $\mathsf{d}_{cr} + \varepsilon\psi$, where $\epsilon << 1$.

Some of the instability phenomena, which are characterized by a rapid transition from stable to unstable equilibrium states, ought to be placed and studied within the framework of dynamics in order to provide the proper interpretation of results. The case in point is the nonlinear instability of shallow truss structure in the first example of this chapter, which is often called snap-through, since the passing of truss through unstable equilibrium states happens so fast that it is often accompanied by a noise.[1] We do not, however, advise to place unless necessary any other instability problem within the framework of dynamics, since this might make it even more difficult to solve, leading to the chaos phenomena (e.g. see Argyris [6] or Gugenheimer and Holmes [94]). The instability in the framework of dynamics can be estimated by the functional of Lyapunov [176], which provides the proper norm on disproportional response increase due to a loading perturbation. For a conservative system where external load derives from a potential, there is an obvious choice of Lyapunov functional in terms of the total energy. The midpoint time-integration scheme with energy conservation or energy dissipation of high-frequency modes, can therefore be applied successfully to this class of problems.

8.2 Material instabilities

A simple example we present for start can be used to illustrate the material instability phenomena, by relying on familiar concepts already presented previously. Namely, we consider an elastic bar of length l, with constitutive behavior described by Saint-Venant–Kirchhoff material model, undergoing a homogeneous compressive deformation under free-end force. The bar is

[1] For the reason of accompanying noise, the same instability phenomenon is called a slap in French.

constrained to move only along its axis, so that any buckling phenomena are excluded. In order to study this problem, we can take the simplest finite element model with a single truss-bar 2-node isoparametric element, with 1 node fixed and the other submitted to the applied displacement d along the bar axis. We can then easily compute the only non-zero component of Green–Lagrange deformation tensor, E_{11}, as well as the corresponding component of the second Piola–Kirchhoff stress, S_{11}:

$$E_{11} = \frac{d}{l} + \frac{1}{2}\frac{d^2}{l^2} \; ; \; S_{11} = EE_{11} \tag{8.28}$$

where E is Young's modulus. The virtual work principle can then be used to obtain the only non-zero component of the internal force in the bar, which can be written:

$$\begin{aligned} wf^{int} &= \Gamma_{11} S_{11} Al \; ; \; \Gamma_{11} := \tfrac{d}{d\varepsilon}[E_{11}(d_\varepsilon)]_{\varepsilon=0} = \tfrac{w}{l} + \tfrac{d}{l}\tfrac{w}{l} \; ; \\ \Longrightarrow f^{int} &= \tfrac{EA}{l}(1 + \tfrac{d}{l})(d + \tfrac{1}{2}\tfrac{d^2}{l}) \end{aligned} \tag{8.29}$$

The derivative of this internal force with respect to displacement will provide the tangent stiffness matrix (or rather its only non-trivial component). We can then obtain the value of displacement d_{cr} resulting with the zero value of tangent stiffness, which reveals the presence of the following critical equilibrium point:

$$\begin{aligned} K &:= \tfrac{\partial f^{int}}{\partial d} = \tfrac{EA}{l}(\tfrac{3}{2l^2}d^2 + \tfrac{3}{l}d + 1) \;\&\; \hat{K}(d_{cr}) = 0 \\ &\Longrightarrow d_{cr}/l = (-1 + 1/\sqrt{3}) \Longrightarrow \lambda_{cr} = \tfrac{1}{\sqrt{3}} \end{aligned} \tag{8.30}$$

We can provide an alternative interpretation of this result from the standpoint of material instability, as produced by the softening constitutive behavior in geometrically linear theory. To that end, we rewrite this constitutive law in terms of the first Piola-Kirchhoff stress that shares in 1D case the same numerical value as the Cauchy or true stress, which would permit us to better relate the results of the model to the physics of the problem. It is easy to see that 1D version of Saint-Venant–Kirchhoff material model will allow to express the first Piola–Kirchhoff stress in terms of the stretch $\lambda = 1 + \frac{\partial d}{\partial x}$, which can be written according to:

$$\sigma \equiv P := E\frac{1}{2}\lambda(\lambda^2 - 1) \tag{8.31}$$

For a compressed bar, this kind of constitutive law will describe a softening behavior reducing the stress once passed its peak value at the critical stretch λ_{cr}, which can easily be verified by direct computations:

$$\mathcal{C}_{cr} = \frac{\partial P}{\partial \lambda}|_{\lambda_{cr}} := C\frac{1}{2}(3\lambda_{cr}^2 - 1) = 0 \Longrightarrow \lambda_{cr} = 1/\sqrt{3} \tag{8.32}$$

The softening behavior produces unstable equilibrium states. We can easily verify that in the neighborhood of the critical stretch λ_{cr}, a small increment of stress will be accompanied by disproportional response with large deformation (or stretch) increment. The latter can thus be interpreted as the material instability phenomenon, which is fully equivalent to the geometric instability phenomena we already discussed. In fact, for this simple, statically determined problem where the stress can be directly obtained from equilibrium and the deformation field remains homogenous, the result obtained for the critical stretch remains identical for either interpretation of instability, geometric or material.
However, it is important to note that, contrary to geometric instabilities, the material instabilities can also occur in small strain problems. For example, the softening constitutive behavior in (8.31) above could be rewritten for small deformation case, with $d \mapsto u$, by using the infinitesimal strain $\epsilon = \frac{du}{dx}$. We can thus obtain an equivalent result for the critical value of deformation revealing the presence of material instability according to:

$$\begin{aligned} &\sigma = E\tfrac{1}{2}(1+\epsilon)[(1+\epsilon)^2 - 1] \implies 0 = \mathcal{C}_{cr} := \tfrac{1}{2}E[3(1+\epsilon_{cr}) - 1] \\ &\implies \epsilon_{cr} = \tfrac{1}{\sqrt{3}} - 1 \end{aligned} \tag{8.33}$$

This law describes the softening behavior at small strain, which can lead to the material instability. For this reason, and even more importantly for special finite element approximation that is needed, the material instability phenomena require a special dedicated approach as opposed to the one already presented for dealing with geometric instabilities for large displacement gradient theories.
The studies of material instability is of a particular interest for softening inelastic behavior. In fact, in such a case, the material instabilities are often accompanied by so-called strain localization phenomena where large inelastic deformations develop within a small zone of the domain (often referred to as the "fracture process zone") forcing the neighboring sub-domains to unload elastically. Our goal in this section is to provide the full understanding of the class of problems of material instability, both in terms of their theoretical formulation and in terms of their finite element implementation. We will start first by the theoretical aspects, and in particular the detection criteria for material instability, for both limit and bifurcation points.

8.2.1 Detection criteria for material instabilities

8.2.1.1 Elastic behavior

The instability phenomena can lead to the loss of uniqueness in the solution of a boundary value problem. If we first consider one such problem in elasticity, we can assume that $u^{(\alpha)}, \alpha = 1, 2$ are two possible solutions (produced by a loss of uniqueness) of the boundary value problem, with each satisfying the same boundary conditions:

$$\dot{u}_i^{(\alpha)} = \dot{\bar{u}}_i \text{ on } \Gamma_{u_i} \; ; \; \dot{\sigma}_{ij} n_j = \dot{\bar{t}}_i \text{ on } \Gamma_{\sigma_i} \tag{8.34}$$

By linearity of elasticity problem, the difference of these two solutions will have to satisfy the boundary value problem with the same governing equations and the homogeneous boundary conditions:

$$\hat{\dot{u}}_i := \dot{u}_i^{(1)} - \dot{u}_i^{(2)} \Rightarrow \begin{cases} \hat{\dot{\epsilon}}_{ij} & = \frac{1}{2}(\hat{\dot{u}}_{i,j} + \hat{\dot{u}}_{j,i}) \\ \hat{\dot{\sigma}}_{ij} & = C_{ijkl} \hat{\dot{\epsilon}}_{kl} \\ \hat{\dot{\sigma}}_{ij,j} & = 0 \end{cases} \text{in } \Omega \tag{8.35}$$

such that:

$$\begin{cases} \hat{\dot{u}}_i = 0 \text{ on } \Gamma_{u_i} \\ \hat{\dot{\sigma}}_{ij} n_j = 0 \text{ on } \Gamma_{\sigma_i} \end{cases} \Longrightarrow \boxed{\hat{\dot{u}}_i \hat{\dot{\sigma}}_{ij} n_j = 0 \text{ on } \Gamma} \tag{8.36}$$

By exploiting the divergence theorem, we can show from the last equation that:

$$\begin{aligned} 0 &= \int_\Gamma \hat{\dot{u}}_i \hat{\dot{\sigma}}_{ij} n_j \, d\Gamma \\ &= \int_\Omega (\hat{\dot{u}}_i \hat{\dot{\sigma}}_{ij})_{,j} \, d\Omega \\ &= \int_\Omega \hat{\dot{u}}_i \underbrace{\hat{\dot{\sigma}}_{ij,j}}_{=0} d\Omega + \int_\Omega \hat{\dot{u}}_{i,j} \hat{\dot{\sigma}}_{ij} \, d\Omega \\ &= \int_\Omega \hat{\dot{\epsilon}}_{ij} \hat{\dot{\sigma}}_{ij} \, d\Omega \end{aligned} \Longrightarrow \hat{\dot{\epsilon}}_{ij} C_{ijkl} \hat{\dot{\epsilon}}_{ij} = 0 \tag{8.37}$$

For any positive definite elasticity tensor, the last result implies the uniqueness of the deformation field, since $\hat{\dot{\epsilon}}_{ij} = 0 \Longrightarrow \dot{\epsilon}_{ij}^{(1)} = \dot{\epsilon}_{ij}^{(2)}$. On the other hand, for a singular elasticity tensor, the result in (8.37) can be interpreted as the instability criterion, with $\hat{\epsilon}_{ij}$ as the corresponding instability deformation mode.

8.2.1.2 Elastoplastic behavior

In this section we consider detection criteria for material instability phenomena in small strain plasticity, which is characterized by: (i) an additive decomposition of the total deformation field into elastic and plastic part, (ii) strain energy density functional defined in terms of elastic deformation and hardening variables, (iii) yield function in terms of dual variables. All the remaining ingredients are obtained from the second principle of thermodynamics and the principle of maximum plastic dissipation, which result with:

$$\begin{aligned}
&\boldsymbol{\epsilon} = \boldsymbol{\epsilon}^e + \boldsymbol{\epsilon}^p & \\
&\psi(\boldsymbol{\epsilon}^e,\zeta) = \underbrace{\frac{1}{2}\boldsymbol{\epsilon}^e \cdot \boldsymbol{\mathcal{C}}\boldsymbol{\epsilon}^e}_{\psi^e(\boldsymbol{\epsilon}^e)} + \underbrace{\frac{1}{2}\zeta K \zeta}_{\Xi(\zeta)} \quad \Longrightarrow \quad & \boldsymbol{\sigma} := \frac{\partial \psi^e}{\partial \boldsymbol{\epsilon}^e} = \boldsymbol{\mathcal{C}}\boldsymbol{\epsilon} \\
& & q := \frac{\partial \Xi}{\partial \zeta} = -K\zeta \\
& & \dot{\boldsymbol{\epsilon}}^p = \dot{\gamma}\frac{\partial \phi}{\partial \boldsymbol{\sigma}} \ ; \ \dot{\zeta} = \dot{\gamma}\frac{\partial \phi}{\partial q} \\
&\phi(\boldsymbol{\sigma},q) := \hat{\phi}(\boldsymbol{\sigma}) - (\sigma_y - q) \leq 0 & \\
&\text{if } d\gamma := \dot{\gamma}\, dt > 0 \Rightarrow 0 = \dot{\phi}\, dt = \underbrace{\frac{\partial \hat{\phi}}{\partial \boldsymbol{\sigma}} d\boldsymbol{\sigma}}_{d\hat{\phi}} - \underbrace{K d\gamma}_{\dot{q}\, dt} \Longrightarrow d\gamma = d\hat{\phi}/K &
\end{aligned} \tag{8.38}$$

We will consider in the subsequent developments, the yield criteria specified with homogeneous function of degree one, which allows us to write that:

$$\hat{\phi}(\alpha\boldsymbol{\sigma}) = \alpha\hat{\phi}(\boldsymbol{\sigma}) \Longrightarrow \frac{\partial \hat{\phi}}{\partial \boldsymbol{\sigma}} \cdot \boldsymbol{\sigma} = \hat{\phi}(\boldsymbol{\sigma}) \Longrightarrow \dot{\gamma} = \frac{\dot{\hat{\phi}}}{K} \tag{8.39}$$

We note that rather a large number of yield criteria of practical interest, from the one proposed by von Mises to those of Hill, fit within this framework.
The solution uniqueness and the absence of instability are checked by the same kind of criterion as the one used for elasticity in (8.37), stating that any perturbation producing two possible solutions $\sigma^{(\alpha)}, \alpha = 1, 2$ at the stable equilibrium state should verify the following:

$$\begin{aligned}
0 &\leq (\dot{\sigma}_{ij}^{(1)} - \dot{\sigma}_{ij}^{(2)})(\dot{\epsilon}_{ij}^{(1)} - \dot{\epsilon}_{ij}^{(2)}) \\
&\leq (\dot{\sigma}_{ij}^{(1)} - \dot{\sigma}_{ij}^{(2)})[\mathcal{C}_{ijkl}^{-1}(\dot{\sigma}_{kl}^{(1)} - \dot{\sigma}_{kl}^{(2)}) + (\dot{\gamma}^{(1)}\frac{\partial \hat{\phi}}{\partial \sigma_{ij}} - \dot{\gamma}^{(2)}\frac{\partial \hat{\phi}}{\sigma_{ij}})] \\
&\leq \| \dot{\boldsymbol{\sigma}}^{(1)} - \dot{\boldsymbol{\sigma}}^{(2)} \|_{\mathcal{C}}^2 + (\dot{\gamma}^{(1)} - \dot{\gamma}^{(2)})(\dot{\hat{\phi}}^{(1)} - \dot{\hat{\phi}}^{(2)})
\end{aligned} \tag{8.40}$$

By further exploiting the result in (8.39) regarding the plastic multiplier, we can further simplify the condition in (8.40) to obtain:

$$0 \leq \| \dot{\boldsymbol{\sigma}}^{(1)} - \dot{\boldsymbol{\sigma}}^{(2)} \|_{\mathcal{C}}^2 + \frac{1}{K}(\dot{\hat{\phi}}^{(1)} - \dot{\hat{\phi}}^{(2)})^2 \tag{8.41}$$

The last result confirms the stability of the equilibrium states produced with hardening ($K > 0$) and also indicates the presence of instability for softening response ($K < 0$).
The stability of elastoplastic behavior can also be checked by Drucker's postulate, which states that the elastoplastic material model provides the stable equilibrium states if the difference between the true stress $\boldsymbol{\sigma}$ with respect to any other plastically admissible stress $\boldsymbol{\sigma}^*$ will verify the following condition:

$$0 \leq (\boldsymbol{\sigma} - \boldsymbol{\sigma}^*) \cdot \underbrace{\dot{\boldsymbol{\epsilon}}\, dt}_{d\boldsymbol{\epsilon}} \tag{8.42}$$

By integrating the last expression over a closed path in deformation space, we obtain the Il'iushin stability postulate, which can be written:

$$0 \leq \oint \underbrace{(\boldsymbol{\sigma} - \boldsymbol{\sigma}^*)}_{\approx \dot{\boldsymbol{\sigma}}\, dt} \cdot \underbrace{d\boldsymbol{\epsilon}}_{=\dot{\boldsymbol{\epsilon}}\, dt} \implies 0 \leq \oint \dot{\boldsymbol{\sigma}} \cdot \dot{\boldsymbol{\epsilon}} \tag{8.43}$$

By finally assuming that $\boldsymbol{\sigma}^*$ is obtained by a small perturbation of the stress $\boldsymbol{\sigma}$ at the equilibrium state, we can write the stability postulates of Drucker and Il'iushin in a unified manner:

$$\begin{aligned} 0 &\leq \dot{\boldsymbol{\sigma}} \cdot \dot{\boldsymbol{\epsilon}} \\ &\leq \underbrace{\dot{\boldsymbol{\sigma}} \cdot \dot{\boldsymbol{\epsilon}}^e}_{\dot{\boldsymbol{\epsilon}}^e \cdot \boldsymbol{C} \dot{\boldsymbol{\epsilon}}^e \geq 0} + \dot{\boldsymbol{\sigma}} \cdot \dot{\boldsymbol{\epsilon}}^p \\ &\leq \dot{\boldsymbol{\sigma}} \cdot \dot{\boldsymbol{\epsilon}}^p \;;\; \forall \dot{\boldsymbol{\epsilon}} \end{aligned} \tag{8.44}$$

The result in (8.44) above is also known as Drucker's inequality. In 1D case, such a criterion will again confirm that unstable equilibrium states are related to softening response in plasticity (see Figure 8.4 for illustration):

$$\dot{\sigma} \cdot \dot{\epsilon}^p \begin{cases} > 0 \text{ stable: hardening} \\ = 0 \text{ stable: perfect plasticity} \\ < 0 \text{ unstable: softening} \end{cases} \tag{8.45}$$

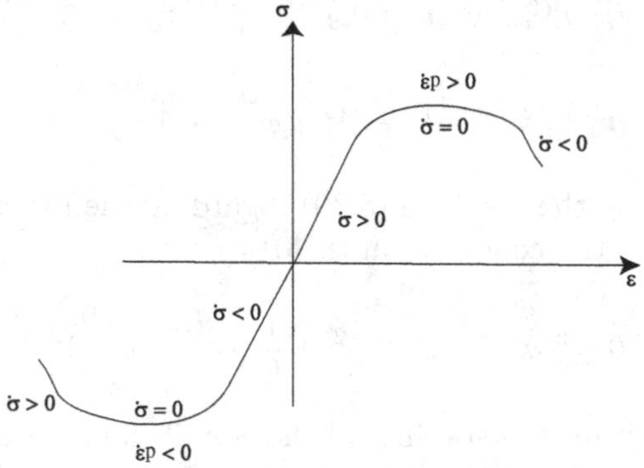

Fig. 8.4 Stability of elastoplastic behavior in 1D case.

By exploiting the results given in (8.42) and (8.44), we can easily show that Drucker's inequality is in agreement with the principle of maximum plastic dissipation; namely, for any plastically admissible stress, we can write:

$$\begin{aligned} \sigma \cdot \dot{\epsilon}^p \geq \sigma^* \cdot \dot{\epsilon}^p \;\Leftrightarrow\; D^p(\sigma) \geq D^p(\sigma^*) \implies \\ (\sigma - \sigma^*) \cdot \dot{\epsilon}^p \geq 0 \Rightarrow D^p(\boldsymbol{\sigma}) = \max_{\sigma^*} D^p(\sigma^*) \end{aligned} \tag{8.46}$$

We can therefore conclude that the softening elastoplastic behavior, when described by (8.38), not only is not stable in the sense of Drucker, but it also no longer obeys the principle of maximum plastic dissipation.
We can show that the softening elastoplastic behavior produces a localized plastic deformation. Several different methods are proposed that are capable of successfully dealing with this kind of localization (or material instability) phenomena. This is done in the next section.
At the end of this section, we discuss how to determine the type of material instability, between a limit point and a bifurcation. For 1D case, with a single non-zero displacement component, there can only be one possibility, which is to find a limit point; the limit point detection criterion for 1D case reduces to:

$$\frac{dq(\zeta)}{d\zeta} =: K = 0 \implies C^{ep} := \frac{E\hat{K}(\zeta)}{E + \hat{K}(\zeta)} = 0 \tag{8.47}$$

For a 3D case, however, we can have more than one material instability mode (e.g. see [25] or [202]). Namely, besides the limit point (equivalent to the peak load in force–displacement diagram for 1D case), we can also find the material instability leading to a bifurcation point. In order to provide the detection criterion for bifurcation points, we follow Rice [227] and consider that a bifurcation point within at homogeneous stress field $\boldsymbol{\sigma}$ breaks the homogeneity of the strain and displacement fields; the latter occurs across the discontinuity surface Γ, which separates the domain Ω into two sub-domains Ω^+ and Ω^- (such that $\Omega^+ \bigcup \Omega^- = \Omega$), each with its own homogeneous strain and displacement fields. It thus follows that the perturbation between these two states ought to be in the form of infinitesimal displacement jump, or rather velocity jump when written in rate-form, across the discontinuity surface Γ. By assuming furthermore that such a velocity jump remains constant along the surface of discontinuity Γ (which is oriented by its exterior normal unit vector $\mathbf{n}$), we can write:

$$\dot{\mathbf{u}}|_\Gamma = cst. \implies [\![\nabla \dot{\mathbf{u}}]\!] = \mathbf{g} \otimes \mathbf{n} = \zeta \mathbf{m} \otimes \mathbf{n} \; ; \; \mathbf{m} = \mathbf{g} / \parallel \mathbf{g} \parallel \tag{8.48}$$

where $[\![\nabla \dot{\mathbf{u}}]\!]$ is the velocity gradient and $\mathbf{g}$ is the vector indicating the direction of the jump (and the type of bifurcation mode). By specifying the corresponding unit vector in that direction $\mathbf{m} = \mathbf{g} / \parallel \mathbf{g} \parallel$, we can fully define the corresponding critical mode for bifurcation. We note that any such mode is placed between mode I with discontinuity in normal direction to Γ and mode II with discontinuity in tangential direction:

$$\begin{aligned} \text{mode I} : \mathbf{m} \parallel \mathbf{n} &\implies \mathbf{m} \cdot \mathbf{n} = 1 \\ \text{mode II} : \mathbf{m} \perp \mathbf{n} &\implies \mathbf{m} \cdot \mathbf{n} = 0 \end{aligned} \tag{8.49}$$

The velocity jump producing the discontinuity in displacement field across the surface Γ does not imply the discontinuity of stress. Namely, the traction vector, which is computed from the Cauchy principle, has to remain continuous across the surface Γ. This can be written:

$$\begin{aligned} \mathbf{0} = [\![\dot{\mathbf{t}}]\!]_\Gamma &:= [\![\dot{\boldsymbol{\sigma}}\mathbf{n} + \boldsymbol{\sigma} \underbrace{\dot{\mathbf{n}}}_{=\mathbf{0}}]\!]_\Gamma \\ &= [\![\dot{\boldsymbol{\sigma}}]\!]_\Gamma \mathbf{n} \end{aligned} \tag{8.50}$$

where $[\![\cdot]\!]$ again denotes the jump. In (8.50) we have taken into account that the discontinuity surface remains fixed in time with $\dot{\mathbf{n}} = 0$, and that it is sufficiently regular to avoid jumps in unit normal vector. If we further introduce in (8.48) the elastoplastic constitutive equations for computing the stress rate $\dot{\boldsymbol{\sigma}}$ as a function of $\nabla\dot{\mathbf{u}}$, we can obtain the following compact result:

$$\begin{aligned} \mathbf{0} &= [\mathcal{C}^{ep}(\mathbf{m} \otimes \mathbf{n})]\mathbf{n} \\ &= \mathbf{A}^{ep}\mathbf{m} \; ; \; A^{ep}_{ik} = n_l \mathcal{C}^{ep}_{ijkl} n_j \end{aligned} \tag{8.51}$$

where $\mathbf{A}^{ep}$ is the elastoplastic acoustic tensor. The result given in (8.51) above represents the detection criterion for material instability of bifurcation type, with the eigenvector $\mathbf{m}$ of the acoustic tensor which specifies the critical mode (placed anywhere between modes I and II). At the critical point, the acoustic tensor is singular, with its determinant equal to zero:

$$\det \mathbf{A}^{ep}(\mathbf{n}) = 0 \tag{8.52}$$

The solution to problem in (8.52) allows us to obtain the orientation $\mathbf{n}$ of the discontinuity surface. In 2D case, where $\mathsf{n} = (n_1, n_2) = (cos\theta, sin\theta)$, we can write the explicit form of the result in (8.52) according to:

$$\begin{aligned} &\det \mathbf{A}^{ep}(\mathbf{n}) := c_0 n_1^4 + c_1 n_1^3 n_2 + c_2 n_1^2 n_2^2 + c_3 n_1 n_2^3 + c_4 n_2^4 = 0 \\ &c_0 = \mathcal{C}^{ep}_{1111}\mathcal{C}^{ep}_{1212} - \mathcal{C}^{ep}_{1112}\mathcal{C}^{ep}_{1211} \\ &c_1 = \mathcal{C}^{ep}_{1111}\mathcal{C}^{ep}_{1222} + \mathcal{C}^{ep}_{1111}\mathcal{C}^{ep}_{2212} - \mathcal{C}^{ep}_{1112}\mathcal{C}^{ep}_{2211} - \mathcal{C}^{ep}_{1122}\mathcal{C}^{ep}_{1211} \\ &c_2 = \mathcal{C}^{ep}_{1111}\mathcal{C}^{ep}_{2222} + \mathcal{C}^{ep}_{1112}\mathcal{C}^{ep}_{1222} + \mathcal{C}^{ep}_{1211}\mathcal{C}^{ep}_{2212} - \mathcal{C}^{ep}_{1122}\mathcal{C}^{ep}_{1212} \\ &\qquad - \mathcal{C}^{ep}_{1122}\mathcal{C}^{ep}_{2211} - \mathcal{C}^{ep}_{1212}\mathcal{C}^{ep}_{2211} \\ &c_3 = \mathcal{C}^{ep}_{1112}\mathcal{C}^{ep}_{2222} + \mathcal{C}^{ep}_{1211}\mathcal{C}^{ep}_{2222} - \mathcal{C}^{ep}_{1122}\mathcal{C}^{ep}_{2212} - \mathcal{C}^{ep}_{1212}\mathcal{C}^{ep}_{2211} \\ &c_4 = \mathcal{C}^{ep}_{1212}\mathcal{C}^{ep}_{2222} - \mathcal{C}^{ep}_{2212}\mathcal{C}^{ep}_{1222} \end{aligned} \tag{8.53}$$

By introducing the change of variable with new unknown $x = tan\theta$, we can rewrite this bifurcation condition in the form of fourth order polynomial:

$$f(x) := c_0 + c_1 x + c_2 x^2 + c_3 x^3 + c_4 x^4 = 0 \; ; \; x = n_2/n_1 \tag{8.54}$$

The minimum value of this function, $df(\bar{x})/dx = 0$, can be obtained explicitly by using Cardan's formulas. It thus only remains to check if such a value is equal to zero, $f(\bar{x}) = 0$, in which case we can conclude that the material instability occurs with the discontinuity surface in the direction $\theta = tan^{-1}\bar{x}$. In 3D case, the explicit solution of the instability condition in (8.52) cannot be obtained, apart the elastic case which has no solution since the determinant of the elastic acoustic tensor always remains positive:

$$\det \mathbf{A}^e(\mathbf{n}) := (\lambda + 2\mu)\mu^2 > 0 \tag{8.55}$$

We thus have to compute a numerical solution:

$$\begin{aligned} & min_{\|\mathbf{n}\|=1}[\det \mathbf{A}^{ep}(\mathbf{n})] \mapsto \bar{\mathbf{n}} \\ & \exists \bar{\mathbf{n}} \text{ such that } \det \mathbf{A}^{ep}(\bar{\mathbf{n}}) = 0 \Longrightarrow \mathbf{A}^{ep}\mathbf{m} = \mathbf{0} \mapsto \bar{\mathbf{m}} \end{aligned} \tag{8.56}$$

where we compute the minimum of $\det \mathbf{A}^{ep}$ for different choices of $\mathbf{n}$, so that all the possible orientations are covered. If the minimum is equal to zero (or below the specified tolerance), we then proceed to computing the critical mode $\mathbf{m}$, which can specify the type of instability.

8.2.2 *Illustration of finite element mesh lack of objectivity for localization problems*

We start this section with a simple illustrative example for loss of objectivity of the computed solution with respect to different choice of finite element mesh in the case of elastoplastic constitutive behavior with softening where the material instability occurs leading to inelastic strain localization. To that end, we consider 1D finite element model of a simple tension test. More precisely, we choose a bar of length l, built-in on one end and submitted on another to a concentrated traction force $\bar{t}$, or rather to an imposed displacement which is more suitable for controlling the softening behavior; see Figure 8.5 for illustration.

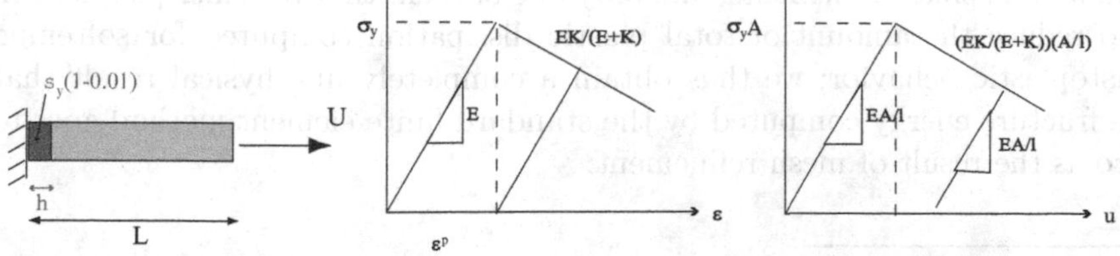

Fig. 8.5 Truss-bar modelling of a simple tension test for elastoplastic softening behavior.

We can obtain from experimental results the energy G_f, yet called the fracture energy, which ought to be introduced in a bar to cause the complete failure that reduces the stress to zero. We would like to obtain the same amount of energy from the numerical solution by integrating the total dissipation in all elements which become plastic:

$$\boxed{D^p_{\Omega^p} = \int_{\Omega^p} (\sigma \dot{\epsilon}^p + q\dot{\zeta})\, dx} \tag{8.57}$$

where Ω^p is the sub-domain of Ω which becomes plastic. In this example, the stress state will remain homogeneous[2] and all the elements will reach the yield stress simultaneously. The latter is only a theoretical result, since in practical computations the round-off numerical errors will then decide which element will become plastic before the others.

In order to be able to better control where instability will occur resulting with localized plastic deformation, we can introduce an imperfection by slightly reducing the yield stress in the chosen element. In this manner we obtain the limit load problem replacing the bifurcation-like problem where the chosen element is the first to become plastic. We note that the imperfection of this kind will not change the physical nature of the problem, since the stress field remains homogeneous along the bar as imposed by the equilibrium equation. Hence, for the case of elastoplastic behavior with hardening, the complete bar will eventually become plastic, starting with the first weakened element. Any finite element mesh will thus lead to the same numerical result regarding the total plastic dissipation (with a minor difference from the weakened element). On the other hand, for the case of elastoplastic behavior with softening (see Figure 8.5) the first weakened element which became plastic will impose the reduction of stress value through softening phase of constitutive behavior in that element. Despite this stress reduction, the equilibrium equation will still require that the stress field remains homogeneous along the bar, with the same value in each element. Therefore, once the softening starts in the weakened element reducing the value of stress, the other elements, which were still elastic, will simply unload and will thus never become plastic. The final result is the plastic zone Ω^p which is limited to a single weakened element. It thus follows that the standard manner for getting a better accuracy of results in finite element analysis by mesh refinement will no longer apply here. In fact, it will have the opposite effect, since the mesh refinement will produce smaller and smaller elements, and only one of them that becomes plastic, will also reduce the amount of total plastic dissipation computed for softening elastoplastic behavior; we thus obtain a completely un-physical result that the fracture energy computed by the standard finite element method goes to zero as the result of mesh refinement:

---

[2] Recall that in a statically determined problem of a truss-bar loaded only at its ends, the stress can be directly computed from equilibrium.

$$G_f^{FEM} := \int_{t_{\epsilon_y}}^{t_{\epsilon_f}} D^p_{\Omega^p}\, dt \implies \Omega^p \mapsto 0 \Leftrightarrow G_f^{FEM} \mapsto 0 \qquad (8.58)$$

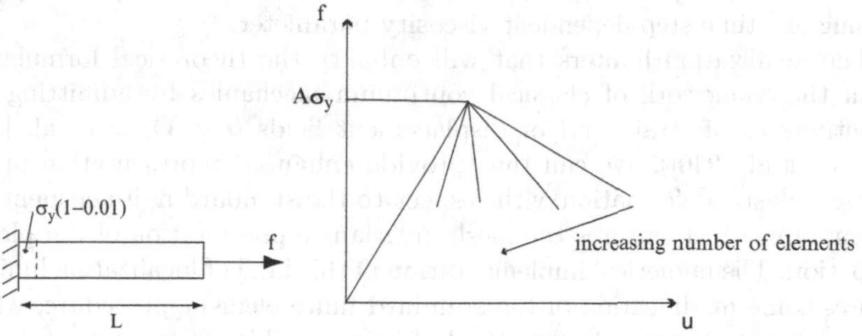

Fig. 8.6 Post-peak response as a function of number of elements for a bar with elasto-plastic softening behavior.

We also obtain that the computed post-peak softening response of the structure is dependent upon the chosen number of elements in the mesh (see Figure 8.6). This lack of objectivity with respect to the chosen mesh is clearly not acceptable, and the standard finite element method cannot be considered reliable in computing the elastoplastic response with softening. The same finding on lack of objectivity of the finite element method for softening problems also applies to 2D and 3D cases.

8.3 Localization limiters

8.3.1 List of localization limiters

In this section we briefly review different methods, often called the localization limiters, proposed to eliminate the finite element mesh lack of objectivity in dealing with softening inelastic behavior. All these methods can be roughly placed within three main groups:
(i) The localization limiters that do not require any modification of the theoretical formulation, which is still placed within the standard continuum mechanics framework. We will only require the corresponding modification of the finite element stress computation with the appropriate, mesh-dependent values of model parameters to ensure the mesh objectivity with respect to computed inelastic dissipation. Among the limiters of this kind, we can first mention the one proposed by Hillerborg et al. [106] or Bicanic et al. [34], where

the finite element mesh objectivity is enforced by using the mesh-dependent softening modulus in order to ensure the invariant amount of plastic dissipation. We can also count herein the localization limiter based on viscoplastic regularization, as proposed in Needleman [198], which ensures mesh independent result for dissipation by replacing the plasticity with viscoplasticity and by using the time step dependent viscosity parameter.
(ii) The localization limiters that will enhance the theoretical formulation within the framework of classical continuum mechanics by admitting the discontinuities of strain and/or displacement fields (e.g. Ortiz et al. [212] or Simo et al. [240]). We can thus provide enhanced representation of the localized plastic deformation with respect to the standard finite element approximation, which ensures the mesh-invariant representation of the plastic dissipation. The numerical implementation of this kind of localization limiters requires some modification of the standard finite element procedure, which is fairly close to the standard method of incompatible modes.
(iii) The localization limiters that require the modifications of the theoretical formulation which no longer fits within the framework of standard continuum mechanics, such as so-called non-local theory of plasticity (Bazant et al. [26]) or non-local damage formulation (Bazant and Pijaudier-Cabot [27]). In non-local theory, the stress value at a point is computed from the (weighted) average of the strains in the chosen neighborhood (e.g. a sphere) of that point, defined by a characteristic length (e.g. the sphere diameter). In this manner we can ensure that the inelastic dissipation remains independent of the chosen finite element mesh, since the non-local neighborhood, as specified by the characteristic length, can also span several elements. Therefore, the stress computation in any element with non-local formulation will impose the communication with all the other elements in the non-local neighborhood, which is not easy to incorporate within the standard finite element code architecture, where we deal with one element at the time. More important disadvantage from this requirement of a more elaborate finite element implementation, is the lack of the physical basis for choosing the appropriate value of the characteristic length in non-local theory. These and other difficulties related to making a proper choice of the non-local neighborhood for a heterogeneous stress field, or to accounting for difference between the volumetric and surface dissipation, were among the reasons which prevented the non-local models from reaching a widespread use in computer codes.
The same conclusion applies to a couple of other localization limiters, which can be derived by an asymptotic expansion from the non-local formulation, such as the gradient plasticity (Coleman [54] or Muhlhaus [192]) and the gradient damage (de Borst et al. [37]). For either of these models, the consistency condition on admissible stress in the sense of plasticity (or damage) criterion is imposed in terms of a differential equation (as opposed to algebraic equation for standard continuum models), which involves the gradients of plastic multiplier and internal variables. This kind of model requires to extend the standard weak form and solve an additional (global) equation

for plastic multiplier, which again employs the characteristic length among the parameters with the same difficulty of defining its sound physical basis. The numerical implementation of gradient plasticity models is fairly complicated by the need to specify the corresponding boundary condition on plastic multiplier, which is in agreement with the evolution of the plastic zone during the given loading program. All this was an important impediment from widespread use of this kind of models in the standard computer codes.
Yet another localization limiter that belongs to this group is given by an extension of the standard continuum model of plasticity (see [38]) by adding the independent rotation field; in the absence of plastic deformation, the latter reduces to the Cosserat continuum, introduced by Cosserat brothers [57]. The addition of an independent rotation field implies the modification of strain measures, with one logical choice being the Biot strain. Moreover, the governing equations imposed by the presence of independent rotations again require that we employ the characteristic length. Finally, another disadvantage of this kind of localization limiter is its limited effectiveness that only pertains to mode II of material instability, with no use for mode I.
We will further elaborate only upon the localization limiters of the first group, which are useful for using in any standard computer code, with the only modification that concerns the choice of parameters dependent upon the finite element mesh or the time-step size. We also present in detail the localization limiters from the second group, which can be handled at the level of a single finite element and thus can easily fit within the standard computer code architecture. The last group of localization limiters cannot be used in the same simple manner for practical problems, and it is not elaborated upon any further.

8.3.2 Localization limiter based on mesh-dependent softening modulus – 1D case

The first localization limiter we present is also the simplest one in the sense that it does not require any modification of the standard plasticity or damage models, or any particular new development in the finite element computer code. It is based on a simple idea to adjust the material softening parameters so that the computed plastic dissipation in the softening regime remains the same regardless of the chosen finite element mesh. For example, for 1D plasticity softening behavior and the finite element model constructed with truss-bar elements representing the simple tension test, we can require that all the plastic dissipation remains concentrated in a single element, the first one to become plastic. If the 2-node truss-bar elements are employed for constructing this mesh, the total dissipation $D^p_{\Omega^p}$ can thus be written:

$$D^p_{\Omega^p} := \int_{l^p} (\sigma \dot{\epsilon}^p + q\dot{\zeta})\, A dx = A l^p [\sigma \dot{\epsilon}^p + q\dot{\zeta}]_{\bar{x}_{l^p}} \tag{8.59}$$

where A is the cross-section of the bar, l^p is the length of the element which is the only one to become plastic and $\bar{x}_{l^p}$ is the Gauss quadrature point of that element. The computed value of the accumulated total dissipation until complete failure (where the stress is reduced to zero) provides the estimate of the fracture energy obtained by this kind of model:

$$G_f^{FEM} := \int_{t_{\epsilon_y}}^{t_{\epsilon_f}} D^p_{\Omega^p}\, dt \; ; \; D^p_{\Omega^p} = Al^p[\sigma\dot{\epsilon}^p + q\dot{\zeta}]_{\bar{x}_{l^p}} \tag{8.60}$$

where t_{ϵ_y} is the pseudo-time value when reaching the limit of elasticity and t_{ϵ_f} is the final time at failure (see Figure 8.6). For the model of associated plasticity with associated hardening, yet called generalized standard material, the evolution equations for all internal variables can be obtained from the yield function potential:

$$\phi(\sigma, q) = |\sigma| - (\sigma_y - q) \leq 0 \, ; \, \dot{\epsilon}^p = \dot{\gamma}\, sign(\sigma) \, ; \, \dot{\zeta} = \dot{\gamma} \tag{8.61}$$

which further allows to write the plastic dissipation in a very compact form directly proportional to the plastic multiplier:

$$\begin{aligned} D^p &= \dot{\gamma}(\underbrace{\sigma sign(\sigma)}_{|\sigma|} + q) \\ &= \underbrace{\dot{\gamma}(|\sigma| - (\sigma_y - q))}_{\dot{\gamma}\phi(\sigma,q)=0} + \dot{\gamma}\sigma_y \\ &= \dot{\gamma}\sigma_y \end{aligned} \tag{8.62}$$

By assuming furthermore the linear softening law, with yet unknown value of the softening modulus K^h to be picked up in accordance with the element size (which is indicated by subscript h), we can obtain the following closed form result for instantaneous plastic dissipation:

$$\dot{\phi} = 0 \,\&\, \dot{q} = K^h\dot{\gamma} \implies D^p := \sigma_y\dot{\gamma} = \sigma_y sign(\sigma)E\dot{\epsilon}/(E + K^h) \tag{8.63}$$

Finally, by integrating the last expression with respect to pseudo-time, we can obtain the finite element estimate of the total dissipation until complete failure, which is given as:

$$\begin{aligned} G_f^{FEM} &= Al^p \int_{\epsilon_y}^{\epsilon_f} \tfrac{\sigma_y sign(\sigma)E}{E+K^h} \overbrace{d\epsilon}^{\dot{\epsilon}\,dt} \\ &= Al^p \tfrac{\sigma_y}{K^h} \underbrace{\frac{EK^h}{E+K^h}(\epsilon_f - \epsilon_y)}_{\sigma_y} \\ &= Al^p \tfrac{\sigma_y^2}{K^h} \end{aligned} \tag{8.64}$$

We will compare this value against the one provided by the classical Griffith criterion of linear fracture mechanics, which defines the fracture according to:

$$G_f = 2Ag_f \tag{8.65}$$

where g_f is the rate of fracture energy per unit area, which can be established experimentally. By requiring that the last two results amount to the same fracture energy, we can obtain the correct value of the softening modulus K^h which is written:

$$K^h = \frac{l^p \sigma_y^2}{2g_f} \tag{8.66}$$

It is obvious that the softening modulus K^h is not only a function of the mechanical properties (σ_y and g_f), but also of the chosen finite element mesh (l^p). This will imply that we need to change this value every time we change the mesh.

8.3.3 Localization limiter based on viscoplastic regularization – 1D case

This localization limiter can easily be exploited within the standard finite element computation, by accepting to replace a softening plasticity model with the corresponding viscoplasticity model. In that sense, the only new parameter to introduce is viscosity, with its numerical value to be chosen in order to ensure the objectivity of results obtained by standard finite element computations for any chosen mesh and time-step. We thus find that such a localization limiter is based upon the viscoplastic regularization, which allows to stabilize the computations with softening plasticity model.
In order to elaborate further upon this kind of localization limiter, we revisit the 1D problem of softening plasticity representing a simple traction test, but with the plasticity law replaced with viscoplasticity. We first start with the perfect plasticity model and its viscoplastic counterpart proposed by Perzyna, which allows to write the corresponding evolution equations according to:

$$\dot{\epsilon}^{vp} = \frac{<\phi(\sigma)>}{\eta^h}\frac{\partial\phi}{\partial\sigma} \; ; \; <\phi> = \begin{cases} \phi \, ; \, \phi > 0 \\ 0 \, ; \, \phi \leq 0 \end{cases} \; ; \; \phi(\sigma) := |\sigma| - \sigma_y \tag{8.67}$$

We can see in (8.67) above that the viscosity η^h is the only new parameter to choose with respect to the perfect plasticity. The superscript 'h' indicates that the viscosity η^h is the parameter that depends on chosen finite element mesh and time-step size, in the manner which will be made explicit subsequently. One way for computing the internal variables in viscoplasticity, which we explained in detail in previous chapters, is by using the computational procedure proposed for plasticity, but with the plastic multiplier replaced with $\dot{\gamma} = <\phi> /\eta$. We will not follow again this kind of approach, but rather turn to the one proposed by Duvaut and Lions [72] in order to show that this kind of viscoplasticity can also be interpreted as the regularization of perfect plasticity model. We thus introduce the period of relaxation $\tau = \frac{\eta}{E}$, which allows to rewrite the evolution equation for viscoplastic strain:

$$\begin{aligned} \dot{\epsilon}^{vp} &= \underbrace{\frac{E^{-1}}{\tau}}_{\frac{1}{\eta}} \underbrace{\{|\sigma| - \sigma_y\}}_{<\phi>} \underbrace{sign(\sigma)}_{\frac{\partial\phi}{\partial\sigma}} \\ &= \frac{E^{-1}}{\tau}\{\sigma - \sigma_y sign(\sigma)\} \end{aligned} \tag{8.68}$$

where $\sigma_y sign(\sigma)$ is the stress value which would have been produced by the perfect plasticity model for the same problem. This value is the asymptotic limit to which the stress produced by viscoplasticity model would converge in a relaxation test, where we apply an initial deformation jump producing the instantaneous stress which can be larger than the yield stress:

$$\begin{aligned} &\epsilon(t) = \epsilon_0 H(t)\,;\; H(t) = \begin{cases} 1;\, t > 0 \\ 0;\, \text{otherwise} \end{cases} \\ &\sigma_0 = E\epsilon_0 \begin{cases} < \sigma_y \Rightarrow \dot{\epsilon}^{vp} = 0 \text{ elastic case} \\ > \sigma_y \Rightarrow \dot{\epsilon}^{vp} \neq 0 \text{ viscoplastic case} \end{cases} \end{aligned} \tag{8.69}$$

The subsequent time-evolution of the stress in the relaxation test can be obtained by solving the following equation:

$$\dot{\sigma} + \frac{1}{\tau}\sigma = E\dot{\epsilon} + \frac{1}{\tau}\sigma_y \text{ with } \epsilon(0) = \epsilon_0 > \frac{\sigma_y}{E} \tag{8.70}$$

By taking into account that the deformation is kept fixed in time in relaxation test $\dot{\epsilon}(t) = 0$, we can obtain the analytic solution of this equation:

$$\left.\begin{aligned} \frac{d\sigma}{\sigma - \sigma_y} &= -\frac{dt}{\tau} \\ \sigma(0) &= E\epsilon_0 \end{aligned}\right\} \Longrightarrow \sigma(t) = (E\epsilon_0 - \sigma_y) exp(-\frac{t}{\tau}) + \sigma_y \tag{8.71}$$

We can thus conclude that the stress produced by this viscoplasticity model in a relaxation test will converge to the yield stress value as set by the perfect plasticity model, $\sigma_\infty = \sigma_y$, with the rate of convergence set by the relaxation period τ.

The same conclusion will apply to the viscoplasticity model with hardening, as well as with softening (with $K < 0$). For one such model, we can write the evolution equation for plastic deformation according to:

$$\dot{\epsilon}^{vp} = \frac{E^{-1}}{\tau}(\sigma - \sigma_\infty)\,;\ \sigma_\infty = \sigma_y + \frac{EK}{E+K}(\epsilon - \epsilon_y)\ ;\ \tau = \frac{\eta}{E+K} \tag{8.72}$$

where σ_∞ is the corresponding value of stress produced by standard plasticity model for the same deformation. This result on relationship between two models is fully equivalent to the one in (8.70) already obtained for perfect plasticity and viscoplasticity, but with σ_∞ replacing σ_y and with the period of relaxation now equal to $\tau = \frac{\eta}{E+K}$.
The stress computation in local phase of the operator split solution procedure for viscoplasticity is carried out in the manner equivalent to the relaxation test, where the total strain to be imposed comes from the elastic trial step. Integrating such an evolution equation by using the implicit backward Euler scheme, for a typical time step $h = t_{n+1} - t_n$, we can obtain the following result:

$$\sigma_{n+1} = \underbrace{\sigma_n + Eh\dot{\epsilon}_{n+1}}_{\sigma_{n+1}^{trial}} + \frac{h}{\tau}(\sigma_{\infty,n+1} - \sigma_{n+1}) \tag{8.73}$$

The last equation can be rewritten in the form which shows that the final value of stress produced by this viscoplasticity model can be written as a linear combination of the trial stress σ_{n+1}^{trial} and the corresponding value of plastically admissible stress $\sigma_{\infty,n+1}$ produced by the classical plasticity for the same deformation:

$$\sigma_{n+1} = \frac{1}{1+\frac{h}{\tau}}\sigma_{n+1}^{trial} + \frac{\frac{h}{\tau}}{1+\frac{h}{\tau}}\sigma_{\infty,n+1} \tag{8.74}$$

The last result allows to easily compute the consistent tangent elasto-viscoplastic modulus according to:

$$C_{n+1}^{vp} := \frac{\partial \sigma_{n+1}}{\partial \epsilon_{n+1}} = \frac{1}{1+\frac{h}{\tau}}E + \frac{\frac{h}{\tau}}{1+\frac{h}{\tau}}C_\infty^{ep}\ ;\ C_\infty^{ep} = \frac{EK}{E+K} \tag{8.75}$$

It is therefore easy to check that the tangent modulus can remain positive, even for softening viscoplasticity (with $K < 0$), until the following limit value is reached:

$$C_{n+1}^{vp} := \frac{E}{1+\frac{h}{\tau}}(1 + \frac{h}{\eta^h}K) \implies C_{n+1}^{vp} \geq 0 \Leftrightarrow (1 + \frac{h}{\eta^h}K) \geq 0 \tag{8.76}$$

It thus follows that for the given values of the softening modulus $K < 0$ and the time step h, we can choose the artificial viscosity parameter η^h for (8.76) leading to a positive value of the viscoplastic tangent modulus, which will eliminate the mesh-dependency:

$$\eta^h \geq h|K| \implies C_{n+1}^{vp} \geq 0 \tag{8.77}$$

In other words, this kind of localization limiter will eliminate the result sensitivity with respect to the chosen finite element mesh by using the viscoplastic regularization of the softening elastoplastic response.

8.3.4 Localization limiter based on displacement or deformation discontinuity – 1D case

In this section we discuss the localization limiter which is based on enhanced strain representation introducing the displacement discontinuity. We will first explain the details for 1D case. We present in particular the new theoretical formulation including a novel interpretation of localized plastic deformation capable of providing a more reliable representation of the strain and displacement fields, as well as the finite element implementation procedure that fits within the framework of incompatible mode method. We also comment on how to accommodate within the same framework a similar localization limiter, where we introduce the strain discontinuities.
We start again with a model problem representing the simple tension test; namely, the bar with elastoplastic softening behavior is built-in on one end and driven on the other by applied traction (or imposed displacement). The governing equation of the problem can be summarized as:

$$\begin{array}{ll}
\text{equilibrium:} & \frac{\partial \sigma}{\partial x} + b = 0 \\
\text{kinematics:} & \epsilon = \frac{\partial u}{\partial x} \\
\text{boundary conditions:} & u(0,t) = 0 \ ; \ u(l,t) = g(t) \\
\text{additive decomposition of def.:} & \epsilon = \epsilon^e + \epsilon^p \\
\text{internal energy:} & \psi(\epsilon, \epsilon^p, \zeta) = \frac{1}{2} E (\epsilon - \epsilon^p)^2 + \frac{1}{2} K \zeta^2 \\
\text{yield function:} & \phi(\sigma, q) = |\sigma| - (\sigma_y - q) \\
\text{stress constitutive eq.:} & \sigma = \frac{\partial \psi}{\partial \epsilon} = E(\epsilon - \epsilon^p) \\
\text{hardening constitutive eq.:} & q = -\frac{\partial \psi}{\partial \zeta} = -K\zeta \\
\text{plastic flow:} & \dot{\epsilon}^p = \dot{\gamma} \frac{\partial \phi}{\partial \sigma} = \dot{\gamma}\, sign(\sigma) \\
\text{hardening/softening flow:} & \dot{\zeta} = \dot{\gamma} \frac{\partial \phi}{\partial q} = \dot{\gamma}
\end{array} \tag{8.78}$$

All these equations are fully justified only for the case of hardening plasticity, where $K > 0$. It is already shown in the previous section that the same equations cannot deliver the right answer for softening plasticity (with $K < 0$), but produces instead a mesh-dependent response for localized plastic deformation which is not in agreement with experimental results. Correcting this deficiency and ensuring the unique value of computed solution for any (sufficiently) refined finite element mesh is the main goal of the localization limiter based on displacement discontinuity. The latter is introduced in order to provide the proper reinterpretation of the fields featuring in all the governing equations in (8.78), in order to be able to apply them to the softening plasticity case.

The stress field is the only one which is not modified when we switch to softening plasticity. Namely, the equilibrium equation, the first of those in (8.78), would still have to remain valid and it would imply that the stress field ought to remain sufficiently regular for its derivative to match the applied loading. For example, in the problem representing the simple tension test where the applied distributed load is equal to zero, the stress remains constant along the bar and equal to the applied traction divided by the cross-section area, even when entering the softening phase of the response.
On the other hand, the localization implies heterogeneities of the displacement and strain fields, which no longer remain regular ever for smooth stress field. In order to account for these heterogeneities, we will introduce the discontinuity of the displacement as the extreme case of localization. More precisely, the displacement field $u(x,t)$ will be written as the sum of a sufficiently smooth, regular part $\bar{u}(x,t)$ and the discontinuous part $\bar{\bar{u}}(x,t)$. The latter can further be represented in terms of product between the Heaviside function $H_{\bar{x}}(x)$, which introduces a unit jump at $\bar{x}$, and the time-dependent displacement jump $[\![\bar{\bar{u}}(t)]\!]$; the displacement field can thus be written:

$$u(x,t) = \bar{u}(x,t) + [\![\bar{\bar{u}}(t)]\!] H_{\bar{x}}(x) \; ; \; H_{\bar{x}}(x) = \begin{cases} 1; \, x > \bar{x} \\ 0; \, x \leq \bar{x} \end{cases} \tag{8.79}$$

This kind of displacement field belongs to the space of functions of bounded variation $u \in BV(\Omega)$ (see Temam [254]). The corresponding deformation field that is produced by the displacement field in (8.79) above, will feature the Dirac delta function leading to:

$$\epsilon(x,t) := \frac{\partial u}{\partial x} = \bar{u}_{,x}(x,t) + [\![\bar{\bar{u}}]\!](t)\delta_{\bar{x}}(x) \; ; \; \delta_{\bar{x}}(x) = \begin{cases} \infty; \, x = \bar{x} \\ 0; \;\; \text{otherwise} \end{cases} \tag{8.80}$$

We can show that such a displacement field will provide a novel interpretation of the (localized) plastic deformation; namely, the last result and those in (8.78) will allow us to express the stress rate according to:

$$E^{-1}\dot{\sigma}(x,t) = (\dot{\bar{u}}_{,x}(x,t) + [\![\dot{\bar{\bar{u}}}(t)]\!]\delta_{\bar{x}}(x)) - \dot{\gamma}(x,t)\;\text{sign}(\sigma(x,t)) \tag{8.81}$$

Since the terms on the left hand side of the last equation always remain bounded, so should be the case with the terms on the right. This can be ensured if the plastic multiplier is proportional to Dirac function with $\dot{\gamma}(x,t) = \dot{\bar{\bar{\gamma}}}(t)\delta_{\bar{x}}(x)$, which allows us to conclude that:

$$\underbrace{E^{-1}\dot{\sigma}}_{\dot{\epsilon}^e} = \dot{\bar{u}}_{,x} + \underbrace{\left[[\![\dot{\bar{\bar{u}}}]\!] - \dot{\bar{\bar{\gamma}}}\;\text{sign}(\sigma) \right] \delta_{\bar{x}}(x)}_{=0} \tag{8.82}$$

We can also conclude that $\text{sign}(\sigma) = \text{sign}[\![\dot{\bar{\bar{u}}}]\!]$. The plastic deformation in softening plasticity will thus remain localized at the discontinuity, and we can restate the corresponding expression in (8.82) according to:

$$\dot{\epsilon}^p := \dot{\bar{\bar{\gamma}}}\, sign(\sigma)\, \delta_{\bar{x}}(x) = [\![\dot{\bar{\bar{u}}}]\!]\, \delta_{\bar{x}}(x) \implies \dot{\bar{\bar{\gamma}}} = |[\![\dot{\bar{\bar{u}}}]\!]| \tag{8.83}$$

The yield function will thus only concern the stress value at the point of discontinuity $\bar{x}$, where the plastic stain is localized:

$$0 = \phi|_{\bar{x}} := [\, |\sigma| - (\sigma_y - q)\,]_{\bar{x}} \tag{8.84}$$

The plastic consistency conditions should then be applied in the sense of distribution as the only appropriate manner imposed by the plastic multiplier interpretation; this further allows to conclude that internal variable q, governing the plasticity threshold evolution due to softening, should also remain bounded:

$$0 = \int_{\Omega} \dot{\gamma}\, \phi\, dx \equiv \dot{\bar{\bar{\gamma}}} \phi|_{\bar{x}} \implies \dot{q}|_{\bar{x}} = -[\dot{\sigma}\, sign(\sigma)]_{\bar{x}} \tag{8.85}$$

In view of proper interpretation of the plastic multiplier in (8.83) and the constitutive relation in (8.78), it further follows that the softening modulus must also be interpreted in the sense of distribution with Dirac delta function:

$$\dot{\bar{\bar{\gamma}}} \delta_x(x) = -K^{-1} \dot{q}|_{\bar{x}} \implies K^{-1} = \bar{\bar{K}}^{-1}\, \delta_{\bar{x}}(x) \tag{8.86}$$

We thus conclude that the plasticity threshold evolution, which is driven by q, will remain directly proportional to the displacement discontinuity with $\bar{\bar{K}}$ as the factor of proportionality:

$$\dot{q}|_{\bar{x}} = -\bar{\bar{K}}\, |[\![\dot{\bar{\bar{u}}}]\!]| \tag{8.87}$$

Furthermore, we conclude that for linear softening law with $\bar{\bar{K}} < 0$ constant, the ultimate value of displacement discontinuity resulting with zero value of the effective stress can be obtained as:

$$0 = \sigma_y + \bar{\bar{K}}\, |[\![\bar{\bar{u}}]\!]_{max}| \implies |[\![\bar{\bar{u}}]\!]_{max}| = \frac{\sigma_y}{|\bar{\bar{K}}|} \tag{8.88}$$

We can compute that the fracture energy G_f, established in a simple tension test, as the equivalent energy of external forces applied to a truss-bar with constitutive behavior of softening plasticity that is needed to reduce the effective stress to zero. This computation is performed by integrating the total dissipation as defined from the Clausius–Duhem inequality for a given loading program, which starts with zero applied traction at time t_0, peaks to maximum value at time t_1 and goes down to zero value again as the result of softening at time t_2; we can thus write:

$$\int_{t_1}^{t_2} \underbrace{[\int_{\Omega} \sigma_y \, \dot{\gamma} \, dx]}_{\sigma_y [\![\dot{\bar{u}}]\!]} \, dt = \underbrace{\int_{t_0}^{t_2} \int_{\Omega} \sigma \dot{\epsilon} \, dx \, dt}_{G_f} - \int_{t_1}^{t_2} \frac{d}{dt} \underbrace{\int_{\Omega} \Xi(\zeta)}_{-(1/2)[\![\bar{u}]\!]|\bar{\bar{K}}|[\![\bar{u}]\!]} \quad dx \, dt \quad (8.89)$$

For the softening plasticity model with linear softening modulus $\bar{\bar{K}} = cst. < 0$, the zero value of the effective stress corresponds to the maximum value of the displacement discontinuity $|[\![\bar{u}]\!]_{max}| = \sigma_y / |\bar{\bar{K}}|$; from (8.89) above, we can further obtain that:

$$\sigma_y \frac{\sigma_y}{|\bar{\bar{K}}|} = G_f + \frac{1}{2} \frac{\sigma_y}{|\bar{\bar{K}}|} |\bar{\bar{K}}| \frac{\sigma_y}{|\bar{\bar{K}}|} \implies |\bar{\bar{K}}| = \frac{1}{2} \frac{\sigma_y^2}{G_f} \quad (8.90)$$

The last result shows that the fracture energy is equal to the area of triangle below the softening part of the response, which starts at the peak value of resistance (where $\sigma = \sigma_y$) and which ends at the final failure state with zero effective stress (where $\sigma = 0$).
It is also possible to obtain the corresponding expression for G_f with any other form of post-peak softening response. For example, we can use the exponential softening that allows any value of displacement discontinuity instead the linear softening law with the maximum authorized value to avoid the stress sign reversal. For the chosen value of fracture energy G_f, the exponential softening law will produce the following plasticity threshold evolution:

$$\begin{array}{l} q(\zeta) = \sigma_y [1 - exp(-\frac{\sigma_y}{G_f} \zeta)] \;\&\; \zeta \equiv |[\![\bar{u}]\!]| \\ \implies \int_0^\infty \sigma_y exp(-\frac{\sigma_y}{G_f} \zeta) \, d\zeta \; = G_f \end{array} \quad (8.91)$$

8.3.4.1 Numerical implementation

The numerical implementation of the softening plasticity model with displacement discontinuity can nicely fit within the framework of the incompatible mode method. We will illustrate the pertinent details of numerical implementation for the simplest finite element model using 2-node isoparametric truss-bar elements. In order to allow for displacement discontinuity representation, without modifying the displacement values at the nodes, we need an additional parameter α^e (which can further be treated in the same manner as the incompatible mode parameter):

$$\begin{array}{l} x(\xi)|_{\Omega^e} = \sum_{a=1}^{2} N_a(\xi) x_a \, ; \; N_a(\xi) = \frac{1}{2}(1 + \xi_a \xi) \\ u^h(x,t)|_{\Omega^e} = \sum_{a=1}^{2} N_a(x) d_a(t) + M^e(x) \alpha^e(t) \, ; \\ M^e(\xi) = \begin{cases} -\frac{1}{2}(1+\xi), & \text{si } \xi \in [-1, 0] \\ \frac{1}{2}(1-\xi), & \text{si } \xi \in [0, 1] \end{cases} \end{array} \quad (8.92)$$

The proposed interpolation will place the displacement discontinuity in the center of this element. The corresponding strain approximation can then be written:

$$\begin{aligned} &\epsilon^h(x,t)|_{\Omega^e} = \textstyle\sum_{a=1}^{2} \underbrace{B_a(x)}_{\frac{dN_a}{dx}} d_a(t) + \underbrace{G^e(x)}_{\frac{dM}{dx}} \alpha^e(t) \\ &B_a^e(\xi) = \tfrac{(-1)^a}{l^e}\,;\; G^e(x) := \begin{cases} -\frac{1}{l^e}, & \text{if } \xi \in [-1,0) \bigcup (0,1] \\ -\frac{1}{l^e} + \delta_0,\ \xi = 0 \end{cases} \end{aligned} \tag{8.93}$$

where δ_0 is the Dirac function centered within the element. In order to ensure the method convergence in the spirit of the patch test, we will make sure that the incompatible mode variation remains orthogonal to the constant stress in each element σ^e; such a work-conjugate couple should thus satisfy:

$$\underbrace{\sigma^e}_{cst.} \int_{\Omega^e} G^e \, d\Omega^e = 0 \Leftrightarrow \int_{\Omega^e} \underbrace{-\frac{1}{l^e}}_{\tilde{G}^e} dx = -\int_{\Omega^e} \delta_0 \, dx \tag{8.94}$$

For the present 1D case, with displacement discontinuity placed in the center of the element, the patch test condition in (8.94) is automatically satisfied leaving G^e intact. We can thus keep exactly the same approximation for the virtual strain field as the one used in (8.93) for real strains, which can be written:

$$\left.\frac{dw(x)}{dx}\right|_{l^e} = \sum_{a=1}^{2} B_a^e(x)\, w_a + G^e(x)\, \beta^e \tag{8.95}$$

The stress field approximation can be obtained from the regular part of the real strain field in (8.93) with no contribution from the singular part representing the plastic strain; we can thus write:

$$\sigma^h(x,t)|_{\Omega^e} = E(\sum_{a=1}^{2} B_a(x)\, d_a(t) + \tilde{G}^e(x)\, \alpha^e(t)) \tag{8.96}$$

By exploiting the approximations for virtual strains and stress, defined in (8.95) and (8.96) respectively, the weak form of equilibrium equations can be recast in the format typical of incompatible mode method:

$$\begin{aligned} &\mathop{\mathbb{A}}_{e=1}^{n_{elem}} (\mathsf{f}^{int,e} - \mathsf{f}^{ext,e}) = 0\,;\; \mathsf{f}^{int,e} = \textstyle\int_{\Omega^e} \mathsf{B}^{e,T} \sigma(d,\alpha^e)\, dx \\ &h^e = 0 \quad \forall\, e \in [1, n_{elem}]\,;\; h^e = \textstyle\int_{\Omega^e} \tilde{G}^T \sigma(d,\alpha^e)\, dx + t(\alpha) \end{aligned} \tag{8.97}$$

where $t = \sigma|_{\bar{x}}$ is the traction force acting at discontinuity. This force, which indirectly depends upon the value of parameter α^e, enters the corresponding plasticity criterion:

$$\phi(t,q) := |t| - (\sigma_y - q) \leq 0 \tag{8.98}$$

where q is the variable defining the current plasticity threshold depending upon the displacement discontinuity.

We will solve this problem by using the operator split solution procedure for finding the solution to equations in (8.97) and (8.98) in an equivalent manner to the one already proposed for standard plasticity model. Namely, we will treat separately the local phase from global phase of computation and solve the former (defined in (8.97)) for total displacement field (and incompatible mode parameter) and then the latter (defined in (8.98)) for the corresponding localized plastic flow. There remains, however, one subtle point to be addressed concerning the role of parameter α^e, which contributes to both local and global phase of computations.

By starting with the local phase of computations, we will assume to be given the best iterative value of displacement $\mathsf{d}_{n+1}^{e,(i)}$, for which we can further perform:

Local computation: displacement discontinuity $\forall e$:

$$\begin{aligned} &\text{Given: } \mathsf{d}_{n+1}^{e,(i)}, \alpha_n^e \ , \ h = t_{n+1} - t_n \\ &\text{Find: } \ \alpha_{n+1}^e, \text{ such that } \phi_{n+1} \leq 0 \end{aligned} \tag{8.99}$$

Between two possibilities for the corresponding value of yield function at time t_{n+1} ($\phi_{n+1} \leq 0$), we will first test for the elastic trial state produced with a zero value of the plastic multiplier associated with discontinuity $\dot{\bar{\bar{\gamma}}}_{n+1}^{trial} = 0$:

$$\left.\begin{aligned} &\bar{\bar{\gamma}}_{n+1}^{trial} := \dot{\bar{\bar{\gamma}}}_{n+1} h = 0 \\ &\alpha_{n+1}^e = \alpha_n^e \end{aligned}\right\} \Rightarrow \begin{cases} t_{n+1}^{trial} = -\int_{\Omega^e} \tilde{G}^T [E(\sum_{a=1}^2 B_a^e \, d_{a,n+1}^{(i)} + \tilde{G}^e \alpha_n^e)] \, dx \\ \phi_{n+1}^{trial} = |t_{n+1}^{trial}| - (\sigma_y - q_n) \end{cases} \tag{8.100}$$

If the elastic trial state does not produce a positive value of the yield function, $\phi_{n+1}^{trial} \leq 0$, we can conclude that the step is indeed elastic and accept the trial values as final with no modification of the plastic strain from the one computed in the previous step. In such a case the incompatible mode parameter α_n^e will also remain intact, and the only change of the traction force can come from a displacement increment; we can thus write:

$$\begin{aligned} &Lin[t_{n+1}]_{\alpha^e} := t_{n+1} + \mathsf{K}_d \, \Delta \mathsf{d}_{n+1} \ ; \ \mathsf{K}_d := \frac{\partial \bar{t}}{\partial \mathsf{d}} = E\mathsf{B}^T \\ &K_{\alpha^e} := \frac{\partial \bar{t}}{\partial \alpha^e} = 0 \end{aligned} \tag{8.101}$$

On the other hand, if the trial value of the yield function is positive, $\phi_{n+1}^{trial} > 0$, the elastic trial step is no longer considered plastically admissible. We are thus sure that the current step is plastic, and that we have to modify the elastic strain and the incompatible mode parameter α_n^e to re-establish the plastic admissibility of the driving traction at discontinuity. Such a computation is carried out in the manner which allows to exploit the results obtained for the trial elastic state. Namely, we can first express the traction at discontinuity:

$$\begin{aligned} \bar{t}_{n+1} &= E(\mathsf{B}^T \mathsf{d}_{n+1}^{(i)} + \tilde{G}^e \alpha_{n+1}^e) \\ &= \underbrace{E(\mathsf{B}^T \mathsf{d}_{n+1}^{(i)} - \frac{1}{l^e}\alpha_n)}_{\bar{t}_{n+1}^{trial}} - \frac{E}{l^e} \underbrace{(\alpha_{n+1}^e - \alpha_n^e)}_{\bar{\bar{\gamma}}_{n+1} sign(\bar{t}_{n+1})} \\ &= \bar{t}_{n+1}^{trial} - \frac{E}{l^e} \bar{\bar{\gamma}}_{n+1} sign(\bar{t}_{n+1}) \end{aligned} \tag{8.102}$$

and then use the last result to write the final, plastically admissible value of the yield function in terms of the corresponding modification of its trial value:

$$\begin{aligned} 0 = \phi(\bar{t}_{n+1}, q_{n+1}) &:= |t_{n+1}| - (\sigma_y - q_{n+1}) \\ &= \underbrace{|\bar{t}_{n+1}^{trial}| - (\sigma_y - q_n)}_{\phi_{n+1}^{trial}} + (q_{n+1} - q_n) - \frac{E}{l^e} \bar{\bar{\gamma}}_{n+1} \end{aligned} \tag{8.103}$$

For a linear softening law, where $q = -\bar{\bar{K}}\,\alpha^e$ with a constant value of $\bar{\bar{K}}$, we can express the last term in (8.103) above in terms of $\bar{\bar{\gamma}}_{n+1}$:

$$q_{n+1} - q_n = -\bar{\bar{K}} \bar{\bar{\gamma}}_{n+1} \tag{8.104}$$

We can thus obtain the final value of the plastic multiplier in this step, which is proportional to the computed trial value of the yield function, and carry out the corresponding update of the incompatible mode parameter:

$$\bar{\bar{\gamma}}_{n+1} = \frac{\phi_{n+1}^{trial}}{(\frac{E}{l^e} + \bar{\bar{K}})} \; ; \; \alpha_{n+1}^e = \alpha_n^e + \bar{\bar{\gamma}}_{n+1} sign(\bar{t}_{n+1}^{trial}) \tag{8.105}$$

For a nonlinear softening law, the result equivalent to (8.105) ought to be obtained iteratively, according to:

$$\begin{aligned} &(j) = 1, 2, \ldots \\ &\text{as long as: } \phi_{n+1}^{(j)} := \phi_{n+1}^{trial} - \frac{E}{l^e} \bar{\bar{\gamma}}_{n+1}^{(j)} + [\hat{q}(\alpha_{n+1}^{e,(j)}) - \hat{q}(\alpha_n^e)] > 0 \; (\approx \text{tol.}) \\ &\Longrightarrow Lin[\phi_{n+1}^{(j)}] := \phi_{n+1}^{(j)} + D_{\bar{\bar{\gamma}}} \phi_{n+1}^{(j)} \, \Delta \bar{\bar{\gamma}}_{n+1}^{(j)} \\ &\bar{\bar{\gamma}}_{n+1}^{(j+1)} = \bar{\bar{\gamma}}_{n+1}^{(j)} + \Delta \bar{\bar{\gamma}}_{n+1}^{(j)} \, ; \; \alpha_{n+1}^{e,(j+1)} = \alpha_{n+1}^{e,(j)} + \Delta \bar{\bar{\gamma}}_{n+1}^{(j)} \, sign(t_{n+1}^{trial}) \\ &(j) \longleftarrow (j+1) \end{aligned} \tag{8.106}$$

In a plastic step, the modification of the discontinuity traction is produced by a change of incompatible mode parameter α^e; hence, the corresponding value of the elastoplastic tangent modulus can be written:

$$K_\alpha := \frac{\partial \bar{t}}{\partial \alpha} = \bar{\bar{K}} sign(t_{n+1}^{trial}) \ ; \ \mathsf{K}_d := \frac{\partial \bar{t}}{\partial \mathsf{d}} = 0 \tag{8.107}$$

Having converged with local computation to the final value of incompatible mode parameter α_{n+1}^e (for either elastic or plastic step), we turn back to the global computation in order to check the convergence of the displacement field; the latter can be written:

Global computation: equilibrium equations

$$\begin{aligned}
&\text{Given:} \quad \mathsf{d}_{n+1}^{(i)}, \alpha_{n+1}^e, \text{ with } h_{n+1}^e(\mathsf{d}_{n+1}^{(i)}, \alpha_{n+1}^e) = 0 \\
&\text{as long as: } \| \mathop{\mathbb{A}}_{e=1}^{n_{elem}} (\hat{\mathsf{f}}^{e,int}(\mathsf{d}_{n+1}^{(i)}, \alpha_{n+1}^e) - \mathsf{f}^{e,ext}) \| > 0 \ (\approx \text{tol.}) \\
&\Longrightarrow \quad 0 = Lin[\mathop{\mathbb{A}}_{e=1}^{n_{elem}} (\hat{\mathsf{f}}^{e,int}(\mathsf{d}_{n+1}^{(i)}, \alpha_{n+1}^e) - \mathsf{f}^{e,ext})] \\
&\qquad := \mathop{\mathbb{A}}_{e=1}^{n_{elem}} (\hat{\mathsf{f}}^{e,int}(\mathsf{d}_{n+1}^{(i)}, \alpha_{n+1}^e) - \mathsf{f}^{e,ext}) + [\mathop{\mathbb{A}}_{e=1}^{n_{elem}} \hat{\mathsf{K}}_{n+1}^{e,(i)}] \Delta \mathsf{d}_{n+1}^{(i)} \\
&\qquad \mathsf{d}_{n+1}^{(i+1)} = \mathsf{d}_{n+1}^{(i)} + \Delta \mathsf{d}_{n+1}^{(i)}
\end{aligned} \tag{8.108}$$

We assumed in (8.108) above that Newton's iterative procedure can be employed to provide an improved guess on best iterative value of displacement field; consequently, the contribution of a typical element to the linearized form of the system in (8.108) is defined as:

$$\begin{aligned}
&Lin[\mathsf{f}^{e,int}(\mathsf{d}_{n+1}^{(i)}, \alpha_{n+1}^e) - \mathsf{f}^{e,ext}] = \mathsf{f}^{e,int}(\mathsf{d}_{n+1}^{(i)}, \alpha_{n+1}^e) - \mathsf{f}^{e,ext} \\
&\quad + \mathsf{K}_{n+1}^{e,(i)} \Delta \mathsf{d}_{n+1}^{e,(i)} + \mathsf{F}_{n+1}^{e,(i)} \Delta \alpha_{n+1}^{e,(i)} \\
&Lin[h_{n+1}^e(\mathsf{d}_{n+1}^{(i)}, \alpha_{n+1}^e)] = \underbrace{h_{n+1}^e(\mathsf{d}_{n+1}^{(i)}, \alpha_{n+1}^e)}_{=0} + \mathsf{F}_{n+1}^{e,(i),T} \Delta \mathsf{d}_{n+1}^e \\
&\quad + \mathsf{H}_{n+1}^{(i),e} \Delta \alpha_{n+1}^e + \mathsf{K}_d \Delta \mathsf{d}_{n+1}^e + \mathsf{K}_\alpha \Delta \alpha_{n+1}^e
\end{aligned} \tag{8.109}$$

where the element matrices can be written explicitly:

$$\begin{aligned}
\mathsf{K}_{n+1}^e &= \int_{\Omega^e} \mathsf{B}^{e,T} \ \mathsf{C}_{n+1}^{ep,(i)} \ \mathsf{B}^e \ dx \\
\mathsf{F}_{n+1}^e &= \int_{\Omega^e} \mathsf{B}^{e,T} \ \mathsf{C}_{n+1}^{ep,(i)} \ \tilde{\mathsf{G}}^e \ dx \\
\mathsf{H}_{n+1}^e &= \int_{\Omega^e} \tilde{\mathsf{G}}^{e,T} \ \mathsf{C}_{n+1}^{ep,(i)} \ \hat{\mathsf{G}}^e \ dx
\end{aligned} \tag{8.110}$$

The right choice of matrices K_d and K_α in (8.109) above is made according to (8.101) for elastic or to (8.107) for plastic step. It is indicated in (8.109) that the residual at discontinuity is equal to zero, since we have obtained the converged value of the incompatible model parameter $\bar{\alpha}^e_{n+1}$. This allows to carry out the static condensation and thus reduce the tangent stiffness matrix to the standard form with the size corresponding to the nodal displacement vector only:

$$\underbrace{[\mathsf{K}^{e,(i)}_{n+1} - \mathsf{F}^e_{n+1}(H^e_{n+1} + K_\alpha)^{-1}(\mathsf{F}^{e,T}_{n+1} + \mathsf{K}_d)]}_{\hat{K}^e_{n+1}} \Delta d^{e,(i)}_{n+1} = \mathsf{f}^{e,ext}_{n+1} - \mathsf{f}^{e,int,(i)}_{n+1} \tag{8.111}$$

We can thus proceed with the standard finite element assembly procedure to account for each element contribution to global equilibrium equations, solve this system and carry out the corresponding displacements update:

$$\hat{\mathsf{K}}^{(i)}_{n+1} \Delta \mathsf{d}^{(i)}_{n+1} = \mathsf{f}^{ext}_{n+1} - \mathsf{f}^{int}_{n+1} \Longrightarrow \mathsf{d}^{(i+1)}_{n+1} = \mathsf{d}^{(i)}_{n+1} + \Delta \mathsf{d}^{(i)}_{n+1} \tag{8.112}$$

With this result in hand, we can restart, if needed, another sweep of the operator split procedure.

8.3.4.2 Deformation discontinuity

In this section we revisit the same localization problem produced with softening plasticity in a simple tension test, in order to briefly illustrate another successful localization limiter. The latter can be interpreted as a modification of the displacement discontinuity localization limiter introduced previously, as the corresponding regularized form where we only introduce the discontinuity of deformation; the latter still allows for the continuity of displacement, and we can write:

$$[\![\frac{\partial u}{\partial x}]\!] := \frac{\partial u}{\partial x}|_{x^+\mapsto\bar{x}} - \frac{\partial u}{\partial x}|_{x^-\mapsto\bar{x}} \neq 0 \; ; \; u|_{x^+\mapsto\bar{x}} - u|_{x^-\mapsto\bar{x}} = 0 \tag{8.113}$$

With the strain discontinuity of this kind we can construct the corresponding finite element approximation of the displacement field that remains suitable for handling the localization problems. For a truss-bar 2-node isoparametric element, this results with:

$$u^h(x,t)|_{\Omega^e} = \textstyle\sum_{a=1}^2 N_a(x) d_a(t) + M^e(x)\alpha^e(t)\,;$$
$$M^e(\xi) = \begin{cases} -\frac{1}{2}(1+\xi),\ \forall\xi \in \ [-1, -\frac{b}{l^e}] \\ \frac{1}{2}(\frac{l^e}{b} - 1)\xi,\ \forall\xi \in \ [-\frac{b}{l^e}, \frac{b}{l^e}] \\ \frac{1}{2}(1-\xi),\ \ \forall\xi \in \ [\frac{b}{l^e}, 1] \end{cases} \tag{8.114}$$

where b is the chosen parameter. The latter will determine the band (or a sub-domain within the element Ω^e) where the plastic deformation is likely to localize, which can easily be seen from the finite element discrete approximation of strain field which can be constructed from (8.114):

$$\epsilon^h(x,t)|_{\Omega^e} = \sum_{a=1}^{2} B_a(x)d_a(t) + G^e(x)\alpha^e(t)\,;$$
$$G^e(x) = \begin{cases} -\frac{1}{l^e}, & \forall\xi \in [-1, -\frac{b}{l^e}] \bigcup [\frac{b}{l^e}, 1] \\ -\frac{1}{l^e} + \frac{1}{b}, & \forall\xi \in [-\frac{b}{l^e}, \frac{b}{l^e}] \end{cases} \tag{8.115}$$

For a small (non-zero) value of parameter b we obtain a regularization of the Dirac delta function employed by the displacement discontinuity localization limiter, which allows us to keep all strain values finite throughout the element. It is important to note, however, that the strain does not remain constant along the element, and that the value inside the band is much larger than the strain value outside; it is thus quite likely that the plastic deformation will remain limited to the chosen band for the case of softening plasticity. The advantage of such a localization limiter is to allow us to keep the standard interpretation of the plastic deformation. Furthermore, with the standard interpretation of the plastic strains, we can also keep the usual procedure for computing the stress and the standard operator split solution method for classical plasticity. However, one should be cautious about the proper choice of the Gauss quadrature rule for the present case where the strain field is not continuous,[3] and realize that even for higher order quadrature rules the final result will remain inaccurate. Therefore, an alternative implementation is preferable where we split the strain field dependent upon α into regular part and the Dirac delta approximation (the latter does not contribute to stress computation again performed as in (8.96)) and use the operator split solution procedure in the same manner as the one explained in previous section for the localization limiter with displacement discontinuity.

8.4 Localization limiter in plasticity for massive structure

8.4.1 *Theoretical formulation of limiter with displacement discontinuity – 2D/3D case*

In this section we will generalize the developments on localization limiter for softening plasticity based upon displacement discontinuity, and make it applicable to practical problem. First, we will extend the theoretical formulation and numerical implementation of this localization limiter to 2D and

[3] Recall that the Gauss quadrature rule implies replacing the true (discontinuous) function with a continuous polynomial.

3D case; second, we will elaborate upon the plasticity model which can account for plasticity threshold evolution both by initial hardening and by final softening regime. The latter is needed for reliable representation of the failure mechanisms for massive structures, where the displacement discontinuity (representing a single macro-crack or a shear-band) is surrounded by the fracture process zone (with a large number of micro-cracks), whose contribution cannot be neglected (as usually done in linear fracture mechanics). These two inelastic mechanisms will co-exist and communicate in each element, with continuum plasticity model used for fracture process zone representation and displacement discontinuity for shear-band.[4]
The 2D/3D continuum plasticity model will be defined by its three principal ingredients: the additive decomposition of total deformation into elastic and plastic part, the strain energy density and the yield function:

$$\begin{aligned} \boldsymbol{\epsilon} &= \boldsymbol{\epsilon}^e + \boldsymbol{\epsilon}^p \\ \psi(\boldsymbol{\epsilon}^e, \zeta) &= \underbrace{\frac{1}{2}\boldsymbol{\epsilon}^e \cdot \boldsymbol{\mathcal{C}}\boldsymbol{\epsilon}^e}_{\psi^e(\boldsymbol{\epsilon}^e)} + \underbrace{\frac{1}{2}\zeta K \zeta}_{\Xi(\zeta)} \\ \phi(\boldsymbol{\sigma}, q) &:= \parallel \boldsymbol{\sigma} \parallel_P - (\sigma_y - q) \end{aligned} \tag{8.116}$$

By the standard procedure using the second principle of thermodynamics and the principle of maximum plastic dissipation, we can then obtain the constitutive equations for this model, the evolution equations and the plastic dissipation in the fracture process zone:

$$\begin{aligned} \boldsymbol{\sigma} &= \frac{\partial \psi^e(\cdot)}{\partial \boldsymbol{\epsilon}^e} \;;\; q = -\frac{\partial \Xi(\cdot)}{\partial \zeta} \\ \dot{\boldsymbol{\epsilon}}^p &= \dot{\gamma}\frac{\partial \phi}{\partial \boldsymbol{\sigma}} \;;\; \dot{\zeta} = \dot{\gamma}\frac{\partial \phi}{\partial q} \\ 0 &\leq D^p := \boldsymbol{\sigma} \cdot \dot{\boldsymbol{\epsilon}}^p + q\dot{\zeta} \end{aligned} \tag{8.117}$$

We will consider the yield criterion defined by a homogeneous function of degree-one, which implies that:

$$\frac{\partial \parallel \boldsymbol{\sigma} \parallel_P}{\partial \boldsymbol{\sigma}} \cdot \boldsymbol{\sigma} \equiv \parallel \boldsymbol{\sigma} \parallel_P \Longrightarrow \frac{\partial \parallel \boldsymbol{\sigma} \parallel_P}{\partial \boldsymbol{\sigma}} \equiv \frac{\partial \phi}{\partial \boldsymbol{\sigma}} \tag{8.118}$$

The plasticity criterion of this kind allows us to express the plastic dissipation as directly proportional to the plastic multiplier:

$$D^p := \underbrace{\boldsymbol{\sigma} \cdot \overbrace{\dot{\boldsymbol{\epsilon}}^p}^{\dot{\gamma}\frac{\partial \|\boldsymbol{\sigma}\|_P}{\partial \boldsymbol{\sigma}}} + q \overbrace{\dot{\zeta}}^{\dot{\gamma}} - \dot{\gamma}\,\sigma_y}_{\dot{\gamma}\phi = 0} + \dot{\gamma}\,\sigma_y \tag{8.119}$$

[4] For macro-crack representation with displacement discontinuity, we refer to anisotropic damage model described in Chapter 3.

All these relations remain valid for the plastic response with hardening. Most of them ought to be modified accordingly in the softening regime. First, in order to provide the appropriate interpretation of the localized plastic deformation, the displacement field is split into a regular part $\bar{\mathbf{u}}$ and displacement discontinuity $\bar{\bar{\mathbf{u}}}$, centered at the surface of discontinuity Γ_s and with the domain of influence limited to a sub-domain $\tilde{\Omega} \in \Omega$; we can thus write the corresponding representation for the displacement field:

$$\begin{aligned} &\mathbf{u}(\mathbf{x},t) = \bar{\mathbf{u}}(\mathbf{x},t) + [\![\bar{\bar{\mathbf{u}}}(t)]\!] M_{\Gamma_s}(\mathbf{x}) \; ; \; M_{\Gamma_s}(\mathbf{x}) = H_{\Gamma_s}(\mathbf{x}) - N_{\tilde{\Omega}}(\mathbf{x}) \; ; \\ &H_{\Gamma_s}(\mathbf{x}) = \begin{cases} 1 \, ; \, \mathbf{x} \in \Omega^+ \\ 0 \, ; \, \mathbf{x} \in \Omega^- \end{cases} \end{aligned} \tag{8.120}$$

where the choice of $N_{\tilde{\Omega}}(\mathbf{x})$ will decide the domain of influence for discontinuity. The deformation field can also be split additively into a regular part, which can have both elastic and plastic component, and a singular part coming from the displacement discontinuity, which will provide an additional contribution to plastic deformation; we can thus write:

$$\begin{aligned} &\boldsymbol{\epsilon}(\mathbf{x},t) = \nabla^s \bar{\mathbf{u}}(\mathbf{x},t) + \tilde{\mathbf{G}}(\mathbf{x}) [\![\bar{\bar{\mathbf{u}}}(t)]\!] + ([\![\bar{\bar{\mathbf{u}}}(t)]\!] \otimes \mathbf{n})^s \delta_\Gamma(\mathbf{x}) \, ; \\ &\tilde{\mathbf{G}}(\mathbf{x}) := -\nabla^s N_{\tilde{\Omega}}(\mathbf{x}) \end{aligned} \tag{8.121}$$

where $\mathbf{n}$ is the unit vector of exterior normal to the surface of discontinuity Γ_s. The last result is obtained by exploiting the following identity from distribution theory (see Stakgold [246]):

$$\int_{\tilde{\Omega}} \varphi(\mathbf{x}) \underbrace{\delta_{\Gamma_s}(\mathbf{x})}_{\nabla H_{\Gamma_s}(\mathbf{x})} \, dV = \int_{\Gamma_s} \varphi(\mathbf{x}) \, \mathbf{n} \, dA \tag{8.122}$$

We will employ subsequently the notation which allows to indicate explicitly the type of discontinuity, according to:

$$[\![\bar{\bar{\mathbf{u}}}]\!] = \alpha \mathbf{m} \; ; \quad \begin{aligned} &\mathbf{m} \perp \mathbf{n} \, (\Leftrightarrow \mathbf{m} \cdot \mathbf{n} = 0) \text{ mode II - sliding} \\ &\mathbf{m} \parallel \mathbf{n} \, (\Leftrightarrow \mathbf{m} \cdot \mathbf{n} = 1) \text{ mode I - openning} \end{aligned} \tag{8.123}$$

We also assume that the discontinuity surface, once activated, will remain fixed, resulting with $\dot{\mathbf{n}} = \mathbf{0}$. This result and the Cauchy principle will then allow us to conclude that the time-derivative of the traction vector acting across the discontinuity can be written:

$$\begin{aligned} \dot{\mathbf{t}} &= \dot{\boldsymbol{\sigma}} \mathbf{n} \\ &= \mathcal{C}(\nabla^s \dot{\bar{\mathbf{u}}} + \tilde{\mathbf{G}} [\![\dot{\bar{\bar{\mathbf{u}}}}]\!] + ([\![\dot{\bar{\bar{\mathbf{u}}}}]\!] \otimes \mathbf{n}) \delta_{\Gamma_s} - \dot{\gamma} \frac{\partial \phi}{\partial \boldsymbol{\sigma}}) \mathbf{n} \\ &= \mathcal{C}(\nabla^s \dot{\bar{\mathbf{u}}} + \tilde{\mathbf{G}} [\![\dot{\bar{\bar{\mathbf{u}}}}]\!]) \mathbf{n} + \mathbf{A}^e \underbrace{[\![\dot{\bar{\bar{\mathbf{u}}}}]\!]}_{\mathbf{m}\dot{\alpha}} \delta_{\Gamma_s} - \dot{\gamma} \mathcal{C} \frac{\partial \phi}{\partial \boldsymbol{\sigma}} \mathbf{n} \end{aligned} \tag{8.124}$$

where $\mathbf{A}^e = \mathbf{n}\boldsymbol{\mathcal{C}}\mathbf{n}$ is the elastic acoustic tensor. We will first follow the development from the previous section for 1D case, and assume that the plastic multiplier can be written:

$$\dot{\gamma} = \bar{\bar{\dot{\gamma}}}\delta_{\Gamma_s} \tag{8.125}$$

With this kind of hypothesis we can keep the traction vector bounded, if the last two terms in (8.124) above would cancel each other implying that:

$$\mathbf{A}^e \mathbf{m}\dot{\alpha} = \bar{\bar{\dot{\gamma}}}\boldsymbol{\mathcal{C}}\frac{\partial \phi}{\partial \boldsymbol{\sigma}}\mathbf{n} \tag{8.126}$$

The value of the plastic multiplier can be obtained from the plastic consistency condition requiring that the time-derivative of the yield function remains equal to zero:

$$0 = \dot{\phi} := \frac{\partial \phi}{\partial \boldsymbol{\sigma}} \cdot \boldsymbol{\mathcal{C}}(\nabla^s \dot{\mathbf{u}} + \tilde{\mathbf{G}}\mathbf{m}\dot{\alpha}) + \underbrace{\frac{\partial \phi}{\partial q}}_{=1}\dot{q} + \delta_{\Gamma_s}\frac{\partial \phi}{\partial \boldsymbol{\sigma}} \cdot [\boldsymbol{\mathcal{C}}(\mathbf{m}\otimes\mathbf{n})^s\dot{\alpha} - \bar{\bar{\dot{\gamma}}}\boldsymbol{\mathcal{C}}\frac{\partial \phi}{\partial \boldsymbol{\sigma}}] \tag{8.127}$$

We require the last term in the expression above to cancel, which leads to:

$$\bar{\bar{\dot{\gamma}}} = \dot{\alpha}\Upsilon \; ; \; \Upsilon = \boldsymbol{\mathcal{C}}\frac{\partial \phi}{\partial \boldsymbol{\sigma}} \cdot (\mathbf{m}\otimes\mathbf{n})^s / (\frac{\partial \phi}{\partial \boldsymbol{\sigma}} \cdot \boldsymbol{\mathcal{C}}\frac{\partial \phi}{\partial \boldsymbol{\sigma}}) \tag{8.128}$$

The last result allows us to restate the condition in (8.124) with the requirements on bounded traction, in the format featuring the elastoplastic acoustic tensor $\mathbf{A}^{ep}$:

$$\begin{aligned} \mathbf{0} &= \{\underbrace{[\boldsymbol{\mathcal{C}} - \boldsymbol{\mathcal{C}}\frac{\partial \phi}{\partial \boldsymbol{\sigma}} \otimes \boldsymbol{\mathcal{C}}\frac{\partial \phi}{\partial \boldsymbol{\sigma}} / (\frac{\partial \phi}{\partial \boldsymbol{\sigma}} \cdot \boldsymbol{\mathcal{C}}\frac{\partial \phi}{\partial \boldsymbol{\sigma}})]}_{\boldsymbol{\mathcal{C}}^{ep}}(\mathbf{m}\otimes\mathbf{n})^s\}\mathbf{n} \\ &= \mathbf{A}^{ep}\mathbf{m}\alpha \; ; \; \mathbf{A}^{ep} = \mathbf{n}\boldsymbol{\mathcal{C}}^{ep}\mathbf{n} \end{aligned} \tag{8.129}$$

We have thus recovered the classical format of the localization condition, such as proposed by Rice [231] or Hill [105]. An alternative interpretation of the localization condition in (8.129) above is possible, indicating that the localization is a material instability phenomenon leading to a critical equilibrium state with the instability mechanism described by the critical mode $\mathbf{m}$; this can be written:

$$[\mathbf{A}^{ep} - \underbrace{\lambda_c}_{=0}\mathbf{1}]\mathbf{m} = \mathbf{0} \tag{8.130}$$

The last result combined with the one in (8.127) can be used to show that the time-derivative of the variable q, which controls the evolution of the plasticity threshold, is proportional to the stress rate:

$$\dot{q} = -\frac{\partial \phi}{\partial \boldsymbol{\sigma}} \cdot \boldsymbol{\mathcal{C}}(\nabla^s \dot{\mathbf{u}} + \tilde{\mathbf{G}}\dot{\alpha}) \equiv -\frac{\partial \phi}{\partial \boldsymbol{\sigma}} \cdot \dot{\boldsymbol{\sigma}} \tag{8.131}$$

Therefore, we can conclude that the stress-like variable q has to remain bounded, in the same way as stress $\boldsymbol{\sigma}$. On the other hand, the internal strain-like softening variable ζ does not remain bounded, since the chosen model enforces its equivalence with the plastic multiplier, which is proportional to the Dirac delta function; we are thus inevitably led to conclusion that the softening modulus should also be written in terms of distribution with:

$$\left.\begin{array}{l}\dot{\zeta} = \dot{\gamma} := \dot{\bar{\bar{\gamma}}}\delta_{\Gamma_s} \\ \dot{\zeta} = -K^{-1}\dot{q}\end{array}\right\} \Longrightarrow K^{-1} = \bar{\bar{K}}^{-1}\delta_{\Gamma_s} \tag{8.132}$$

By combining the last two results and exploiting the explicit expression for $\dot{\bar{\bar{\gamma}}}$ in (8.128), we obtain the corresponding evolution equation for displacement jump:

$$\dot{\alpha} = \frac{\bar{\bar{K}}^{-1}}{\boldsymbol{C}\frac{\partial\phi}{\partial\boldsymbol{\sigma}} \cdot (\mathbf{m} \otimes \mathbf{n})^s / (\frac{\partial\phi}{\partial\boldsymbol{\sigma}} \cdot \boldsymbol{C}\frac{\partial\phi}{\partial\boldsymbol{\sigma}})}(\frac{\partial\phi}{\partial\boldsymbol{\sigma}} \cdot \dot{\boldsymbol{\sigma}}) \tag{8.133}$$

The second case we consider herein is the localized failure in a massive structure, where the dissipation along the discontinuity surface should be accompanied with the dissipation in the fracture process zone. The strain energy density in this case can thus be written:

$$\psi(\bar{\boldsymbol{\epsilon}}^e, \bar{\zeta}, \bar{\bar{\zeta}}) = \underbrace{\psi^e(\bar{\boldsymbol{\epsilon}}^e) + \bar{\Xi}(\bar{\zeta})}_{\text{regular}} + \bar{\bar{\Xi}}(\bar{\bar{\zeta}})\delta_{\Gamma_s} \tag{8.134}$$

where the first term is the elastic energy, whereas the second and the third are the contributions of hardening and softening mechanisms, respectively. In order to be able to handle these two mechanisms independently, the plastic multiplier ought to have a more general form than the one given in (8.125), with two components: one defined on the discontinuity surface and the other in the surrounding volume:

$$\dot{\gamma} = \dot{\bar{\gamma}} + \dot{\bar{\bar{\gamma}}}\delta_{\Gamma_s} \tag{8.135}$$

With last two results in hand, we can express the total plastic dissipation as the sum of the dissipation from fracture process zone in $\tilde{\Omega}$ and from the discontinuity on Γ_s, which can be written:

$$\begin{aligned} 0 \leq D^p_{\tilde{\Omega}} &:= \textstyle\int_{\tilde{\Omega}} \boldsymbol{\sigma} \cdot \dot{\boldsymbol{\epsilon}} - \dot{\psi}(\bar{\boldsymbol{\epsilon}}^e, \bar{\zeta}, \bar{\bar{\zeta}})\, dV \\ &= \textstyle\int_{\tilde{\Omega}} [(\boldsymbol{\sigma} - \frac{\partial\psi^e}{\partial\boldsymbol{\epsilon}^e}) \cdot \dot{\bar{\boldsymbol{\epsilon}}}^e + \boldsymbol{\sigma} \cdot \dot{\boldsymbol{\epsilon}}^p + (-\frac{d\bar{\Xi}}{d\zeta})\dot{\bar{\zeta}}]\, dV + \int_{\Gamma_s} (-\frac{d\bar{\bar{\Xi}}}{d\bar{\bar{\zeta}}})\dot{\bar{\bar{\zeta}}}\, dS \\ &+ \textstyle\int_{\tilde{\Omega}} \boldsymbol{\sigma} \cdot \tilde{\mathbf{G}}\mathbf{m}\dot{\alpha}\, dV + \int_{\Gamma_s} (\mathbf{t} \cdot \mathbf{m})\dot{\alpha}\, dA \end{aligned} \tag{8.136}$$

where $\mathbf{t} := (\boldsymbol{\sigma}\mathbf{n})|_{\Gamma_s}$ is the traction vector acting on discontinuity. For the case of elastic process, the last result will allow us to define the constitutive equations for the stress, and the hardening and softening stress-like variables controlling the evolution of the plasticity threshold:

$$\boldsymbol{\sigma} = \frac{\partial \psi^e}{\partial \bar{\boldsymbol{\epsilon}}^e} \; ; \; \bar{q} = -\frac{d\bar{\Xi}}{d\bar{\zeta}} \; ; \; \bar{\bar{q}} = -\frac{d\bar{\bar{\Xi}}}{d\bar{\bar{\zeta}}} \tag{8.137}$$

By assuming these equations to remain valid for the case of the most general plastic process, we can than obtain the additive decomposition of the total plastic dissipation into a regular and a singular part:

$$D^p_{\tilde{\Omega}} = \int_{\tilde{\Omega}} (\boldsymbol{\sigma} \cdot \dot{\bar{\boldsymbol{\epsilon}}}^p + \bar{q}\dot{\bar{\zeta}})\, dV + \int_{\Gamma_s} \bar{\bar{q}}\dot{\bar{\bar{\zeta}}}\, dA \tag{8.138}$$

For the last result to be valid, we ought to eliminate the last term in (8.136), resulting with the corresponding condition connecting the driving forces of two dissipation mechanisms:

$$\int_{\tilde{\Omega}} \boldsymbol{\sigma} \cdot \tilde{\mathbf{G}}\mathbf{m}\dot{\alpha}\, dV + \int_{\Gamma_s} (\mathbf{t} \cdot \mathbf{m})\dot{\alpha}\, dA = 0 \tag{8.139}$$

Each plastic dissipation mechanism activation is controlled by the corresponding plasticity criterion. First, we require that the stress tensor $\boldsymbol{\sigma}$ remains plastically admissible in domain $\tilde{\Omega}$ in the sense of the following plasticity criterion:

$$0 \geq \bar{\phi}(\boldsymbol{\sigma}, \bar{q}) := \| \boldsymbol{\sigma} \|_P - (\sigma_y - \bar{q}) \tag{8.140}$$

The second plastic admissibility condition is imposed upon the traction vector component t_m acting upon the surface of discontinuity Γ_s, as specified by the following plasticity criterion:

$$0 \geq \bar{\bar{\phi}}(t_m \bar{\bar{q}}) := |\underbrace{\mathbf{t} \cdot \mathbf{m}}_{t_m}| - (t_y - \bar{\bar{q}}) \tag{8.141}$$

The plastic multiplier computation is now done in two steps. We first compute the regular part of the plastic multiplier from the plastic consistency condition related to yield function $\bar{\phi}$ according to:

$$\left. \begin{array}{l} \bar{\phi} = 0 \\ \dot{\bar{\boldsymbol{\epsilon}}}^p = \dot{\bar{\gamma}} \frac{\partial \bar{\phi}}{\partial \boldsymbol{\sigma}} \\ \dot{\bar{\xi}} = \dot{\bar{\gamma}} \end{array} \right\} \Longrightarrow \dot{\bar{\gamma}} = \frac{1}{\frac{\partial \phi}{\partial \boldsymbol{\sigma}} \cdot \mathcal{C} \frac{\partial \phi}{\partial \boldsymbol{\sigma}} + \bar{K}} (\frac{\partial \phi}{\partial \boldsymbol{\sigma}} \cdot \mathcal{C}\dot{\bar{\epsilon}}) \tag{8.142}$$

For the computed plastically admissible value of stress, we can then carry on to compute the singular part of the plastic multiplier from the plastic consistency condition related to the yield function $\bar{\bar{\phi}}$, which results with:

$$\left.\begin{array}{l}\bar{\bar{\phi}} = 0 \\ \dot{\bar{\bar{\xi}}} = \dot{\bar{\bar{\gamma}}}\end{array}\right\} \Longrightarrow \int_{\Gamma_s} \dot{\bar{\bar{\gamma}}}\, dA = \frac{1}{\bar{\bar{K}}} \int_{\Omega} \boldsymbol{\sigma} \cdot \tilde{\mathbf{G}}\mathbf{m}\, dV \tag{8.143}$$

8.4.2 Numerical implementation within framework of incompatible mode method

In order to illustrate the finite element approximation and solution procedure for this kind of plasticity models, we choose the one first proposed in Ibrahimbegovic and Brancherie [117]. This plasticity model employs the von Mises plasticity criterion (typical of metallic materials) for controlling the fracture process zone dissipation, along with the saturation hardening of exponential type:

$$\begin{array}{c} 0 \geq \bar{\phi}(\boldsymbol{\sigma}, \bar{q}) := \sqrt{\frac{3}{2}} \parallel dev[\boldsymbol{\sigma}] \parallel - (\sigma_y - \bar{q}) \\ \bar{q} = K\bar{\zeta} - (\sigma_\infty - \sigma_y)\,(1 - exp[-\beta \zeta]) \end{array} \tag{8.144}$$

For this kind of criterion, we can obtain the corresponding form of the acoustic tensor:

$$\mathbf{A}^{ep} = \mu \mathbf{1} - \frac{2\mu}{(1 + \frac{\bar{K}}{3\mu}) \parallel dev[\boldsymbol{\sigma}] \parallel} dev[\boldsymbol{\sigma}]\mathbf{n} \otimes dev[\boldsymbol{\sigma}]\mathbf{n} \tag{8.145}$$

The localization condition can be written explicitly in terms of the principal stress values, expressing the critical value of the hardening modulus $\bar{K}$ and the corresponding instability mode defined by angle θ with respect to the principal axis of the largest stress:

$$\bar{K} = 3\mu[(\frac{\sqrt{2}\tau}{\parallel dev[\boldsymbol{\sigma}] \parallel} sin2\theta)^2 - 1] \; ; \; \tau = \frac{1}{2}(\sigma_1 - \sigma_2) \; ; \; \mathbf{n}^T = [cos\theta, sin\theta] \tag{8.146}$$

If the elastic response is considered as quasi-incompressible, with $\nu \approx 0.5$, we can find the limit point at $\bar{K} = 0$ with the critical mode producing the shear band at 45° with respect to the maximum principal stress:

$$\theta = \frac{1}{2} \sin^{-1}(\frac{\parallel dev[\boldsymbol{\sigma}] \parallel}{\sqrt{2}\tau}) \Rightarrow \theta = \frac{\pi}{4} \text{ (if } \nu = 0.5) \tag{8.147}$$

This direction is also the direction of the maximum shear stress, which confirms the classical result on shear band localization and yield line theory.

We next address the discrete approximation issues. The quasi-incompressible elastic response, as well as the choice of von Mises criterion enforcing incompressible plastic response, can lead to the serious locking problem with difficulties for standard isoparametric finite element approximation in representing quasi-incompressible deformation modes. In order to solve this locking problem, we employ the assumed strain method, often referred to as $\bar{B}$ (e.g. see [110] or [242]). The latter provides the special finite element interpolation for the strain field of a 4-node isoparametric element:

$$\bar{\boldsymbol{\epsilon}}|_{\Omega^e} = \sum_{a=1}^{4} \bar{\mathbf{B}}_a^e \mathbf{d}_a^e \; ; \; \bar{\mathbf{B}}_a^e = \mathbf{B}_a^{dev,e} + \bar{\mathbf{B}}_a^{vol,e} \tag{8.148}$$

where only the average value of the spherical part of deformation tensor is used within the element:

$$\bar{\mathbf{B}}_a^{vol,e} = \frac{1}{3}\mathbf{1} \otimes \bar{\mathbf{b}}_a \; ; \; \bar{\mathbf{b}}_a = \int_{\Omega^e} \mathbf{b}_a \, dV \; ; \; tr[\boldsymbol{\epsilon}]|_{\Omega^e} = \sum_{a=1}^{4} \mathbf{b}_a^T \mathbf{d}_a^e \tag{8.149}$$

By assuming that the influence domain of the discontinuity is limited to a single element, with $\tilde{\Omega} \equiv \Omega^e$, we can further construct the finite element approximation of the displacement field including the discontinuity represented by the Heaviside function H_{Γ_s}; the strain field should then combine both $\bar{B}$-approximation and the derivatives of displacement discontinuity, leading to:

$$\begin{aligned} &\bar{\bar{\boldsymbol{\epsilon}}}|_{\Omega^e}(\mathbf{x}, t) = \tilde{\mathbf{G}}^e(\mathbf{x})\bar{\bar{\alpha}}(t) + \bar{\bar{\alpha}}(t)(\mathbf{m} \otimes \mathbf{n})^s \delta_{\Gamma_s^e} \\ &\tilde{\mathbf{G}}^e = -\textstyle\sum_{b \in \Omega^{e,+}} \bar{\mathbf{B}}_b^e \mathbf{m} + \frac{1}{\Omega^e} \int_{\Omega^e} \sum_{b \in \Omega^{e,+}} \bar{\mathbf{B}}_b^e \mathbf{m} \, dV - \frac{l_{\Gamma_s^e}}{\Omega^e} (\mathbf{m} \otimes \mathbf{n})^s \end{aligned} \tag{8.150}$$

where $\delta_{\Gamma_s^e}$ is the Dirac function in domain Ω^e and $l_{\Gamma_s^e}$ the total length of discontinuity in the element. By using the same interpolation for the virtual strain field, we can write the corresponding discrete approximation of the weak form of equilibrium equations in the format typical of incompatible mode method:

$$\begin{aligned} &\mathop{\mathbb{A}}_{e=1}^{n_{elem}} \left(\mathbf{f}_{n+1}^{int,e} - \mathbf{f}_{n+1}^{ext,e} \right) = \mathbf{0} \; ; \; \mathbf{f}_{n+1}^{int,e} = \textstyle\int_{\Omega^e} \bar{\mathbf{B}}^e \cdot \boldsymbol{\sigma}_{n+1} \, dV \\ &h_{n+1}^e := \textstyle\int_{\Omega^e} \tilde{\mathbf{G}}^e \cdot \boldsymbol{\sigma}_{n+1} \, dV + \int_{\Gamma_s^e} t_{m,n+1} \, dA \; ; \; \forall e \in [1, n_{el}] \end{aligned} \tag{8.151}$$

The stress tensor $\boldsymbol{\sigma}_{n+1}$ and the traction force at discontinuity in the last equation have to remain plastically admissible in the sense which is specified by the chosen plasticity criteria:

$$\bar{\gamma}_{n+1}\bar{\phi}(\boldsymbol{\sigma}_{n+1}, \bar{q}_{n+1}) = 0 \; ; \; \bar{\bar{\gamma}}_{n+1}\bar{\bar{\phi}}(t_{m,n+1}, \bar{\bar{q}}_{n+1}) = 0 \tag{8.152}$$

The computation of the internal variables, which ensures the plastic admissibility of stress and traction at discontinuity, remains independent in each element. The regular part of plastic multiplier will have the corresponding value at each of four Gauss quadrature points, whereas the value of the singular part of the plastic multiplier is computed only in the center of the element where discontinuity will always pass regardless of its orientation. The displacement and displacement jump increments are computed from the linearized form of the equilibrium equations, which can be written:

$$\begin{aligned}
&\mathop{\mathbb{A}}_{e=1}^{n_{elem}} (\mathbf{K}^{e,(i)}_{n+1} \Delta \mathbf{d}^{(i)}_{n+1} + \mathbf{F}^{e,(i)}_{n+1} \Delta\alpha^{(i)}_{n+1} = \mathbf{f}^{ext,e}_{n+1} - \mathbf{f}^{int,e,(i)}_{n+1}) \\
&[\mathbf{F}^{e,(i)}_{n+1} + K^{e,(i)}_{d,n+1}]\Delta \mathbf{d}^{(i)}_{n+1} + [\mathbf{H}^{e,(i)}_{n+1} + K^{e,(i)}_{\alpha,n+1}]\Delta\alpha^{(i)}_{n+1} = -h^{e,(i)}_{n+1} \ ; \ \forall e \\
&\qquad \mathbf{d}^{(i+1)}_{n+1} = \mathbf{d}^{(i)}_{n+1} + \Delta \mathbf{d}^{(i)}_{n+1} \\
&\Longrightarrow \\
&\qquad \alpha^{(i+1)}_{n+1} = \alpha^{(i)}_{n+1} + \Delta\alpha^{(i)}_{n+1}
\end{aligned} \tag{8.153}$$

where:

$$\begin{aligned}
&\mathbf{K}^{e,(i)}_{n+1} = \int_{\Omega^e} \bar{\mathbf{B}}^{e,T} \mathcal{C}^{ep,(i)}_{n+1} \bar{\mathbf{B}}^e \, dV \ ; \ \mathbf{F}^{e,(i)}_{n+1} = \int_{\Omega^e} \bar{\mathbf{B}}^{e,T} \mathcal{C}^{ep,(i)}_{n+1} \tilde{\mathbf{G}} \mathbf{m} \, dV \ ; \\
&H^{e,(i)}_{n+1} = \int_{\Omega^e} \mathbf{m}^T \tilde{\mathbf{G}}^{e,T} \bar{\bar{\mathcal{C}}}^{ep,(i)} \tilde{\mathbf{G}}^e \mathbf{m} \, dV \ ; \\
&k^{(i)}_{d,n+1} = l_{\Gamma^e_s} \frac{\partial t_m}{\partial \mathbf{d}^e}\Big|^{(i)}_{n+1} ; \ k^{(i)}_{\alpha,n+1} = l_{\Gamma^e_s} \bar{\bar{\mathcal{C}}}^{ep,(i)}_{n+1} \ ; \ \bar{\bar{\mathcal{C}}}^{ep,(i)}_{n+1} = \frac{\partial t_m}{\partial \alpha}\Big|^{(i)}_{n+1}
\end{aligned} \tag{8.154}$$

The global iterative method of this kind using the linearized form in (8.153) is not necessarily the most efficient nor the most robust. We can rather employ the operator split solution procedure, which will separate solution for new iterative values of displacements from computing the incompatible mode parameter by enforcing the plasticity criterion at the discontinuity; the latter is equivalent to the Gauss-point local computation of the plastic strain, which ought to be reinterpreted in accordance with an alternative definition of α as the localized plastic deformation at discontinuity. More precisely, for the best iterative guess of displacement $\mathbf{d}^{(i)}_{n+1}$ and the incompatible mode parameter $\alpha^{(i)}_{n+1}$ providing the value of plastically admissible stress, we will obtain the trial value of driving traction acting at discontinuity, which can be written:

$$0 = h^{(i),trial}_{n+1} := \int_{\Omega^e} \tilde{\mathbf{G}}^e \cdot \boldsymbol{\sigma}_{n+1} \, dV + \int_{\Gamma^e_s} t^{trial}_{m,n+1} \, dA \implies t^{trial}_{m,n+1} \tag{8.155}$$

If the corresponding trial value of the plasticity criteria remains negative or zero, we need not modify the incompatible mode parameter by this local computation:

$$\bar{\bar{\phi}}^{trial}_{n+1} \leq 0 \implies \alpha_{n+1} = \alpha^{(i)}_{n+1} \tag{8.156}$$

On the contrary, if the trial value is positive $\bar{\bar{\phi}}_{n+1}^{trial} > 0$, this implies the shear band is activated and we need to compute the corresponding localized plastic deformation that will re-establish the plastic admissibility of the traction force acting on Γ_s^e:

$$
\begin{aligned}
&(j) = 1, 2, \ldots \\
&\mathrm{IF}(\bar{\bar{\phi}}_{n+1}^{(j)} > 0 \, (\approx \mathrm{tol.}))\mathrm{THEN} \\
&\frac{\partial \bar{\bar{\phi}}_{n+1}^{(j)}}{\partial \alpha_{n+1}^{(j)}} \Delta\alpha_{n+1}^{(j)} = -\bar{\bar{\phi}}_{n+1}^{(j)} \\
&\qquad\qquad \alpha_{n+1}^{(j+1)} = \alpha_{n+1}^{(j)} + \Delta\alpha_{n+1}^{(j)} \quad \Longrightarrow \; \mathrm{next}(j) \\
&\mathrm{ELSE} \\
&\alpha_{n+1} = \alpha_{n+1}^{(j)} \quad \Longrightarrow \; \mathrm{next}(i)
\end{aligned}
\tag{8.157}
$$

Having reached convergence in local computations, we carry on with the next iterative sweep of the global computation, which provides (if needed) an improved iterative guess of the displacement $\mathbf{d}_{n+1}^{(i+1)}$; we can thus write:

$$
\begin{aligned}
&\mathbb{A}_{e=1}^{n_{elem}} \{\hat{\mathbf{K}}_{n+1}^{e,(i)} \Delta\mathbf{d}_{n+1}^{e,(i)} = \mathbf{f}_{n+1}^{ext,e} - \mathbf{f}_{n+1}^{int,e,(i)} - \mathbf{F}_{n+1}^{e,(i)} [\mathbf{H}_{n+1}^{e,(i)} \\
&\qquad\qquad + K_{\alpha,n+1}^{(i)}]^{-1} h_{n+1}^{e}\} \\
&\hat{\mathbf{K}}_{n+1}^{e,(i)} = \mathbf{K}_{n+1}^{e,(i)} - \mathbf{F}_{n+1}^{e,(i),T} [\mathbf{H}_{n+1}^{e,(i)} + K_{\alpha,n+1}^{(i)}]^{-1} [\mathbf{F}_{n+1}^{e,(i)} + K_{d,n+1}^{e,(i)}] \\
&\Longrightarrow \; \mathbf{d}_{n+1}^{(i+1)} = \mathbf{d}_{n+1}^{(i)} + \Delta\mathbf{d}_{n+1}^{(i)}
\end{aligned}
\tag{8.158}
$$

We can note that the proposed operator split solution procedure will allow us to reduce the element stiffness matrix to the standard format for an isoparametric element (without incompatible modes), which can directly fit within the standard finite element assembly procedure. The components of the tangent stiffness are computed by using the results of local computations with either $K_{\alpha,n+1} = 0$ for elastic step or $K_{d,n+1} = 0$ for plastic step with an active shear band. We also note that such a global computation will need the results for stress values $\boldsymbol{\sigma}_{n+1}$ in the fracture process zone which are plastically admissible in the sense of the von Mises plasticity criterion $\bar{\phi}_{n+1} \leq 0$, along with the corresponding plastic deformation $\bar{\boldsymbol{\epsilon}}^p$ and the consistent tangent elastoplasticity tensor.

8.4.3 Numerical examples for localization problems

In this section we will briefly present the results of a couple of numerical simulations carried out with the proposed localization limiter for hardening/

softening plasticity. We will show that the proposed model can eliminate completely the sensitivity of computed results with respect to the choice of finite element mesh.

8.4.3.1 Simple shear test

In the first example we study a simple shear test. The specimen has a rectangular form, with length equal 20 cm, height equal 10 cm and unit thickness (see Figure 8.7). The plane strain conditions are imposed. The constitutive behavior is described by the proposed hardening/softening plasticity model, which combines the von Mises plasticity criterion with saturation hardening for representing the fracture process zone with the linear softening law for controlling an active shear band. The chosen values of mechanical parameters are given in Table 8.1.

Table 8.1 Simple shear test: material parameters.

Plasticity model with hardening/softening	
Young's modulus	210 GPa
Poisson's coefficient	0.4999
Limit of elasticity σ_y	0.55 GPa
Saturation limit σ_∞	0.75 GPa
Exp. coef. $\bar{\beta}$	200
Softening modulus $\bar{\bar{K}}$	−0.1 GPa/mm

The numerical analysis is performed by using the imposed horizontal displacement on the top boundary, while keeping the bottom boundary fixed. Zero traction boundary conditions are used on left and right boundaries. Two different meshes are employed in computations, the coarse mesh with 8×13 and the fine mesh with 14×21 enhanced finite elements $Q4$ with $\bar{B}$ assumed strain interpolation and shear band discontinuity. In each mesh, we weaken slightly one element in order to better control where the shear band will develop.

The objectivity of the computed response is ensured by using the proposed approach based upon hardening/softening plasticity model and the elements with embedded discontinuity, as shown in Figure 8.8, with the same force-displacement diagram obtained with either coarse or fine mesh. We also note that the quasi-incompressibility locking problem is alleviated by using the assumed strain $\bar{B}$ interpolation for strain filed, with the better quality results than those obtained by the standard isoparametric interpolations.

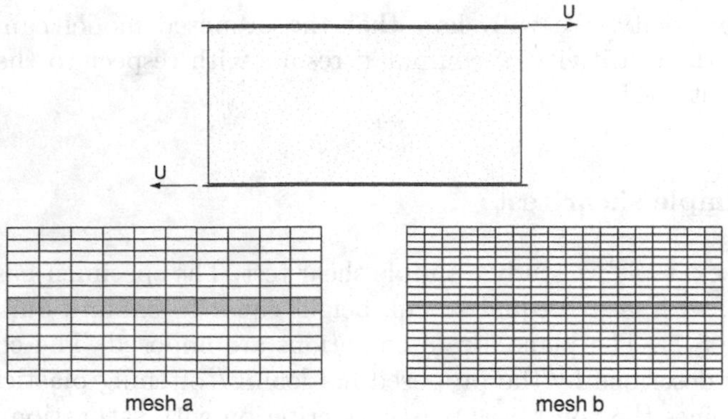

Fig. 8.7 Simple shear test: Model and finite element mesh.

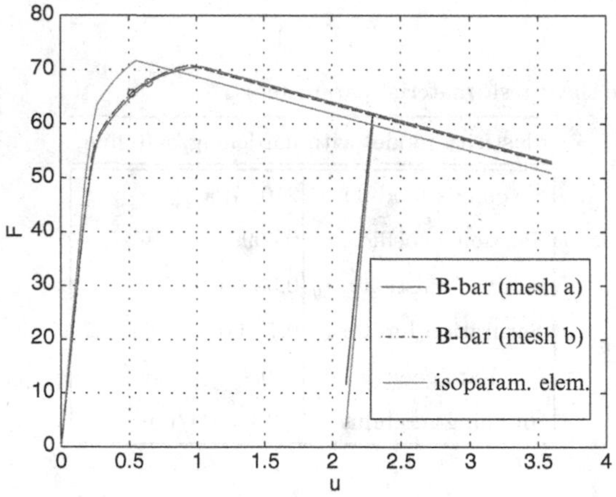

Fig. 8.8 Simple shear test: force–displacement diagram.

8.4.3.2 Simple tension test

In this example, we present the results of numerical simulation of the simple tension test with a double-edge notched specimen shown in Figure 8.9. The specimen has the length equal to 20 cm, width equal to 10 cm and a unit thickness, with plane strain conditions imposed. The material properties are chosen the same as presented in previous example (see Table 8.1).
The computation is carried out under displacement control applied at the right end of the specimen, with other hand kept fixed. The presence of the notches will perturb the homogeneity of the stress field, guiding the fracture process zone creation followed eventually by the shear band appearance. Even at this final stage both plastic mechanisms remain active, as shown in Figure 8.10.

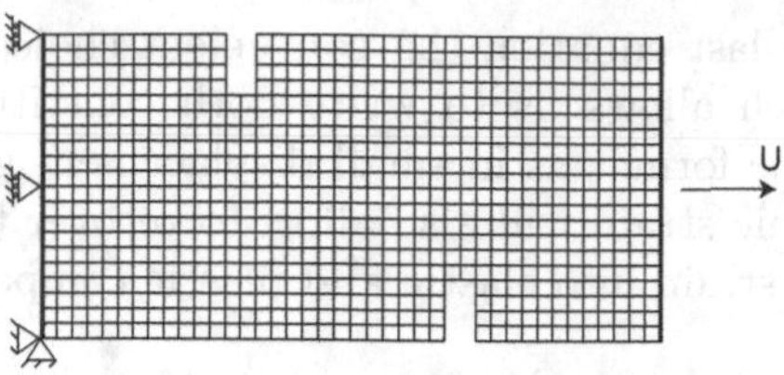

Fig. 8.9 Simple tension test: chosen specimen and boundary conditions.

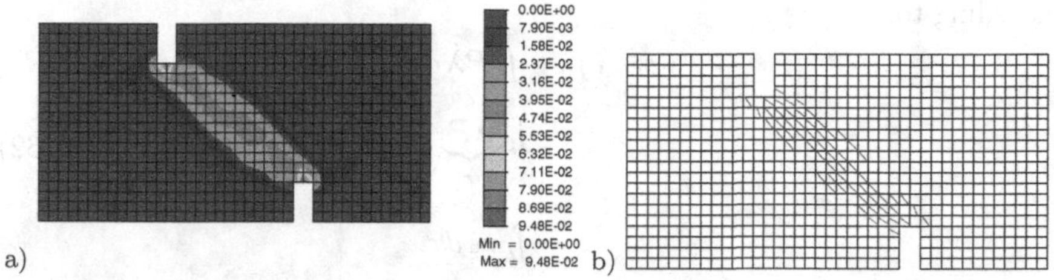

Fig. 8.10 Simple tension test: (**a**) equivalent plastic strain in fracture process zone (**b**) plastic shear bands.

8.5 Localization problem in large strain plasticity

The limiter of localization presented in the previous section can be generalized to the large strain softening plasticity problems. We will illustrate all the modifications to provide in 1D case where all the criteria reduce to very much the same statement, leaving to readers to choose from wide variety of criteria of interest for 3D problems (e.g. see [7], [162] or [210]). For 1D hardening plasticity in large strains, we can define the main ingredients defining the multiplicative decomposition of strain, the strain energy and the yield criterion:

$$\begin{aligned} &\frac{\partial \overbrace{(x+d)}^{\varphi}}{\partial x} =: \lambda = \lambda^e \lambda^p \\ &\phi(\tau, q) := |\tau| - (\sigma_y - q) \\ &\psi(\lambda^e, \zeta) := \tfrac{1}{2} E (ln\lambda^e)^2 + \tfrac{1}{2} K \zeta^2 \end{aligned} \tag{8.159}$$

By the standard developments based upon the second principle of thermodynamics and the principle of maximum plastic dissipation, we can further obtain the constitutive equations for stress and the evolution equations for plastic strain and hardening variable:

$$\begin{aligned} &\tau = E\epsilon^e \; ; \; \epsilon^e = ln\lambda^e \; ; \; q = -K\zeta \\ &\dot{\epsilon}^p = \dot{\gamma}\tfrac{\partial \phi}{\partial \tau} \; ; \; \epsilon^p = ln\lambda^p \; ; \; \dot{\zeta} = \dot{\gamma} \end{aligned} \tag{8.160}$$

We note that in the last equation the convenient choice of the logarithmic strain measure, which allows us to write both constitutive and evolution equations in the same format as in small displacement gradient case. Introducing the logarithmic strain will also allow to recover the additive decomposition of the total strain into elastic and plastic component with:

$$ln\lambda = ln(\lambda^e \lambda^p) \implies \epsilon = \epsilon^e + \epsilon^p \tag{8.161}$$

The weak form of equilibrium equations for this model can then be written in terms of the given strain and the work-conjugate Kirchhoff stress according to:

$$\begin{aligned} G^{int}(\varphi, \cdot; v) &:= \int_l P\dot{\lambda}\, dx \\ &= \int_l \underbrace{\lambda P}_{\tau} \underbrace{\frac{\dot{\lambda}}{\lambda}}_{\dot{\epsilon}} dx \\ &= \int_l \tau\dot{\epsilon}\, dx \end{aligned} \tag{8.162}$$

We now turn to the softening plasticity model at finite strains, revisiting and eventually modifying each of the equations of the hardening plasticity model. Namely, the localization phenomena will impose that the plastic deformation remains localized, which will enforce the change in its representation along with the corresponding changes of all other fields affected by the plastic deformation. We will consider only the extreme case where the plastic deformation will localize in a point, leading to the displacement discontinuity representation introduced in the deformed configuration:

$$\begin{gathered} \varphi(x,t) = \bar{\varphi}(x,t) + [\![\bar{\varphi}(t)]\!] H_{\bar{x}}(x) \;;\; \bar{\varphi}(x,t) = x + \bar{u}(x,t) \;; \\ H_{\bar{x}} := \begin{cases} 1 \;;\; x > \bar{x} \\ 0 \;;\; \text{otherwise} \end{cases} \Rightarrow \tfrac{d}{dx} H_{\bar{x}} = \delta_{\bar{x}} := \begin{cases} \infty \;;\; x = \bar{x} \\ 0 \;;\; \text{otherwise} \end{cases} \end{gathered} \tag{8.163}$$

For this kind of displacement field, we can then obtain the deformation gradient:

$$\lambda(x,t) := \bar{\lambda}(x,t) + [\![\bar{u}(t)]\!]\delta_{\bar{x}}(x) \;;\; \lambda := \frac{\partial\varphi(x,t)}{\partial x} \;;\; \bar{\lambda} = \frac{\partial\varphi(x,t)}{\partial x} \;; \tag{8.164}$$

as well as its time derivative:

$$\dot{\lambda}(x,t) = \dot{\bar{\lambda}}(x,t) + [\![\dot{\bar{u}}(t)]\!]\delta_{\bar{x}}(x) \tag{8.165}$$

The last equation allows us to obtain the strain rate decomposition valid for the case when the material instability phenomenon occurs leading to strain localization, which can be written:

$$\dot{\lambda}\bar{\lambda}^{-1}|_{\lambda=\bar{\lambda}} = \underbrace{\dot{\bar{\lambda}}\bar{\lambda}^{-1}}_{\dot{\bar{\epsilon}}} + \underbrace{[\![\dot{\bar{u}}]\!]\bar{\lambda}^{-1}}_{\dot{\bar{\bar{\epsilon}}}}\delta_{\bar{x}} \tag{8.166}$$

Even in the presence of localized plastic deformation, the stress ought to remain bounded, which implies that the plastic multiplier should also be interpreted in the sense of distribution:

$$E^{-1}\dot{\tau} = \dot{\bar{\epsilon}} + \dot{\bar{\bar{\epsilon}}}\delta_{\bar{x}} - \dot{\gamma}sign(\tau) \implies \dot{\gamma} = \dot{\bar{\bar{\gamma}}}\delta_{\bar{x}} \tag{8.167}$$

The corresponding value of the plastic multiplier can be computed from the plastic consistency condition, which can be written:

$$\begin{aligned} 0 = \dot{\phi} &:= \tfrac{\partial \phi}{\partial \tau}\dot{\tau}|_{\bar{x}} + \tfrac{\partial \phi}{\partial q}\dot{q} \\ &= \tfrac{\partial \phi}{\partial \tau}E\dot{\bar{\epsilon}}|_{\bar{x}} + \dot{q} + [\tfrac{\partial \phi}{\partial \tau}\dot{\bar{\bar{\epsilon}}} - \dot{\bar{\bar{\gamma}}}]\delta_{\bar{x}} \end{aligned} \tag{8.168}$$

The last expression will allow us to obtain two important conclusions: first that the plastic multiplier is proportional to the absolute value of jump at discontinuity and second that the internal variable q that controls the plasticity threshold should also remain bounded; in other words, we can write:

$$\dot{\bar{\bar{\gamma}}} = |\dot{\bar{\bar{\epsilon}}}| \ ; \ \dot{q} = -sign(\tau)E\dot{\bar{\epsilon}}|_{\bar{x}} \tag{8.169}$$

The last result will further allow to conclude that the softening modulus should also be interpreted in terms of distribution:

$$\left.\begin{aligned} \dot{\zeta} &= \dot{\bar{\bar{\gamma}}}\delta_{\bar{x}} \\ \dot{\zeta} &= -\tfrac{1}{K}\dot{q} \end{aligned}\right\} \implies \frac{1}{K} = \frac{1}{\bar{\bar{K}}}\delta_{\bar{x}} \tag{8.170}$$

For the constant value of the softening modulus, the latter can be obtained from the corresponding fracture energy G_f (the energy needed to induce the failure and reduce the stress to zero):

$$\bar{\bar{K}} = -\frac{\sigma_y^2}{2G_f} \tag{8.171}$$

The simple tension test in large strain is often accompanied by so-called necking phenomena where the specimen will finally have a significantly reduced cross-section in the failure regime, which leads to the large strains and large rotations in the fracture process zone and imposes the use of the 2D or 3D plasticity models for proper interpretation. We do not elaborate further on these models, but only give the illustrative result obtained by using them in modelling the localized failure (see Figure 8.11) showing the mesh objectivity of computed results.

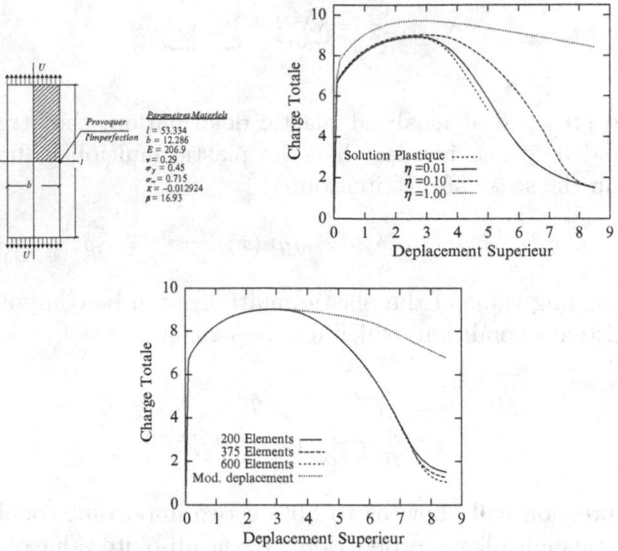

Fig. 8.11 Material instability of necking phenomena: model, computed response with respect to chosen value of viscosity η and with respect to number of elements.

Chapter 9
Multi-scale modelling of inelastic behavior

In this chapter we will briefly explore the multi-scale modelling of inelastic behavior, the topic of much interest for the current research (e.g. see [118] for the state-of-the-art contributions). With this kind of approach, the phenomenological models are replaced by refined models of inelastic behavior constructed at two scales: macro-scale that represents the homogenized behavior of material for computing the global structural response, and micro-scale that allows us to capture the fine details of microstructure for heterogeneous multi-phase materials. The latter provides the main advantage of this kind of approach in offering a more reliable interpretation of inelastic behavior mechanisms, failure and/or plasticity modes. In trying to keep our developments within the most general framework, we will again employ the finite element method at the level of micro-scale.

The chapter outline is as follows. In the first part we study the coupling conditions between the macro and micro scales. We start by weak coupling of scales leading to the classical problem of homogenization, where the macro and micro scales need not have the permanent communication, so that we can carry out a separate analysis to establish the averaged properties at the micro-scale as the only one needed for macro-scale analysis. We then turn to studying the strong coupling of scales, where the two scales communicate throughout the analysis, and the computations on both scales have to be advanced simultaneously.

In the second part of the chapter we study different possibilities for microstructure representation by using the finite element method. The vast majority of early works (e.g. see [95]) have used the exact finite element representation of the microstructure, where the element frontiers are positioned at the phase-interface so that each finite element domain contains only one phase. Since the mesh preparation for this kind of finite element representation of the microstructure can become very costly for a heterogenous, multi-phase material with complex microstructure, we turn towards so-called structured mesh representation where each finite element will keep a regular form. Therefore, the inter-element frontiers will no longer be placed

A. Ibrahimbegovic, *Nonlinear Solid Mechanics: Theoretical Formulations and Finite Element Solution Methods*, Solid Mechanics and its Applications 160,

along the phase-interface, and each element domain can contain more than one phase. We show that the isoparametric finite element interpolations are no longer suitable for representing the strain field of the elements crossed by the phase-interface, and we also show how to provide the corresponding strain field enhancements. Some recent developments in multi-scale nonlinear analysis of heterogeneous materials are also discussed, along with the current research goals.

9.1 Scale coupling for inelastic behavior in quasi-static problems

In principle, we can distinguish between two classes of problems with inelastic behavior: the first that concerns the failure of (massive) structures and the second that deals with elaboration of new materials and studies of its inelastic behavior. For the first class of problems in nonlinear analysis of failure of massive structures, the micro–macro approach is currently employed mostly in the sense of homogenization. In other words, we first carry out the analysis at micro-scale by studying so-called representative volume element in order to identify the best suitable phenomenological model of constitutive behavior capable of representing all pertinent details of inelastic behavior of given material. We then directly employ such a phenomenological model within the nonlinear structural analysis at macro scale. In this case, we talk about the weak scale coupling, where the solution can be constructed in two separate steps, the first one dealing only with micro-scale and the second with only macro-scale computations. The second class of problems concerns the micro–macro nonlinear analysis of inelastic behavior with strong coupling, where the two scales constantly communicate in the course of analysis and the computations advance simultaneously at each scale. Such a nonlinear analysis is currently used in order to obtain the most reliable interpretation of inelastic behavior of a given material, or to complete the experimental results and insufficient measurements for the case where the specimen is of much smaller size than the complete structure. In the last case it is possible that the characteristic size of heterogeneities becomes too big with respect to the size of the specimen and no longer allows to easily identify the representative volume element within the specimen nor to accept only the weak scale coupling. One example of this kind is the concrete with large aggregate size, where the required size of the specimen allowing for scale separation would have been too large for any standard testing machine.
We will present in this section how to handle the coupling between the scales micro and macro, both for the case of weak and strong coupling. In order to keep the highest level of generality, we will assume that the finite element method is employed at both scales, micro and macro.

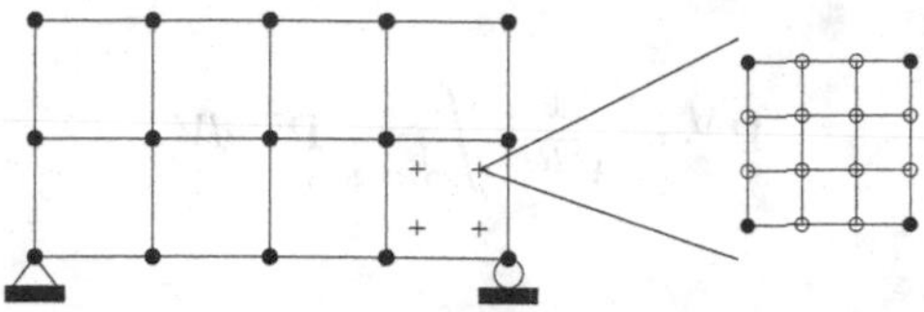

Fig. 9.1 Model FEM^2 for a three-point bending test, with a finite element mesh composed of $Q4$ isoparametric elements at macro-scale and another mesh of $Q4$ elements constructed at each Gauss point at micro-scale.

9.1.1 Weak coupling: nonlinear homogenization

Weak coupling between micro and macro scales concerns the class of problems equivalent to the classical homogenization procedure (e.g. see [36], [170] or [232]), but applied to nonlinear inelastic constitutive behavior (e.g. see [79], [158] or [187]). The tacit hypothesis which is at the origin of the homogenization approach pertains to a sufficient separation of micro and macro scales, which allows to define a representative volume element for quantifying the inelastic behavior of material. With respect to macro scale, the representative volume element corresponds to a single point, here denoted with $\mathbf{x}^M$, which is typically chosen as the numerical integration point of the macro-scale element (see Figure 9.1). Therefore, we can define the material response by a work-conjugate couple of tensors, with on one side the first Piola-Kirchhoff stress tensor $\mathbf{P}^M$ and on the other the deformation gradient $\mathbf{F}^M$, which derives from the macro-scale displacement field $\mathbf{u}^M$. These two tensors can thus replace the corresponding stress–strain couple, which was defined until now from phenomenological constitutive model of inelastic behavior. For the finite element mesh of macro-elements, the points where we ought to define this kind of model are set again by the rules of numerical integration, such as Gauss quadrature, that we use in the standard finite element computations. As the only difference with respect to the classical homogenization that provides the analytic (linear) response, the nonlinear response obtained by numerical homogenization is obtained by numerical procedure using the finite element mesh of micro-scale elements placed at the Gauss point. The micro-scale finite element mesh ought to be sufficiently refined to be able to provide the reliable description of the inelastic behavior mechanisms in accordance with the corresponding criteria featuring the stress at micro-scale and work-conjugate strain measures. For example, by choosing the work-conjugate couple of the first Piola–Kirchhoff stress tensor $\mathbf{P}^m$ and the deformation gradient $\mathbf{F}^m$ at micro-scale, we can construct from the average values the corresponding constitutive relation at the macro-scale:

$$\mathbf{P}^M = \widehat{\mathbf{P}}(\mathbf{F}^M, \cdot) \tag{9.1}$$

with:

$$\mathbf{P}^M = \frac{1}{V^{RE}} \int_{\Omega^{RVE}} \mathbf{P}^m \, dV \tag{9.2}$$

as well as:

$$\mathbf{F}^M = \frac{1}{V^{RE}} \int_{\Omega^{RVE}} \mathbf{F}^m \, dV \tag{9.3}$$

where V^{RE} is the volume of the representative element and Ω^{RVE} is its domain of integration. By taking into account that the stress at the micro-scale $\mathbf{P}^m$ ought to satisfy the corresponding equilibrium equations $div[\mathbf{P}^m] = 0$ and by exploiting the divergence theorem, we can express the result in (9.2) in an equivalent form in terms of the boundary integrals:

$$\mathbf{P}^M = \frac{1}{V^{RE}} \int_{\partial\Omega^{RVE}} \mathbf{t}^m \otimes \mathbf{x}^m \, dA; \qquad \mathbf{t}^n = \mathbf{P}^m \mathbf{n}^m; \tag{9.4}$$

The divergence theorem will also allow us to obtain an equivalent form of the result in (9.3), which can be written:

$$\mathbf{F}^M = \frac{1}{V^{RE}} \int_{\partial\Omega^{RVE}} \boldsymbol{\varphi}^m \otimes \mathbf{n}^m \, dA; \tag{9.5}$$

The definitions of macro-scale stress and deformation gradient given in (9.4) and (9.5) above, can also be employed for a very general case where several phases or voids are contained in the representative volume element.
By using the following identity: $\mathbf{P}^m \cdot \dot{\mathbf{F}}^m = div[\mathbf{P}^m \dot{\boldsymbol{\varphi}}^m] - div[\mathbf{P}^m] \cdot \dot{\boldsymbol{\varphi}}^m$, and by exploiting equilibrium equations, we can write the corresponding expression for the stress power:

$$\mathbf{P}^M \cdot \dot{\mathbf{F}}^M = \frac{1}{V^{RE}} \int_{\Omega^{RVE}} \mathbf{P}^m \cdot \dot{\mathbf{F}}^m \, dV = \frac{1}{V^{RE}} \int_{\partial\Omega^{RVE}} \mathbf{t}^m \cdot \dot{\boldsymbol{\varphi}}^m \, dA \tag{9.6}$$

The last result will allow us to construct the consistent basis for including the computational results obtained at micro-scale within the macro-scale framework. More precisely, we can define the stress at the micro-scale through a constitutive equation deriving from a potential:

$$\mathbf{P}^M = \frac{\partial W(\mathbf{F}^M, \cdot)}{\partial \mathbf{F}^M}; \tag{9.7}$$

where $W(\mathbf{F}^M, \cdot)$ is the strain energy density defined as a function of deformation gradient. This kind of potential can be defined by computations at the micro-scale. Any such analysis would require the corresponding choice of the boundary conditions, to be made among three following possibilities:

(i) Imposed displacements and/or deformations:

$$W(\mathbf{F}^M) = \min_{\boldsymbol{\varphi}^m} \max_{\mathbf{t}^m} \left\{ \tfrac{1}{V^{RE}} \int_{\Omega^{RVE}} W(\mathbf{F}^m) \, dV \right. \\ \left. \tfrac{1}{V^{RE}} \int_{\partial\Omega^{RVE}} \mathbf{t}^m \cdot (\boldsymbol{\varphi}^m - \mathbf{F}^M \mathbf{x}^m) \, dA \right\} ; \tag{9.8}$$

(ii) Imposed stress:

$$W(\mathbf{F}^M) = \min_{\boldsymbol{\varphi}^m} \max_{\mathbf{t}^m} \left\{ \tfrac{1}{V^{RE}} \int_{\Omega^{RVE}} W(\mathbf{F}^m) \, dV \right. \\ \left. -\tfrac{1}{V^{RE}} \int_{\partial\Omega^{RVE}} \underbrace{(\mathbf{P}^m \mathbf{n}^m)}_{\mathbf{t}^m} \cdot \boldsymbol{\varphi}^m \, dA + \mathbf{P}^M \cdot \mathbf{F}^M \right\} ; \tag{9.9}$$

(iii) Periodicity conditions:

$$W(\mathbf{F}^M) = \min_{\boldsymbol{\varphi}^m} \max_{\mathbf{t}^m} \left\{ \tfrac{1}{V^{RE}} \int_{\Omega^{RVE}} W(\mathbf{F}^m) \, dV - \right. \\ \left. \tfrac{1}{V^{RE}} \int_{\partial\Omega^{RVE}} \mathbf{t}^m \cdot ([\![\boldsymbol{\varphi}^m]\!] - \mathbf{F}^M [\![\mathbf{x}^m]\!]) \, dA \right\} ; \tag{9.10}$$

The first choice for boundary conditions on imposed displacements leads to the upper bound (Voigt bound) and the second choice with imposed stress leads to the lower (Reuss) bound for the homogenized mechanical properties. The periodicity boundary conditions provide an equivalent hypothesis to the classical homogenization in linear analysis only; this equivalence will not necessarily hold for nonlinear analysis, especially for the inelastic material behavior with different regimes (e.g. hardening versus softening).
We can obtain the same kind of results for weak scale coupling for small displacement gradient theory, where we will use the Cauchy stress $\boldsymbol{\sigma}$ instead of the first Piola-Kirchhoff stress accompanied by the infinitesimal strains $\boldsymbol{\varepsilon}$; we can thus write:

$$\boldsymbol{\sigma}^M = \frac{1}{V^{RE}} \int_{\Omega^{RVE}} \boldsymbol{\sigma}^m \, dV; \qquad \boldsymbol{\varepsilon}^M = \frac{1}{V^{RE}} \int_{\Omega^{RVE}} \boldsymbol{\varepsilon}^m \, dV; \tag{9.11}$$

and make the appropriate choice of the boundary conditions in order to obtain the lower or the upper bound values for homogenized material properties.

9.1.2 Strong coupling micro–macro

The strong coupling of micro and macro scales concerns the problems of inelastic behavior of material, where the characteristic size of heterogeneities is not sufficiently small compared against the structure size at macro-scale. For that reason, we employ the micro–macro model presented in Figure 9.2.

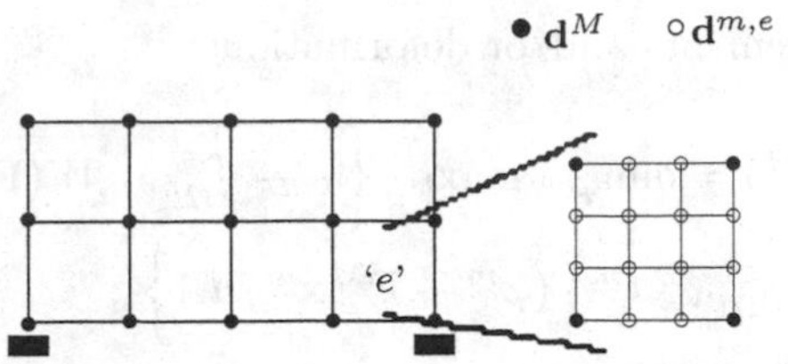

Fig. 9.2 Model micro–macro of a three-point bending test, where each finite elements at macro-scale is built from the corresponding finite element mesh at micro-scale.

The variational formulation for a multi-scale problem of this kind can be cast at the fixed value of the internal variables $\overline{\zeta}_k$ according to (e.g. see [134]):

$$\Pi(\mathbf{u}^M, \mathbf{u}^m, \boldsymbol{\lambda}, \bar{\zeta}_k) = \Pi^M + \Pi^m + \Pi_{\Gamma^{Mm}}; \tag{9.12}$$

where Π^m is the strain energy at micro-scale, Π^M strain-energy at macro-scale and $\Pi_{\Gamma^{Mm}}$ is the energy at interface for coupling two scales. Without loss of generality for the developments to follow, we assume that the energy of external forces is only defined at the macro-scale (e.g. the dead load of the specimen obtained by homogenization), which then reduces to:

$$\Pi^M = -\int_{\Omega^M} \mathbf{u}^M \cdot \mathbf{b}^M \, dV - \int_{\Gamma^M_\sigma} \mathbf{u}^M \cdot \bar{\mathbf{t}} \, dA \tag{9.13}$$

We can also accept the hypothesis that the mechanisms of inelastic behavior and the internal variables are only defined at the micro-scale, which allows us to define the corresponding potential from the strain energy density at that scale:

$$\Pi^m = \int_{\Omega^m} \Psi^m(\mathbf{u}^m, \overline{\zeta}_k) \, dV \tag{9.14}$$

Finally, the last terms in (9.12) will allow us to ensure the scale coupling by introducing the Lagrange multiplier $\boldsymbol{\lambda}$ with:

$$\Pi^{\Gamma^{Mm}} = \int_{\Gamma^{Mm}} \boldsymbol{\lambda} \cdot (\mathbf{u}^M - \mathbf{u}^m) \, dA \tag{9.15}$$

In summary, the state variables in the present variational problem of micro–macro analysis are the displacement field at macro-scale $\mathbf{u}^M$, the displacement field at micro-scale $\mathbf{u}^m$, the Lagrange multipliers $\boldsymbol{\lambda}$ for scale coupling, as well as the internal variables ζ_k. The latter will impose that we perform incremental/iterative solution procedure starting with the initial configuration. A typical step of such a procedure can be presented in terms of:

Central problem of micro–macro incremental analysis (strong scale coupling)

Given: $\mathbf{u}^M_n = \mathbf{u}^M(\mathbf{x}^M, t_n), \mathbf{u}^m_n = \mathbf{u}^m(\mathbf{x}^m, t_n), \boldsymbol{\lambda}_n = \lambda(\mathbf{x}^M, t_n),\ \zeta_{k,n} = \zeta_k(\mathbf{x}^m, t_n),\ h = t_{n+1} - t_n$

Find: $\mathbf{u}_{n+1}^M, \mathbf{u}_{n+1}^m, \boldsymbol{\lambda}_{n+1}, \zeta_{k,n+1}$
such that:

$$\begin{aligned}
0 &= \frac{d}{d\varepsilon}\Big|_{\varepsilon=0} \Pi^m(\mathbf{u}_{n+1}^m + \varepsilon \mathbf{w}^m, \cdot) \equiv G^m(\mathbf{u}_{n+1}^m, \boldsymbol{\lambda}_{n+1}, \zeta_{k,n+1}; \mathbf{w}^m) \\
&= \int_{\Omega^m} \nabla^s \mathbf{w}^m \cdot \hat{\boldsymbol{\sigma}}^m(\mathbf{u}_{n+1}^m, \zeta_{k,n+1})\, dV - \int_{\Gamma^{Mm}} \mathbf{w}^m \cdot \boldsymbol{\lambda}\, dA \\
0 &= \frac{d}{d\varepsilon}\Big|_{\varepsilon=0} \Pi^M(\mathbf{u}_{n+1}^M + \varepsilon \mathbf{w}^M, \cdot) \equiv G^M(\mathbf{u}_{n+1}^M, \boldsymbol{\lambda}_{n+1}; \mathbf{w}^M) \\
&= \int_{\Gamma^{Mm}} \mathbf{w}^M \cdot \boldsymbol{\lambda}\, dA - \int_{\Omega^M} \mathbf{w}^M \cdot \mathbf{b}\, dV - \int_{\Gamma_\sigma^M} \mathbf{w}^M \cdot \bar{\mathbf{t}}\, dA \\
0 &= \frac{d}{d\varepsilon}\Big|_{\varepsilon=0} \Pi^{Mm}(\boldsymbol{\lambda}_{n+1} + \varepsilon \boldsymbol{\nu}, \cdot) \equiv G^{Mm}(\mathbf{u}_{n+1}^M, \mathbf{u}_{n+1}^m, \boldsymbol{\nu}) \\
&= \int_{\Gamma^{Mm}} \boldsymbol{\nu} \cdot (\mathbf{u}_{n+1}^M - \mathbf{u}_{n+1}^m)\, dA
\end{aligned} \tag{9.16}$$

with:

$$\zeta_{k,n+1} = \zeta_{k,n} + h\, \hat{f}(\mathbf{u}_{n+1}^M, \mathbf{u}_{n+1}^m, \boldsymbol{\lambda}_{n+1}, \zeta_{k,n+1}) \tag{9.17}$$

We have indicated in (9.16) above that the governing variational equations are obtained as the directional derivatives of the total energy potential in (9.12) with respect to the variations of displacements at the macro-scale, $\mathbf{w}^M$, displacements at the micro-scale, $\mathbf{w}^m$, and Lagrange multipliers, $\boldsymbol{\nu}$. We have also indicated that these variational equations ought to be accompanied by the results of time-integration providing the evolution of the internal variables $\zeta_{k,n+1}$. In constructing the solution by the finite element method, we ought to provide the finite dimensional approximations for all unknown fields. First, we choose a local reference frame for each macro-scale element for placing all the micro-scale elements on the same basis with a particular macro-scale element as the corresponding envelope; in other words, by using $\mathbf{x}^{M,E} \equiv \mathbf{x}^m$, we can construct the micro-scale displacement field approximation according to:

$$\mathbf{u}_{n+1}^m\big|_{\Omega^{m,e}}(\mathbf{x}^m) = \sum_{a=1}^{n_{el}^m} \mathbf{N}_a^{m,e}(\mathbf{x}^m)\mathbf{d}_{a,n+1}^m \quad ; \forall \Omega^{m,e} \subset \Omega^{M,e} \tag{9.18}$$

where $\mathbf{N}_a^{m,e}(\mathbf{x}^m)$ are the shape function and $\mathbf{d}_{a,n+1}^m$ are nodal values of micro-displacements at time t_{n+1}.

We can thus limit the internal variable definition only to numerical integration points of all micro-scale elements. In order to define the scale coupling, we also have to choose all the interpolations along the macro-scale element interface $\Gamma^{Mm,E}$ according to:

$$\boldsymbol{\lambda}_{n+1}\big|_{\Gamma^{Mm,E}}(\mathbf{x}^m) = \sum_{a \in \Gamma^{Mm,E}} \mathbf{P}_a^{M,E}(\mathbf{x}^m)\boldsymbol{\beta}_{a,n+1} \tag{9.19}$$

$$\mathbf{u}_{n+1}^M\big|_{\Gamma^{Mm,E}}(\mathbf{x}^m) = \sum_{a \in \Gamma^{Mm,E}} \mathbf{N}_a^{M,E}(\mathbf{x}^m)\mathbf{d}_{n+1}^{M,E} \tag{9.20}$$

With these approximations in hand, we can restate the central problem of micro-macro incremental analysis with:

Central problem of micro–macro incremental analysis in discrete approximation

Given: $\mathbf{d}_n^{M,E}, \quad \mathbf{d}_n^m, \quad \boldsymbol{\beta}_n, \quad \zeta_{k,n}, \quad ; \qquad \forall \quad \Omega^{M,E}$

Find: $\mathbf{d}_{n+1}^{M,E}, \quad \mathbf{d}_{n+1}^m, \quad \boldsymbol{\beta}_{n+1}, \quad \zeta_{k,n+1},$

such that: $(\forall e \in [1, n_{el}^m] \; ; \; \forall E \in [1, n_{el}^M])$

$$
\begin{aligned}
\mathbf{0} &= \mathbf{r}^m(\mathbf{d}_{n+1}^m, \boldsymbol{\beta}_{n+1}, \zeta_{k,n+1}) \\
&= \mathop{\mathbb{A}}_{e=1}^{n_{elem}^m} \left[\int_{\Omega^{m,e}} \mathbf{B}^{mT} \cdot \hat{\boldsymbol{\sigma}}^m(\mathbf{d}_{n+1}^m, \zeta_{k,n+1})\, dV \right] \\
&\quad - \int_{\Gamma^{Mm,E}} \mathbf{N}^{m,eT} \mathbf{P}^{M,E} \boldsymbol{\beta}_{n+1}\, dA \,; \\
\mathbf{0} &= \mathbf{r}^M(\mathbf{d}_{n+1}^M, \boldsymbol{\beta}_{n+1}) \\
&= \mathop{\mathbb{A}}_{E=1}^{n_{elem}^M} \left[- \int_{\Omega^{M,E}} \mathbf{N}^{M,E^T} \mathbf{b}\, dV - \int_{\Gamma^{M,E}} \mathbf{N}^{M,E^T} \bar{\mathbf{t}}\, dA \right. \\
&\quad \left. + \int_{\Gamma^{M,E}} \mathbf{N}^{M,E^T} \mathbf{P}^{M,E} \boldsymbol{\beta}_{n+1}\, dA \right] \\
\mathbf{0} &= \mathbf{p}^{M,E}(\mathbf{d}_{n+1}^{M,E}, \mathbf{d}_{n+1}^m, \boldsymbol{\beta}_{n+1}) \\
&= \int_{\Gamma^{M,F}} \mathbf{P}^{E,T} (\mathbf{N}^{M,E} \mathbf{d}^{M,E} - \mathbf{N}^{m,e} \mathbf{d}^m)\, dA \,;
\end{aligned}
\tag{9.21}
$$

with:

$$
\zeta_{k,n+1} = \zeta_{k,n} + h\, \hat{f}(\mathbf{d}_{n+1}^{M,E}, \mathbf{d}_{n+1}^m, \boldsymbol{\beta}_{n+1}, \zeta_{k,n+1}) \tag{9.22}
$$

We have indicated in (9.21) above that the equilibrium equations at micro-scale are solved independently in each macro-scale element, with macro-scale displacement kept fixed. The finite element assembly of those contributions will then provide the corresponding equilibrium equations to be solved at the macro-scale for computing the unknown values of displacement. The details of such computation remain dependent upon the chosen coupling between the scales, with a couple of different possibilities discussed next.

(i) Scale-coupling in displacements

For the coupling between the scales in terms of displacements, we assume that the chosen variation of the displacement at the macro-scale will impose the corresponding variation of micro-scale displacement. Namely, in order to ensure compatibility between the scales, the same displacement variation is enforced at two scales along the macro-scale element boundaries (see Figure 9.3).

This kind of displacement coupling is in agreement with the discrete approximation presented previously, if the Lagrange multipliers' interpolations are

Table 9.1 Flow-chart for micro-macro incremental/iterative solution procedure: displacement coupling.

$$
\left[\begin{array}{l}
n = 0,1,2,... \\
\left[\begin{array}{l}
(i) = 1,2,...\ \text{given: } \mathbf{d}_{n+1}^{M(i)} (\text{ with } \mathbf{d}_{n+1}^{M(1)} = \mathbf{d}_{n}^{M}) \\
\left[\begin{array}{l}
E = 1,2,...,n_{el}^{M} \\
\bar{\mathbf{d}}_{n+1}^{m,e(1,i)} = \mathbf{T}^{E}\mathbf{d}_{n+1}^{M,E,(i)} \\
\left[\begin{array}{l}
(j) = 1,2,... \\
\mathop{\mathbb{A}}_{e=1}^{n_{elem}^{m}} \left[\mathbf{K}_{n+1}^{m,e(j,i)} \left(\mathbf{d}_{n+1}^{m,e(j+1,i)} - \mathbf{d}_{n+1}^{m,e(j,i)} \right) = -\mathbf{r}_{n+1}^{m,e(j,i)} \right] \\
\text{IF: } \| \mathbf{r}_{n+1}^{m,e(j+1,i)} \| > tol \\
(j) \longleftarrow (j+1)
\end{array}\right. \\
\text{ELSE IF: } \| \mathbf{r}_{n+1}^{m,e(j+1,i)} \| \leq tol \\
\mathbf{r}_{n+1}^{M,E(i)} = \mathbf{T}^{E,T}\widetilde{\mathbf{r}}_{n+1}^{m}; \qquad \mathbf{K}_{n+1}^{M,E(i)} = \mathbf{T}^{E,T}\widetilde{\mathbf{K}}_{n+1}^{m}\mathbf{T}^{E} \\
E \longleftarrow E+1
\end{array}\right. \\
\mathop{\mathbb{A}}_{E=1}^{n_{elem}^{M}} \left[\mathbf{K}_{n+1}^{M,E(i)} \left(\mathbf{d}_{n+1}^{M,E(i+1)} - \mathbf{d}_{n+1}^{M,E(i)} \right) = -\mathbf{r}_{n+1}^{M,E(i)} \right] \\
\text{IF: } \| \mathbf{r}_{n+1}^{M,E(i+1)} \| > tol \\
(i) \longleftarrow (i+1)
\end{array}\right. \\
\text{ELSE IF: } \| r_{n+1}^{M,E(i+1)} \| \leq tol \\
n \longleftarrow n+1
\end{array}\right.
$$

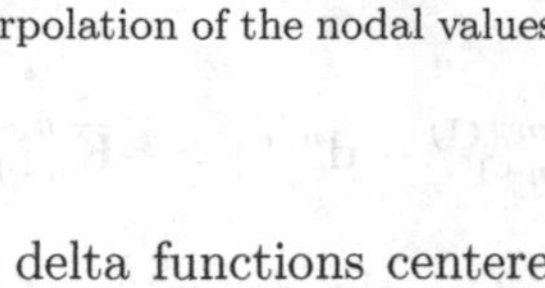

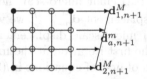

Fig. 9.3 Linear constraint imposed by 4-node macro-scale element, which allows to obtain connectivity matrix $\mathbf{T}^{e}$, enforcing that the nodal displacements at micro-scale elements $\mathbf{d}_{a,n+1}^{m}$ are computed as linear interpolation of the nodal values of displacements at macro-scale $\mathbf{d}_{1,n+1}^{M,E}$ and $\mathbf{d}_{2,n+1}^{M,E}$.

chosen in terms of the Dirac delta functions centered upon the micro-scale nodes placed on the interface $\Gamma^{M,E}$, which can be written: $\mathbf{P}_a \mapsto \delta(\mathbf{x}-\mathbf{x}_a)$. In this case, the scale-coupling can be defined explicitly, since the last equations in (9.21) will reduce to:

$$\mathbf{p}_a(\cdot) := \bar{\mathbf{d}}_{a,n+1}^{m} - \underbrace{\sum_{b} N_b^{M,E}(\mathbf{x}_a)}_{T_{ab}^{E}}\mathbf{d}_{b,n+1}^{M} = \mathbf{0}\ ; \qquad \forall a \tag{9.23}$$

Table 9.2 Quadratic convergence rate for micro–macro iterative solution procedure in a typical time step.

Iter. macro	Macro resid. energy	Iter. micro	Micro resid. energy
1	1.96×10^6	1	4.34×10^4
		2	6.70×10^0
		3	3.16×10^{-7}
		4	8.19×10^{-22}
2	2.11×10^4	1	1.55×10^6
		2	2.30×10^0
		3	7.53×10^{-3}
		4	1.86×10^{-14}
3	8.61×10^0	1	9.34×10^3
		2	2.61×10^{-3}
		3	6.94×10^{-14}
4	7.75×10^{-6}	1	6.06×10^0
		2	1.93×10^{-10}
5	7.45×10^{-17}	1	6.22×10^{-3}
		2	2.80×10^{-6}
		3	5.54×10^{-20}

which allows us to constrain all the micro-scale nodes on the interface $\Gamma^{M,E}$ to the displacement of macro-scale corner nodes; the final result can be written:

$$\left.\overline{\mathbf{d}}^{m}_{n+1}\right|_{\Gamma^{M,E}} = \mathbf{T}^{E}\mathbf{d}^{M,E}_{n+1} \tag{9.24}$$

where the components of matrix $\mathbf{T}$ are placed between 0 and 1, depending upon the position of a particular micro-scale node. With the best iterative value of macro-scale displacement $\mathbf{d}^{M,E,(i)}_{n+1}$, we will recover by (9.23) all the corresponding values of micro-scale displacements for the nodes placed at the boundary $\Gamma^{M,E}$. The first iteration at the micro-scale is thus performed with the linearized form of first of equations in (9.21) driven by the imposed boundary displacements:

$$\begin{aligned} &Lin[\mathbf{r}^{m}_{n+1}] = \mathbf{0} \Rightarrow \\ &\mathop{\mathbb{A}}_{e=1}^{n^{m}_{elem}} \left[\mathbf{K}^{m,e,(1)}_{n+1}(\mathbf{d}^{m,e(1)}_{n+1} - \mathbf{d}^{m,e}_{n}) = -\overline{\mathbf{K}}^{m,e}_{n+1}(\overline{\mathbf{d}}^{m,e(1)}_{n+1} - \overline{\mathbf{d}}^{m,e}_{n})\right] \end{aligned} \tag{9.25}$$

where $\mathbf{K}^{m,e}_{n+1}$ is computed as the assembly of the stiffness matrices of the micro-scale elements in the interior of macro-scale element, whereas $\overline{\mathbf{K}}^{m,e}_{n+1}$ is the corresponding assembly of stiffness matrices of the micro-scale elements along the boundary $\Gamma^{M,E}$. The solution of equation (9.25) will provide the first iterative guess for the interior nodal displacements at the micro-scale. The latter should then be tested and eventually corrected in order to enforce the equilibrium at the micro-scale:

$(j) = 1, 2, ...$ store internal variables: $\zeta^{m,e(j+1)}_{n+1} \longleftarrow \zeta^{m,e(j)}_{n+1}$

$$Lin[\mathbf{r}_{n+1}^{m,(j)}] = \mathbf{0} \Rightarrow \mathop{\mathbb{A}}_{e=1}^{n_{elem}^{m}} \left[\mathbf{K}_{n+1}^{m,e(j)} (\mathbf{d}_{n+1}^{m,e,(j+1)} - \mathbf{d}_{n+1}^{m,e,(j)}) = -\mathbf{r}_{n+1}^{m,e,(j)} \right] \quad (9.26)$$

If $\| \mathbf{r}_{n+1}^{m,e,(j+1)} \| > tol$, we carry on iterating with $(j) \longleftarrow (j+1)$; or else if $\| \mathbf{r}_{n+1}^{m,e,(j+1)} \| < tol$, we finish with local iterations, and proceed to the static condensation (see [263]) in order to obtain the corresponding stiffness matrix of the macro-scale element for the solution of global equilibrium equations:

$$\begin{aligned} \overbrace{\left[\overline{\overline{\mathbf{K}}}_{n+1}^{m} - \overline{\mathbf{K}}_{n+1}^{m\;T} (\mathbf{K}_{n+1}^{m})^{-1} \overline{\mathbf{K}}_{n+1}^{m} \right]}^{\widetilde{\mathbf{K}}_{n+1}^{m}} (\overline{\mathbf{d}}_{n+1}^{m} - \overline{\mathbf{d}}_{n}^{m}) \\ = \underbrace{-\overline{\mathbf{r}}_{n+1}^{m} + \overline{\mathbf{K}}_{n+1}^{m\;T} (\mathbf{K}_{n+1}^{m})^{-1} \overbrace{\mathbf{r}_{n+1}^{m}}^{=\mathbf{0}}}_{\widetilde{\mathbf{r}}_{n+1}^{m}} \end{aligned} \quad (9.27)$$

In the final step, we will exploit the result in (9.24) in order to obtain the standard format of macro-scale element arrays by keeping only the macro-scale nodal displacements:

$$\mathbf{r}_{n+1}^{M,E,(i)} = \mathbf{T}^{E,T} \widetilde{\mathbf{r}}_{n+1}^{m} \; ; \qquad \mathbf{K}_{n+1}^{M,E,(i)} = \mathbf{T}^{E,T} \widetilde{\mathbf{K}}_{n+1}^{m} \mathbf{T}^{E} \quad (9.28)$$

which allows us to obtain the internal force vector and tangent stiffness matrix for each macro-scale element. We can then pass these contributions to the standard finite element assembly procedure, leading to a linearized form of the macro-scale equilibrium equations:

$$\mathop{\mathbb{A}}_{E=1}^{n_{elem}^{M}} \left[\mathbf{K}_{n+1}^{M,E(i)} \left(\mathbf{d}_{n+1}^{M,E(i+1)} - \mathbf{d}_{n+1}^{M,E(i)} \right) = -\mathbf{r}_{n+1}^{M,E(i)} \right] \quad (9.29)$$

With the computed iterative contribution to macro-displacement, we can carry out the corresponding update in order to obtain an improved iterative guess $\mathbf{d}_{n+1}^{M,E(i+1)}$, and restart the iterative procedure described in (9.26). The computations in (9.29) is continued until the equilibrium equations at macro-scale are finally satisfied, resulting with $\| \mathbf{r}_{n+1}^{M} \| = 0$ within the specified tolerance. For more clarity, we present in Table 9.1 the flow-chart of the complete incremental/iterative solution procedure for the case of scale-coupling in displacements.

By using Newton's iterative scheme at micro and macro scales, we can indeed obtain with the proposed micro–macro solution procedure a quadratic convergence rate at each scale; an illustrative result for the quadratic convergence in a typical time step is given in Table 9.2.

(ii) Scale-coupling in forces

In the case of micro-macro scale-coupling in forces, we will seek to impose the variation of stress at the micro-scale that is in agreement with the chosen stress variation at macro-scale (see [179]). The latter is motivated by the

Table 9.3 Flow-chart of incremental/iterative micro–macro solution procedure: force coupling.

$$
\begin{array}{l}
n = 0, 1, 2, \ldots \\
\quad (i) = 1, 2, \ldots \\
\quad \text{given: } \mathbf{d}_{n+1}^{M,(i)} (\text{with}(\mathbf{d}_{n+1}^{M,(1)} = \mathbf{d}_n^M)) \\
\quad\quad E = 1, 2, \ldots, n_{el}^M \\
\quad\quad\quad (j) = 1, 2, \ldots \\
\quad\quad\quad \text{given: } \boldsymbol{\beta}_{n+1}^{E,(j)} \qquad (\text{ avec } (\boldsymbol{\beta}_{n+1}^{E(1)} = \boldsymbol{\beta}_n^E) \\
\quad\quad\quad\quad (k) = 1, 2, \ldots \\
\quad\quad\quad\quad \mathbb{A}_{e=1}^{n_{elem}^m} \left[\mathbf{D}_{n+1}^{m,e,(k,j,i)} \left(\mathbf{d}_{n+1}^{m,e,(k+1)} - \mathbf{d}_{n+1}^{m,e(k)} \right) = -\mathbf{r}_{n+1}^{m,e,(k,j,i)} \right] \\
\quad\quad\quad\quad \text{avec: } \widetilde{\mathbf{r}}_{n+1}^{m,(k,j,i)} = \mathbf{r}_{n+1}^{m,(k,j,i)} + \mathbf{E}(\boldsymbol{\beta}_{n+1}^{E,(j)} - \boldsymbol{\beta}_{n+1}^{E,(j)}) \\
\quad\quad\quad\quad \text{IF: } \| \widetilde{\mathbf{r}}_{n+1}^{m(k,j,i)} \| > tol \\
\quad\quad\quad\quad\quad (k) \longleftarrow (k+1) \\
\quad\quad\quad \text{ELSE IF: } \| \widetilde{\mathbf{r}}_{n+1}^{m(k,j,i)} \| \leq tol \\
\quad\quad\quad \mathbf{H}^{E,(j,i)^{-1}} (\boldsymbol{\beta}_{n+1}^{E,(j+1)} - \boldsymbol{\beta}_{n+1}^{E,(j)}) = -\widetilde{\mathbf{p}}_{n+1}^{E,(j,i)}; \\
\quad\quad\quad \text{with: } \mathbf{H}^{E,(j,i)} = \mathbf{E}^T \mathbf{D}^{m(j,i)} \mathbf{E} \\
\quad\quad\quad \widetilde{\mathbf{p}}_{n+1}^{(j,i)} = -\mathbf{p}_{n+1}^{(j,i)} + \mathbf{E}^T \mathbf{D}^{m-1} \mathbf{r}_{n+1}^{m,(i,j)} + \mathbf{F}(\mathbf{d}_{n+1}^{M,(k+1)} - \mathbf{d}_{n+1}^{m,e,(k)}) \\
\quad\quad\quad \text{IF: } \| \widetilde{\mathbf{p}}_{n+1}^{(j+1,i)} \| > tol \\
\quad\quad\quad (j) \longleftarrow (j+1) \\
\quad\quad \text{ELSE IF: } \| \widetilde{\mathbf{p}}_{n+1}^{(j+1,i)} \| \leq tol \\
\quad\quad E \longleftarrow E+1 \\
\quad \mathbb{A}_{E=1}^{n_{elem}^M} \left[\mathbf{K}_{n+1}^{M,E,(i)} \left(\mathbf{d}_{n+1}^{M,E(i+1)} - \mathbf{d}_{n+1}^{M,E,(i)} \right) = -\mathbf{r}_{n+1}^{M,E(i)} \right] \\
\quad \text{with: } \mathbf{K}_{n+1}^{M,E(i)} = \mathbf{F}^T \mathbf{H}^{E,(i)^{-1}} \mathbf{F} \\
\quad \widetilde{\mathbf{r}}_{n+1}^{M,E,(i)} = \mathbf{r}_{n+1}^{M,E,(i)} + \mathbf{F}^T \mathbf{H}^{E(i)^{-1}} \mathbf{p}_{n+1}^{(i)} - \mathbf{F}^T \mathbf{H}^{E,(i)^{-1}} \mathbf{E} \mathbf{D}^{(i)^{-1}} \mathbf{r}_{n+1}^{m,(i)} \\
\quad \text{IF: } \| \widetilde{\mathbf{r}}_{n+1}^{M(i)} \| > tol \\
\quad (i) \longleftarrow (i+1) \\
\text{ELSE IF: } \| \widetilde{\mathbf{r}}_{n+1}^{M,(i)} \| \leq tol \\
n \longleftarrow n+1
\end{array}
$$

optimal choice for 2D element with 4-nodes in bending dominated deformation modes, which is proposed by Pian and Sumihara (see [223]):

$$
\begin{aligned}
\sigma_{xx} &= \beta_1 + \beta_2 y \\
\sigma_{yy} &= \beta_3 + \beta_4 x \\
\sigma_{xy} &= \beta_5
\end{aligned}
\tag{9.30}
$$

We thus choose the corresponding interpolations for Lagrange multipliers in agreement with such a stress approximation, which can be constructed by appealing to the Cauchy principle for each edge of macro-scale element:

$$
\boldsymbol{\lambda}(x, y) := \left[\boldsymbol{\sigma}(x, y) \mathbf{n} \right]_{\Gamma^{Mm,E}} = \mathbf{P}\boldsymbol{\beta} \tag{9.31}
$$

The result in (9.31) is also in agreement with the physical interpretation of Lagrange multipliers as the components of the Cauchy stress vector acting along the boundaries of the macro-scale elements. The stress field of this kind ought to be self-equilibrated, and represented with five independent parameters. When accompanied by three rigid body modes, the latter can provide the correct range of the tangent stiffness matrix for 4-node macro-scale element, represent any deformation pattern and reconstruct the displacement field for such element. In practical computations we choose a minimum of local supports for eliminating the rigid body modes. The corresponding computational procedure is somewhat more involved than one employed for the previous case of scale-coupling in displacements, since the coupling equation can no longer be solved explicitly. The contribution of one micro-scale element to linearized form of equilibrium equations in (9.21) can now be written:

$$\begin{bmatrix} \mathbf{D} & \mathbf{E} & \mathbf{0} \\ \mathbf{E}^T & \mathbf{0} & -\mathbf{F} \\ \mathbf{0} & -\mathbf{F}^T & \mathbf{0} \end{bmatrix} \begin{bmatrix} \Delta \mathbf{d}^m \\ \Delta \boldsymbol{\beta}^E \\ \Delta \mathbf{d}^{M,E} \end{bmatrix} = \begin{bmatrix} -\mathbf{r}^m \\ \mathbf{p}^{M,E} \\ -\mathbf{r}^{M,E} \end{bmatrix} \tag{9.32}$$

with:

$$\mathbf{D} = \mathop{\mathbb{A}}_{e=1}^{n^m_{elem}} \int_{\Omega^{m,e}} \mathbf{B}^{mT} \mathbf{C}^m \mathbf{B}^m \, dV \; ; \qquad \mathbf{C}^m = \frac{\partial \boldsymbol{\sigma}^m}{\partial \boldsymbol{\epsilon}^m}\Big|_{\overline{\zeta_k}} \tag{9.33}$$

$$\mathbf{E} = \int_{\Gamma^{M,E}} \mathbf{N}^{mT} \mathbf{P} \, dA \; ; \qquad \mathbf{F} = \int_{\Gamma^{M,E}} \mathbf{P}^T \mathbf{N}^M \, dA \tag{9.34}$$

The computation of the tangent modulus $\mathbf{C}^m$ in (9.33) above requires plastically admissible stress and the results of computations of internal variables in (9.22). The set of linear equations in (9.32) will allow us to express $\Delta \mathbf{d}^m$ as well as $\Delta \boldsymbol{\beta}$ in terms of $\Delta \mathbf{d}^M$. We can thus provide the tangent stiffness matrix and residual vector by the static condensation of the system in (9.32), where we only keep $\Delta \mathbf{d}^{M,E}$ as unknowns. In order to increase the robustness of this computational procedure, we can choose the operator split procedure, where we iterate independently on each of equations in (9.32). The flow chart of proposed incremental/iterative micro–macro procedure for scale-coupling in force is given in Table 9.3.

9.2 Microstructure representation

The main advantage of the micro–macro approach for nonlinear analysis of inelastic behavior proposed herein is its ability to provide the most reliable representation of the mechanisms governing the inelastic behavior. The latter is constructed at the micro scale, where we can properly account for heterogeneities of real materials and its microstructure. Needless to say, none of the phenomenological models can provide the same level of accuracy and reliability in interpreting the inelastic behavior.

9.2.1 *Microstructure representation by structured mesh with isoparametric finite elements*

The first possibility which we propose is to represent the details of heterogeneities by using the isoparametric finite elements at the micro-scale. In such a case, we can adapt the inter-element boundaries so that each element sub-domain will only contain one phase. We can obtain in this manner quite reliable representation of the strain field, since the mechanical properties as well as the corresponding criteria of inelastic behavior remain constant (and continuous) within each element. However, the complexities of real materials can easily lead to practically intractable difficulties for preparing the finite element mesh for this kind of exact representation of their microstructure (e.g. see [81] for some practical methods of mesh preparation), as well as the significant reduction of results quality because of excessive element distortion in such a mesh. For that reason, we consider herein the use of structured finite element mesh, where each element keeps a regular form (e.g. square or rectangular), which is compatible with the form of the domain of interest (e.g. typical specimen for standard tests). With the structured mesh representation, the phase-interface is not limited to inter-element boundaries, but it can be placed inside the element. Therefore, each element sub-domain can contain more than one phase. The resulting strain field representation for such a case does not possess the same regularity as for the case with elements with homogeneous material properties. Two potential solutions are proposed and tested herein for dealing with the heterogeneous element sub-domains of this kind: one that employs an increased number of numerical integration points with respect to the minimum required for homogeneous element sub-domains, and the other that uses the corresponding incompatible mode method to enhance the strain approximation of isoparametric finite elements. The proposed methods are illustrated for a material model with two-phases; see Figure 9.4 for the exact mesh and structured mesh representations in a typical macro-scale element.

The first method for microstructure representation by structured finite element mesh of isoparametric elements considers an increase of total number of numerical integration points (with respect to the minimum requirement). For example, for a 4-node micro-scale element, instead of usual quadrature rule $2{\times}2 = 4$, we can employ $5{\times}5 = 25$ Gauss quadrature points. We thus hope to provide better sampling of the complex microstructure throughout the element and corresponding participation of each phase. However, the results presented in Figure 9.5, specifying the convergence in energy norm with

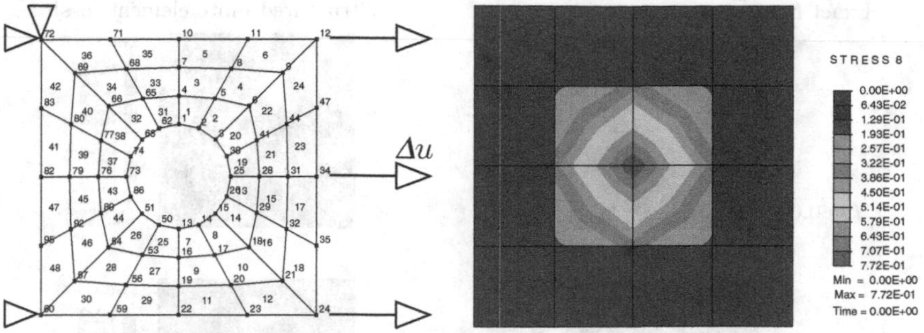

a) Exact finite element mesh b) Structured finite element mesh

Fig. 9.4 Microstructure representation within macro-element by using: (**a**) exact finite element mesh; (**b**) structured finite element mesh.

respect to the chosen mesh, confirm that it is better to increase the number of elements than the number of Gauss points in order to increase the accuracy.

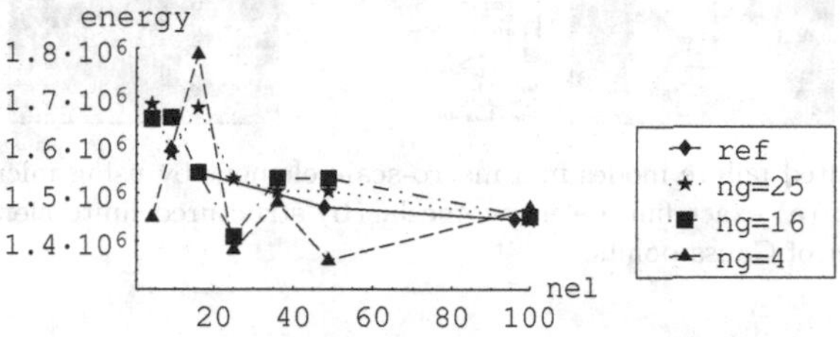

Fig. 9.5 Computed energy in a simple tension test for a single macro-scale element by using structured mesh representation of microstructure, with respect to number of micro-scale elements and number of Gauss quadrature points.

This kind of results also indicates that a more precise representation of strain and stress fields, achieved by using more micro-scale elements or enhancing their performance, provides far better accuracy than the more precise information (from using increased order of Gauss quadrature) from the representation of inferior quality. In fact, increasing the number of Gauss points will only change the "filter" that provides observation of inelastic behavior mechanisms, which remain represented (not sufficiently well) by micro-scale elements. For that reason the result in Figure 9.6 is very important to confirm that one should rather change the element which is not capable of representing the inelastic behavior mechanisms; the case in point, is the failure mode observed in plastic matrix with exact mesh as a typical mode II failure, which is interpreted by the structured mesh representation as a diffuse mode I failure.

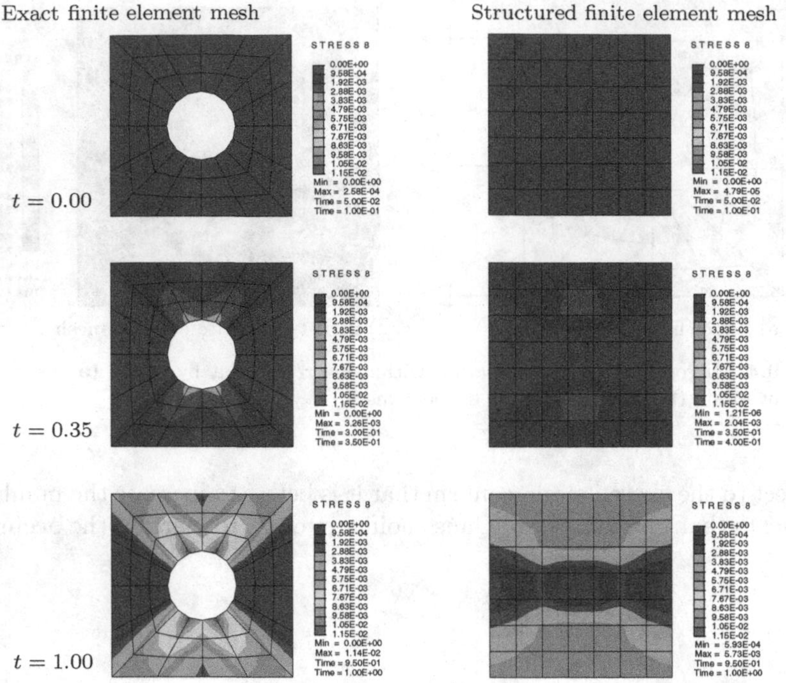

Fig. 9.6 Computed failure modes in a macro-scale element by using microstructure representation with: (**a**) exact finite element mesh, (**b**) structured finite element mesh with increased number of Gauss points.

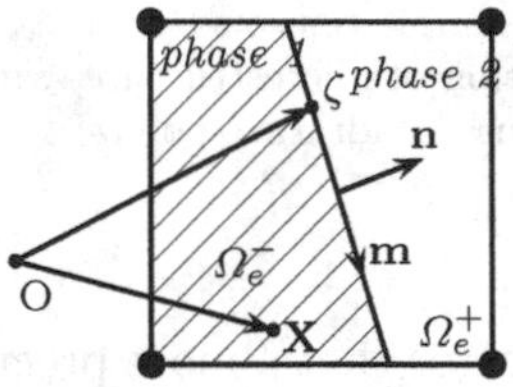

Fig. 9.7 Two-phase micro-scale element cut with the phase-interface in two sub-domains.

9.2.2 *Microstructure representation by structured mesh with incompatible mode elements*

The incompatible mode method can be used to provide the corresponding enhancement of the strain field for the element which contains two or more different phases. We will consider for illustration a micro-scale element in Figure 9.7 crossed by the phase interface. For simplicity, and with no loss of generality, we will consider the zero volume force. The condition of stress vector continuity across the phase interface will impose:

$$\mathbf{t}^+ = \mathbf{t}^- \Rightarrow \begin{cases} \sigma^+_{nn} = \sigma^-_{nn} \\ \sigma^+_{nm} = \sigma^-_{nm} \end{cases} \tag{9.35}$$

where $(\cdot)^+$ and $(\cdot)^-$ are the values on each side of the interface. The result in (9.35) also contains the component form of the stress continuity condition, which is obtained by appealing to the Cauchy principle to express the stress vector in terms of stress tensor components. If the mechanical properties (e.g. Young's modulus) in each phase are not the same, we can conclude from (9.35) that the corresponding strain filed components are discontinuous across the interface; this can be written:

$$\epsilon^+_{nn} \neq \epsilon^-_{nn} \quad ; \quad \epsilon^+_{nm} \neq \epsilon^-_{nm} \tag{9.36}$$

Such a discontinuity of strain field cannot be represented by an isoparametric element, which always results with a smooth approximation of strain field within the element. The corresponding approximation can only be constructed by enhancing the strain field of the isoparametric element by the judicious choice of incompatible modes (see [134]); in present case, we can choose:

$$\mathbf{u}(\mathbf{x})|_{\Omega^{m,e}} = \sum_{a=1}^{4} N_a^{m,e}(\mathbf{x})\,\mathbf{d}_a^{m,e} + \alpha\widehat{\mathbf{n}}(\mathbf{x}) + \beta\widehat{\mathbf{m}}(\mathbf{x}) \tag{9.37}$$

with:

$$\widehat{n}_i(\mathbf{x}) = \begin{cases} \theta n_i[\mathbf{n}\cdot(\mathbf{x}-\bar{\boldsymbol{\xi}})]; & \mathbf{n}\cdot(\mathbf{x}-\bar{\boldsymbol{\xi}}) \geq 0 \\ (1-\theta)n_i[\mathbf{n}\cdot(\mathbf{x}-\bar{\boldsymbol{\xi}})]; & \mathbf{n}\cdot(\mathbf{x}-\bar{\boldsymbol{\xi}}) < 0 \end{cases} \tag{9.38}$$

$$\widehat{m}_i(\mathbf{x}) = \begin{cases} \theta m_i[\mathbf{n}\cdot(\mathbf{x}-\bar{\boldsymbol{\xi}})]; & \mathbf{n}\cdot(\mathbf{x}-\bar{\boldsymbol{\xi}}) \geq 0 \\ (1-\theta)m_i[\mathbf{n}\cdot(\mathbf{x}-\bar{\boldsymbol{\xi}})]; & \mathbf{n}\cdot(\mathbf{x}-\bar{\boldsymbol{\xi}}) < 0 \end{cases} \tag{9.39}$$

where $\bar{\boldsymbol{\xi}}$ is the position vector at the interface, $\mathbf{x}$ is the position vector in the sub-domain Ω^{e+} or Ω^{e-}, and θ is the parameter to be determined with respect to the element shape. The displacement field of this kind belongs to the incompatible mode method, which provides the strain field approximation:

$$\boldsymbol{\epsilon}^m(\mathbf{x})|_{\Omega^{m,e}} = \nabla^s\mathbf{u}^m(\mathbf{x}) = \sum_{a=1}^{4} \mathbf{B}_a^{m,e}(\mathbf{x})\mathbf{d}_a^{m,e} + \widetilde{\epsilon}(\mathbf{x}) \tag{9.40}$$

where the enhanced strains $\widetilde{\epsilon}$ can be defined explicitly:

$$\widetilde{\epsilon}(\mathbf{x}) = \alpha \begin{cases} \theta \mathbf{n} \otimes \mathbf{n}; & \mathbf{n} \cdot (\mathbf{x} - \bar{\boldsymbol{\xi}}) \geq 0 \\ (1-\theta)\mathbf{n} \otimes \mathbf{n}; & \mathbf{n} \cdot (\mathbf{x} - \bar{\boldsymbol{\xi}}) < 0 \end{cases}$$
$$+\beta \begin{cases} \theta(\mathbf{n} \otimes \mathbf{m} + \mathbf{m} \otimes \mathbf{n}); & \mathbf{n} \cdot (\mathbf{x} - \bar{\boldsymbol{\xi}}) \geq 0 \\ (1-\theta)(\mathbf{n} \otimes \mathbf{m} + \mathbf{m} \otimes \mathbf{n}); & \mathbf{n} \cdot (\mathbf{x} - \bar{\boldsymbol{\xi}}) < 0 \end{cases} \tag{9.41}$$

The patch test convergence condition imposed upon the incompatible mode method, will thus require the value of parameter θ equal to:

$$\int_{\Omega^{m,e}} \widehat{\epsilon}(\mathbf{x})\, dV = 0 \quad \Rightarrow \quad \theta = \frac{\Omega^{e-}}{\Omega^{e}}; \qquad \Omega^{e-} + \Omega^{e+} = \Omega^{e} \tag{9.42}$$

An illustrative result with force–displacement diagram obtained for a simple tension test by using the proposed enhancement is given in Figure 9.8 to confirm that the results are close to the reference value obtained by the exact finite element mesh representation of the microstructure, and also very much improved with respect to those obtained by an equivalent mesh constructed with isoparametric elements, even if an increased number of numerical integration points is used.

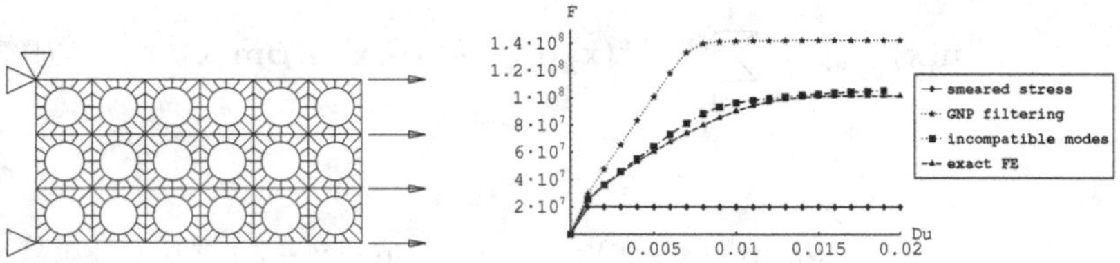

Fig. 9.8 Simple tension test for material with microstructure: force–displacement diagram computed with three different models.

9.2.3 Microstructure representation with uncertain geometry and probabilistic interpretation of size effect for dominant failure mechanism

The structured mesh representation of the microstructure can also be employed for modelling the most general situation with different inelastic behavior in each phase within a single element domain, as well as the localized failure at the phase interface resulting with softening behavior. In this case, we will need additional kinematics enhancement in terms of the displacement discontinuity for the corresponding element of the structured mesh. We will

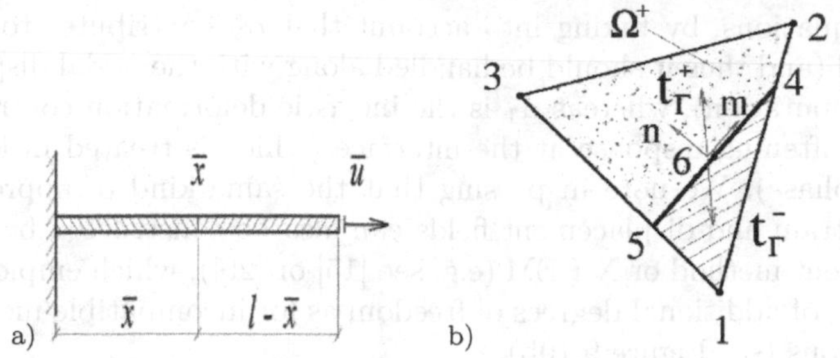

Fig. 9.9 Two-phase elements with failure mode at phase interface: (**a**) 1D 2-node truss-bar element, (**b**) 2D 3-node triangular element with stress vector continuity enforced across phase interface.

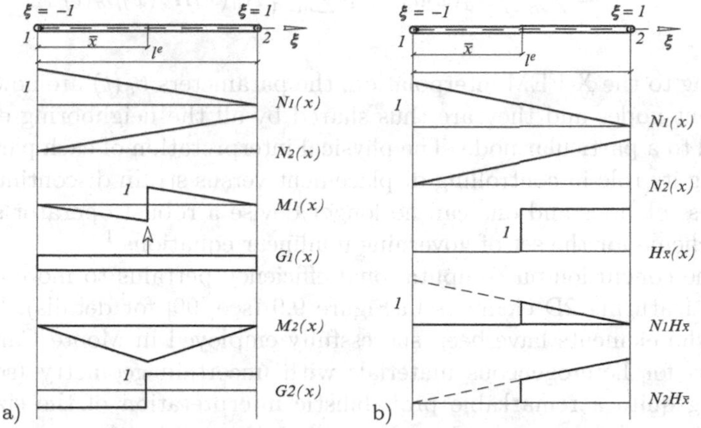

Fig. 9.10 Two-node bar element with: (**a**) embedded strain and displacement discontinuities placed at $\bar{x}$, where α_1 and α_2 are element parameters scaling $M_1(x)$ and $M_2(x)$ (**b**) equivalent X-FEM displacement interpolation, where β_1, β_2 are nodal parameters scaling $N_1(x)H_{\bar{x}}(x)$ and $N_2(x)H_{\bar{x}}(x)$.

illustrate this approach for the simplest 1D case with 2-node truss-bar element (see [136]) and for 2D case with 3-node triangular element (see [99]), both presented in Figure 9.9. The truss-bar element in Figure 9.9 consists of two different phases with inelastic behavior and the possibility of localized failure at phase interface placed at $\bar{x}$.

For the choice of the discontinuity influence domain corresponding to a single element $\widetilde{\Omega} = \Omega^e$, the displacement field interpolation can be written:

$$u(x,t)\Big|_{\Omega_i^e} = \sum_{a=1}^{2} N_a(x) d_a(t) + M_1(x)\alpha_1(t) + M_2(x)\alpha_2(t) \; ; \; x \in \Omega_i^e \quad (9.43)$$

where different shape functions are illustrated in Figure 9.10. A very robust operator split solution procedure can be developed (see [136]) for such a set of equations, by taking into account that α_2 contributes to the total strain field (and thus it should be handled along with the nodal displacement global computation), whereas α_1 is the inelastic deformation controlling the localized softening response at the interface (which is treated in local computation phase). We note in passing that the same kind of approximation for total strain and displacement fields can also be constructed by extended finite element method or X-FEM (e.g. see [15] or [29]), which employs identical number of additional degrees of freedom as for incompatible modes based interpolations (see Figure 9.10b):

$$\begin{aligned} u(x,t)\Big|_{\Omega^e} &= \textstyle\sum_{a=1}^{2} N_a(x)\left[d_a(t) + H_{\bar{x}}(x)\beta_a(t)\right] \\ &= \textstyle\sum_{a=1}^{2} N_a(x)d_a(t) + \sum_{a=1}^{2} N_a(x)H_{\bar{x}}(x)\beta_a(t) \end{aligned} \qquad (9.44)$$

According to the X-FEM interpolation, the parameters $\beta_a(t)$ are now placed at element nodes and they are thus shared by all the neighboring elements attached to a particular node. The physical interpretation of each parameter, regarding its role in controlling displacement versus strain discontinuity, becomes less obvious and one can no longer devise a robust operator split solution scheme for the set of governing nonlinear equations.[1]

The same conclusion on computational efficiency pertains to more complex problem featuring 2D elements in Figure 9.9 (see [99] for details). For that reason, the elements have been successfully employed in Monte Carlo computations for heterogeneous materials with uncertain geometry (see [55]), providing quite a remarkable probabilistic interpretation of the size effect for dominant failure mechanism. We will elaborate shortly upon this idea through the following example. We consider a two-phase specimen as the microstructure representation of a rectangular domain corresponding to the representative volume element. The mechanical properties of each phase are deterministic (in particular the circular inclusions are considered elastic, and the surrounding matrix is elastoplastic obeying the Drucker-Prager perfect plasticity criterion), but the exact internal geometry of each phase distribution is not known with certainty. The phase distribution is handled by the Gibbs point process, which results in different realizations for the position of inclusions and their size. The structured mesh representation of a single realization of the microstructure is shown in Figure 9.11. The results of the numerical simulation for each realization in terms of imposed displacement versus failure load are also shown in Figure 9.11.

---

[1] The X-FEM interpolation is still needed for non-local representation of the corresponding field; the case in point is the analysis of crack spacing in reinforced concrete; see [116].

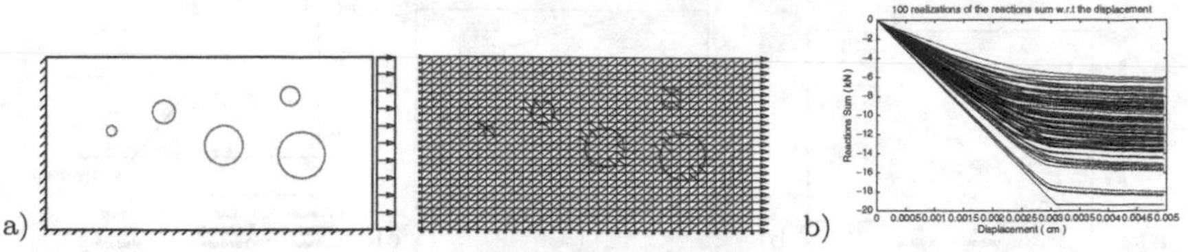

Fig. 9.11 Representative volume element computations: (**a**) two-phase microstructure with uncertain geometry and structured finite element mesh representation, (**b**) force-displacement diagrams for different realizations.

We will further interpret the results computed by this 2D model for a simple tension test in terms of the corresponding 1D model for failure of massive structures quite similar to the one described in the previous chapter (see Figure 9.12), apart the fact that the yield stress and the ultimate failure stress are no longer considered as deterministic material properties but rather the correlated random fields. The proper interpretation of the results of 2D model (see [99]) can provide typical realizations of yield and ultimate failure stress variations along the bar as shown in Figure 9.12), where the length of the bar is qualified as short, medium or long with respect to the correlation length.

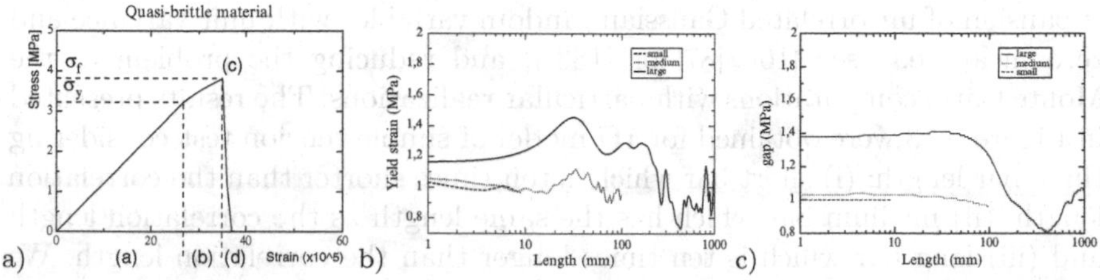

Fig. 9.12 Diffuse-localized failure computation for 1D truss-bar for short, medium and long bar: (**a**) stress–strain diagram for failure model of massive structures with yield stress and ultimate stress as random fields, (**b**) normalized realizations of yield stress σ_y, (**c**) normalized realizations of difference between ultimate and yield stress e_f.

In other words, we have thus developed a 1D model for diffuse-localized failure of quasi-brittle materials, where we consider that the yield stress σ_y and the difference e_f between the ultimate failure stress σ_f and the yield stress σ_y are uncertain, whereas Young's modulus and fracture energy are kept deterministic, for simplicity. The key point of stochastic description pertains to choosing the random fields σ_f and e_f as correlated,[2] since the start of yielding

[2] If we were following Weibull theory of weakest link for explaining the failure, we would choose to model this set of material properties by two uncorrelated random fields (or an infinite set of random variables indexed by the geometrical domain; see [171]) amounting to modelling the white noise along the geometrical domain.

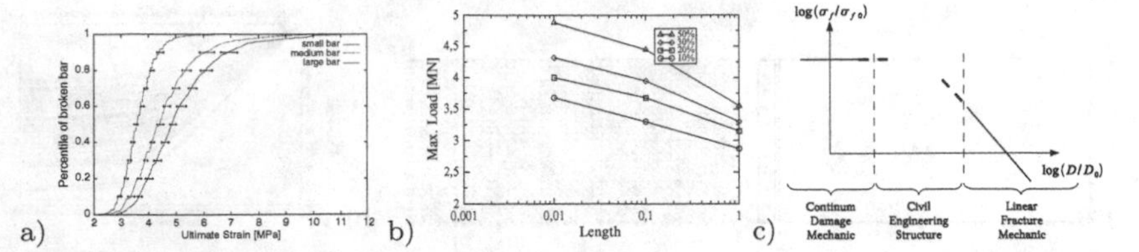

Fig. 9.13 Diffuse-localized failure computation for 1D truss-bar for short, medium and long bar: (**a**) Cumulative density function of the failure load obtained by Monte Carlo simulation (with 99% error bars), (**b**) Different percentile for each bar (short, medium and long), (**c**) size-effect explanation from [25]

or ultimate stress related localized failure in one point will inevitably influence the values of these fields within the whole neighborhood defined by the correlation length. Both random fields are considered stochastically homogeneous, which implies they can be characterized by their marginal distribution and their covariance function; moreover, by considering that both σ_y and e_f are positive and supposed to have a finite known variance, we can choose lognormal marginal distribution and consider for convenience that these two random fields are defined as non-linear transformations of Gaussian random fields γ_1 and γ_2. The latter is helpful for constructing an effective computational representation of the random fields in terms of the Karhunen-Loève expansion of uncorrelated Gaussian random variables with unit variance and zero mean (e.g. see [16], [87] or [183]), and reducing the problem to the Monte Carlo computations with particular realizations. The results presented in Figure 9.13 were obtained for 1D model of simple tension test considering three bar length: (i) short bar which is ten times shorter than the correlation length, (ii) medium bar which has the same length as the correlation length and (iii) long bar which is ten times longer than the correlation length. We show in particular the corresponding results for the maximum load cumulative density functions of these three bars in Figure 9.13a, along with the error bars showing the 99% confidence interval, which never overlap allowing us to conclude that the Monte Carlo simulations were accurate enough. In Figure 9.13b, we gather the ultimate load values for a chosen percentile of broken bars; for example, with 10% we find ultimate strength of 3.68 MP for the small bar, 3.3 MP for the medium one and 2.87 MP for the long bar. We can conclude from those results that not only the long bar breaks the first with the smallest ultimate failure stress, but also that it has the smallest fracture process zone with the smallest difference between the ultimate and the yield stress. We can also conclude that the short bar has the largest fracture process zone. The results of this kind thus provide the probability based interpretation of the size effect specifying the typical failure mechanism for each structure size based upon material heterogeneities and thus more intrinsically related to microstructure than the previous interpretations (e.g. see [25]).

9.3 Conclusions and remarks on current research works

Recent reviews on current research interests (e.g. see [206]) are very much keen on multi-physics applications, since it has been recognized that the computational mechanics offers practically unlimited potential to enhance the understanding of a number of practical problems and further advance the technological level currently available. Some of the examples of this kind are: providing better understanding of advanced composite material in aerospace (e.g. [157]), multi-physics problems related to durability of construction materials under extreme loading (e.g. see [178] or [233]), numerical simulations of impact, blast and nonlinear dynamics problems in industrial applications (e.g. see [76], [213] or [220]), better interpretation of experimental results obtained in tests on structures (e.g. see [188]) or development of advanced design of complex structures with better understanding of the most appropriate model of inelastic structural behavior (e.g. see [216] or [247]). In each of these studies, one will provide an enhanced interpretation of the corresponding multi-physics problems interacting with the solid mechanics component, which thus calls for equally enhanced interpretation such as provided by multi-scale models.
The multi-scale models of inelastic behavior provide an important advantage with respect to the classical phenomenological models with their capability to provide a more reliable criteria for initiation and propagation of inelastic mechanisms. The ever increasing computational resources and tools allow us to tackle successfully the multi-scale models of large complexity, and to make them more pertinent to the nonlinear inelastic analysis of real-life structures. If we carry on this line of developments, we can soon arrive at the widespread use of novel representation of the nonlinear inelastic material behavior taking into account its microstructure. There, we can encounter a number of open problems, which are currently waiting for solutions. For example, the details of inelastic mechanism typical of a given microstructure are not always well represented by the finite element mesh, but rather by discrete elements which can provide a reliable representation of cohesive forces (e.g. [225]). One model of this kind is proposed by Ibrahimbegovic and Delaplace [121] for representing the brittle failure mechanisms at micro and meso-scale, by using the Voronoi polygon-cell representation of microstructure and the cohesive force model based upon the geometrically exact Reissner beam (e.g. see Ibrahimbegovic and Taylor [138]). The latter allows to take into account the large displacements and rotations that can be produced locally within the fracture process zone, as well as the fracture process where groups of cells detach from the main structure; see Figure 1.1. Some other discrete elements based upon the Voronoi polygons (e.g. see [142]) are also suitable for problems with large shape change, such as fluid mechanics, as well as for enhancing the computational efficiency brought by a quick mesh construction. The optimal approach for constructing the best finite element model is very likely to turn towards the adaptive modelling, where one can change the mesh a number

of times in the course of a given loading program requiring with each mesh change the corresponding field transfer capable to ensure the conservation of inelastic dissipation, admissibility of stress etc. (see [39] or [139]).
With the multi-scale models providing a detailed description of the microstructure, we can further explore how to make the optimal use of (composite) materials with respect to their inelastic behavior. One such illustrative example is presented by Ibrahimbegovic et al. [126] for a model two-phase material, where one phase is described with elastoplastic and the other with damage model. The optimization problem considered herein concerns the search for optimal shape of phase interface. The cost function is chosen either to render the amount of inelastic dissipation (which is of interest for durability problems), or the maximum dissipation (which is of interest for design of energy-dissipation devices trying to limit the spread of damage within a structure).
Control of mechanical tests on structures, or tests under heterogeneous stress field for heterogeneous materials with microstructure, where one tries to achieve the desired stress and strain states (e.g. see Ibrahimbegovic et al. [129] or Kucerova et al. [155]) is also a research goal of great current interest. The successful solution to this kind of problems would allow to reduce the number of tests, since each one of them would be able to provide much more information from standard test under homogeneous stress field. Additional concern for this kind of tests in the sense of multi-scale modelling is to identify the scale where identification of inelastic behavior mechanisms will be the most reliable. Finally, in view of results dispersion for the real materials, such an identification procedure should be placed within the probabilistic bounds for the corresponding mechanical properties of interest.
We can think that the multi-scale modelling of inelastic mechanical behavior will next be followed by the multi-scale modelling of accompanying problem, such as heat transfer or physical-chemistry, which should allow even more reliable interpretation of evolution of inelastic mechanisms. In other words, the multi-scale models for multi-physics problems will probably increase their field of practical applications. The multi-scale analysis for all the problems of this kind will require that we should deal not only with multi-scale problems in space, but also with multi-scale problems in time (e.g. see Ibrahimbegovic et al. [120] or Kassiotis et al. [147]) as an additional complexity.
With the increase in problem complexity that stems from multi-physics and multi-scale problems, as well as from coupling of nonlinear mechanics with the corresponding tools for optimization, control or probability computations, we believe that the software development will certainly remain a very important issue. In that sense, quite an interesting development in Niekamp et al. [204] should be of interest for all those who would like to integrate their favorite piece of software within a more general software product.
One can wonder if the current developments on micro-macro models will eventually leave any place to the phenomenological models. There is no danger that the phenomenological models will phase out; we should rather imagine

that the vast majority of problems in computing the ultimate limit load of complex structures will call for a judicious combination of phenomenological and multi-scale models for computing nonlinear inelastic behavior, where the latter is limited to the local zones where more detailed information (such as those important for durability problems) is required, which cannot be obtained by the phenomenological models. Within this context, the quality of a phenomenological model will be measured with respect to its capability to ensure the smooth transition and accept the wealth of information coming from micro-macro models.[3] It is thus quite likely that the phenomenological models and corresponding methods will remain practical computational tools for all those who are interested in nonlinear solid mechanics and its applications.

[3] In that sense, the model reduction is currently a very active research field, especially for coupled and interaction problems; see [180] or [218].

References

1. R. Abraham and J.E. Marsden. *Foundations of Mechanics.* Addison-Wesley, Reading, MA, 1985.
2. P. Alart and A. Curnier. A mixed formulation for frictional contact problems prone to Newton like solution methods. *Computer Methods in Applied Mechanics and Engineering*, 92:353–375, 1991.
3. O. Allix and F. Hild, editors. *Continuum Damage Mechanics of Materials and Structures.* Elsevier, Amsterdam, 2002.
4. S.S. Antmann. *Problems in Nonlinear Elasticity.* Springer, New York, 1995.
5. J.H. Argyris and J. St. Doltsinis. On the natural formulation and analysis of large deformation coupled thermomechanical problems. *Computer Methods in Applied Mechanics and Engineering*, 25:195–253, 1981.
6. J.H. Argyris, G. Faust, and M. Haase. *An Exploration of Chaos.* North-Holland, Amsterdam, 1994.
7. F. Armero and K. Garikipati. An analysis of strong discontinuities in multiplicative finite strain plasticity and their relation with the numerical simulation of strain localization in solids. *International Journal of Solids and Structures*, 33:2863–2885, 1996.
8. F. Armero and E. Petocz. A new dissipative time-stepping algorithm for frictional contact problems: formulation and analysis. *Computer Methods in Applied Mechanics and Engineering*, 179:151–178, 1999.
9. F. Armero and J.C. Simo. A priori stability estimates and unconditionally stable product formula algorithms for nonlinear coupled thermoplasticity. *International Journal of Plasticity*, 9:749–782, 1993.
10. P.J. Armstrong and C.O. Frederick. A mathematical representation of the multiaxial Bauschinger, effect. Technical Report RD/B/N731, C.E.G.B., Berkeley Nuclear Laboratories, R&D Department, 1966.
11. M. Arnold and M. Gunther. Preconditioned dynamic iteration for coupled differential-algebraic systems. *BIT Numerical Mathematics*, 41:1–25, 2001.
12. V.I. Arnold. *Mathematical Mehods in Classical Mechanics.* Springer, New York, 1980.
13. R. Asaro. Micromechanics of crystals and polycrystals. *Advances in Applied Mechanics*, 23:1–115, 1983.
14. P.W. Atkins. *Physical Chemistry.* W.H. Freeman, New York, 1998.
15. I. Babuška and J.M. Melenk. The partition of unity method. *International Journal for Numerical Methods in Engineering*, 40:727–758, 1997.
16. I. Babuška, R. Tampone, and G.E. Zouraris. Solving elliptic boundaray value problem with uncertain coefficients by the finite element method: the stochastic formulation. *Computer Methods in Applied Mechanics and Engineering*, 194:1251–1294, 2005.

17. J. Ball. Polyconvexity conditions. *Archives Rational Mechanics and Analysis*, 2:206–235, 1977.
18. J. Barlow. Optimal stress locations in finite element models. *International Journal for Numerical Methods in Engineering*, 10:243–251, 1976.
19. K.J. Bathe. *Finite Element Procedures*. Prentice-Hall, Englewood Cliffs, NJ, 1996.
20. K.J. Bathe and E.L. Wilson. *Numerical Methods in Finite Element Analysis*. Prentice-Hall, Englewood Cliffs, NJ, 1976.
21. K.J. Bathe, E.L. Wilson, and E. Ramm. Finite element formulations for large deformation dynamic analysis. *International Journal for Numerical Methods in Engineering*, 9:353–86, 1975.
22. J.L. Batoz and G. Dhatt. Incremental displacement algorithms for nonlinear problems. *International Journal for Numerical Methods in Engineering*, 14:1261–1266, 1979.
23. J.L. Batoz and G. Dhatt. *Moélisation des structures par éléments finis. Vol. 1: solides élastiques*. Hermes Science, Paris, 1990.
24. O.A. Bauchau, G. Damilano, and N.J. Theron. Numerical integration of nonlinear elastic multibody systems. *International Journal for Numerical Methods in Engineering*, 38:2727–2751, 1995.
25. Z.P. Bazant and L. Cedolin. *Stability of Structures: Elastic, Inelastic, Fracture and Damage Theories*. Oxford University Press, Oxford, 1991.
26. Z.P. Bazant and F.B. Lin. Non-local yield limit degradation. *International Journal for Numerical Methods in Engineering*, 26:1805–1823, 1988.
27. Z.P. Bazant and G. Pijaudier-Cabot. Non linear continuous damage, localization instability and convergence. *Journal of Applied Mechanics*, 55:287–293, 1988.
28. T. Belytschko, W.K. Liu, and B. Moran. *Nonlinear Finite Elements for Continua and Structures*. Wiley, Chichester, 2000.
29. T. Belytschko, N. Moes, S. Usui, and C. Parimi. Arbitrary discontinuities in finite elements. *International Journal for Numerical Methods in Engineering*, 50:993–1013, 2001.
30. T. Belytschko and D.F. Schoeberle. On the unconditional stability of an implicit algorithm for nonlinear structural dynamics. *Journal Applied Mechanics, ASME*, 97:865–869, 1975.
31. A. Benallal, editor. *Continuum Damage and Fracture*. Elsevier, Amsterdam, 2000.
32. A. Benallal, R. Billardon, and L. Doghri. An integration algorithm and the corresponding tangent operator for fully coupled elastoplastic and damage equations. *Communications in Numerical Methods in Engineering*, 4:731–740, 1988.
33. C. Bernardi, Y. Maday, and A.T. Patera. A new non conforming approach to domain decomposition - the mortar method. Paris, 1994. Collège de France.
34. N. Bicanic, E. Pramono, S. Sture, and K.J. Willam. On numerical prediction of concrete fracture localizations. In *Proceedings of the NUMETA Conference*, pages 385–392. Balkema, 1985.
35. M. Bischoff, E. Ramm, and D. Braess. A class of equivalent enhanced assumed strain and hybrid stress finite elements. *Computational Mechanics*, 22:443–449, 1999.
36. M. Bornert, T. Bretheau, and P. Gilormini, editors. *Homogenization in Mechanics of Materials, Vol. I & II*. Hermes-Science, Paris, 2001.
37. R. De Borst, M. Geers, and H. Petereekns. Fundamental issues in finite element analysis of localization of deformation. *International Journal for Numerical Methods in Engineering*, 10:99–121, 1995.
38. R. De Borst, L.J. Sluys, H.B. Hühlhaus, and J. Pamin. Fundamental issues in finite element analysis of localization of deformation. *Engineering Computations*, 10:99–121, 1993.
39. D. Brancherie, P. Villon, and A. Ibrahimbegovic. On a consistent field trnafer in nonlinear inelastic analysis and ultimate load computation. *Computational Mechanics*, 42:213–226, 2008.

40. H. Brezis. *Analyse fonctionnelle: théorie et applications.* Masson, Paris, 1983.
41. F. Brezzi and M. Fortin. *Mixed and Hybrid Finite Element Methods.* Springer, New York, 1991.
42. C.G. Broyden. Quasi-Newton methods and their application to function minimization. *Mathematics and Computation*, 21:368–381, 1967.
43. J. Casey and P.M. Naghdi. A remark on the use of the decomposition $F = F_e F_p$ in plasticity. *Journal of Applied Mechanics, ASME*, 47:672–675, 1980.
44. J.L. Chaboche. On some modifications of kinematic hardening to improve the description of ratcheting effects. *International Journal of Plasticity*, 7:661–678, 1991.
45. J.L. Chaboche and G. Cailletaud. On the calculation of structures in cyclic plasticity and viscoplasticity. *Computers and Structures*, 23(1):23–31, 1994.
46. P. Chadwick. *Continuum Mechanics.* Wiley, New York, 1976.
47. Y. Choquet-Bruhat and C. deWitt Morette. *Analysis, Manifolds and Physics.* North-Holland, Amsterdam, 1989.
48. L. Chorfi and A. Ibrahimbegovic. Modèle de plasticité et viscoplastcité pour le chargement cyclique. *Revue eurpéenne des éléments finis*, 10:77–97, 1991.
49. P.G. Ciarlet. *The Finite Element Method for Elliptic Problems.* North-Holland, Amsterdam, 1978.
50. P.G. Ciarlet. *Mathematical Elasticity. Vol. 1: Three-Dimensional Elasticity.* North-Holland, Amsterdam, 1988.
51. P.G. Ciarlet. *Introduction à l'analyse numérique matricielle et à l'optimisation.* Masson, Paris, 1990.
52. F.H. Clarke. *Optimization and Nonsmooth Analysis*, Vol. 9. SIAM, Philadelphia, PA, 1990.
53. R.W. Clough and J. Penzien. *Dynamics of Structures.* McGraw-Hill, New York, 2 edition, 1993.
54. B.D. Coleman and H.D. Hodgdon. On shear band in ductile materials. *Archives in Rational Mechanics and Analysis*, 90:219–247, 1985.
55. J.B. Colliat, M. Hautefeuille, A. Ibrahimbegovic, and H. Matthies. Stochastic approach to size effect in quasi-brittle materials. *Comptes Rendus de l'Académie des Sciences. Part B. Mécanique*, 335:430–436, 2007.
56. J.B. Colliat, A. Ibrahimbegovic, and L. Davenne. Saint-venant multi-surface plasticity model in strain space and in stress resultants. *Engineering Computations*, 41:536–557, 2005.
57. F. Cosserat and E. Cosserat. *Corps déformables.* Librarie Scientifique A. Hermann et fils, Paris, 1909.
58. R. Courant. Variational methods for the solution of problems of equilibrium and vibration. *Bulletin of the American Math Society*, 49:1–61, 1943.
59. M.A. Crisfield. A fast incremental/iterative solution procedure that handles "snap through". *Computers and Structures*, 13:55–62, 1981.
60. M.A. Crisfield. *Non-linear Finite Element Analysis of Solids and Structures*, Vol. 2. Wiley, Chichester, 1997.
61. C.M. Dafermos. *Hyperbolic Conservation Laws in Continuum Physics.* Springer, Berlin, 2000.
62. G. Dahlquist and A. Björck. *Numerical Methods.* Prentice-Hall, Englewood Cliffs, NJ, 1974.
63. P.A. Dashner. Invariance considerations in large strain elasto-plasticity. *Journal of Applied Mechanics, ASME*, 53:55–60, 1986.
64. C. Agelet de Saracibar. Numerical analysis of coupled thermomechanical frictional contact: Computational model and applications. *Archives of Computational Methods in Engineering*, 5(3):243–301, 1998.
65. J.E. Dennis and R.B. Schnabel. *Numerical Methods for Unconstrained Optimization and Nonlinear Equations.* Society for Industrial and Applied Mathematics, Philadelphia, PA, 1996.

66. G. Dhatt and G. Touzot. *Une présentation de la méthode des éléments finis.* Maloine S.A., Paris, 1984.
67. F.L. DiMaggio and I.S. Sandler. Material models for granular soils. *Journal of Applied Mechanics, ASME*, 97:935–970, 1971.
68. S. Dolarevic and A. Ibrahimbegovic. A modified three-surface elasto-plastic model and its numerical implementation. *Computers and Structures*, 85:419–430, 2007.
69. T.C. Doyle and J. Ericksen. Non-linear elasticity. *Advances in Applied Mechanics*, 4:53–116, 1956.
70. I.S. Duff, A.M. Erisman, and J.K. Reid. *Direct Methods for Sparse Matrices.* Clarenden Press, Oxford, 1986.
71. G. Duvaut. *Mécanique des milieux continus.* Masson, Paris, 1990.
72. G. Duvaut and J.-L. Lions. *Les inéquations en mécanique et physique.* Dunod, Paris, 1972.
73. I. Ekeland and R. Temam. *Analyse convexe et problèmes variationnels.* Gauthier-Villars, Paris, 1974.
74. J.L. Ericksen. *Introduction to the Thermodynamics of Solids.* Springer, Berlin, 1998.
75. C.A. Felippa. Iterative procedure for improving penalty function solutions of algebraic systems. *International Journal for Numerical Methods in Engineering*, 12:165–185, 1978.
76. C.A. Felippa and K.C. Park. Synthesis tools for structural dynamics and partitioned analysis of coupled systems. In A. Ibrahimbegovic and B. Brank, editors, *Engineering Structures under Extreme Conditions: Multi-physics and Multi-scale Computer Models in Non-linear Analysis and Optimal Design*, pages 50–110. IOS Press, Amsterdam, 2005.
77. R.M. Ferencz and T.J.R. Hughes. *Iterative Finite Element Solutions in Nonlinear Solid Mechanics.* Handbook of Numerical Analysis. *Vol. VI.* Elsevier, Amsterdam, 1998.
78. G. Fonder and R.W. Clough. Explicit addition of rigid body motion in curved finite elements. *Journal of AIAA*, 11:305–315, 1970.
79. S. Forest, G. Cailletaud, D. Jeulien, F. Feyel, I. Gaillet, V. Mounoury, and S. Quilici. Elements of microstructural mechanics. *Mechanique et industrie*, 3:439–456, 2002.
80. D. François, A. Pineau, and A. Zaoui. *Comportement mécanique des matériaux : viscoplasticité, endommagement, mécanique de la rupture, mécanique du contact.* Hermes Science, Paris, 1993.
81. P.J. Frey and P.L. Georges. *Maillages: applications aux éléments finis.* Hermes Science, Paris, 1999.
82. B.G. Galerkin. Series solution of some problems in elastic equilibrium of rods and plates. *Vestn. Inzh. Tech.*, 19:897–908, 1915.
83. G.W. Gear. *Numerical Initial Value Problems in Ordinary Differential Equations.* Prentice-Hall, Englewood Cliffs, NJ, 1971.
84. M. Géradin and D. Rixen. *Théorie des vibrations: Applications à la dynamique des structures.* Masson, Paris, 1992.
85. P. Germain. *Cours de mécanique des milieux continus. Vol. 1.* Masson, Paris, 1973.
86. P. Germain, Q.S. Nguyen, and P. Suquet. Continuum thermodynamics. *Journal of Applied Mechanics, ASME*, 105:1010–1020, 1983.
87. R.G. Ghanem and P.D. Spanos. *Stochastic Finite Elements: A Spectral Approach.* Springer, Berlin, 1991.
88. F. Gharzeddine and A. Ibrahimbegovic. Incompatible mode method for finite deformation quasi-incompressible elasticity. *Computational Mechanics*, 24:419–425, 2000.
89. R. Glowinski and P. Le Tallec. *Augmented Lagrangian and Operator-Splitting Methods in Nonlinear Mechanics*, Vol. 9. SIAM, Philadelphia, PA, 1989.
90. H. Goldstein. *Classical Mechanics.* Addison-Wesley, Reading, MA, 2nd edition, 1980.
91. G.H. Golub and C.F. Van Loan. *Matrix Computations.* The Johns Hopkins University Press, Baltimore MD, 3rd edition, 1996.

92. A.E. Green and P.M. Naghdi. A note on invariance under superposed rigid body motion. *Journal of Elasticity*, 9:1–8, 1979.
93. A.E. Green and W. Zerna. *Theoretical Elasticity*. Dover, New York, 1992.
94. J. Guckenheimer and P. Holmes. *Nonlinear Oscillations, Dynamical Systems, and Bifurcation of Vector Fields*. Springer, Berlin, 1983.
95. A.L. Gurson. Continuum theory of ductile rupture by void nucleation and growth: Part i - yield criteria and flow rules for porous ductile media. *Journal of Engineering Materials Technology*, 99:2–15, 1977.
96. M.E. Gurtin. *An Introduction to Continuum Mechanics*. Academic, New York, 1981.
97. E. Hairer and G. Wanner. *Solving Ordinary Differential Equations II: Stiff Equations*. Springer, Berlin, 1991.
98. B. Halphen and Q.S. Nguyen. Sur les matériaux standars généralisés. *J. de Mécanique*, 14:39–63, 1975.
99. M. Hautefeuille, S. Melnyk, J.B. Colliat, and A. Ibrahimbegovic. Failure model for heterogeneous structures using structured meshes and accounting for probability aspects. *Engineering Computations*, 26(1-2):166–184, 2009.
100. G. Herve, F. Gatuingt, and A. Ibrahimbegovic. On numerical implementation of a coupled rate dependent damage-plasticity constitutive model for concrete in application to high-rate dynamics. *Engineering Computations*, 22(5/6):583–604, 2005.
101. H. Hilber, T.J.R. Hughes, and R.L. Taylor. Improved numerical dissipation for the time integration algorithms in structural dynamics. *Earthquake Engineering and Structural Dynamics*, 5:283–292, 1977.
102. F.B. Hildebrand. *Methods of Applied Mathematics*. Prentice-Hall, Upper Saddle River, NJ (reprinted by Dover Publishers, 1992), 2 edition, 1965.
103. R. Hill. *The Mathematical Theory of Plasticity*. Clarendon Press, Oxford, 1950.
104. R. Hill. On constitutive inequalities for simple materials. *Journal Mechanics and Physics of Solids*, 16:544–555, 1968.
105. R. Hill. Aspects of invariance in solid mechanics. *Advances in Applied Mechanics*, 18:1–75, 1978.
106. A. Hillerborg, M. Modeer, and P.E. Petersson. Analysis of crack formation and crack groth in concrete by means of fracture mechanics and finite elements. *Cement and Concrete Research*, 6:773–782, 1976.
107. M.W. Hirsch and S. Smale. *Differential Equations, Dynamical Systems and Linear Algebra*. Academic, Boston, MA, 1974.
108. G. Hofstetter, J.C. Simo, and R.L. Taylor. A modified cap model: closest point solution algorithms. *Computers and Structures*, 48:203–214, 1993.
109. T.J.R. Hughes. A note on the stability of newmark algorithm in nonlinear structural dynamics. *International Journal for Numerical Methods in Engineering*, 11:383–386, 1977.
110. T.J.R. Hughes. Generalization of selective integration procedures to anisotropic and non-linear media. *International Journal for Numerical Methods in Engineering*, 15:1413–1418, 1980.
111. T.J.R. Hughes. *The Finite Element Method: Linear Static and Dynamic Analysis*. Prentice-Hall, Englewood Cliffs, NJ, 1987.
112. A. Ibrahimbegovic. A consistent finite element formulation of nonlinear elastic cables. *Communications in Numerical Methods in Engineering*, 8:547–556, 1992.
113. A. Ibrahimbegovic. Equivalent spatial and material descriptions of finite deformation elastoplasticity in principal axes. *International Journal of Solids and Structures*, 31:3027–3040, 1994.
114. A. Ibrahimbegovic. Finite elastoplastic deformations of space-curved membranes. *Computer Methods in Applied Mechanics and Engineering*, 119:371–394, 1994.
115. A. Ibrahimbegovic. *Mécanique non linéaire des solides déformables: Formulation théorique et résolution numérique par éléments finis*. Hermes-Science, Lavoisier, Paris, 2006.

116. A. Ibrahimbegovic, A. Boulkertous, and L. Davenne. Fe modelling of rc structures providing details of constitutive ingredients behavior and novel operator split solution procedure based on x-fem. *International Journal for Numerical Methods in Engineering*, 73:in press, 2009.
117. A. Ibrahimbegovic and D. Brancherie. Combined hardening and softening constitutive model of plasticity: precoursor to shear slip line failure. *Computational Mechanics*, 31:88–100, 2003.
118. A. Ibrahimbegovic and B. Brank, editors. *Engineering Structures under Extreme Conditions: Multi-physics and Multi-scale Computer Models in Non-Linear Analysis and Optimal Design*. IOS Press, Amsterdam, 2005.
119. A. Ibrahimbegovic, L. Chorfi, and F. Gharzeddine. Thermomechanical coupling at finite elastic strain : Covariant formulation and numerical implementation. *Communications in Numerical Methods in Engineering*, 17:275–289, 2001.
120. A. Ibrahimbegovic, J.B. Colliat, and L. Davenne. Thermomechanical coupling in folded plates and non-smooth shells. *Computer Methods in Applied Mechanics and Engineering*, 194:2686–2707, 2005.
121. A. Ibrahimbegovic and A. Delaplace. Microscale and mesoscale discrete models for dynamics fracture of structures built of brittle material. *Computers and Structures*, 81:1255–1265, 2003.
122. A. Ibrahimbegovic and F. Frey. Geometrically nonlinear method of incompatible modes in application to finite elasticity with independent rotation field. *International Journal for Numerical Methods in Engineering*, 36:4185–4200, 1993.
123. A. Ibrahimbegovic and F. Frey. Stress resultant geometrically nonlinear shell theory with drilling rotations. part iii: Linearized kinematics. *International Journal for Numerical Methods in Engineering*, 37:3659–3683, 1994.
124. A. Ibrahimbegovic and F. Gharzeddine. Finite deformation plasticity in principal axes: from a manifold to the euclidean setting. *Computer Methods in Applied Mechanics and Engineering*, 171:341–369, 1999.
125. A. Ibrahimbegovic, F. Gharzeddine, and L. Chorfi. Classical plasticity and viscoplasticity models reformulated: Theoretical basis and numerical implementation. *International Journal for Numerical Methods in Engineering*, 42:1499–1535, 1998.
126. A. Ibrahimbegovic, I. Gresovnik, D. Markovic, S. Melnyk, and T. Rodic. Shape optimizatoin of two-phase inelastic material with microstructure. *Engineering Computations*, 22(5/6):605–645, 2005.
127. A. Ibrahimbegovic and F. Gruttmann. A consistent finite element formulation of nonlinear membrane shell theory with particular reference to elastic rubberlike material. *Finite Elements in Analysis and Design*, 12:75–86, 1993.
128. A. Ibrahimbegovic, P. Jehel, and L. Davenne. Coupled damage-plasticity model and direct stress interpolation. *Computational Mechanics*, 42:1–11, 2008.
129. A. Ibrahimbegovic, C. Knopf-Lenoir, A. Kucerova, and P. Villon. Optimal design and optimal control of structures undergoing finite rotations and elastic deformations. *International Journal for Numerical Methods in Engineering*, 61:2428–2460, 2004.
130. A. Ibrahimbegovic and I. Kozar. Nonlinear wilson's brick element for finite elastic deformation of three-dimensional solids. *Communications in Numerical Methods in Engineering*, 11:655–664, 1995.
131. A. Ibrahimbegovic and I. Kozar, editors. *Extreme Man-Made and Natural Hazards in Dynamics of Structures*. Springer, Berlin, 2007.
132. A. Ibrahimbegovic and S. Mamouri. Nonlinear dynamics of flexible beams in planar motion: Formulation and time-stepping scheme for stiff problem. *Computers and Structures*, 70:1–22, 1999.
133. A. Ibrahimbegovic and S. Mamouri. Energy conserving/decaying implicit time-stepping scheme for nonlinear dynamics of three-dimensional beams undergoing finite rotation. *Computer Methods in Applied Mechanics and Engineering*, 191:4241–4258, 2002.

134. A. Ibrahimbegovic and D. Markovic. Strong coupling methods in multi-phase and multi-scale modeling of inelastic behavior of heterogeneous structures. *Computer Methods in Applied Mechanics and Engineering*, 192:3089–3107, 2003.
135. A. Ibrahimbegovic, D. Markovic, and F. Gatuingt. Constitutive model of coupled damage-plasticity and its finite element implementation. *Revue eurpéenne des éléments finis*, 12:381–405, 2003.
136. A. Ibrahimbegovic and S. Melnyk. Embedded discontinuity finite element method for modeling of localized failure in heterogeneous materials with structured mesh: an alternative to extended finite element method. *Computational Mechanics*, 40:149–155, 2007.
137. A. Ibrahimbegovic, H. Shakourzadeh, J.L. Batoz, M. Al Mikdad, and Y.Q. Guo. On the role of geometrically exact and second order theories in buckling and post-buckling analysis of three-dimensional beam structures. *Computers and Structures*, 61:1101–1114, 1996.
138. A. Ibrahimbegovic and R.L. Taylor. On the role of frame-invariance in structural mechanics models at finite rotations. *Computer Methods in Applied Mechanics and Engineering*, 191:5159–5176, 2002.
139. A. Ibrahimbegovic, P. Villon, and G. Herve. Nonlinear impact dynamics and field tranfer suitable for parametric design studies. *Engineering Computations*, 26(1-2): 185–204, 2009.
140. A. Ibrahimbegovic and E.L. Wilson. A modified method of incompatible modes. *Communications in Numerical Methods in Engineering*, 7:187–194, 1991.
141. A. Ibrahimbegovic and E.L. Wilson. Unified computational model for static and dynamic frictional contact analysis. *International Journal for Numerical Methods in Engineering*, 34:233–247, 1991.
142. S. Idelsohn, E. Oñate, N. Calvo, and F. Del Pin. The meshless finite element method. *International Journal for Numerical Methods in Engineering*, 58:893–912, 2003.
143. B.M. Irons. Engineering applications of numerical integration in stiffness methods. *Journal of AIAA*, 4:2035–2037, 1966.
144. C. Johnson. *Numerical Solutions of Partial Differential Equations by the Finite Element Method.* Cambridge University Press, Cambridge, 1987.
145. G.C. Johnson and D.J. Bammann. A discussion of stress rates in finite deformation problems. *International Journal of Solids and Structures*, 20:725–737, 1984.
146. W. Ju. On energy-based coupled elastoplastic damage theories: Constitutive modeling and computational aspects. *International Journal of Solids and Structures*, 25:803–833, 1989.
147. C. Kassiotis, J.B. Colliat, A. Ibrahimbegovic, and H. Matthies. Multiscale in time and stability analysis of operator split solution procedure applied to thermomechanical problems. *Engineering Computations*, 26(1-2):205–223, 2009.
148. C.T. Kelley. *Iterative Methods for Linear and Nonlinear Equations.* SIAM, Philadelphia, PA, 1995.
149. J. Kestin and J.M. Rice. Paradoxes in the applications of thermodynamics to strain solids. In B. Gal'Or E.B. Stuart and A.J. Brainard, editors, *A Critical Review of Thermodynamics*, pages 275–298. Mono Book Corp., Baltimore, MD, 1970.
150. W.T. Koiter. Stress-strain relations, uniqueness and variational theorems for elastic-plastic materials with a singular yield surface. *Quarterly Journal of Applied Mathematics*, 11:350–354, 1953.
151. M. Kojic and K.J. Bathe. Studies of finite element procedures - stress solution of closed elastic strain path with stretching and shearing using the updated lagrangian jaumann formulation. *Computers and Structures*, 26:175–179, 1987.
152. D. Kondepudi and I. Prigogine. *Modern Thermodynamics: From Heat Engines to Dissipative Structures.* Wiley, New York, 1998.
153. D. Krajcinovic. *Damage Mechanics.* North-Holland, Amsterdam, 1996.

154. R.D. Krieg and D.N. Krieg. Accuracy of numerical solution methods for the elastic, perfectly plastic model. *Journal of Pressure Vessel Technology-Transactions of the ASME*, 99:510–515, 1977.
155. A. Kucerova, D. Brancherie, A. Ibrahimbegovic, J. Zeman, and Z. Bittnar. Novel anisotropic continuum-discrete damage model capable of representing localized failure of massive structures. part II: identification from tests under heterogeneous stress field. *Engineering Computations*, 26(1-2):128–144, 2009.
156. P. Ladevèze. *Mécanique non linéaire des structures*. Hermes Science, Paris, 1996.
157. P. Ladeveze. On computational strategies for multiscale, time-dependent, nonlinear structural mechanics problems. In A. Ibrahimbegovic and B. Brank, editors, *Engineering Structures under Extreme Conditions: Multi-physics and Multi-scale Computer Models in Non-linear Analysis and Optimal Design*, pages 139–169. IOS Press, Amsterdam, 2005.
158. P. Ladevèze, O. Loiseau, and D. Dureisseix. A micro-macro and parallel computational strategy for highly heterogeneous structures. *International Journal for Numerical Methods in Engineering*, 52:121–138, 2001.
159. P. Ladevèze and J.P. Pelle. *La maitrise du calcul en mécanique linéaire et non linéaire*. Lavoisier, Paris, 2001.
160. C. Lancoz. *The Variational Principles of Mechanics*. Dover, New York, 1986.
161. S. Lang. *Differential Manifolds*. Springer, New York, 1985.
162. R. Larsson, P. Steinmann, and K. Runesson. Finite element embedded localization band for finite strain plasticity based on a regularized strong discontinuities. *Mechanics of Cohesive-Frictional Material*, 4:171–194, 1998.
163. T.A. Laursen. On development of thermodynamically consistent algorithm for thermomechanical frictional contact. *Computer Methods in Applied Mechanics and Engineering*, 177:273–287, 1999.
164. T.A. Laursen. *Computational Contact and Impact Mechanics: Fundamentals of Modeling Inferfacial Phenomena in Nonlinear Finite Element Analysis*. Springer, Berlin, 2003.
165. E.H. Lee. Elastic-plastic deformations at finite strains. *Journal of Applied Mechanics, ASME*, 36:1–6, 1969.
166. E.H. Lee. Some comments on elasto-plastic analysis. *International Journal of Solids and Structures*, 17:859–872, 1980.
167. J. Lemaître. Coupled elasto-plasticity and damage constitutive equations. *Computer Methods in Applied Mechanics and Engineering*, 51:31–49, 1985.
168. J. Lemaitre and J.L. Chaboche. *Mécanique des matériaux solides*. Dunod, Paris, 1988.
169. R.W. Lewis and B.A. Schrefler. *The Finite Element method in the Deformation and Consolidation of Porous Media*. Wiley, Chichester, 1987.
170. J.L. Lions. *Asymptotic calculus of variations*. Singular perturbation and asymptotics. Academic, New York, 1980.
171. M. Loève. *Probability Theory*. Springer, Berlin, 1977.
172. J. Lubliner. On the thermodynamic foundations of non-linear solid mechanics. *International Journal of Non-linear Mechanics*, 7:237–254, 1972.
173. J. Lubliner. A maximum-dissipation principle in generalized plasticity. *Acta Mechanica*, 52:225–237, 1984.
174. J. Lubliner. *Plasticity Theory*. Macmillan, New York, 1990.
175. D.G. Luenberger. *Linear and Nonlinear Programming*. Addison-Wesley, Reading, MA, 1984.
176. A.M. Lyapunov. *Stability of motion*. Taylor & Francis, Bristol, 1992.
177. J. Mandel. Equations constitutives et directeurs dans les milieux plastiques et viscoplastiques. *International Journal of Solids and Structures*, 9:725–740, 1973.
178. H. Mang, R. Lackner, and C. Pichler. Thermochemomechanics of cement-based materials at finer scales of observation: Application to hybrid analysis of shortcrete tunnel

linings. In A. Ibrahimbegovic and B. Brank, editors, *Engineering Structures under Extreme Conditions: Multi-physics and Multi-scale Computer Models in Non-linear Analysis and Optimal Design*, pages 170–199. IOS Press, Amsterdam, 2005.

179. D. Markovic and A. Ibrahimbegovic. On micro-macro interface conditions for micro-scale based fem for in elastic behavior of heterogeneous materials. *Computer Methods in Applied Mechanics and Engineering*, 193:5503–5523, 2004.
180. D. Markovic, A. Ibrahimbegovic, and K.C. Park. Partition based reduced order modelling approach for transient analysis of large structures. *Engineering Computations*, 256(1-2):46–68, 2009.
181. J.E. Marsden and M.J. Hoffman. *Elementary Classical Analysis.* W.H. Freeman, New York, 1993.
182. J.E. Marsden and T.J.R. Hughes. *Mathematical Foundations of Elasticity.* Dover, New York, 1994.
183. H.G. Matthies. Quantifying uncertainty: Modern computational representation of probability and applications. In A. Ibrahimbegovic and I. Kozar, editors, *Extreme Man-Made and Natural Hazards in Dynamics of Structures*, pages 105–136. Springer, Berlin, 2007.
184. H.G. Matthies and G. Strang. The solution of nonlinear finite element equations. *International Journal for Numerical Methods in Engineering*, 14:1613–1626, 1979.
185. G.A. Maugin. *The Thermomechanics of Plasticity and Fracture.* Cambridge University Press, Cambridge, 1992.
186. J. Mazars. A description of micro and macroscale damage of concrete structures. *Journal of Engineering Fracture Mechanics*, 25:729–737, 1986.
187. C. Miehe. Computational micro-to-macro transitions for discretized micro-structures of heterogeneous materials at finite strains based on the minimization of averaged incremental energy. *International Journal for Numerical Methods in Engineering*, 192:559–591, 2003.
188. F.J. Molina and M. Geradin. Earthquake engineering experimental research at jrc-elsa. In A. Ibrahimbegovic and I. Kozar, editors, *Extreme Man-Made and Natural Hazards in Dynamics of Structures*, pages 311–353. Springer, Berlin, 2007.
189. M. Mooney. A theory of large elastic deformation. *Journal of Applied Physics*, 1:582–592, 1940.
190. B. Moran, M. Ortiz, and C.F. Shi. Formulation of implicit finite element methods for multiplicative finite deformation plasticity. *International Journal for Numerical Methods in Engineering*, 29:483–514, 1990.
191. J.J. Moreau. On unilateral constraints, friction and plasticity. In *New Variational Techniques in Mathematical Physics*, pages 173–322, Rome, 1973. C.I.M.E., Cremonese.
192. H.B. Muhlhaus and E.C. Aifantis. A variational principle for gradient plasticity. *International Journal of Solids and Structures*, 28:845–857, 1991.
193. P.M. Naghdi. *The Theory of Shells.* Springer, Berlin, 1972.
194. P.M. Naghdi. A critical review of the state of finite plasticity. *Journal of Applied Mechanics and Physics*, 41:315–394, 1990.
195. P.M. Naghdi and P. Trapp. Plasticity in strain space. *ASME Journal of Applied Mechanics*, 21:259–265, 1975.
196. J.C. Nagtegaal and J.C. DeJong. Some aspects on nonisotropic work-hardening in finite strain plasticity. In E.H. Lee and L. Mallett, editors, *Plasticity of Metals at Finite Deformations*, CA, 1982. Stanford University.
197. J.C. Nagtegaal, D.M. Parks, and J.R. Rice. On numerical accurate finite element solutions in the fully plastic range. *Computer Methods in Applied Mechanics and Engineering*, 4:153–177, 1974.
198. A. Needleman. Material rate dependence and mesh sensitivity in localization problems. *Computer Methods in Applied Mechanics and Engineering*, 67:69–85, 1988.
199. A. Needleman and V. Tvergaard. An analysis of ductile rupture in notched bar. *J. Mechanics and Physics of Solids*, 32:461–469, 1984.

200. S. Nemat-Nasser. On finite deformation elasto-plasticity. *International Journal of Solids and Structures*, 18:857–872, 1982.
201. Q.S. Nguyen. On elastic plastic initial-boundary value problem and its numerical implementation. *International Journal for Numerical Methods in Engineering*, 11:817–832, 1977.
202. Q.S. Nguyen. *Stabilité en mécanique non linéaire.* Lavoisier, Paris, 2000.
203. Q.S. Nguyen and J. Zarka. Quelques méthodes de resolution numérique en elastoplasticité classique et en elasto-viscoplasticité. *Sciences et Technique de l'Armement*, 47:407–436, 1973.
204. R. Niekamp, D. Markovic, A. Ibrahimbegovic, H.G. Matthies, and R.L. Taylor. Multi-scale modelling of heterogeneous structures with inelastic constitutive behavior. part II: software coupling implementation aspects. *Engineering Computations*, 26(1-2): 6–28, 2009.
205. J.T. Oden. *Qualitative Methods in Nonlinear Mechanics.* Prentice-Hall, Englewood Cliffs, NJ, 1986.
206. J.T. Oden, T. Belytschko, I. Babuska, and T.J.R. Hughes. Research directions in computational mechanics. *Computer Methods in Applied Mechanics and Engineering*, 192:913–922, 2003.
207. J.T. Oden and L.F. Demkowicz. *Applied Functional Analysis.* CRC Press, Boca Raton, FL, 1996.
208. J.T. Oden and R.B. Fost. Convergence, accuracy and stability of finite element approximations of a class of non-linear hyperbolic equations. *International Journal for Numerical Methods in Engineering*, 6:357–365, 1973.
209. R.W. Ogden. *Non-linear Elastic Deformations.* Ellis Horwood, Limited (reprinted by Dover, 1997), Chichester, England, 1984.
210. J. Oliver, A.E. Huespe, M.D.G. Pulido, and E. Samaniego. On the strong discontinuity approach in finite deformation setting. *International Journal for Numerical Methods in Engineering*, 4:171–194, 1998.
211. M. Ortiz. A constitutive theory for inelastic behavior of concrete. *Mechanics and Materials*, 4:67–93, 1985.
212. M. Ortiz, Y. Leroy, and A. Needleman. A finite element method for localized failure analysis. *Computer Methods in Applied Mechanics and Engineering*, 61:189–214, 1987.
213. D.R.J. Owen, Y.T. Feng, M.G. Cottrell, and J. Yu. Computational issues in the simulation of blast and impact problems: an industrial perspective. In A. Ibrahimbegovic and I. Kozar, editors, *Extreme Man-Made and Natural Hazards in Dynamics of Structures*, pages 3–36. Springer, Berlin, 2007.
214. D.R.J. Owen and E. Hinton. *Finite Element Method in Plasticity.* Pineridge Press, Swansea, 1980.
215. P. Papadopoulos and R.L. Taylor. A mixed formulation for the finite element solution of contact problems. *Computer Methods in Applied Mechanics and Engineering*, 94:373–389, 1992.
216. M. Papadrakakis, M. Fragiadakis, and N. Lagaros. Optimum performance-based reliability of structures. In A. Ibrahimbegovic and I. Kozar, editors, *Extreme Man-Made and Natural Hazards in Dynamics of Structures*, pages 137–162. Springer, Berlin, 2007.
217. K.C. Park, C. Felippa, and G. Rebel. A simple algorithme for localized construction of non-matching structural interfaces. *International Journal for Numerical Methods in Engineering*, 53:2117–2142, 2002.
218. K.C. Park, C.A. Felippa, and R. Ohayon. Reduced-order partitioned modelling of coupled systems: Formulation and computational algorithms. In A. Ibrahimbegovic and B. Brank, editors, *Engineering Structures under Extreme Conditions: Multi-physics and Multi-scale Computer Models in Non-linear Analysis and Optimal Design*, pages 139–169. IOS Press, Amsterdam, 2005.

219. B.N. Parlett. *The Symmetric Eigenvalue Problem.* Prentice-Hall, Englewood Cliffs, NJ, 1980.
220. D. Peric, W.D. Dettmer, and P.H. Saksono. Modelling fluid induced structural vibrations: Reducing the structural risk for stormy winds. In A. Ibrahimbegovic and I. Kozar, editors, *Extreme Man-Made and Natural Hazards in Dynamics of Structures*, pages 225–256. Springer, Berlin, 2007.
221. D. Peric, D.R.J. Owen, and E. Honnor. A model for finite strain elasto-plasticity based on logarithmic strains: Computational implications. *Computer Methods in Applied Mechanics and Engineering*, 94:35–61, 1992.
222. P. Perzyna. Fundamental problems in viscoplasticity. *Advances in Applied Mechanics*, 9:243–377, 1966.
223. T.H.H. Pian and K. Sumihara. Rational approach for assumed stress finite elements. *International Journal for Numerical Methods in Engineering*, 20:1685–1695, 1985.
224. W. Prager. *An Introduction to Plasticity.* Addison-Wesley, Reading, MA, 1959.
225. E. Ramm and G.A. d'Addetta. Discrete models for geomaterials. In A. Ibrahimbegovic and B. Brank, editors, *Engineering Structures under Extreme Conditions: Multi-physics and Multi-scale Computer Models in Non-linear Analysis and Optimal Design*, pages 383–406. IOS Press, Amsterdam, 2005.
226. S. Reese and W. Dettmer. On the theoretical and numerical modelling of armstrong-frederics kinematic hardening in the finite strain regime. *Computer Methods in Applied Mechanics and Engineering*, 193:87–116, 2004.
227. J.R. Rice. The localization of plastic deformation. In W.T. Koiter, editor, *Theoretical and Applied Mechanics*, pages 207–220. North-Holland, Amsterdam, 1977.
228. E. Riks. An incremental approach to the solution of snapping and buckling problems. *International Journal of Solids and Structures*, 15:529–551, 1979.
229. W. Ritz. Über eine neue Methode zur Lösung gewisser variationsproblem der mathematischen physik. *Journal für die reine und angewandte Mathematik*, 135:1–61, 1908.
230. P. Rougée. *Mécanique des grandes transformations.* Springer, New York, 1997.
231. J.W. Rudnicki and J.R. Rice. Conditions for the localization of deformations in pressure sensitive dilatant materials. *Journal of Mechanics and Physics of Solids*, 23:371–394, 1975.
232. E. Sanchez-Palencia. *Non-homogeneous media and vibration theory.* Springer, Berlin, 1980.
233. B.A. Schrefler, F. Pesavento, L. Sanavia, and D. Gawin. Multi-physics problems in thermo-hydro-mechanical analysis of partially saturated geomaterials. In A. Ibrahimbegovic and B. Brank, editors, *Engineering Structures under Extreme Conditions: Multi-physics and Multi-scale Computer Models in Non-linear Analysis and Optimal Design*, pages 351–382. IOS Press, Amsterdam, 2005.
234. J.G. Simmonds. *A Brief on Tensor Analysis.* Undergraduate Texts in Mathematics. Springer, Berlin, 1982.
235. J.C. Simo. A framework for finite strain elastoplasticity based on maximum plastic dissipation and multiplicative decomposition. part i: Continuum formulation. *Computer Methods in Applied Mechanics and Engineering*, 66:199–219, 1988.
236. J.C. Simo. Nonlinear stability of the time-discrete variational problem of evolution in nonlinear heat conduction, plasticity and viscoplasticity. *Computer Methods in Applied Mechanics and Engineering*, 88:111–131, 1991.
237. J.C. Simo, F. Armero, and R.L. Taylor. Improved versions of assumed enhanced strain tri-linear elements for 3d finite deformation problems. *Computer Methods in Applied Mechanics and Engineering*, 110:359–386, 1993.
238. J.C. Simo and T.J.R. Hughes. *Computational Inelasticity*, volume 7 of *Interdisciplinary Applied Mathematics.* Springer, Berlin, 1998.
239. J.C. Simo and W. Ju. Stress and strain based continuum damage model-i. formulation, ii. computational aspects. *International Journal of Solids and Structures*, 23:821–869, 1987.

240. J.C. Simo, J. Oliver, and F. Armero. An analysis of strong discontinuities induced by strain-softening in rate-independent inelastic solids. *Computational Mechanics*, 12:277–296, 1993.
241. J.C. Simo and M.S. Rifai. A class of mixed assumed strain methods and the method of incompatible modes. *International Journal for Numerical Methods in Engineering*, 29:1595–1638, 1990.
242. J.C. Simo, R.L. Taylor, and K.S. Pister. Variational and projection methods for the volume constraint in finite deformation plasticity. *Computer Methods in Applied Mechanics and Engineering*, 51:177–208, 1985.
243. I.S. Sokolnikoff. *The Mathematical Theory of Elasticity.* McGraw-Hill, New York, 2 edition, 1956.
244. I.S. Sokolnikoff. *Tensor Analysis: Theory and Applications to Geometry and Mechanics of Continua.* Wiley, New York, 1964.
245. E. Souza-Neto, D. Peric, M. Dutko, and D.R.J. Owen. Design of simple low order finite elements for large strain analysis of nearly incompressible solids. *International Journal of Solids and Structures*, 33:3277–3296, 1996.
246. I. Stakgold. *Green's Function and Boundary Value Problems.* Wiley, New York, 2 edition, 1979.
247. E. Stein and M. Ruter. Computational mechanics with model adaptivity in analysis and design: Current research and perspectives. In A. Ibrahimbegovic and B. Brank, editors, *Engineering Structures under Extreme Conditions: Multi-physics and Multi-scale Computer Models in Non-linear Analysis and Optimal Design*, pages 383–406. IOS Press, Amsterdam, 2005.
248. G. Strang. *Introduction to Applied Mathematics.* Wellesley Cambridge Press, Wellesley, MA, 1986.
249. G. Strang and G.J. Fix. *An Analysis of the Finite Element Method.* Prentice-Hall, Englewood Cliffs, NJ, 1971.
250. R.L. Taylor, P.J. Beresford, and E.L. Wilson. A non-conforming element for stress analysis. *International Journal for Numerical Methods in Engineering*, 10:1211–1219, 1976.
251. R.L. Taylor and P. Papadopoulos. A patch test for contact problems in two dimensions. In P. Wriggers and W. Wagner, editors, *Nonlinear Computational Mechanics*, pages 690–702. Springer, Berlin, 1991.
252. R.L. Taylor and P. Papadopoulus. On a finite element method for dynamic contact impact problem. *International Journal for Numerical Methods in Engineering*, 36:2123–2140, 1993.
253. R.L. Taylor, O.C. Zienkiewicz, J.C. Simo, and A.H.C. Chan. The patch test – a condition for assessing FEM convergence. *International Journal for Numerical Methods in Engineering*, 22:39–62, 1986.
254. R. Temam. *Problèmes mathématiques en plasticité.* Gauthier-Villars, Paris, 1983.
255. S.P. Timoshenko and J.N. Goodier. *Theory of Elasticity.* McGraw-Hill, New York, 3 edition, 1969.
256. J.L. Troutman. *Variational Calculus and Optimal Control.* Springer, New York, 1995.
257. C. Truesdell and W. Noll. *The Non-Linear Field Theories of Mechanics.* Springer, Berlin, 1965.
258. C. Truesdell and R.A. Toupin. *Principles of Classicale Mechanics and Field Theory.* Springer, Berlin, 1960.
259. M.J. Turner, R.W. Clough, H.C. Martin, and L.J. Topp. Stiffness and deflection analysis of complex structures. *Journal of the Aeronautical Sciences*, 23:805–823, 1956.
260. K. Washizu. *Variational Methods in Elasticity and Plasticity.* Pergamon, New York, 3 edition, 1982.
261. M.L. Wilkins. Calculation of elastic-plastic flow. In B. Alder, editor, *Methods in Computational Physics*, volume 3, pages 211–263. Academic, New York, 1964.

262. E.L. Wilson. SAP - a general structural analysis program for linear systems. *Nucl. Engr. Des.*, 25:257–274, 1973.
263. E.L. Wilson. The static condensation algorithm. *International Journal for Numerical Methods in Engineering*, 8:1974, 199-203.
264. P. Wriggers. *Computational Contact Mechanics.* Wiley, London, 2000.
265. P. Wriggers and C. Miehe. Contact constraints within coupled thermomechanical analysis - a finite element model. *Computer Methods in Applied Mechanics and Engineering*, 113:301–319, 1994.
266. P. Wriggers and J.C. Simo. A note on tangent stiffness for fully nonlinear contact problems. *Communications in Numerical Methods in Engineering*, 1:199–203, 1985.
267. N.N. Yanenko. *The Method of Fractional Steps.* Springer, Berlin, 1971.
268. H. Ziegler. A modification of Prager's hardening rule. *Quarterly of Applied Mathematics*, 17:55–65, 1959.
269. H. Ziegler. *An Introduction to Thermomechanics.* North-Holland, Amsterdam, 1983.
270. O.C. Zienkiewicz and R.L. Taylor. *The Finite Element Method*, volume 1. McGraw-Hill, London, 4th edition, 1989.
271. O.C. Zienkiewicz and R.L. Taylor. *The Finite Element Method: The Basis*, *Vol. 1.* Butterworth-Heinemann, Oxford, 5th edition, 2000.
272. O.C. Zienkiewicz, J. Too, and R.L. Taylor. Reduced integration technique in general analysis of plates and shells. *International Journal for Numerical Methods in Engineering*, 3:275–290, 1971.

Index